Air Pollution and Community Health

A Critical Review and Data Sourcebook

FOG ALONG THE MEUSE VALLEY

WEDNESDAY, SEPTEMBER 1, 1976

20 DEAD IN DONORA; RAIN CLEARING AIR

Belgium's Mystery Fog

'Smog' Linked to 18 Deaths in Day And Hospital Jam in Donora, Pa.

CINCINNATI SMOG

LONDON'S NEW FOG SUFFUSED BY GASES

Medical Journal Confirms Warning of Lethal Content From Smokeless Fuels

URVIVORS OF SMOG 'LEE TO HIGH LAND

remen Praised for Making ygen Tents With Sheets— 'Odor' Detected in Plane

The DECEMBER SMOG

RAIN MAY WASH OUT 4-DAY SMOG TONIGHT

THE NEW YORK TIMES, SATURDAY, NOVEMBER 21, 1953.

Heavy Hangs Over: Smaze, a Blend of Smoke and Haze, Smothers City Area With a Grayish Pall

3-Day Smog Torments Entire East As Cold 'Lid' Traps Irritants in Air

Four-Day Concentration of Smoke and Haze Causes Optical Illusions and Discomfort --Two Airports Close as Fog Is Added

Air Pollution and Community Health

A Critical Review and Data Sourcebook

Frederick W. Lipfert

VNR VAN NOSTRAND REINHOLD
New York

Library of Congress Catalog Card Number 93-1364
ISBN 0-442-01444-9

I(T)P Van Nostrand Reinhold is an International Thomson Publishing company. ITP logo is a trademark under license.

Printed in the United States of America

Van Nostrand Reinhold
115 Fifth Avenue
New York, NY 10003

International Thomson Publishing GmbH
Königswinterer Str. 518
5300 Bonn 3
Germany

International Thomson Publishing
Berkshire House, 168–173
High Holborn, London WC1V 7AA
England

International Thomson Publishing Asia
38 Kim Tian Rd., #0105
Kim Tian Plaza
Singapore 0316

Thomas Nelson Australia
102 Dodds Street
South Melbourne 3205
Victoria, Australia

International Thomson Publishing Japan
Kyowa building, 3F
2-2-1 Hirakawacho
Chiyada-Ku, Tokyo 102
Japan

Nelson Canada
1120 Birchmount Road
Scarborough, Ontario
M1K 5G4, Canada

ARCKP 16 15 14 13 12 11 10 9 8 7 6 5 4 3 2 1

Library of Congress Cataloging in Publication Data

Lipfert, F.W.
Air pollution and community health: a critical review and data sourcebook/Frederick W. Lipfert.
p. cm.
Includes bibliographical references and index.
ISBN 0-442-01444-9
1. Air—Pollution—Toxicology—Statistics. 2. Air—Pollution—Toxicology—Research. I. Title.
RA576.L596 1994
363.73′92—dc20 93-1364
CIP

To Charlotte, who has at long last regained a husband and a dining room table

Contents

Preface

For the past 20 years or so, I have studied and pondered reports of the health effects of air pollution, especially those dealing with mortality and other irreversible endpoints. This book represents a culmination of that effort. Publication in book form is necessary in order to cover the breadth of the subject comprehensively. In addition, I believe that it is necessary to consider the interactions among the three supporting disciplines: physiology, statistics, and the various aspects of air pollution. As Ziman so aptly put it (as summarized by Thomas [1974]):

> A regular journal carries from one research worker to another the various . . . observations which are of common interest. . . . A typical scientific paper has never pretended to be more than another little piece in a larger jigsaw—not significant in itself but as an element in a grander scheme. This technique, of soliciting many modest contributions of the store of human knowledge, has been the secret of Western science since the seventeenth century, for it achieves a corporate collective power that is far greater than one individual can exert.

I agree with these thoughts and note further that journal publication has expanded substantially in the intervening twenty-plus years since these words were written. Not only are there many more journal titles, but they have grown more specialized. The opportunities for specialists to isolate themselves from the broader implications of their work seem to have increased and may constitute one of the adverse impacts of the growth of science. This book is my attempt to transform the work of many researchers into a coherent picture of studies of the effects of air pollution on populations.

In selecting articles for review in this book, I have tried to emphasize those which genuinely strive to communicate and inform through presentation and analysis of observed data and thus appear to add to the "store of human knowledge." Articles with a predominately methodological or theoretical focus have been given lower priority. It is my goal to assemble and analyze enough of the factual "pieces" so that at least the outline of the "grander scheme" comes into view.

This book is based primarily on reviews of previously published work. However, in my reviews I have replotted and reanalyzed some of the key data sets, as a check on the validity and completeness of the original authors' conclusions and to provide a comprehensive account, emphasizing graphical presentations. These augmented analyses are clearly identified by use of the first person pronoun. In addition, there are some new analyses that had not been published elsewhere by the time this text was completed. This synthesis process follows Ziman's (1969) admonition: "It is not enough to observe, experiment, theorize, calculate, and communicate; we must also argue, criticize, debate, expound, summarize, and otherwise transform the information that we have obtained individually into reliable, well established public knowledge."

Portions of this work have appeared elsewhere, in slightly different forms. A portion of Chapter 3 first appeared in a Brookhaven National Laboratory (BNL) informal report. Chapters 5, 6, and 7 began life as "Mortality and Air Pollution," in the *Handbook of Environmental Chemistry* (Part 4C, Springer-Verlag, Heidelberg, 1991) but have since been revised extensively. A portion of a recent BNL report has been included in Chapter 7. Portions of Chapter 8 were presented to the American Meteorological Society, Tenth Conference on Biometeorology and Aerobiology (Salt Lake City, 1991), and to the Ninth World Clean Air Conference (Montreal, September 1992), and appear in those proceedings. Chapters 9 and 10 form the nucleus of a report to the Electric Power Research Institute, "A Review of Studies of the Association between Hospital Utilization and Air Pollution" (1991), which has since been published (1993).

Many authors whose work I reviewed freely shared data, advance copies of manuscripts, and additional insights, for which I am grateful. I am particularly indebted to the authors of the 1960s and earlier, who often included detailed plots and listings of their data in their publications, making them thereby freely available for the sort of *a posteriori* analysis featured in this book. I would also like to acknowledge the contributions of three anonymous reviewers and those made (often unwittingly) by colleagues at BNL and elsewhere, through hundreds of conversations and "brain picking" sessions over the years. Dr. Robert Frank provided helpful comments on Chapter 11. Financial support for certain portions of this effort was provided by the Electric Power Research Institute, and I thank Ron Wyzga for his encouragement and criticism. Much of the library research was supported by BNL. However, nothing published in this book should be construed as a policy statement on the part of any institution with which I have been associated. I thank those who have reviewed various portions of the text but take full responsibility for any errors or oversights that may remain.

Frederick W. Lipfert
Northport, NY

REFERENCES

Lipfert, F.W. (1993), A Critical Review of Studies of the Association between Demands for Hospital Services and Air Pollution, *Env. Health Persp. Suppl. 101* (Suppl. 2): 229–68.

Thomas, L. (1974), *The Lives of a Cell*, Viking Press, New York, p. 16.

Ziman, J.M. (1969), Information, Communication, Knowledge, *Nature* 224:318–24.

PART I

Introduction

The impact of community air pollution on health is one of the most complicated health problems of the day because of the extensive range of scientific disciplines needed to describe the essential elements of the problem as well as because of the interaction of subjective and objective effects in need of investigation.

Not only are all the medical sciences and specialties required to describe the health effects of pollution, but meteorology, chemistry, physics, sociology, mathematics, psychology, and statistics are involved. The conventional manner of organizing and reporting scientific research in professional societies and specialty journals has some disadvantages in an effective attack on the air pollution problem since the audience may be inappropriate.

M.H. Merrill,
Director, California State
Department of Public Health
1964

1
Purpose and Scope

The driving force behind the control of air pollution in the United States is the protection of public health. This goal has been overwhelmingly supported by public surveys and is reflected in the language of the Clean Air Act and its various amendments. This statute mandates the achievement of health-related ambient air quality standards, without regard to the costs of pollution controls required to do so. In the 1970s, annual costs of air pollution control in the United States were estimated at $500 million (Eisenbud, 1970). The annual cost was more recently estimated to be about $33 billion; when the 1990 Amendments to the Clean Air Act are fully implemented, this figure may rise to over $60 billion (Portney, 1990; O'Neal, personal communication, 1991). In view of the stated emphasis on public health, air pollution control costs should be viewed in the context of health care expenditures. The total costs of health care reached $620 billion in 1989, or more than 11% of the gross national product (GNP) (Ginzburg, 1990). Using another measure, 1990 health care costs to workers (presumably the healthiest adult group in our society) reached over $3200 per worker (Freudenheim, 1991). This book is intended to explore the relationships between health and environment, as observed in the United States and elsewhere, and to examine the range of public health benefits that environmental control expenditures are likely to purchase.

There have been many previous books on environmental epidemiology. This one is concerned largely with irreversible health effects, including mortality, hospitalization, and impaired lung function, and is intended to emphasize several themes, most of which have not been specifically stressed before:

1. *Observational epidemiological studies are featured*, in which existing information on health effects is matched retrospectively with corresponding environmental data. Most of these studies are "ecological" in nature, in that the behavior of groups is studied rather than that of specific individuals (Robinson,

1950). This study design has often been denigrated for its lack of specificity (see Ware *et al.*, 1981, for example), yet Kleinbaum *et al.* (1982) characterized the observational genre as "central to epidemiologic research." Many of the studies proposed in the first federal examination of community air pollution and health in 1955 were of this design (Mancuso and Mordell, 1969).

2. *Close attention is given to the environmental data*, in this case air pollution measurements, as demanded by the nature of observational studies (Lodge, 1969). In an observational study, it may be unavoidable to rely on air quality measurements made by others, but since the conclusions may depend heavily on the reliability and limitations of the data describing the independent variables, one must consider carefully if all of the candidate variables have been given an equal chance to be significant predictors. This topic is discussed in detail in Chapter 3. Too often, air quality data have been implicitly treated as a set of error-free values (Lipfert, 1980). Since most people, and particularly sick people, spend the overwhelming majority of their time indoors, it is important to consider the interactions of indoor and outdoor environments. Almost all of the studies reviewed in this book were based on outdoor air quality measurements.

3. *At current levels, community air pollution is not treated as a toxicant in the classical sense but is assumed to be one of several factors that can exacerbate existing medical conditions*. As a case in point, cancer has seldom been (robustly) associated with community air pollution (Lipfert, 1978; Buffler *et al.*, 1988; Shannon *et al.*, 1988); multifactorial diseases involving the cardiopulmonary system are of primary interest. Thus, in today's environment an exacerbation model is more appropriate than a toxicological or disease model, and this requires that the nature of the population at risk be considered when assessing health effects (U.S. DHHS, 1984). Lead in the environment is an important exception to this generalization, as are acute pollutant exposures at industrial concentration levels, which have been shown to result in chronic disease, disability, and death (Bates *et al.*, 1971).

4. *Several different types of models are considered, since the dose–response functions, including their shapes, are unknown*. The concept of an air quality standard below which the health risk is defined as zero is based on toxicology, for which the "dose defines the poison." According to this concept, a substance may be an essential nutrient at low doses and toxic at high doses; examples include water, some metals, and sodium chloride. This principle or "toxicological model" clearly applies to individuals and to genetically similar groups of test animals; it is less likely to apply to heterogeneous community populations that include individuals having a wide range of previous impairments and sensitivities. History has steadily pushed the perceived thresholds of "no effect" downward for environmental agents; examples include lead, carbon monoxide, tobacco smoke. Thus, a linear dose-response function with no threshold may not be unreasonable for an observational study, although several of the recent studies of daily mortality have used log-linear models (the logarithm of response is regressed against a linear measure of air quality).

5. *The literature is examined by type of endpoint rather than by pollutant* (i.e., the focus is on risks rather than on agents). The criteria documents required by the Clean Air Act and published by the U.S. Environmental Protection Agency (EPA), which form the cornerstone of environmental health protection in the United States, are structured to find threshold effects uniquely associated with specific pollutants (see Chapter 2). Below these thresholds, health risks are postulated to be minimal or zero. These documents try to address the question: how much air pollution is acceptable? Studies that do not fit this model have often been rejected from consideration in criteria documents. Since in the United States, most community populations are quite diverse, more than one air pollutant may be associated with a given health response or endpoint.

6. *Studies that purport to find chronic health effects are evaluated in the light of the comparable acute effects* (for the same endpoints), and vice versa. While time-series analysis methods may be devised to exclude the long-term (annual or seasonal) components of change, the reverse is not true. The annual rates of illness or death include the sum of the year's short-term (acute) effects, by definition. If such a persistent change is not apparent, one may question whether the acute effects are transitory or are canceled by opposing effects a few days or weeks later. It is one of the purposes of this book to make the appropriate comparisons among these fundamentally different types of studies, as part of the process of reconciliation and "closing the loop."

Frank eloquently suggested several of these ideas in 1975. He wrote that "the effects of urban air pollutants on health . . . are neither discrete nor readily quantified" and noted that urban air pollution may act in combination with other forms of stress. He cited Lawther's (1966) observation that urban air pollution may act to "worsen rather than initiate new disease." Frank also called for the use of large data banks on a national scale as a source of epidemiological information. Many of the ecological studies reviewed in this book did make use of such data banks, and their results will be seen to support Frank's original ideas, with the addition of quantitative data on risks.

If the concept that current levels of air pollution cause exacerbation rather than disease *per se* is accepted, it may be appropriate to reexamine the design of the National Ambient Air Quality Standards (Chapter 2) and their accompanying regulatory mechanisms. The 1967 SO_2 Criteria Document (U.S. DHEW, 1967) stated that guidelines for air quality criteria should be strict enough that "the health of even sensitive or susceptible segments of the population would not be adversely affected. . . ." The authors apparently failed to realize that this may be essentially a prescription for zero emissions (no threshold). Ferris and Speizer (1980) pointed out that, even if a standard protected 99.99% of the people, over 20,000 could still be affected; they recommended that the no-effect threshold concept be abandoned in favor of a risk analysis approach based on well-defined health endpoints that should clearly be regarded as "adverse." Such a risk analysis model, in which specific endpoints are evaluated, should be based on the actual health experiences encountered by heterogeneous populations.

MORTALITY RATES AS A STUDY TOOL

Farr's emphasis on mortality as the appropriate measure of public health was stated forcefully in his memorial volume (1885, p. 111): "It may confidently be assumed that the most important branch of vital statistics is that which deals with deaths and rates of mortality. This is not only the most complex branch of the subject, therefore demanding the more careful study, but the influence of health on the human race is so powerful for good or evil that statistics of deaths and rates of mortality acquire their greatest value from their acceptation as trustworthy indications of public health."

To examine whether this statement is still true more than a century later, one need only consider the actuarial value of a life (several million dollars) in relation to the cost of illness. The costs of hospital treatment may be a useful paradigm for the relative role of morbidity. About two million deaths and about 35 million hospital discharges occur in the United States each year (all diagnoses). Thus, for the aggregate economic value to be equivalent to the "value" of mortality, each hospitalization would have to be valued at about $100,000, which seems excessive. Even if the comparison were limited to respiratory diagnoses, mortality "costs" outweigh morbidity costs. However, the degree of prematurity of the "excess" mortality is one of the elements missing from this crude comparison and is considered in Section II of this book, along with other measures of morbidity.

Russell (1924, 1926) was the first to find a quantitative link between daily increases in mortality and the smoky fogs of London, using correlation and regression analyses; the great episode of 1952 in which 4000 perished rammed this lesson home with a vengeance. However, the analyses of these events were still focused on a toxicological model (Firket, 1936; Shrenk *et al.*, 1949; Ministry of Health, 1954). The general procedure in the analyses of these episodes was to compare the estimated air pollutant concentrations to the levels thought to be "safe" based on toxicological experiments. Since the exposures appeared to be well within these limits, the investigators were left with the conclusion that the synergism of pollutant mixtures might have been responsible for the observed health effects. However, Firket (1936), with remarkable prescience, estimated that if an episode similar to that of the Meuse Valley were to occur in a city as large as London, as many as 3200 people might succumb. This constituted one of the first rudimentary air pollution risk analyses, although its message was not heeded at the time.

At a conference in 1979, Robert Waller said, "The evidence concerning effects of sulphur dioxide and associated urban air pollutants on health has been derived mainly from epidemiological work. There are, however, major problems in the interpretation of findings from the diverse range of studies that have been done, largely because there is no effect on health that can be uniquely related to pollution, and no single component of the pollution component can be indicted rather than another." More than a decade later, and after the expenditure of many millions of dollars on research, these statements still ring true.

RISK ANALYSIS VERSUS NO-EFFECT THRESHOLD CONCENTRATIONS

As will be shown by the studies reviewed in this book, the health effects of community air pollution are now seen largely as contributory rather than as primary causal factors for lung or heart disease, although there may be some evidence for chronic effects at the higher air pollution levels of the past. Air pollution is known to exacerbate these diseases and their symptoms, rather than to create them *de novo*. The implications of the primary ambient air quality standard provisions of the Clean Air Act are that achievement of these standards essentially eliminates the risks due to air pollution. Thus, the EPA sets its highest priorities on the achievement of ambient standards and a much lower priority on estimating the actual health risks involved under various conditions.

It is the purpose of the analyses in Sections II and III of this book to examine the validity of the exacerbation model when applied to heterogeneous populations. Such populations will inevitably include individuals with varying degrees of sensitivity to the various constituents of ambient air. The numbers of sensitive subjects are expected to decrease as the threshold concentration levels are reduced, and thus testing of randomly selected groups of volunteers is unlikely to locate and include the most sensitive individuals. Many of the epidemiological population studies reviewed below found linear, no-threshold, dose-response models, which is consistent with a heterogeneous community population and health effects that are limited to the "tail" of the distribution of individual responses.

Because it is physically impossible to eliminate all air pollution and hence all risks, a more appropriate regulatory model may now be one in which the average population risks are quantified, based on epidemiological studies of specified types of health outcomes (death, hospitalization, for example) associated with exposure to defined air pollution doses. Informed judgments could then be made of the acceptable levels of such risks, for example, by comparison to suitably rare natural events such as meteorite impacts, earthquakes, or lightning strikes. However, Portney (1990) cites a 1980 court decision in which the balancing of costs and (presumably health) benefits was found to be "inconsistent with the law," so that judgments about the degree of acceptable risk are not likely to be based solely on economics. In a democratic society, such judgments will involve the concepts of risk perception and risk communication. Careful consideration of these arguments, however simplistic, may call for a renewed emphasis on epidemiological research; epidemiological efforts devoted to air pollution in the United States were funded at only about $2 million in 1990, which is a tiny fraction of the sums involved in either health care or air pollution control costs.

CONTENT OF THE BOOK

The remainder of Part I presents additional background material: Chapter 2 offers a brief primer on air pollution in general and a description of the U.S. experience; this material is intended to serve as a resource in interpreting the studies that are reviewed in subsequent chapters. Chapter 3 presents some basic statistical information relevant to the two fundamental epidemiological designs used in most of the studies reviewed. These are: studies of the coincidence in timing of pollution and various health-related events, which is largely an acute consideration, and studies of the spatial variability in prevalence rates, the so-called cross-sectional design, which has usually been intended to deal with chronic illness. Chapter 4 presents some of the elementary physiological considerations for air pollution health effects and the principles that apply to acute and chronic responses.

After this introductory material, Section II examines mortality studies of both designs (Chapters 5 through 8), and Section III follows suit with respect to selected morbidity measures: hospital use and admissions in Chapters 9 and 10, lung function studies in Chapter 11. The last topic has been included to examine whether a physiological basis exists for the observational studies. Finally, Section IV is intended to place all of these findings in perspective; Chapter 12 examines whether lung function *per se* can serve as a predictor of premature mortality and thus close the gap between individual and group studies, and Chapter 13 summarizes and derives the final conclusions. I have chosen not to include studies of respiratory symptoms in part because this literature is voluminous and in part because of the often subjective and variable nature of the data.

REFERENCES

Bates, D.V., Macklem, P.T., and Christie, R.V. (1971), *Respiratory Function in Disease*, 2nd ed. Saunders, Philadelphia, Chapt. 17.

Buffler, P.A., Cooper, S.P., Stinnett, S., *et al.* (1988), Air Pollution and Lung Cancer Mortality in Harris County, Texas, 1979–1981, *Amer. J. Epidem.* 128:683–99.

Eisenbud, M. (1970), Environmental Protection in the City of New York, *Science* 170:706–12.

Farr, W. (1885), *Vital Statistics: A Memorial Volume of Selections from the Reports and Writings of William Farr*, N. Humphreys, ed., Sanitary Institute of Great Britain, London, p. 164.

Ferris, B.G., Jr., and Speizer, F. (1980), The Business Roundtable Air Quality Project, The Business Roundtable, Washington, DC.

Firket, J. (1936), Fog along the Meuse Valley, *Trans. Faraday Soc.* 32:1192–97.

Frank, R. (1975), Biologic Effects of Air Pollution, in *Energy and Human Welfare—A Critical Analysis. Vol I. The Social Costs of Power Production*, B. Commoner, H. Boksenbaum, and M. Corr, eds., Macmillan, New York, pp. 17–27.

Freudenheim, M. (1991), "Health Care a Growing Burden," *New York Times*, Jan. 29, 1991, p. D-1.

Ginzburg, E. (1990), *The Medical Triangle, Physicians, Politicians, and the Public*, Harvard University Press, Cambridge, MA, p. 257.

Kleinbaum, D.G., Kupper, L.L., and Morgenstern, H. (1982), *Epidemiologic Research, Principles and Quantitative Methods*, Wadsworth, Belmont, CA.

Lawther, P.J. (1966), Air Pollution, Bronchitis and Lung Cancer, *Postgrad. Med. J.* 42:703 (as cited by Frank, 1975).

Lipfert, F.W. (1978), Statistical Studies of Geographical Factors in Cancer Mortality: Association with Community Particulate Air Pollution, Paper 78–6.6, presented at the seventy-first Air Pollution Control Association (APCA) Annual Meeting, June 1978.

Lipfert, F.W. (1980), Differential Mortality and the Environment: The Challenge of Multicollinearity in Cross-Sectional Studies, *Energy Systems and Policy* 3:367–400.

Lodge, J.P., Jr. (1969), Air Pollutant Monitoring, *Env. Res.* 2:134–36.

Mancuso, T.F., and Mordell, J.S. (1969), Proposed Initial Studies of the Relationship of Community Air Pollution to Health, *Env. Res.* 2:108–33.

Ministry of Health (1954), *Mortality and Morbidity during the London Fog of December 1952*, Reports on Public Health and Medical Subjects No. 95., HMSO, London.

Portney, P.R. (1990), Economics and the Clean Air Act, *J. Econ. Perspectives* 4:173–79.

Robinson, W.S. (1950), Ecological Correlations and the Behavior of Individuals, *Am. Sociol. Rev.* 15:351–57.

Russell, W.T. (1924), The Influence of Fog on the Mortality from Respiratory Diseases, *Lancet* 2:335–39.

Russell, W.T. (1926), The Relative Influence of Fog and Low Temperature on the Mortality from Respiratory Disease, *Lancet* 2:1128–30.

Shannon, H.S., Hertzman, C., Julian, J.A., *et al.* (1988), Lung Cancer and Air Pollution in an Industrial City—A Geographical Analysis, *Can. J. Public Health* 79:255–59.

Shrenk, H.H., Heimann, H., Clayton, G.D., Gafafer, W.M., and Wexler, H. (1949), *Air Pollution in Donora, PA*, Public Health Bulletin No. 306, Public Health Service, Washington, DC.

U.S. Dept. of Health and Human Services (U.S. DHHS) (1984), *The Health Consequences of Smoking, Chronic Obstructive Lung Disease*, Office on Smoking and Health, Rockville, MD.

U.S. Dept. of Health, Education, and Welfare (U.S. DHEW) (1967), *Air Quality Criteria for Sulfur Oxides*, PHS Publ. 1619, Washington, DC.

Waller, R.E. (1979), The Effect of Sulphur Dioxide and Related Urban Air Pollutants on Health, presented at the International Symposium on Sulphur Emissions and The Environment, Society of Chemical Industry, London.

Ware, J.H., Thibodeau, L.A., Speizer, F.E., Colome, S., and Ferris, B.G., Jr. (1981), Assessment of the Health Effects of Ambient Sulfur Oxides and Particulate Matter: Evidence from Observational Studies, *Environ. Health Perspect.* 41:255–76.

2

A Primer on Air Pollution, Past and Present

This most excellent canopy, the air . . . why it appears no other thing to me than a foul and pestilent congregation of vapours.

William Shakespeare, *Hamlet*, Act II, Scene II

As mentioned in Chapter 1, one of the themes of this book is the need to understand the nature of the air pollution dose. This chapter is intended to present some basic information toward this end; those readers who are thoroughly familiar with the topic should feel free to proceed to Chapter 3, although the historical material may be of interest to even the experienced air quality analyst.

Observational studies deal with gradients in either space (cross-sectional studies) or time (time-series analyses). Since cross-sectional studies usually purport to analyze chronic effects, it is important to understand the long-term air pollution exposures of the areas studied. Similarly, time-series analyses deal with the temporal variability of such areas as cities; thus, it is necessary to understand the spatial variability under the conditions of interest. The relative roles of exposures to indoor air pollutants adds a layer of complexity to both types of epidemiological studies.

This book emphasizes the United States, Canada, and Great Britain (although there are studies from other locations as well). Hence, this chapter emphasizes U.S. air quality, with some additional information on London. Table 2-1 presents a chronology of significant events with respect to air pollution control and health effects concerns in these three countries.

EARLY ATTITUDES TOWARD AIR POLLUTION

The desire for clean air dates back virtually to antiquity. Air was one of the four basic elements of the ancient world, along with fire, water, and earth.

TABLE 2-1 Significant events in the development of air pollution control

Year	Event
1863	Alkali Act establishes industrial smoke inspectors in Britain.
1881	Local smoke ordinances passed in Cincinnati and Chicago.
1891	Society for the Prevention of Smoke established in Chicago.
1907	Association of municipal smoke inspectors formed in Milwaukee (forerunner of AWMA).
1912	U.S. Bureau of Mines publishes 3 bulletins on causes and prevention of smoke.
1914	Sootfall collection begun in 22 English cities.
1915	U.S. Bureau of Mines publishes report on the Selby smelter air pollution study.
1930	5-day episode strikes the Meuse Valley in Belgium. 60 die.
1935	Agreement to compensate U.S. farmers for crop loss from Canadian smelter emissions.
1937	15-month mobile surveys of SO_2 levels made in 5 U.S. cities; spot surveys in 20 more.
1943	First severe smog episode hits Los Angeles (July 26).
1947	California establishes first state regulations; air sampling begins in Los Angeles.
1948	4-day episode strikes Donora, PA. 20 die.
1949	First National Air Pollution Symposium (Pasadena, CA).
1950	Resolutions in U.S. Congress calling for research on health effects. No action taken.
1950	U.S. Technical Conference on Air Pollution held in Washington (requested by President Truman).
1952	Resolution on health research passes House, fails in Senate.
1952	The "Great Fog" episode strikes London for 4 days. 4000 die.
1953	Five-day episode in New York City; 200 die (not reported until 1962).
1953	National air monitoring begun for particulates using high-volume samplers.
1954	*Ad hoc* committee on air pollution formed in Department of HEW.
1955	President Eisenhower recommends a program of research on air pollution health effects.
1955	First national legislation passes (PL 84-159), emphasizing state and local control.
1956	Britain passes Clean Air Act regulating smoke, but not SO_2.
1957	National Advisory Committee on Air Pollution formed.
1958	First National Conference (Washington, DC).
1958	U.S. Public Health Service opposes Federal enforcement role.
1959	Study of automobile exhaust begun.
1962	Second National Conference (Washington, DC).
1963	President Kennedy backs federal enforcement role; supported by AMA. PHS changes position.
1963	First U.S. Clean Air Act passed, calling for federal-state cooperation in enforcement.
1967	President Johnson recommends federal emissions standards; Air Quality Act of 1967 passed.
1968	British Clean Air Act extended to industry.
1970	Last of the 50 States establishes regulations.
1970	U.S. EPA created as part of federal reorganization.
1970	Clean Air Act Amendments call for national ambient standards and federal enforcement.
1971	Canadian Clean Air Act enacted.
1972	Last AMA Air Pollution Medical Research Conference (series begun in 1958).
1977	U.S. Clean Air Act Amendments protect sensitive land areas.
1980	United States and Canada sign a Memorandum of Intent (MOI) to develop transboundary air pollution policy.
1982	United States rejects Canadian proposal for 50% emissions reduction; MOI negotiations cease.
1985	5-day episode strikes Central Europe; 6% excess deaths reported in Germany.
1988	Canadian Environmental Protection Act passed.
1990	Clean Air Act Amendments mandate SO_2 reductions to reduce acid rain, regulate toxic emissions, reduce volatile organic compounds emissions.

Sources: Ripley, 1969; Stern, 1982; Brydges, 1987; Milburn-Hopwood, 1989.

Before the spread of infectious disease was known to be caused by germs, "bad air" in crowded cities was thought to be responsible for a whole host of calamities. Even as late as the mid-nineteenth century, the mechanism by which polluted water spread disease was thought by some to be through the air. Evelyn's somewhat polemical 1661 tract, *Fumifugium*, singled out industrial coal burners ("brewers, diers, lime-burners, salt, and sope-boylers") as the primary culprits in fouling the air of London and cited adverse effects due to the coal smoke on health, vegetation, and building materials. He made clear that both black smoke and sulfur were offensive by favorable reference to the atmosphere of Paris, where wood was then the primary fuel. Brimblecomb (1987) provides additional evidence on this aspect of life in medieval cities and some fascinating accounts of early London. Coal smoke may have been singled out as the most noxious of the common air contaminants because of its distinctive odor.

Another line of reasoning may have operated with respect to the effects of sulfureous air pollutants. Since the corrosive properties of sulfuric acid were well-known in chemical laboratories, it was relatively easy to ascribe the shorter lifetimes of metals and stonework exposed in urban areas, relative to those in rural areas, to polluted urban air. The question was then posed, "if these visible effects are due to polluted air, surely our lungs must be adversely affected as well."

Quantitative linkage between public health and air pollution was realized only in the latter half of this century, after the advent of routine air quality measurements. In the preface to the 1772 edition of Evelyn's *Fumifugium*, reference is made to 10,000 excess deaths annually in London, apparently based on urban-rural gradients in mortality rates. In the absence of knowledge of the spread of contagious diseases, air pollution was blamed for all of this excess mortality (Brimblecomb, 1977). However, by the nineteenth century, it was speculated that air pollution was only one of the factors that led to increased annual mortality rates in locations of higher population density. During this era, public health concerns focused on infectious diseases and water sanitation problems and Farr (1885) stated, "Now we have never observed any connexion between the increase of the mortality and the London fogs. The diseases, again, caused by smoke must be of a mechanical nature, and affect the lungs and air-passages; it may increase the pulmonary diseases, but will assuredly not produce scarlatina, measles, typhus, and other diseases which prevail in towns." Bowler and Brimblecomb (1991) cite problems of vegetation and materials damage in conjunction with controversy over early power station construction in England, but health is not mentioned. The great air pollution episode of 1952 caught the public's attention, however, and the subsequent air pollution cleanup campaign in London (beginning with the Clean Air Act of 1956) emphasized domestic coal burning sources, in keeping with the notion of the importance of population density (Brimblecomb, 1982). See Appendix 2A for an analysis of associations between air pollution, population density, and mortality.

In contrast, in the United States, prior to the mid-twentieth century, air pollution was linked to soiling, corrosion, and damage to vegetation, and smoke control efforts in industrial cities such as Pittsburgh and St. Louis were primarily focused on such economic factors under "nuisance law." Factory smoke was considered a sign of a healthy economy, even if there might be occasional "nuisance" incidents (Tarr, 1984). In one of the earliest environmental impact investigations, the effects of operations of a metallurgical smelter near San Francisco were studied (Holmes *et al.*, 1915). The report concluded that the effects of visible emissions constituted a nuisance with respect to damage done to "certain horses" but that sulfur dioxide (SO_2) impacts above the odor threshold—about 0.5 parts per million (ppm)—did not constitute a nuisance to humans. Swain's (1950) account of air pollution from smelters emphasizes vegetation damage; the copper smelter at Ducktown, Tennessee, near the Georgia border, left a bare spot large enough to show on a satellite photo mosaic map of the United States.

The first U.S. smoke control ordinances, in the early twentieth century, were aimed at industrial sources and vehicles but had little effect. In 1940, St. Louis included homeowners in their smoke enforcement efforts and began to show progress (Tarr, 1984). The 1948 episode at Donora, PA, was the first tangible sign in the United States that air pollution might constitute a serious health hazard, but according to Stern (1982), the public soon forgot about Donora, and attention swung to the smog problems of Southern California where the predominant syndrome was eye irritation (Magill, 1949). There was virtually no mention of permanent health damage problems at the 1949 American Chemical Society Symposium on "Atmospheric Contamination and Purification" (Swain, 1950).

The history of air pollution and its abatement in Pittsburgh has been reviewed by Davidson (1979), beginning with complaints about coal smoke in 1804. However, the common attitude in Pittsburgh was that smoke was a necessary evil and abatement did not get under way until the late 1940s, when over 300 hours of "heavy smoke" were recorded per year. By 1958, the annual number of hours was down to zero, as a result of switches to natural gas (especially for home heating) and to "smokeless" coal (Davidson, 1979). Unfortunately, few if any studies have been done on the long-term public health benefits resulting from this or other air pollution cleanups.

BACKGROUND INFORMATION ON COMMUNITY AIR POLLUTION

The earth's atmosphere is roughly 79% nitrogen, 21% oxygen, varying amounts of water vapor, and 0.036% carbon dioxide (CO_2), plus a host of minor trace substances. Those substances we deem to have adverse effects we call "pollution," after the Latin root for "dirt." As analytical chemistry and sampling improve, new compounds and radicals continue to be identified in the atmosphere. For regulatory purposes, air pollutants can be grouped into two classifications: *Criteria* pollutants are those substances deemed to present a general risk to public health and for which National Ambient Air Quality Standards (NAAQSs) have been issued. A *criteria document* is intended to lay out the scientific basis for the derivation of standards for a specific air pollutant, including its properties, typical ambient concentrations, and its adverse effects on human health and welfare (U.S. DHEW 1969b, 1969c, 1970a, 1970b, 1971; U.S. EPA, 1978, 1982a, 1982b, 1986, 1990). *Hazardous* air pollutants (sometimes called "toxics") are those substances identified with cancer, birth defects, and neurotoxicity and for which ambient standards are neither appropriate nor practical. In general, criteria pollutants are virtually ubiquitous in urban atmospheres (hence the name, "community air pollution"), whereas most hazardous species tend to be identified with industrial atmospheres and the workplace. We are concerned only with community air (mainly criteria pollutants), which includes particulate matter (smoke, dust, total suspended particulates [TSP], inhalable particulates [IP or PM_{10}, sometimes referred to as *thoracic* particles], soiling or coefficient of haze [COH]), SO_2, nitrogen dioxide (NO_2), ozone (O_3), carbon monoxide (CO), and lead (Pb). Volatile organic compounds (nonmethane hydrocarbons) are an important precursor of ozone and thus are regulated, even though no direct health effects are involved. Particulate matter includes many substances, including soil dusts, carbon, trace metals, organic compounds, and water-soluble compounds including hydrogen (H^+), sulfate (SO_4^{-2}), nitrate (NO_3^-), and ammonium (NH_4^+) ions. The hydrogen ion content is also referred to as the *acidity* of the aerosol. In the 1950s, there was concern about the effects of fluoride on human and animal health (U.S. DHEW, 1958); these concerns seem to have since abated.

One of the early methods of classifying the pollution characteristics of urban atmospheres was concerned with the level of oxidants present. Los Angeles was said to have an "oxidizing" atmosphere, while the air of London and New York was said to be "reducing" (Lodge, 1956). This was also implicitly a seasonal characterization, since oxidants peak in summer and the "reducing" species were associated with space heating emissions. However, this characterization may have been mostly due to lack of good information; although oxidant levels in Manhattan in the mid-1950s were usually well below those of Southern California, average levels in the outlying parts of the New York metropolitan area were similar to California levels (Interstate Sanitation Commission, 1958). Current air quality in New York, London, and Los Angeles shows the presence of both "oxidizing" and "reducing" species, albeit at lower concentrations than were found during the period when this terminology was coined.

The main effects of these pollutants on human health are as follows (for details, see the respective Criteria Documents [U.S. DHEW, 1969b, c, 1970a, b, 1971; U.S. EPA, 1978, 1982a, 1982b, 1986, 1990] and see Chapter 4 for elementary information on lung physiology).

Sulfur Dioxide (SO_2). Sulfur dioxide is a respiratory irritant that can cause impaired breathing in experimental animals and man. Because it is a soluble gas, SO_2 is removed in the mouth and pharynx and primarily affects the upper respiratory tract. However, SO_2 can also attach to particle surfaces and may form acidic coatings. In such cases, the loci of response will follow those of the inhaled particles.

Carbon Monoxide (CO). Carbon monoxide has a chemical affinity for hemoglobin and displaces oxygen (O_2) in the blood; it thus reduces O_2 delivery to compromised tissue and can augment angina (chest pain). High concentrations (>1000 ppm) and long exposures (>8 hours) can be fatal; chronic exposure to low concentrations may accelerate atherosclerosis or precipitate coronary vessel spasm. Carbon monoxide poisoning can also cause gastrointestinal symptoms (Mitchell *et al.*, 1974).

Ozone (O_3). Ozone is irritating to the upper respiratory tract and can cause impaired breathing and reduced athletic performance. Since it is not soluble, ozone can penetrate to the deeper reaches of the lung; high concentrations can cause edema. Chronic effects of ozone are less well defined.

Nitrogen Dioxide (NO_2). Nitrogen dioxide has been shown to cause pulmonary disease in experimental animals. It is also thought to increase susceptibility to respiratory infections and to lower respiratory tract illness (Hasselblad *et al.*, 1992).

Particulate Matter (TSP, PM_{10}). The sizes of inhaled particles determines where they may deposit in the respiratory system; since smaller particles can penetrate deeper into the lung, the basis of the primary standard was changed from total suspended particles (sizes up to 50 μm) to inhalable particles (PM_{10}), 50% of which are 10 μm or less. Airborne particles vary widely in composition and toxicity and can include organic matter, carbon, mineral dusts, metal oxides, and soluble compounds such as sulfate and nitrate salts. Particles may also act as carriers for adsorbed gases such as SO_2. Inhaled particles can affect respiratory mechanics and cause irritation and edema. Although the normal respiratory defense mechanisms act to clear foreign matter from the lungs, the half-time to clear acute exposures is a matter of several days (American Petroleum Institute, 1969). Friedlander (1977) presents basic data on the properties of aerosols and the distinctions between dusts (particles formed by disintegration of solids), mists (liquid particles), and smoke (solid particles formed by condensation of gases). Particles formed in the atmosphere are referred to as "secondary."

Lead (Pb). Lead affects the central nervous system, especially in young children. The primary source of atmospheric lead in urban areas was formerly leaded gasoline, which has effectively been phased out of the marketplace. The current concern is with indoor particles containing lead from flaking paint. Since airborne lead is no longer a problem, it will not be discussed further in this chapter.

Acidity. Acidity is one of the properties of airborne substances that raises environmental concerns, and "acid aerosol" is the most recent candidate for criteria pollutant status in the United States. Breathing acidic substances can lead to respiratory irritation if the body's natural defenses are overwhelmed; sulfuric acid was one of the agents suspected of contributing to mortality during the severe air pollution episodes of the 1930s to 1960s, which are discussed in Chapter 5. Following a request by the Clean Air Scientific Advisory Committee (CASAC) of the U.S. Environmental Protection Agency (EPA) relating to concerns about possible adverse health effects, that agency is now considering adding acid aerosols to the present list of criteria air pollutants (CASAC, 1988).

An aerosol is defined as a suspension of solid or liquid particles in a gaseous medium. The acidity of an aerosol normally refers to the particles, not to the suspension medium, although it is not uncommon to find acidic gases and acidic particles together. Commonly found acidic gases include SO_2, nitric (HNO_3) and hydrochloric (HCl) acids, and a variety of organic acids such as formic ($HCOOH$) and acetic (CH_3COOH) acids. Acidic particles include sulfuric acid, ammonium bisulfate (NH_4HSO_4), and certain organic compounds. Most of the common organic acids are in the gaseous phase. In the

TABLE 2-2 U.S. National (Primary) Ambient Air Quality Standards

Pollutant	Units	Averaging Time	Concentration	Statistic
Sulfur dioxide	μg/m³	Annual	80	Annual mean
		24 hours	365	Maximum*
Carbon monoxide	mg/m³	8 hours	10	Maximum
		1 hour	40	Maximum
Ozone	μg/m³	1 hour	235	Maximum†
Nitrogen dioxide	μg/m³	Annual	100	Annual mean
Inhalable particulate	μg/m³	Annual	50	Annual mean
(PM_{10})		24 hours	150	Maximum
Lead	μg/m³	3 months	1.5	Quarterly average

* Maximum values may be exceeded once per year.
† Not to be exceeded more than three times in 3 years.

TABLE 2-3 World Health Organization Ambient Air Quality Guidelines (1977–1979)

Pollutant	Units	Averaging Time	Concentration	Statistic
SO_2 plus	μg/m³	Annual	40–60	Annual mean
British smoke	μg/m³	Annual	40–60	Annual mean
SO_2 plus	μg/m³	24 hours	100–150	Maximum*
British smoke	μg/m³	24 hours	100–150	Maximum
Carbon monoxide	Ambient limits to be derived from 2.5–3% maximum carboxyhemoglobin in blood (air concentrations will vary)			
Ozone	μg/m³	1 hour	120	Maximum†
Nitrogen dioxide	μg/m³	1 hour	190–320	Maximum†
Total suspended	μg/m³	24 hours	150–230	Maximum
particulate	μg/m³	Annual	60–90	Annual mean

* Maximum value should not be exceeded more than 7 days per year.
† Maximum value should not be exceeded more than once per month.

Eastern United States, most of the acidic particles are sulfuric acid and its salts of ammonia, collectively referred to as "sulfates" (Lipfert, 1988).

The current health-related (primary) U.S. NAAQSs are listed in Table 2-2. Another source of information on the general health effects of air pollutants, including many trace substances, is the World Health Organization (WHO) series on Environmental Health Criteria, which also gives guidelines on human exposure limits (Table 2-3). Appendix 2B to this chapter discusses some of the other air pollutants that have been studied as part of community health investigations; Appendix 2C gives some of the factors commonly used to convert from volumetric concentration units (ppm) to mass concentration units (μg/m³).

The WHO guidelines are seen to be more conservative than the U.S. standards in some cases, although where a greater number of 24-hour exceedances is allowed, it may be difficult to make this determination. The WHO tried to specify safety factors from the lowest known levels at which adverse effects have been observed. Other noteworthy differences between the two tables are the specification of a short-term standard for NO_2, rather than annual, and the use of combined smoke and SO_2 standards by the WHO. Since the U.S. standards are largely based on the studies used by the WHO, the question arises as to why the United States does not recommend joint consideration of both pollutants. The WHO guideline for TSP is based on an approximate extrapolation from British smoke data (see Appendix 2D).

MEASUREMENT OF AIR POLLUTION IN THE UNITED STATES

Although the air pollutants of interest in urban areas are well defined chemically and physically in their own right, for the purpose of ambient monitoring and demonstrating compliance with federal ambient standards, they are defined by official EPA *reference methods*, which are published in the Code of Federal Regulations. According to this regulatory protocol, each criteria pollutant is defined as the material that is collected when the official (i.e., "reference") sampling and analysis protocol is followed. As new and better methods are developed, the reference methods may change.

Particulate Matter. Early measurements of air pollution in cities focused on two pollutants: sulfur dioxide (SO_2) and particulate matter. Characterization of the latter has progressed from simple dustfall measurements (gravimetric catch in a glass jar), to smokeshade (degree of staining of a filter paper), to total suspended particles (TSP) (gravimetric catch of particles $<50\,\mu m$ diameter on a glass-fiber filter), to size-classified particle mass. The rationale behind these changes was concern that smaller particles are potentially more injurious to health, since they can penetrate deeper into the lung. In addition, the technology of particle sampling has been influenced by the need for subsequent chemical analysis, for example, to determine the content of sulfates and other ions or trace elements. Several methods are now available for continuous determination of atmospheric particle mass loading, although these innovations do not provide data on particle composition.

The method of choice for determining the contributions of water-soluble compounds (SO_4^{-2}, NO_3^-, NH_4^+, etc.) to total particulate mass is now ion chromatography. Care must be taken to avoid interferences from the filters used to collect particles. When aerosol acidity is required, care must also be taken to preclude neutralization by ambient ammonia or by co-collected alkaline particles (Brauer *et al.*, 1989).

Sulfur Dioxide. Sulfur dioxide measurements have progressed from lead peroxide sulfation plates or candles (sulfur deposition gauges), to simple wet-chemical bubblers and their automated counterparts, to automated gas-phase detectors that telemeter results to a central data acquisition system. Data from SO_2 bubblers can be suspect, particularly in hot weather when evaporation of solution can result in overstatement of concentrations. For both SO_2 and particle measurements, averaging times for routine monitoring have progressed from monthly to hourly, although 24-hour samples normally are taken for particle mass. Some of the early publications on air pollution (Shaw and Owens, 1925, for example) made the assumption that all of the fuel sulfur would eventually become sulfuric acid in the atmosphere. We now know that, while almost all of the sulfur in fuel is emitted as SO_2, only a portion of the airborne SO_2 is oxidized to sulfuric acid (H_2SO_4). Additional oxidation takes place in the aqueous phase, and the balance of the SO_2 is deposited on surfaces.

Nitrogen Dioxide. The early (pre-1970) bubbler measurements of NO_2 are particularly suspect in the United States. In 1973, EPA withdrew the then-reference method, known as the Jacobs-Hochheiser method, because of excessively variable collection efficiencies. Thus, NO_2 data from health studies employing this method should be disregarded. The current reference method for NO_2 employs the chemiluminescence method of measurement.

Oxidants. Attention first focused on photochemical smog in Southern California after World War II. "Total oxidants" were measured by using a buffered potassium iodide solution but were reported as ozone. Electrochemical methods have since been developed that are specific to ozone and are used in most of the United States. Peroxyacetylnitrate (PAN) is one of the oxidant compounds sometimes associated with eye irritation (although conclusive data are lacking); typical PAN concentrations are an order of magnitude lower than ozone. The two compounds tend to be proportional during afternoon hours (Roberts *et al.*, 1990).

Carbon Monoxide. Carbon monoxide is usually determined by nondispersive infrared analysis (NDIR), an automated method designated as the EPA reference method. It is based on the absorption of infrared radiation by the CO molecule and is one of the most reliable methods available for any pollutant. It is perhaps ironic that the pollutant that can be measured most accurately has one of the least adequate data bases for population exposure assessment, because in part of regulatory agencies' primary focus on determining compliance with ambient standards.

Sampling Locations. The early air sampling networks were designed to gather information for research (as opposed to regulatory) purposes. Unfortunately, the spatial coverage of urban air monitoring has decreased in the 1980s. In many cities, the extensive networks formerly present have been reduced, so that population exposures are now less well defined. In addition, much of the air monitoring in the United States is devoted to checking compliance with standards around large point sources, rather than determining the impact on populations. Since urban concentrations were higher in the early years (before 1970), epidemiological studies of this period are often of special interest. Stocks (1960) cites work in England that determined that the coefficient of variation for annual averages of the mean of a network of eight smoke samplers was about 10%, increasing to about 28% for a single site. This study was conducted in an area of 62 square miles, but with both topographic relief and strong gradients in population density, so that Stocks felt that these estimates of variability might be on the high side. However, many of the U.S. epidemiological studies reviewed in Sections II and III attempted to characterize entire Standard Metropolitan Statistical Areas (SMSAs), sometimes comprising thousands of square miles, with a single air monitoring station.

The height of sampling can also make a difference. Air quality at the top of the World Trade Center (411 m) in New York City is quite different from street level, to cite an extreme example. Barratt (1989) found that the vertical gradients (6–35 m) in Birmingham, England, were changing with time; smoke was getting more uniform, but for SO_2, the effect of sampling height on concentration was increasing.

A more recent realization of importance to health studies is the role of indoor air quality.* Since populations in temperate climates spend most of their time indoors, their total exposures to air pollution can be dominated by indoor air quality. While some pollutants have important indoor sources (such as NO_2, formaldehyde, and small particles), indoor concentrations of others may be greatly attenuated from outdoor levels (notably SO_2 and aerosol acidity). The relationships between indoor and outdoor exposures are discussed in more detail later in the chapter.

METHODS OF MEASUREMENT USED IN EARLY STUDIES IN GREAT BRITAIN

Since British epidemiological studies are of considerable interest in subsequent chapters, some attention should be given to their air quality data. According to Commins and Waller (1967), the following measurement methods were in early use in London, at least up until 1964.

British smoke (BS). The British smoke sampler operates at flow rates much lower than the U.S. high-volume sampler, and thus it is thought that the particles sampled are mostly in the respirable range (Holland *et al.*, 1979). Lawther *et al.* (1968) report mass median diameters for the particles in London air to lie between 0.7 and 0.9 μm but do not state the method of sampling or size determination. After 1954, BS data were based on reflectometer measurements of the degree of dark staining of Whatman No. 1 filter paper, calibrated to read in gravimetric units ($\mu g/m^3$). Prior to that, visual comparisons of the shade of filter staining had been used. It has been noted (IERE, 1981) that the actual smoke readings during the 1952 London episode are uncertain and may have been biased low, because of saturation of the filters (difficulty in distinguishing among shades of black).

Several calibrations of BS have been performed against U.S. high-volume samplers (TSP); the results may have changed over time because of changes in the mix of particle sizes: 1955–62 (based on regression of annual means): TSP = 75 + 1.21 BS (72 < BS < 180):

$$\text{Summer 1970: TSP} = 6 + 3.1\ \text{BS}$$
$$\text{Winter 1970: TSP} = 35 + 1.4\ \text{BS}$$

$$\text{Summer 1975: TSP} = 22 + 1.6\ \text{BS}$$
$$\text{Winter 1975–76: TSP} = 37 + 1.2\ \text{BS}$$

Holland *et al.* (1979) report an unpublished correlation by Waller,

$$\text{TSP} = 119 + 0.824\ \text{BS}$$

* Actually, concern about indoor air quality dates back to Victorian times (Williams-Freeman, 1892), when the lack of proper ventilation was a sanitary concern with respect to the spread of infectious diseases.

and calibrations of coefficient of haze (COH) versus TSP ranging from 69 to 200 $\mu g/m^3$ per COH unit.

Lee *et al.* (1972) compared several different types of particulate samplers at 11 British sites in 1970 and found that smokeshade tended to read higher than TSP at urban sites during the heating season, but lower at other times. When inverted to provide estimates of TSP based on BS measurements, this difference was reflected primarily as a higher TSP intercept in summer. Their empirical equations for London were:

$$\text{Heating season: TSP} = 57 + 0.48\ \text{BS}$$
$$\text{Nonheating season: TSP} = 72 + 0.56\ \text{BS}$$

They also determined that the differences related to particle color, rather than to size.

Coefficient of haze measurements were found to correlate better with smokeshade than with TSP; one COH unit was found to be equivalent to $125 \pm 110\ \mu g/m^3$ TSP and $100 \pm 85\ \mu g/m^3$ of smoke. Drawing on data collected by Ingram in New York City, Lee *et al.* (1972) compared equivalent smokeshade there and at the 11 British sites, as shown in Figure 2-1. The approximation used elsewhere in this book (in Sections II and III), that one COH unit represents about 100 $\mu g/m^3$ on a mass basis, is thus seen to be reasonable, but the variability in these relationships suggests that they be regarded as rules of thumb. The differences between TSP and BS can become large in atmospheres containing only small amounts of carbonaceous material (rural areas, for example).

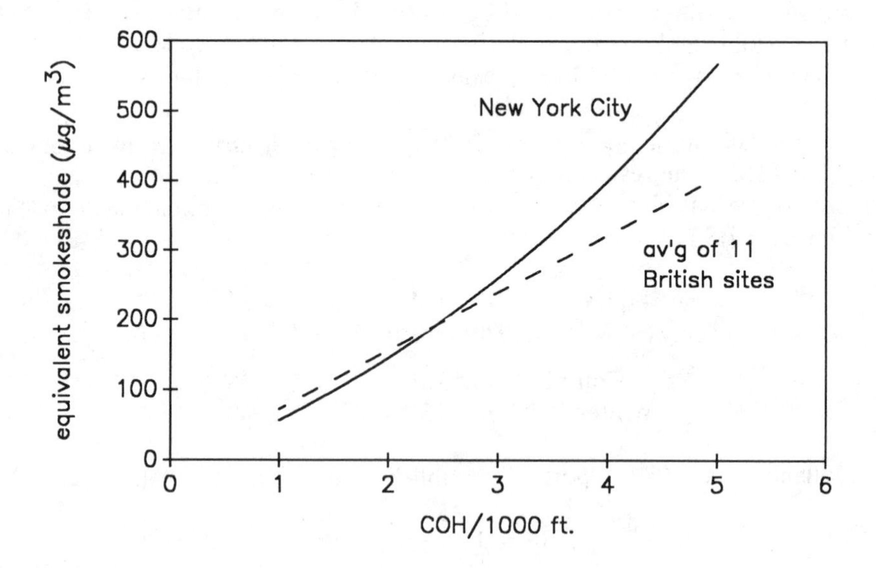

FIGURE 2-1. Relationship between British smoke and coefficient of haze in New York (averaged over 11 British sites). *Source: Lee* et al. *(1972).*

The American Iron and Steel Institute (Pashel and Egner, 1981) evaluated the two methods, using side-by-side sampling at 16 industrial locations in the United States, some of which were sites where fugitive dust might have been expected to predominate. Based on sample collection every sixth day for 1 year, the average BS value was 13 $\mu g/m^3$, the average TSP was 118 $\mu g/m^3$. This example illustrates how different the two measures can be, especially in industrial locations. However, the site-site correlation was 0.69 ($p < 0.01$), so that either method could still be used to separate "clean" from "dirty" locations.

Sulfur Dioxide. With the apparatus used in Britain in the 1960s and before, SO_2 is determined by the hydrogen peroxide method from the same air sample that passed through the filter. Thus there is a possibility that some SO_2 was adsorbed onto the particles collected by the filter. Such a bias would be more important for low SO_2 levels than high, because of saturation of the filter adsorption sites. Because of the titration method used with this method, it is sensitive to all acid gases, not just SO_2, and thus readings may include HCl and HNO_3. This method was also used in New York City prior to the mid-1970s.

Aerosol Acidity. Aerosol acidity (hydrogen ion, H^+) was determined from the British smoke filter catch by titration (Commins, 1963). Some of the acid is neutralized by contact with insoluble bases in the particulate catch. Commins reports errors of around 10% at times of high pollution and 30% at times of low pollution. Lawther *et al.* (1968) report that the particle collection efficiency of Whatman No. 1 filters drops for particles below about 1 μm, which could be a further source of downward bias for the London H_2SO_4 measurements. The reported annual mean for 1964–65 was 5.4 $\mu g/m^3$, with a mean in winter of 7.2 and summer of 3.5 $\mu g/m^3$. Ito and Thurston (1987) show mean values in London declining from 6.8 $\mu g/m^3$ in 1964 to 3.2 $\mu g/m^3$ in 1972 and report that the measurements were taken at the Medical College in central London. They also point out that this parameter represents the net free acidity of the particulate catch and thus may include both H_2SO_4 and NH_4HSO_4, as well as other acidic particles. Some authors have used the label "sulfuric acid" for this data set, which describes the units of reporting, not the actual composition.

Particulate sulfate. Particulate sulfate (sulfate ion, SO_4^{-2}) was determined by precipitating barium sulfate from aqueous extracts of the particulate catch of glass-fiber filters (high volume samplers). This method has known artifacts due to conversion of SO_2 on the filters. Due to the high levels of SO_2 in London, this error could be appreciable. Commins (1963) points out that there is a poor relationship between total water-soluble sulfate concentrations and acidity in London, the latter accounting for from 20 to 80% of the measured sulfate.

Carbon Monoxide. Carbon monoxide is determined by NDIR in Britain, as in the United States.

AIR POLLUTION DATA AND TRENDS FOR THE UNITED STATES

Emissions

Trends in air pollution in the United States follow mainly from the history and patterns of fuel use. Table 2-4 gives typical values (from various sources)

TABLE 2-4 Typical (Uncontrolled) Air Pollutant Emissions from Fuel Combustion

		Emission Rate (lb/million Btu)*			
Fuel	Application	SO_2	NO_x	CO	Particulates
Wood	Wood stove	0.04	0.4	20.0	1.2–10.0
Natural gas	Home furnace	~0.0	0.1	0.02	0.01
Distillate oil (2% S)	Home furnace	0.20	0.12	0.04	0.07
Residual oil (2% S)	Industrial boiler	2.2	0.40	0.03	0.16
Coal (bituminous, 2% S)	Home stove	2.7	0.12	3.6	0.60–6.0
	Power plant†	3.0	0.70	0.04	0.10

* Engineering units are customarily used for these parameters; to convert from lb/10^6 British thermal unit (Btu) to g/10^6 cal, multiply by 1.8.

† Fitted with electrostatic precipitator with 98% efficiency.

of pollutant emissions per unit of fuel used in various external combustion devices, for comparison; national estimates could be obtained by summing over the total fuel use of each example. The table shows that coal and residual (heavy) fuel oil are the major sources of sulfur oxides (transportation sources and industrial processes, which are not shown, make only minor contributions). Note that all fuel combustion produces oxides of nitrogen (NO_x); however, transportation sources contribute about half of the national NO_x total. Among these combustion devices, the wood stove is the most prodigious emitter of CO, since it operates most efficiently with restricted combustion airflow. However, on a national total basis, transportation sources are the major emitter of CO, by far. Particulates are emitted from all fuel combustion processes except those using gaseous fuels. In addition to the processes mentioned above, sulfur oxides and metals are emitted from smelters, and there are various natural sources of certain atmospheric compounds such as ammonia and chlorides.

In Colonial times, wood was the primary fuel, even to the extent of being pyrolized to provide charcoal for the early iron furnaces. An important exception was the early use of coal in Pennsylvania. As transportation became mechanized, coal was used for railroads and steamships, eventually giving way to diesel oil. This transformation was completed in the 1950s; the number of coal-burning locomotives dropped from over 30,000 to fewer than 400 by 1960 (Tarr, 1984). The choices for home heating fuels involve trade-offs between economy and convenience; natural gas has become the heating fuel of choice (59% of homes in 1980), initially because it was so much more convenient than coal. In 1980, fewer than 5% of U.S. homes used solid heating fuels or were unheated.

The period following World War II saw greatly changing patterns in air pollution, especially for home heating sources (Table 2-5), which can have important effects on urban air quality. Most of these changes were the results of the availability of utility gas from pipelines constructed in the late 1940s and 1950s and the advent of cheap electricity from hydropower in the 1960s. Changes in heating fuels since then have been modest, involving a minor resurgence in wood fuel in the mid-1970s and some conversions from oil to gas in the East.

TABLE 2-5 Trends in Predominant Home Heating Fuels

Region	1950	1960	1970
New England	Oil, wood	Oil	Oil
Mid-Atlantic	Coal, oil	Oil, gas, coal	Oil, gas
East N. Central	Coal	Gas, oil	Gas, oil
West N. Central	Oil, coal	Oil, gas	Gas, oil
South Atlantic	Wood	Oil, wood	Oil, gas
East S. Central	Wood, coal	Wood, gas, coal	Gas, elect.
West S. Central	Gas, wood	Gas	Gas
Mountain	Coal, gas, wood	Gas, oil	Gas, oil
Pacific	Gas, oil, wood	Gas, oil, wood	Gas, electricity, oil

Source: Bureau of Census (Map GE-70, No. 3).

Trends in air pollution emissions since the turn of the century are charted in Figure 2-2a,b and since 1970 in Figures 2-2b–e, based on data from EPA annual reports and the analysis of Gschwandtner *et al.* (1985). The long-term trends for SO_2 (Figure 2-2a) show a steep rise from 1900 to about 1925, a period of industrial and population expansion, followed by fluctuations that reflect economic cycles. The fraction of SO_2 attributable to electric utilities began to rise sharply following World War II and has since leveled off. Although residential SO_2 emissions may have been more important because of their proximity to populations and the reduced dispersion resulting from ground-level release, they have always been a small fraction of the U.S. total. The maximum estimated annual contribution from residential and commercial SO_2 sources was only about 5 million (metric) tons in 1945. The bulk of SO_2 emissions before about 1960 thus came from industrial sources, including smelters. Emissions of NO_x are divided between transportation sources, utilities, and others (mainly industrial). The decline in utility SO_2 while NO_x continues to increase (Figure 2-2b) reflects the imposition of SO_2 emission controls, including the use of fuels with lower sulfur content. Transportation NO_x reached its peak around 1980 and reflects a trade-off between better controls on new vehicles and the steady increase in the number of miles driven. Particulate emissions (Figure 2-2d) have declined drastically since 1970 and earlier (the 1950 estimate was about 25 million tons); these figures do not include fugitive emission sources. The transportation components of CO and volatile organic compounds (VOCs) are compared in Figure 2-2e; most of the reduction in both pollutants came from transportation sources.

Transportation fuel use in the United States is now dominated by gasoline (61% of energy input in 1979 [U.S. DOE, 1991]), followed by diesel and jet fuels. The number of vehicles on the road continues to grow every year, so that in some locales, the air quality improvements gained by tighter emission controls on new cars are overcome by the growth in usage. Transportation fuel use has been growing at about the rate of 3% per year since about 1950, with the fastest growth during the late 1960s and early 1970s. This growth was temporarily interrupted by the oil price shock of the mid-1970s but effectively doubled from 1960 to 1979, nevertheless. Growth has been slower since 1980, about 1.3% per year.

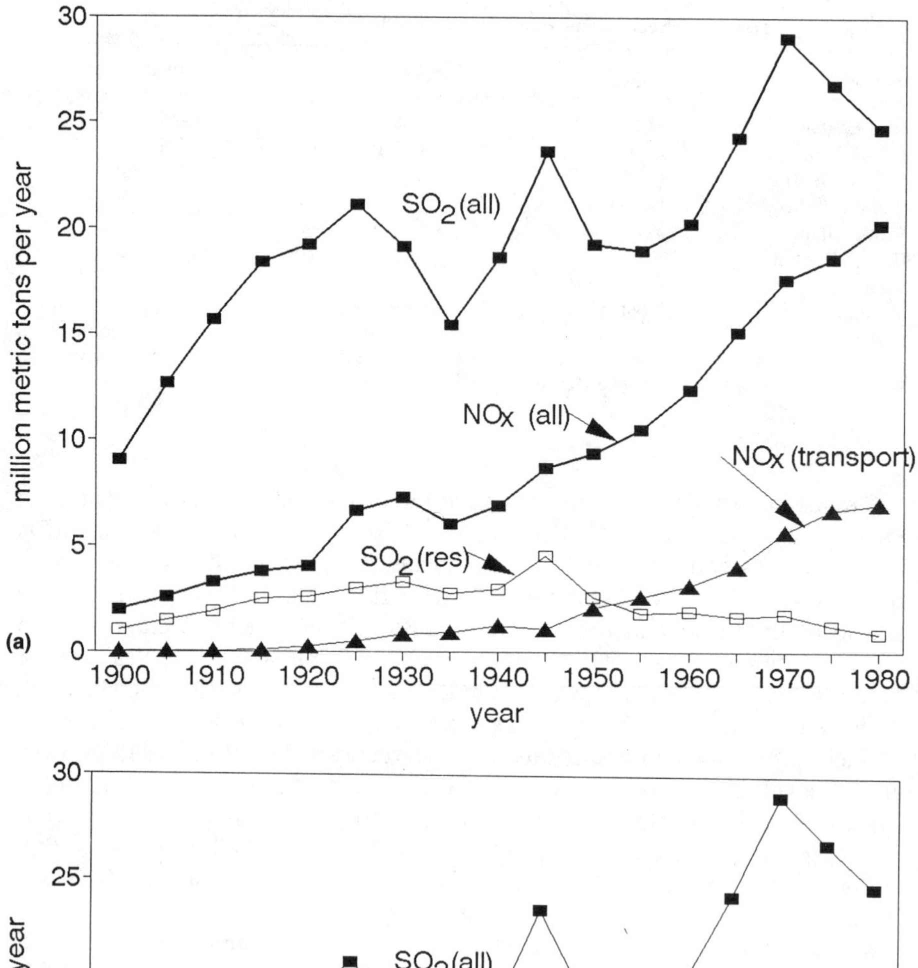

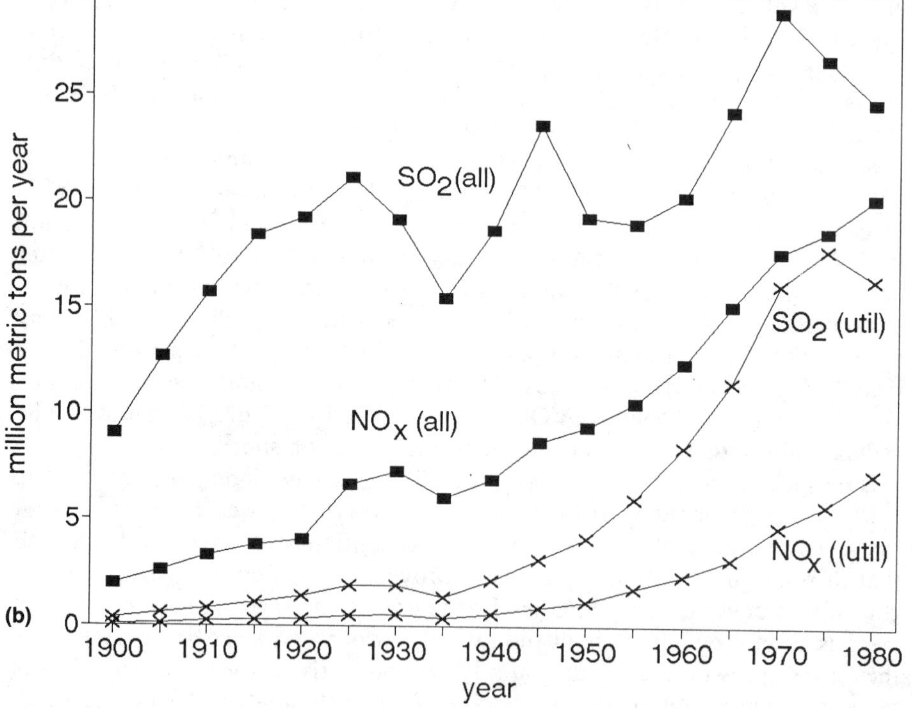

FIGURE 2-2. Trends in U.S. national emissions of air pollutants. (a) SO_2 and NO_x, 1900–1980, all sources, residential SO_2 and transportation NO_x. (b) SO_2 and NO_x, 1900–1980, all sources and electric utility sources. (c) SO_2 and NO_x, 1970–1990, all sources, transportation and electric utility sources. (d) SO_2, NO_x, and particulates, 1970–1990. (e) CO and VOCs, 1970–1990. *Data from EPA annual reports; Gschwandtner* et al. *(1985).*

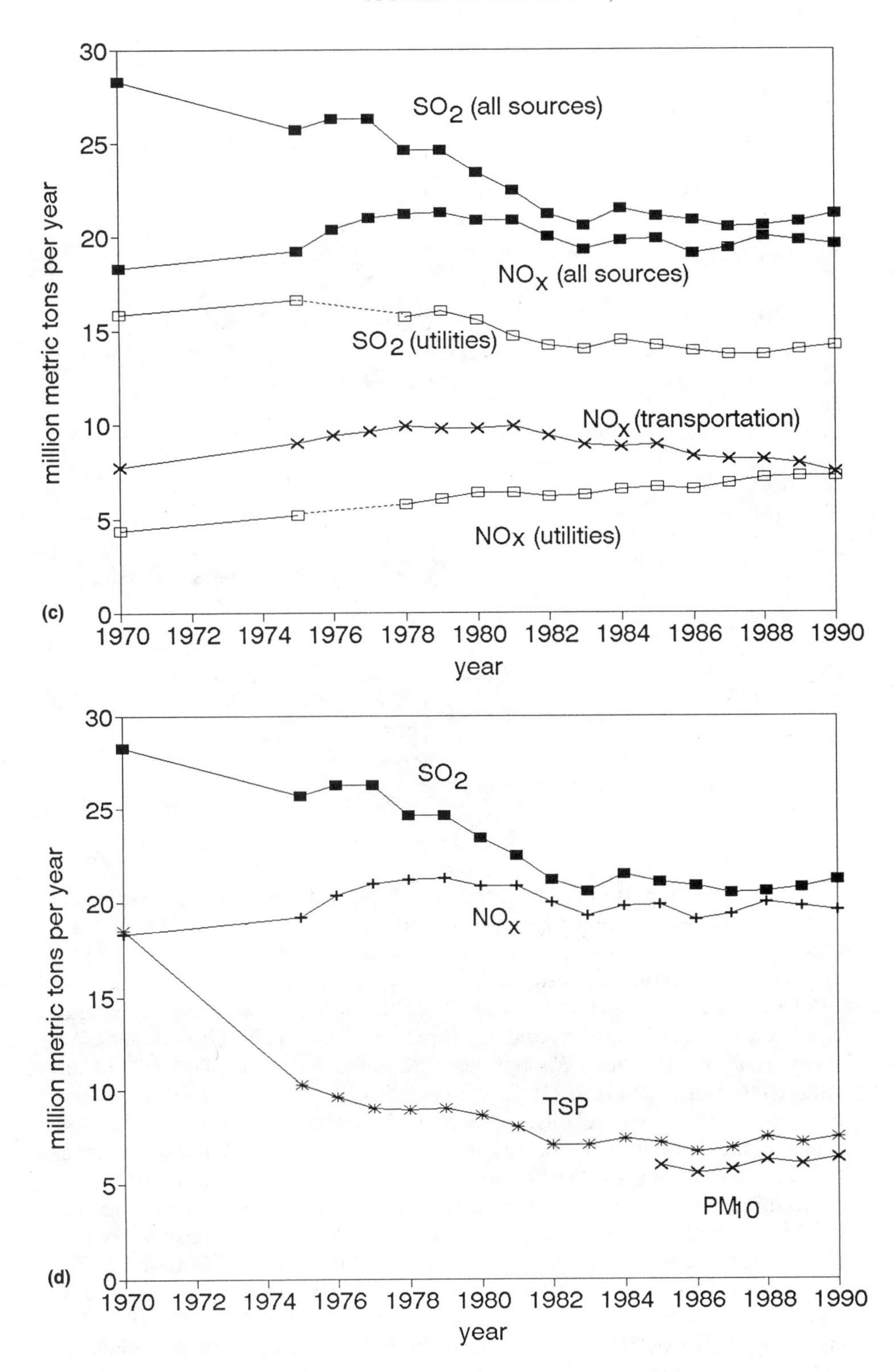

FIGURE 2-2. *Continued*

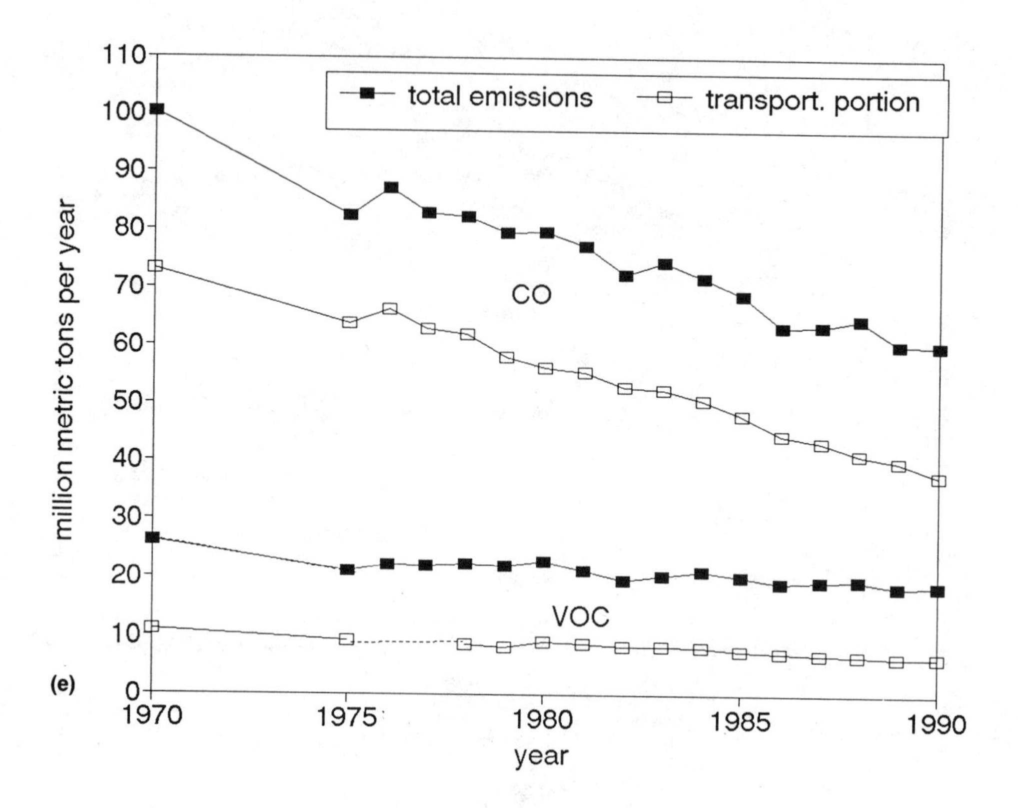

FIGURE 2-2. *Continued*

Industrial energy use has increased at a slower rate, less than 1% per year, and is dominated by natural gas (42%) and petroleum products (25%). Only about 18% of industrial fuel use would be considered important sources of SO_2 and particulates (coal and residual fuel oil).

The electric utility sector has been one of the fastest growing users of energy and fossil fuels, increasing at about 6% per year since 1950. Growth has been slower since 1980, about 1.3% per year. Bituminous coal supplied 55% of the energy input in 1979 (U.S. DOE, 1991), followed by nuclear energy, natural gas, hydro power, and fuel oils. In the early days of the electric utility industry, power plants tended to be located in urban areas along with other industrial facilities. These choices were often dictated by minimizing transmission and distribution power losses as well as tapping into urban fuel supplies. Stack heights were often modest (under 150 m) by today's standards, but usually much higher than those of most of the other urban fuel users. Since the 1960s, new power plants have been built at more suburban and remote locations, some with stack heights reaching 1000 ft (300 m). Tarr (1984) reports that 178 stacks in excess of 500 ft (150 m) were built from 1970 to 1979, mainly by utilities. The effect of increasing the height of air pollution release is to decrease the average ground-level air concentrations and to spread the impacts over a larger area. One of the regions with a concentration of power plants is the Ohio River Valley, especially at the eastern end. As a result, the regional annual background SO_2 concentration there has increased to levels normally found in urban areas (around 25–40 $\mu g/m^3$).

These trends in fuel use must be overlain with concomitant population shifts in order to understand the resulting changes in urban ambient air quality. Some generalizations follow.

1. Home heating and transportation impacts on air quality are related to population density, modified by the types of fuels available in specific locales.
2. The locations of industrial facilities have been reasonably constant over the years, being fixed either by the sources of labor or of raw materials (such as ore smelters). The largest air quality impacts have been from the so-called smokestack industries (primary metals, auto manufacturing, chemical plants, refineries), many of which have declined in recent years due to competition from imports.
3. Power plant locations tend to be geared to fuel supplies and access to cooling water, as well as to reasonable proximity to load (population) centers. Since under the terms of the 1970 Clean Air Act, new facilities must meet stricter pollution control standards than existing facilities (see the latter part of the chapter), most of the locations of highest air pollutant emissions have been fixed since the late 1960s and early 1970s.

Trends in Ambient Air Quality

For the purpose of discussing air quality trends, it is convenient to divide the United States into four regional sections:

1. Northeast corridor: urban areas stretching from Washington, DC, to Portland, ME.
2. Midwest industrial: the region bounded roughly by Chicago-Buffalo-Charleston, WV-St. Louis.
3. Southeast: the region east of the Mississippi and south of Kentucky-Virginia
4. Southern California: that part of the state south of Santa Barbara (containing about 30% of the population of the Pacific Census Division).

The approximate air quality trends since 1940 in these four regions are estimated in Table 2-6, by decade. The estimates suggest that TSP has improved in most areas while NO_x has gotten worse. In the large urban centers of the Northeast and Midwest SO_2 improved greatly, largely because of substitutions in space heating fuels; in contrast, sulfate aerosol has declined only slightly. Ozone improved in Southern California but has not changed much in recent years elsewhere. In most cases it has not been possible to estimate changes in aerosol acidity (H^+), because of the complex interplay between changes in SO_2, oxidants, and neutralizing agents (including TSP). Changes in CO result from changes in space heating fuels in the early years, institution of controls on cars in the 1970s, and the steady growth in traffic.

TABLE 2-6 Estimated Urban Air Quality Trends in Selected U.S. Regions

Decade	1940–50	1950–60	1960–70	1970–80
Northeast				
SO_2	0	–	– –	–
TSP	–	– –	– –	–
NO_x	0	+	+	0
CO	0	+	–	– –
O_3			–	0
Aerosol H^+			–	0
Midwest				
SO_2	–	– –	+	0
TSP	– –	– –	–	0
No_x	0	+	++	+
CO	–	+	–	– –
O_3		–	–	0
Aerosol H^+			0	0
Southeast				
SO_2	0	–	+	0
TSP	–	–	+	0
NO_x	0	+	+	+
CO	–	0	–	0
O_3		+	+	0
Aerosol H^+			–	0
Southern California				
SO_2	+	+	+	–
TSP	–	+	– –	0
NO_x	+	++	+	0
CO	+	++	– –	0
O_3	++	+	0	–
Aerosol H^+				0

Key: ++ = much worse, + = slightly worse, – slightly better, – – = much better, 0 = no change, blank = no estimate possible.

VARIABILITY OF AIR POLLUTION

Epidemiology is the study of variation in the incidences of disease or in states of well-being. With respect to air pollution, variances may exist in either space or time; if air concentrations were always constant everywhere, epidemiological studies of its effects would be impossible. The study of long-term spatial variability is often called cross-sectional analysis; spatial gradients in air pollution may exist because of the presence of specific pollution sources, such as industrial facilities, or because of gradients in population density. The study of short-term temporal variability at a given location is termed time-series analysis. Only a few cases of long-term temporal variability, such as pollution source cleanup, have been analyzed. A further example of temporal variability involves contrasts in seasonal changes in air pollution and mortality (see Chapter 8).

Variability in air concentrations is created by the distribution of pollution sources and by changes in meteorological dispersion. Sources are conveniently divided into stationary and mobile (vehicular) types. Knowledge of their distribution in space is required to predict or understand the resulting impacts on ambient air quality. It is convenient to classify pollution sources as either area or point source types. Area sources are agglomerations of small individual sources, such as home heating plants or vehicles (which may also be modeled as line sources under some circumstances), and are characterized by the areal density of their emissions (g/s/m^2). It is normally assumed that area source emissions emanate essentially from ground elevation. Large point sources are often considered individually and must be characterized by their rates of emissions and their effective heights of pollutant release (physical stack height plus thermal plume rise).

Current Regional Variability in Air Quality

Data from the 1990 EPA report (U.S. EPA, 1991) on air quality and emissions trends were used to construct Table 2-7, which presents composite averages for the ten federal administrative regions for the years 1988–90. These composite averages are not based on population weighting. Both TSP and PM_{10} data (1990 only) are shown, in order to provide some context with previous data on TSP. The relationship between the two particulate measures was consistent for all but Regions VII and IX, for which the ratio PM_{10} was somewhat lower. The relatively higher TSP levels there were probably due to windblown dust.

In general, many of the composite regional averages were similar. The highest SO_2 and NO_2 levels were in the Northeast (Regions I, II, III, and V). Levels of NO_2 were also high in Region IX (California). Carbon monoxide was highest in the Mountain states (Region VIII), probably because of the altitude effect on vehicle performance. Surprisingly, average maximum l-hour ozone was nearly as high in the Northeast as it was in Region IX.

In order to provide a single air quality measure for each region, the composite averages for each pollutant were divided by the corresponding NAAQS level and summed (Region X was omitted because of the lack of valid NO_2 data there). The rankings according to these sums are shown in Table 2-7, using TSP or PM_{10} alternatively as the measure of particulate pollution. The rankings were quite similar on either basis and showed Region IX to have the worst overall air quality, followed by Region II. Air quality in the Northeastern states is essentially a consequence of high population densities and industrial production. In the West, population growth is also important, but natural factors are more important: stagnations in the Los Angeles basin with high photochemical activity, windblown dust in the arid regions, and high-altitude effects on internal combustion engines. Aerosols and precipitation tend to be the most acid in the Northeast; fog acidity is highest in the Los Angeles basin.

Temporal Variability

Temporal variability in ambient air quality results from changes in both emissions and the weather. Area (surface releases) and point (elevated releases) sources of pollution differ in the ways in which meteorological variability affects ambient concentrations. Under low-wind conditions, surface releases may stagnate and build up high concentration levels, while elevated releases may stay aloft to be transported out of the area. Under high-wind conditions, low-level releases will be dispersed while elevated plumes may be brought

TABLE 2-7 Composite Air Quality Averages, 1988–1990 (μg/m³)

		Pollutant	TSP	SO_2	NO_2	CO (ppm)	O_3	PM_{10} (1990)	Rank based on pollutant standard index (TSP)	Rank based on pollutant standard index (PM_{10})
Region	States	NAAQS Average	75 Annual	80 Annual	100 Annual	9 8 hours	235 1 hour	50 Annual		
I	CT, MA, ME, NH, RI, VT		37.3	24.1	43.9	6.1	265	22	4	4
II	NY, NJ		39.3	26.7	47.7	6.2	261	26	2	2
III	DC, DE, MD, PA, VA, WV		47.3	34.1	42.4	5.4	241	30	3	3
IV	AL, FL, GA, KY, MS, NC, SC, TN		43.8	17.9	30.8	5.6	221	32	9	8
V	IL, IN, OH, MI, WI		49.2	23.5	37.7	5.4	219	31	5	5
VI	AR, LA, NM, OK, TX		43.5	15.0	29.2	6.2	246	26	7	7
VII	IO, KS, MO, NE		56.5	19.0	30.0	5.2	194	30	8	9
VIII	CO, MT, ND, SD, UT, WY		40.7	16.9	23.7	8.5	213	26	6	6
IX	AZ, CA, HI, NV		67.3	7.2	49.8	6.7	274	39	1	1
X	AK, ID, OR, WA		47.8	19.5	n/a	8.7	201	30		

Source: U.S. EPA, 1991.

down to the ground in localized areas. The thermal structure of the atmosphere is also important in this regard; a temperature inversion can trap low-lying emissions vertically, while preventing elevated releases from contacting the surface. These situations occur most often at night but can persist for days at a time under certain conditions. When averaged over a season and the same area, emissions from a ground-level area source have over 30 times the impact (concentration averaged over the same area) of those from an elevated stack, in this case 180 m in height (Lipfert *et al.*, 1973).

Seasonal and Long-Term Variability

Examples of seasonal variability of the criteria air pollutants in the 1950s and 1960s, before air pollution control was undertaken at the federal level, are given in Figure 2-3a–f for six of the cities of the Continuous Air Monitoring Project (CAMP) (U.S. HEW, 1969a). Chicago, which still used some coal for residential space heating at that time, shows high TSP and SO_2 concentrations in winter, but high CO and NO_2 in summer. All the cities except San Francisco showed the traditional summer peak in oxidants. For most of the other cities, SO_2, NO_2, and CO showed little seasonal behavior. Even though sulfur-bearing fuels were widely used for space heating before about 1950, winter emissions were only about one-third of the annual total. The 1957–61 TSP data (Figure 2-3f) were only available on a seasonal basis and show that minimum TSP levels were recorded in summer, as they were in the earlier period (Figure 2-3e).

Long-term TSP trends are shown by season for several cities in Figure 2-4a through 2-4f. These plots are based on monthly data obtained from the EPA AIRS data bank; winter (December–March) and summer (June–September) seasons were defined to correspond to the peak and minimum periods in the typical annual mortality cycle. Peak TSP levels were similar in all six cities and occurred in winter during the 1960s. By the mid- to late 1970s, peak TSP levels had dropped substantially and switched to summer, apparently because of the use of cleaner fuels for space heating. Since the seasonal difference in average atmospheric dispersion is small in these locations (Holzworth, 1972), one must conclude that the increase in summer particulates is the result of photochemical conversion processes.

Table 2-8 gives more recent trends in statewide SO_2 ambient concentrations (Pollack and Burton, 1985). All locations peak in the winter, and most of them show a downward trend since 1975, which parallels the trends seen for TSP in Figure 2-4.

Local Scale Air Quality Variability

The air monitoring data presented by Heimann (1970) for Boston during some stagnation episodes in 1966 affords comparison of spatial and temporal variability between species (Figure 2-5a–b). The comparison between TSP and COH is of particular interest, since time-varying epidemiological studies tend to use COH while spatial studies tend to use TSP. Figure 2-5a compares the temporal variability of the averages of about 20 measuring stations. The relationship between TSP and COH was given by TSP $= (96 \pm 5) \times$ COH

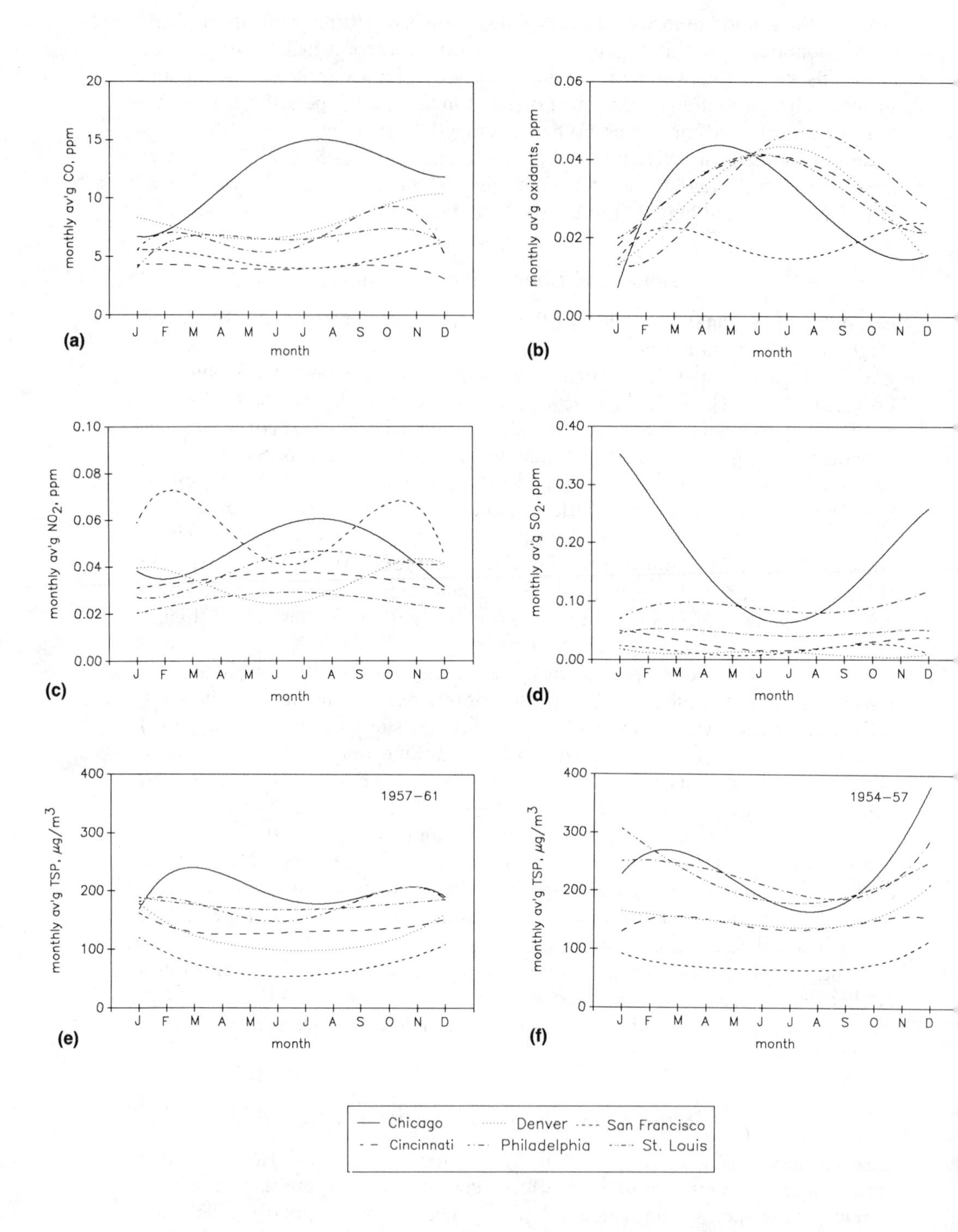

FIGURE 2-3. Seasonal variations in ambient air quality in six U.S. cities. Lines represent best-fit 4th-order polynomial regressions through monthly averages. (a) CO, 1962–1967. (b) oxidants, 1962–1967. (c) NO_2, 1962–1967. (d) SO_2, 1962–1967. (e) TSP, 1957–1961. (f) TSP, 1954–1957. *Data from U.S. DHEW (1969a).*

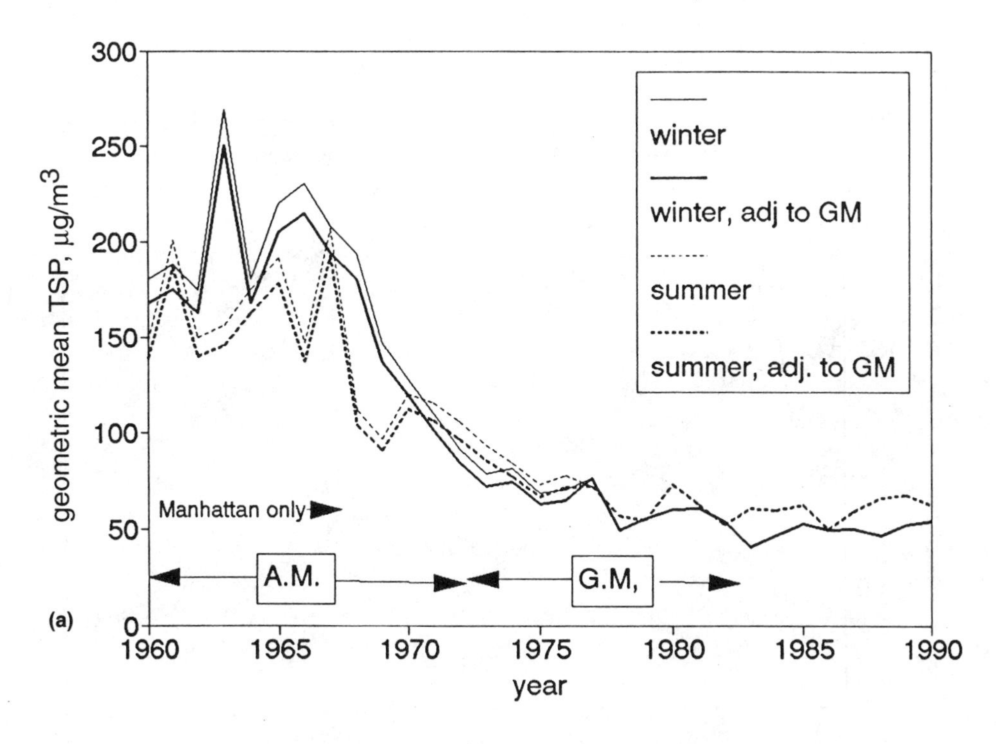

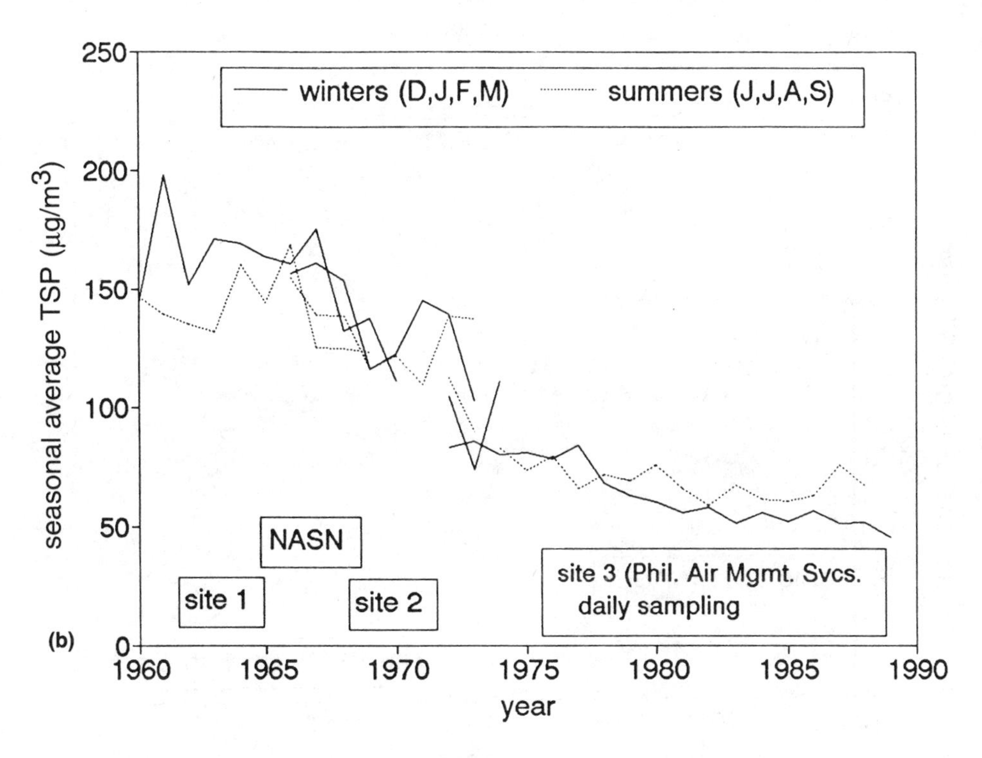

FIGURE 2-4. TSP trends in U.S. cities, by season. (a) New York. (b) Philadelphia. (c) Pittsburgh. (d) Cleveland. (e) Cincinnati. (f) Los Angeles. *Data from U.S. EPA Airs data bank.*

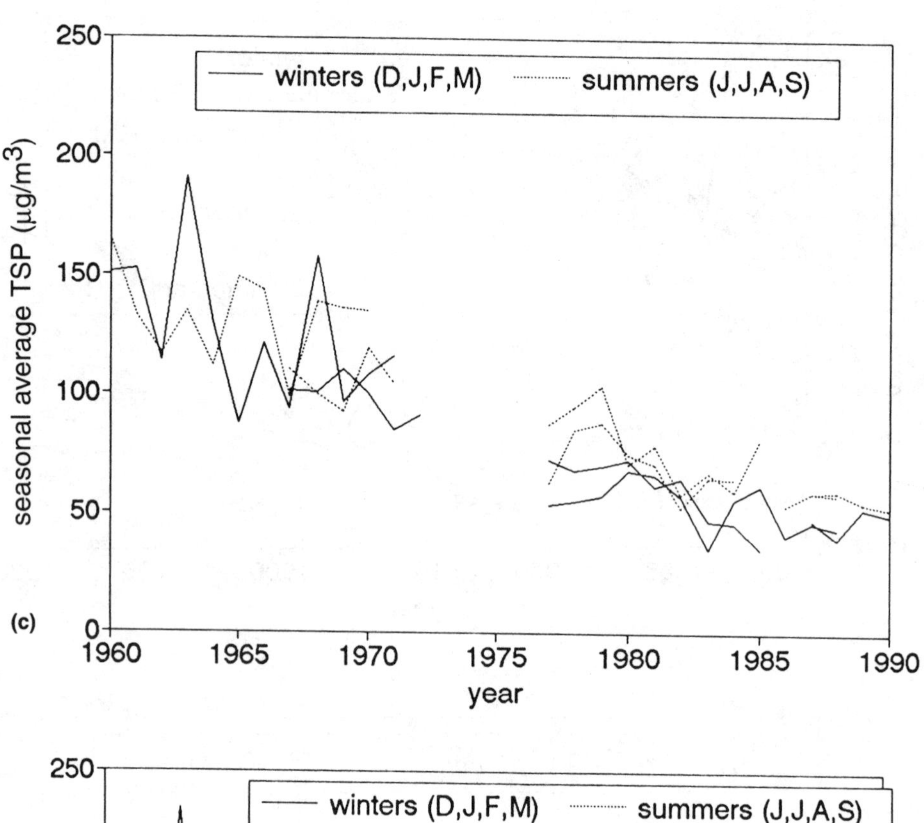

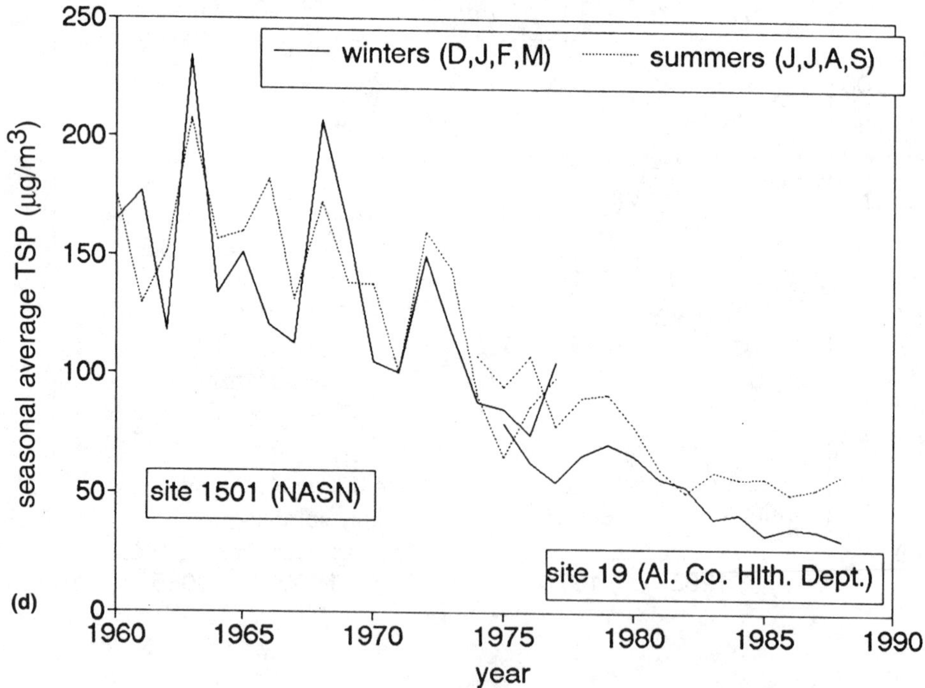

FIGURE 2-4. *Continued*

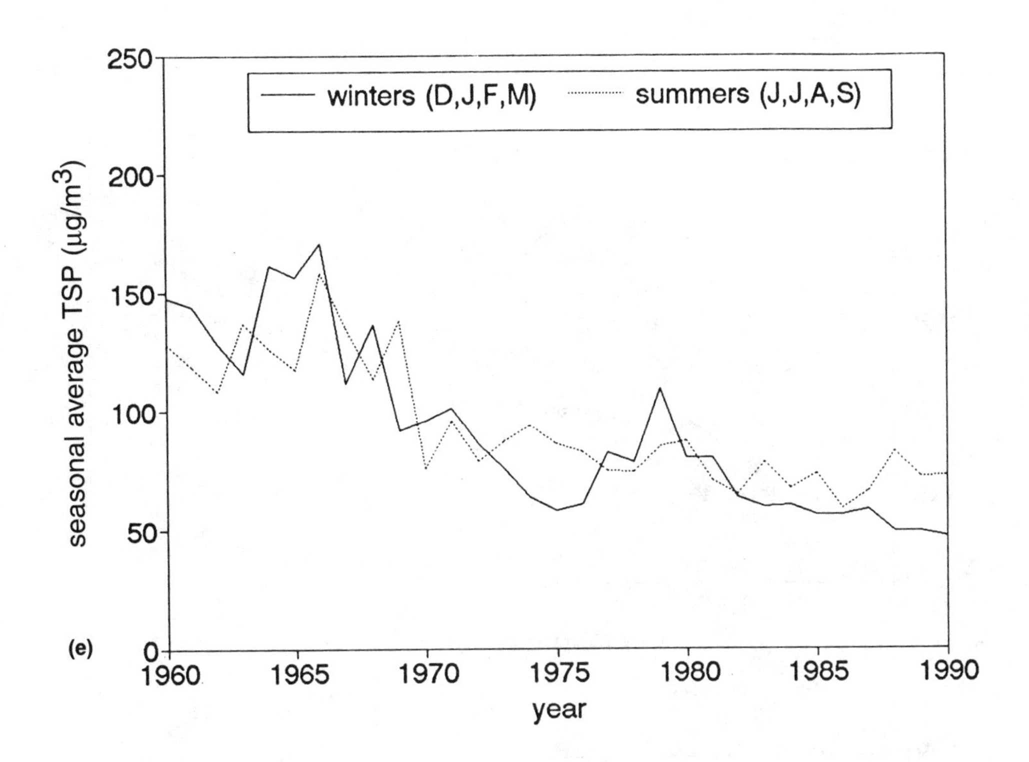

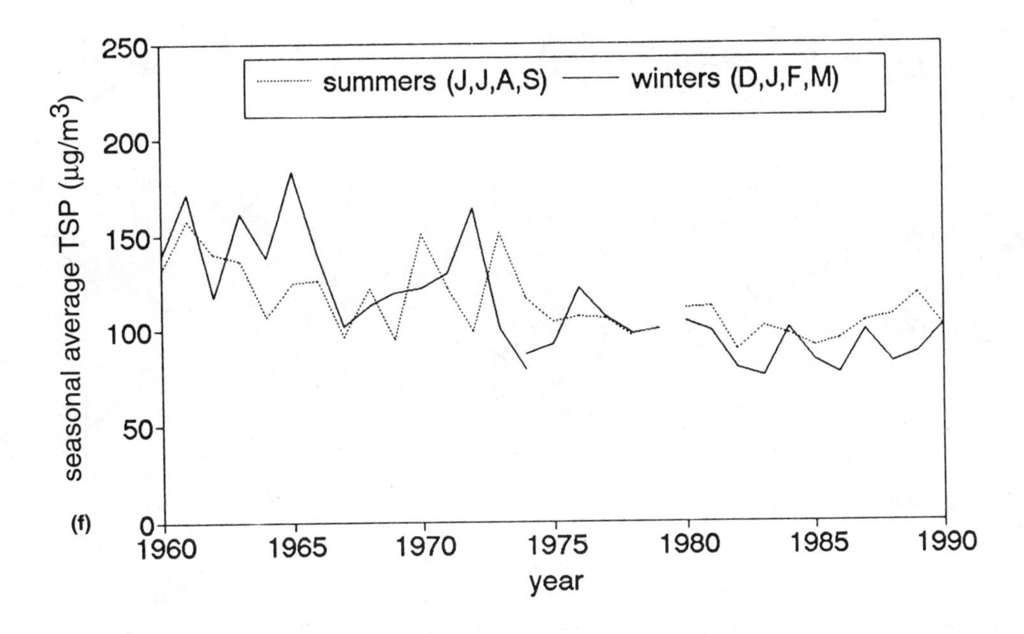

FIGURE 2-4. *Continued*

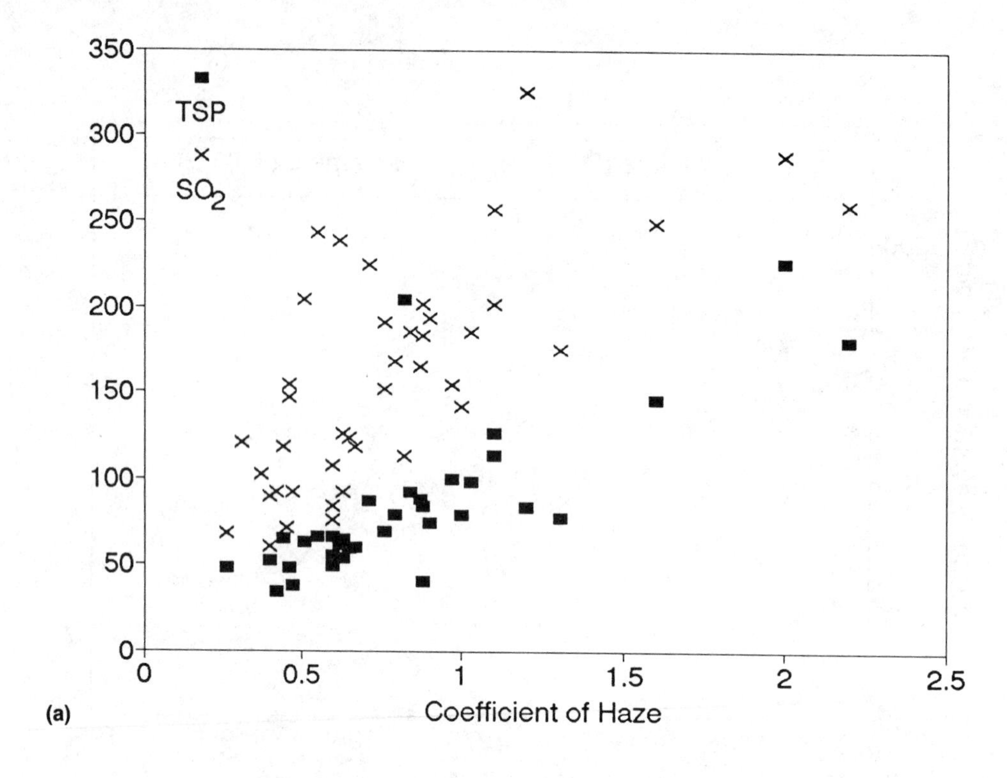

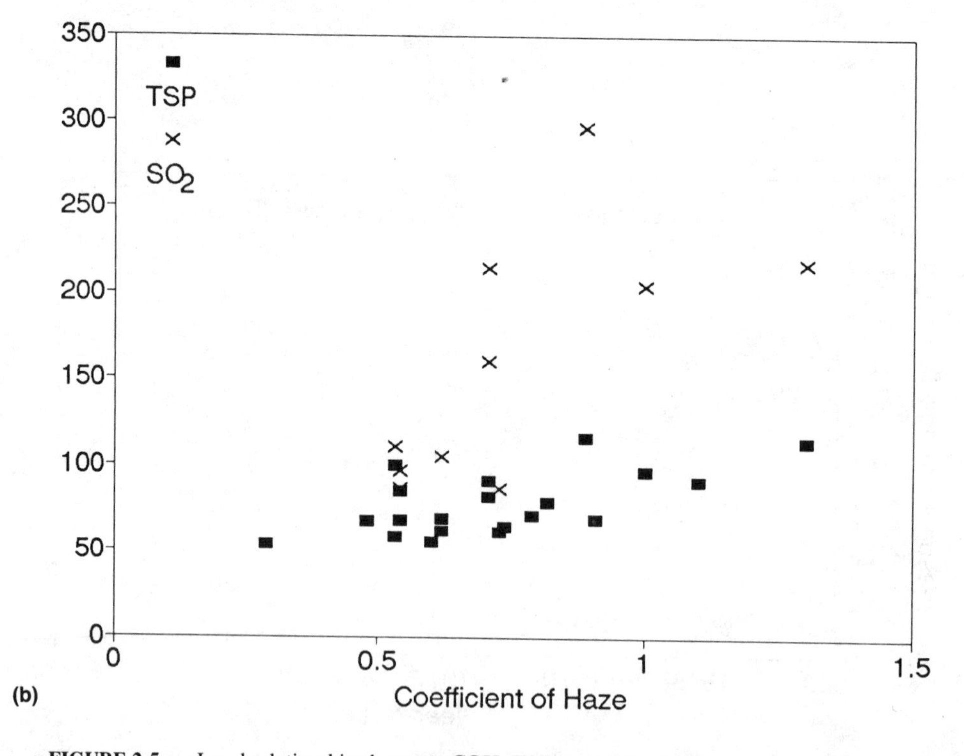

FIGURE 2-5. Local relationships between COH, TSPs, and SO_2. (a) Temporal variability at about 20 stations. (b) Spatial variability averaged over 2 months. *Data from Heimann (1970).*

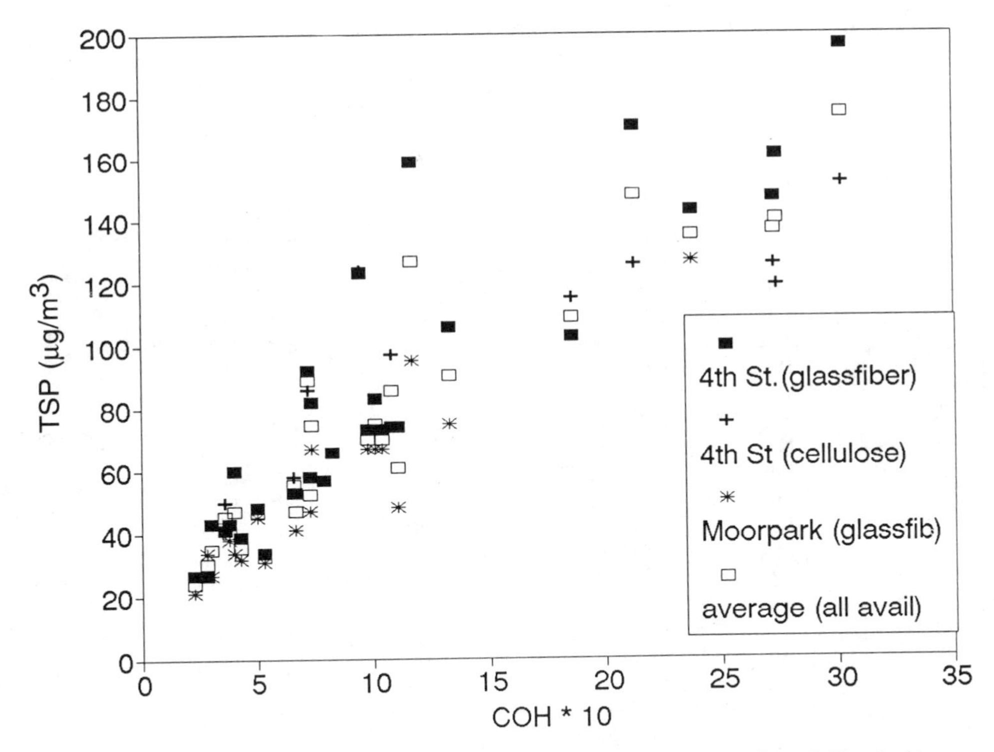

FIGURE 2-6. Comparison of TSPs and the COH at San Jose, CA. *Data from California Air Resources Board air monitoring reports.*

[$R = 0.78$], which checks our previous rule of thumb quite well. This figure also shows how well SO_2 tracks the particulate measures; there were a few days with high SO_2 at medium smoke levels, but not vice versa. The situation with spatial variability (Figure 2-5b) is similar, but the correlation between TSP and COH is poor. However, the average slope, obtained by forcing the regression through the origin, was 102 μg/m^3 TSP per COH unit, also a very good check.

The relationship between TSP and COH at San Jose, CA, is shown in Figure 2-6. At the Fourth Street monitoring station, two types of filters were used in the hi-vol (TSP) sampler; the glass-fiber data are generally higher. The COH was only available at the Fourth Street location, but TSP measurements at both locations seem well correlated with the COH data.

Ambient Air Quality in Relation to Emissions

It is also of interest to examine the average ratios* of ambient air quality to emissions for various pollutants (Table 2-9). The order of magnitude variation in these ratios for the different pollutants reflects several factors, some of them competing. The high ambient/emission ratio for CO reflects the concentration of measurements in urban areas, where the emission density is much higher and the heights of release are low. It also reflects the insolubility of CO and its

---

* This ratio is often denoted as χ/Q, where χ represents an ambient air quality concentration in μg/m^3, and Q represents emissions in g/s. The ratio then has units of s/m^3.

TABLE 2-8 Monthly Average SO_2 Data, 1975–1982 (ppm)

State	Long-term trend	Peak season	Monthly range
WV	0.025–0.018	All	0.010–0.040
PA	0.022–0.017	W	0.011–0.042
OH	0.021–0.015	W	0.010–0.030
RI	0.024–0.012	W	0.005–0.040
DC	0.016	W	0.006–0.036
NY	0.015–0.015	W	0.008–0.032
ME	0.020–0.008	W	0.005–0.044
CT	0.015–0.013	W	0.006–0.024
NH	0.015–0.013	W	0.005–0.035
IL	0.018–0.010	W	0.009–0.028
IN	0.016–0.012	W	0.008–0.030
MD	0.012–0.015	W	0.002–0.026
NJ	0.013–0.013	W	0.006–0.030
DE	0.018–0.008	W	0.003–0.027
MI	0.014–0.010	W	0.006–0.019
NC	0.017–0.006	W	0.005–0.032
MA	0.012–0.011	W	0.005–0.022
VI	0.016–0.009	W	0.005–0.024
VT	0.012–0.010	W	0.001–0.024
KY	0.012–0.010	W	0.007–0.026
WI	0.014–0.008	All	0.005–0.022
TN	0.008–0.008	W–S	0.005–0.017

Source: Pollack and Burton, 1985.

TABLE 2-9 Ratios of Ambient Air Quality to Emissions (national annual averages, 1990)

Pollutant	Annual Emissions $10^{12}\,g/y$	Annual Average Concentrations (approx.) $\mu g/m^3$	Ratio $10^{-12}\,s/m^3$
CO	60.1	~800	420
SO_2	21.2	21	31
NO_x	19.6	36	58
TSP	7.5 (+40.8 fugitive)	48	200 (31)*
PM_{10}	6.4 (+15 SO_4^{-2})	30	150 (44)*

* After adjustment (see text).

lower rates of deposition. The low ambient/emission ratio for SO_2 reflects the opposite extremes: high elevation releases, high solubility, and high rates of dry deposition; NO_x enters into photochemical reactions, is released from both high and low sources, is relatively insoluble, and shows an intermediate value for ambient/emission ratio. For particulates, sources other than direct emissions must be considered. Adding the EPA estimate for national fugitive emissions brings the ambient/emission ratio for TSP into line with SO_2, which is appropriate given the relatively high rates of deposition for particles larger than 5 μm. For PM_{10}, adding an estimate for the amount of SO_2 that ends up as airborne SO_4^{-2} brings its estimate into line, given the low rate of dry deposition for smaller particles.

These ambient/emission ratios can be placed in context by comparing with typical annual average values computed using dispersion models. Values downwind of tall stacks are about 2000×10^{-12} s/m^3 at the worst location (Lipfert and Dupuis, 1985); averaged over a large city, about 4000×10^{-12} s/m^3 (Holzworth, 1972). The fact that the values in Table 2-9 are much lower than these local estimates reflects the dispersion of emissions on the national scale relative to the locations where ambient air quality is monitored.

THE METEOROLOGY AND PHYSICS OF AIR POLLUTION EPISODES

Elevated concentrations of air pollutants can occur for two reasons: exceptional releases of toxic materials (such as occurred in Bhopal, India), or changes in the dispersion characteristics of the atmosphere such that the "normal" emissions become trapped and result in substantially higher ground-level concentrations. The episodes that relate to community air pollution fall into the second group and result from prolonged stagnations during which elevated concentrations are experienced for several days. These conditions would apply to all the air pollutants normally present in a given area; elevation of concentrations of secondary pollutants would also depend on whether reaction rates were affected by the meteorological conditions of the episode. In general, the historical analyses and modern perceptions of these episodes have tended to focus on the maximum concentrations reached, rather than on their durations or total doses.

The physical nature of the pollution sources is an important consideration in this regard; releases from tall stacks, say taller than 150 m, may be trapped aloft with little or no effect at ground level in the immediate vicinity. If the stagnation is of regional extent, such emissions may still contribute to regional pollution levels, however. Ground-level emissions are thus of primary concern during trapping conditions. During the time periods in question, prior to about 1960, the bulk of sulfur oxide emissions came from stacks less than 150 m; particulates and carbon monoxide have always been associated with low-elevation sources.

One simple way to consider the consequences of trapping of emissions from surface sources is through the use of dispersion models for area sources such as the Air Quality Display Model (AQDM) (TRW, 1969) or its successor the Climatological Dispersion Model (CDM) (U.S. EPA, 1973). Using meteorological data from the eastern United States, Lipfert and Dupuis (1985) found that the overall level of annual average concentrations computed from area sources could be represented in terms of two parameters: the emission density and the physical size of the area source. This simple model estimates the spatial average of annual average concentrations (μg/m^3) χ_a as:

$$\chi_a = kQ/A \qquad [2\text{-}1]$$

where k is a function of the dimensions of the area source, as shown in Figure 2-7.

However, in a trapping situation, a more appropriate model is the "box" model, which estimates the concentration by dividing the mass of material emitted by the volume of the box (which is given by the ground area of the area source and the height of the trapping layer). This model assumes perfect

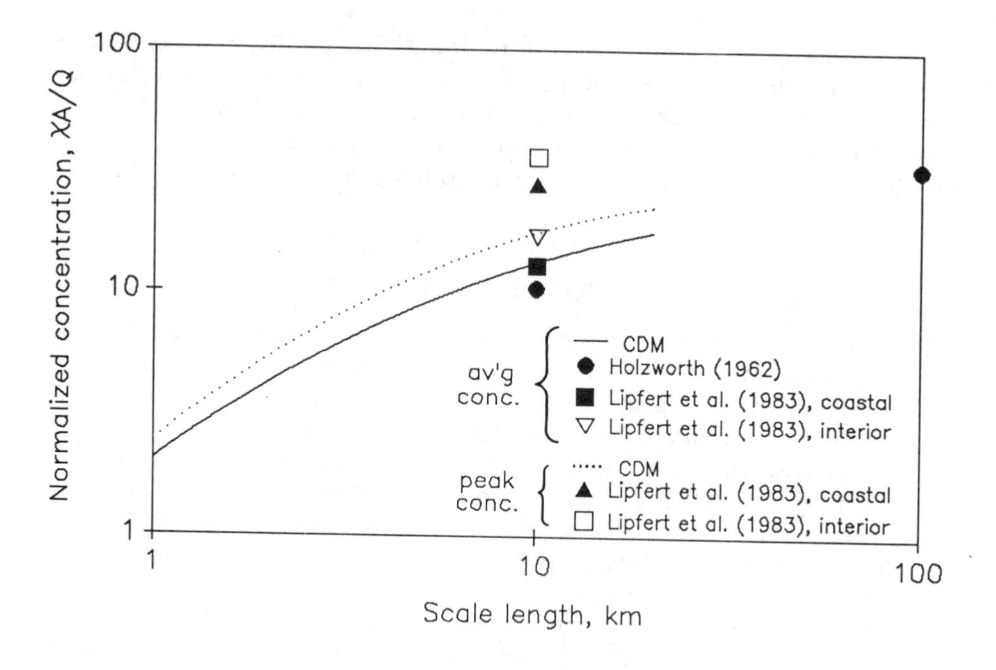

FIGURE 2-7. Scale effect on area source dispersion. *From Lipfert and Dupuis (1985).*

mixing within the box and projects a steady increase in concentration over time as long as the emissions and the trapping situation persists:

$$\chi_t - \chi_0 = Q/Ah\ dt \qquad [2\text{-}2]$$

where h is the height of the box, given by the inversion or mixing height or by the physical dimensions of the valley, if appropriate. This is similar to the model considered by Jensen and Petersen (1979) who found that a quasi-stationary model gave essentially the same results as a time-dependent version, as judged by a time-dependent relationship with mixing heights observed by acoustic sounder.

Thus, according to Eq. 2-2, the average concentration reached in one day will be given by 86.4 times the emission density divided by the trapping height in kilometers. For an area source of 20 km diameter and a 1 km mixing height, the average concentration would thus approach about six times the "normal" level (given by Figure 2-7) on the first day of trapping, with higher levels on subsequent days. Concentrations of a factor of about six over normal were reported during the 1952 London episode (Wilkins, 1954), and a model similar to Eq. 2-2 was used by Firket (1936) to estimate maximum levels during the Meuse Valley episode. A time step dt of one day is consistent with the air sampling measurements used at the times of the major episodes and avoids dealing with diurnal variations in the various parameters.

However, one must also account for advection out of the box (if trapping is not complete) and for the sinks that remove pollutants from the air by deposition or by chemical transformations. For SO_2, these sinks include oxidation to sulfates, washout by precipitation, and "dry" deposition to surfaces. The last process is greatly enhanced when the surface is wet and reactive (Lipfert, 1989). Particulates are removed by washout and dry deposition. Because of their low solubilities in water, carbon monoxide and nitrogen oxides are removed primarily by chemical reactions through photochemical processes.

A more complete box model is given by

$$\chi_t = \frac{\chi_0 - \chi_w + Qdt/Ah}{1 + u(V_d/h + u/a)dt} \qquad [2\text{-}3]$$

where χ_0 is the initial concentration at $t = 0$, χ_w is the concentration lost to washout and chemical transformations, V_d is the dry deposition velocity, and u is the advection wind speed averaged over the mixing height and oriented parallel to the box dimension a. Note that since the empirical area source representation (Eq. 2-1) developed by Lipfert and Dupuis (1985) was based on use of a half-life for SO_2 of a few hours in the original CDM modeling, losses due to transformations and deposition have already been accounted for in Figure 2-7.

An important element of the three most severe community air pollution episodes of this century (in terms of mortality) was the presence of fog. Fog can be a result of both the meteorological situation and the presence of particulates to provide condensation nuclei, and thus a feedback mechanism could operate during stagnation periods (Neuberger and Gutnick, 1949). There are several consequences of fog: Photochemical processes near the surface are essentially turned off because of the absence of sunlight, fog droplets become a sink for water-soluble gases (such as SO_2) and small particles, and ground surfaces are wetted and become more effective sinks for dry deposition of water-soluble gases. Since fog droplets are relatively large and have appreciable gravitational settling velocities, fog can actually clean the air by removing those pollutants that are captured in its droplets (Lipfert, 1992). However, if no sunlight reaches the ground, the normal daytime increases in turbulence may not occur, extending the time period of reduced mixing rates.

Indoor-Outdoor Air Quality Relationships*

Since most people spend the majority of their time indoors, the quality of indoor environments can have important health consequences. Much of the literature on indoor air quality is concerned with sources of air pollution that are found primarily indoors, including tobacco smoke, other types of combustion products, outgassing from building materials, and pathogens (in the case of health care facilities). The advent of increased emphasis on energy conservation has led to reduced infiltration of outside air in certain types of buildings, which can exacerbate the effects of indoor sources; however, the quality of outdoor air is a baseline on which the characteristics of a building and its occupants act to determine the total exposure to various air pollutants.

With respect to studies of population responses to community (outdoor) air pollution, indoor air pollutants can either add or subtract from the total

* Much of this material also appears in Wyzga and Lipfert, 1992.

exposure that would have been implied by the use of outdoor air quality as the sole index of exposure. Additions result from either the presence of additional pollutant species not normally found outdoors at appreciable concentrations (such as radon or formaldehyde) or from indoor sources of common outdoor pollutants (such as particles, CO, or NO_2). Subtractions result from the confined indoor environment, which provides additional pollutant sinks beyond those normally found outdoors. Air filtration systems can also remove particles from indoor air.

Mass Balance Relationships

Sinclair *et al.* (1988) developed a comprehensive mass balance relationship for application to commercial buildings with air handling and ventilation systems, for a steady-state situation in the absence of indoor sources (expressed as an indoor-outdoor ratio [I/O]):

$$\mathrm{I/O} = \frac{v_l(l - F_l) + v^*(l - F_s)f}{k_d A_d + v^* F_s(l - f) + v^* f} \qquad [2\text{-}4]$$

where v_l is the fraction of air leaking into and out of the building, F_l is the fractional filter efficiency of leakage paths, k_d is the pollutant deposition velocity, A_d is the interior surface area for deposition, f is the fraction of circulating air from the outside, v^* is the volume of flow in the air handling system, and F_s is the fraction of particles in a given size range removed by filtration. In the absence of air handling systems and filtration, this equation may be simplified as (Petersen and Sabersky, 1975; Weschler *et al.*, 1989):

$$\mathrm{I/O} = ACH/(ACH + \{k_d A_d/V\}) \qquad [2\text{-}5]$$

where ACH is the outside air infiltration rate (air changes per hour) and V is the building volume. The term $k_d A_d/V$ represents a net rate of loss (fraction per hour) and could be modified to account for source terms, if necessary. The geometric ratio A_d/V has been estimated at about $3\,\mathrm{m}^{-1}$ for omnidirectional deposition to surfaces within residences (Weschler *et al.*, 1989); however, coarse particles large enough to be influenced primarily by gravitational settling would act only on horizontal surfaces, so that the ratio A_d/V would be about $1\,\mathrm{m}^{-1}$ in this case.

Deposition Velocities

The deposition of air pollutants is a mass transfer phenomenon governed by the laws of fluid mechanics. In the absence of air motion, only gravitational forces and random (Brownian) motion would act, so that gases and small particles would remain airborne for very long periods. Air motion is greatly reduced indoors relative to outdoors, and deposition velocities (k_d) are correspondingly lower. Appropriate values have been determined empirically by measuring rates of decay in sealed environments (Petersen and Sabersky, 1975) and by comparing accumulations on indoor surfaces with average air concentrations (Sinclair *et al.*, 1988).

For particles, the density and aerodynamic diameter are the most important parameters, although surface roughness and relative temperature can also play a role (Sehmel, 1980). Minimum k_d values occur in the particle diameter range of 0.05 to 0.5 μm, which accounts for the long lifetime of fine particles in the atmosphere. Sinclair *et al.* (1990a) found indoor k_d values from 0.004 to 0.005 cm/s for sulfate particles in telephone equipment buildings in four different cities; k_d values for other fine particle ions (<2.5 μm) were up to an order of magnitude larger and k_d for coarse particles (2.5–15 μm) was estimated at 0.7 cm/s.

The deposition of gases also depends on their reactivity with the surfaces in question; for example, SO_2 deposits much faster on wet reactive surfaces (Lipfert, 1989) and ozone deposits more readily than CO. A value for ozone deposition inside an automobile was found to be 0.086 cm/s, whereas CO absorption was assumed to be negligible (Petersen and Sabersky, 1975). For ozone depositing in an office building, a somewhat lower k_d value was reported: 0.036 cm/s (Wechsler *et al.*, 1989).

Parametric Calculations

Figure 2-8 is a plot of Eq. 2-5 for a range of air exchange and pollutant loss rates. Typical air exchange rates are given in Table 2-10; for environments where susceptible individuals might be found, the extremes are given by energy-efficient residences (<0.5 ACH) and hospital operating rooms (15 ACH, although not all of this may be outside air). Residential ACH values with open

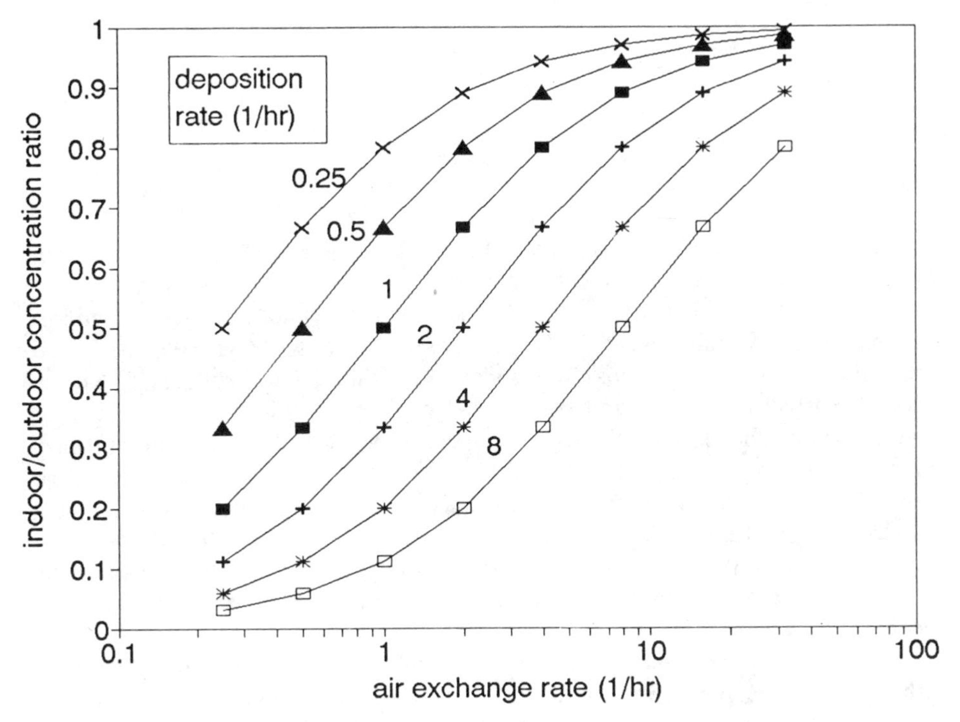

FIGURE 2-8. Relationship between indoor-outdoor concentration ratio and air exchange rate, for various rates of pollutant deposition.

TABLE 2-10 Typical Building Air Exchange Rates (air changes per hour, ACH)

Type of Facility	ACH	Outdoor Air (ft^3/min·person)
Restaurants	8–12	
Cafeterias	12–15	
Kitchens	12–15	
Bars	15–20	
Nightclubs	20–30	
Office buildings	4–10	5
Libraries and museums	8–12	
Warehouses	1–4	
Transport terminals	8–12	
Garages	4–6	
Classrooms	5	15
Hospitals		
Patient rooms	2*	
Operating rooms	5–15*	
Recovery rooms	2*	
Intensive care	2*	
Nursery suite	5*	
Trauma room	5*	
Nursing homes		
Patient care	2*	
Examination room	2*	
Physical therapy	2*	
Residences		
Windows sealed	0.5†	
Windows open	5†	
Mobile homes	0.5††	

* Outdoor air supply.
† Weschler *et al.*, 1989.
†† Mumford *et al.*, 1991.

windows can exceed 5; a median value of 1.5 was found for 600 Southern California homes in July (Wechsler *et al.*, 1989), which sample probably included many with air conditioning. Values inside automobiles (with air conditioning at maximum) were in the range of 18 to 40 ACH (Wechsler and Shields, 1988). Data on loss rates may be estimated for sulfates ($0.5\,h^{-1}$), coarse particles ($25\,h^{-1}$), and ozone ($4–9\,h^{-1}$). From Figure 2-8 or Eq. 2-5, sulfate (and other fine particles) I/O concentration ratios would range upward from about 0.5, coarse particles would be less than 0.4, and O_3 would be in the range of 0.05 to 0.8. These predicted values are generally consistent with actual measured I/O ratios (Dockery and Spengler, 1981). Sulfur dioxide tends to deposit rapidly on many indoor surfaces, and airborne acidity tends to be neutralized by the ammonia typically present in indoor environments (Brauer *et al.*, 1990).

These estimates neglect the effects of indoor sources, which will add to the burden received from outdoor contamination. Indoor sources may be characterized in terms of the species emitted and their frequencies of operation. Tobacco and cooking sources may be thought of as quasi-continuous, while space heating sources are seasonal. Spengler *et al.* (1981) found that each smoker in a home added about 20 μg/m^3 to the long-term average respirable particulate concentration, for example. There are many other sources of indoor particles in residences, including carpeting, vacuum cleaners, and electric resistance heaters. Dockery and Spengler (1981) postulated a linear model relating indoor air concentrations in terms of outdoor levels, in which the slope was the transfer coefficient or penetration, p (from Figure 2-8, for example), and the intercept represented the sum of indoor source terms (S/q):

$$C_i = pC_o + S/q. \qquad [2\text{-}6]$$

The correlation between C_i and C_o is one measure of the applicability of this model; values in the literature range from 0.25 to 0.43 for pooled summer and winter data (2 weeks each) in Toronto (Hosein *et al.*, 1990) to 0.85 for a 9-month study of a factory building (Sinclair *et al.*, 1990b). By pollutant species, Brauer *et al.* (1990) found high indoor/outdoor correlations in 11 non-air-conditioned homes in the Boston area in both summer and winter for SO_2 and SO_4^{-2} (0.7–0.9) and lower values for NO_2, HNO_3, NH_4^+, and H^+ (0.4–0.6). Correlations between long-term average indoor and outdoor NO_2 concentrations across communities were also successful (Butler *et al.*, 1990): for a sample of eight cities including variable mixes of gas and electric cooking with pooled summer and winter data, the correlation was 0.97.

Two special cases are appropriate for this model: (a) the use of air conditioning, which reduces the slope, and (b), the presence of smokers, which increases the intercept. A third special case could consist of (a) + (b). Note that, according to this model, for a mixture of situations typical of a community, average indoor and total exposure will still be linearly related to outdoor air quality, although the individual exposures will vary from house to house. For a linear health-effects dose–response relationship with no threshold, this mix of exposure situations will only add to the natural variability of individual susceptibilities within the community and thus would not obscure the detection of linear effects. However, the converse is also true, that it will be difficult to detect health responses characterized by thresholds on the basis of outdoor measurements alone.

BASIS FOR AIR POLLUTION REGULATION IN THE UNITED STATES

Stern (1982) presents a comprehensive account of the history of air pollution regulation in the United States, beginning with smoke control ordinances in 1881 in St. Louis and Cincinnati. The first county law was passed in 1913, for Albany County, NY. Oregon passed the first comprehensive state legislation in 1952; by 1970, all states had air pollution control laws on the books. In the first half of the twentieth century, the emphasis was on smoke control, based primarily on the appearance of stack plumes, judged against shades of gray. According to Stern (1982), in the 1940s and 1950s, most of the local regulations allowed smoke as dense as 60% black or opaque; by 1975, most of them prohibited more than 20% opacity. See Appendix 2D for data relating plume opacity or blackness to mass emission rates. California established statewide authority for automotive emissions in 1957 and still maintains regulations stricter than the federal counterparts.

The first federal air pollution legislation was a 1955 bill (PL 84-159) authorizing research and education (Stern, 1982). The legislative history of the Clean Air Act and its associated air quality standards dates back to 1963. As a result of this first national Clean Air Act (PL 88-206), the first SO_2 criteria document (U.S. DHEW, 1967) was issued "as guides for municipal, state, and interstate air pollution control authorities." It was followed by the better-known criteria documents for particulate matter (PM) and sulfur oxides (U.S. DHEW 1969a, b) issued in 1969 pursuant to the Air Quality Act of 1967 (PL 90-148), to assist the States in "taking responsible technological, social, and political action to protect the public from the adverse effects of air pollution."

The Clean Air Act amendments of 1970 (PL 91-604) placed the regulatory authority in federal hands through the newly established Environmental Protection Agency (EPA) and defined the (NAAQSs) concept. Primary standards were to be set to protect human health (regardless of cost), and secondary standards are intended to protect other aspects of our environment (e.g., vegetation, materials, visibility), with a consideration of cost-benefit ratios. Under the Clean Air Act, states have the responsibility for establishing emissions limits for existing air pollution sources that would ensure that primary standards were satisfied within their boundaries; the satisfaction of secondary standards is left as a desirable goal. States are also free to set their own (stricter) air quality standards.

The other regulatory mechanism established by the 1970 Amendments was that of the New Source Performance Standard (NSPS). These regulations established maximum rates of emissions of criteria air pollutant for various classes of sources, ranging from vehicles to power plants. A source that predated these regulations was deemed an existing or "old" source and was free to continue its operations and emissions as long as no "major modifications" were made (for example, an increase in output or emissions). A "new" source, on the other hand, had to meet prescribed emissions limits based on the best available control technology (BACT), as determined through testing according to defined protocols. The framers of this legislation thought that this regulatory mechanism would eventually bring emissions down to minimal levels through turnover of the industrial plant and vehicle fleets. This has been the case for vehicles, but less so for industrial facilities. The costs of compliance with the NSPS for SO_2 proved so high that, in many cases, it was more economical to try to keep old plants operating beyond their normal lifetimes than to build new ones. As a result SO_2 emissions have not declined as much as particulates or CO. Furthermore, the law provides that NSPS emission limits are to be reviewed and lowered as technology improves, which creates a sort of BACT "moving target" for industry.

The 1977 amendments to the Clean Air Act (PL 95-95) established the concept of Prevention of Significant Deterioration (PSD) in order to maintain air quality at levels better than the NAAQS where it had previously so existed. This was intended to be a regulatory device to preclude all of the nation's air resources from eventually degrading to the point where the NAAQS were just met everywhere. The amendments classified land areas according to the degree of future air quality degradation that would be allowed as a result of industrial or population growth, which is primarily a "new source" consideration. "Class I" areas, which include named National Parks and Wilderness Areas, are severely limited in terms of the allowable incremental changes in ambient air quality. If a proposed new source or modification is projected to exceed the allowable PSD air quality increments for one pollutant, a review is required for all pollutants. The 1977 amendments also precluded the use of "excessive" stack heights (greater than necessary to prevent aerodynamic downwash) as a means of meeting the NAAQS around a specific point source.

The 1977 amendments also established the concept of "Non-Attainment Areas" (NAAs), defined as those locations where one or more NAAQSs are not being met. Unlike PSD, non-attainment is considered one pollutant at a time. Non-attainment area status can preclude any further growth in the area until the NAAQS are achieved. In 1990, NAAs for SO_2 involved about 1.7 million people; altogether, in 1990 about 74 million people lived in counties where at least one ambient air quality standard was violated (U.S. EPA, 1991). Non-attainment area issues include the method of determining NAA status and the geographic extent of the affected area. Regulatory agencies have considerable latitude here; NAA status can be based on either models or on monitored data, and the area can be as big as an entire Air Quality Control Region (several counties) or limited to a small area around a specific source.

The 1990 amendments (PL 101-549) consolidated federal authority even further and introduced some innovations to the regulation of SO_2. In Title IV, "Acid Deposition Control," 110 power plants in the eastern United States are specifically identified and mandated to reduce SO_2 emissions according to a schedule of "allowances," regardless of their "old source" status or the level of ambient air quality nearby. However, these sources are allowed to "trade" emissions rights, so that those sources that can reduce emissions more cheaply can sell some of this capability to others who are not so fortunate. Sulfur dioxide emissions reductions are to begin in the year 2000, reaching an aggregate of about 10 million tons per year by the year 2010. This section also mandates a program of NO_x emissions reductions intended to require the use of low-NO_x combustion technology, by 1995. The purpose of these blanket emissions reduction programs is to reduce acidic deposition over a large region by reducing the average emission rates while allowing the maximum flexibility as to how this overall goal is to be achieved.

AIR POLLUTION REGULATION IN BRITAIN, CANADA, AND JAPAN

Britain. The British approach to regulation of air pollution differs substantially from that of the United States, having begun with the Alkali Works Regulation Act of 1863. Under this regulation, source owners must show to the satisfaction of the chief inspector that the "best practicable means" have been provided for "preventing the escape of noxious or offensive gases to the atmosphere and for rendering such gases harmless and inoffensive." In addition, for certain processes, limits are established on the acidity of effluent gases. This is essentially regulation on a case-by-case basis, and, under this policy, British electric utilities increased stack heights to alleviate local air concentrations. The Clean Air Act of 1956 emphasized control of particulates by establishing "smokeless" zones in urban areas, in which only better quality fuels could be used for space heating. The act was extended to industry in 1968. Concerns about acid rain impacts on the continent due to uncontrolled British SO_2 emissions from tall-stack utilities led to new environmental control pressures on the United Kingdom in the 1970s and 1980s.

Canada. In Canada, there has been interplay between federal and provincial responsibilities, just as in the United States prior to 1970. The 1971 Canadian Clean Air Act was intended to assist the provinces in developing their own control programs but gave the federal government the authority to protect human health, regulate fuels, and acquire data on sources. The largest individual pollution sources in Canada are the nickel and copper smelters; the INCO smelter at Sudbury, Ontario, was the largest SO_2 source in North America and was fitted with the world's tallest stack in the 1970s in order to reduce local air concentrations. Since that time, emissions have been substantiality reduced as well. The 1988 Canadian Environmental Protection Act consolidated federal pollution control authority in all media and defined 44 substances with the potential to harm human health and the environment, regardless of the medium of transport (Hilborn and Still, 1990).

Japan. Japan may be unique among the industrialized countries in having a law that provides for compensation of the victims of adverse environmental health impacts. The first case concerned methyl mercury poisoning and paid about $3.8 million to 45 victims (Namekata, 1986; Namekata and du Florey, 1987). Other cases have concerned cadmium discharges and air pollution involving sulfur oxides; more recently, there has been heavy debate about an ambient standard for NO_2 that is about half the level of the U.S. standard (Awaji and Tsukatani, 1988). The law distinguishes the specific diseases due to heavy metal discharges from the more general problems of air pollution. Claims under the latter must meet two criteria: annual average SO_2 over 0.05 ppm, and prevalence of respiratory disease of at least two to three times the "natural" prevalence of persistent cough and phlegm (i.e., the British definition of chronic bronchitis). There are also criteria for the duration of residence in areas that meet these conditions and the existence of disease. As of the end of 1984, there were over 90,000 "certified respiratory patients" receiving compensation under this law, and the annual payout exceeded 100 billion yen (about $1 billion) in 1986 (Awaji and Tsukatani, 1988). However, Namekata also pointed out that there was no epidemiological basis for compensation on the basis of SO_2, especially since the law made no distinctions for smokers, and no account has been taken of the 75% drop in SO_2 levels since the peaks of the 1960s.

REFERENCES

American Petroleum Institute (1969), *Particulates: Air Quality Criteria Based on Health Effects*, API Air Quality Monographs, Washington, DC.

Awaji, T., and Tsukatani, T. (1988), Current Problems and Prospects with the Japanese Compensation System for Pollution-Related Health Damage, *Soc. Sci. Med.* 27: 1053–59.

Barratt, R.S. (1989), Characteristics of Air Pollution in Birmingham, England. II. On the Significance of Sampling Height, *Sci. Total Environ.* 84:149–57.

Bowler, C., and Brimblecomb, P. (1991), Battersea Power Station and Environmental Issues 1929–1989, *Atm. Environ.* 25B:143–52.

Brauer, M., Koutrakis, P., and Spengler, J.D. (1989), Personal Exposures to Acidic Aerosols and Gases, *Envir. Sci. Tech.* 23:1408–12.

Brauer, M., Koutrakis, P., Keeler, G.J., and Spengler, J.D. (1990), Indoor and Outdoor Concentrations of Acidic Aerosols and Gases, *Proc. IAQ/90*, Vol. 3, Toronto, pp. 447–52.

Brydges, T.G. (1987), Some Observations on the Public Response to Acid Rain in Canada, paper prepared for the Svante Oden Commemorative Symposium, Skokloster, Sweden.

Brimblecomb, P. (1977), London Air Pollution, 1500–1900, *Atm. Environ.* 11:1157–62.

Brimblecomb, P. (1982), Long Term Trends in London Fog, *Science Tot. Envir.* 22:19–29.

Brimblecomb, P. (1987), *The Big Smoke*, Methuen, London.

Butler, D.A., Ozkaynak, H., Billick, I.H., and Spengler, J.D. (1990), Predicting Indoor NO_2 Concentrations as a Function of Home Characteristics and Ambient NO_2 Levels, *Proc. IAQ/90*, Vol. 3, Toronto, pp. 519–24.

Clean Air Scientific Advisory Committee (CASAC) (1988), Subcommittee on Acid Aerosols, *Report on Acid Aerosol Research Needs*, EPA-SAB/CASAC-89-002, U.S. Environmental Protection Agency, Washington, DC.

Commins, B.T. (1963), Determination of Particulate Acid in Town Air, *Analyst* 88: 364–67.

Commins, B.T., and Waller, R.E. (1967), Observations from a Ten-Year Study of Pollution at a Site in the City of London, *Atm. Eniron.* 1:49–68.

Davidson, C.I. (1979), Air Pollution in Pittsburgh: A Historical Perspective, *J. APCA* 29:1035–41.

Dockery, D.W., and Spengler, J.D. (1981), Indoor-Outdoor Relationships of Respirable Sulfates and Particles, *Atm. Environ.* 15:335.

Evelyn, J. (1661), *Fumifugium: Or the Inconvenience of the Aer and Smoake of London Dissipated*, reprinted by Maxwell Reprint Co., Elmsford, NY.

Farr, W. (1885), *Vital Statistics: A memorial Volume of Selections from the Reports and Writings of William Farr*, N. Humphreys, ed., Sanitary Institute of Great Britain, London, p. 164.

Firket, J. (1936), Fog along the Meuse Valley, *Trans. Faraday Soc.* 32:1192–97.

Friedlander, S.K. (1977), *Smoke, Dust and Haze*, Wiley, New York.

Gschwandtner, G., Gschwandtner, K.C., and Eldridge, K. (1985), *Historic Emissions of Sulfur and Nitrogen Oxides in the United States from 1900 to 1980*, EPA-600/7-85-009a, U.S. Environmental Protection Agency, Washington, DC.

Hasselblad, V., Kotchmar, D.J., and Eddy, D.M. (1992), *Synthesis of Environmental Evidence: Nitrogen Dioxide Epidemiology Studies*, EPA/600/8-91/049A, U.S. Environmental Protection Agency, Research Triangle Park, NC.

Heimann, H. (1970), Episodic Air Pollution in Metropolitan Boston, *Arch. Environ. Health* 20:239–51.

Hilborn, J., and Still, M. (1990), *Canadian Perspectives on Air Pollution*, SOE Report No. 90–1, Environment Canada, Ottawa.

Holmes, J.A., Franklin, E.C., and Gould, R.A. (1915), *Report of the Selby Smelter Commission*, Bureau of Mines Bull. 98, Washington, DC.

Holzworth, G.C. (1972), *Mixing Heights, Wind Speeds, and Potential for Urban Air Pollution throughout the Contiguous United States*, AP-101, U.S. Environmental Protection Agency, Research Triangle Park, NC.

Hosein, H.R., Corey, P., and Silverman, F. (1990), Air Pollution Models Based on

Personal, Indoor and Outdoor Exposure, *Proc. IAQ/90*, Vol. 3, Toronto, pp. 423–28.

International Electric Research Exchange (IERE) (Nov. 1981), *Effects of SO_2 and Its Derivatives on Health and Ecology. Vol. 1. Human Health*, available from EPRI Research Reports Center, P.O. Box 50490, Palo Alto, CA 94303.

Interstate Sanitation Commission (1958), *Smoke and Air Pollution, New York–New Jersey*, Interstate Sanitation Commission, New York.

Ito, K., and Thurston, G.D. (1987), The Estimation of London, England, Aerosol Exposures from Historical Visibility Records, Paper 87-47.2, presented at the 80th Annual Meeting of the Air Pollution Control Association, New York.

Jensen, N.O., and Petersen, E.L. (1979), The Box Model and the Acoustic Sounder, a Case Study, *Atm. Environ.* 13:717–20.

Lawther, P.J., Ellison, J. McK., and Waller, R.E. (1968), Some Medical Aspects of Aerosol Research, *Proc. Roy. Soc. A* 307:223–34.

Lee, R.E., Jr., Caldwell, J.S., and Morgan, G.B. (1972), The Evaluation of Methods for Measuring Suspended Particulates in Air, *Atm. Environ.* 6:593–622.

Lipfert, F.W. (1988), *Exposure to Acidic Sulfates in the Atmosphere: A Review and Assessment*, Electric Power Research Institute Report EA-6150, Palo Alto, CA.

Lipfert, F.W. (1989), Dry Deposition Velocity as an Indicator for SO_2 Damage to Materials, *J. APCA* 39:446–52.

Lipfert, F.W. (1992), An Assessment of Acid Fog, presented at the 9th World Clean Air Conference, Montreal, Canada.

Lipfert, F.W., and Dupuis, L.R. (1985), *Methods for Mesoscale Modeling for Materials Damage Assessment*, BNL 37508, Brookhaven National Laboratory, Upton, NY.

Lipfert, F.W., Kroetz, C.A., and Mahoney, J.R. (1973), A Comparison of Air Pollution Impacts Associated with Electric vs. Oil-Fired Space and Water Heating, Proc. 8th Intersociety Energy Conversion Engineering Conference, Philadelphia, PA pp. 623–30.

Lodge, J.P. (1956), discussion in *Proc. Air Pollution Research Planning Seminar*, Cincinnati, OH, U.S. HEW, p. 32.

Magill, P.L. (1949), The Los Angeles Smog Problem, *Ind. Eng. Chem.* 41:2476–86.

Milburn-Hopwood, S. (1989), The Role of Science in Environmental Policy Making: A Case Study of the Canadian Acid Rain Policy, M.S. Thesis, University of Toronto.

Mitchell, D.J., Ireland, R.B., Hume, A., McVey, E., and Blakey, D.L. (1974), Epidemiologic Notes and Reports, Carbon Monoxide Poisoning–Mississippi, *Morbidity and Mortality Weekly Report* 23:1.

Mumford, J.L., *et al.* (1991), Indoor Air Pollutants from Unvented Kerosene Heater Emissions in Mobile Homes: Studies on Particles, Semivolatile Organics, Carbon Monoxide, and Mutagenicity, *Environ. Sci. Tech.* 25:1732.

Namekata, T. (1986), The Japanese Compensation System for Respiratory Disease Patients: Background, Status and Trend, and Review of Epidemiological Basis, in Aerosols: Research, Risk Assessment and Control Strategies, *Proc. 2nd US-Dutch International Symposium*, Lee, Schneider, Grant, and Verkek, eds., Williamsburg, VA, pp. 1159–69.

Namekata, T., and Florey, C., du, eds. (1987), *Health Effects of Air Pollution and the Japanese Compensation Law*, Battelle Press, Columbus, OH.

Neuberger, H., and Gutnick, M. (1949), Experimental Study of the Effect of Air Pollution on the Persistence of Fog, in *Proc. First National Air Pollution Symposium*, Pasadena, CA, Stanford Research Institute, Menlo Park, CA, pp. 90–96.

Pashel, G.E., and Egner, D.R. (1981), A Comparison of Ambient Suspended Particulate Matter Concentrations as Measured by the British Smoke Sampler and the High Volume Sampler at 16 Sites in the United States, *Atm. Environ.* 15:919–27.

Petersen, G.A., and Sabersky, R.H. (1975), Measurements of Pollutants inside an Automobile, *J. APCA* 25:1028.

Pollack, A.L., and Burton, C.S. (1985), *Trends in Sulfur Dioxide Emissions from The Electric Utility Industry and Ambient Sulfur Dioxide Concentrations in the Northeastern United States, 1975 to 1982*, EPA/600/3-85/035, U.S. Environmental Protection Agency, Research Triangle Park, NC.

Ripley, R.B. (1969), Congress and Clean Air, in *Congress and Urban Problems*, F.N. Cleaveland and Associates, eds., The Brookings Institute, Washington, DC, p. 224.

Roberts, J.M., *et al.* (Aug. 1990), *Relationships between PAN and Ozone at Sites in Eastern North America*, BNL informal report, Brookhaven National Laboratory, Upton, NY.

Sehmel, G.A. (1980), Model Predictions and a Summary of Dry Deposition Velocity Data, in *Atmospheric Sulfur Deposition, Environmental Impact and Health Effects*, D.S. Shriner, C.R. Richmond, and S.E. Lindberg, eds., Ann Arbor Science, Ann Arbor, MI, pp. 223–35.

Shaw, N., and Owens, J.S. (1925), *The Smoke Problem of Great Cities*, Constable & Co., London, p. 42.

Sinclair, J.D., Psota-Kelty, L.A., and Weschler, C.J. (1988), Indoor/Outdoor Ratios and Indoor Surface Accumulations of Ionic Substances at Newark, New Jersey, *Atm. Environ.* 22:461.

Sinclair, J.D., Psota-Kelty, L.A., Ibidunni, A.O., and Peins, G.A. (1990a), Indoor/Outdoor Relationships of Airborne Ionic Contaminants: Comparison of Electronic Equipment Offices and a Factory Environment, *Proc. IAQ/90*, Vol. 3, Toronto, pp. 601–05.

Sinclair, J.D., Psota-Kelty, L.A., Weschler, C.J., and Shields, H.C. (1990b), Measurement and Modeling of Airborne Concentrations and Indoor Surface Accumulations of Ionic Substances at Neenah, Wisconsin, *Atm. Environ.* 24A:627.

Spengler, J.D., Dockery, D.W., Turner, W.A., Wolfson, J.M., and Ferris, B.G., Jr. (1981), Long-Term Measurements of Respirable Sulfates and Particles inside and outside Homes, *Atm. Environ.* 15:23.

Stern, A.C. (1982), History of Air Pollution Legislation in the United States, *J. APCA* 32:44–61.

Stocks, P. (1960), On the Relations between Atmospheric Pollution in Urban and Rural Localities and Mortality from Lung Cancer, Bronchitis, and Pneumonia, with Particular Reference to 3:4 Benzopyrene, Beryllium, Molybdenum, Vanadium, and Arsenic, *Brit. J. Cancer* 14:397–418.

Swain, R.E. (1950), Smoke and Fume Investigations, *Ind. Eng. Chem.* 41:2384–88.

Tarr, J.A. (1984), The Search for the Ultimate Sink: Urban Air, Land and Water Pollution in Historical Perspective, *Records of the Columbia Historical Society of Washington, DC*, University of Virginia Press, Charlottesville, VA.

TRW Systems Group (1969), Air Quality Display Model, prepared for U.S. Dept. of Health, Education and Welfare, available from NTIS, U.S. Dept. Of Commerce, Springfield, VA. PB-189194.

U.S. Department of Energy (1991), *State Energy Data Report, Consumption Estimates 1960–1989*, DOE/EIA-0214(89), Energy Information Administration, Washington, DC.

U.S. Department of Health, Education, and Welfare (1969a), *1962–1967 Summary of Monthly Means and Maximums—Continuous Air Monitoring Projects*, National Air Pollution Control Administration Publication APTD 69-1, Washington, DC.

U.S. Department of Health, Education, and Welfare (1969b), *Air Quality Criteria for Particulate Matter*, National Air Pollution Control Administration Publication AP-49, Washington, DC.

U.S. Department of Health, Education, and Welfare (1969c), *Air Quality Criteria for Sulfur Oxides*, National Air Pollution Control Administration Publication AP-50, Washington, DC.

U.S. Department of Health, Education, and Welfare (1970a), *Air Quality Criteria for*

Carbon Monoxide, National Air Pollution Control Administration Publication AP-62, Washington, DC.

U.S. Department of Health, Education, and Welfare (1970b), *Air Quality Criteria for Photochemical Oxidants*, National Air Pollution Control Administration Publication AP-63, Washington, DC.

U.S. Department of Health, Education, and Welfare (1971), *Air Quality Criteria for Nitrogen Oxides*, National Air Pollution Control Administration Publication AP-84, Washington, DC.

U.S. Environmental Protection Agency (1973), *User's Guide for the Climatological Display Model*, EPA-R4-73-024, Research Triangle Park, NC.

U.S. Environmental Protection Agency (1978), *Air Quality Criteria for Ozone and Other Photochemical Oxidants*, Report EPA-600/8-78-004, Research Triangle Park, NC.

U.S. Environmental Protection Agency (1982a), *Air Quality Criteria for Particulate Matter and Sulfur Oxides*, Report EPA-600/8-82-029a (in 5 volumes), Research Triangle Park, NC.

U.S. Environmental Protection Agency (1982b), *Air Quality Criteria for Oxides of Nitrogen*, Report EPA-600/8-82-026F, Research Triangle Park, NC.

U.S. Environmental Protection Agency (1986), *Air Quality Criteria for Ozone and Other Photochemical Oxidants*, Report EPA-600/8-84-020aF, 5 vols., Research Triangle Park, NC.

U.S. Environmental Protection Agency (1989a), *An Acid Aerosols Issue Paper*, Report EPA-600/8-88-005F, Washington, DC.

U.S. Environmental Protection Agency (1989b), *National Air Pollution Emission Estimates, 1940–1987*, EPA-450/4-88-022, Research Triangle Park, NC.

U.S. Environmental Protection Agency (1990), *Air Quality Criteria for Carbon Monoxide*, Draft Report EPA-600/8-90-045A, Research Triangle Park, NC.

U.S. Environmental Protection Agency (1991), *National Air Quality and Emissions Trends Report, 1990*, EPA-450/4-91-023, Office of Air Quality Planning and Standards, Research Triangle Park, NC.

Wechsler, C.J., and Shields, H.C. (1988), The Influence of HVAC Operation on the Concentrations of Indoor Airborne Particles, in *Engineering Solutions to Indoor Air Problems: IAQ/88*, American Society of Heating, Refrigerating, and Air-Conditioning Engineers, Atlanta, pp. 166–81.

Weschler, C.J., Shields, H.C., and Nalk, D.V. (1989), Indoor Ozone Exposures, *J. APCA* 39:1562.

Wilkins, E.T. (1954), Air Pollution in a London Smog, *Mech. Eng.* 76:426–28.

Williams-Freeman, J.P. (1892), On the Importance of More Actively Enforcing the Ventilation of Public and other Buildings, Suggested as a Standard of Impurity of Air as a Basis of Prosecutions, in *Trans. 7th Int. Congress of Hygiene and Demography*, Vol. I, Eyre and Spottiswoode, London, pp. 393–97.

Wyzga, R.E., and Lipfert, F.W. (1992), Exposure Assessment for Particulates: A Pilot Study of a Group of Susceptibles, Paper 92-65.01, presented at the Annual Meeting of the Air and Waste Management Association, Kansas City, MO.

APPENDIX 2A EFFECTS OF POPULATION DENSITY ON MORTALITY

Farr (1885) found that population density was a strong predictor of general mortality in nineteenth-century England. He discusses how personal life-styles differed little between rural and urban areas, and how hard life was even in uncrowded surroundings. He then concluded that the striking increase in mortality that was seen with increasing population density (Figure 2A-1) must be due to the community environment, especially sanitary conditions. He calculated that, on average, mortality rate was proportional to population density to the 0.12 power, a figure that is strikingly close to the logarithmic regression coefficient that may be computed from his tabulated data. This relationship could be attributed to the spread of communicable diseases and to air pollution from space heating, since in an area source (i.e., city), the air concentration is proportional to the areal density of emissions (Eq. 2-1). Communicable diseases have been conquered in modern industrial societies, but a certain amount of air pollution remains proportional to population density.

Figure 2A-1 compares Farr's data with regression results for U.S. SMSAs in 1969 (Lipfert, 1984) and cities in 1980 (Lipfert *et al.*, 1988). The (positive) effect of population density on mortality has all but disappeared; in some of the other regression analyses of the spatial variation of U.S. mortality, the association is negative, presumably because of better medical care in larger

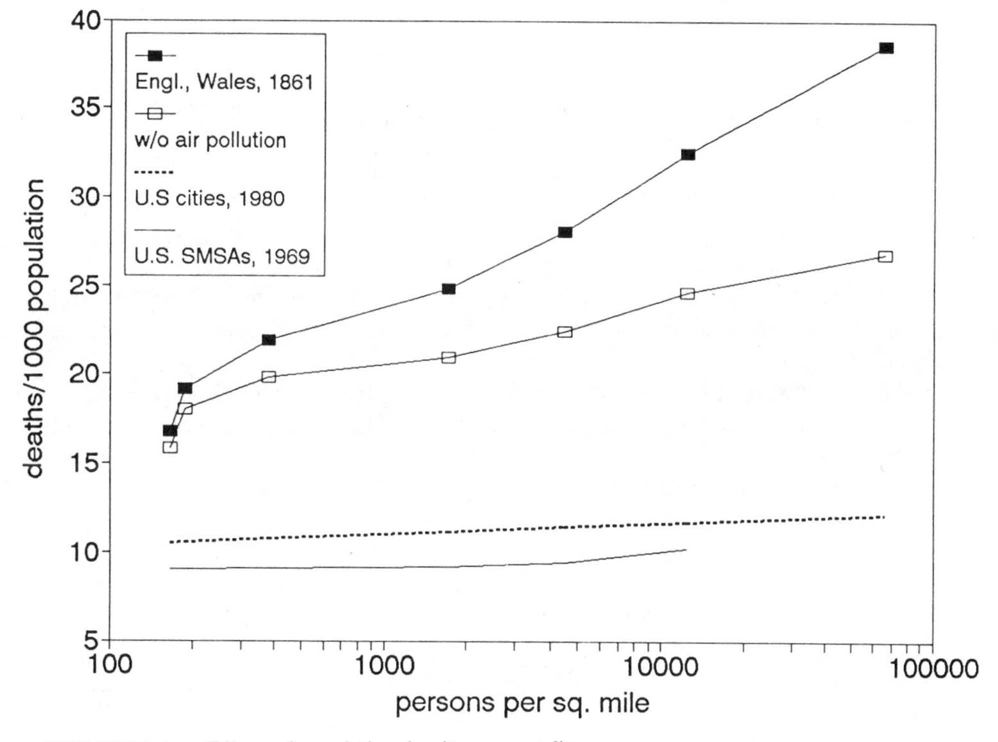

FIGURE 2A-1. Effects of population density on mortality.

cities. Note that the current population density in Manhattan is about the same as the densest districts in nineteenth-century England; however, the use of the vertical dimension in modern cities allows this utilization of land to be achieved without the severe overcrowded living conditions that must have been present in low-rise housing. The U.S. data in Figure 2A-1 are from multiple regressions that allow the effects of air pollution and poverty to be accounted for separately, which is not the case with Farr's data for England. The plot for 1861 England and Wales "without air pollution" was created by estimating the air pollution levels based on coal use and adjusting the mortality rates according to cross-sectional regression results. This admittedly crude exercise suggests that about half the incremental effect of population density on mortality in Victorian England might have been due to air pollution. The other half was presumably due to increased spreading of infectious diseases and other socioeconomic factors.

REFERENCES

Farr, W. (1885), *Vital Statistics: A Memorial Volume of Selections from the Reports and Writings of William Farr*, N. Humphreys, ed., Sanitary Institute of Great Britain, London, p. 164.

Lipfert, F.W. (1984), Air Pollution and Mortality: Specification Searches Using SMSA-Based Data, *J. Environ. Econ. & Mgmt.* 11:208–43.

Lipfert, F.W. Malone, R.G., Daum, M.L., Mendell, N.R., and Yang, C.-C., (1988), *A Statistical Study of the Macroepidemiology of Air Pollution and Total Mortality*, BNL Report 52122, Brookhaven National Laboratory, Upton, NY.

APPENDIX 2B OTHER AIR POLLUTANTS USED IN HEALTH EFFECTS STUDIES

The preceding discussion has emphasized the criteria pollutants that have been defined in the United States as having the potential for adverse effects on human health. There are many other substances present in ambient air that could be classified as pollutants, and some of these have been considered in health studies.

Most of these miscellaneous compounds are particles, and their presence is identified by analyzing the material collected on filters. There are three general classifications used: water-soluble ions, principally SO_4^{-2}, NO_3^-, and NH_4^+, organic compounds, and trace metals.

Of the water-soluble ions, only sulfate measurements are considered reasonably reliable in historical data bases, and even these data have often been compromised by conversion of (gaseous) SO_2 to SO_4^{-2} on alkaline filters (such as glass-fiber). This positive artifact tends to overstate the true particulate sulfate levels in locations with high SO_2 by about 2 to 25% of the SO_2. A similar phenomenon affects nitrate measurements, by absorption of gaseous nitric acid on filters. However, ammonium nitrate is an unstable compound and can decompose into ammonia (NH_3) and HNO_3, both of which may volatilize,

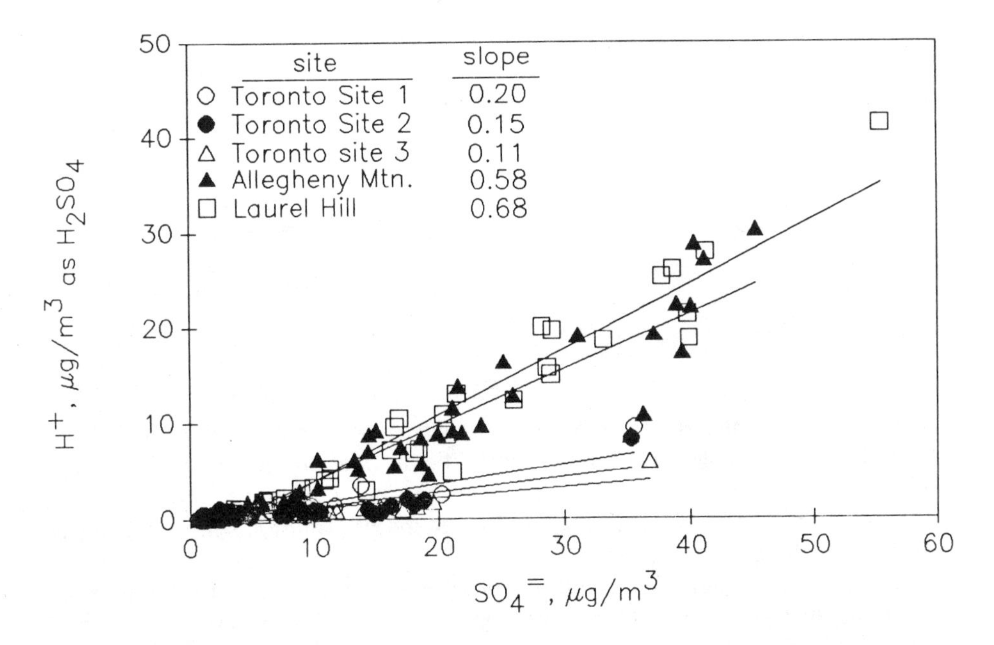

FIGURE 2B-1. Aerosol acidity as a function of SO_4^{-2}. *Data from Waldman* et al. *(1990) and Pierson* et al. *(1989).*

creating a negative artifact. Particulate nitrate data from routine glass-fiber filters are thus generally considered unreliable, as are ammonium ion (NH_4^+) data (due to volatilization over time and loss of NH_3).

Some of the more recent studies of health effects have been concerned with the *acidity* (H^+) of the aerosol, which is usually determined by measuring the pH of an aqueous extract of collected particles. It is important that such samples not be neutralized by ambient ammonia after collection. Occasionally there is confusion between the net acidity of an aerosol mixture, the content of H_2SO_4, and the concentration of all sulfates (SO_4^{-2}). Although most of the acidity in ambient aerosols is derived from sulfate compounds, the amount of atmospheric neutralization can be quite variable (Figure 2B-1). Thus, a clear distinction must be maintained between studies involving acidity *per se* (H^+) and those involving SO_4^{-2}.

The organic fraction of particulate matter most often reported in the older data bases was obtained by extraction with benzene and was reported as "benzene soluble organics" (BSO). The only specific organic compound that has received much attention is benzo(alpha)pyrene, B(a)P or BaP, which is a carcinogen emitted from incomplete combustion, especially from coke ovens.

Trace metal determinations were usually made by ashing the filters and using spectroscopic methods to compare metal concentrations to those of an unexposed ("blank") filter. Metals of interest include lead, iron, manganese, cadmium, vanadium, and arsenic. Mercury is a toxic metal of concern but is usually emitted as a vapor in very low concentrations and thus cannot be determined from a routine examination of filter deposits. Iron and manganese are "signatures" for ferrous metal industries, although iron is also an important component of soil. Lead (which is now a criteria pollutant) is emitted primarily

TABLE 2B-1 Selected Historical TSP Data

City	Period	Maximum TSP	Average TSP
Buffalo	1961–63	206 (as TSP) 161 (from dustfall)	112
Chicago	1946 (winter)	539	177
Detroit	1941–2 (winter)	550	299
Pittsburgh	1938–40	728	313
Cincinnati	1930–40	442	144

from leaded gasoline and from primary ore smelters. Vanadium is a "signature" for heavy residual fuel oil from certain oil fields. The interactions of these and other various characteristic source signatures have been used in statistical manipulations to identify the contributions of major pollution source categories to the overall particle loading.

Many of the older studies used crude integral measures of air pollution, including dustfall or sootfall and sulfation rate. Dustfall is simply the amount of material collected in an open container over a specified time period, usually a month. Precipitation was included in these collections, but the total catch was analyzed for insoluble material and some ions, such as SO_4^{-2} and chloride ion (Cl^-). There is a rough relationship between dustfall and TSPs, which can vary with location and average particle size. In the Erie County (NY) Respiratory Study (Winkelstein *et al.*, 1967), the correlation between TSP and dustfall was 0.84, and the relationship between the two was given by ($TSP[\mu g/m^3] = 40.7 + 80.6 * Dustfall\ [mg/cm^2 mo]$). This provides a rough basis for estimates of TSP levels in other locations based on dustfall measurements for the early years. Such estimates are useful in comparing environmental conditions during the older studies versus the more recent ones. According to this formula, the long-term average TSP values ($\mu g/m^3$), as shown on Table 2B-1, would pertain.

Sulfation rate is an integral measure of the amount of lead dioxide (PbO_2) converted to sulfate over some period of time, usually a month. Since the SO_2 in the air must deposit to the surface, this metric measures dry deposition, not just air concentration, and is thus sensitive to wind speed and turbulence. As a rule of thumb, 1 part per billion (ppb) of SO_2 will deposit about 1 mg sulfur trioxide (SO_3) on 100 cm^2 per month, under ideal conditions.

REFERENCE

Winkelstein, W. Jr., Kantor, S., Davis, E.W., Maneri, C.S., and Mosher, W.E., (1967), The Relationship of Air Pollution and Economic Status to Total Mortality and Selected Respiratory System Mortality in Men, *Arch. Env. Health* 14:162–71.

APPENDIX 2C CONVERSION FACTORS FOR CONCENTRATION UNITS

To convert from ppm units (volumetric) to mass concentration units (μg/m^3) at 25°C and one atmosphere, multiply by the following factors:

Species	*Factor*
SO_2	2620
NO_2	1880
NO	1230
CO	1150
O_3	1960
CH_4	655

To refer a mass concentration χ to volumetric units at a temperature (T) other than 25°C, use the formula

$$C_T = \chi/(\text{factor}) \times (T + 273.15)/298.15$$

To convert from ppm to ppb, divide by 1000. To convert from μg/m^3 to mg/m^3, divide by 1000.

APPENDIX 2D THE RELATIONSHIP BETWEEN PLUME OPACITY OR DARKNESS AND PARTICULATE MASS EMISSION RATE

The primary tool for air pollution control for many years was the smoke inspector's judgment as to the color or opacity of the smoke issuing from the chimney, as compared to a standard chart known as the Ringelmann chart. "Ringelmann numbers" were graded on an arbitrary scale from zero (invisible) to 5 (solid black). Intermediate Ringelmann numbers are also given as percentage opacity figures: Ringelmann 1 = 20% opacity, and so on. Unfortunately, this measure was not only subjective, but it also was influenced by the diameter of the chimney and by the position of the sun relative to the observer, as well as by steam condensation. Nevertheless, this standard gave the authorities a rudimentary tool with which to pursue at least the most egregious offenders. Modern facilities may now be equipped with in-stack transmissometers that provide boiler operators with instant information on combustion or control equipment conditions.

The crudeness of the Ringelmann number may be the primary reason why so little information is found in the technical literature on its relationship to other parameters of combustion conditions and smoke emission rates. Experiments were conducted in the 1940s in Britain on a hand-fired boiler burning

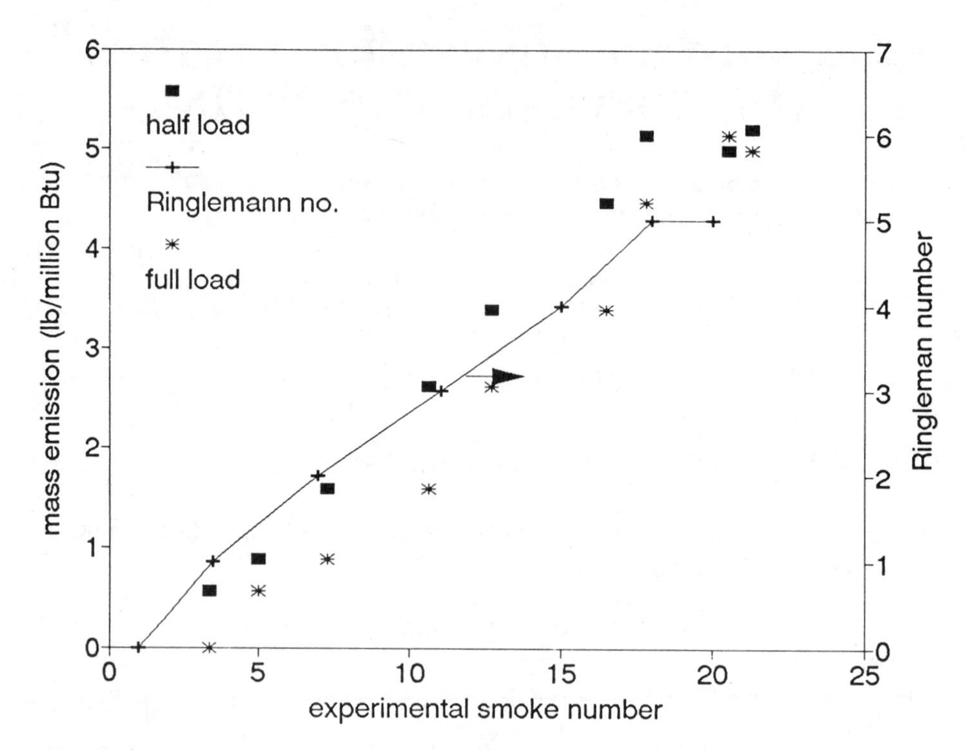

FIGURE 2D-1. Relationship between plume opacity and particle mass emission rate for experimental coal combustion. *Data from DSIR (1946).*

medium sulfur, low-ash coal (DSIR, 1946). The full load of the boiler was 5000 lb of steam per hour; tests were made at full and half loads. Measurements consisted of fuel analyses, stack gas composition, optical light transmission in the stack, and particulate mass and its composition. The authors point out that the mass emitted depends on the size of the unit and the load it carried and thus must be expressed in concentration units in order to be transferable to other situations. I transformed the data from g/1000 ft^3 at stack exit conditions to lb per unit of heat input, which is the unit used in the United States for emissions standards. The results are given in Figure 2D-1.

The figure shows a nearly proportional relationship between Ringelmann number and mass emission rate, which is essentially the same for both load conditions. Whether this relationship holds for other facilities cannot be determined. It is interesting to note that Table 2-4 gives particulate emission rates for coal stoves from 0.6 to 6.0 lb/million Btu, which is very close to the range given in Figure 2D-1. The U.S. particulate emission standard for new power plants is 0.03 lb/million Btu, by way of comparison.

These results are reasonably consistent with laboratory tests reported by Conner and Hodkinson (1972), who used artificially generated smoke. Their relationship was of the form

$$\log(\%\ \text{light transmission}) = 2 - 0.77\,M$$

where M is the mass density in the plume in g/m^3. The mass loadings according to this relationship are lighter than those shown in Figure 2D-1 at low opacities but agree very well for dense plumes.

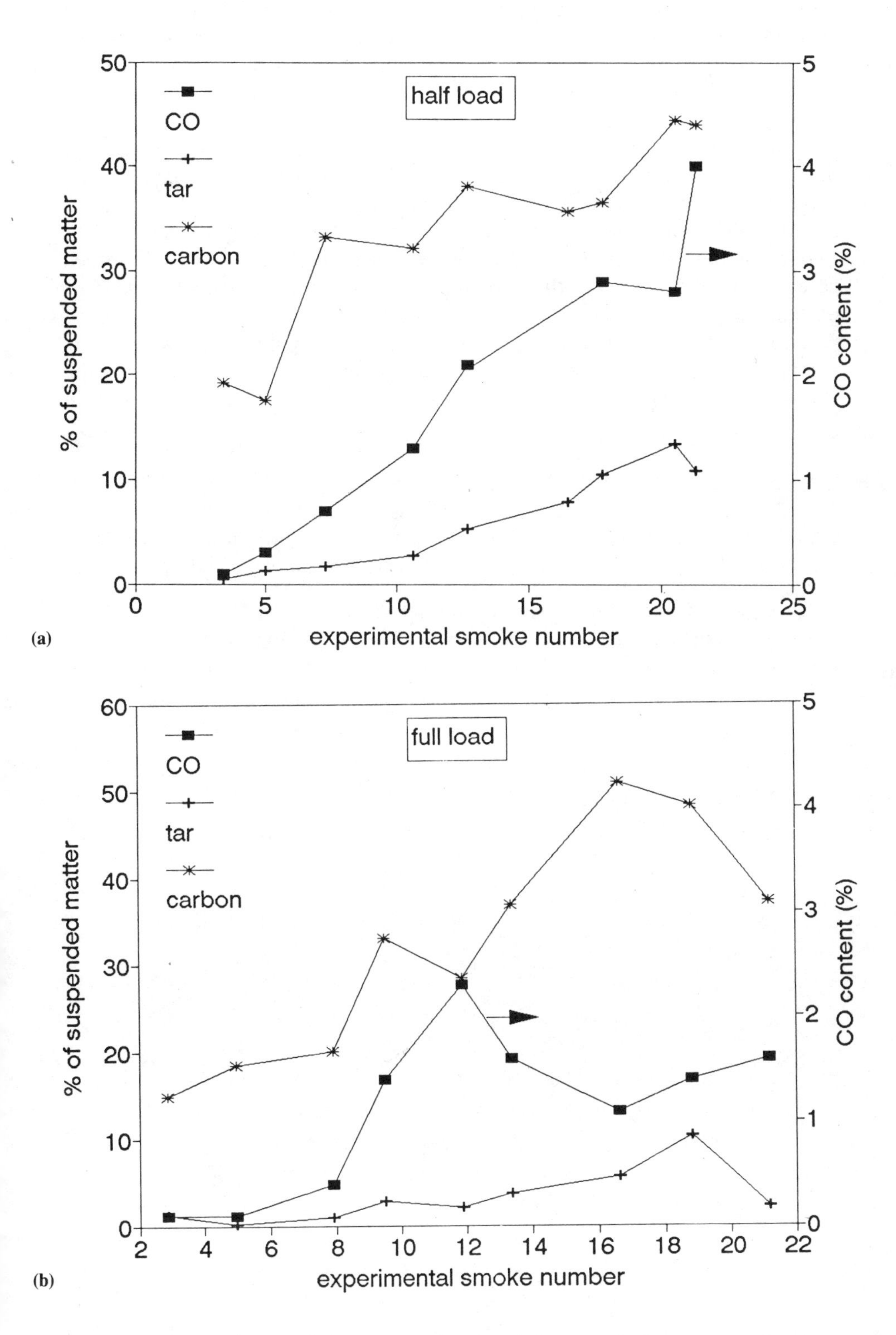

FIGURE 2D-2. Pollutant emission rates as a function of plume opacity for experimental coal combustion. (a) Half load. (b) Full load. *Data from DSIR (1946).*

In London, before smoke control was widely adopted, ambient smoke and SO_2 levels were comparable, so we may assume that emission rates were also roughly comparable. This might place the average smoke emission rate at about 2 lb/million Btu (smoke number about 8, Ringelmann number about 2 (40% opacity)).

Figures 2D-2a, b present the available data from the DSIR report on pollutant emission rates. "Tar" and carbon were determined from separate aliquots, the former by extraction with carbon disulfide. The pollutants associated with incomplete combustion are seen to increase dramatically with smoke number. At a smoke number of 8, we might expect 1% CO in the stack gases, 20 to 30% carbon in the particulate matter, but only a small percentage of "tar."

These data are presented here in part because they may be unique in the technical literature but also to aid in estimating ambient conditions during the severe air pollution episodes that occurred before air pollution controls were in place.

REFERENCES

Conner, W.D., and Hodkinson, J.R. (1972), *Optical Properties and Visual Effects of Smoke-Stack Plumes*, U.S. Environmental Protection Agency, Research Triangle Park, NC.

Department of Scientific and Industrial Research (DSIR) (1946), *Smoke and Its Measurement*, Fuel Research Technical Paper no. 53, HMSO, London.

3

Methodological and Statistical Considerations

The glitter of the t *table diverts attention from the inadequacies of the fare.*

Sir A.B. Hill, 1965

Statistics is the branch of mathematics dealing with collecting data, analyzing them, and drawing conclusions from them (Snedecor and Cochran, 1967). We also use the principles of statistics to test hypotheses, that is, to decide whether a particular finding may reasonably have resulted solely from the vagaries of chance. For studies of air pollution, we use statistics to define exposures and risks and to compare the contributions of the various terms of multivariate relationships. However, statistics alone cannot make the final determination of whether an apparent association is causal or circumstantial.

EPIDEMIOLOGICAL METHODS

Epidemiology differs from clinical medicine or biomedical research by virtue of its study of populations rather than individual cases or specimens. Often, this emphasis stems from a fundamental objective of epidemiology: to improve public health (Kleinbaum *et al.*, 1982). However, the health effects of air pollution are usually subtle (i.e., associations are weak) and can only be observed in large populations, for which consideration of individual cases is clearly impractical.

To consider a numerical example, the daily mortality rate in a typical U.S. city of one million people is about 20 deaths per day. If this rate were to double for a few days due to an air pollution disaster, for example, only about 0.005% of the population would have been affected. Within a city of this size, considerable variation in air pollution exposure would be expected, even during a stagnation episode. Since those individuals most at risk cannot be identified precisely *a priori*, a large number of people would have to be monitored in order to determine the individual exposures of the decedents.

Szklo (1987) points out the problems of extrapolation from experimental panel studies to large populations and warns that fundamentally weak relationships are difficult to detect in all kinds of study designs.

Population Considerations and the Ecological Fallacy

Studies of population health responses to air pollution are thus necessarily observational, that is, involving naturally occurring rather than manipulated environmental conditions (Kleinbaum *et al.*, 1982). Since the characterization of individual environmental exposures is clearly impractical, such an epidemiological study is likely to be ecological* as well as observational, involving the study of group attributes rather than those of individuals (Piantidosi *et al.*, 1988). According to Kleinbaum *et al.* (1982), the primary feature of an ecological study is the lack of knowledge of the joint distribution of the study factor (i.e., exposure to air pollution) and the disease within each group. The primary weakness in ecological regression relates to the lack of specificity of the affected individuals and the exposed individuals, because groups are used in the regression analysis. This problem is most critical when the pollutant is very localized (such as emissions from a toxic waste dump) or when the disease is relatively rare (such as leukemia). However, the problem diminishes for broadly distributed regional pollutants, such as fine particles or sulfates, and for mortality from all causes or from very common causes (such as heart disease). In such instances, it is reasonable to assume that all the cases were exposed at least to some degree, although the particulars of their exposures will vary because of indoor-outdoor pollutant relationships and other local factors.

Greenland and Morgenstern (1989) present some examples in which ecological analyses clearly lead to incorrect results. Their first example introduces the concept of "effect modification," in which the effect (i.e., slope of the dose-response function) depends on the proportion of exposed subjects in each of the regions used for ecological regression. In this case, the esophageal cancer rate for nonsmokers was constant across all regions, but the rates for smokers increased as the proportion of smokers in the region decreased. Such a situation might occur due to selective survival: A low proportion of smokers could result from smokers' higher rates of cancer and heart disease and subsequent demise from these conditions. In this example, the numbers of smokers' cancer deaths was constant across regions, even though the absolute numbers of smokers varied by a factor of 1.67. It is difficult to conceive a physiological rationale for such a situation, since mortality rates are much too low to deplete the population, even for smokers. Greenland and Morgenstern's other two examples also had somewhat special situations involving two compet-

* Perhaps the most ecological of published ecological studies is that of Kagamimori *et al.* (1990), who showed a positive correlation between the prevalence of respiratory symptoms and annual tree ring growth. The authors reasoned that both trends were the result of air pollution changes as a result of upgrading a nearby power station.

ing cancer risks, smoking being the "confounding" variable. In both examples, the regional rates for the two risk factors were strongly negatively correlated ($r = -1$ and -0.7). One of the most important checks to be made a multifactor ecological analysis is for the existence of such confounding situations. If the desired risk factor is strongly correlated (+ or −) with other variables, disentanglement will be difficult if not impossible, in all cases, ecological or not. These examples also involved use of a linear regression model in situations where the effects had been established as multiplicative. The confounding factor (smoking) increased the risk by factors of 5 and 10, respectively. In such extreme situations, it should not be surprising to find anomalous results. In Greenland and Morgenstern's examples, the difficulties arose from the assumed properties of the regions, not from differences between group and individual behavior. One might thus argue that the chances of such situations seriously affecting an analysis will diminish as the size of the groups decreases and their numbers and diversity increase (Robinson, 1950).

However, Piantidosi *et al.* (1988) compared regression coefficients for a group of physiological and dietary variables, contrasting the "true" values obtained from individuals with those obtained from various aggregations. In four out of the 13 cases, the true values lay outside the confidence limits for the aggregated values, with errors in both directions.

Time-Series Studies

For a time-series analysis, the group is the single city or other geographic entity whose temporal responses are being studied, and the "within-group" variation is temporal. The time intervals used in such studies range from daily to monthly periods. Since for each day, month, and so on, a different subgroup is likely to respond (die, be admitted to hospital, etc.), the ecological hypothesis is that the same set of air monitoring locations faithfully represents the actual exposures of these different subgroups, for all intervals throughout the period studied. The term "ecological fallacy" refers to a situation where this hypothesis is not supported. The likelihood of such support depends strongly on the size of the area being studied and the spatial coverage of the air monitoring network, in addition to the duration of the time periods being studied. It follows that daily weather corrections will not be spatially dependent, but this is not necessarily true for air pollution because of the presence of local sources. (During the infamous London fogs, for example, for which we are fortunate to have multistation air quality data, the location of maximum air pollution varied. For calm situations, it was in central London, as expected, but with light winds, it was found downwind in the more outlying areas.) Given a sufficiently large data set, the net result of random error in the independent variable is to bias the regression coefficient downward, as will be discussed in more detail. Some insights into the validity of the ecological hypothesis might be gained for time-series analyses by subdividing the population into several subgroups, for example according to the availability of air monitoring data or proximity of monitors to residences. One would then examine whether the regression results for each of the subgroups corresponded to their exposures. Such a procedure conforms to Cohen's (1990) call for data sets large enough to be stratified.

Cross-Sectional Studies

For cross-sectional analyses, spatial variation constitutes the within-group variance at issue with respect to the ecological fallacy. We desire that each of the cities or locations we are studying have the same within-city spatial distribution of air quality (assuming that adequate monitoring networks are not always available) and the same within-city distributions of potential confounding variables, such as age, race, poverty neighborhoods, and so on. This is not likely to be true in general, but it can easily be seen that these considerations favor the use of the smallest possible units for geographic analysis. As larger geographic units are used for analysis, for example, Standard Metropolitan Statistical Areas (SMSAs),* the representativeness of air monitoring is quite likely to diminish. In some of the studies reviewed in Chapter 7, only one monitoring station was available in a given metropolitan area.

There can be important regional biases in the spatial distributions within SMSAs or counties. The large urban centers of the Northeast and West Coast often contain contiguous SMSAs that may be more homogeneous than isolated SMSAs in other parts of the country. These characteristics are not independent of air pollution, which varies both regionally (more sulfur in the East, more ozone in Southern California) and according to the activities of the area. Industrial SMSAs may have centrally located poor neighborhoods, while in the South, poverty pockets are often found in the outskirts of cities. Some pollutants are higher in central cities (CO, particulates), while some may be higher in the suburbs (ozone, aerosol acidity). Use of successively larger geographic units of analysis surrounding an air monitoring station creates a bias, since the population characteristics are averaged over the area, but the air pollution data used in the analysis remains unchanged (Lipfert, 1980a, b). Thus the nature of the central city with respect to its suburbs is an important parameter to consider when selecting the geographic unit of analysis.

It is thus not possible to predict all the biases involved when large geographic areas are used in cross-sectional analysis, but it can safely be stated that smaller areas are preferable, as long as the populations are large enough for stable health statistics. Morgenstern (1982) calls for smaller and more numerous units of analysis to improve the precision of estimates, and Cohen (1990) argues that there is safety in numbers, that using large numbers of observations in a geographic study reduces the chance for serious ecological bias. The same claim may be made for confounding (see that section).

Combined Spatial-Temporal Analyses

This mode of analysis is relatively rare in the literature but can provide information not otherwise available (see Chapter 8). However, the potential pitfalls of both time-series and cross-sectional analyses apply, and such studies must be carefully designed to avoid them.

* Groups of counties surrounding a central city of 50,000 or more, now called Metropolitan Statistical Areas (MSAs).

Interactions between Air Pollution and the Size of Geographic Unit

The accuracy with which exposure to air pollution can be estimated will also vary with the nature of the pollutant. Some primary pollutants, such as TSP, CO, and SO_2, tend to be very local, and concentrations may vary substantially within a few city blocks. Secondary pollutants, such as NO_2, oxidants, and sulfate particles, may exhibit less spatial variability, although ozone can be strongly attenuated locally by the presence of NO_x sources. Most authors of cross-sectional studies have had to work with data from a few air pollution monitors and have made arbitrary assumptions about the size of the area that each monitor represents. The lack of true representation of the air pollution exposure of the population constitutes an important source of error in the independent variables.

This source of error is also associated with the choice of the type of political subdivision for the observational unit, since the larger its area, the larger the chances for errors in estimating true population exposures (assuming a fixed number of monitors and that local pollution sources are present). For example, assume that there is a true relationship between particle concentration and mortality (this need not be a causal relationship, since there may be other aspects of the pollution source to consider, such as occupational factors). Often, two measures of particle concentration have been available: total suspended particles (TSP), which tend to be somewhat local, because they may include particles up to 50 μm in diameter, and the sulfate portion of the particulate catch, which is usually distributed regionally, since the particles are much smaller and travel further. (Recently in the United States, particulate monitoring has led to the distinction between fine and coarse particles.) When relatively small areas (such as cities or portions of cities) are used as the observational units, TSP exposures may be reasonably well represented. On the other hand, if larger units are used with the same monitoring network, such as entire counties or metropolitan conurbations, any "true" TSP effect on mortality is likely to be masked by the exposure error, since many of the people "assigned" to the TSP monitor live so far away that they are not actually exposed to the pollution measured there. Now, if at the same time there is a regional trend toward higher mortality, any broadly distributed pollutants characterizing the region (such as fine particles or ozone) will become the significant variables. This result may appear to be a health-based causal finding, since small particles and ozone can penetrate deeper into the lung, but, in this case, the result appeared as a statistical artifact because a regionally distributed pollutant was matched with a regionally distributed mortality trend. An analysis based on large geographic units is unlikely to capture local pollution effects, only regional ones, but a city-based analysis should be able to detect either type. This distinction is similar to separating the high-frequency (short-term) effects from the seasonal effects in a time-series analysis. Richardson *et al.* (1987) recommends checking the stability of results from ecological analyses in relation to geographic scale.

However, mortality rates may be statistically unstable if the population base is too small. One solution to this problem is to use small geographic areas (i.e., central cities) with data averaged over several years, which will improve the stability of estimates of both mortality and air pollution exposure.

Confounding

The term "confounding" refers to the incorrect assignment of an effect to an agent when in fact a third variable (the confounder) is responsible. Such a situation requires that the confounder have an effect on the outcome variable and be correlated with the first agent. In other words, a confounder must have the property of different distributions for exposed and nonexposed subjects (Miettinen and Cook, 1981). A hypothetical example might be a situation in which smokers are more likely to be exposed to certain air pollutants because they work outdoors, or more recently, because they are required to smoke outdoors (as discussed in Chapter 2, concentrations of some, but not all, air pollutants are higher outdoors than indoors). According to Stellman (1987), confounding is the "cause of great angst among epidemiologists." In ecological case-control studies of environmental factors, in which an exposed city is compared to an unexposed city, the opportunity for confounding is very large, since there are many other ways in which two population groups may differ. As the number of locations or time periods increases and regression methods become appropriate, the opportunities for serious confounding are diminished. It is possible to concoct examples that feature serious confounding (Greenland and Morgenstern 1989, for example), but according to Stellman, "rarely, however, does confounding itself, especially from unidentified sources, live up to its reputation by introducing seriously spurious associations" (1987, p. 165).

Population migration patterns can cause errors in estimated pollution exposures, as well as confounding of regression results. Confounding results from either selective migration of sick people or of the more economically advantaged. In either case, current (local) air quality may not represent the true long-term exposures of current residents. Polissar (1980) gives some examples where migration biases the estimation of cancer risk based on geographic comparisons.

Other problems can arise when unadjusted total mortality data are used (all causes, ages, races, both sexes). For smaller geographic subdivisions, this is often the only type of data available. Age adjustment is the most important correction to make, since the probability of dying in a given year increases exponentially for persons over 35. If mortality rates are available for detailed age groups, they can be combined into one age-adjusted total rate by reference to the age distribution of a standard population. If, on the other hand, only total deaths are available but details are available on the population's age distribution, then the expected total number of deaths may be computed on the same basis. In many of the cross-sectional studies reviewed in Chapter 7, neither procedure was followed, but surrogate age adjustments were attempted by using a population age descriptor variable as an independent variable in the multiple regression. "Percentage of population aged 65 and over" is a common choice. If all populations have similar age distributions, such a choice may be acceptable, but simple algebra shows, for example, that the regression coefficient for "% $\geq$ 65" should be numerically equal to the mortality rate for this age group minus the rate for the under 65 group (Goodman, 1953; Lipfert *et al.*, 1988). Many studies do not meet this simple test. Similar considerations apply to other explanatory variables employing percentages of the population, such as "% nonwhite" or "% poverty." Such checks are tantamount to comparing the ecological regression results with individual studies.

Autocorrelation

Because of the persistence of weather patterns (yesterday's weather is a good predictor for today's), daily time-series involving atmospheric phenomena tend to exhibit serial (auto) correlation; that is, observations on successive days are not independent. This violates the required conditions for the basic regression model, Eq. 3-1 (see p. 69), that the errors u be uncorrelated. Although serial correlation does not affect the magnitude of a regression coefficient, it inflates the statistical significance and thus should be accounted for. Various methods have been devised for dealing with serial correlation; for example, see Box and Jenkins (1976). "Local" air pollutants like CO and SO_2 (and perhaps TSP) are expected to exhibit less daily autocorrelation than regional pollutants like SO_4^{-2} aerosol or ozone.

Spatial data can also exhibit autocorrelation, in that adjacent observations may be similar and the number of "independent" observations is thus reduced. Cliff and Ord (1981) give tests for the presence of spatial autocorrelation.

STATISTICAL MODELS

Some studies of air pollution health effects have been content to identify the existence of associations, primarily by means of calculating correlation coefficients. In general, bivariate correlations are not only inadequate to define the relationships that are ultimately of interest, they can be misleading because of confounding variables (Hammerstrom *et al.*, 1991). Furthermore, at this stage of our knowledge of air pollution health effects, in many cases the *existence* of associations is no longer an important issue. This book is largely concerned with establishing consistency or coherence and in estimating the relative magnitudes of the important relationships. Unfortunately, there is very little theoretical guidance available on the structure of appropriate statistical models; according to the NAS Committee on the Epidemiology of Air Pollutants (NRC, 1985, Appendix C), "most model formulations have no biologic basis other than intuitive plausibility."

When temporal variability is at issue, weather, seasonal, and day-of-week patterns must be taken into account in order to derive the true associations with air pollution. Meteorological factors can confound, because they can affect both health status and air quality. For example, breathing cold air can precipitate respiratory distress; lower outside temperatures call for increased space heating and pollutant emissions. However, during adverse weather conditions, people, especially those already ill, may choose to spend more of their time indoors. In summer, similar confounding can occur between heat wave distress and the effects of increased ozone concentrations. Seasonal and day-of-week effects can exert independent influences on health (viral outbreaks) and on the reporting of health-based events (availability of clinics and physicians). When air pollution patterns correspond to these exogenous temporal patterns, spurious correlations result. Note that emissions of certain air pollutants may be reduced on weekends and holidays.

For spatial or cross-sectional analysis, there are more opportunities for confounding, since the same sources that create more air pollution in a given location can have many other effects on the population. Industrial neighborhoods are generally less desirable for residential purposes, hence their populations may be less well educated or economically advantaged. Many other life-style differences accompany such socioeconomic gradients, including smoking, alcohol consumption, diet, access to medical care, and so on. On the other hand, industrial workers *per se* are often healthier than the general population because of self-selection. It should thus be evident that identification of air pollution health effects by means of spatial gradients must account for many factors in addition to the obvious demographic adjustments (age, sex, and race).

The ways in which researchers choose to deal with the need for multivariate analysis constitutes their statistical "models." The literature varies greatly with regard to these methods and models, and some data sets have been subjected to several different types of analysis. One of the first decisions to make is whether to preadjust for a confounding variable (this may be thought of as two-stage analysis) or to perform a multivariate analysis that allows the confounding variable to interact with the air pollution variables. This dichotomy occurs most often with time-series analyses and the need to account for simultaneous weather effects. If the data are preadjusted without recourse to exogenous data to define the adjustments, there is a risk that some portion of the pollution effect will have been assigned to the weather effect. We may have more confidence in such procedures if the weather "adjustments" are consistent with known physiological responses.

For cross-sectional data, we must distinguish between the process of trying to define a model and that of estimating its coefficients. These two processes have often been implicitly combined, and it should be obvious that two independent data sets are required to do justice to both tasks. This, of course, is one of the motivations for quantitative comparisons of independent data analyses. Since we have no basis for a "true" model of the spatial variability of health indices (especially for mortality) and the data available for analysis are always limited, we must resort to empirical "specifications" of the important terms. It follows that there can be any number of such models, and prudent researchers will investigate whether their findings of effects due to air pollution are robust against plausible variations in these models. Further, they may wish to test the distributions of residuals to determine whether similar models result in statistically significant differences in their assignments of pollutant effects.

Researchers also differ in the types of multivariate analyses conducted. Two-way contingency tables were used to display the interactions of variables in some of the earlier studies, but multiple regressions seem to be the current method of choice. Some researchers use stepwise variable selection methods; some of these are sensitive to the order of variable entry. Others have predefined their models and used forced variable entry. In cross-sectional studies, 10-variable models are not uncommon and collinearity can be very important as the last few variables are entered. Suffice it to say that the burden of proof remains with the researchers to show that their findings *vis-à-vis* air pollution and health are robust against changes in model specifications and that the assumptions of normality and independence have been met, as required.

MEASURES OF RISK

Risk can be quantified as the probability of an event occurring within a given time. If 10 members of a group of 1000 die within a year, the observed annual mortality rate is 10 per thousand population, which is a statement that each person in that group had a 1% risk of dying that year. Of course, we also know that the individual risk increases exponentially with age, above about age 35. The annual risk to those aged 65 and over is about 6%, for example (Lipfert, 1978). In this book, we are primarily interested in whether exposure to air pollution also increases the risk within such a group.

For contributory factors such as air pollution, we are interested in the incremental or "excess" risk associated with given levels of ambient air concentrations. The fundamentals of excess risk must be developed from various statistical measures of association, such as correlation or regression coefficients. The classical regression equation is given by

$$y = a_0 + \Sigma_i b_i x_i + u \qquad [3\text{-}1]$$

where the b_i are the regression coefficients for the independent variables x_i and u is the residual error. For a linear dose-response model such as Eq. 3-1, which is the simplest form, the excess risk $b_i x_i$,* (where the index i refers to air pollution variables) may be expressed per unit of air concentration, regardless of concentration level. For example, some time-series analyses have derived daily risk factors for smoke exposure (b_{smoke}), of about 4% excess deaths per 100 $\mu g/m^3$ of smoke (Schwartz and Dockery, 1992). Thus, if the normal risk of dying is 6% per year (0.0164% per day), in a population of 125,000 persons aged 65 and over, the number of deaths expected each day would be 20.

On a day with a smoke concentration of 125 $\mu g/m^3$, this risk would be increased by 5%, so that one "excess" death would be expected on that day. (Such studies also have often shown a "persistence effect" carrying over for several days, so that the risk of an excess death on any of several days following the pollution event could also be elevated.) This analysis methodology presumes that the agents and exposures of concern have been identified (in this case, smoke).

Much of the epidemiological literature deals with risk factors of a categorical nature, such as being overweight or having serum cholesterol above some concentration. For such risk factors, the concepts of *relative risk*, *attributable risk*, or *logistic odds ratio* are useful (Kleinbaum *et al.*, 1982). For exposure to a risk factor of varying strength, such as air pollution, concentration level must be specified to use these concepts. In the example above for exposure to 125 $\mu g/m^3$ of smoke, the relative risk (exposed/unexposed) would be 21/20 or 1.05. The risk attributable to that exposure would be 1/20 or 0.05. Note that when causality is an issue (see below), categorical analysis is not likely to be useful, since increased exposure to air pollution usually entails some other confounding variables, such as poverty, socioeconomic status, or industrial

* The product $b_i x_i$ is also referred to as the "effect" of variable i. The use of the term "effect" in this context is a measure of the relative importance of the statistical association and should not necessarily be construed as conveying that a causal relationship has been established.

employment and attendant occupational exposures. For this reason, the presence of a consistent linear (or at least monotonic) dose-response relationship is an essential ingredient of air pollution epidemiology, which requires at least three independent observations.

The example we gave pertains to a group or population. For the study of individuals, some fraction of which develop a disease or condition, *logistic regression* is useful (Kleinbaum *et al.*, 1982):

$$P(x) = 1/(1 + e^{[-\{a+\Sigma_i b_i x_i\}]}) \quad [3\text{-}2]$$

where the x_i are independent risk factors. For example, $P(x)$ could be the dichotomous variable (0,1) representing dead or alive states, and x could be various levels of air pollution exposure. This analysis protocol may have the advantage of greater statistical power, since it deals with individuals. However, individual data on air pollution exposures are not often available. The logistic odds ratio (OR), which is the measure of effect for the multiplicative model, is defined as

$$\text{OR} = e^{[b(x^*-x)]} \quad [3\text{-}3]$$

where x^* and x represent two different levels of risk factor x (or air pollution exposure) and b is the regression coefficient analogous to a linear model.

Since the regression coefficients in Eq. 3-1 must be expressed in units consistent with the dependent and independent variables, it is often difficult to assess their practical importance based on numerical values. A useful concept is that of the *elasticity* (at the mean), a term taken from economics defining a nondimensional regression coefficient as

$$e_i = b_i \bar{x}_i / \bar{y} \quad [3\text{-}4]$$

Elasticities can be expressed as decimals or in percent and offer another measure of attributable risk, based on the mean values of the x_i. Comparison of two elasticities may be misleading if the mean values differ widely or if one represents a "perturbation" variable from which the mean has been subtracted. Note that when the "effect" of a variable ($b_i x_i$) is expressed as a percentage of the mean total response, "effect" and elasticity are synonymous.

When only the bivariate or partial correlation coefficient (r_i) is available as the measure of association, the elasticity may be estimated if the coefficients of variation (CV = standard deviation/mean) are known or may be estimated:

$$e_1 = r_1 \text{CV}_y / \text{CV}_{xi}$$

The coefficient of variation for daily air quality measurements is often around unity, and relatively rare health outcomes, such as mortality or admission to hospital, are often distributed binomially (variance = mean; CV = $[\text{mean}]^{-0.5}$). Elasticities for nonlinear models are discussed later.

The absolute excess risk in the example we use is seen to be 1:125,000, but this figure depends on the baseline level, since the fundamental dose-response relationship was expressed as a percentage increase. Obviously, the absolute risk from air pollution is much less for a group of healthy teenagers than for a group of senior citizens.

COMPARISON OF MODELS

According to the exacerbation model of air pollution effects on health, air pollution is seldom, if ever, the only factor contributing to the prevalence of a health effect. Thus, in the multiple regression model given,

$$y = a_0 + \Sigma_i b_i x_i + u \quad [3\text{-}1]$$

air pollution variables will account for only some of the x_i.

If we desire to evaluate Eq. 3-1 for alternative pollutant species that are highly correlated, such as smoke versus SO_2, the only practical method is to evaluate the model for each species separately, which may give rise to models that may differ very little from one another. There is always a temptation to declare the model with the highest adjusted correlation coefficient (R) value or the highest t statistic for the pollution variable as "best," however close its competitors might be. This practice ignores the fact that a given data set represents only one realization from the universe of possible data sets, and that its regression statistics thus all carry confidence limits. When alternative models are independent, the conventional confidence limits for R may be used as a guide toward defining statistically significant differences between models. However, in the cases of interest here, models generally only differ in the pollution variables chosen and thus are not independent, and special techniques are required in order to test the differences for statistical significance.

Snedecor and Cochran (1967) give such a method, which requires the correlation between sets of residuals from alternative models, R_r, in order to compare two correlated variances (in this case the residuals from competing models for predicting the same set of y's). Their procedure appears to be similar to that of Wolfe (1976). Cohen (1989) offers a model comparison technique for the case of highly correlated alternative predictors. The development is given in Appendix 3A.

In particular, when air pollution is responsible for only a small part of a multivariate relationship, R_r is likely to be large, and it will be difficult to distinguish among competing models with certainty. One of the important determinations in this regard is whether the same events or locations constitute influential observations for both models.

DOSE-RESPONSE FUNCTIONS

When quantitative estimates of the effect of an independent variable are required, the regression equation or some portion thereof becomes in effect a dose-response function (DRF). The mathematical form of such a function can be very important, especially when one extrapolates beyond the range of the original data (which is always dangerous). The implications of the derived shapes of various empirically defined DRFs are discussed in Chapter 7.

For a simple linear regression model, there are two parameters, the slope and the intercept. If the x intercept is positive (negative y intercept), the function is said to have a threshold, which, in the case of ambient air pollution, can be interpreted as the basis for an air quality standard (such DRFs have been called "hockey-stick" functions, because of their shape). Such a function has a constant slope, but the elasticity is conventionally defined at the mean. Obviously, the function

$$e = \frac{dy}{dx}\frac{x}{y}$$

takes on different values along the curve of $y = mx + b$ if b is not zero. Thus, two different DRFs having the same slope may have very different elasticities if the ranges of the x values are greatly different.

Some investigators have found that logarithmic transforms provide a better fit to their data. For the model

$$\ln(y) = m \ln(x) \qquad [3\text{-}5]$$

the elasticity is simply

$$e = m = \frac{dy}{dx}\frac{x}{y} = \frac{d[\ln(y)]}{d[\ln(x)]}$$

and is constant along the entire length of the DRF, which is curvilinear in Cartesian coordinates (concave downward). A model that fits this definition provides the same percentage response, regardless of the absolute value of x, and implies increased toxicity at low doses, which seems physiologically implausible. However, when dealing with heterogeneous populations, the appropriate application of toxicological data derived from relatively uniform populations may not be immediately obvious.

Some recent time-series analyses of relatively rare events, such as daily mortality in smaller cities, have used the *Poisson* regression model,

$$y = \exp(\Sigma_i B_i x_i) \qquad [3\text{-}6]$$

This model is a special case of the logistic model (Eq. 3-2) for small values of p; the regression model used to fit Eq. 3-6 is

$$\ln y = a_0 + \Sigma_i b_i x_i + u \qquad [3\text{-}7]$$

For this model, the elasticity at the mean is given by

$$e_i = B_I \bar{x}_i \qquad [3\text{-}8]$$

In Cartesian coordinates, Eq. 3-6 is concave upward and thus could be said to resemble the hockey stick for large ranges in x.

The fourth model paradigm considered here is the semilog model

$$y = m_1 \ln(x_1) + m_2 x_2 \cdots, \qquad [3\text{-}9]$$

in which only the pollution terms have been transformed to logarithms. The elasticity of this model is given by

$$\frac{dy}{dx}\frac{x}{y} = \frac{mx}{xy} \qquad [3\text{-}10]$$

or $m/\bar{y}$ when evaluated at the mean.

This model is concave downward and could represent a physiological situation involving either adaptation (individuals) or depletion of the supply of susceptibles (populations).

One of the problems with using logarithmic transforms for the pollution variables is the inability to estimate the risk at zero pollution. Instead, comparisons must be made at some baseline pollution level, which should be within the range of the original data. Obviously, when models or studies are compared, a common baseline must be used for all.

For data sets of limited range in x and small values of e, these four types of models may appear to be essentially equivalent. Figure 3-1a is based on a "true" log-log relationship with $e = 0.05$ and $e = 0.25$ for Figure 3-1b. Linear and semilog models were fit to the log-log data to examine the degree of approximation. In both cases, the log-log and semilog curves are practically identical, but the linear fits are significantly different. For real data sets with substantial variability, plots of the regression residuals may be required to establish the best form of model.

USE OF SCATTER PLOTS AND GRAPHICAL ANALYSIS

Statistics alone cannot tell the whole story about the relationship between two variables.* Anscombe (1973) presented a clever series of scatter plots to illustrate possible analytical pitfalls that are not obvious from simple correlations and regression statistics. Benarie (1980) then amplified this presentation, which I further modified, as shown in Figure 3-2.

Figures 3-2a through 3-2d all have the same mean values for x and y, are fitted by the same regression equation ($y = 3 + 0.5x$), and have the same correlation coefficient ($R = 0.817$) and standard error of the regression coefficient (0.118). It is thus possible that all four data sets are different realizations of the same model using samples drawn from the same universe. However, visual inspection of the graphs conveys additional information with different implications.

Figure 3-2a shows points scattered about the regression line in apparent random order. This is the "base case" and represents the kind of situation the investigator would like to see. The slope of the line (dy/dx) is 0.5, and there is less than a 1% chance that the true value of the slope is zero. For the case shown in Figure 3-2b, the data imply that the true relationship is not linear and

*The essence of this material originally appeared in a Brookhaven National Laboratory memorandum (Lipfert, 1989).

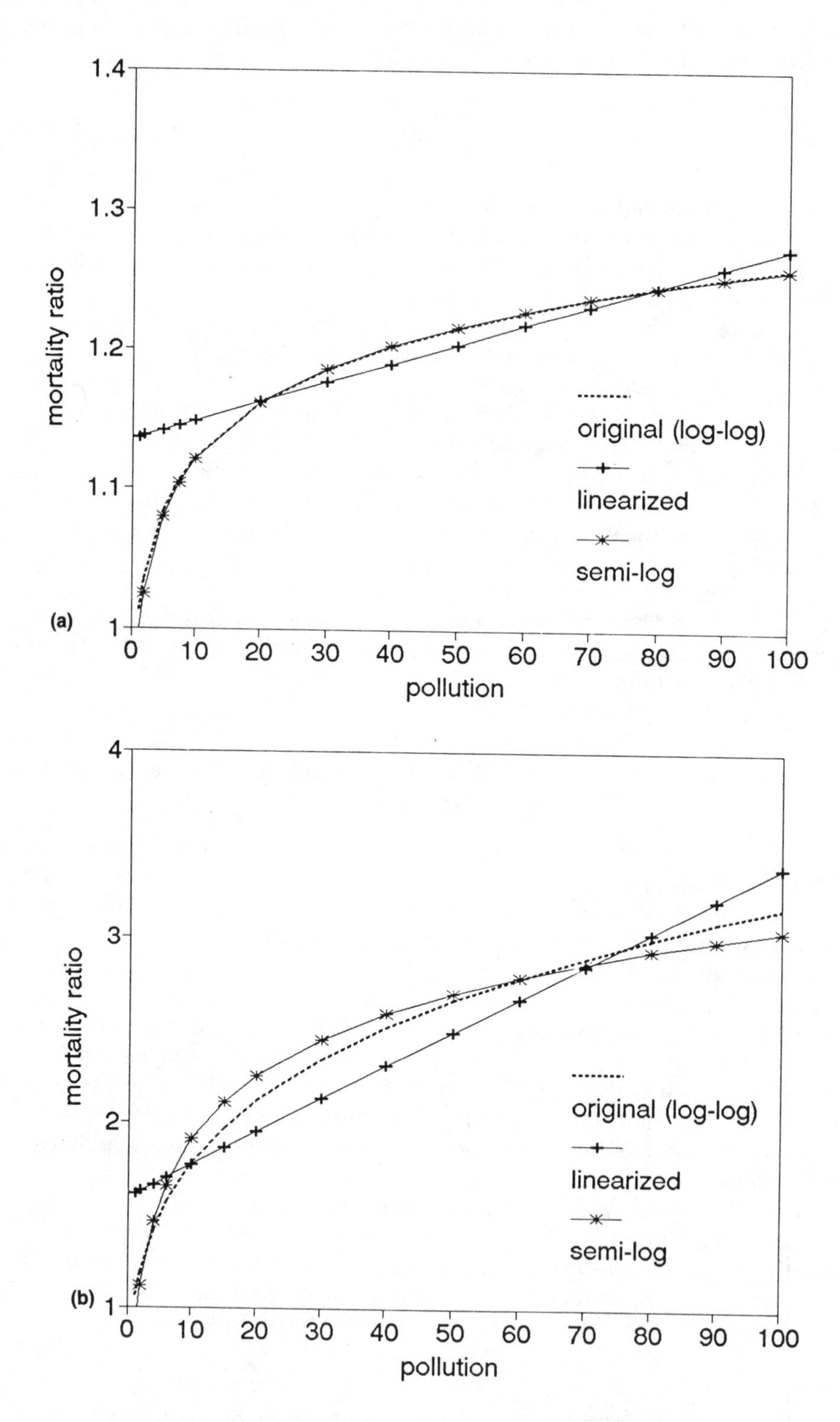

FIGURE 3-1. Log-log dose-response relationships. (a) $e = 0.05$. (b) $e = 0.25$.

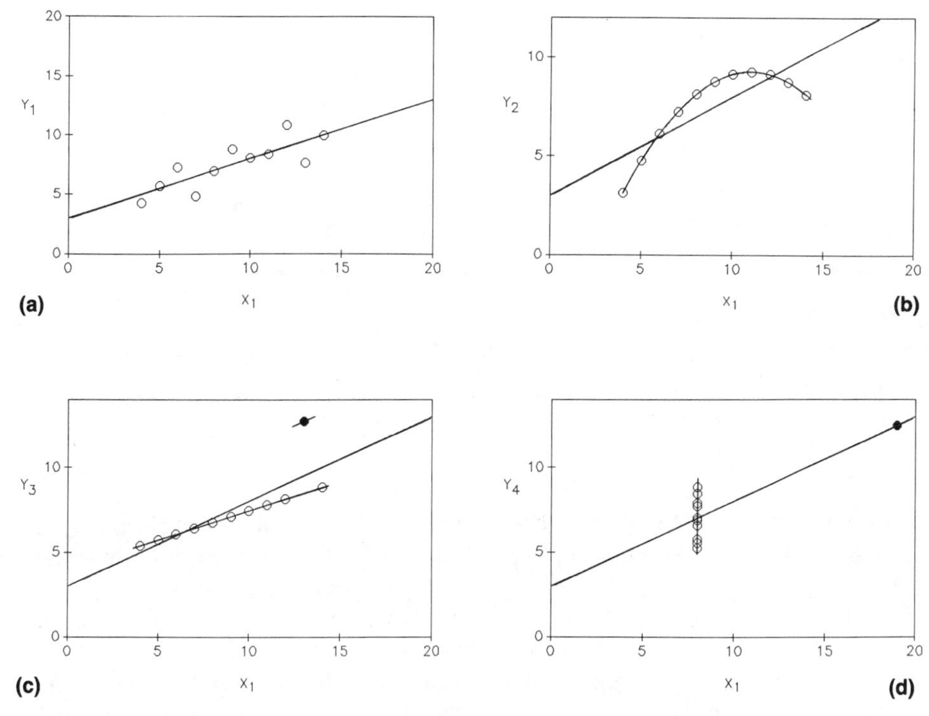

FIGURE 3-2. Different data sets with the same slope and correlation. (a) Random data. (b) Quadratic data. (c) and (d) Missing independent variables. Reprinted with permission. *Source: Anscombe (1973).*

that the model should have a quadratic term in x:

$$y = -6 + 2.78x - 0.126x^2 \quad [3\text{-}11]$$

The correlation coefficient for the quadratic equation is essentially unity, and there is virtually no chance that either coefficient is truly zero. This is a case where physical insight should play a role; is there a physiological reason to expect a quadratic relationship? What about the negative intercept? If we have solid evidence for a linear model, then $y = 3 + 0.5x$ is still our best estimate. However, a data set like the one displayed in Figure 3-2b provides strong motivation to investigate further, including the influence of additional (perhaps unmeasured) variables.

Figures 3-2c and 3-2d present data that appear to be from two different populations, represented by the open and solid symbols, respectively. In Figure 3-2c, the solid point may be an outlier; if the true relationship were given by the open points, the outlier would be hundreds of standard deviations away from the expected value. More likely, the solid point is from a different population; perhaps a different measurement instrument was used or the experimental situation differed in some way. Such unknown factors may be represented for statistical analysis purposes by a dummy variable; the open points would take on a value of 0, and the solid points, 1. However, note that the "true" slope given by the open points, 0.345, is still within the uncertainty of the original slope, 0.5 ± 0.267, for a 95% confidence interval. Thus, the

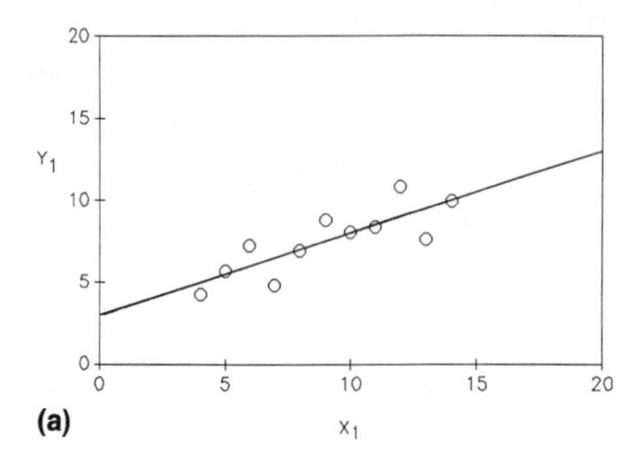

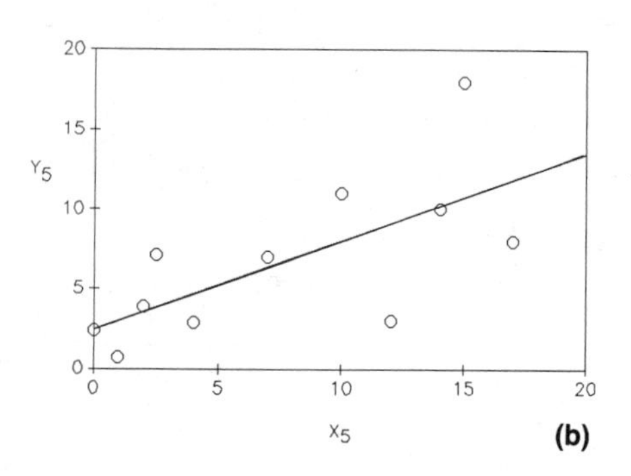

FIGURE 3-3. Data sets with the same underlying relationship but different amounts of scatter. (a) $R = 0.817$. (b) $R = 0.68$. *Sources: (a) Anscombe (1973); (b) Benarie (1980). Reprinted with permission.*

presence of this outlier has not seriously affected the ability to derive a valid estimate of dy/dx. When there are only a few outliers, using a dummy variable is essentially the same as deleting the outliers; the slope will be given by the main body of data. However, ignoring outliers in this way may represent a loss of information about the underlying physical model.

Figure 3-2d depicts a different situation. Here the solid point represents an "influential observation," since it conveys all of the available information about the effect of x on y. If this point differed from the others in any way other than its value of x, the situation would become indeterminate, because the dummy variable separating the two populations would be collinear with x. If we arbitrarily assign any one of the open points to the solid point population, we can then estimate dy/dx taking the dummy variable into account. These slopes range from 0.33 to 0.66, which are still within the confidence limits of Figure 3-2a. Thus, interference from an unmeasured variable is not necessarily a serious problem unless it is collinear with one of the variables of interest.

Figures 3-3a and 3-3b are intended to compare data sets with the same underlying relationship but with different amounts of scatter. Figure 3-3b is taken from Benarie (1980); the correlation coefficient is 0.68; slope, 0.55; standard error of slope, 0.197. Figure 3-3a is the same as 3-2a, but it is plotted on an expanded scale as in Figure 3-3b. Although the scatter was increased by a factor of 3, Figure 3-3b still has a statistically significant slope (2% chance that the true value is zero). If we remove the largest y value, the slope drops to 0.37, which is still within the original confidence limits, and the probability of the true value being zero increases to about 4%. This example thus illustrates the robustness of randomly distributed data.

The final example (Figure 3-4) is also from Benarie (1980) and illustrates a "dumbbell" distribution. If both groups of data were from the same population, the correlation coefficient would be 0.982, and the probability of a zero slope, less than 0.1%. However, if they are from different populations and we enter a dummy variable to that effect, the multiple correlation remains at over 0.98, but neither x nor the dummy variable is statistically significant (dy/dx = 0.001). This situation arises from the high correlation between x and the dummy variable (0.996). As expected, if we drop x from the model and use

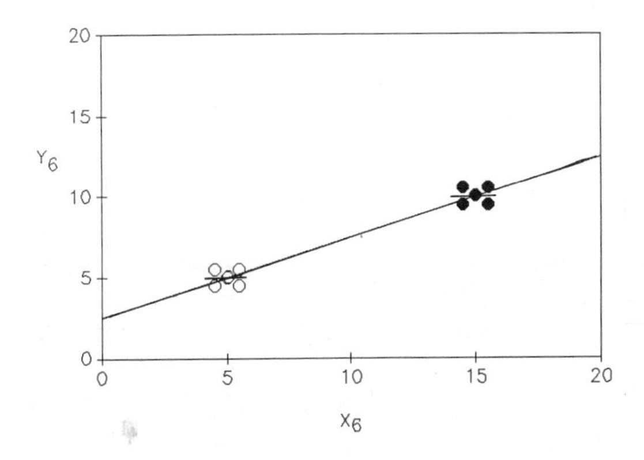

FIGURE 3-4. "Dumbbell" distribution. *After Benarie (1980).*

only the dummy variable, the correlation remains at over 0.98. The message here is the importance of the distribution of x values, since in many practical situations it will be difficult to guarantee that there are not some other differences between the two groups of data points that would justify the use of a dummy variable.

These simple examples were intended to reinforce the use of scatter plots in examining the likely viability of statistical relationships, particularly candidate dose-response functions. Plots of the residuals from alternative regression models can convey much of the same information and are also recommended. Sections II and III contain many such plots and the reader should remember to compare them to this set of examples.

THE ISSUE OF CAUSALITY

As has been noted many times, statistical analysis can reveal associations but can never establish cause and effect. Perhaps the classic paper on this topic is the address by Hill (1965), in which he proposed nine criteria to be considered in judging whether a statistical association might be causal:

1. *Strength of the association.* Large effects are more likely to be causal than weak ones. The example given was cigarette smoking and lung cancer. Either elasticity or odds ratio can be a measure of strength. The effects of community air pollution are usually small, relative to other factors; their importance derives from the potential to affect large numbers of people.
2. *Consistency of the association.* Is it robust over space, time, and investigator? If so, the likelihood of spurious association is greatly reduced. According to multiple inference theory (Miller, 1981), the probability of, say, five studies, each with $p = 0.05$, to all be false is approximately 0.01.
3. *Specificity of the association.* This criterion applies to physiologically linked pairs of causes and effects (asbestos and lung cancer, for example) but could also apply to a more general situation in which controls are analyzed in expectation of finding no effect. One of the more common findings from ecological regressions, an association between mortality from all causes and

total suspended particulates, is not very specific in this regard. Specificity also refers to the uniqueness of the independent variables (can the same effect be explained by more than one independent factor?).

4. *Temporality*. Does the dose precede the response? In an extreme example put forth by Lave and Seskin (1978), does mortality cause air pollution (crematorium plumes excepted)? This criterion is difficult to apply to slowly developing diseases or to mortality from chronic diseases.
5. *Biological gradient* (dose-response relationships). According to toxicology, a larger dose should lead to a stronger response.
6. *Biological plausibility*. There must be a physiological pathway between dose and response.
7. *Coherence*. Concordance with the prevailing medical wisdom, or more specifically, support among various endpoints: If an agent is associated with excess mortality, it should also be associated with excess morbidity.
8. *Experimental manipulation*. Does reduction of the dose weaken or eliminate the effect? Case studies of pollution cleanup campaigns may be useful in this regard.
9. *Analogy*. Is the association consistent with a similar, previously estabished cause-and-effect relationship?

Hill goes on to lament the folly of slavishly following the criteria of statistical significance, as opposed, for example, to practical significance. A small effect may be only of academic interest regardless of the t value, while a larger effect on public health may be worthy of further study even if $p = 0.10$. This philosophy was recently reemphasized by Rose (1989), who also opined that, "it is hard to think of any instance where a relative risk below 1.5 has been shown convincingly to be causal, for at these low levels it is always possible that the association has some other explanation."

Wherever possible, the risks estimated by the various factors in an ecological regression should be compared to their counterparts from studies of individual risks (Lipfert *et al.*, 1988). For example, the effects of age, race, education, smoking, cholesterol, and poverty have been estimated for individuals, and these results should be used to validate the appropriate coefficients derived from ecological regressions that also include pollution variables. If an ecological regression derives "correct" values for the nonpollution variables, it will be difficult to argue that the results for the pollution variables are spurious.

Causal Issues for Ecological Regressions

Many ecological regression methods often appear to be rather *ad hoc* and have been called "data mining" or "fishing expeditions" (Rencher and Pun, 1980). The basic problem is lack of a theoretical *a priori* "model" for human mortality and morbidity, as well as lack of appropriate data bases for the known risk factors. For example, for heart disease, which accounts for a large fraction of all deaths, the known risk factors include smoking, diet, exercise, lung function, and perhaps drinking-water hardness and ethnic background. A few of the well-known cross-sectional studies have attempted to represent some of these risk factors; most have not.

The types of associations sought in this book are likely to be classified as "weak associations," that is, relative risks less than 3 (Wynder, 1987, 1990). The reader should be forewarned that some of the studies reviewed in this book find relative risks of the order of 1.05; since they involve community air pollution and thus exposures of millions of people, the absolute risk is nontrivial, and the burden of trying to establish causality is worth considering (Rose's comment above notwithstanding).

In spite of the inherent difficulties in distinguishing causality from association in all statistical studies, there are some useful "numerical experiments" that can be performed with ecological data sets to test some of the hypotheses relating to the issue of causality or its converse, circumstantial association.

Environmental Exposures. For bivariate relationships or in the absence of covariance, the statistical consequence of (randomly) poorly characterized environmental exposures is to bias the regression coefficients and their significance downward, making a true effect harder to detect. Thus, improving the characterization of exposure should increase the coefficients of "true" responses. Such improvements are accomplished by considering the data sets with better monitoring coverage, using mathematical models to smooth between stations, or increasing the averaging times. Note that for a pollutant with large hourly variations (which is the case for most "local" pollutants, due to wind movements), the spatial representativeness of the hourly maximum will be worse than for the 24-hour average. The health effects response would thus have to be highly nonlinear (for all respondents) in order for the hourly maximum concentration to provide a better measure of the population-averaged dose than the 24-hour average. See Chapter 9 for some examples of this phenomenon.

Biases in exposures can occur when source-oriented monitoring station data are used to characterize entire communities (which can be a problem for SO_2 and CO, but less so for NO_2, SO_4^{-2} aerosol, and ozone). Additional biases can occur when smoking effects or occupational exposures are neglected. Smoking can increase indoor particulate levels substantially, so that the outdoor measurements represent only part of the actual exposure, even to nonsmokers. These effects can often be discerned by considering specific population subsets, for example, by age or gender. A less obvious bias can occur when the monitoring coverage is better (or worse) for a segment of the population that is more likely to be susceptible to the effect.

Lagged Responses. A common device in studying time-series phenomena is the analysis of lagged responses. It is unreasonable to expect an instant response to an air pollution insult at community concentration levels, so that some degree of (positive) lag is usually found. A finding of statistically significant negative lag can be interpreted as an indicator of spurious association (Hill's fourth criterion). As the period of the (positive) lag is increased, the cumulative response should also increase, since different portions of the population may respond differently. When the lag begins to exceed the actual response time, the significance for a true association should begin to decrease. These findings should be robust for a given pollutant and endpoint across different subsets of the population, assuming equally well-characterized exposures. As exposure characterization improves, the temporal structure of responses should become better defined if the relationship is causal.

Specificity of Responses. Considering specific diagnoses often involves a trade-off between accuracy of diagnosis and the introduction of sampling (Poissonian) variance as the numbers of relevant cases become smaller. As diagnoses unrelated to pollution are eliminated from the data set, we would expect the coefficient to remain constant in absolute terms (cases per dose) but its statistical significance to increase. Consideration of the nonsignificance of nonpollution-related diagnoses must be keyed to the magnitude of the sampling variances.

Collinearity Issues. Collinearity is often an issue in ecological regression analysis and must be considered on two levels: collinearity among pollutant species and collinearity between pollutant and other explanatory variables, such as weather or socioeconomic life-styles. In the first instance, we must recognize that pollutants do not occur in the natural atmosphere one at a time and that different people may be sensitive to different species. There may also be interactions between species, such as the adsorption of gases onto particles. Thus it may be truly impossible to separate the effects of different pollutant species. Forcing several related pollutants into the same model (or one pollutant with several different lags) can result in no single variable achieving statistical significance, even in cases where each variable may be significant on its own (NRC, 1985, Appendix C). Appendix 3B discusses the interactions between collinearity and measurement error, which can have serious effects.

Collinearity between pollutants and other variables is a more serious matter with respect to causality. If pollutant x_1 is highly correlated with nonpollutant variable x_2, we must look to exogenous evidence to separate the two variables, and it does not automatically follow that such a situation precludes finding a significant relationship for the pollutant in question. For example, sulfate aerosol tends to be highly correlated with the age of housing stock (Lipfert, 1978), since in the U.S. concentrations tend to be higher in the parts of the country that were settled first. Since living in an old house would not generally be considered a health risk factor *per se*, there would be no reason to include a housing age variable in a regression model. Similar considerations may apply to certain weather variables; weather effects might be separated from pollution effects by performing time-series analyses in locations for which the pollutant-weather relationships differ.

There are no standard tests for causality, and Lave (1982) even suggested that causality is only a theoretical construct, based on a consensus of scientific opinion. For long-term health studies, establishing causality requires that the known clinical effects of the pollutant are consistent with the health endpoint in question, that associations based on spatial relationships are consistent with those based on temporal relationships, and that all reasonably plausible confounding variables are accounted for.

Feinstein (1988) postulates five "scientific standards" for valid epidemiological studies: (1) a stipulated research hypothesis (as opposed to a "fishing expedition"), (2) a well-specified cohort, (3) high-quality data, (4) analysis of attributable actions, and (5) avoidance of detection bias. Although not all of these "standards" apply directly to the study of the effects of community air pollution, these requirements should be kept in mind in reviewing the specific studies in Parts II and III.

Decisions on Causality

Causality tends to be cast in "shades of gray," rather than as well-defined, black-and-white issues. The general case involves selecting from alternative regression models, and, as the numbers of independent variables are increased, the differences in overall fit tend to become small. It is important to realize that the statistics used to compare models and coefficients themselves have distributions and that undue importance should not be attached to a given realization (i.e., data set) or to arbitrary levels of statistical significance. Statistical tests should be used to decide whether the alternative models are really different (Cohen, 1989); such tests often involve comparing the distributions of residuals (Lipfert *et al.*, 1988). Although causality may really be defined only by a consensus of scientists, it behooves the conscientious investigator to seek consensus from the data first. According to Wynder (1990), "a hallmark for establishing causation is the internal and external consistency of findings."

REFERENCES

Anscombe, F.J. (1973), Graphs in Statistical Analysis, *Amer. Statistician* 27:17–21.

Benarie, M.M. (1980), *Urban Air Pollution Modelling*, MIT Press, Cambridge, MA, pp. 24–25.

Box, G.E.P., and Jenkins, G.M. (1976), *Time Series Analysis, Forecasting and Control*, Holden-Day, San Francisco.

Cliff, A.D., and Ord, J.K. (1981), *Spatial Processes: Models and Applications*, Pion, London.

Cohen, A. (1989), Comparison of Correlated Correlations, *Statistics in Medicine* 8:-1485–95.

Cohen, B.L. (1990), Ecological versus Case-Control Studies for Testing a Linear-No Threshold Dose-Response Relationship, *Int. J. Epidemiology* 19:680–84.

Feinstein, A.R. (1988), Scientific Standards in Epidemiologic Studies of the Menace of Daily Life, *Science* 242:1257–63.

Goodman, L.A. (1953), Ecological Regressions and Behavior of Individuals, *Amer. Sociol. Rev.* 13:663–64.

Greenland, S., and Morgenstern, H. (1989), Ecological Bias, Confounding, and Effect Modification, *Int. J. Epidemiology* 18:269–74.

Hammerstrom, T., Silvers, A., Roth, N., and Lipfert, F. (1991), Statistical Issues in the Analysis of Risks of Acute Air Pollution Exposure, manuscript prepared for the Electric Power Research Institute.

Hill, A.B. (1965), The Environment and Disease: Association or Causation? *Proc. Roy. Soc. Med., Occupational Medicine* 58:295–300.

Kagamimori, S., *et al.* (1990), An Ecological Study on Air Pollution: Changes in Annual Ring Growth of the Japanese Cedar and Prevalence of Respiratory Symptoms in Japanese Schoolchildren in Japanese Rural Districts, *Env. Res.* 52:47–61.

Kleinbaum, D.G., Kupper, L.L., and Morgenstern, H. (1982), *Epidemiologic Research*, Wadsworth, Belmont, CA.

Lave, L.B. (1982), *Quantitative Risk Assessment in Regulation*, Brookings Institution, Washington, DC.

Lave, L.B., and Seskin, E.P. (1978), *Air Pollution and Human Health*, Johns Hopkins University Press, Baltimore.

Lipfert, F.W. (1978), The Association of Human Mortality with Air Pollution: Statistical Analyses by Region, by Age, and by Cause of Death, Ph.D. Dissertation,

Union Graduate School, Cincinnati, Ohio. Available from University Microfilms.

Lipfert, F.W. (1980a), Differential Mortality and the Environment: The Challenge of Multicollinearity in Cross-Sectional Studies, *Energy Systems and Policy* 3:367–400.

Lipfert, F.W. (1980b), Sulfur Oxides, Particulates, and Human Mortality: Synopses of Statistical Correlations, *J. APCA* 30:366–71.

Lipfert, F.W. (1989), *Guidelines for Statistical Analysis*, Brookhaven National Laboratory memorandum, Upton, NY.

Lipfert, F.W., Malone, R.G., Daum, M.L., Mendell, N.R., and Yang, C.-C. (1988), *A Statistical Study of the Macroepidemiology of Air Pollution and Total Mortality*, BNL Report 52122 to U.S. Dept. of Energy, Brookhaven National Laboratory, Upton, NY.

Miettinen, O.S., and Cook, E.F. (1981), Confounding: Essence and Detection, *Am. J. Epidemiology* 114:593–603.

Miller, R.G., Jr. (1981), *Simultaneous Statistical Inference*, Springer-Verlag, New York.

Morgenstern, H. (1982), Uses of Ecologic Analysis in Epidemiologic Research, *Am. J. Public Health* 72:1336–44.

National Research Council (NRC) (1985), *Epidemiology and Air Pollution*, Committee on the Epidemiology of Air Pollutants, National Academy Press, Washington, DC.

Piantidosi, S., Byar, D.P., and Green, S.B. (1988), The Ecological Fallacy, *Am. J. Epidemiology* 127:893

Polissar, L. (1980), The Effect of Migration on Comparison of Disease Rates in Geographic Studies in The United States, *Am. J. Epidemiology* 111:175–82.

Rencher, A.C., and Pun, F.C. (1980), Inflation of R^2 in Best Regression, *Technometrics* 22:49–53.

Richardson, S., Stucker, I., and Hemon, D. (1987), Comparison of Relative Risks Obtained in Ecological and Individual Studies: Some Methodological Considerations, *Int. J. Epidemiology* 16:111–20.

Robinson, W.S. (1950), Ecological Correlations and the Behavior of Individuals, *Am. Sociol. Rev.* 15:351–57.

Rose, G. (1989), Science, Ethics, and Public Policy, in *Assessment of Inhalation Hazards*, D.V. Bates *et al.*, eds., Springer-Verlag, Berlin, 1989, pp. 349–56.

Schwartz, J., and Dockery, D.W. (1992), Particulate Air Pollution and Daily Mortality in Steubenville, Ohio, *Am. J. Epidemiology* 135:12–19.

Snedecor, G.W., and Cochran, W.G. (1967), *Statistical Methods*, 6th ed., The Iowa State University Press, Ames, pp. 195–97.

Szklo, M. (1987), Design and Conduct of Epidemiologic Studies, *Preventive Medicine* 16:142–49.

Wolfe, D.A. (1976), On Testing Equality of Related Correlation Coefficients, *Biometrika* 63:214–15.

Wynder, E.L. (1987), Workshop on Guidelines to the Epidemiology of Weak Associations, Introduction, *Preventive Medicine* 16:139–41.

Wynder, E.L. (1990), Epidemiological Issues in Weak Associations, *Int. J. Epidemiology* 19(Suppl. 1):S5–S7.

APPENDIX 3A A METHODOLOGY FOR TESTING THE RESIDUALS FROM ALTERNATIVE REGRESSION MODELS

Snedecor and Cochran (1967) pose this problem in terms of a variance ratio (see also Lipfert *et al.*, 1988)

$$\phi = \sigma_1^2/\sigma_2^2 \qquad [3A\text{-}1]$$

where σ is the (true) standard error of estimate from a regression. For the case in which the models have the same numbers of explanatory variables and thus degrees of freedom for error,

$$\phi = (1 - R_1^2)/(1 - R_2^2) \qquad [3A\text{-}2]$$

For any single data set purporting to be a realization of the true relationship, we have a regression standard error of estimate s that is an estimate of σ. Let $F = s_1^2/s_2^2$; then confidence limits for ϕ are given by

$$\phi = F(K \pm [K^2 - 1]^{0.5}) \qquad [3A\text{-}3]$$

where $K = 1 + 2(1 - R_r)t^2/(n - 2)$. R_r is the correlation between sets of residuals for competing models 1 and 2, and t corresponds to the desired confidence interval for $n - 2$ degrees of freedom.

In order to test the hypothesis that $\phi = 1$, we are interested in confidence limits that include unity. The above expressions can be solved to yield the value of F that would have to be observed in a given realization in order to find a statistically significant difference between models 1 and 2. The parameters are the correlation between sets of residuals R and the number of degrees of freedom ($\cong$ the number of observations). The resulting expression for $K = 1$ is:

$$\phi = F(1 \pm 2t)((1 - R_r^2)/(n - 2))^{0.5} \qquad [3A\text{-}4]$$

Thus, for $R_r = 0$ (uncorrelated residuals),

$$\phi = F(1 \pm 2t)(n - 2)^{-0.5} \qquad [3A\text{-}5]$$

which provides very nearly the same confidence limits for the regression correlation coefficients as does the standard approach using the z transform (Snedecor and Cochran, 1967, p. 185). However, for the case of $R \neq 0$, especially $R \cong 1.0$, the expression for ϕ provides much tighter confidence limits. The correlation R_r between alternative sets of residuals will be high when the residuals are controlled by either random factors or explanatory factors not included in the model, as opposed to the explanatory factors that distinguish model 1 from model 2.

APPENDIX 3B THE EFFECT OF MEASUREMENT ERROR ON MULTIPLE REGRESSION STABILITY

INTRODUCTION AND BACKGROUND

A common problem in air pollution epidemiology is the presence of collinear pollutants. Such collinearity can result when a dominant pollutant source tends to emit a constant mixture of pollutants. Examples include NO and CO from mobile sources, SO_2 and NO_x from power plants and space heating sources. Day-to-day variability in concentration levels may be driven by meteorological variability, but unless the source characteristics also change from day-to-day, the pollutants comprising the mix will tend to be highly correlated. This phenomenon also applies to spatial distributions, in which common source types are seen at many locations.

The purpose of air pollution epidemiology is to derive relationships between population health status and air pollution exposure. Regression techniques are often used for this purpose. Unfortunately, in many cases, air concentrations measured at a limited set of fixed monitoring sites must serve as surrogates for actual exposure, which is impractical to measure directly in large populations. The difference between measured concentrations and actual exposures may be expected to vary from day-to-day and by pollutant species. Some pollutants are distributed more uniformly throughout a community than others, and some penetrate to indoor environments more readily than others. Thus, we should expect that the error in exposure will vary by pollutant species.

The effect of error in the independent variable of a (bi-variate) regression is to bias the regression coefficient and its significance toward the null (Snedecor and Cochran, 1967). In a multiple regression in which the various independent variables may have different magnitudes of measurement errors, the results may be less predictable (Fleiss and Shrout, 1977). The effect of collinearity among independent variables entered simultaneously into a multiple regression is to inflate the variance of the regression coefficients (Theil, 1978), with attendant reductions in statistical significance. It thus follows that measurement error serves to obscure the extent of the actual (structural) collinearity that may be present among pollutants.

This note explores the consequences of differing measurement errors in collinear pollutants, making use of rudimentary simulation techniques. Quasi-random variables were generated using the random number function of Quattro-Pro® (Borland International, 1992) and near-normal distributions were derived by adding independent sets of random values. A time-series of 101 days was simulated and 20 trials were performed in order to generate a range of results.

THE SIMULATION MODEL

We postulate a daily mortality model in which daily deaths are proportional to two pollutants:

$$D = \beta_0 + \beta_1 P_1 + \beta_2 P_2 + \varepsilon_D \qquad [1]$$

We solve this relationship for the regression coefficients and their standard errors using ordinary least squares. In the simulations performed, the assumed values of β_1 and β_2 are unity. We assume that the pollutants P_1 and P_2 emanate from a common source or group of sources and are thus highly correlated. However, the relationships between monitored concentrations (the observed values in this case) and actual population exposure differ; additional (random) variance is present in the distribution of the values of P_2. We designate the unobservable value of this pollutant as P_2 (true) and the actual measurement as P_2 (measured).

Simulations were performed for seven different combinations of relative measurement error and assumed collinearity between P_1 and the true value of P_2, using 20 trials each. This was deemed to be a number sufficient to define average regression coefficients and t values, but was undoubtedly insufficient to define their distributions. The first of these simulations ("baseline") is described in detail, including plots of the results. Summary statistics are then given for the remaining simulation cases.

BASELINE SIMULATION RESULTS

The means and standard deviations for the baseline simulation are given in Table 3B-1.

TABLE 3B-1 Means and Standard Deviations (Baseline)

Variable	Mean	Std. Dev.
D	38.6	4.7
P_1	6.1	0.93
P_2 (true)	7.6	0.94
P_2 (meas.)	7.6	1.18

The additional exposure measurement error corresponds to an increase in variance of 56%. The correlations between P_1 and P_2 (true) and P_2 (measured) were 0.96 and 0.75, respectively. Note that the average correlation between observed smoke and SO_2 in London during winters from 1958 to 1972 was about 0.89 (Roth Associates, Inc., 1986). The common source for these pollutants in winter was largely space heating.

Figure 3B-1 plots the simulated daily values of the three pollutant variables. Since air pollution is often distributed log normally, one might assume that the logs of concentrations have been simulated rather than the actual values; this entails no loss in generality. The correlations between P_1 and the two alternative values for P_2 are shown on the figure. The imperfect relationship between P_1 and the "true" P_2 is intended to reflect measurement error in P_1, or perhaps daily variations in pollutant mix.

Each of three pollutants was regressed against the simulated mortality variable, one at a time. The results for 20 trials are plotted in Figure 3B-2, in terms of the regression coefficient t values. As expected, we see that P_1 and P_2 are essentially equivalent when the "true" P_2 values are used, but that use of the simulated P_2 measurements results in lower t values. Note that the parameters

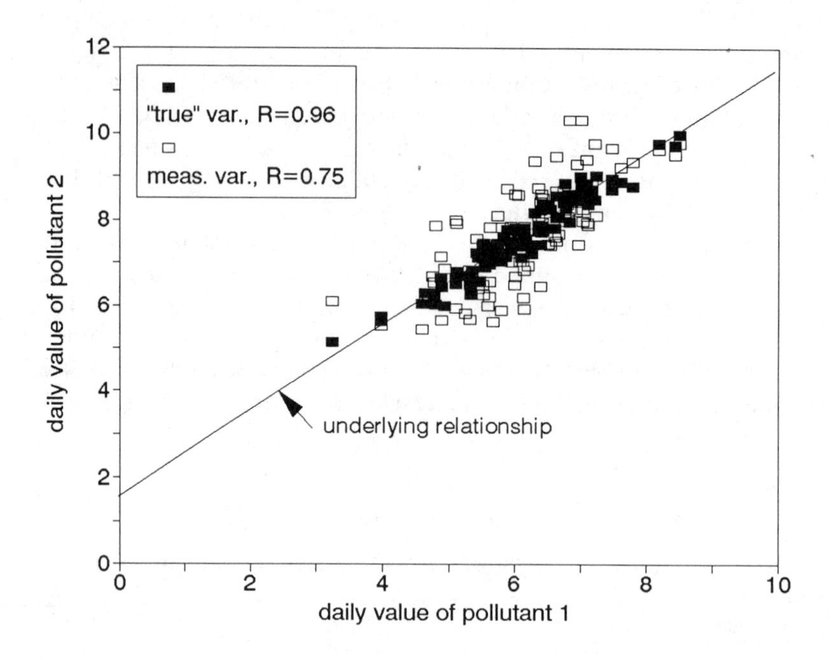

FIGURE 3B-1. Relationships among pollutants for the 101 day simulation. (a) Case 1. (b) Case 2.

of this simulation were chosen to yield highly significant regression coefficients for bivariate regressions.

Figure 3B-3 compares the *t* values for P_1, for two alternative multiple regressions using the "true" and measured values for P_2, respectively. When regressed jointly with the "true" P_2, P_1 loses significance, on average. However, when regressed jointly with P_2 in the presence of increased measurement error, the loss of significance is much less and the regression coefficients are virtually unchanged (Figure 3B-4).

The corresponding t-value information for P_2 is given in Figure 3B-5. In both cases, P_2 lost significance in most of the 20 trials. The relationship between regression coefficients for P_1 and P_2 in multiple regressions is shown in Figure 3B-6, and they are seen to be negatively correlated. Thus a negative result for one variable contributes to a more strongly positive value for the other.

The average regression statistics for the 20 trials are given in Table 3B-2, for the five different regression formats.

Note that only regression no. 4 provides the "correct" coefficients, which were defined as 1.0 in the formulation of the model; because of the collinearity, these coefficients are not significant, and thus are likely to be disregarded in the absence of a priori information on the "true" model. In addition, the regression coefficients derived in the bivariate case (regressions nos. 1 and 2, one pollutant at a time) are inflated to essentially twice the expected value,

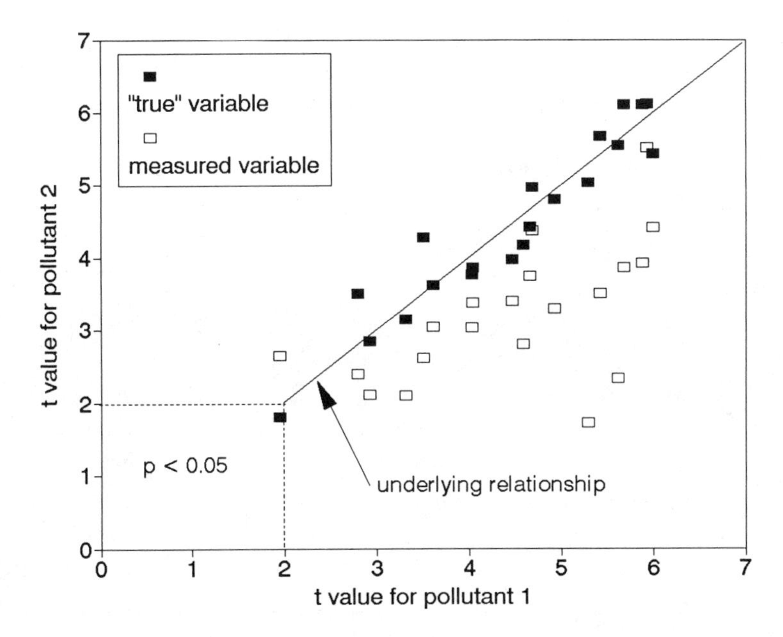

FIGURE 3B-2. Relationships among the coefficient t values for the 20 trials of simulation Case 1, bivariate regressions. Coefficients from regressions 2 and 3 are plotted against the coefficient for regression 1, for each trial.

TABLE 3B-2 Baseline Regression Statistics

Regression #	1	2	3	4	5
Statistic	D vs. P_1	D vs. P_2 true	D vs. P_2 meas.	D vs. $P_{1,2}$ true	D vs. $P_{1,2}$ meas.
R^2	0.17	0.17	0.098	0.17	0.185
SEE	4.18	4.18	4.37	4.18	4.17
β_1	1.88			1.06	1.88
$\sigma\ \beta_1$	0.45			2.1	0.78
t	4.47			0.6	2.94
β_2		1.81	1.07	0.85	−0.01
$\sigma\ \beta_2$		0.46	0.27	2.1	0.5
t		4.46	3.21	0.48	0.0

since the effect of both pollutants is picked up by the single pollutant entered into the model. With measurement error, the coefficient for P_2 (meas) is biased low with respect to that for P_2 (true), as expected.

We see that the regression coefficient for P_1 is essentially unchanged from the bivariate case when regressed jointly with P_2 in the presence of measurement error (regression no. 5), while the joint coefficient for P_2 has become nil, even though the basic underlying relationship assigns equal weight to both

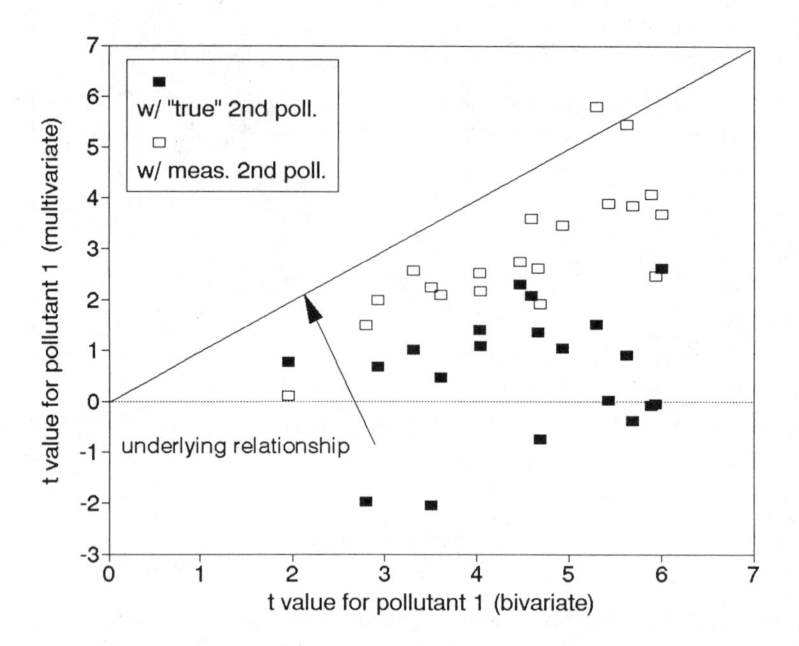

FIGURE 3B-3. Relationships among the coefficient t values for the 20 trials of simulation Case 1. t values from regressions 4 and 5 are plotted against the t value for regression 1, for each trial.

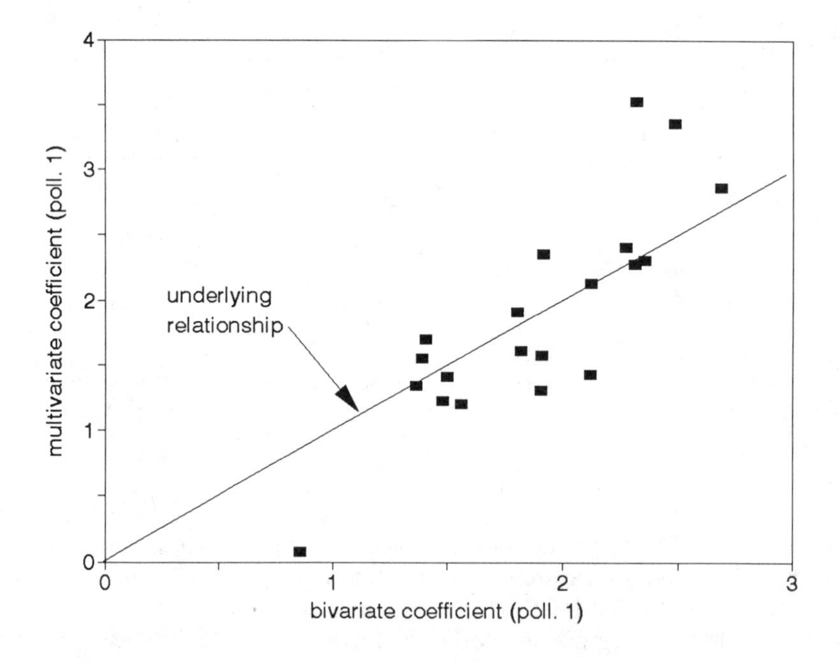

FIGURE 3B-4. Relationships among the coefficient t values for the 20 trials of simulation Case 1. Coefficients from regression 5 are plotted against the coefficient for regression 1, for each trial.

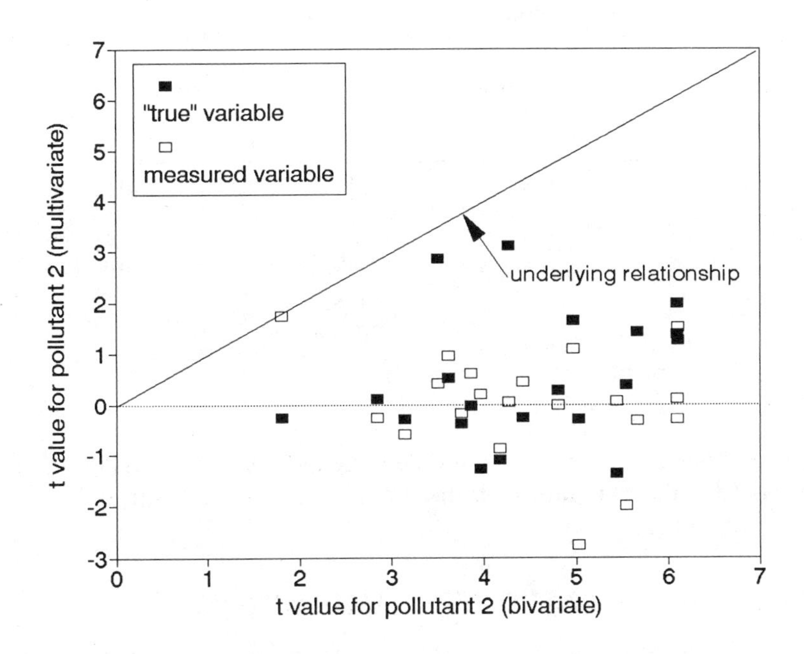

FIGURE 3B-5. Relationships among the coefficient t values for the 20 trials of simulation Case 1. t values from regressions 4 and 5 are plotted against the t value for regression 2, for each trial.

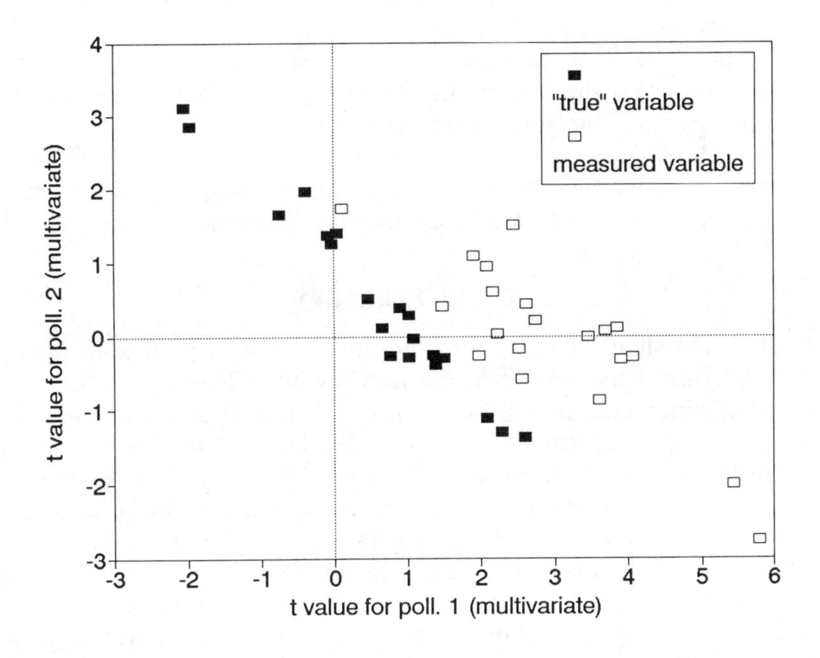

FIGURE 3B-6. Relationships among the coefficient t values for the 20 trials of simulation Case 1. t values for P_2 from regressions 4 and 5 are plotted against the corresponding t value for P_1, for each trial.

TABLE 3B-3 Summary of Simulation Results for the Joint Regression of Mortality on P_1 and measured P_2

	Pollutant Correlations		Average Regression Coefficients	
Case #	P_1, P_2 true	P_1, P_2 meas.	P_1	P_2 meas.
1.	0.96	0.75	1.88	−0.01
2.	0.84	0.75	1.71	0.33
3.	0.90	0.84	1.75	0.36
4.	0.96	0.91	2.04	−0.08
5.	0.72	0.65	1.08	0.87
6.	0.66	0.29	1.76	0.18
7.	0.50	0.27	1.52	0.56

variables. Thus the "standard" multiple regression procedure yields an inflated coefficient for the pollutant with the lower measurement error and essentially zero for the other.

SENSITIVITY SIMULATIONS

Additional simulations were performed in the same way to explore the robustness of these findings. The results from regression model 5 (joint regression of mortality on P_1 and P_2 meas) are given in Table 3B-3, along with the previous result. There were little or no changes in the bivariate regression coefficients for these simulations, except that large amounts of measurement error result in loss of significance even in a bivariate regression.

Note that the regression coefficients tend to sum to the correct value of 2.0 in all cases, but that the proper split (1:1) is approached only in simulation case 5. In this case, the introduced measurement error is small and the regression on P_1, P_2 measured is essentially the same as on P_1, P_2 true. Note that simulation cases feature very modest amounts of pollutant collinearity, and that biased coefficients still result from the joint regression.

CONCLUSIONS

This simulation should sound a warning for the use of joint regression as a means of trying to separate the effects of highly collinear variables, in the presence of differences in exposure error. When measurement error is considerable, this warning extends to only modest levels of observed collinearity, since the "true" degree of collinearity cannot be observed. Note that in the absence of differential exposure error, neither pollutant can be identified as "causal" on the basis of statistical significance. While the simulations are admittedly crude, they serve to illustrate this important point: When independent variables are highly correlated, it will not be possible to separate their effects and the only conclusion possible is that both are associated with the dependent variable.

With respect to decisions regarding pollution controls, this outcome may not be as bleak as it sounds, since control of the source of one pollutant should also yield control of the other, assuming that commonality of sources is responsible for their collinearity. However, control of only one pollutant may change the existing pollutant mix and thus invalidate the conditions under which the analysis was performed.

REFERENCES

Assessment of the London Mortality Data. Draft report prepared by Roth Associates, Inc., Rockville, MD., for the Electric Power Research Institute, Palo Alto, CA, March 1986.

Borland International, Inc. (1992), Quattro-Pro 4.0®, Scotts Valley, CA.

Fleiss, J.L., and Shrout, P.E. (1977), The Effects of Measurement Errors on Some Multivariate Procedures, *Am. J. Public Health* 67:1188–91.

Snedecor, G.W., and Cochran, W.G. (1967), *Statistical Methods*, 6th ed., The Iowa State University Press, Ames, Iowa, pp. 164–66.

Theil, H. (1978), *Introduction to Econometrics*, Prentice-Hall, Englewood Cliffs, NJ, p. 137.

4

A Primer on Physiological Effects of Air Pollution and Measurement of Respiratory Function

The ultimate test animal is man.

W.S. Spicer, Jr. (1963)

Since the lung is the primary target organ for the health effects of air pollution, we turn to the literature on respiratory disease and mechanics for guidance.* As Spicer (1963) pointed out early on, the relationship between respiratory health and air pollution is complex, and the concept of respiratory health *per se* is not well defined. Respiratory health is commonly assessed either by the study of symptoms (presence of cough, wheezing, shortness of breath [dyspnea]) or by means of measurements of the mechanics of breathing, referred to collectively as *lung function*. (The general terms *lung function*, *pulmonary function*, and *respiratory function* are used interchangeably in this discussion. Specific measures are defined in the following text.) Since data on symptoms often contain an element of subjectivity, measures of lung function are preferred, although many studies have found that the two are correlated. There are also many biochemical measurements available for use in specific clinical situations (see Bates *et al.* (1971), for example). A more recent approach toward assessing respiratory health involves imputation of subclinical disease based on autopsies of accident and homicide victims (Sherwin and Richter, 1991).

A number of studies, discussed in Chapter 12, have identified impaired lung function as an independent risk factor for heart disease morbidity and

---

* In preparing the material on the physiology and testing of respiration, I have drawn heavily on material originally presented by the American Lung Association (1973); Comroe (1965); Comroe *et al.* (1955); Cotes (1979); Frank (1975); Kittredge (1989); and O'Brien and Drizd (1981).

mortality, as well as for all-cause mortality. The degree to which air pollution causes reductions in lung function (Chapter 11) can thus provide an independent estimate of the plausible effects of air pollution on mortality, by combining the two relationships.

THE HUMAN RESPIRATORY SYSTEM

The primary purpose of the respiratory system is to supply the body with oxygen and to remove CO_2. The system also has a number of important defense mechanisms by which infections and foreign bodies are resisted. The uppermost part of the system (Figure 4–1) is the pharynx, which includes the mouth and nose and extends to the trachea (windpipe). Inspired air is warmed to body temperature and humidified here to essentially 100% relative humidity, to prevent drying the lower respiratory tissues. The nasal mucous membrane, which comprises an area of about $160\,cm^2$, is covered with glands that secrete the fluid necessary for this humidification.

The airway consists of the trachea, bronchi, and bronchioles. The trachea divides into the two major bronchi, which then divide into smaller bronchi. The subdivision into sequentially smaller passages continues down to the bronchioles. The bronchioles divide further, and each finally ends in 10 to 20 small air sacs or alveoli, in whose thin membranes a dense network of fine capillary blood vessels is embedded. There are 300 to 400 million alveoli in each lung. Oxygen and CO_2 exchange takes place here by means of diffusion through the thin ($0.1\,\mu m$) membrane, as the red blood cells are forced through the narrow blood vessels. The right ventricle of the heart supplies the pumping energy for this process.

The larger airways have rigid cartilage rings that keep them from collapsing as air is inspired. The smaller branches have a lining of strong muscles that can contract and relax, thus regulating the air flow to the lower regions.

The interiors of the trachea and bronchi are covered with mucous glands and ciliary cells, each of which carries several hundred very small fibers, called cilia. The cilia beat 1000 to 1500 times per minute, sweeping mucous fluid and captured foreign particles outward to where they are expectorated or swallowed. Mucus that has collected in the bronchi as a reaction to infection or to inhaled substances is expelled by coughing and by ciliary action. Sneezing is a way of eliminating particles and irritating substances from the nasal passages. The interior of the alveoli is covered with a thin layer of surfactant fluid that keeps the surface tension low, allowing air to be drawn in without collapsing the tiny air sacs. The elastic fibers in the space between alveoli create a tendency of the alveoli to collapse during exhalation.

The human cardiorespiratory system functions as a coupled system in order to perform its task, exchanging oxygen and carbon dioxide across the alveolar membranes. This process is driven by concentration gradients across the membrane in the two fluids involved: air and blood (plasma plus red corpuscles). The pumps providing these two fluids operate independently but are coupled through the involuntary control of respiration. Cardiac insufficiency (reduced blood flow) results in more rapid breathing. This is also a symptom of carbon monoxide poisoning, since the loss of hemoglobin from binding CO to form carboxyhemoglobin reduces the ability of the blood to carry oxygen. Thus,

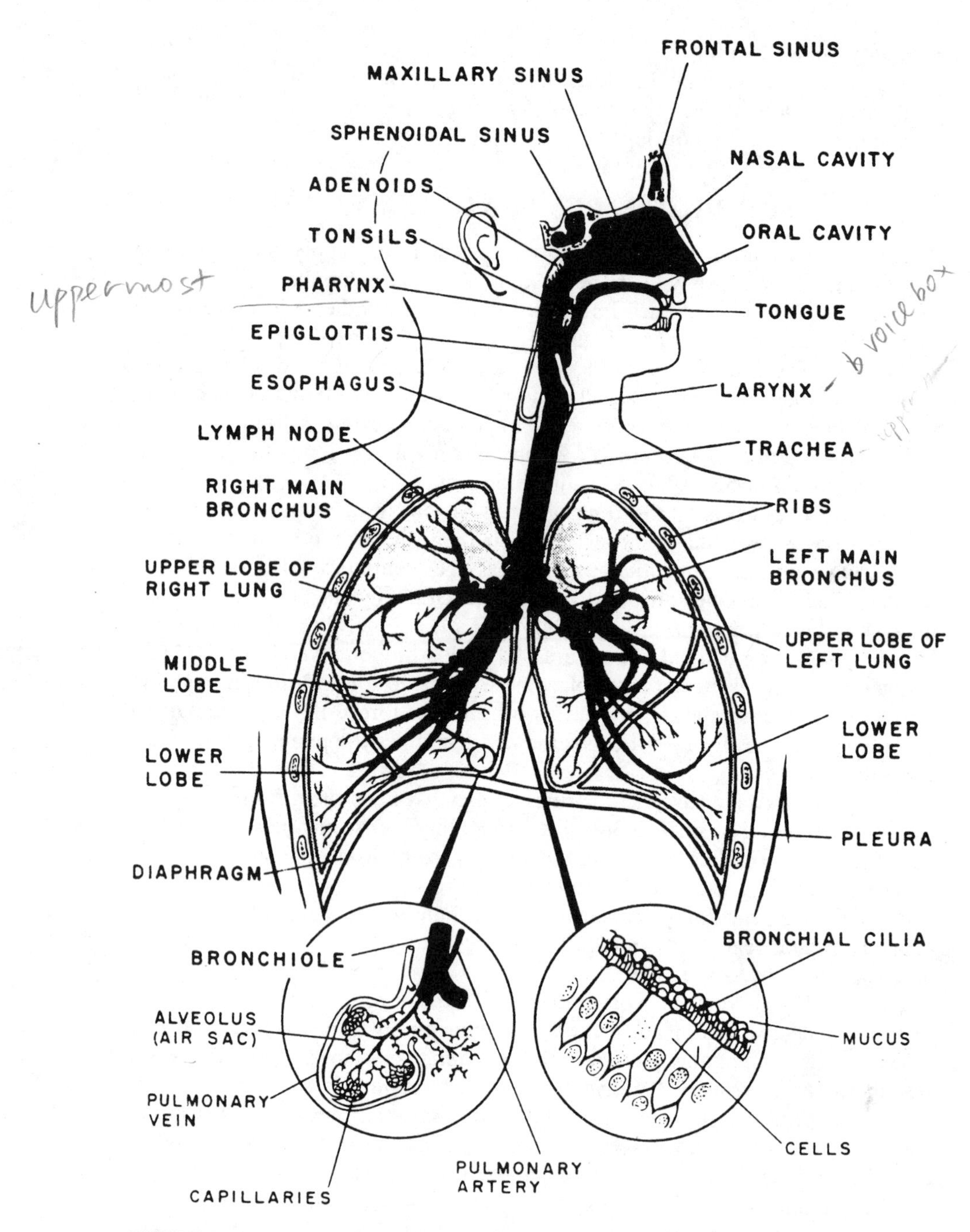

FIGURE 4-1. The human respiratory system. *Reprinted with permission of the American Lung Association.*

exposure to a mixture of CO and other pollutants could cause higher breathing rates and increased intake of the air mixture. Similarly, reduced respiration forces the heart to work harder; emphysema is an example, and its victims may suffer heart damage. Changes in vital capacity were noted as an indicator of cardiac patients' prognosis early in this century (McClure and Peabody, 1917).

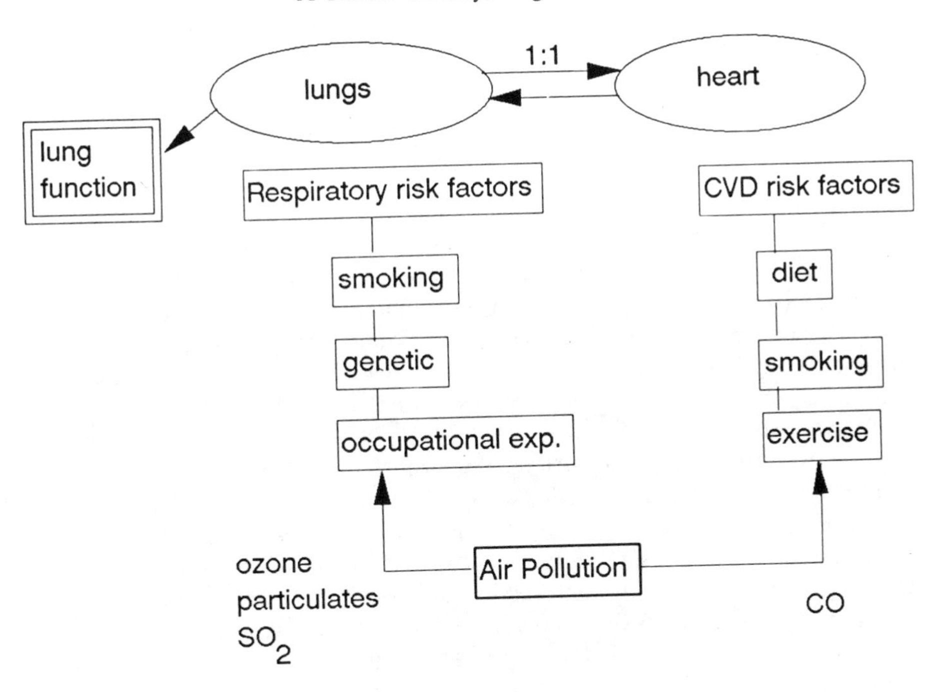

FIGURE 4-2. Interrelationships in the cardiorespiratory system.

A simplified "engineering model" of the cardiorespiratory system and some of the agents that can result in deteriorated performance is depicted in Figure 4-2. Well-known respiratory risk factors include smoking, genetic factors, and occupational exposures to fibers, dusts, metal fumes, and bacterial agents. Factors that can cause cardiovascular deterioration include diet (cholesterol can create plaques that choke off arteries) and smoking; exercise is generally regarded as beneficial, acting to strengthen the heart muscle. Air pollution can act on both systems, either through toxic agents in the blood, asphyxiants (CO), or deterioration of lung performance. The "1:1" notation symbolizes the findings from Chapter 12 that a given percentage loss in respiratory performance is reflected as that same percentage of excess risk due to heart disease. Lung function measurements are the simplest manifestations of the performance of this coupled system, although heart rate, blood pressure, and pulse data are useful indicators for cardiac performance *per se*.

The first step in considering the potential effects of air pollution on respiratory functioning is to determine the fate of these substances after inspiration. Water-soluble gases are likely to be scrubbed out by means of contact with the moist surfaces of the upper airways, since diffusion to the passage walls is faster for gases than for small particles. These gases include SO_2, HCl, HNO_3, NH_3, and HCOOH. Insoluble gases, including CO, O_3, and NO_2, will reach the deep lung (alveoli), where they may interact with oxygen transfer and/or enter the bloodstream.

Large particles (>10 μm) will be caught by the upper respiratory defenses and are unlikely to penetrate beyond the trachea. Ninety percent of particles greater than 2 μm are caught by the mucous layer and are expelled through the action of the cilia. Some of these larger particles will be swallowed, which is a means of transferring their toxicity into the digestive system, a possible pathway for an effect of air pollution on stomach cancer. Small particles (<2 μm) can penetrate deeper into the lung, where some of them will be deposited. Their subsequent fate depends on solubility and chemical interactions; for example, acid particles will reduce the pH of the airway fluids as they dissolve. Particles of all sizes can be removed by macrophages (scavenger cells).

The work of breathing is used to overcome the pulmonary resistance of the airways and the elastic recoil of the lungs and thorax. Under normal conditions, the latter dominates, but when the smaller airways become obstructed, pulmonary resistance can increase markedly. Under the laws of Poiseuille flow, resistance to a given air flow increases with the fourth power of the radius reduction. According to Comroe (1965, pp. 118–119), obstruction of smaller airways can come from "constriction of bronchiolar smooth muscle, mucosal congestion or inflammation, edema (swelling due to accumulation of fluids) of bronchiolar tissues, plugging of the lumen (open space) by mucus, edema fluid, exudate or foreign bodies, cohesion of mucosal surfaces by surface tension forces, infiltration, compression or fibrosis of bronchioles, or collapse or kinking of bronchioles due to loss of the normal pull of alveolar elastic fibers on bronchiolar walls or to loss of structural, supporting tissues of the bronchial walls ('weak walls')."

THE EFFECTS OF AIR POLLUTION

Volumes have been written about the physiological effects of air pollution on animals and man; the references to Chapter 2 include the EPA Criteria Documents, which are intended to define these effects specifically for each criteria pollutant. However, an excellent general description of the physiology involved was given by Frank (1975), which forms the basis for the following section.

Altered Respiratory Mechanics. Irritating air pollutants can cause localized constriction of the airways, which results in an uneven distribution of air and increases the work of breathing. This can cause shortness of breath and insufficient gas exchange. Exercise can increase these effects, which also tend to be more severe for persons with already compromised respiration, such as asthmatics.

Reduced Supply of Oxygen. Carbon monoxide can interfere with the oxygen supply to tissue by virtue of its strong ability to replace oxygen in hemoglobin. The brain and the heart are especially sensitive to these effects. Other mechanisms for this effect include damage to the alveolar-capillary membrane, which is the pathway for CO_2 exchange with oxygen. Thickening (swelling) or scarring of this membrane can impair its diffusion performance. Ozone is one of the pollutants that have been associated with such effects. In addition, the maldistribution of air supply can impair oxygen supply.

Reduced Resistance to Infection. Macrophages and ciliary action are critical to the ability of the lung to clean itself and repel invaders. Nitrogen dioxide and ozone have been shown to cause such effects in laboratory animals, but such evidence has been lacking in humans. Studies of respiratory clearance mechanisms in a laboratory setting have shown that alterations are caused by sulfuric acid (Lippmann, 1987), but the interpretation of these results is not straightforward, since the concentrations required to retard clearance are much higher than those experienced in community air.

Aging and Chronic Disease of the Lung. High concentrations of NO_2 and O_3 have been shown to contribute to structural changes resembling increased rates of aging and development of emphysema in laboratory animals. The extension to humans (and to lower concentrations) is unknown.

Lung Cancer. The primary agent for respiratory cancer is tobacco smoke. Industrial pollutants such as benzo(alpha)pyrene and asbestos also play a role. Although lung cancer death rates were formerly higher in urban than in rural areas, most investigators discount the role of urban air pollution at present community air levels. It is certainly possible that compromised respiratory defenses could contribute to lung cancer in combination with other agents, but the effects of smoking are so strong that it is difficult to make such a determination based on epidemiological data.

THE MEASUREMENT OF PULMONARY FUNCTION

Types of Measurements

Measures of human pulmonary function date back to the mid-nineteenth century; the early terminology was "vital capacity," a measure of anatomical lung size given by the volume of gas that can be expelled from the lungs, with maximum effort following a maximal inspiration. Since there is a residual volume which cannot be expelled, this measure is smaller than the actual total lung volume. Changes in vital capacity have long had diagnostic value for disease and pulmonary disorders (McClure and Peabody, 1917; Wilson and Edwards, 1922). Vital capacity is generally now referred to as *forced vital capacity* (FVC). It was also recognized early on that lung function measurements have little value unless compared to some sort of standards, either values from a "normal" population or changes in values for an individual or group over time (Miller and Thornton, 1980).

The *nitrogen washout test* is used to measure the distribution of inspired gas to the alveoli by recording the uniformity of decay of nitrogen in the breath following inspiration of pure oxygen. The diffusing capacity of the lungs may be measured by breathing a known (and safe) concentration of CO and measuring the difference in inspired and exhaled concentrations. Maldistribution of ventilation is an important parameter in the analysis of respiratory performance (Bates *et al.*, 1971).

If we think of the respiratory system as an air handling system, the forced expiratory maneuvers provide some indices of performance under maximal conditions. More sensitive measurements of *airway resistance*, essentially the pressure drop of the system under more ordinary conditions, can often provide additional diagnostic information. These data are obtained from a *whole-body plethysmograph*, a closed chamber in which the subject's respiratory efforts are recorded on sensitive pressure and flow transducers. The pressure drop of the system is given by the difference between atmospheric pressure at the mouth and some lower value in the alveoli. A measurement of the latter is obtained when the flow into the chamber is closed off abruptly; under these conditions of zero flow, the alveolar pressure instantaneously becomes the chamber pressure. A whole-body plethysmograph yields different types of information, based on pressure, flow, and volume relationships. These parameters include thoracic gas volume (TGV), which includes gas in those regions of the lung that are not ventilated, and functional residual capacity (FRC), defined as the volume of gas in the lung during quiet breathing when the subject is relaxed and sitting upright. Since airway resistance (R_{aw}) varies inversely with the volume at which measurements are made (Lawther *et al.*, 1973), the specific airway resistance (SR_{aw}) is defined as the product of resistance times volume. For obvious reasons, there are many more spirometric studies in the literature than those of whole-body plethysmography.

Spirometry is used to provide a permanent record of breathing patterns, which may be used for sequential comparisons and for comparison with population norms for various features of the volume versus time tracing (Figure 4–3). These include:

- Timed vital capacities (volume expired in a given time). The commonly used times are 1 second (FEV_1) and 0.75 seconds ($FEV_{0.75}$).
- Forced vital capacity, the asymptotic volume reached at the end of the test (FVC).
- The ratio of FEV_1/FVC, the fraction of the total capacity reached in a given time. When expressed as a percentage, this parameter is often denoted as FEV_1%.
- The peak instantaneous expiratory flow rate (*L*/min), that is, the maximum slope of the curve (PEFR).
- Various instantaneous forced expiratory flow rates defined as the slopes at defined increments of the FVC ($FEF_{25\%}$, for example).
- Various average forced expiratory flow rates defined as the average slopes between defined increments of the FVC ($FEF_{25-75\%}$, for example, which is sometimes also called maximum midexpiratory flow [MMEF]).

All of these lung function parameters* change with aging of the subject. In addition, there are known average differences by sex, race, physical stature, altitude (Schoenberg *et al.*, 1978), and smoking status; no significant effect was found due to alcohol consumption (Sparrow *et al.*, 1983). The position of the subject can make a difference (many years ago, tests were administered to seated patients; the current practice requires them to stand).

Peak flow (PEFR) can also be measured with a different, specific apparatus that is portable and can be self-administered (Lawther *et al.*, 1974).

---

* Lung Function Acronyms and Abbreviations

FVC	forced vital capacity
R_{aw}	airway resistance
SR_{aw}	specific airway resistance
FEV_1	forced expiratory volume in 1 second
$FEV_{0.75}$	forced expiratory volume in 0.75 seconds
PERF, PFR	peak expiratory flow rate
$FEF_{(n)\%}$, $MEF_{(n)\%}$, $\dot{V}_{max(n)}$	forced expiratory flow at (n)% vital capacity
$FEF_{25-75\%}$, MMEF, or MMFR	maximum midexpiratory flow (average slope between defined increments of the FVC)

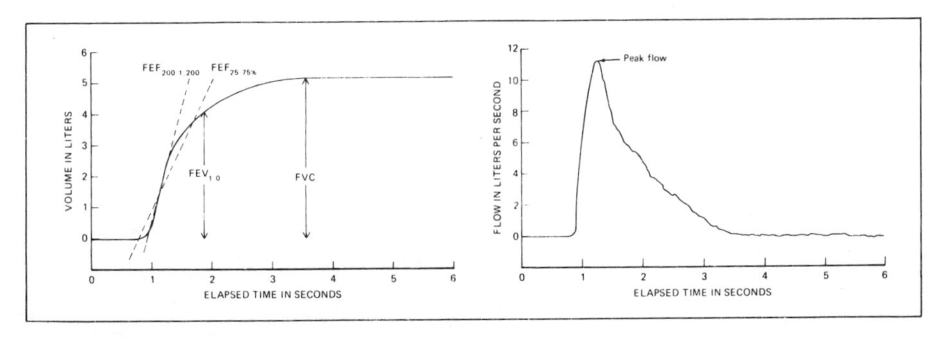

FIGURE 4-3. Sample spirometric time-volume and time-flow tracings. *Source: O'Brien and Drizd (1981).*

Unlike studies of mortality, hospital usage, or respiratory symptoms obtained from diaries or questionnaires, the collection of lung function data may be limited by the resources of the investigative team. A certain amount of laboratory equipment is required, each subject must be personally attended to, and the recorded tracings must be analyzed to extract the desired statistical parameters. The simple apparatus (peak flow meter, for example) tends to yield cruder, less sensitive measurements than the sophisticated laboratory-scale apparatus (plethysmograph). Thus, when a large number of subjects are examined at one location, an extensive period of time will be required. Recording the temporal variation of patients' lung function in order to generate a time series will be limited to a relatively small number of subjects. Measurements other than simple spirometry will generally be limited to laboratory settings (Lippmann, 1987). Epidemiological studies may contain elements of both spatial and temporal variability, both of which must be addressed if valid estimates of the environmental factors *per se* are to be obtained.

Sources of Variation in Lung Function

There are other sources of systematic variations in lung function that must be considered in making comparisons, especially for grouped averages. For example, altitude can increase FVC in young adults, about 0.3 L per 1000 m. (Schoenberg *et al.*, 1978). Diurnal variations (Kerr, 1973; Reinberg and Gervais, 1972) have been found in lung function parameters. The amplitudes of these variations are generally of the order of a few percent for lung volume measurements and up to about 20% for airway resistance measurements. These fluctuations are the same order of magnitude as the effects of air pollution. This means that longitudinal studies must be organized so as to obtain repeated measurements at the same time of day.

Seasonal factors have been studied by several groups of investigators. McKerrow and Rossiter (1968) followed a group of 28 ex-miners with pneumoconiosis in Wales for 3 years, taking monthly measurements of $FEV_{0.75}$. The linear decrease over this time period (about 0.2 L) was much greater than the cyclical perturbations, which had an average peak-to-trough amplitude of 0.044 L with the trough occurring in February. They also studied 17 younger normal subjects for 13 months, who exhibited an average linear increase in

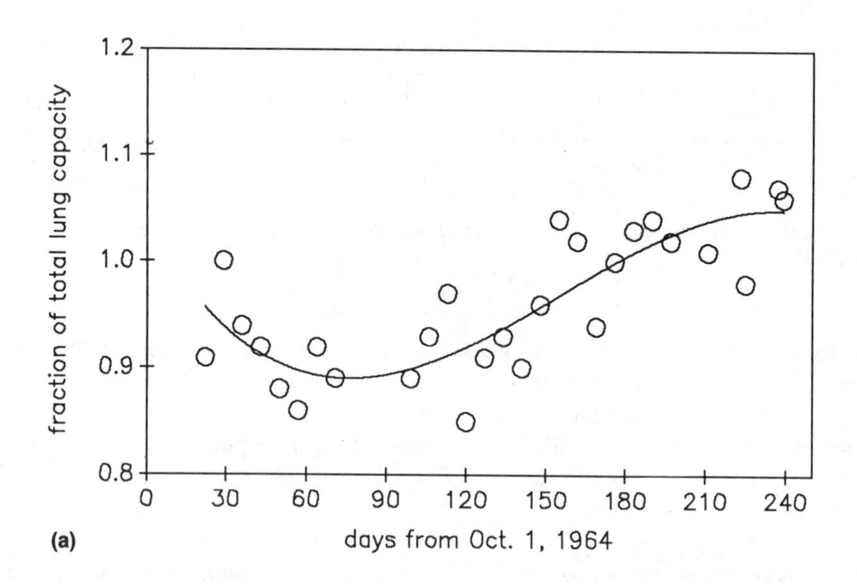

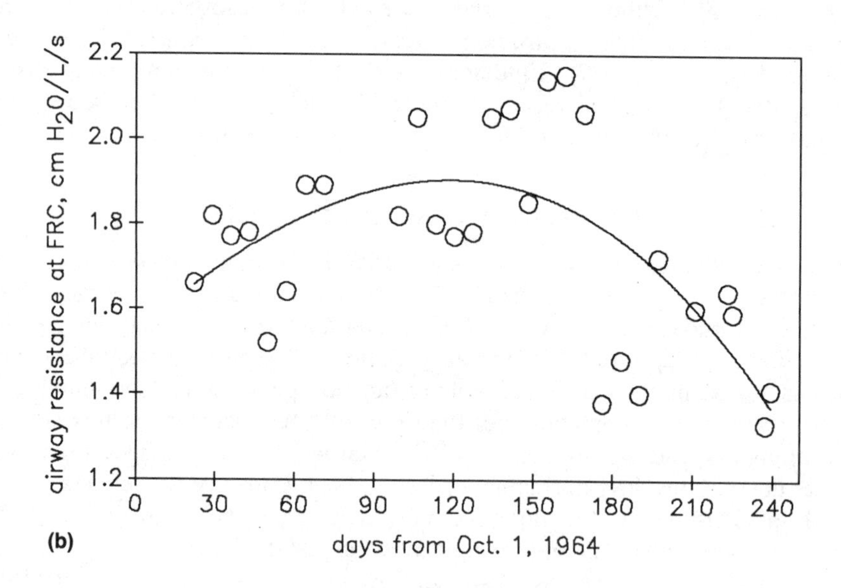

FIGURE 4-4. Seasonal changes in selected lung function measures for a group of white male college students. (a) Total lung capacity. (b) Airway resistance at FRC. (c) Upper and lower airway resistances. *Data from Spodnik* et al. *(1966).*

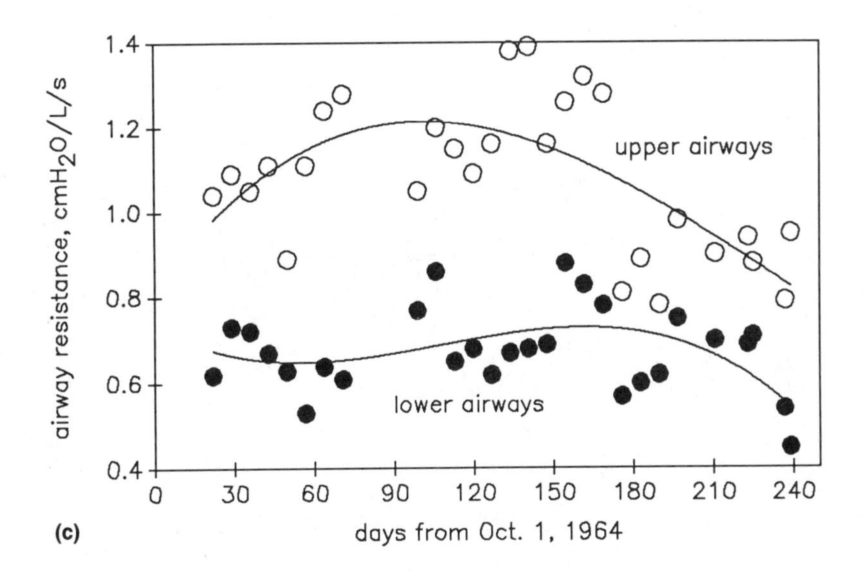

FIGURE 4-4. *Continued*

$FEV_{0.75}$ over time of about 0.1 L and had a cyclical perturbation of about 0.2 L (this finding was reported as 0.1 L on the basis of smoothing by McKerrow and Rossiter; this description has been repeated by others [Cotes, 1979]). The cycle for normal subjects, based on only 13 observations, peaked in June with a minimum in November–December.

Lawther *et al.* reported daily spirometry measurements on four colleagues for 5 years (1974) and measurements of airway resistance for himself and two other staff members over a period of 6 years (1977), in central London. They reported no difference between summer and winter average FVC, 1.5% lower FEV_1 in winter, and 4.6% lower MMEF in winter. Some of the difference may have been due to differences in air pollution between summer and winter (see Chapter 11). Only one of three subjects exhibited cyclical variation in R_{aw}, and this was seen only during the first year.

Spodnik *et al.* (1966) studied 100 white male college students in Baltimore for about 9 months. They reported data on total lung capacity and several measures of airway resistance, stratified by seven different groups of subjects. These results are plotted in Figures 4-4a through 4-4c for the entire cohort; each measure tends to have its own cycle, and some of the groups of subjects were shown to be significantly different from the others.

The data of Spicer *et al.* (1962) for chronic obstructive pulmonary disease (COPD) patients show seasonal dependencies for airway resistance and MMEF, but not for vital capacity. The first two measures also showed a dependency on TSP. Rokaw and Massey (1962) reported no cyclical variations among a group of COPD patients in Los Angeles; this could be an indication that ambient temperature is the operative variable (as suggested by Spodnik *et al.*, 1966),

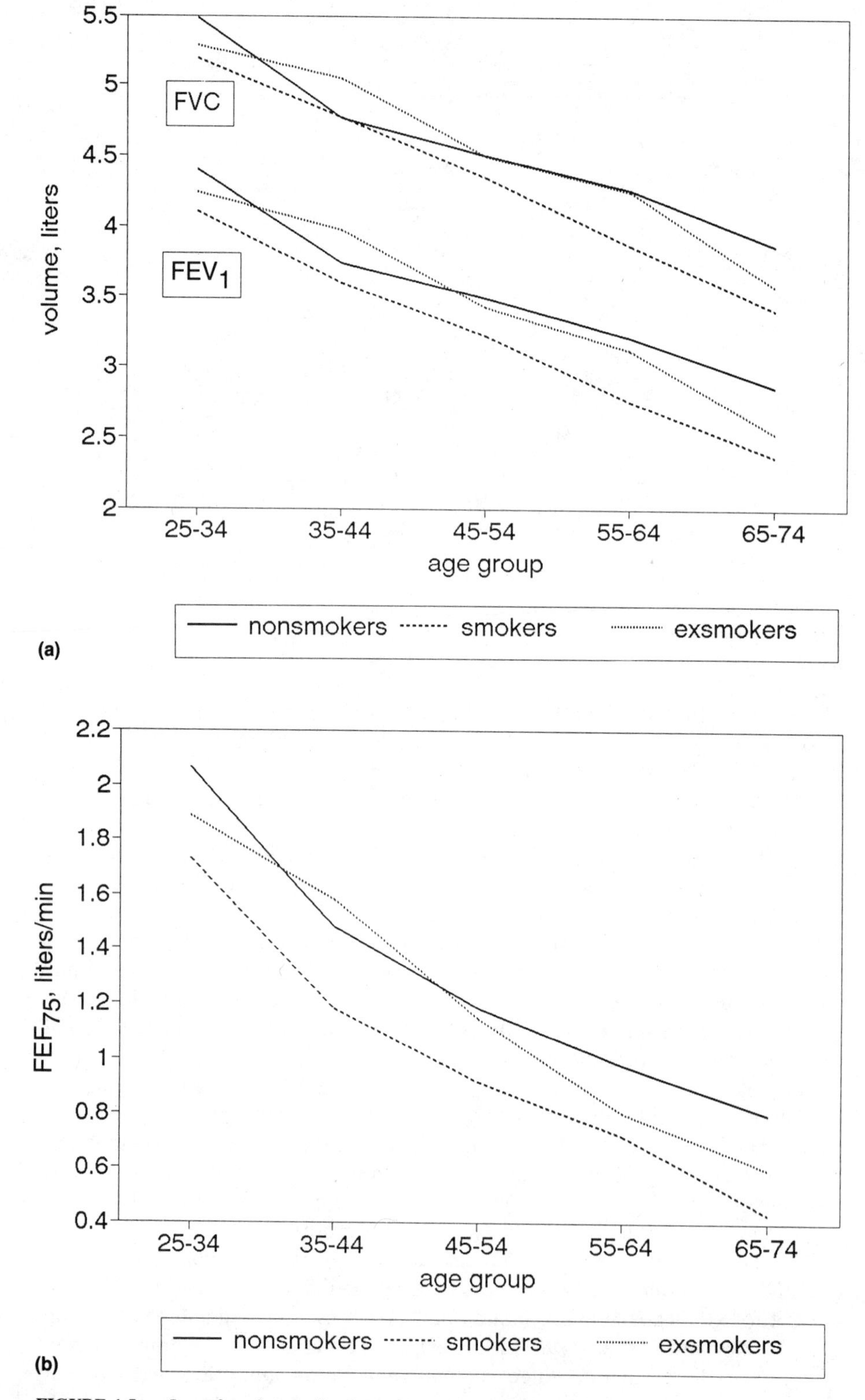

FIGURE 4-5. Lung function decline by smoking for white males. (a) FVC and FEV_1. (b) $FEV_{0.75}$. (c) Ratio of FEV_1/FVC. *Data from O'Brien and Drizd (1981).*

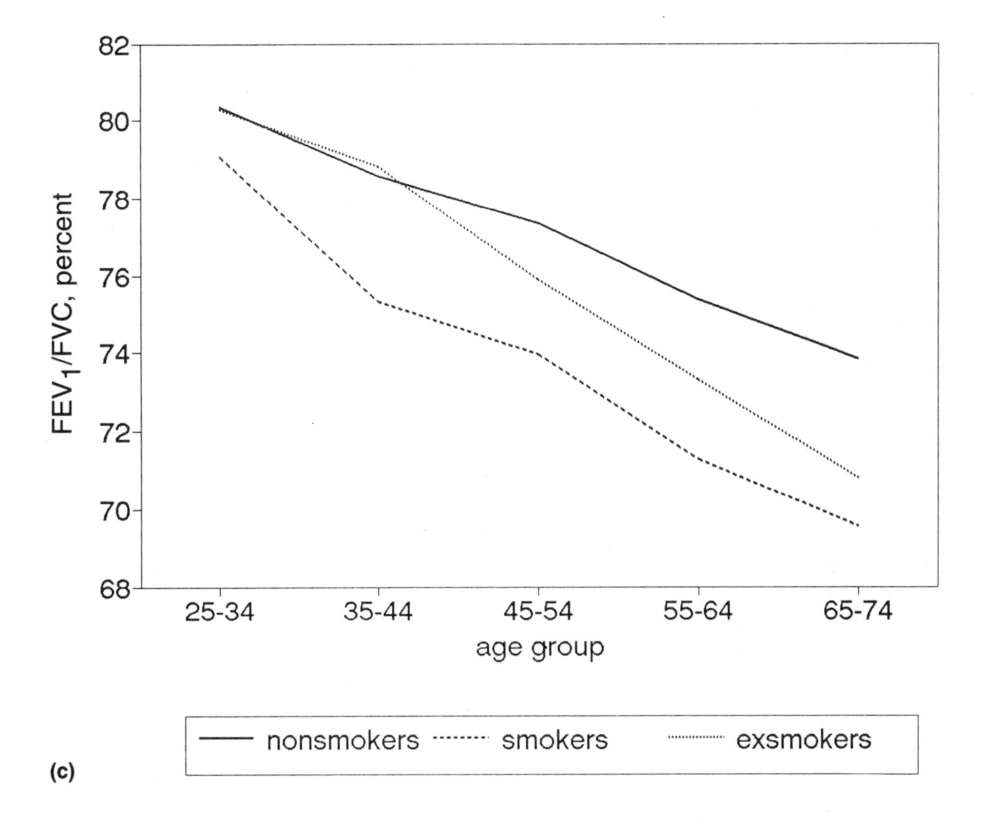

FIGURE 4-5. *Continued*

rather than some intrinsic biological rhythm. This hypothesis could also help explain the variations in timing of lung function cycles that have been observed.

Dampness and molds in the home have been reported to affect children's respiratory symptoms and, to a lesser extent, lung function (Brunekreef *et al.*, 1989). The average effects on $FEF_{25-75\%}$ were 1% for dampness and 1.6% for molds.

Elevated levels of copper in drinking water have been shown to be associated with an improvement of about 10% in FEV_1 in never-smokers only (Sparrow *et al.*, 1982). These results were rationalized physiologically in terms of the role of copper in certain enzymatic reactions. (This finding could be interpreted as a benefit of acidified drinking water supplies, since softer water tends to leach metals from the supply pipes. However, such leaching could also create a hazard because of leaching of lead from the soldered joints, for example.)

Medical Significance of Lung Function Measures

Peto *et al.* (1983) pointed out that none of the available techniques could provide a perfect measurement of the degree of airflow obstruction. Various physiological interpretations have been given to deviations from normal values for each of these parameters. For example, FEV_1, $FEV_1\%$, and peak flow are

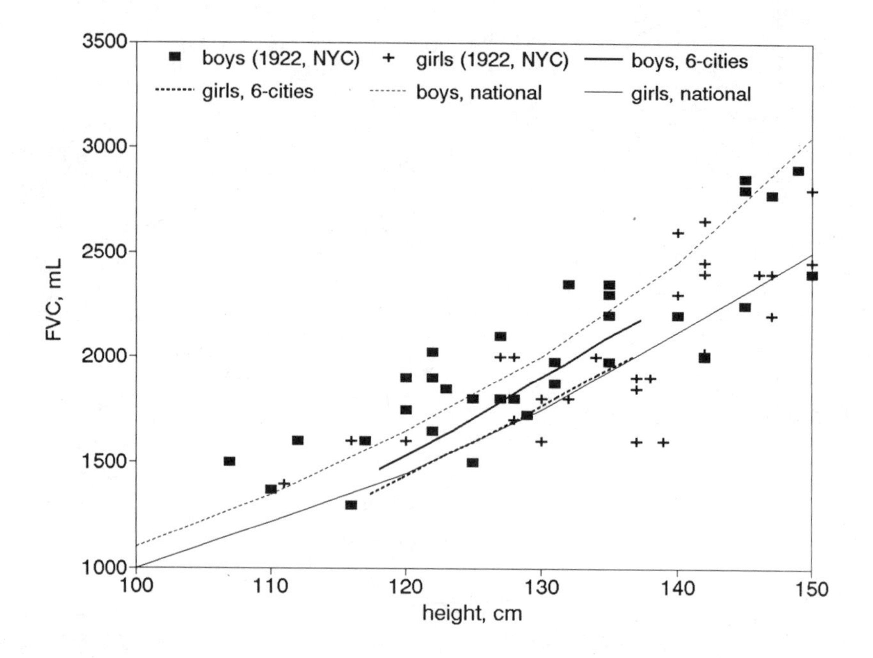

FIGURE 4-6. Comparison of children's lung function data, 1922 versus contemporary. *Data from Wilson and Edwards (1922) and Ware* et al. *(1986).*

thought to be inversely related to the amount of obstruction in larger airways; $FEF_{50\%}$ and $FEF_{25-75\%}$ are thought to be related to obstruction in both large and small airways; and reductions in $FEF_{75\%}$ are thought to be related to airflow resistance in the smaller airways. Sobol *et al.* (1974) found MMEF to be the most sensitive measure for detecting pulmonary abnormalities, although FEV_1 was almost as good. For illustrative purposes, Peto *et al.* (1983) showed a diagram in which an FEV_1 value of about 30% of the value obtained by an individual at age 25 was defined as "disability," and about 10%, as "death." Respiratory impairment is often defined as a ratio of FEV_1 to FVC that is less than 0.70.

Lippmann (1987) discussed the types of reversible spirometric effects commonly seen as a result of exposure to common air pollutants, including SO_2, NO_2, O_3, and H_2SO_4, and noted the difficulty of defining which responses might be regarded as "adverse" effects, in the context of the Clean Air Act. For example, he cited a 1985 statement by the American Thoracic Society that a "statistically significant" increase in the rate of decline of lung function with age should be considered as adverse. However, it can be seen that this definition presents two practical difficulties: first, impaired individuals often display poor repeatability of spirometric measurements, and second, if group averages are used, their statistical significance can depend on the numbers of observations or patients. Lippmann went on to point out that in the

TABLE 4-1 Regression Results from U.S. National Sample, 1971–75

Group	Smoking	Measure	a	b (year)$^{-1}$	Std error of slope	R^2	Std error of estimate	Predicted Value Age 25	Predicted Value Age 60
White male	No	FVC (L)	6.437	−0.0372	0.0048	0.952	0.153	5.51	4.21
		FEV_1 (L)	5.352	−0.0362	0.0041	0.963	0.130	4.45	3.18
		Ratio (%)	85.230	−0.1620	0.0059	0.996	0.185	81.20	75.50
		FEF_{75} (liters/min)	2.820	−0.0304	0.0047	0.934	0.147	2.06	1.00
White female	No	FVC (L)	4.686	−0.0290	0.0032	0.964	0.100	3.96	2.95
		FEV_1 (L)	3.986	−0.0280	0.0015	0.991	0.048	3.29	2.31
		Ratio (%)	87.190	−0.1520	0.0240	0.928	0.770	83.40	78.10
		FEF_{75} (liters/min)	2.285	−0.0260	0.0029	0.964	0.091	1.64	0.73
Black male	No	FVC (L)	5.439	−0.0314	0.0120	0.693*	0.380	4.65	3.56
		FEV_1 (L)	4.388	−0.0270	0.0110	0.655*	0.360	3.71	2.77
		Ratio (%)	80.740	−0.0500	0.1000	0.070*	3.330	79.50	77.70
		FEF_{75} (liters/min)	2.045	−0.0190	0.0100	0.534*	0.330	1.57	0.91
Black female	No	FVC (L)	4.147	−0.0270	0.0036	0.947	0.115	3.47	2.53
		FEV_1 (L)	3.465	−0.0240	0.0023	0.974	0.071	2.87	2.03
		Ratio (%)	84.560	−0.0830	0.0670	0.335*	2.130	82.50	79.60
		FEF_{75} (liters/min)	1.732	−0.0190	0.0055	0.803	0.173	1.26	0.59
White male	Ex	FVC (L)	6.637	−0.0421	0.0041	0.972	0.130	5.58	4.11
		FEV_1 (L)	5.597	−0.0428	0.0030	0.985	0.096	4.53	3.03
		Ratio (%)	88.040	−0.2440	0.0136	0.991	0.430	81.90	73.40
		FEF_{75} (liters/min)	2.887	−0.0337	0.0021	0.988	0.067	2.04	0.87
White male	Yes	FVC (L)	6.554	−0.0447	0.0008	0.999	0.026	5.44	3.87
		FEV_1 (L)	5.363	−0.0431	0.0012	0.998	0.037	4.29	2.78
		Ratio (%)	85.390	−0.2310	0.0200	0.978	0.640	79.60	71.50
		FEF_{75} (liters/min)	2.527	−0.0306	0.0036	0.960	0.114	1.76	0.69

Source: O'Brien and Drizd, 1981.
*$p > 0.05$.

absence of any better guidance, regulatory agencies have developed *ad hoc* guidelines for defining adverse effects.

Average Values

A national sample of spirometry data was obtained in 1971–75 by the National Center for Health Statistics as part of the National Health and Nutrition Examination Survey (O'Brien and Drizd, 1981), based on the examination of 6913 persons. Results were reported by age, smoking status, and the success of the measurements; some of the data are shown here in Figures 4-5a through 4-5c; results of my regressions against age are given in Table 4-1, together with predicted values for ages 25 and 60 based on these linear regressions (value = $a + b * \text{age}$). The values predicted for age 25 presumably reflect conditions before chronic exposure to cigarette smoke or occupational air pollution have had time to exert a substantial effect. However, Miller and Thornton (1980) point out that the age curve is not strictly linear in this range, tending to be flat from age 24 to 35, and thus use of a linear age term may result in the overprediction of values for young adults. Such a tendency is not apparent from Figures 4-4 and 4-5. All of these regressions were statistically significant, except those for black males and the ratio of FEV_1/FVC for black females (presumably because of the smaller numbers of subjects in these subgroups).

The data show that the annual rates of decline for FVC and FEV_1 are about the same in each subgroup but are accelerated by smoking. As expected, ex-smokers are intermediate between smokers and nonsmokers, presumably reflecting the ages at which smoking ceased. The ratio of FEV_1 to FVC is thought to be one of the best indications of large airway obstruction; we see that this statistic is highest for females of all ages and both races. The differences between predicted values for nonsmoking black and white males is not statistically significant. The forced expiratory flow rate at 75% of FVC ($FEF_{75\%}$), which is thought to be a good measure of small airway obstruction, also declines substantially with smoking, even at age 25. Racial differences for this statistic have largely disappeared by age 60.

Lung function data for children are compared in Figure 4-6. The data points were taken in the early 1920s in New York City; they compare quite well with contemporary data (in spite of the high air pollution levels undoubtedly present then).

REFERENCES

American Lung Association (1973), *Introduction to Lung Diseases*, Washington, DC.

Bates, D.V., Macklem, P.T., and Christie, R.V. (1971), *Respiratory Function in Disease*, Saunders, New York, pp. 25–26.

Brunekreef, B., *et al.* (1989), Home Dampness and Respiratory Morbidity in Children, *Am. Rev. Respir. Dis.* 140:1363–67.

Comroe, J.H., Jr. (1965), *Physiology of Respiration*, Year Book Publishers, Chicago.

Comroe, J.H., Jr., Forster, R.E., Dubois, A.B., Briscoe, W.A., and Carlsen, E. (1955), *The Lung–Clinical Physiology and Pulmonary Function Tests*, Year Book Publishers, Chicago.

Cotes, J.E. (1979), *Lung Function*, 4th ed., Blackwell, Oxford.

Frank, R. (1975), Biologic Effects of Air Pollution, in *Energy and Human Welfare—A Critical Analysis. Vol I. The Social Costs of Power Production*, B. Commoner, H. Boksenbaum, and M. Corr, eds., Macmillan, New York, pp. 17–27.

Kerr, H.D. (1973), Diurnal Variation of Respiratory Function Independent of Air Quality, *Arch. Environ. Health* 26:144–52.

Kittredge, M. (1989), *The Respiratory System*, Chelsea House, New York.

Lawther, P.J., Brooks, A.G.F., Lord, P.W., and Waller, R.E. (1974), Day-to-Day Changes in Ventilatory Function in Relation to the Environment. Part I. Spirometric Values, *Env. Res.* 7:37–40.

Lawther, P.J., Lord, P.W., Brooks, A.G.F., and Waller, R.E. (1977), Air Pollution and Pulmonary Airways Resistance: A 6-Year Study with Three Individuals, *Env. Res.* 13:478–92.

Lippmann, M. (1987), Health Significance of Pulmonary Function Tests, paper 87-32.3 presented at the Eightieth Annual Meeting of the Air Pollution Control Association, New York.

McClure, C.W., and Peabody, F.W. (1917), Relation of Vital Capacity of Lungs to Clinical Condition of Patients with Heart Disease, *JAMA* 69:1954–59.

McKerrow, C.B., and Rossiter, C.E. (1968), An Annual Cycle in the Ventilatory Capacity of Men with Pneumoconiosis and of Normal Subjects, *Thorax* 23:340–49.

Miller, A., and Thornton, J.C. (1980), The Interpretation of Spirometric Measurements in Epidemiologic Surveys, *Env. Res.* 23:444–68.

O'Brien, R.J., and Drizd, T.A. (1981), Basic Data on Spirometry in Adults 25–74 Years of Age, U.S. Dept. of Health and Human Services, National Center for Health Statistics, DHHS Publ. No. (PHS) 81–1672.

Peto, R., *et al.* (1983), Relevance in Adults of Air-Flow Obstruction, but Not of Mucus Hypersecretion, to Mortality from Chronic Lung Disease, *Am. Rev. Resp. Dis.* 128:491–500.

Reinberg, A., and Gervais, P. (1972), Circadian Rhythms in Respiratory Functions, with Special Reference to Human Chronophysiology and Chronopharmacology, *Bull. Physio-Path. Resp.* 8:663–75.

Rokaw, S.N., and Massey, F. (1962), Air Pollution and Chronic Respiratory Disease, *Am. Rev. Respir. Dis.* 86:703–4.

Schoenberg, J.B., Beck, G.J., and Bouhuys, A. (1978), Growth and Decay of Pulmonary Function in Healthy Blacks and Whites, *Resp. Physiology* 33:367–93.

Sherwin, R.P., and Richters, V. (1991), Chronic Bronchitis in Youths (Coroner Cases): Glandular Inflammation and Alterations, presented at the Annual Conference, Society for Occupational and Environmental Health, Crystal City, VA.

Sobol, B.J., Herbert, W.H., and Emirgil, C. (1974), The High Incidence of Pulmonary Functional Abnormalities in Patients with Coronary Artery Disease, *Chest* 65: 148–51.

Sparrow, D., Silbert, J.E., and Weiss, S.T. (1982), The Relationship of Pulmonary Function to Copper Concentrations in Drinking Water, *Am. Rev. Respir. Dis.* 126:312–15.

Sparrow, D., Rosner, B., Cohen, M., and Weiss, S.T. (1983), Alcohol Consumption and Pulmonary Function, *Am. Rev. Respir. Dis.* 127:735–38.

Spicer, W.S., Jr. (1963), Health Considerations, Report of the Chairman, Panel C., in *Proc. National Conference on Air Pollution*, U.S. Dept. of Health, Education, and Welfare, Washington, DC, pp. 387–89.

Spicer, W.S., Jr., Storey, P.B., Morgan, W.K.C., Kerr, H.D., and Standiford, N.E. (1962), Variations in Respiratory Function in Selected Patients and Its Relation to Air Pollution, *Am. Rev. Respir. Dis.* 86:705–19.

Spodnik, M.J., Jr., Cushman, G.D., Kerr, D.H., Blide, R.W., and Spicer, W.S., Jr. (1966), Effects of Environment on Respiratory Function, *Arch. Env. Health* 13:243–54.

Wilson, M.G., and Edwards, D.J. (1922), Diagnostic Value of Determining Vital Capacity of Lungs of Children, *JAMA* 78:1107–10.

PART II

Mortality Studies

And therefore the Empoysoning of Aer, *was ever esteemed no lesse fatall than the poysoning of Water or Meate itself, and forborn even amongst* Barbarians.

John Evelyn, *Fumifugium*, London, 1661

5

The Air Pollution Disasters

The inhabitants groped in a world of unreality, in which everything was shadowy and assumed fantastic shape. Judgment of the reasoning faculties was betrayed by the false witness of misled perception. The fog alone was materially in evidence, and it blinded, choked and chilled them. The pervasive phenomenon penetrated their thought. There wasn't anything else to engage it. Nothing is more contagious than hysteria.

Louisville Courier-Journal, describing the Meuse Valley air pollution disaster

In this chapter I emphasize short-term incidents in which simultaneous increases in air pollution and in mortality were noted. Concurrent increases in morbidity (hospitalization) are discussed in Chapter 9. These incidents are important, because they laid the foundations for modern concerns about the health effects of air pollution, by virtue of the nearly incontrovertible evidence they provided.

The published reports of these disasters are first reviewed, and then effects on mortality are considered in the context of dose-response functions.

REPORTS OF EXCESS MORTALITY DURING AIR POLLUTION DISASTERS

Ground-level concentrations of air pollutants can increase for either of two reasons: increased rates of emission, or reduced rates of atmospheric dispersion. The 1950 hydrogen sulfide (H_2S) disaster at Poza Rica, Mexico (McCabe and Clayton, 1952), and the 1984 toxic gas release at Bhopal, India, belong to the first category. A more recent (December 1990) subway fire in New York City also belongs to this group of events, which are random accidents about which little can be said except to express regrets for the conditions that allowed the excess emissions to occur. However, nature provides a range of atmospheric dispersion conditions, and thus extreme events of the second kind may be expected to occur with some regularity, even in the absence of human error. This latter category is the subject of this chapter.

The common factors associated with the earliest and best known air pollution disasters (1930–1962) included stagnant meteorology, fog, and cool or cold weather, which resulted in high levels of pollutant emissions from space heating as well as reduced dispersion. Industrial emissions were a common factor in the first two disasters. Stagnation conditions persisted for several days, causing people and animals alike to sicken and die at rates far above those expected for that time of the year. More recent stagnation episodes have generally not involved fog, perhaps because air pollution levels were lower (Neuberger and Gutnick, 1949).

The Meuse Valley, Belgium, December 1930

The Meuse Valley episode occurred along a narrow stretch of the Meuse River about 20 km long, between Liege and Huy, that contained 27 factories producing coke, steel, glass, phosphate, sulfuric acid, and zinc (Batta *et al.*, 1933; Firket, 1936; Roholm, 1937; Ashe, 1959). Given the year, we may assume that vehicle emissions were minor. Little is known about this episode (even the population at risk was uncertain [McDonald *et al.*, 1951]), but 63 "excess" deaths were reported in two days, about 10 times the estimated normal number. Ashe (1959) reported that around 6000 people became ill. Firket (1936) also reported that "many head of cattle had to be slaughtered." Meteorological conditions consisted of a high-pressure zone and a strong inversion in the narrow valley. Temperatures were estimated at slightly below freezing (Batta *et al.*, 1933, p. 329).

Fifteen (human) autopsies were performed, and "inhalation of fine particles of soot even as far as the pulmonary alveoli" were noted (Firket, 1936). Interviews with survivors revealed that most people were affected at about the same time along the entire 20 km stretch, thus ruling out a traveling cloud of toxic gas. Severe coughing and breathlessness were common syndromes.

In the community of Engis (3500 inhabitants) in the center of the valley, for example, the fog began on December 1, became somewhat intense on December 2, and very intense on December 3 and 4 (Batta *et al.*, 1933). Eight people died on the 4th and six more on the 5th, after the worst fog intensity had passed. There were no deaths on December 6. The death toll during these 6 days was thus about 14 times the normal rate in this town.

Firket (1936) used dispersion calculations to estimate air concentration levels, since there was no air monitoring in operation; he estimated about 10 ppm (26,000 $\mu g/m^3$) for SO_2. The fog would have provided an aqueous medium for the transformation of SO_2 to H_2SO_4 and would have greatly enhanced rates of deposition; Firket based his estimates of H_2SO_4 on 100% conversion of SO_2 to H_2SO_4, an extremely unlikely assumption (Lipfert, 1992). Estimates of particulate levels were not made, since no dust or smoke emissions were estimated for the various pollution sources. Firket mentions the deposition of fine soot particles in the fog droplets. Fluoride was also implicated in this episode because of industrial emissions (Roholm, 1937). Given the industrial mix present, there must have been a variety of metal and organic particles present as well. Zinc oxide in Engis was mentioned, for example, and Mills (1954) reported a zinc smelter in the valley.

There is some controversy about wind conditions in the reports. Although stagnation conditions were reported, Batta *et al.* (1933) also mention "feeble" winds along the valley (<3 km/h). This area represented essentially a line-source of pollution; a light wind along its axis with inversion trapping could represent a worst condition, since the effluents would be transported and accumulated along the valley with minimal dispersion. Firket's diffusion estimates were based on a "box" model in which pollutants built up because of continuous emissions with no sinks or advection out of the system. This is clearly inappropriate to the actual situation, in which advection was present (however "feeble") and in which the fog would have naturally removed airborne pollution by means of absorption and deposition in droplets. Such wetting of surfaces would have also enhanced the rates of "dry" deposition of SO_2 (Lipfert, 1989). Firket's SO_2 concentration calculation thus seems a gross overestimate. Based on an area source model, I estimated that "normal" SO_2 levels would have been around 0.1 ppm; it seems difficult to imagine an increase of more than a factor of 10, given the SO_2 sinks present. However, deposition rates may have been lower for other pollutants such as soot or CO, and since the sulfur content of the coal was only around 1% (British coals were higher, for example), it is quite likely that the rates of soot* and CO emission were also high in comparison to those of SO_2.

Batta *et al.* (1933, p. 320 [based on a rough translation]) concluded: "After having examined which substances were susceptible of explaining the toxicity of the fog, and having eliminated most of them, we concluded that, above all, the sulfurous elements that came principally from the burning of coal and exerted their harmful action either in the form of SO_2 or H_2SO_3 or H_2SO_4, whose production reached a sufficient level by the exceptional meteorological conditions." Ashe's (1959) paraphrased conclusions of the official report by Dehalu *et al.* (1931) cited acid gases absorbed onto particles, probably based on the observations of soot particles deep in the lungs of victims.

Donora, PA, October 1948

Donora, an industrial community of about 12,500 people near Pittsburgh, was afflicted by a heavy, stagnating fog for four days in late October 1948. Fifteen people died on the third day, two on the fourth day, and one more on the twelfth day (two additional deaths were later attributed to the episode). The expected death rate was about one in two days, so that the death rate was inflated by about a factor of 10 (similar to the Meuse disaster). Excess deaths were also reported in two nearby areas (Shrenk *et al.*, 1949, p. 74). Most of the area residents reported being affected by the fog, and there were reports of many animal deaths. A high-pressure system was in place over the area, but, in

*For example, Wilkins (1954a) cites an emission factor of 0.025 g smoke/g coal in London; the SO_2 emission factor for 1% sulfur coal would be from 0.015–0.02 g SO_2/g coal, depending on how it was burned.

contrast to many of the other episodes, temperatures were not exceptionally cold (Pittsburgh highs were in the upper 60s and lows in the high 30s). Upper air soundings taken at Pittsburgh showed the atmosphere to be very stable.

The official report of this disaster was produced by the Public Health Service (PHS) (Shrenk *et al.*, 1949). Two interesting and more personal accounts are those by Mills (1954) and by Roueché (1984). Mills addresses the apparent culpability of the pollution sources, an issue that most of the other accounts of this incident seemed to eschew. Ashe (1952) reported on a study sponsored by the plant's owner, the American Steel and Wire Company, a subsidiary of U.S. Steel.

The event began on Wednesday, October 27, 1948. About 60 people (later) reported feeling ill effects on the first two days, 750 on the next two, and 1600 on Friday. The fatalities began at 2 A.M. on Saturday and totaled 15 for that day. Of the 18 fatalities, 17 were ill not more than 60 hours (Ashe, 1952). The acute nature of these attacks was shown by the fact that two seamen became ill on river boats passing through the area. The PHS surveys were conducted from December to the following March; because of the elapsed time, some recall bias may have entered the survey. Deane (1965) observed that the nonwhite mortality rate during the Donora episode was substantially higher than the white rate and suggested that "some groups receive more sophisticated medical attention than others."

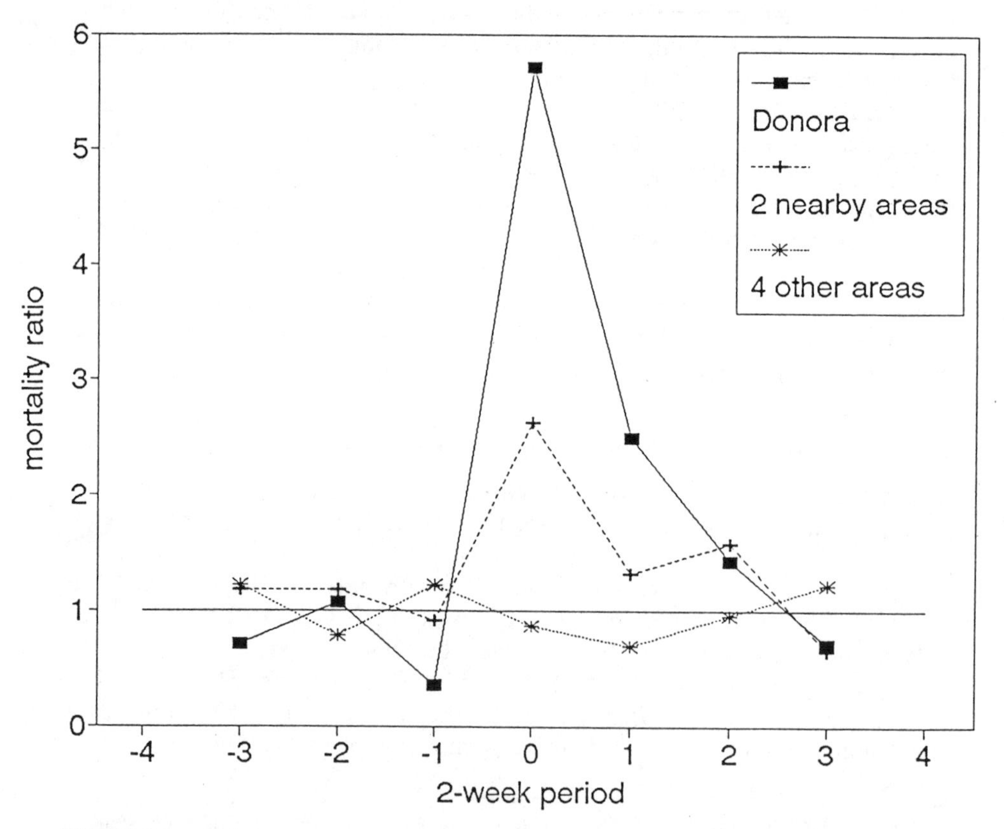

FIGURE 5-1. Relative mortality responses in and near Donora, PA, October 1948. *Data from Shrenk* et al. *(1949).*

The mortality data presented by Shrenk *et al.* (1949) allow some interesting comparisons. Figure 5-1 shows the time histories of deaths in Donora, two nearby areas, and four areas somewhat further away. The 2-week death counts have been divided by the respective averages not affected by the episode. In Donora, it appears that effects of the episode may have been felt up to 6 weeks after the fog began (period "0" began October 24). This was true to a lesser extent in the two nearby areas, but no perturbations were seen elsewhere. This appears to rule out the simultaneous presence of disease outbreaks throughout the region.

Questions also arose as to the baseline health of the affected community, given the presence of major pollution sources since about 1913. By cause of death, Donora had more cancer deaths than expected on the basis of the rates for the state of Pennsylvania, but rates for respiratory causes were as expected, and heart disease mortality rates were low. Figure 5-2 compares Donora's mortality experience for 1945–48 with Pittsburgh's, on the basis of unadjusted death rates. The rate in Donora averaged about 70% of Pittsburgh, which does not suggest a compromised baseline health status in Donora (at least on the basis of unadjusted mortality rates).

Three periods of high rates in Donora stand out: April and September 1945, and October–November 1948. The confidence limits shown were computed after excluding the three highest observations (n = 45). We would thus expect two or three outliers from the 95% limits to be just due to chance, but the two

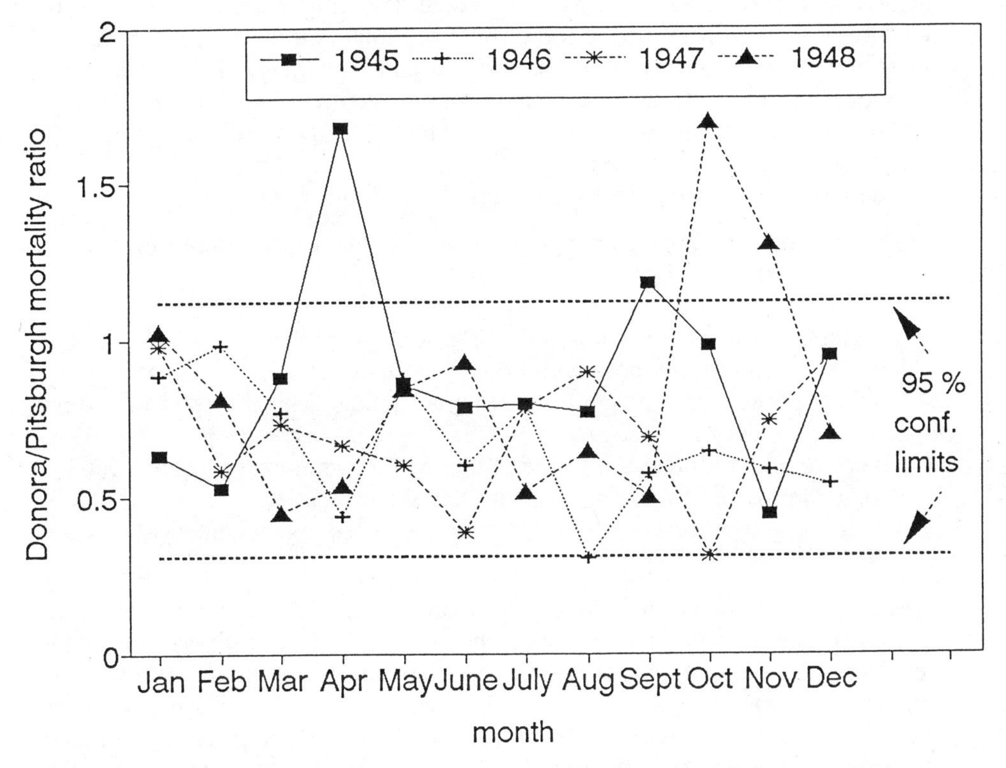

FIGURE 5-2. Mortality in Donora relative to Pittsburgh's, by month, 1945–1948. *Data from Shrenk* et al. *(1949).*

highest points have t values greater than 4 and thus are not likely to be random occurrences. Shrenk *et al.* (1949) report that meteorological conditions in September 1945 were "not conducive to the development of smog." Figure 5-2 also suggests that the effects of October 1948 were felt during November. It is also of interest that there were no seasonal differences between Donora and Pittsburgh, on average. This implies that the air pollution effects of space heating (if any) were about the same in both locations.

Mills (1954) visited Donora 6 days after the event and monthly thereafter and reported on his own survey and findings. He provided details on animal deaths, including 14 sheep, 427 chickens, 2 pigs, 12 colts, 6 cows, and numerous canaries. Shrenk *et al.* (1949) also reported animal deaths, including rabbits, but in general their account of effects on animals minimized this aspect of the episode. Livestock would have been farther away from the pollution source on farms rather than in the town where the human deaths occurred. According to Mills, after the disaster, the Humane Society forbade the importation of animals into the area! The death rate for house cats was about the same as for humans, but for dogs, it was three times as high. Mills also reported on a survey of about 87% of the area's population about their reactions to the fog. Among the respondents, 41% reported "serious effects," 805 were treated by physicians, and 338 unsuccessfully tried to get medical treatment. The percentage of affected people over age 50 was about double that of under age 10.

Roueche's account, which was apparently first published in the *New Yorker* magazine in 1950, tells the Donora story from the perspective of the medical emergency workers. As in the Meuse episode, the really severe effects were all felt within a relatively short period of time beginning on the fourth day, as if a toxicity threshold had been crossed, and also ended suddenly. The fact that not everyone was affected was demonstrated by the fact that the annual Halloween parade was held as usual and was "well attended." Some of the emergency service workers "felt fine" during the incident.

There are a number of interesting anecdotes in this account:

- The inversion was strong enough to prevent smoke from rising from a steam locomotive stack.
- The smoky fog penetrated indoors.
- Visibility was down to a few feet for a long period of time, but this had happened before in Donora without (apparent) ill effects.
- Smokers found that they could not tolerate the additional respiratory stress of cigarettes during the fog.
- Severe coughing to the point of vomiting was a common syndrome.
- People improvised "smog masks' with handkerchiefs.
- The fog was still present when the demands for emergency medical services stopped.

One of the main pollution sources was a zinc smelter, which emitted SO_2 and H_2SO_4 as by-products from the zinc sulfide ore. Mills (1954) reports a ground-level ambient measurement of H_2SO_4 near the plant of 700 mg/m^3 (this value is higher than measurements made in stack gases and thus does not seem credible; perhaps Mills meant µg/m^3). According to Mills, the company reported no occupational illness on Saturday, the day when 15 deaths occurred, but the workmen themselves told a far different story—an estimated 200 cases treated in the plant clinic. It is perhaps not surprising that half as many of the plant workers reported ill effects as did the general population. This could either have been self-selection of the work force or fear of reprisals. In 262 adult males examined, Shrenk *et al.* (1949) reported finding no effects of dental erosion that might have been attributed to acid aerosols. Such dental problems have been observed in lead-acid battery plants (Gamble *et al.*, 1984).*

Again, no air monitoring system was in place at the time, and estimates were based on subsequent measurements that showed 10-week averages of 0.12 ppm SO_2, 0.15 ppm total sulfur (assuming that "total sulfur" includes particles, the difference could be sulfates), 740 $\mu g/m^3$ particulate matter, and 85 $\mu g/m^3$ zinc. Cadmium and lead were also found. Maximum values observed during this subsequent sampling period were about 0.6 ppm SO_2 and 3000 $\mu g/m^3$ TSP. However, Mills reports that improvements were made to the zinc plant processes so that the PHS measurements represented neither maximum emissions nor worst-case dispersion. Air quality during the October fog could have been 10 times worse than the average values.

The official PHS report concluded that none of these substances could have been concentrated enough to have caused the observed effects alone, and therefore a combination of pollutants must have been responsible. The health effects observed seem to point to two different kinds of responses: acute and immediate respiratory irritation causing severe coughing, and delayed effects that resulted in death some days or weeks later. It thus follows that more than one agent may have been involved.

A follow-up study 10 years later (Ciocco and Thompson, 1961) indicated higher mortality rates for persons exposed to the episode and affected by it, but could not distinguish whether these individuals had preexisting conditions that might also have played a role in their survival rates.

London, England, December 1952 and After

London had been notorious for its "pea soup" fogs for many years and suffered deadly air pollution disasters on several occasions before and after the major episode that occurred in December 1952 (McDonald *et al.*, 1951). The statistics of seven such episodes are reviewed by Brasser *et al.* (1967); the numbers of excess deaths ranged from 200 to 3900, according to their account. This group of events is examined quantitatively. Air quality was characterized by the peroxide method for SO_2 and by the darkness of filter stains for particulates ("British smoke" or just "smoke" when referring to British data). As discussed in Chapter 2, the latter measurement is not gravimetric, although

*The modern experience of battery workers could be used to estimate maximum H_2SO_4 exposures in Donora. Gamble *et al.* (1984) report that the lowest average exposure (i.e., dose) in which tooth etching was observed was 4 months at 230 $\mu g/m^3$. Erosion was first seen at an exposure of 30 months. If we assume an exposure period of 20 years in Donora, we can solve for the equivalent continuous concentration levels by dividing the minimum doses by 20 years. This yields estimates of about 4 $\mu g/m^3$ and 30 $\mu g/m^3$ for the "normal" conditions in Donora, based on tooth etching and erosion, respectively. Using the figure for erosion and a concentration multiplier of 10 to account for stagnation conditions would give an estimate of 300 $\mu g/m^3$ for maximum H_2SO_4 levels in Donora. This estimate is reasonably consistent with actual measurements of aerosol acidity in London (Waller and Commins, 1967).

the results are reported in units of μg/m^3. Comparisons with other particulate sampling methods are discussed in Chapter 2.

In the official report, over 4000 excess deaths were attributed to the December 1952 disaster, which affected all age groups, especially infants and those over age 45 (Ministry of Health [MOH], 1954). Both central London and the outlying districts (as far away as 50 km [Logan, 1953]) recorded sharp increases in weekly mortality; the largest increases were for bronchitis, pneumonia, and respiratory tuberculosis, although large increases were also noted for deaths due to coronary heart disease and myocardial degeneration. Local levels of air pollution reached about 4000 μg/m^3 for both SO_2 and smoke (48-hour averages), although there is some question as to the maximum smoke level because of saturation of the filters and the use of visual colorimetric evaluation. Since pollution levels at the height of the episode were averaged over 48 hours, the peaks were probably higher. The principal source of air pollution was the burning of soft coal for home heating, mainly in inefficient open grates; as a result, both residential and commercial portions of the city were affected. The 1952 event is also analyzed in more detail in the following section; a time line of mortality and air pollution is shown in Figure 5-3.

This disaster looms large in the entire history of air pollution control, and much has been written about it. Wise's popularized account (1968) provides some of the human dimensions. For example, the fog was so dense that traffic

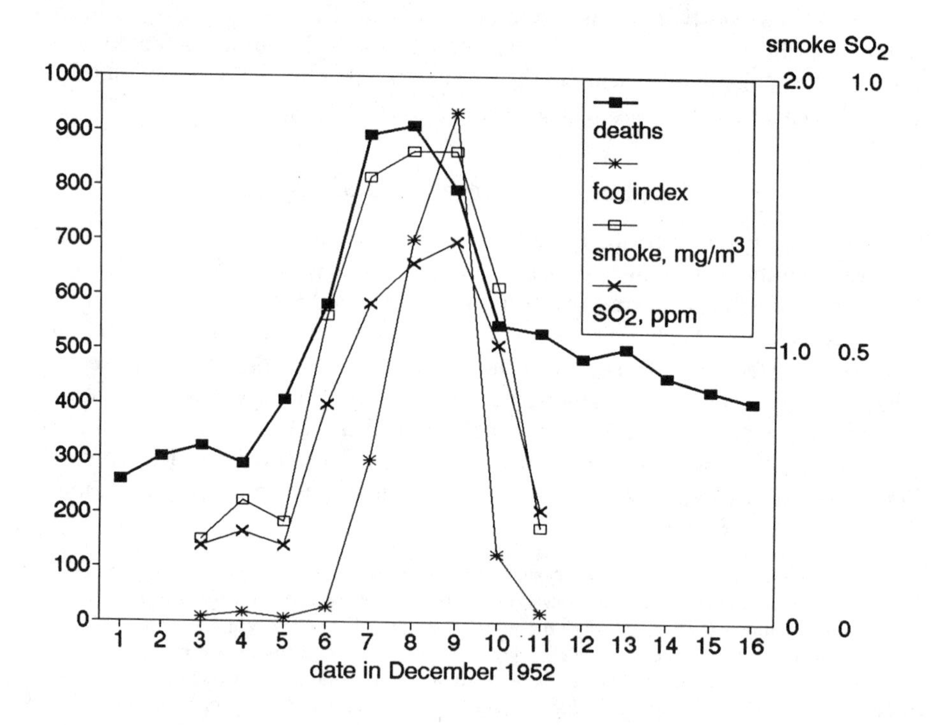

FIGURE 5-3. Time line for the 1952 air pollution disaster in London. *Data from Ministry of Health (1954).*

was brought to a standstill and eventually stopped altogether. (According to Baker [1989], London traffic in 1952 was minimal in any event, especially at night, because there were relatively few private cars by current standards.) People got lost in the fog, even on foot. Ambulances could not get through to bring people to hospital. The fog penetrated indoors, so that theatrical performances were affected. It tarnished metals and permanently soiled fabrics, which implies acidity. The fog droplets contained soot particles that soiled skin and clothing, even indoors. A common symptom was nausea, often with vomiting; a sulfur smell was apparent, and the fog was yellowish in color. Wise's personalized accounts of sickness and death would probably be described as "docudrama" today, since he combined details from more than one interviewee. Nevertheless, he described deaths of people with no previously known symptoms, as well as of those with preexisting conditions. He also ascribed the *onset* of chronic conditions as a result of exposure to the fog. The debate over imposition of controls after the full dimensions of the disaster became evident is reminiscent of contemporary environmental disputes (Ashby, 1974).

A common notion of the severe air pollution episodes is that of people dropping dead on the street. While this may have indeed happened in London (1952), most of the casualties occurred at home, in bed. The fatalities in the street or at work in London were overwhelmingly male heart cases. Another notion is that most or all of the cases suffered from prior conditions, which was one of the conclusions of the MOH report; this conclusion was contested by Kjellstrom (1989).* However, according to the combined MOH data (which constitute a small sample), 70% of the casualties had definite histories of prior pulmonary or cardiac problems. In one sample of fatalities, 36% were suffering from either an acute or subacute infection before the fog. The MOH report gives the average numbers of "conditions" noted on death certificates, which could be regarded as a measure of prior illness. This statistic increased slightly after the fog (from 1.8 to 2.1), suggesting that the patients with more preexisting conditions died later. The cardiac death rate dropped markedly as soon as the fog cleared (December 9–12), but deaths due to respiratory systems continued at a high rate for another week.

The age distribution of dead people showed the highest ratios of excess deaths for those 45 and older and under 1 year. There was no gradient within the over-45 group (i.e., the elderly did not seem at particularly higher risk during the fog, although there is some suggestion that their risks were higher immediately after the fog). This episode differed markedly from those at Donora and the Meuse in the persistence of excess mortality long after the fog had cleared, but the excess mortality ratio was also much lower in London (2.6 versus 10).

The MOH report also gives data on the spatial distributions of deaths and of pollution; it concluded that no spatial proximity of excess fatalities to

---

*The MOH report concluded "so far as can be ascertained, there were no death attributable to fog among previously healthy people." This seems a bit of an overstatement, since the Medical Officer of St. Pancras reported that 10% of his sample had no prior history and that this information was not available for another 40%. In addition, the report stated that a large number of fatalities occurred in bed and that "the nature of these sudden deaths remains a matter for speculation since no specific cause was found at autopsy."

areas of high SO_2 was noted but that there was some association with local sources of excessive smoke. One of the outcomes of the 1952 episode was the recognition that daily death data would be required for a sensitive analysis of the effects of air pollution (MacFarlane, 1977).

A number of lesser air pollution episodes were experienced in London throughout the 1950s. The last major episode was December 1962; it is believed that this weather system also affected other parts of the world at that time, including New York City on November 29 (Greenburg *et al.*, 1963). In London, SO_2 levels reached 2 ppm on December 5, 1962, and aerosol acidity was measured at 678 μg/m^3 (as H_2SO_4) (Waller and Commins, 1967). Both of these readings were maximum hourly averages; they represent 8% conversion of the SO_2. Fog was also present in London on this occasion (Marsh, 1963). According to Wichmann *et al.* (1989), 24-hour SO_2 reached about 2 ppm in the Ruhr district in December 1962, which was probably a result of the same meteorological disturbance.

Although London air quality improved by an order of magnitude through the application of smoke controls during the 1960s, relatively high levels of pollution can still be approached under adverse meteorological conditions. In December 1975, smoke reached about 550 μg/m^3 and SO_2 about 1000 μg/m^3 (24-hour averages); excess mortality for the week was of the order of 6–11% (Holland *et al.*, 1979).

New York City, 1950s and 1960s

Like London, New York had several episodes in the 1950s and 1960s. They have been analyzed by several authors (Greenburg *et al.*, 1962; McCarroll and Bradley, 1966), and some similarities to the London experiences were noted. Notable exceptions included warmer temperatures and the absence of heavy fog. During the episode of November 17–21, 1953, SO_2 reached 0.86 ppm and smokeshade, 8.5 COHs (about 850 μg/m^3 equivalent TSP, Greenburg *et al.*, 1962). Winds were calm up to 5000-ft altitude at the height of the episode. Eye irritation and respiratory complaints were received from all parts of the city, indicating that a large geographic area was affected. Mortality for all ages above 45 was affected. Air pollution data were reported only from a central station, and SO_2 monitoring was sporadic. Compared to its rate of incidence in London, bronchitis was not a major factor in New York, the percentage of excess deaths was lower, and SO_2 levels were somewhat lower while particulates were much lower (reflecting the generally cleaner fuels used in New York). The SO_2 measuring system used in New York City at that time is believed to have been comparable to that of London, but the tape sampler used for "smoke" was slightly different, and data were reported in units of "coefficient of haze" (COH). For the quantitative analysis and data plots, COH units were converted to equivalent gravimetric units as given in Lipfert, 1978 [approximately 100 μg/m^3 per COH unit]. As discussed in Chapter 2, this conversion factor agrees best with actual calibration data at the higher concentration levels (IERE, 1981; Holland *et al.*, 1979).

Figure 5-4 plots data from each day of this event, based on a 2-day lag. There is no suggestion that excess mortality continued beyond the episode period. The slope of a linear regression through these data was 0.00027 (0.027%

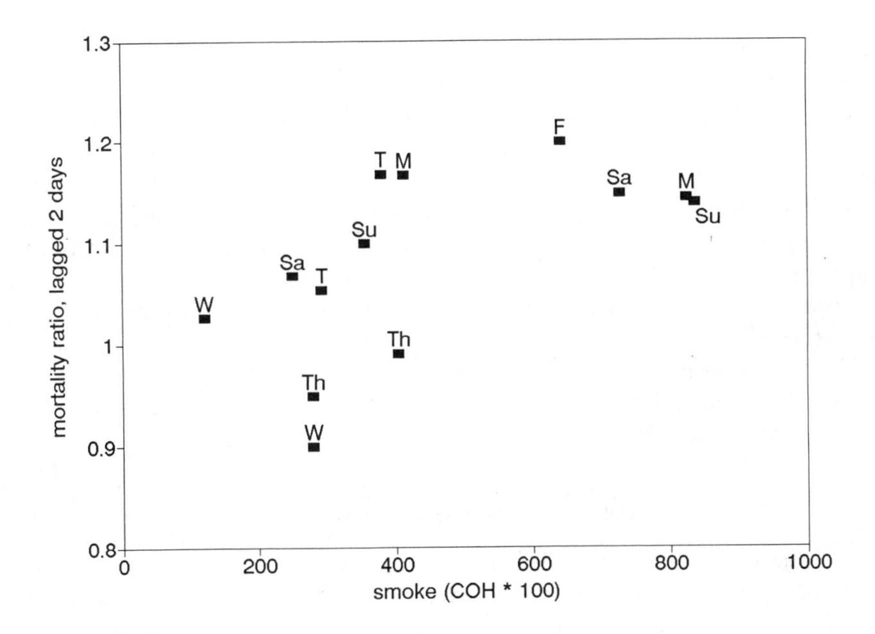

FIGURE 5-4. Mortality ratios for New York City, November 1953. *Data from Greenburg* et al. *(1962).*

excess deaths per μg/m^3 of smoke), with an R^2 of 0.45 ($p < 0.01$). The days of the week are indicated, since the weekend had the worst air quality during this episode, and, in general, Monday death registrations are sometimes inflated because of delays from the weekend. There is no indication of this phenomenon from Figure 5-4, however.

In November–December 1962, a potential for air pollution stagnation was forecast for the entire Eastern United States; SO_2 in New York City reached 1.4 ppm on November 30 (Greenburg *et al.*, 1963). Only one day showed appreciable excess mortality, which was followed by a drop. The 7-day yield was 33 excess deaths, or 13% per episode day. As discussed, the same period of restrictive dispersion seems to have swept across the Northern Hemisphere.

In 1963, an air pollution episode was confounded with an influenza outbreak (Greenburg *et al.*, 1967). A 15-day "critical" period was identified with intermittent high air pollution, but a clear mortality response could not be identified from the data given. Data from this paper also suggested that influenza might have been a factor during the 1962 episode.

McCarroll and Bradley (1966) analyzed five separate incidents of mortality "spikes" in New York from November 1962 to March 1964. Formal time-series methods were not used, but time line plots were used to display the sequences of the "episodes." Four of the five events featured sharp rises in mortality following simultaneous peaks in SO_2 and smoke, usually caused by reductions in wind speed. In one case, a second mortality peak was shown 4 days later, which corresponded to a peak in SO_2 but not smoke. One of the events, which occurred in April 1963, showed a very sharp peak in mortality with no accompanying rise in air pollution (or drop in wind speed). Crude estimates of the dose-response functions for four of the episodes were in the range of 0.0004 to 0.0006 per $\mu g/m^3$ smoke, or 0.0001 to 0.00013 per $\mu g/m^3$ SO_2.

Data from the 3-day Thanksgiving 1966 episode in New York, during which smoke reached about 600 $\mu g/m^3$ and SO_2, 0.5 ppm (1300 $\mu g/m^3$), are also difficult to interpret. The total "yield" was 23.5% over 7 days, which corresponds to a slope of 0.039% per $\mu g/m^3$, but there were other large swings in mortality before and after the event. No excess deaths were reported in Boston from this stagnation episode (Heimann, 1970), but my analysis of a longer period there showed a mortality increase in response to SO_2 similar to that found in other cities (see Chapter 6).

The conclusion from analyzing these relatively "mild" episodes was that continuous time-series methods would be more appropriate. This subject is taken up in Chapter 6.

Paris, December 1972–January 1973

Loewenstein and colleagues (1983) described two episodes of excess mortality in Paris: December 1979, with 45% excess mortality for the month, attributed mainly to an epidemic of influenza, and December 25, 1972–January 14, 1973, with 22% excess mortality reported for the 20-day period. Pollution was near the normal monthly average in the first case and reached 260 $\mu g/m^3$ for black smoke and 660 $\mu g/m^3$ for "acidité forte" (a measure of SO_2 believed to be similar to the methods used in England and in New York City prior to the mid-1970s) for the second case. Loewenstein's analysis does not clearly separate the various contributing factors: influenza, a holiday period, cold temperatures, and air pollution. I was unable to find a day-to-day correspondence between air pollution and mortality for any lag or averaging period. If the SO_2 measure was comparable to that for London and New York, one would conclude that the observed excess mortality in Paris was somewhat higher than expected. The mortality coefficient was 0.00085 per $\mu g/m^3$ based on particulates and 0.00033 per $\mu g/m^3$ based on SO_2. However, like in New York, pollution was only measured at one station and thus may not be representative of citywide exposures.

Sao Paulo, Brazil, July–August 1973

Mendes and Wakamatsu (1977) present a discussion of air pollution episodes in an industrialized city in the Southern Hemisphere, which offers an interesting comparison by virtue of the differences in seasons and climate. The high-mortality months in Sao Paulo were May and August; the event for which data were given occurred July 25–August 8, 1973. The original data have been cross-plotted to provide dose-response curves, shown in Figure 5-5. Since the time trend showed persistent mortality elevation after the peak pollution had passed, a 3-day average was used in the plot, which displays the average deaths for lags 0, 1, 2 versus SO_2 or suspended particulate matter (measurement method not stated). The best fit seems to be with SO_2 alone (Figure 5-5a), but the slope of this relationship seems very high (0.124% per $\mu g/m^3$). Also, the

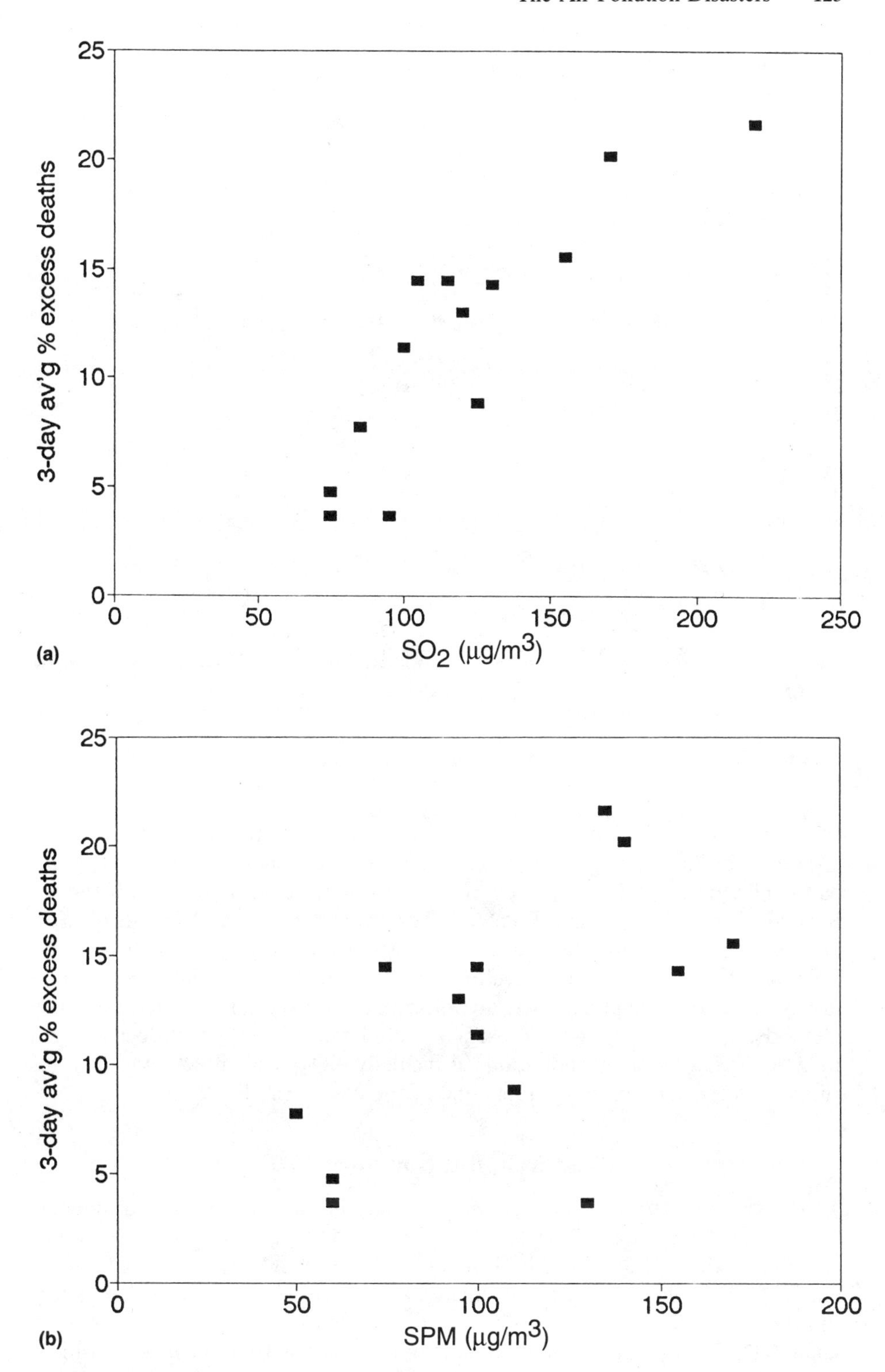

FIGURE 5-5. Excess mortality in Sao Paulo, July 25–August 8, 1973. (a) Plotted versus SO_2. (b) Plotted versus suspended particulate matter. (c) Plotted versus the sum of SO_2 and particulate matter. *Data from Mendes and Wakamatsu (1977).*

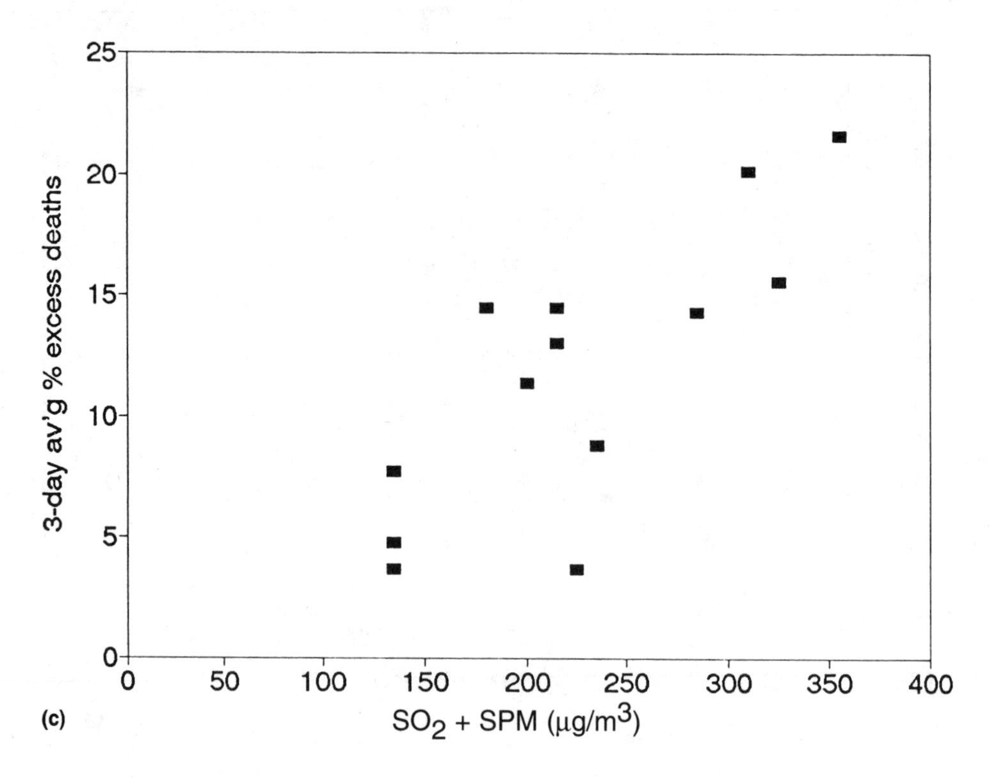

FIGURE 5-5. *Continued*

SO_2 plot shows a tendency toward saturation at $SO_2 > 150\,\mu g/m^3$, which does not seem credible. The average slope of Figure 5-5c, using the sum of SO_2 and particulate matter (0.063% per $\mu g/m^3$), seems more reasonable, even if the fit is worse than for SO_2 alone. Poor fit may be a consequence of the air quality measurements or simply the randomness associated with considering only an isolated incident. It may be difficult to compare this incident with other events discussed in this chapter, since no information was given on measurement methods, meteorology, or the presence of epidemics. However, it does appear to demonstrate that air pollution (principally SO_2) can affect mortality in climates other than the temperate zone of the Northern Hemisphere.

Pittsburgh, PA, November 1975

This event is classified as an "episode" mainly because industrial curtailment was ordered by the U.S. Environmental Protection Agency (EPA) in response to deteriorating air quality. The levels of SO_2 reached $200\,\mu g/m^3$ (0.076 ppm) and TSP, $770\,\mu g/m^3$. The EPA estimated that there were 14 excess deaths for this event (Riggan *et al.*, 1976). This corresponds to a mortality coefficient of about 0.0003 per $\mu g/m^3$. Changes in pulmonary function for a subset of sensitive school children were also reported (Stebbings and Fogelman, 1979). A more extensive study of acute health effects in Pittsburgh is discussed in Chapter 6.

Central and Western Europe, January 1985

The most recent air pollution episode may have affected several countries (EURASAP, 1986), including Eastern Europe, but health effects have been analyzed in only a few. In West Germany, SO_2 levels reached 800 $\mu g/m^3$ (0.3 ppm) and suspended particulates, 600 $\mu g/m^3$ (24-hour averages, Wichman *et al.*, 1989). The reported adverse health effects included 6% excess mortality, 12% excess hospital admissions, and 7% excess outpatients. "Excesses" were determined by comparison with a control area in the same region but with low levels of air pollution; evidently the meteorological conditions creating the episode extended over much of Europe, but levels of air pollution were governed mainly by local emissions. The mortality coefficient for particulates for this event would be 0.0001 per $\mu g/m^3$, which is substantially lower than the effects seen in the previous episodes.

The Schweizerhalle Chemical Disaster

On November 1, 1986, a chemical warehouse near Basel, Switzerland, caught fire and burned, exposing parts of the city to odors and unknown chemicals. The worst environmental impact seems to have been to aquatic resources, since the water used to fight the fire became contaminated in the process and flowed back into the Rhine River and thence downstream (Ackermann-Liebrich *et al.*, 1992). The public health impacts of this event seem to have been minimal; no impacts were observed directly at local hospitals, and subsequent analysis of daily mortality found only one peak, one week later, which could not be attributed to the fire. There was a slight excess in respiratory symptoms reported, but it could not be determined whether psychological factors may have played a role. The presence of a defensive attitude on the part of those whose jobs depended on the chemical industry was also reported.

COMPARATIVE ANALYSIS OF AIR POLLUTION EPISODES

In this section, I attempt to "explain" these events by deriving dose-response functions and examining other contributing factors. The London experience is considered separately because of the commonality of the population at risk and the air monitoring methods used. The analysis is then broadened to include other locations.

Analysis of Eight London Episodes

By far, the most scientifically interesting of the older episodes are the eight events that occurred in London from 1948 to 1962, in part because of the high pollution levels experienced. Air monitoring for SO_2 and British smoke was in place at a network of stations, and weather observations were published for several of these events. This set includes the 1952 disaster, in which 4000 people were reported to have died prematurely in Greater London, and the 1962 episode, in which the toll was only about 700.

TABLE 5-1 London Air Pollution Episodes, 1948–1962 (London Administrative County)

Dates	References	Duration	Excess Deaths*	Maximum Smoke	Maximum SO_2	Mortality Lag from Peak Pollution*
Nov. 26–Dec. 1, 1948	Logan, 1949; MOH, 1954; Bradley *et al.*, 1958	6 days	414	2.19	2.20	n/a
Dec. 5–8, 1952	MOH, 1954; Wilkins, 1954b	5 days	2153	2.38	2.65	0
Nov. 15–19, 1954	Gore & Shaddick, 1958	5 days	227	1.18	0.68	4–6 days
Jan. 12–22, 1955		11 days	383	1.80	0.86	6 days
Jan. 3–7, 1956	Bradley *et al.*, 1958; Gore & Shaddick, 1958	5 days	429	3.20	1.41	1 day
Dec. 18–26, 1956	Gore & Shaddick, 1958	9 days	295	1.25	0.94	7 days
Dec. 2–5, 1957	Bradley *et al.*, 1958	5 days	522	2.30	1.75	0
Dec. 3–7, 1962	Scott, 1963; Marsh, 1963; Waller *et al.* 1969	5 days	340	2.05	3.35	0

* Air monitoring data usually refer to the 24 hours ending at 9 A.M. on the day in question; thus an indicated lag of 0 corresponds to nearly one day.

This analysis is based on data from the published literature, as shown in Table 5-1. "Excess deaths" refers to mortality above the expected rate, for a period of 17 days after the onset of each event, for the London Administrative County (as opposed to Greater London). This is the geographic entity corresponding to the 7-station air monitoring network described by Gore and Shaddick (1958).

Figure 5-6 profiles these episodes in terms of their time histories of smoke and SO_2, averaged over the central city network; the diversity among these events is striking. (For the 1948 event, only the average concentrations were reported; hence the time histories could not be displayed.) Although the 1952 episode affected the most people, it does not exhibit the highest air concentrations for either pollutant. The meteorological data (Bradley *et al.*, 1958) provide some insights: the 1952 episode had 54 consecutive hours of calm winds, as measured on the roof of the Air Ministry building at Kingsway (118 ft above ground); the next most stagnant episode was the 1948 event, which had two periods of calms, 18 and 30 hours, respectively. The fog was worse in 1952; 48 consecutive hours of visibility of 20 yards or less were recorded. None of the other events came close to this. I devised an index of fog intensity based on the inverse of visibility. Daily values consisted of the average of the (inverse) of the four readings reported. For this purpose, the visibility report of "nil" on December 7 was assigned a value of 2 meters. However, visibility data were available for only four of the eight events, which precluded a more quantitative analysis.

Figures 5-7a through 5-7d and 5-8a through 5-8d present my attempts to develop a consistent dose-response relationship for these data, trying to re-

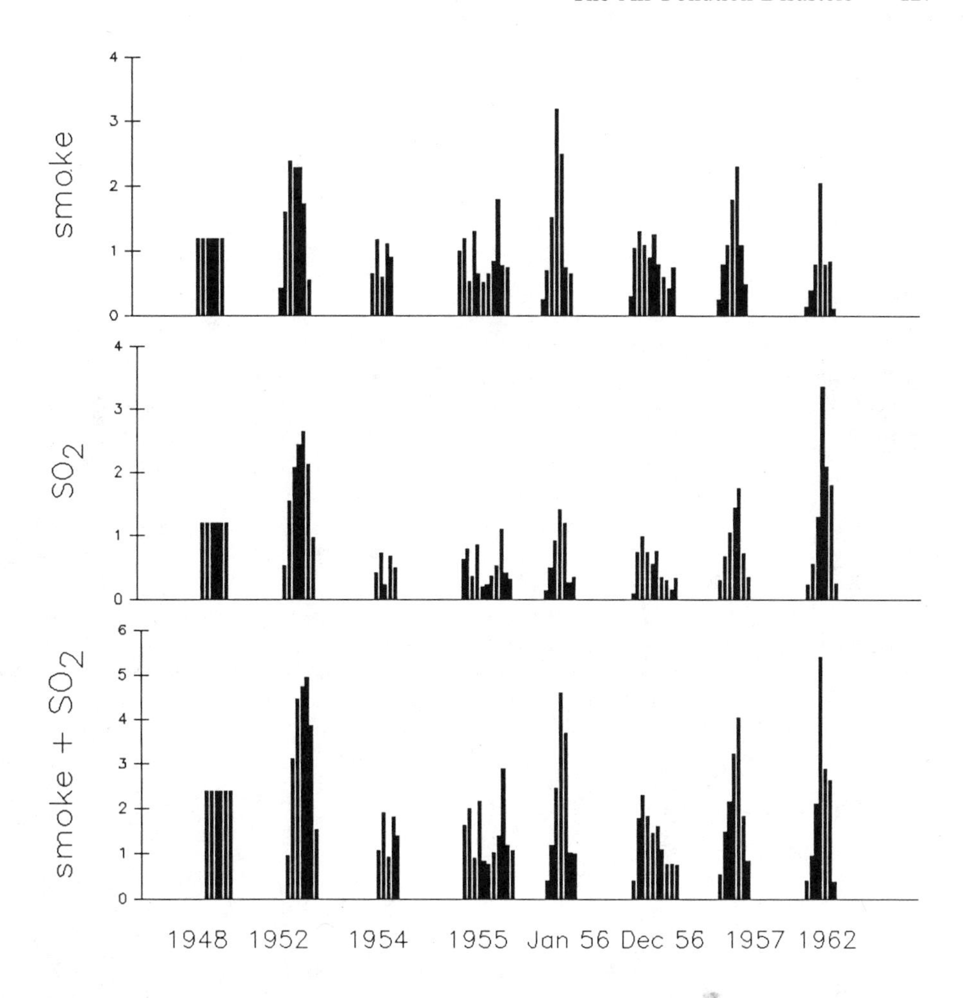

FIGURE 5-6. Air pollution profiles of eight episodes in London, 1948–1962; 24-hour average values are shown for each day of each episode, except for 1948, for which the averages over the entire period are shown.

present the cumulative mortality assigned to each event in terms of some combination of air quality parameters. Two ways of representing the mortality data were considered:

- The *total* number of excess deaths in 17 days and its logarithm (Figure 5-7).
- The *6-day average* number of excess deaths and its logarithm (Figure 5-8).

The log-linear model assumes that the effect of air pollution rises exponentially with concentration. The 17-day mortality accumulation period was an arbitrary choice based on the data available, but it captured almost all of even the worst episode (1952). In all cases, the numbers of deaths were referenced to the "baseline" daily rate at the onset of the episode; these reference values varied from 104 in mid-November to 146 in late January. Using the baseline rate as a reference accounts for the normal seasonal mortality cycle and may help account for the presence of any infectious epidemics. In all of the plots, the

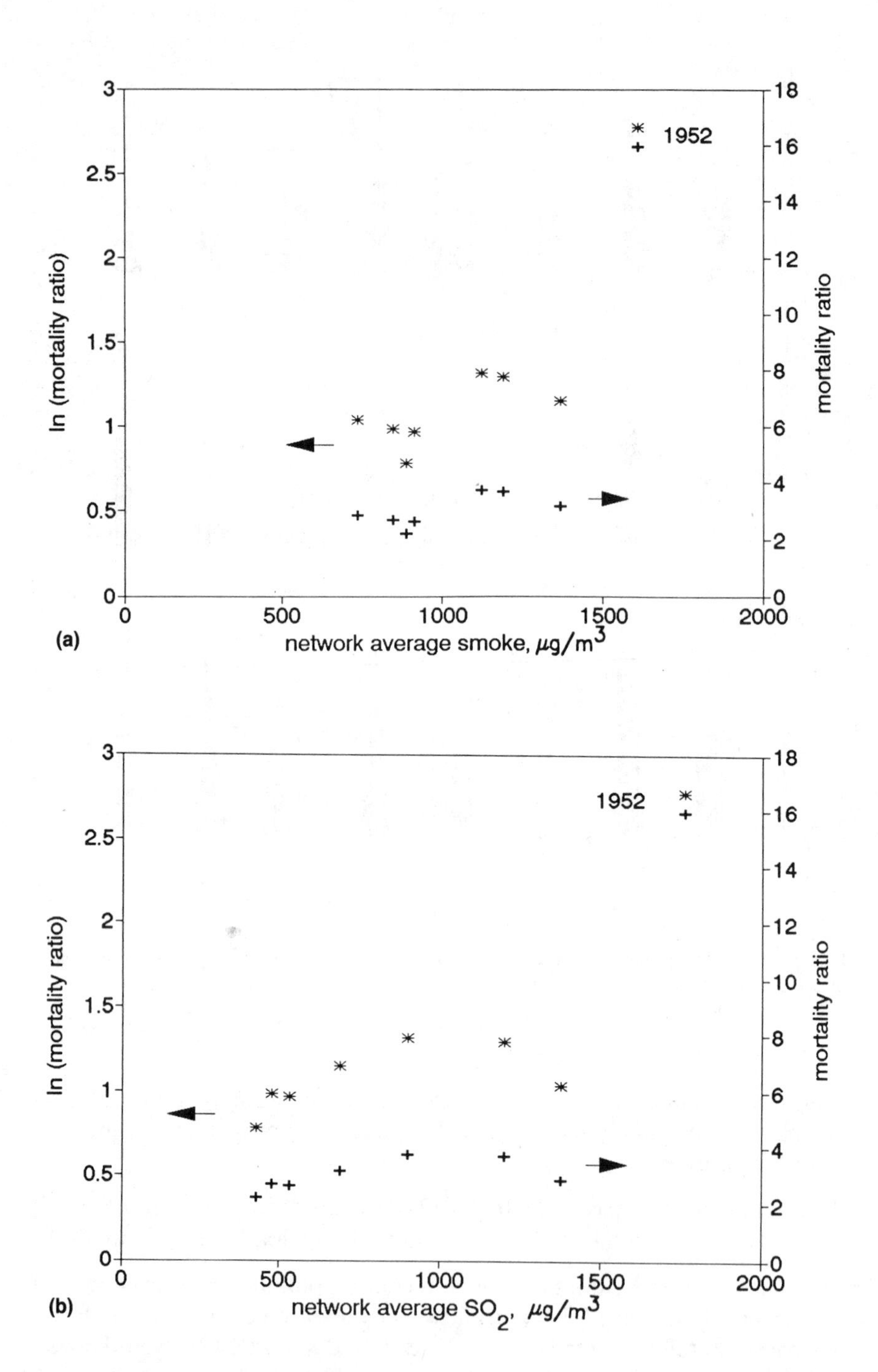

FIGURE 5-7. Dose-response functions for linear and log-linear models, based on the 17-day total mortality response for each of eight London episodes. (a) Plotted versus smoke. (b) Plotted versus SO_2. (c) Plotted versus the sum of smoke and SO_2. (d) Plotted versus the product of SO_2 and smoke. *Data from Ministry of Health (1954), Gore and Shaddick (1958), Bradley* et al. *(1958), and Scott (1963).*

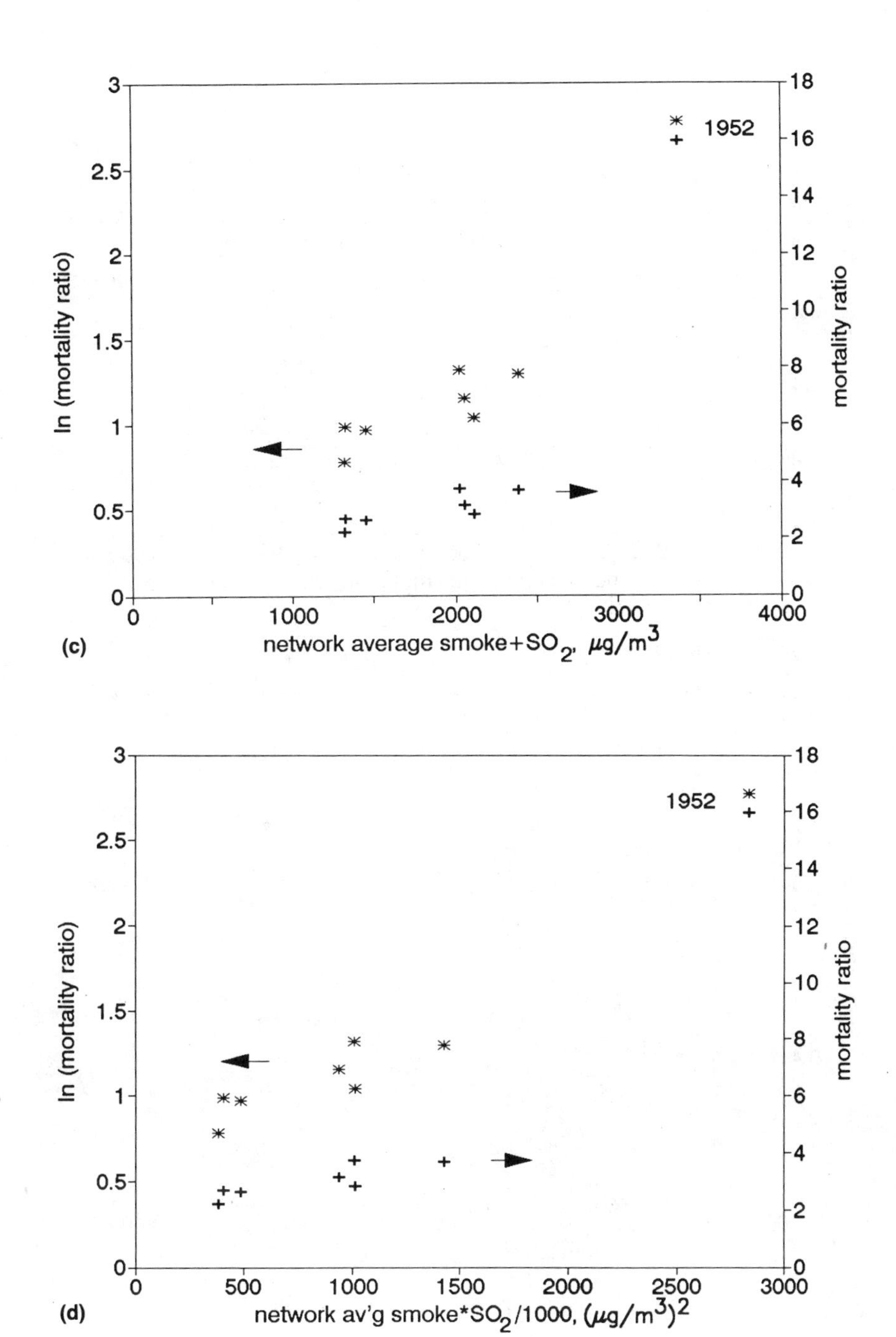

FIGURE 5-7. *Continued*

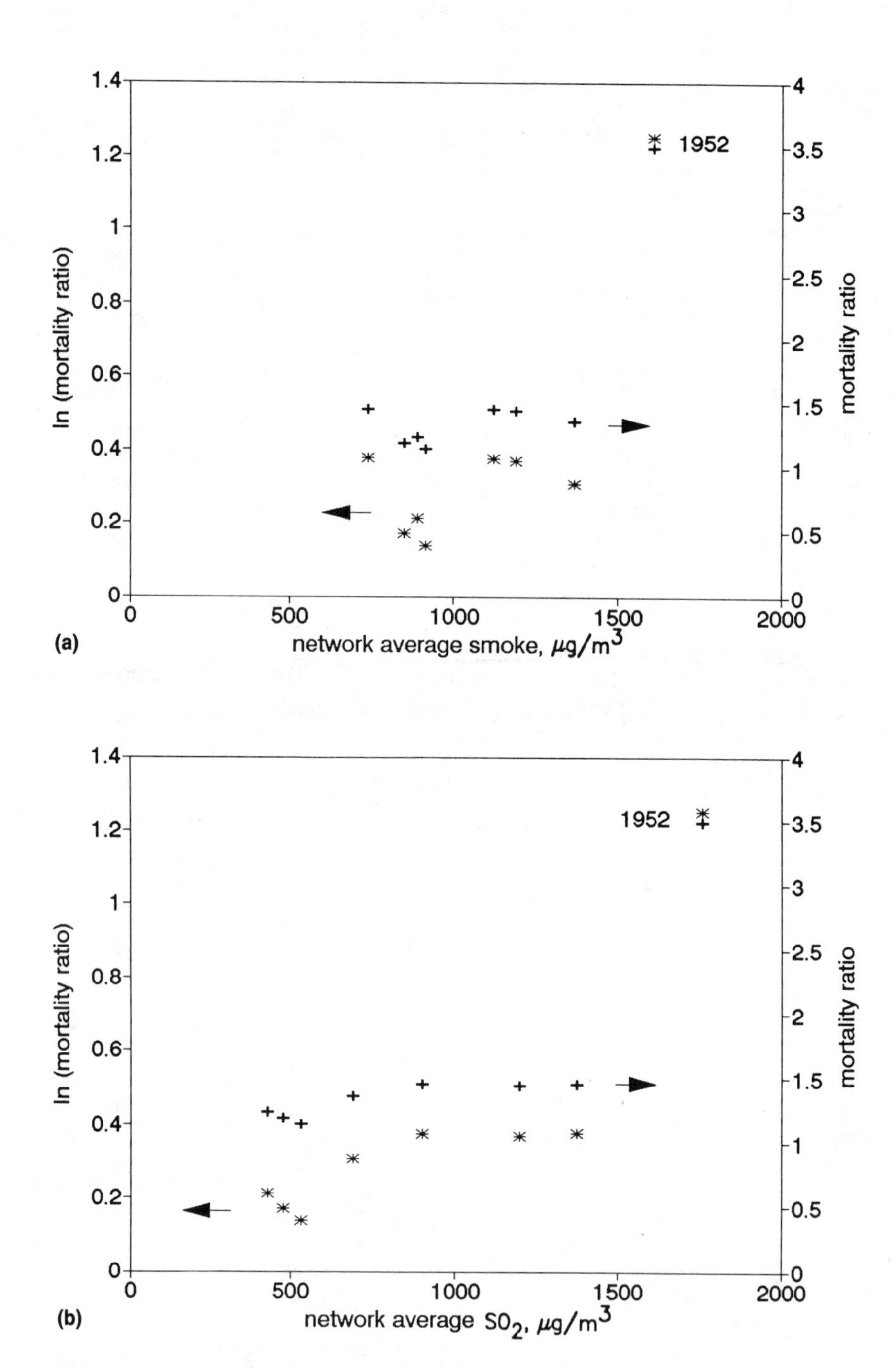

FIGURE 5-8. Dose-response functions for linear and log-linear models, based on the 6-day average mortality response for each of eight London episodes. (a) Plotted versus smoke. (b) Plotted versus SO_2. (c) Plotted versus the sum of smoke and SO_2. (d) Plotted versus the product of SO_2 and smoke. *Data from Ministry of Health (1954), Gore and Shaddick (1958), Bradley* et al. *(1958), and Scott (1963).*

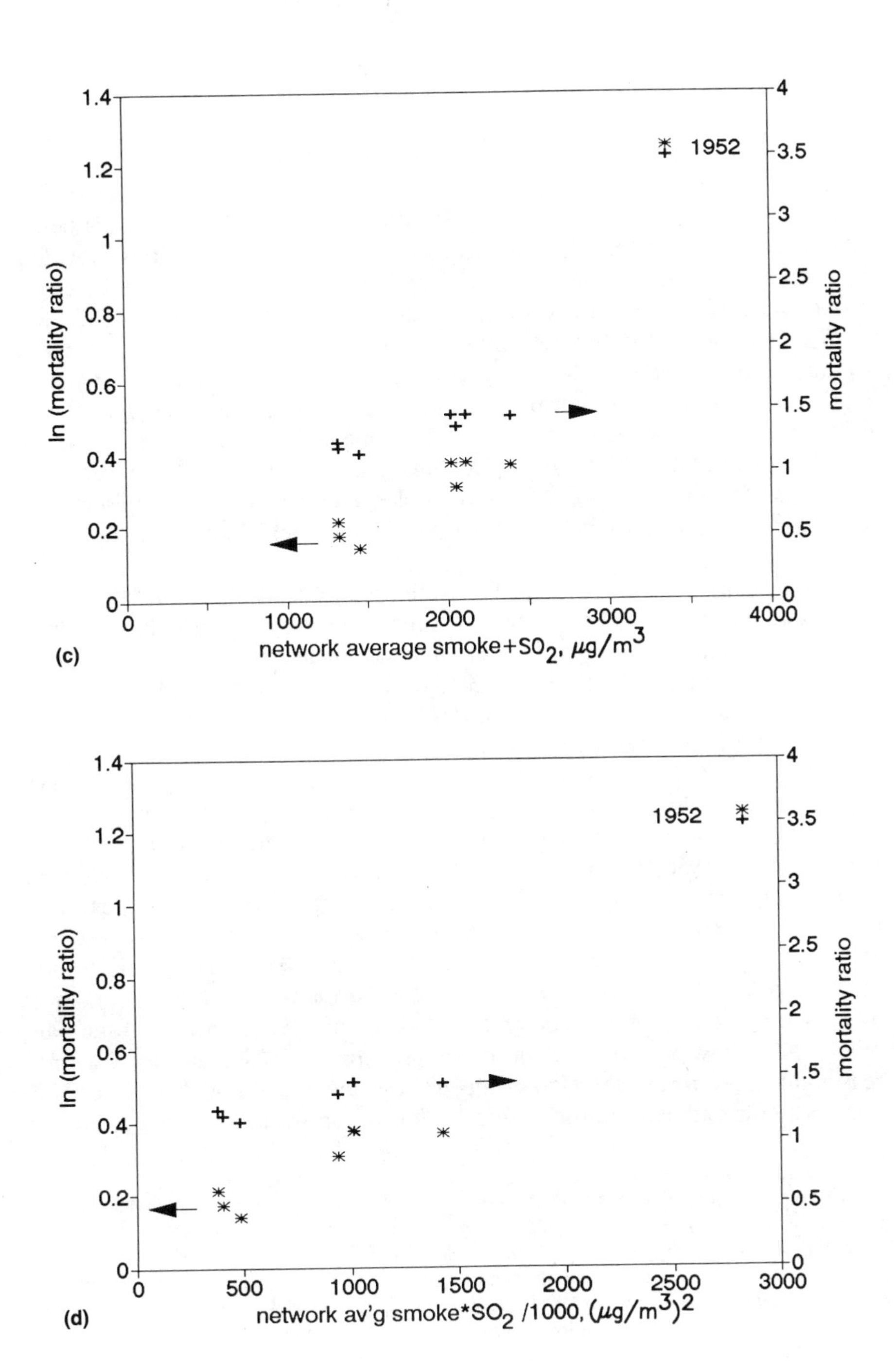

FIGURE 5-8. *Continued*

1952 episode stands out as an anomaly (shown in the upper right-hand corner), even with a log-linear model. However, the other seven events formed consistent dose-response relationships when plotted against the sum or product of smoke and SO_2 (Figures 5-7c and 5-7d, 5-8c and 5-8d). The best fit of all eight episodes was obtained by plotting the log mortality ratio against the product of smoke and SO_2 (regardless of which mortality accumulation period was used).

I confirmed these relationships by regression analysis, in part to estimate the regression coefficients for comparison with the time-series analyses discussed in Chapter 6. The log-linear models provided the best fits, since they were better able to accommodate the 1952 "outlier." The pollutant combinations providing the best fit were the product of smoke and SO_2, their sum, followed by either smoke or SO_2 alone. The coefficient for the 17-day episode "yield" was 0.075% per $\mu g/m^3$ for $SO_2 \times$ smoke, or 0.043% per $\mu g/m^3$ for the 6-day average of daily deaths. The average slope for all eight episodes considered individually, assuming a "normal" smoke level of 300 $\mu g/m^3$, was 0.048% per $\mu g/m^3$ of smoke. For the linear model, the slope of Figure 5-8c corresponds to 0.027% $\pm$ 0.014% (95% confidence limits) excess deaths per episode day per $\mu g/m^3$ for the sum* of SO_2 and smoke averaged over the episode and over the London County network; the correlation coefficient was 0.91 for this relationship.

Consideration of the effects of fog were limited by the lack of data for the less severe episodes. In the absence of comparable series of visibility readings, anecdotal evidence may provide some help. Douglas and Stewart (1953) confirm the characteristics of the 1952 fog and report that the 1948 fog was "closely comparable." They describe the 1948 fog as "more extensive and with higher temperature and water content." The MOH (1954) report stated that the 1948 fog lasted for 114 continuous hours in parts of London. The January 1955 event was described as "dry"—no water droplets or H_2SO_4 were present (Waller and Lawther, 1957). Visibility was reduced to 70 yards, however. Bradley *et al.* (1958) report that the 1957 fog was thick enough to have been blamed for a major railway accident (the resulting 87 deaths were excluded from this analysis). Fog was present during the 1962 event, but it was described as less dense and persistent than in 1952 (Marsh, 1963). However, Prindle (1963) reported his personal observation that visibility was down to about 15 feet on the evening of December 6 and that the episode was "substantially over" the following evening. From these data, it is difficult to discern a reasonable dose-response relationship for fog *per se*, although this certainly does not rule out contributions due to fog for individual episodes. For ex-

---

* Use of this sum is admittedly crude; however it does not exclude considering that some other attribute of SO_2, such as H_2SO_4, could be considered as part of the mix, which would imply attributing a smaller contribution to SO_2 *per se*. Data from London show that H_2SO_4 tends to be proportional to SO_2, for example (Waller and Commins, 1967). The assumption that smoke and SO_2 may be equally weighted when forming sums or products was checked with multiple regressions of the logarithm of the mortality ratios on smoke and SO_2, for the eight episodes. The coefficients for mortality summed over 17 days were 1.17 $\pm$ 0.42 for smoke and 0.66 $\pm$ 0.26 for SO_2 (concentrations in mg/m^3). For mortality summed over 6 days, these values were 0.54 $\pm$ 0.24 and 0.45 $\pm$ 0.15, respectively. These values suggest slightly stronger effects for smoke, particularly for the delayed effect but do not rule out uniform weighting. Similar results were achieved by regressing on the logarithms of smoke and SO_2.

ample, it is possible that the disruptions to normal traffic flow created delays for some affected persons in reaching medical care, especially during the 1952 episode.

Another interesting finding from this analysis is the lack of a temporal trend among the seven less severe episodes. The literature refers to the widespread publicity given the 1952 event and the use of "smog masks" and ammonia to help neutralize acids (Marsh, 1963; Prindle, 1963). However, the 1948 and 1962 events fall quite near to each other on Figure 5-8c, suggesting that the mortality difference is largely explained by the ambient air quality (i.e., that there is nothing special about the 1962 event). The 1962 episode was high in terms of the peak SO_2 concentration reached, relative to all the others; it has been suggested that its modest mortality response exonerated SO_2 as the primary causal agent. However, the same claim could be made for the peak smoke level reached during the January 1956 episode. It appears that the *duration* of these peak levels must be accounted for; the use of averages over the duration of the episode does this, although there remains some degree of arbitrariness in deciding when an episode begins and ends.

Further Consideration of the 1952 London Disaster

The preceding analysis "explains" the eight London episodes, but the statistical analysis is highly dependent on the high mortality experienced in 1952, which was similar to that seen during previous cholera and influenza epidemics but not during other fog episodes. Depending on the pollution metric chosen, 1952 may be seen as an outlier with respect to the other seven episodes. Three possibilities come to mind: unmeasured air pollutants, errors in the estimates of average air concentrations and exposures, and other influences on mortality besides air pollution.

The 1952 event was marked by an extended period of calm winds, during which other (unmeasured) pollutants could have built up. Among the prime candidates are H_2SO_4 and CO. The main factor promoting the formation of H_2SO_4 is the fog itself, which provides an aqueous reaction medium for the conversion of SO_2 to H_2SO_4. Fog also acts to remove both smoke and SO_2 from the air as the relatively large fog particles are deposited. The parallel trends for both pollutants, even during the fog, appear to rule out preferential removal and absorption of SO_2.*

Carbon monoxide was suggested in *The Lancet* (1953, editorial, p. 765), both directly from traffic sources, since CO emissions may have increased because of the slowing of traffic, and indirectly by a "new and popular type of slow-burning (domestic) fire, which is commonly stoked up to keep it going all

---

*I demonstrated this by regressing the ratio of SO_2 to smoke for the 11 measuring stations and the 9 days of data, with the fog index and the distance from Charing Cross as independent variables. Neither was significant; if anything, the effect of fog was positive (to indicate SO_2 absorption in the fog, it would have to be negative). However, it is possible that the highest smoke measurements were biased low because of filter saturation, in which case the corresponding SO_2 readings could suggest losses to the fog. In the absence of data, however, this scenario must remain speculation.

night." Routine CO measurements were later made in the vicinity of traffic in London and showed a continuous average of about 8 ppm, rising to 17 ppm during the business day (Waller *et al.*, 1965). During the fog, CO emissions may have been higher and dispersion would have been much lower, so that the multiplying factor of 6 applicable to SO_2 and smoke could have been higher for CO. However, anecdotal evidence suggests that traffic-related emissions may have been negligible at night. In addition, air concentrations of SO_2 and smoke may have been reduced by deposition due to the fog, but CO would have been relatively unaffected because of its low solubility in water. A multiplying factor of 10 or more does not seem out of line for CO, which would give concentrations of 100 to 200 ppm (0.01–0.02%) in traffic zones. Nighttime concentrations from domestic heating could also have been high in residential areas, and the duration of these high levels was of the order of 100 hours during the 1952 episode.

Such a dose could push blood carboxyhemoglobin (COHb) levels in smokers to around 20%, which is high enough to cause adverse effects in cardiovascular patients (Ministry of Health, 1954; National Academy of Sciences, 1977). In a recent case of fatal CO poisoning in three children, the COHb levels were reported as 15 to 20%, 23 to 28%, and 31 to 36% (Brown *et al.*, 1991). In addition, elevated CO can increase heart and breathing rates, so that the intake of other air pollutants would have increased.

The differing behaviors of smoke and SO_2 in London over the years from 1952 to 1962 may be seen from the annual average data of Craxford *et al.* (1967). For smoke, the ratio of annual average concentration to emissions was 77×10^{-9} sec/m^3 in 1952 and 70×10^{-9} sec/m^3 in 1962, which provides an excellent check. For SO_2, the corresponding values were 37×10^{-9} and 22×10^{-9} sec/m^3. The lower values for SO_2, especially in 1962, suggest more losses of SO_2 to deposition, oxidation, and so on, relative to smoke and that these losses have increased over time, probably because an increasing share of SO_2 was released from tall stacks. Such a change in source distribution implies that the ability of fixed monitoring networks to estimate population exposure will also have changed.

Wilkins (1954b) reported that mortality in London continued to rise during January 1953 because of an influenza outbreak. His graph, reproduced below as Figure 5-9, shows that this increase, relative to the previous year, began before the 1952 fog, approximately on Nov. 20. A possible explanation for the severity of the air pollution effects may lie in the combined effects of a disease outbreak and the fog. The MOH report (1954) points out that pneumonia was 50% above normal in London on December 6, for example, and rose to 1.6 times normal throughout England and Wales on December 20. In an exchange of correspondence in *The Lancet*, MacDonald (March 14, 1953, p. 547) pointed out that demands for emergency hospital admissions began to rise in September of 1952 and were 40% above normal at the beginning of the fog. J.A. Scott responded in the next issue of *The Lancet* that emergency bed applications were an unduly sensitive indicator and that his estimate of the weekly deaths during the fog, not attributed to the fog, was 1100. Using this figure would change the 17-day excess percentage from 1493 to 1130, which would still leave the 1952 event as an outlier as seen on Figures 5-7 and 5-8, depending on the pollution metric used.

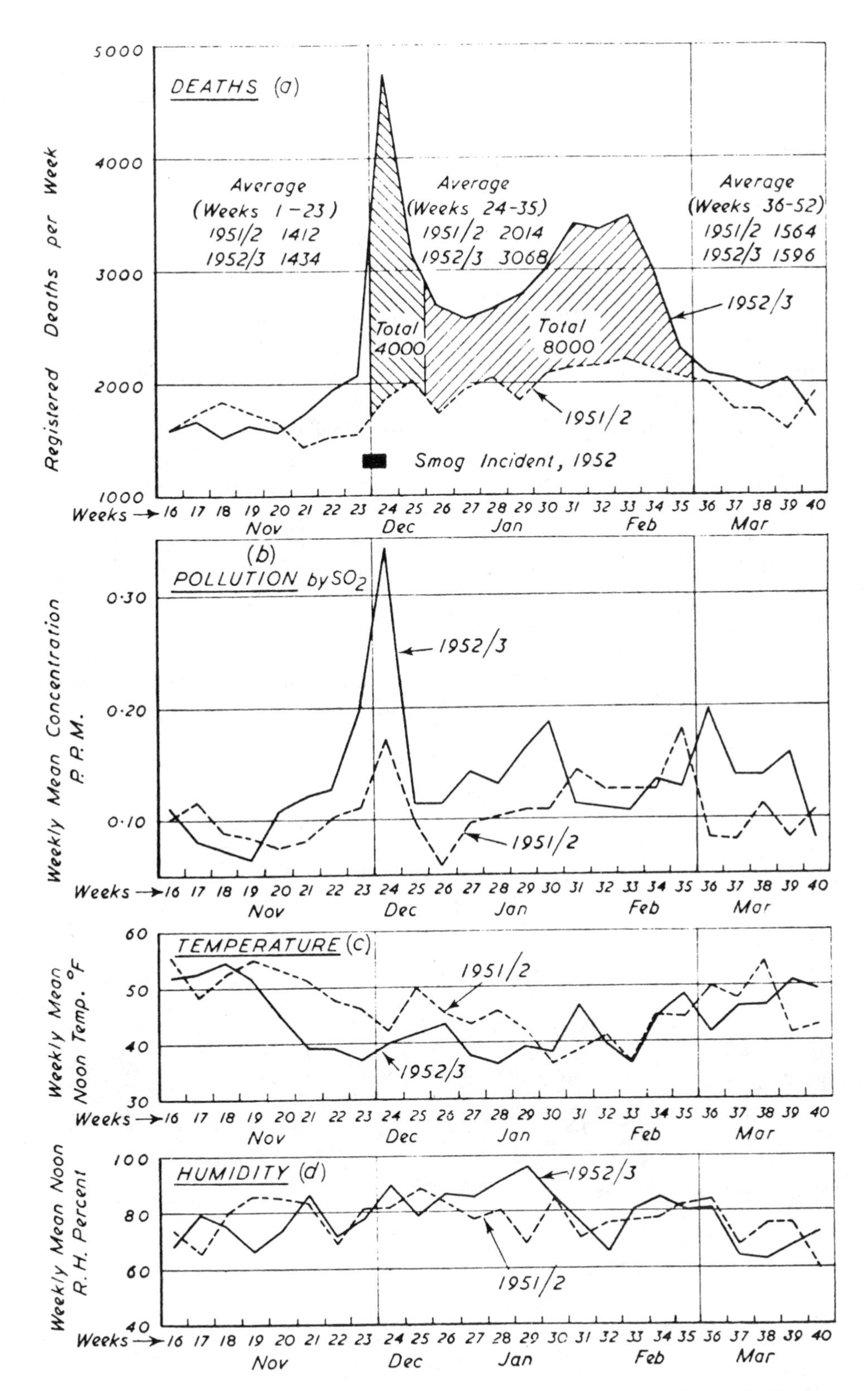

FIGURE 5-9. Weekly registered deaths, SO_2, temperature, and humidity for Greater London, winters of 1951–52 and 1952–53. *Source: Reproduced from Wilkins (1954b) with permission.*

To explore this topic further, I used the weekly mortality counts from the "160 Great Towns of England and Wales," less Greater London, as a baseline (Ministry of Health, 1954). Mortality generally increased during this period in the 160 towns and also showed a 4% jump during the week of the London fog. The 7-week time trends relative to this baseline are shown in Figure 5-10, for London Administrative County and the "outer ring." For additional clarity, these data were also normalized with respect to the average values during the 2 weeks before the fog. As expected, the effects of the episode were slightly less in the outer ring, but if strict proportionality with the change in air quality were expected, the difference in the two peaks should have been larger. There is also a dip after Christmas followed by an increase, suggesting that death registrations in London may have been delayed more than elsewhere during the holidays. The upward secular trend is virtually identical for both plots, which suggests that the seasonal trend was also larger in London. This plot does not suggest any subsequent drop in mortality after the fog, due to premature "harvesting" of particularly susceptible individuals. It also suggests that appreciable excess mortality from the fog did not persist beyond December 27, since such an excess should have been more obvious in the central district (a similar conclusion was reached by the MOH report). Using a gradually increasing baseline to account for the secular trend, the number of excess deaths from the 1952 London episode in Greater London would be just under 3000. Using this figure would bring the 1952 data point into line with the other seven episodes as depicted on Figure 5-7 and 5-8.

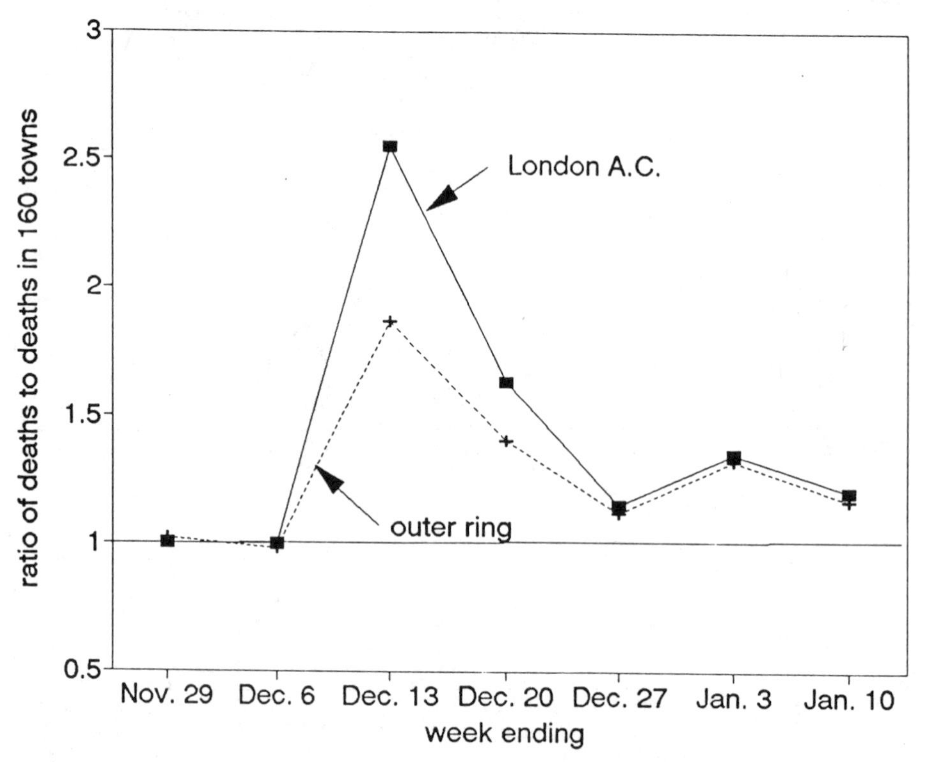

FIGURE 5-10. Mortality in London and suburbs relative to 160 large cities, winter 1952–1953. *Data from Ministry of Health (1954).*

This analysis does not address the observation that the number of 1952 deaths in London was running well ahead of those of 1951 beginning in mid-November, as shown in Figure 5-9. That figure also shows a spike in SO_2 in 1951, with the suggestion of a small subsequent rise in mortality. I used data for 7 weeks from the end of November in both years to examine the joint effects of SO_2 (smoke data were not available), temperature, season (secular trend), and the difference between the 2 years. This regression analysis was done for mortality in the same week and for the average of that week and the week after the pollution and temperature measurements were made. The latter case yielded the following "model" for weekly deaths in Greater London:

$$\text{Deaths} = 87 + 771\ \text{year} + 111\ \text{week} + 26.6\ \text{temperature} + 7009\ SO_2,$$

with temperature in degrees Fahrenheit and SO_2 levels in ppm.

All factors except temperature were significant. This result suggests that there were 771 more deaths per week in 1952 relative to 1951, on average, after controlling for the other factors (including SO_2 but not smoke), and that during the episode and the following 2 weeks, 333 additional deaths would have occurred due to seasonal variation. The residuals from this model, not including the SO_2 term, are plotted in Figure 5-11, and we see that the two 1952 fog points still stand out, perhaps since the plot only accounts for SO_2. If we take the cluster of data points in the lower left-hand corner as the baseline, the number of excess deaths drops to about 2100.

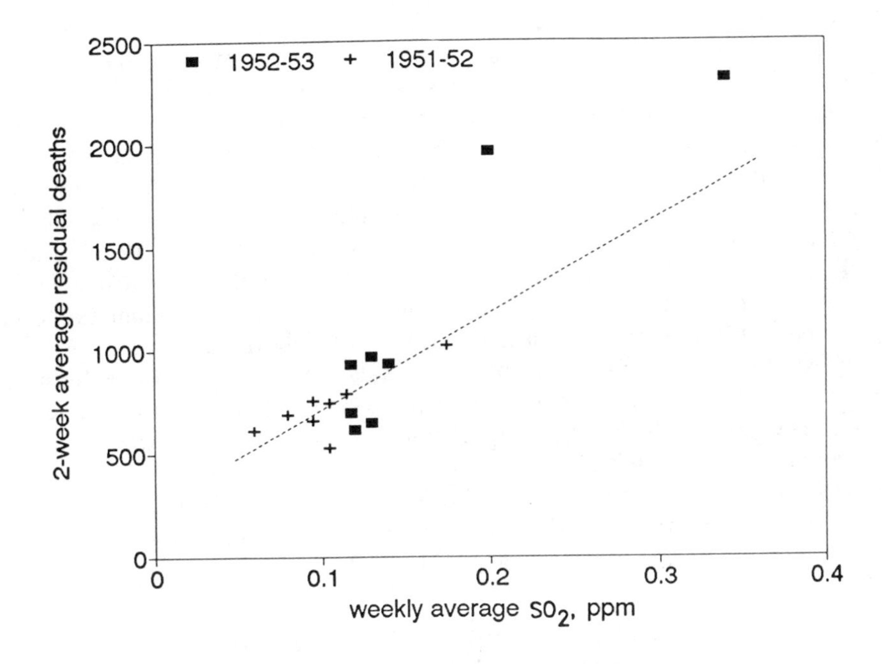

FIGURE 5-11. Two-week average London mortality versus SO_2, for the winters of 1951–52 and 1952–53. *Source: Wilkins (1954b).*

In summary, we can account for the large number of excess deaths during the 1952 London fog, relative to previous fogs, only by a combination of factors:

- The presence of high concentrations of other pollutants that were not measured, including CO and (probably) H_2SO_4.
- Probable underestimation of the actual population exposures to smoke and SO_2.
- Attribution to the fog of a large number of deaths that may have been due to secular trends or to the presence of a disease outbreak.
- Additional numbers of susceptible persons because of this disease outbreak.

None of this rather labored analytical treatment should be taken to diminish the importance of this event. Rather, the intended message is that there are many other factors to consider besides air pollution before a consistent dose-response relationship can be defined with certainty.

Animal Deaths

Firket (1936) mentioned cattle deaths in the Meuse Valley. Mills (1954) discussed animal deaths in Donora extensively and cited a large number of affected species, both domestic and livestock. Ashe (1959) seemed to discount citizens' reports of animal casualties and reported that, during the 1952 London episode, there were no deaths at the London Zoo, at horse farms, or at small animal hospitals (this information was confirmed by Lawther, 1955). The only animal deaths known from this episode were the prize cattle, who had arrived by train to participate in the Smithfield Club Show. Sheep and pigs at the show were unaffected. Lawther (1955) interpreted the deaths of the prize cattle as *prima facie* evidence of the toxic effects of airborne acids, since more of the manure was removed from the stalls of the prize cattle than from the others. Ammonia is released from cattle droppings, and Lawther thus reasoned that the ordinary animals may have been protected from the airborne acids. However, one would think that the cages at the Zoo would also be reasonably free from droppings and Meetham (1981) reported that during the fog of 1875, "cattle died in their byres where conditions were probably not clean." Further, in an indoor environment, one would speculate that the general ammonia level would be high throughout, regardless of the local state of cleanliness.

In fact, there is a simpler, alternative explanation, available from veterinary experts (D. Joel, personal communication, 1991). All cattle tend to suffer respiratory distress from shipping, a syndrome known as "shipping fever" (Siegmund *et al.*, 1973). The fat cattle tend to suffer more, since their breathing is more difficult. Appendix A to the MOH report discussed the cattle deaths and confirmed the presence of fever. Furthermore, that report mentions cattle deaths in previous fogs with ordinary stall maintenance.

CONCLUSIONS FROM EPISODE MORTALITY STUDIES

Meteorological conditions that severely limit atmospheric dispersion can increase ground-level concentrations by as much as a factor 10. The resulting effects on health depend on the nature of the population at risk, the contaminants involved, and the duration of these adverse conditions. The return period for such severe meteorological conditions appears to be of the order of 10 years.

The published historical records of these events affords a reasonable understanding of what happened, although much of it is based on circumstantial and some on anecdotal evidence. Industrial emissions dominated the episodes of the Meuse Valley and Donora and built up to toxic levels after several days of stagnation. These events were marked by acute responses, which diminished quickly as these situations eased. However, there was also an indication at Donora of delayed responses in a few of the victims. These events were also characterized by a high ratio of morbidity to mortality, two orders of magnitude or more. Many species of animals were affected.

The remaining events can be characterized as "community" air pollution, consisting of emissions from space heating, power generation, and traffic, with industrial emissions playing a lesser role (this may not be the case for the Ruhr Valley, however). In many cases, peak mortality responses lagged the peak air pollution periods by several days or more, and in some cases excess deaths continued for up to 2 weeks. Respiratory cases lagged longer than cardiac cases. Persons with (known) existing debilitating conditions were most at risk but were not the exclusive victims. The slopes of mortality dose-response functions were in the range 0.0001 to 0.0008 per $\mu g/m^3$ for smoke and 0.0003 to 0.0005 for SO_2. It was difficult to separate the effects of these two pollutants, and the evidence from London suggested that both of them played a role in the mortality responses.

Morbidity responses to these events are less well documented but appear to be of the same order of magnitude as the deaths. The only documented case of animal deaths in response to air pollution appears to be a group of prize cattle in London; they may have been compromised by being shipped and by being overweight. There is some suggestive evidence that responses to acute air pollution are heightened by the presence of infectious disease outbreaks. The evidence for the role of responses to temperature gradients in conjunction with air pollution is equivocal, since the U.S. events were marked by winter temperatures warmer than usual.

The London episodes, in terms of the total mortality "yield," may be predicted on the basis of the air concentrations averaged over the duration of the event. The total dose thus appears to be more important than the peak daily values. Air quality should be averaged over the city in order to provide a representative estimate of population exposure; peak local air pollution concentrations are not representative of the entire city. There is suggestive evidence that the number of excess deaths attributed to the 1952 London disaster may have been overestimated.

REFERENCES

Ackermann-Liebrich, U.A., Braun, C., and Rapp, R.C. (1992), Epidemiologic Analysis of an Environmental Disaster: The Schweizerhalle Experience, *Envir. Res.* 58:1–14.

Anon. (Feb. 23, 1963), The Latest London Fog, *Brit. Med. J.* 1:489–90.

Anon. (Apr. 27, 1963), Smog Deaths in December, 1962, *Brit. Med. J.* 1:1171.

Lord Ashby (Dec. 1974), Clean Air over London, The Second Sir Hugh Beaver Memorial Lecture, National Society for Clean Air.

Ashe, W.F. (1952), Acute Effects of Air Pollution in Donora, Pennsylvania, in *Proc. U.S. Technical Conference on Air Pollution*, L.C. McCabe, ed. McGraw-Hill, New York.

Ashe, W.F. (1959), Exposure to High Concentrations of Air Pollution. 1. Health Effects of Acute Episodes, *Proc. of the National Conference on Air Pollution*, Washington, DC: GPO, 1958, pp. 188–95.

Baker, R. (1989), *The Good Times*, Morrow. New York.

Batta, G., Firket, J., and Leclerc, E. (1933), *Les Problèmes de pollution de L'atmosphère*, Masson, Paris, pp. 260–35 (in French).

Bradley, W.H., Logan, W.P.D., and Martin, A.E. (1958), The London Fog of December 2nd–5th, 1957, *Monthly Bulletin of the Ministry of Health* 17:156–65.

Brasser, L.J., Joosting, P.E., and Van Zuilen, D. (1967), *Sulphur Dioxide—To What Level Is It Acceptable?* Report G300, Dutch Research Institute for Public Health Engineering.

Brown, J. *et al.* (1991), Fatal Carbon Monoxide Poisoning in a Camper-Truck—Georgia, *Morbidity and Mortality Weekly Report* 40:154.

Ciocco, A., and Thompson, D.J. (1961), A Follow-Up of Donora Ten Years After: Methodology and Findings, *Am. J. Public Health* 51:155–64.

Craxford, S.R., Clifton, M., and Weatherley, M.-L.P.M. (1967), Smoke and Sulphur Dioxide in Great Britain: Distribution and Changes, *Proc. 1966 Int. Clean Air Congress*, London, pp. 213–6.

Deane, M. (1965), Epidemiology of Chronic Bronchitis and Emphysema in the United States II. The Interpretation of Mortality Data, *Med. Thorac.* 22:24–37.

Dehalu, M. *et al.* (1931), The Causes of the Symptoms Found in the Meuse Valley during the Fogs of 1930, *Bul. Acad. Royale de Med. de Belgique* 11:683. [As cited by Ashe, 1959.]

Douglas, C.K.M., and Stewart, K.H. (1953), London Fog of December 5–8, 1952, *Meteorological Magazine* 82:67–71.

Firket, J. (1936), Fog along the Meuse Valley, *Trans. Faraday Soc.* 32:1192–97.

Gamble, J., Jones, W., Hancock, J., and Meckstroth, R.L. (1984), Epidemiological-Environmental Study of Lead Acid Battery Workers. III. Chronic Effects of Sulfuric Acid on the Respiratory System and Teeth, *Env. Res.* 35:30–52.

Glasser, M., Greenburg, L., and Field, F. (1967), Mortality and Morbidity during a Period of High Levels of Air Pollution, *Arch. Env. Health* 15:684–94.

Greenburg, L., Erhardt, C., Field, F., Reed, J.L., and Serif, N.S. (1963), Intermittent Air Pollution Episode in New York City, 1962, *Public Health Reports* 78:1061–64.

Greenburg, L., Jacobs, M.B., Drolette, B.M., Field, F., and Braverman, M.M. (1962), Report of an Air Pollution Incident in New York City, November, 1953, *Public Health Reports* 77:7–16.

Greenburg, L., Field, F., Erhardt, C., Glasser, M., and Reed, J.L. (1967), Air Pollution, Influenza, and Mortality in New York City, *Arch. Env, Health* 15:430–38.

Heimann, H. (1970), Episodic Air Pollution in Metropolitan Boston, *Arch. Env. Health* 20:230–47.

Hoffmann, M.R., and Jacob, D.J. (1984), Kinetics and Mechanisms of the Catalytic Oxidation of Dissolved Sulfur Dioxide in Aqueous Solution: An Application to Nighttime Fog Water Chemistry, in *SO_2, NO, and NO_2 Oxidation Mechanisms: Atmospheric Considerations*, J.G. Calvert, ed. Butterworth, Boston, pp. 101–72.

Kjellstrom, T. (Sept. 1989), London Fog Revisited: An Analysis of Severe Health Effects in a 35-Year Perspective, presented at the International Epidemiology Symposium, Brookhaven National Laboratory, Upton, NY.

Lawther, P.J. (Apr. 1955), Some Clinical Aspects of the Atmospheric Pollution Problem in London, *Proc. 3rd National Air Pollution Symposium*, Standford Research Institute, Pasadena, CA.

Lipfert, F.W. (1989), Dry Deposition Velocity as an Indicator for SO_2 Damage to Materials, *J. APCA* 39:446–52.
Lipfert, F.W. (Sept. 1992), An Assessment of Acid Fog, Paper IU-8.09, presented at the 9th World Clean Air Conference, Montreal, Canada. Also presented at the 1992 Air and Waste Management Association Symposium on Measurement of Toxic and Related Air Pollutants, Durham, NC, May 1992.
Logan, W.P.D. (1949), Fog and Mortality, *Lancet* 1:78.
Logan, W.P.D. (1956), Mortality from Fog in London, January, 1956, *Brit. Med. J.* 1:722–25.
Macfarlane, A. (1977), Daily mortality and environment in English conurbations 1: Air Pollution, low temperature, and influenza in Greater London, *Brit. J. Prev. Soc. Med.* 31:54–61.
Marsh, A. (1963), The December Smog, a First Survey, *J. APCA* 13:384–87.
McCabe, L.C., and Clayton, G.D. (1952), Air Pollution by Hydrogen Sulfide in Poza Rica, Mexico, *AMA Arch. Ind. Hyg. Occup. Med.* 6:199–213.
McCarroll, J., and Bradley, W. (1966), Excess Mortality as an Indicator of Health Effects of Air Pollution, *Amer. J. Public Health* 56:1933–42.
McDonald, J.C., Drinker, P., and Gordon, J.E. (1951), *Am. J. Med. Sci.* 221:325–41.
Meetham, A.R. (1981), *Atmospheric Pollution*, 4th ed., Pergamon Press, London.
Mendes, R., and Wakamatsu, C. (1977), A Study of the Relation of Air Pollution to Daily Mortality in Sao Paulo, Brazil—1973, *Proc. 4th Int. Clean Air Congress*, Japanese Union of Air Pollution Prevention Associations, Tokyo, pp. 76–80.
Mills, C.A. (1954), *Air Pollution and Community Health*, Christopher Publishing, Boston.
Ministry of Health (MOH) (1954), *Mortality and Morbidity during the London Fog of December 1952*, Reports on Public Health and Medical Subjects No. 95, HMSO, London.
National Academy of Sciences (1977), *Carbon Monoxide*, Washington, DC.
Neuberger, H., and Gutnick, M. (1949), Experimental Study of the Effect of Air Pollution on the Persistence of Fog, in *Proc. First National Air Pollution Symposium*, Stanford Research Institute, Pasadena, CA.
Prindle, R.A. (1963), Notes Made during the London Smog in December, 1962, *Arch. Env. Health* 7:493–96.
Roholm, K. (1937), The Fog Disaster in the Meuse Valley, 1930: A Fluorine Intoxication. *J. Ind. Hyg. & Toxic.* 19:126–37.
Roueche, B. (1984), *The Medical Detectives*, Vol. II, Dutton, New York.
Scott, J.A. (Apr. 26, 1963), The London Fog of December, 1962, *The Medical Officer*, pp. 250–53.
Shrenk, H.H., Heimann, H., Clayton, G.D., Gafafer, W.M., and Wexler, H. (1949), *Air Pollution in Donora, PA*, Public Health Bulletin No. 306, Public Health Service, Washington, DC.
Siegmund, O.H. *et al.*, eds. (1973), *The Merck Veterinary Manual*, Merck, Rahway, NJ.
Waller, R.E., and Commins, B.T. (1967), Episodes of High Pollution in London, 1952–1966, *Proc. Int. Clean Air Congress*, London, pp. 228–31.
Wallar, R.E., and Lawther, P.J. (1957), Further Observations on the London Fog, *Brit. Med. J.* 2:1473–75.
Waller, R.E., Commins, B.T., and Lawther, P.J. (1965), Air Pollution in a City Street, *Brit. J. Ind. Med.* 22:128–38.
Wilkins, E.T. (May 1954a), Air Pollution in a London Smog, *Mech. Eng.* 76:426–28.
Wilkins, E.T. (1954b), Air Pollution and the London Fog of December, 1952, *J. Roy. Sanitary Institute* 74:1–21.
Wise, W. (1968), *Killer Smog, the World's Worst Air Pollution Disaster*, Rand McNally, New York.

6

Time-Series Studies of Mortality

I do not think it is wrong to say that we do not even know what disease or diseases are caused by every day pollution of our urban air . . . we have a cause but no disease to go with it.

E.J. Cassell

Chapter 5 provided evidence of the *existence* of effects of air pollution on human mortality; the similarity of temporal mortality responses from several different times and places seems to rule out substantial confounding by meteorological variables or by faulty reporting (but not by covarying pollutants, which may be similar in various cities). These studies of isolated air pollution events constitute a special case in the general study of the *timing of death*. This chapter deals with this general case, in which continuous temporal patterns of mortality are compared with weather and air pollution patterns in search of consistent relationships.

Since these studies deal with all levels of daily air pollution, as opposed to isolated incidents of high values, they can address the question of threshold concentrations for the effects of air pollution on health. This question is at the heart of the setting of air quality standards and of the estimation of the cost-benefit ratios of proposed air pollution controls. With severe air pollution episodes, which occur quite rarely, the mortality responses may be obvious in relation to expected patterns. However, as will be seen in this chapter, modest perturbations in air quality produce responses of the same order of magnitude as the random day-to-day mortality fluctuations, so that longer time periods must be considered to determine whether such associations are statistically significant. This problem requires the use of the formal methods of time-series analysis, as outlined in any number of texts on econometrics and statistical methods.

The chapter begins with a discussion of statistical methods. Time-series studies of mortality in London and New York City are reviewed extensively, in part because of the rich bodies of literature available and in part because of the importance of these two case studies of air pollution abatement. Studies in California are similarly important and offer the additional attribute of a different mix of air pollutants: an "oxidizing" atmosphere as opposed to the "reducing" atmospheres of New York and London. A group of studies in various diverse U.S. and foreign cities provides additional evidence of the universality of the phenomenon. Finally, a group of recent studies of various U.S. cities is reviewed; these present examples of a common analysis methodology applied to a variety of situations. Appendix 6A discusses a related but different method of analyzing temporal concordance between mortality and the environment: the stimulus-response methodology developed by Lebowitz (1973).

TIME-SERIES METHODS

Several methodological issues arise with time-series analyses. Some of these relate to the "ecological" nature of such observational studies, that is, the attempt to infer effects on individuals from observations of the behavior of groups. These issues were discussed in Chapter 3. Other statistical problems arising with time-series studies include lag effects, serial correlation, confounding by exogenous long-term trends, and collinearity among independent variables (also discussed in Chapter 3).

Confounding Variables

All time-series studies of air pollution effects must take into account the "natural" temporal patterns in air quality, which can confound statistical analysis of response variables with similar patterns. These include both long- and short-term patterns. On a long-term basis, seasonal trends include higher ozone levels in summer and higher levels of pollutants associated with space heating in winter. Such seasonal cycles may be confounded with seasonal weather cycles, which can also affect health responses (Rogot and Blackwelder, 1970; Bull, 1973; Bull and Morton, 1975). For example, respiratory illness shows a seasonal pattern, higher in winter, in part because of periodic outbreaks of viral infections (Hall, 1981). There may also be longer term air quality trends associated with community growth or with air pollution abatement efforts. Weekly cycles may be expected in traffic-related air pollutants or in those pollutants related to local industrial sources that operate with reduced emissions on weekends.

Since weather affects air pollution and seems to affect health, it is important to account for these interactions correctly, especially at the lower pollution levels where effects may be more subtle. During the past decade or so, much of the literature on the short-term health effects of air pollution has been concerned with this topic.

Figure 6-1 is a flowchart that may be useful in depicting the complexity of the temporal influences involved in the context of conceptual path analysis. Beginning at the top left of the chart, weather patterns are seen to influence both the emissions of air pollution and their dispersion, both of which are

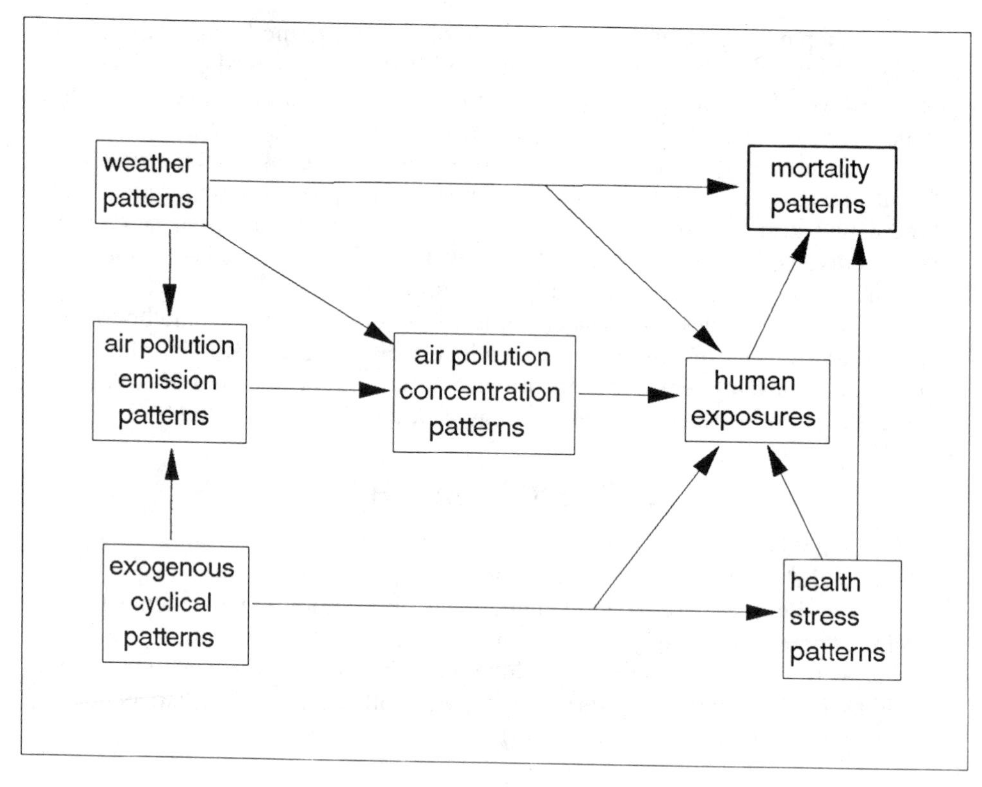

FIGURE 6-1. Temporal influences on mortality.

responsible for daily fluctuations in concentrations of air pollutants. Research in biometeorology (Hodge and Nicodemus, 1980) suggests that weather patterns can have a direct influence on health and mortality and can also influence the patterns of human activities that control exposure to outdoor air pollution. There are also cyclical patterns in health stresses that are independent of weather *per se*, since they occur in all climates; viral activity is an example. The weekly cycle of normal activities could provide additional health stress patterns and also affect human exposures to outdoor air pollution. Thus, to adequately estimate the importance of the path from air concentration to exposure to mortality, which is the goal of all of the analyses considered in this chapter, one must take into account both of the other paths that influence mortality patterns. The latter includes not only the well-known seasonal patterns, which affect mortality regardless of the weather or climate but also effects of holidays and the day of the week that may influence the way in which deaths are reported. The various studies of mortality–air pollution time series vary greatly in the ways they account for these potentially confounding cycles.

One of the traditional methods of avoiding or reducing temporal confounding is the use of a "deviation" variable, that is, the difference between the observed value and the value expected for that day of week, season, year, and so on. This practice is essential for the dependent variables; its use for the air pollution variables is tantamount to assuming linear responses, which are often assumed by the use of linear regression models in any event. Some studies have used deviations from 15-day moving averages; however, as seen in Chapter 5, mortality responses to the most severe episodes persisted for periods of about this length. Thus, caution must be exercised when the mortality perturbations are no longer "small."

Autocorrelation

Tests of significance of regression coefficients require that the residuals be normally distributed and uncorrelated. This condition is frequently violated in both time-series and spatial analyses, with the result that significance levels can be overstated. In time-series analysis, the phenomenon is called serial correlation, and formal methods have been developed to deal with it. Hammerstrom *et al.* (1991) showed that confidence limits on bivariate correlation coefficients were on average about 7% wider when based on bootstrapping, as opposed to the classical method based on the z-transform. If serial correlation has not been accounted for, the alpha level for hypothesis testing should be decreased, say, from $p < 0.05$ to $p < 0.01$.

Many of the more recent time-series analyses have adopted the methods of Liang and Zeger (1986) for dealing with serial correlation. According to Schwartz and Dockery (1992a), this approach "allows the calculation of robust variance estimates, which give unbiased hypothesis tests even if the covariance is misspecified. The Liang and Zeger approach incorporates the covariance structure in the estimation of the regression coefficients as well as their variances, giving more efficient estimates of the parameters."

Selection of Averaging Times and Lag Periods

Most time-series studies of mortality use daily data, the shortest practical time period. Aggregating to longer periods is one way of reducing the effects of Poissonian variability in the dependent variable, at the risk of losing the details of the time-response patterns. Some studies have used weekly or monthly data, for example.

When daily data are used, it is important to account for lags between stimulus and response. Even persons experiencing acute industrial exposures to air pollutants have shown delays of a half day or more before exhibiting severe symptoms and seeking treatment (Bates *et al.*, 1971). Studies that do not consider a range of lags are thus likely to underestimate the total response. One way of capturing the total response to an air pollution event is to sum the responses for some number (n) of succeeding days. For a stationary time series, this is equivalent to summing or averaging the pollution over n days prior to death. This summation accounts for the "harvesting" effect sometimes seen when the higher lags take on the opposite sign, as well as the tendency for the total response to be split over the various lag periods, which reduces the significance of any individual lag variable. An approximate estimate of the significance (t value) of the total effect over several lags may be obtained from a multiple regression by adding the coefficients for each lag and dividing by the average standard error.

Since air quality data for many species are often available on an hourly basis, the analyst must decide whether to use daily peaks or daily averages as the measure of the effective dose. If a nonlinear or threshold-type response is expected, daily peak values would be appropriate for a physiological model. This would also be the case if the peaks occur during a time of day when outdoor exposure is more likely. However, there are other factors to consider in making this choice, in order to distinguish between statistically effective and physiologically appropriate doses. First, the peak value may only apply to a localized area around the monitoring station and thus may not be representative of the entire community. Daily averages tend to be smoother spatially and thus more representative of community levels. Note that the average dose gave more consistent dose-response relationships during the London episodes (Chapter 5) than the peak concentrations. Second, when one wants to compare regression or correlation results among various pollutants, it is essential that data for all pollutant species be based on the same averaging time, in order to compare on the basis of equivalent random noise levels and spatial representativeness. Differences in the "noise" in the independent variables can bias the comparison of regression statistics. Since most particulate mass data are only available for 24-hour samples, daily averages may be the "least common denominator" for the purpose of comparing models employing alternative pollutants.

REVIEWS OF TIME-SERIES STUDIES OF DAILY MORTALITY IN LONDON

Early Studies

The effects of air pollution and fogs on health in London have been studied for over 60 years. Russell (1924, 1926) analyzed a 21-year record beginning in the nineteenth century, using regression and correlation analyses of weekly deaths on fog and temperature variables. His technique involved numbering the weeks of each year and examining the relationship of respiratory mortality for each week across the 21 years, for two residential boroughs of London. He concluded that fog and cold weather had a larger effect than fog alone, and that usually only adults were affected. Because emissions from space heating are increased during colder temperatures and atmospheric dispersion is often reduced, Russell's studies could be interpreted as also implying an air pollution effect.

Logan (1949) analyzed the causes of death attributed to polluted fogs, with reference to episodes occurring during the winter of 1948. He also found that the excess deaths occurred in people aged 45 and older, primarily for bronchitis and pneumonia with some contributions from cancer and myocardial degeneration, but not "old age."

Early mortality analyses in London that directly incorporated air pollution data include the work of Martin and Bradley (1960) and Martin (1964). All of the major studies of London mortality have used an air quality data base comprising the averages of seven monitoring stations for British smoke and SO_2, beginning with Gore and Shaddick (1958). These data are reported for the 24-hour period ending at 9:00 A.M., and thus represent nearly a 1-day lag with respect to the mortality counts reported for the same day. The report of Martin (1964) dealing with the winter of 1958–59 does not specify the source of air quality data, described as "indices" of smoke and SO_2 in Greater London. On an episodic basis, there can be substantial differences in air quality between central London (i.e., London Administrative County) and Greater London, which includes the outer ring. As discussed in Chapter 3, such differences can affect the validity of the ecological hypothesis.

Martin and Bradley (1960) analyzed the mortality data from Greater London for the winter of 1958–59, using mortality deviations from a 15-day moving average as the dependent variable. This technique removes the effects of long-term variations, due, for example, to weather or season. Pollution data were based on the seven-station network for London County, although reference was also made to a 12-station network for Greater London that usually had lower readings. (This study thus apparently established the precedent of using the London County pollution data in conjunction with Greater London mortality data, even though the latter area was about five times bigger and had 2.6 times the population.) For 1958–59, Martin and Bradley found statistically significant correlations between either smoke or SO_2 and mortality from all causes and from bronchitis, but apparently without jointly considering the effects of daily temperature changes (temperature and relative humidity were only weakly correlated with mortality). The relationship was slightly stronger for smoke and for a model using the logarithm of the pollutants, but was not noticeably improved by considering smoke and SO_2 jointly.

Martin (1964) extended the technique to the winters of 1959–60. Regression analysis of the tabulated excess mortality data suggested a threshold for smoke at 450 $\mu g/m^3$ and 330 $\mu g/m^3$ for SO_2 (considered separately without temperature effects), although Martin commented on the difficulty of defining "safe" levels of pollution. The slopes of these regressions were 0.013% excess mortality per $\mu g/m^3$ smoke, or 0.016% excess mortality per $\mu g/m^3$ SO_2, which are similar to those derived from the analysis of episodes in Chapter 5.

Macfarlane (1977) discussed the importance of proper consideration of influenza epidemics and concluded on the basis of visual inspection that there was no longer an association between daily mortality and suspended particulates, based on the seven-station network (ca. 1970). She also commented that the study area was changed slightly and data collection was begun on a year-round basis in 1965, and that the codings for causes of death were changed in 1968 in accordance with changes in the International Classification of Diseases (ICD). Macfarlane and White (1977) considered the weekly mortality cycle in England and Wales and showed a 3% excess in cardiac causes on Mondays, especially for males; there was also a deficit on Sundays. They discounted the hypothesis that the weekly cycle is an artifact of the death registration system.

Studies by the Schimmel, Mazumdar, and Higgins Teams

Mazumdar *et al.*'s analysis (1982) was the first comprehensive study of a long time-series of Greater London mortality (14 winters, 1958–72). The environmental data were daily mean values of SO_2 and British (black) smoke, averaged over the seven London County measuring stations, and readings of temperature and relative humidity taken at 9:00 A.M. at a single site (Heathrow Airport). The 15-day moving average was used, dummy variables were included for days of the week, and variables were divided by their mean (winter) values to account for the long-term trends (which were considerable in part because of the air pollution controls that had been instituted; year-to-year changes such as discussed by Macfarlane (1977) would also be accounted for by this procedure). Weather effects were handled through a two-stage procedure: first, mortality and pollution variables were regressed on temperature and humidity to provide "corrections." The corrected variables were then analyzed for the pollution-mortality relationships. The authors reported that the results of this two-stage procedure did not differ substantially from those obtained using weather variables jointly in the same regression with pollution and mortality. This finding is consistent with the more casual observations in Chapter 5 that temperature did not seem to be an important factor in explaining the differences among episodes. This study is particularly useful for this review, since it features several different analytic approaches, including nonlinear pollution models, and since some of their data were available for reanalysis.

Mazumdar *et al.* (1982) made several types of analyses. First, they provided regressions for each winter, for pooled data for the first and second groups of seven winters, and for all fourteen winters as a group. Smoke and SO_2 were regressed against mortality separately and jointly; a separate set of mortality regressions was also provided for $(smoke)^2$, $(SO_2)^2$, and (smoke $\times$ SO_2), to assess the linearity of responses and interactions. Note that such nonlinear pollution models allow a check on the applicability of the log-linear model used elsewhere. Additional yearly regression results were obtained from the their contract progress report (Mazumdar *et al.*, 1980). Twelve of the 14 individual SO_2 regressions and 13 of the individual smoke regressions were statistically significant. Six of the 14 joint regressions had significant (positive) smoke coefficients; of the remaining winters, 2 had significant (positive) SO_2 coefficients. The 1964–65 winter had a significant positive smoke coefficient and a significant negative SO_2 coefficient. For the first seven winters (mean smoke from 130 to 550 $\mu g/m^3$; mean SO_2 from 270 to 420 $\mu g/m^3$), the SO_2 coefficient for the mortality ratio was -0.000033 and the smoke coefficient, 0.00022 (per $\mu g/m^3$). For the second group of winters (mean smoke from 60 to 110 $\mu g/m^3$; mean SO_2 from 180 to 240 $\mu g/m^3$), the overall smoke coefficient was 0.00028 per $\mu g/m^3$ and the SO_2 coefficient, 0.000056 per $\mu g/m^3$. For all 14 winters, the smoke coefficient was 0.00025 and the SO_2 coefficient, 0.00012 (per $\mu g/m^3$). None of these three grouped SO_2 coefficients was statistically significant; all of the smoke coefficients were significant. Note that the smoke coefficients were robust but slightly lower than those found from the London episode studies (Chapter 5).

Mazumdar *et al.* (1982) identified two subsets of the data: the "episodic" days consist of those days with smoke in excess of 500 $\mu g/m^3$ and 7 days before and after. The remaining days constitute the "nonepisodic" days. Lag effects were not discussed in Mazumdar *et al.*, but some results are available from their 1980 progress report. In this exploratory analysis, five different pollutant combinations were considered individually, and then jointly, for two different subsets of total mortality: the episodic period, and the nonepisodic period, against total mortality. The five combinations were SO_2, $(SO_2)^2$, smoke, $smoke^2$, and SO_2 $\times$ smoke. There were only minor differences in bivariate correlations (after corrections for weather, day-of-week, etc.) between SO_2 and smoke

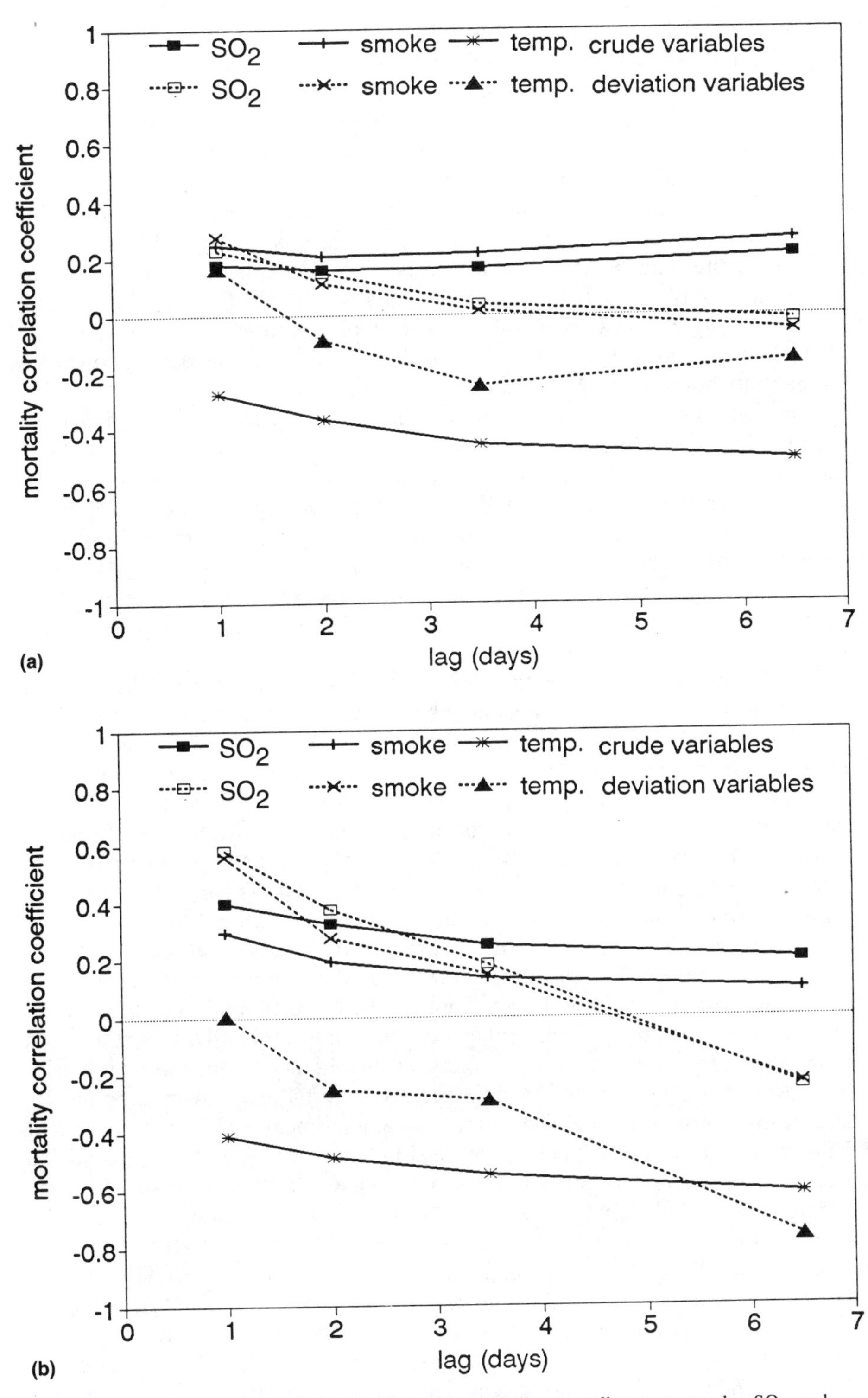

FIGURE 6-2. Bivariate correlation coefficients for London mortality versus smoke, SO_2, and temperature, for various lag periods. (a) Episodic period. (b) Nonepisodic period.
Source: Mazumdar et al. (1982).

(smoke was higher in the nonepisodic period, and SO_2 was higher in the episodic and combined periods). Correlations for the higher-order variables were always lower, but the cross-product $SO_2 \times$ smoke was the best of these and was only slightly inferior to the individual species correlations for the episodic period. As lags were added to the multiple regressions, there was little change in the pollution effect (sum of coefficients over all lags) for the individual species but the cross-product term showed a 48% increase when lags of 1, 2, and 3 days were summed. Note that there is a nearly 1-day lag incorporated in the data by virtue of the respective data collection times, so that a nominal 1-day lag actually represents nearly 2 days. For the nonepisodic period, the lag effects were smaller. For the joint regression, the net effect of all pollutants summed decreased for all lags, but there was a shift toward a larger contribution from SO_2 as lag time increased.

Some insights into the behavior of these variables may be obtained from Figure 6-2a, b, which presents bivariate correlation coefficients for total mortality versus SO_2, smoke, and temperature. Correlations are presented for the same day (actually a nearly 1-day lag), the previous day, the average of lags 2 and 3, and the average of lags 4, 5, 6, and 7. On each graph, "crude" variables (raw data as measured) and "deviation" variables, which consist of differences from 15-day moving averages, are correlated. The latter correlations represent the "high-frequency" components since the 15-day averaging and difference process removes the seasonal trend.

For the episodic period (Figure 6-2a), correlations change only slightly with lag and the temperature effect actually becomes stronger. The SO_2 correlations are slightly higher than the values for smoke. When the seasonal effect is removed (deviation variables), more of the remaining variance is explained by pollution, but, at longer lags, the pollution effect dies out and temperature takes over. For the lower air concentrations of the nonepisodic period (Figure 6-2b), the trends are qualitatively the same except that the pollutant correlations are lower (and smoke is slightly higher than SO_2). The fact that the crude temperature correlations differ according to the air pollution level would seem to demonstrate the confounding effect present in the crude variables.

Mazumdar *et al.* (1982) introduced the parameter of *mortality effect*, obtained by multiplying the regression coefficient by the corresponding mean value for pollution, expressed as a percentage of mean mortality. This parameter is numerically equivalent to the elasticity evaluated at the mean. Figure 6-3a compares the effect on mortality as expressed, alternatively, by both pollutants and by each separately, plotted against the annual mean smoke concentration. Best-fit second-order polynomial regression lines are also shown (since the trend is not linear, contrary to expectations) and although there are some differences for individual years, the best-fit lines are virtually identical. The inability to distinguish between the two pollutants stems from their high correlations for the daily values, which ranged from 0.79 to 0.96. The analysis indicates a pollution effect on mortality of about 3% at the lowest pollution levels, rising to almost 9% for the worst year (1958–59). This type of response is consistent with the exponential relationship assumed by a log-linear model. In Figure 6-3b, the mortality effect data are plotted against their respective mean pollutant concentration levels: SO_2 effect versus SO_2 concentration, smoke effect versus smoke concentration, and the joint effect versus the sum of

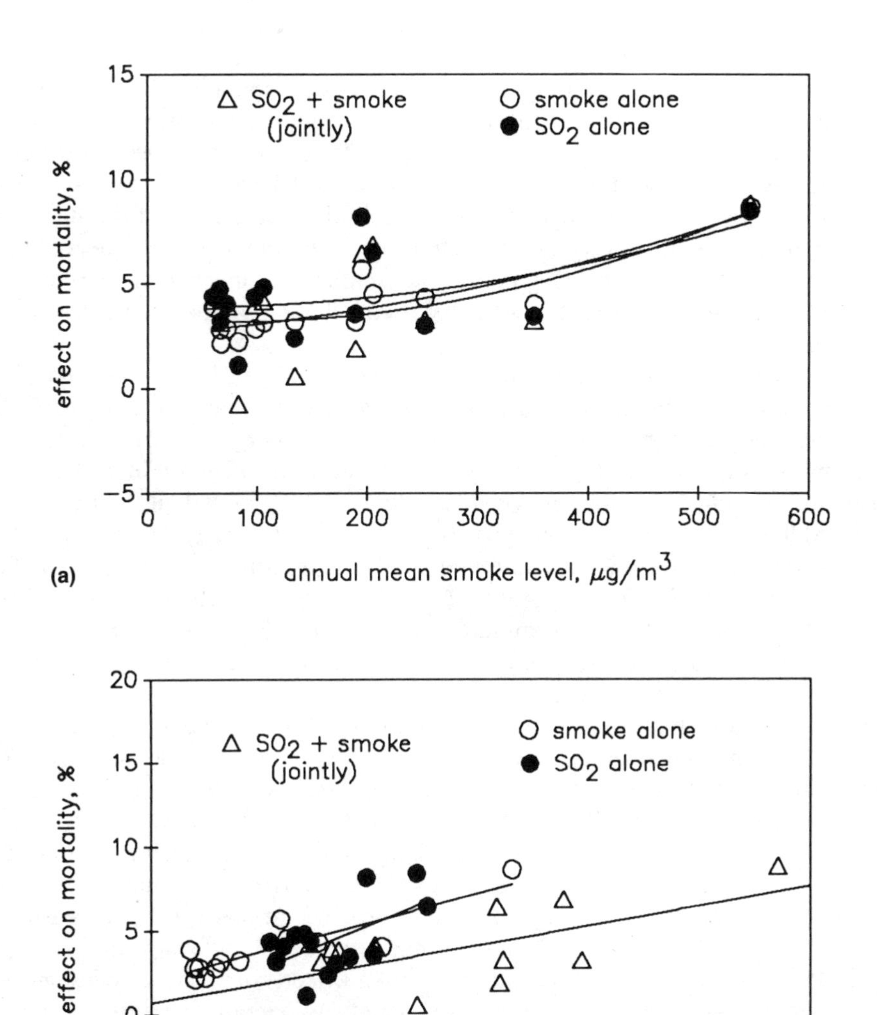

FIGURE 6-3. Comparison of mean pollution effects on London mortality. (a) Plotted versus smoke. (b) Plotted versus various combinations of smoke and SO_2. *Source: Mazumdar et al. (1982).*

SO_2 plus smoke concentrations. The figure suggests that the effect of pollution on mortality passes through the origin only when SO_2 is considered as a contributing pollutant, which implies that SO_2 has an effect, even though it is not statistically significant when regressed jointly with smoke.

Mazumdar *et al.* (1982) addressed the question of separation of the effects of the two pollutants in three different ways: the joint regression, for which SO_2 was usually not significant; a nested quartile analysis; and a case study of high-pollution days. Based on the nested quartile analysis, the authors report a mortality ratio coefficient of 0.00022 per $\mu g/m^3$ for smoke and 0.00006 for SO_2. The analysis of high-pollution days yielded a model with both positive and negative terms for the linear and squared pollution terms, such that SO_2 was important at the higher concentration levels. On the basis of these three analyses, Mazumdar *et al.* concluded that smoke was much more important than SO_2 in London mortality and that it likely had a nonlinear dose-response relationship.

This conclusion was challenged by Goldstein *et al.* (1983), who felt that the methodology used by Mazumdar *et al.* was overly complicated and unjustified. They also criticized the neglect of possible confounding by influenza epidemics, the emphasis on crude total mortality, and the arbitrary deletion of outliers. In their response to this criticism, Mazumdar *et al.* (1983) amplified their findings with additional tables of year-by-year regression results. First, they supplied R^2 values for the regressions based on two-stage temperature corrections. For the joint regressions on SO_2 and smoke, these R^2 values were all in the range 0.16 to 0.38, except for the two winters (0.57 for 1958–59 and 0.70 for 1962–63) that had statistically significant coefficients for SO_2 but not for smoke. The year with a significant negative SO_2 coefficient had an R^2 of 0.29. Next, they presented results for regressions of mortality on SO_2 and smoke, as deviations from 15-day moving averages, without considering weather or day of the week. The SO_2 effects were decreased slightly and the smoke effects increased; R^2's decreased for all years. The smoke coefficient increased for the later years, which had low mean values of smoke. The estimated average mean effect of smoke on mortality increased slightly over the "corrected" values. The final table of regression results was based on "crude" variables with no seasonal adjustments; these results were quite similar to their previous results, except that R^2's were uniformly low and the smoke effects were increased slightly. The year-to-year variation in smoke coefficients was greatly increased and had little resemblance to the other results for specific years; however, the average for the 14 years was not greatly different. These additional data appeared to effectively rebut the criticisms of Goldstein *et al.*

In their discussion of "outliers," Mazumdar *et al.* (1983) presented mortality and pollution data for the 5 days of the December 1962 episode and pointed out how their model greatly overpredicts the maximum mortality actually observed during this event. However, the slope of mortality on pollution for these five points is significant for SO_2 with a value close to that obtained for the entire winter, while smoke is nonsignificant. This finding suggests that SO_2 played a role in London mortality, in spite of the consistent tendency for multiple regressions to indicate only smoke.

Mazumdar *et al.* (1983) tabulated the nested quartile data by using temperature corrections (MacFarlane, 1977) and on several different bases (Mazumdar *et al.*, 1980). Since each of the 16 quartiles contains the same number of original observations, these data provide a convenient means of reanalyzing the London data set for the purposes of this chapter. The quartiles were intended to separate the effects of SO_2 and smoke, by alternatively holding each constant (nesting). This procedure was only partially successful, since the correlation between the two was still 0.89 for the 16 quartiles. However, if only the quartiles are used for which one pollutant is held reasonably constant while the other is varied, the resulting smoke coefficients range from 0.00012 to 0.00036 and the SO_2 coefficients from -0.00009 to 0.00019 per $\mu g/m^3$.

A joint regression of the tabulated quartile data from Mazumdar *et al.* (1980) with corrections for temperature yielded the model (standard errors of regression coefficients in parentheses):

$$\begin{array}{ll} \%\ \text{excess mortality} = 0.0064 + \underset{(0.0043)}{0.008\ SO_2} + \underset{(0.0048)}{0.012\ \text{smoke}}. & [6\text{-}1] \\ R^2 = 0.85 & \end{array}$$

Dropping the highest smoke quartiles resulted in:

$$\begin{array}{ll} \%\ \text{excess mortality} = -0.013 + \underset{(0.0040)}{0.012\ SO_2} + \underset{(0.0068)}{0.0068\ \text{smoke}}. & [6\text{-}2] \\ R^2 = 0.74 & \end{array}$$

Plots of the residuals from these two models showed that the differences between the two were quite small, which illustrates the difficulty in partitioning the effects of the two pollutants. The data at about 250 $\mu g/m^3$ smoke with a residual of about 2% have a large influence on this analysis. Figure 6-4 shows

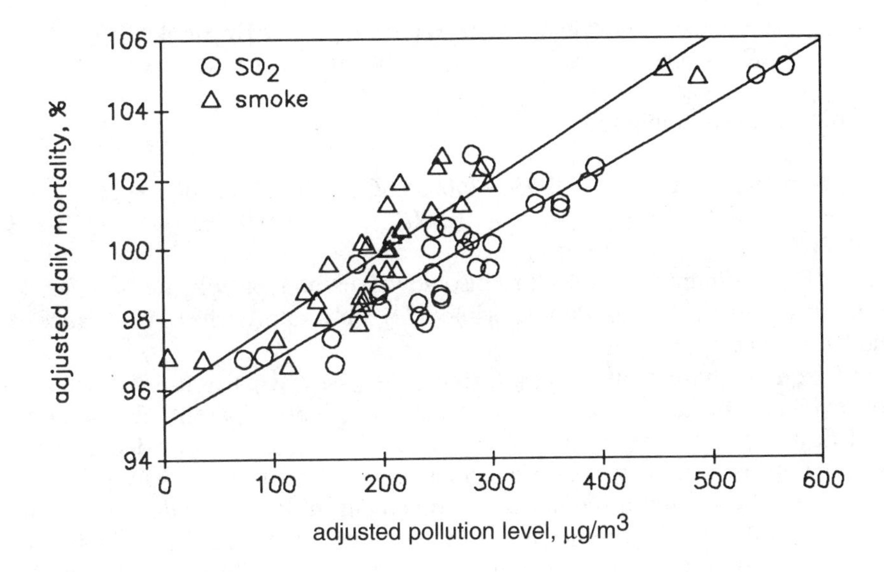

FIGURE 6-4. Dose-response relationships for London mortality quartiles, winters, 1958–72. *Source: Mazumdar et al. (1982).*

the dose-response relationships of each pollutant alone. These plots suggest a threshold at about 100 μg/m^3, but these aggregated data do not permit statistical testing of thresholds.

Mazumdar *et al.*'s 1980 progress report presented tabulations of nested smoke and SO_2 quartiles based on deviations from 15-day moving averages, for both "episodic" and "nonepisodic" periods. To combine these two data sets, mortality data must be converted back from percentages to daily death counts, since the two periods were normalized by different mean values. Figures 6-5 and 6-6 present these data. Figure 6-5a plots the dose-response relationship for smoke as the sole pollutant. The nonepisodic data have a significantly higher slope (0.088 versus 0.057 deaths/μg/m^3), giving rise to the logarithmic or quadratic relationship that others have noted. Figure 6-5b shows the same data plotted against SO_2. The slopes of the two data sets are almost the same, at 0.052 and 0.056 deaths/μg/m^3. The difference between the two lines is due to the difference in average smoke level (249 μg/m^3), plus all other differences between episodic and nonepisodic periods (such as fog and cold). A detailed view of the nonepisodic period is given in Figure 6-6. Again, there is the suggestion of a threshold at about 100 μg/m^3. Using linear models, the joint regressions for these data are (standard errors of regression coefficients in parentheses):

For episodic periods:

$$\begin{aligned} &\%\text{ excess mortality} = -0.066 = \underset{(0.0036)}{0.0039\ SO_2} + \underset{(0.0045)}{0.0146\text{ smoke}}. \\ &R^2 = 0.88 \end{aligned} \quad [6\text{-}3]$$

For nonepisodic periods:

$$\begin{aligned} &\%\text{ excess mortality} = 0.03 + \underset{(0.0036)}{0.0\ SO_2} + \underset{(0.0102)}{0.0313\text{ smoke}}. \\ &R^2 = 0.68 \end{aligned} \quad [6\text{-}4]$$

For both periods combined:

$$\begin{aligned} &\%\text{ excess mortality} = -0.05 + \underset{(0.0036)}{0.0055\ SO_2} + \underset{(0.0036)}{0.0146\text{ smoke}} + \underset{(0.57)}{3.6\ E} \\ &R^2 = 0.93 \end{aligned} \quad [6\text{-}5]$$

where E is a dummy variable for episodic versus nonepisodic periods. Smoke is always statistically significant, while SO_2 only approaches significance in model (6-5) ($p > 0.10$).

To explore these relationships further, the residual patterns were checked for evidence of heteroscedasticity, which was not apparent. In addition, statistical tests were performed to see if the degrees of fit provided by using one or both pollutants (6-1) were significantly different. This analysis showed that adding SO_2 made a significant improvement in fit to a model with smoke alone, even though the coefficient for SO_2 was not quite statistically significant.

The final attempt at rederiving dose-response relationships from Mazumdar *et al.*'s (1980, 1982, 1983) studies of the London mortality data involves their tabulated means of the data by day of the week, for episodic and nonepisodic periods. Again, the best regressions were all based on smoke; adding SO_2

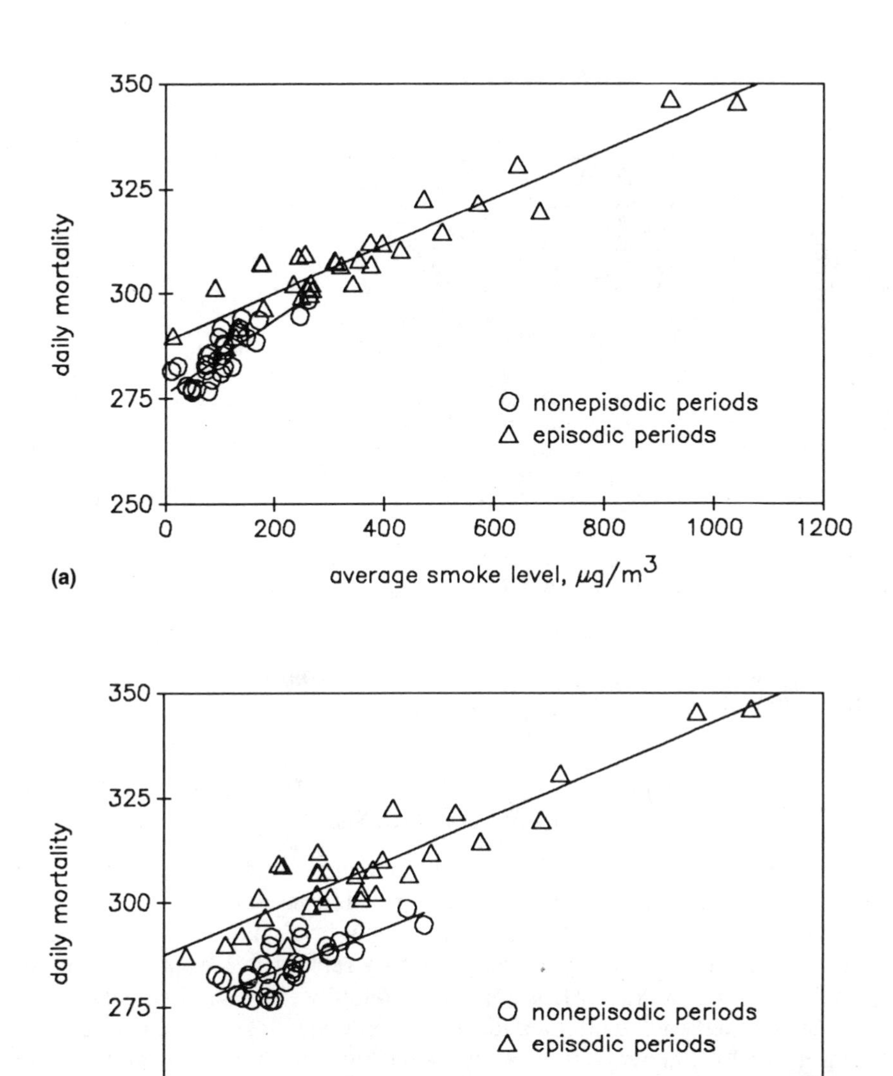

FIGURE 6-5. Dose-response relationships for London mortality quartiles, winters, 1958–72. (a) Plotted versus average smoke concentration. (b) Plotted versus average SO_2 concentration. *Source: Mazumdar et al. (1982).*

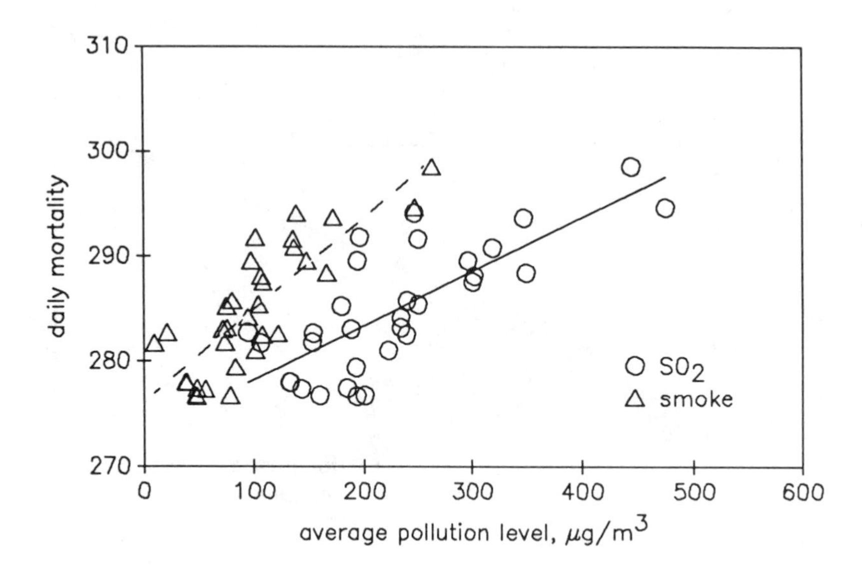

FIGURE 6-6. Dose-response relationships for nonepisodic London mortality quartiles, winters, 1958–72. *Source: Mazumdar et al. (1982).*

made little difference. Smoke coefficients for the combined period were in the range of 0.00027 to 0.00032 per μg/m³.

Subsequent Studies of the 1958–1972 London Mortality Data Set

Ostro (1984) used deviations from 15-day moving averages with temperature and humidity entered directly into the regression model and found "no evidence to support the existence of a no-effects level. Further, the reanalysis suggests that the estimated pollution-mortality relationship exists even in nonepisodic winters, when British smoke readings were less than 500 μg/m³." Ostro subdivided each winter's data into days with smoke greater or less than 150 μg/m³ and estimated a regression coefficient for each winter and subset. I calculated group average regression coefficients from Ostro's tabulated data for the high- and low-smoke subsets, including the winters of 1960–72 for the low values (the minimum daily smoke values in 1958–60 were only slightly below the 150 μg/m³ cutoff), and for the winters of 1958–68 and 1969–70 for the high-smoke values (there were insufficient high values in the other three winters). For days with smoke levels over 150 μg/m³, the average coefficient was 0.044 ± 0.042 excess deaths/μg/m³ (2-sigma confidence limits); for values less than 150, the average was 0.14 ± 0.13 excess deaths/μg/m³. Such a dose-response function has a counterintuitive shape (the lower the pollution concentration, the more lethal it is!) and implies a constant elasticity of about 0.04. An alternative explanation is that important factors were omitted from the analysis and that at low levels, smoke is behaving at least in part as a surrogate for some other factor, possibly SO_2 or CO, among others. Ball and Hume (1977) showed a strong correlation ($R = 0.98$) between black smoke and particulate lead at a nontraffic site in London, raising the possibility that other motor vehicle air pollutants may be similarly correlated. Additional criticisms of the Ostro study include the omission of variables for flu epidemics, holidays, and days of the week. In addition, more rigorous statistical methods for searching for thresholds employing dummy variables could have been used. It should also be noted from the confidence limits given that there was a great deal of year-to-year variability in both subsets, which reinforces the need to examine a relatively long period of record in a time-series mortality study.

The next analysis of this London mortality data set was performed by Schwartz and Marcus (1986), who used both autoregressive models (allowing for correlations between data on successive days) and deviations from a 15-day moving average. They also performed regressions with and without temperature and humidity variables in the regression model and explored subsets defined by smoke concentration level. The authors found that "multiple regression models for daily mortality and British smoke do indeed reflect a relationship that cannot be attributed to time-series effects, temperature, SO_2, or functional misspecification. . . . There does appear to be some tendency for higher slopes in later years when both British smoke and SO_2 reflect more nearly contemporary conditions." This study established the autocorrelation structure of the data and showed that deaths on day t are correlated with deaths on day $t - 1$, even when deviations from a 15-day moving average are used. The dose-response findings generally confirmed previous ones: smoke coefficients increasing in the later years as mean levels dropped, with an average range of 0.00024 to 0.00027 excess deaths/$\mu g/m^3$. In joint regressions, smoke was positive with SO_2 negative. However, Schwartz and Marcus also showed that including temperature and humidity resulted in larger and more significant pollution coefficients, especially for SO_2. When SO_2 was used as the sole pollutant in a random effects model across all years, the result was substantially more significant than when smoke was used alone ($t = 21$ versus $t = 6$). Schwartz and Marcus concluded that an independent SO_2 effect on mortality could not be excluded, but that the effects of smoke were independent of SO_2. Further debate on this point was joined by Fleisher and Nayeri (1991), who called for tests of the interaction between smoke and SO_2 as a way of assessing possible physical and chemical interactions between the two species, which could imply the presence of sulfuric acid on the smoke particles. Schwartz (1991a) agreed with the premise but reported that additional regressions incorporating such an interaction term failed to be significant, and thus that the London data did not support the premise.

Shumway *et al.* (1983) emphasized spectral analysis in their treatment of the London data, using a multiple regression approach for the daily counts of total, cardiovascular, and respiratory mortality, separately. The authors concluded that the best models used lagged temperature in conjunction with the logarithm of either smoke or SO_2 (same day) and that the two pollutants appeared be "acting identically in all respects." No threshold effects were evident. Relative humidity was not found to be important, but a delayed effect was found due to daily temperature variations (negative). Their regression analysis used detrended (filtered) variables in the frequency domain, which allowed them to examine lags of arbitrary length. The seasonal and epidemic effects on mortality were removed by this filter. The resulting regression equations were then transformed back into the physical domain (although the units were not given). They found that lag effects were minimal for total, cardiovascular, and respiratory mortality, but their tabulated results indicate that respiratory effects persisted for at least 3 days. Including these lags added 60% to the pollution effect on mortality. Thus, the pollution effect on respiratory mortality was proportionately larger than on total or cardiovascular mortality. None of the three mortality measures showed a subsequent drop at longer lags, and none showed appreciable negative lag effects. The results were thus compatible with a causal model.

The spectral analysis examined the coherence between the time series, which was found to be strongest for frequencies corresponding to periods of 7 to 21 days, implying that pollution or temperature episodes have more effect on mortality when averaged over periods of this length. This finding may reflect day-of-week patterns, which were not specifically removed by this analysis. Shumway *et al.* (1983) showed that temperature effects were more important than pollution effects for cardiovascular mortality, but not for respiratory mortality. They warned that these results should not be extrapolated to the United States, a point that was reinforced by Roth *et al.* (1986) on the basis of differences in indoor air quality and of the chemical composition of particulate matter.

Using Eq. 3-7 to estimate the SO_2 elasticities at the mean yields values of 0.076 for total mortality, 0.071 for heart disease, and 0.094 for respiratory causes, based on Shumway *et al.*'s tabulated semilog regression equations. For example, for total mortality,

$$\% \text{ excess deaths} = 7.6 \ln(SO_2) + 0.072 \text{ temp} - 0.051 \text{ temp}_{t-2}, \quad [6\text{-}6]$$

with SO_2 in $\mu g/m^3$ and temperatures in degrees celsius (standard errors were not given).

When additional lag days are included, the respiratory elasticity increased to about 0.15. These values are consistent with those of other studies of London mortality and imply contributions to total mortality from causes other than cardiovascular and respiratory. If we convert these nondimensional coefficient values to linearized regression slopes by dividing by the mean pollution value, we find a value of 0.00028 excess deaths (total) per $\mu g/m^3$ based on SO_2. If we further assume perfect correlation between smoke and SO_2, the value based on the sum would be 0.00017 excess deaths per $\mu g/m^3$.

With respect to the threshold question, Shumway *et al.*'s analysis (1983) is silent on year-to-year variations during the period of improving air quality, since the data for each year were normalized with respect to the mean values for that year. They used logarithmic transforms for the pollution variables, and thus effects of the temporal changes in mean pollution values that occurred over the years would tend to be suppressed. Therefore, the authors' conclusion that "no threshold effect was evident" should not be interpreted as providing evidence that thresholds do not exist, since the study was not specifically structured to find them.

The difference between linear and logarithmic models can be seen from

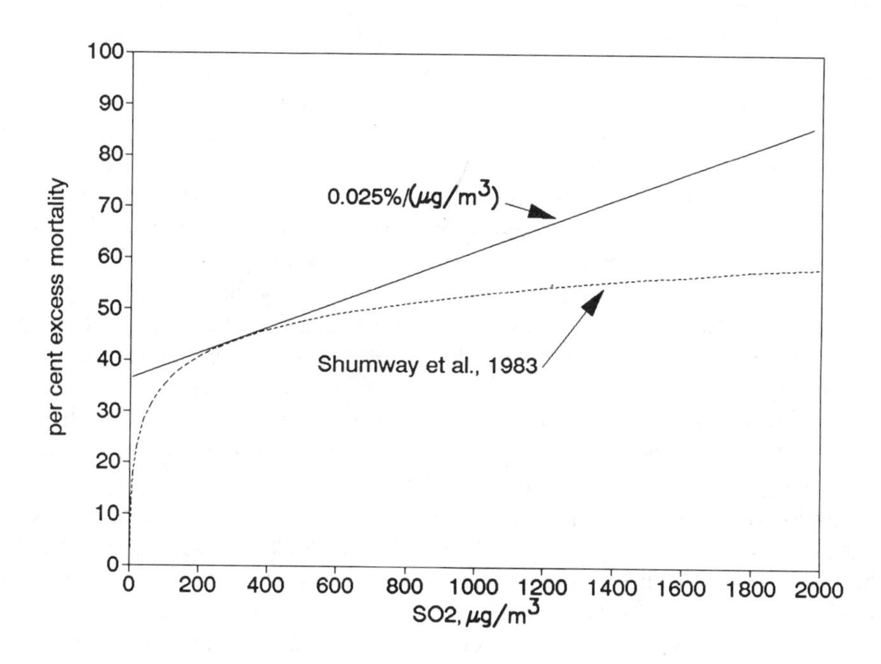

FIGURE 6-7. Comparison of linear and logarithmic dose-response functions for London mortality data.

Figure 6-7. Both of these dose-response functions have about the same effective slope at the mean and about the same elasticity. Near the origin, the differences are dramatic, but one should keep in mind that the lowest mean value of SO_2 in any winter was about 180 μg/m^3. It is possible that the very high values of SO_2 experienced during the 1962 episode (about 3500 μg/m^3) had an undue influence on the analysis (see Chapter 5); for example, Mazumdar *et al.* excluded this value. However, the analysis of Mazumdar *et al.*, using individual years, also found increasing (linear) coefficients as the mean pollution values decreased, which is consistent with a logarithmic relationship.*

Consideration of Aerosol Acidity

Ito and Thurston (1989) added data on aerosol acidity to their time-series analyses of London mortality. These measurements have often been referred to as "sulfuric acid," and H_2SO_4 is undoubtedly a major constituent. However, the actual measurement consists of a titration of the filter extract (Commins, 1963), and thus it is not specific to H_2SO_4. The early literature mentions an excess of SO_4^{-2} relative to H^+, indicating partial neutralization. Also, Junge

* A possible physical explanation of this phenomenon could involve SO_2 adsorption onto the surfaces of particles as the mechanism for transport into the lung. If the number of particles were limited, higher levels of SO_2 could adsorb less effectively and thus have a lesser health impact.

and Scheich (1971) compared methods of H^+ determination in London and some German cities and concluded that acidity was higher in London but that the British methodology gave higher results (probably because the titration endpoint included organic acids as well). The London acidity measurements were taken at a single site in central London and thus may not be strictly compatible with the SO_2 and smoke data from the seven-station network. Ito and Thurston show that the mean smoke levels are lower in central London (1963–71 winters) and that the mean SO_2 levels are higher than the network averages. Although fog is expected to promote the oxidation of SO_2 to H_2SO_4 and has historically been associated with high concentrations, the correlation between acidity and relative humidity (RH) was low in these data. Using multiple regression techniques, Ito and Thurston concluded that more than half of the variance in daily aerosol acidity could be explained in terms of the variability of smoke, SO_2, and other meteorological parameters.

In a companion paper, Thurston *et al.* (1989) examined the relationships of these acidity measurements with mortality. They combined the data into a single data set after filtering to remove the long-term variations. The bivariate correlation of acidity with mortality was about the same as the smoke and SO_2, for same-day and next-day deaths. (Note that the acidity measurements were in phase with the mortality counts, while the seven-station network data preceded by 12 hours.) For all species, correlations improved when logarithmic transforms were used, suggesting an attenuation of the effect at high concentrations. The best correlation was for raw (unfiltered) mortality with acidity, but it is not clear how much other variables (such as temperature) may have contributed to this finding. No multiple regressions were considered in this analysis.

In an updated report on the subject, Ito *et al.* (1991) reported that the association between mortality and acidity in London was "not statistically different from that found for British smoke or SO_2 in the 'short-term' effects based models, with or without control for weather variables."

Comparison and Reconciliation of the London Studies

The most important issues remaining from the various analyses of the London daily mortality data are concerned with the shape of the dose-response relationship(s) and the separation of smoke effects from those of SO_2 and acidity. In the analysis that follows, these two issues and their interrelationship will be examined.

The changing pollution levels over the 14 winters provide an opportunity to examine dose-response relationships on an aggregated basis. First, since the daily mortality rate decreased by 25% during period of study, for reasons that are only partly related to air pollution, it is important to calculate regression coefficients as *percent* excess deaths per unit of pollution (or as mortality ratios), as opposed to death counts. An alternative would be to account for the temporal change separately, for example, by pooling the data and including an appropriate independent variable in the regression. It is also useful to consider some theoretical linear dose-response models, as shown schematically in Figure 6-8, in terms of both a regression coefficient and the corresponding effect on mortality.

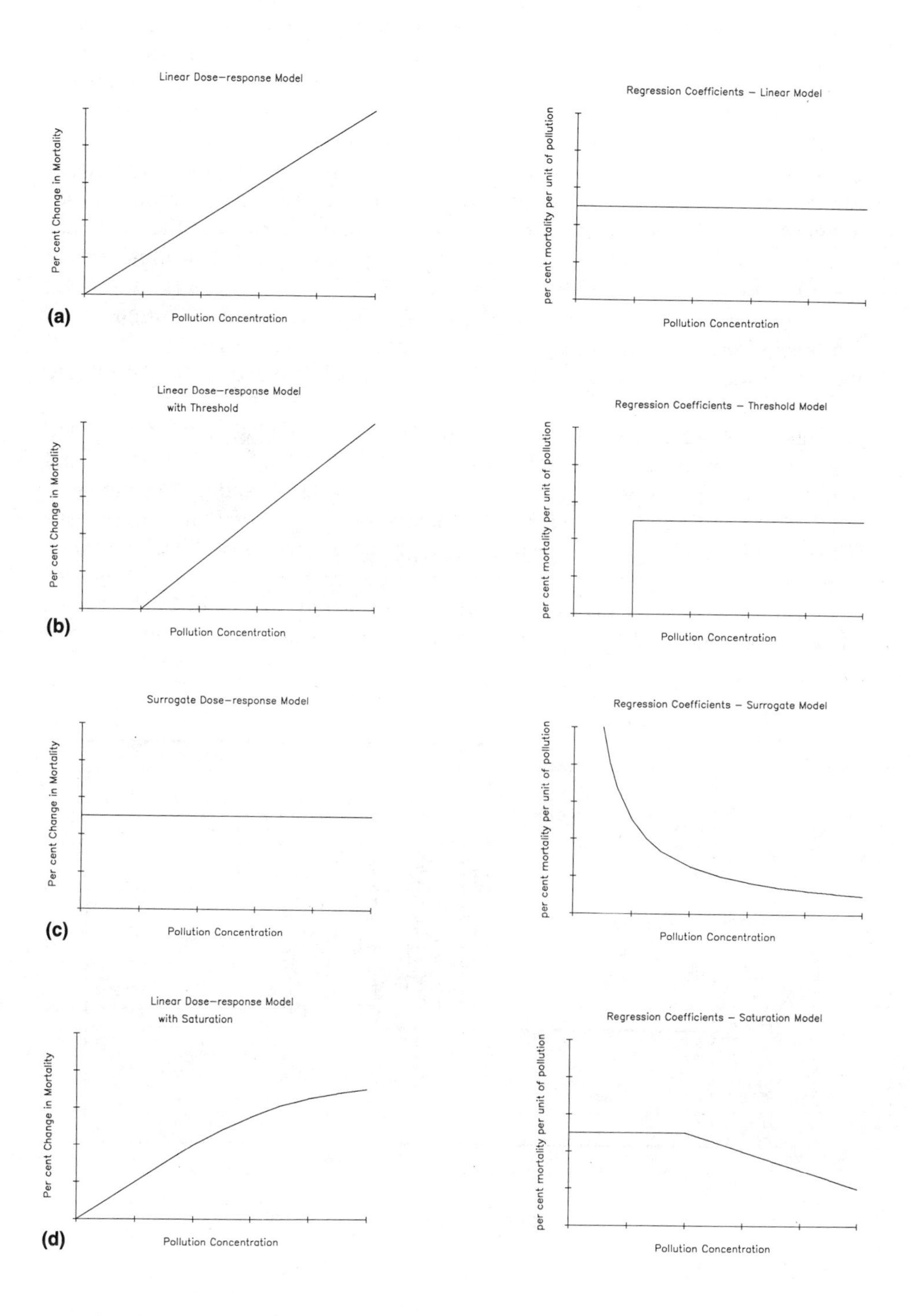

FIGURE 6-8. Theoretical dose-response models. (a) Linear, no-threshold model. (b) Linear, threshold model. (c) Surrogate model. (d) Linear model with saturation. The corresponding regression coefficients (slopes) are shown to the right of each model.

A linear, no-threshold, model (Figure 6-8a) has a constant coefficient (C) over the entire range of pollution (P) and an effect ($C \times P$) on mortality that is linear and passes through the origin. A linear threshold model (Figure 6-8b) is similar, except that the coefficient becomes indeterminate between zero and the threshold pollution level (P_0), and the mortality effect has an x intercept at P_0. Log-linear models have these same characteristics if one substitutes the logarithm of the response variable. The next possibility (Figure 6-8c) is the surrogate model, which is characterized by a constant effect (or an effect not influenced by the pollutant used for plotting and analysis), and therefore has a coefficient that increases with decreasing pollutant concentration as the (constant) effect is divided by smaller values. Nonlinear models (Figure 6-8d) in theory may either have coefficients that increase with pollution level, because effects become more severe at higher exposure levels, or decrease, because saturation is reached and the dose loses effectiveness. The latter case is shown in the figure.

Figure 6-9 is a plot of SO_2 coefficients (as the sole pollutant) as determined from the London data by Schwartz and Marcus (1986), with and without consideration of temperature and humidity effects (t&h), and by Mazumdar *et al.* (1982) and Shumway *et al.* (1983). Below about 300 $\mu g/m^3$, the coefficients generally are no longer significant, although the estimated coefficients remain remarkably constant over the range of annual mean SO_2 levels. The datum for Shumway *et al.* represents my estimate of a linearized coefficient for the whole data set, plotted at the mean. The linear regression line is intended to depict

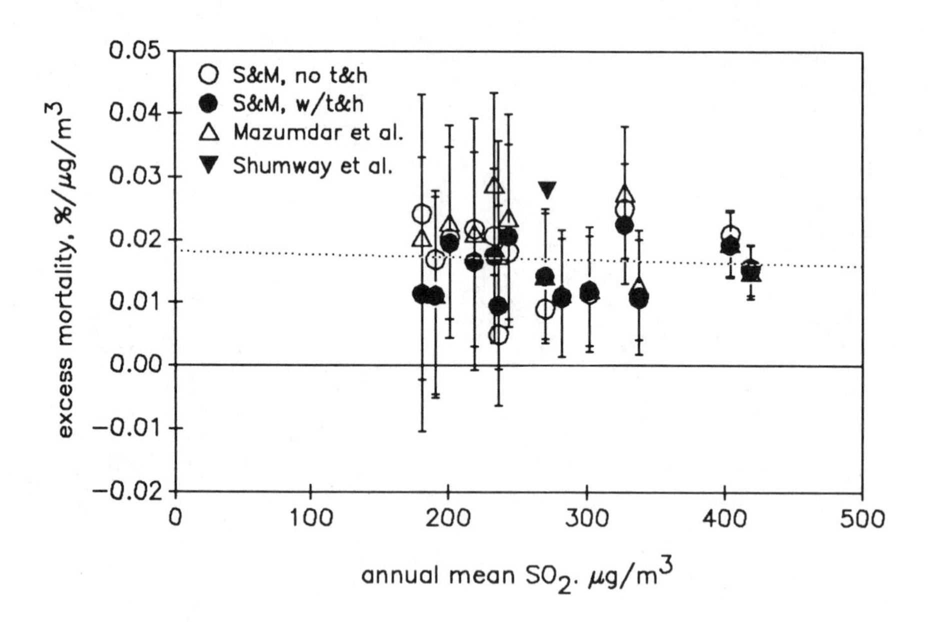

FIGURE 6-9. Comparison of SO_2 regression coefficients for London mortality for the winters of 1958–72. The dotted lines represents a linear regression for the entire plot versus SO_2 level; error bars are for 2-σ confidence limits on the individual regression coefficients. *Source: Mazumdar et al. (1982); Schwartz and Marcus (1986), and Shumway et al. (1983).*

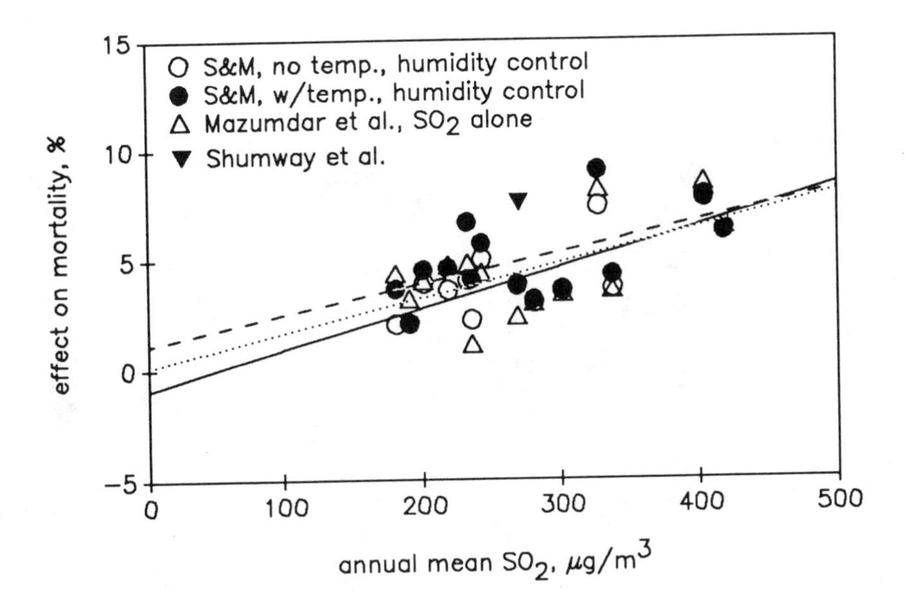

FIGURE 6-10. Comparison of mean SO_2 effects (regression coefficient x mean SO_2 level) on London mortality, winters, 1958–72. Lines represent linear regressions of the individual effects versus SO_2 level. *Source: Mazumdar et al. (1982); Schwartz and Marcus (1986).*

the relationships between coefficients and mean SO_2 levels; its slope is not significantly different from zero, suggesting that the regression coefficients are independent of the mean pollutant level (Figure 6-8a). Since all three of these year-by-year analyses produced virtually indistinguishable results, it appears that t&h corrections are not critical. Given its distance from the plot regression lines, the result of Shumway *et al.* stands out from the ensemble, perhaps because they used a log transform and pooled all the years.

The corresponding plot of mortality effects, Figure 6-10, passes through the origin, as in Figure 6-8a. However, if the lower confidence limits on Figure 6-9 had been used to compute the mortality effect, a threshold of effect would have been shown as in Figure 6-8b. Such a threshold is suggested from the joint regression for SO_2 and smoke; the SO_2 contribution approaches zero at $SO_2 = 260\,\mu g/m^3$.

Figure 6-11 compares the regression coefficients for smoke. Six different regression models are compared in Figure 6-11a; the results are quite consistent above about $200\,\mu g/m^3$ and correspond closely with the results obtained from episodes in several cities (Chapter 5). Below this level, the curve looks like a surrogate model (Figure 6-8c). Note that such a surrogate is not likely to be aerosol acidity, since this parameter is highly correlated with smoke and SO_2. However, carbon monoxide is a possible surrogate not known to be so correlated. Confidence limits are shown on Figure 6-11b for two of the models

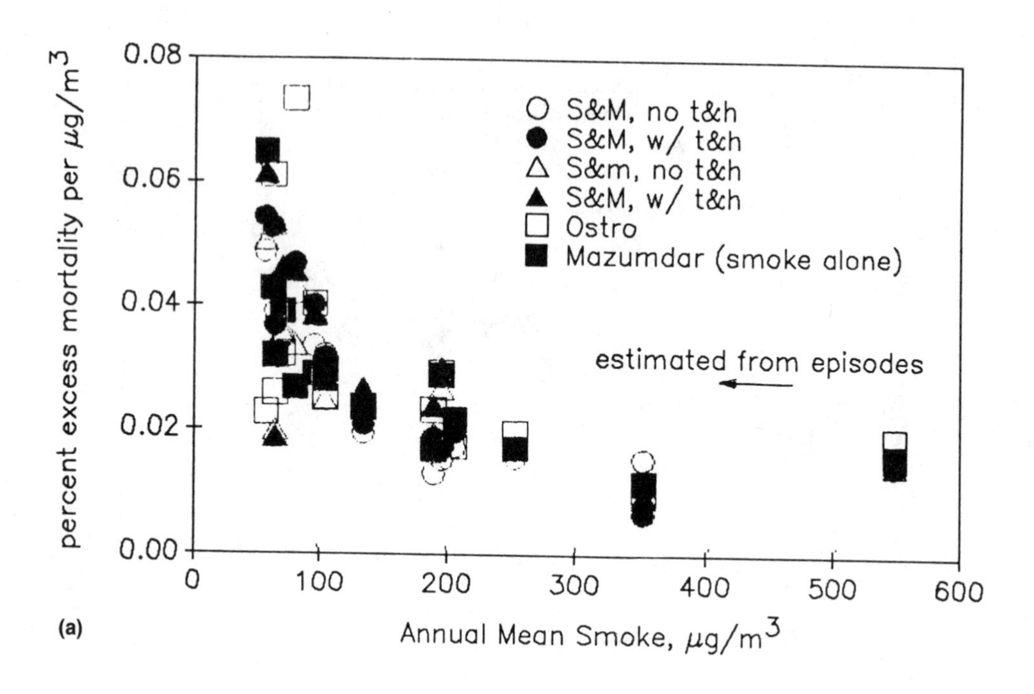

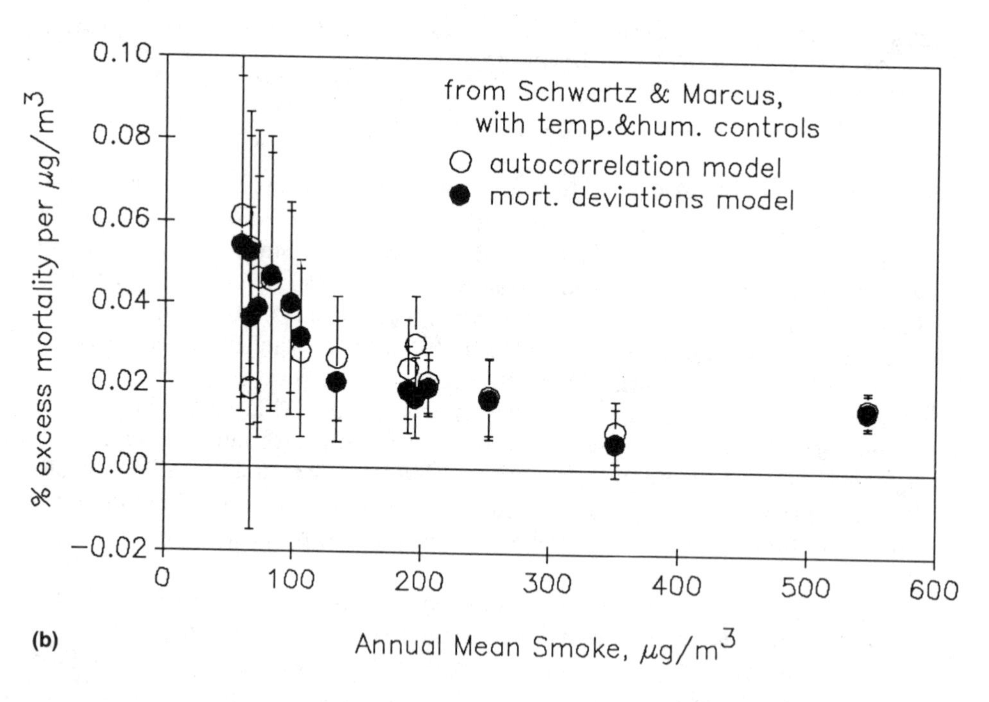

FIGURE 6-11. Comparison of smoke regression coefficients for London mortality, winters, 1958–72. (a) Various models. (b) Models with temperature and humidity controls, Schwartz and Marcus (1986). Error bars are for 2-σ confidence limits on the individual regression coefficients. *Source: Mazumdar et al. (1982); Schwartz and Marcus (1986); and Ostro (1984).*

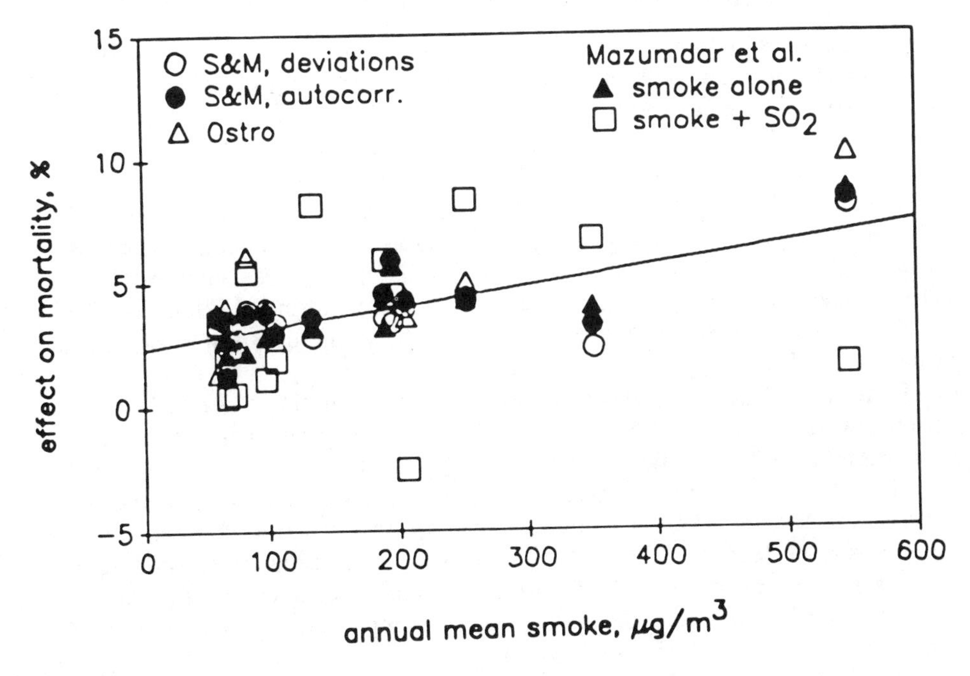

FIGURE 6-12. Comparison of mean smoke effects (regression coefficient x mean SO_2 level) on London mortality, winters, 1958–72. *Source: Mazumdar et al. (1982), Schwartz and Marcus (1986), and Ostro (1984).*

presented by Schwartz and Marcus (1986); most of the lower confidence limits would allow a constant coefficient that is reasonably consistent with the episode data (0.00011 per $\mu g/m^3$). Figure 6-12 shows the corresponding smoke effects on mortality, and additional data for Mazumdar *et al.*'s (1982) regression for smoke and SO_2 together. Linear regressions through these data indicate a statistically significant mortality intercept of about 2.5 to 3%, which is a symptom of a surrogate model (Figure 6-8c).

Hammerstrom (personal communication, 1988) also analyzed the London data but emphasized the 1964–72 time period, seasonal variations, and included data on aerosol acidity (H^+). Results were presented by sex and by season. For both sexes combined, the overall effect found by Hammerstrom in winter was about 2% (combined elasticity for smoke and $H^+ = 0.02$), which compares well with the results from other investigators. It also implies that adding H^+ as an additional pollutant did not significantly increase the overall effect. Including SO_2 in addition also made little difference. However, seasonal differences were considerable: The total effects on mortality were the largest in spring and nonsignificant in summer and fall. Given the much higher air pollutant concentrations in winter and the larger probability of outdoor exposure in summer, these results suggest that the effects of air pollution on London mortality are still not completely understood, especially at low concentrations. Similarly, when mortality was broken down by sex, females showed larger total effects, even though it may be hypothesized that their outdoor exposure was probably less.

Conclusions from Studies of London Mortality

From a statistical point of view, the various studies are quite consistent. Smoke is the most important pollutant; however, its effect appeared to become more severe per unit as concentrations decreased when clean air controls were imposed. The interpretations of this finding differ considerably. Ostro (1984) argues that the lack of a threshold has been demonstrated (which is also the case when a log transform is used); however, Ostro's analysis ignores the simultaneous presence of over 180 μg/m^3 SO_2 (and the accompanying aerosol acidity) and levels of CO probably up to 15 to 20 ppm, just as Shumway *et al.*'s (1983) analysis ignores the simultaneous presence of smoke. None of these studies specifically tested threshold-type models, so the question must be considered as still open.

Joint regression is the classical method of apportioning the effects of two contributing independent variables. All of the analyses seem to exhibit a consistent lack of statistical significance of SO_2 when regressed jointly with smoke. It is clear that this situation results at least in part from the high correlation between the two pollutants in the London data base. If this correlation were perfect ($R = 1$), we would be justified in using either variable but would have to specify that the resulting dose-response function was valid only for this special situation and would have to regard the pollutant as being measured in some sort of combined units.

However, it is also clear that the spatial distributions of the two pollutants are quite different (Gore and Shaddick, 1958; Ito and Thurston, 1989), and thus the high correlation may be due partly to averaging across the seven stations of the network. When two independent variables are highly correlated, the variable with the least measurement error will dominate the regression. Since the sources of SO_2 in London shifted over the years from distributed area sources to discrete point sources (Ellison and Waller, 1978), it is possible that the seven monitoring stations represent smoke exposures better than SO_2 exposures, especially since the mortality data are for Greater London, a much larger area. This would be consistent with the possible existence of a physiological effect of SO_2 in London, despite the lack of statistical significance when regressed jointly with smoke. Effects due to carbon monoxide are another possibility, but lacking data, we cannot explored this hypothesis.

The British philosophy of air pollution control in the 1960s was centered on the control of smoke, primarily from domestic heating sources. This may explain the almost universal emphasis on the analysis of winter data, to the exclusion of the other seasons. However, the correlations between pollutants is expected to change with season, and other species such as ozone may be more important in summer (Derwent and Stewart, 1973). Also, the relationship between acidity and SO_2 tends to vary seasonally (Lipfert, 1988). Therefore, it may be useful to analyze the London data on a continuous year-round basis, rather than just during the winters.

TIMES-SERIES STUDIES OF MORTALITY IN NEW YORK CITY

As in London, studies of the effects of air pollution on mortality in New York began with descriptions of isolated incidents (Chapter 5) and progressed through a series of more formal time-series analyses. A useful summary was given by Shils and Skolnick (1978). Unlike London, where only winters have been analyzed, most of the New York mortality studies dealt with the entire year.

Routine air pollution monitoring in New York City began in the 1950s at the 121st Street Laboratory station in upper Manhattan, where SO_2 was measured using the peroxide method and smokeshade by the tape sampler, reporting the readings as coefficient of haze (COH). Depending on the situation, one COH unit could correspond to a range of mass units, from 69 to 200 $\mu g/m^3$ (Holland *et al.*, 1979). At high levels, a conversion factor of 100 $\mu g/m^3$ per COH unit appears reasonable and is convenient (Lipfert, 1978; see Chapter 2). In 1962, Cornell University Medical College established an air monitoring station about 10 km farther south on the lower east side of Manhattan, data from which were used in several studies (McCarroll and Bradley, 1966; Hodgson, 1970). By the 1970s, an extensive network was in operation in all five boroughs of the city (about 40 stations), which has since been scaled down considerably. Goldstein (1976) describes some of the details and interrelationships among the stations of this network.

Peak air pollution levels were recorded in New York during the 1960s, a period when residual fuel oil with high (2–3%) sulfur content was used extensively in large commercial and residential buildings. Also, it was the practice to require on-site incineration of municipal refuse, which added to the particulate burden (Simon, 1978). Hourly values of SO_2 as high as 1 ppm and smokeshade as high as 8.0 COHs were recorded then during adverse meteorological conditions (Greenburg *el al.*, 1962). Carbon monoxide readings in the mid-1970s ranged from 3 to 15 ppm annual average, with peak 8-hour averages as high as 30 ppm (Simon, 1978). CO readings were probably higher in the 1960s before automotive pollution controls came into use.

Early Studies

The early studies of episodes focused on mortality by cause and age (McCarroll and Bradley, 1966; Glasser and Greenburg, 1967; Greenburg *el al.*, 1967), rather than on developing dose-response relationships. Causes that were affected include influenza-pneumonia, vascular lesions of the central nervous system, diseases of the heart, and lung cancer. All age groups were affected, except perhaps infants. The peak air pollution levels cited above, which occurred in November 1966, were associated with about 10% total excess mortality, for the entire city (Glasser and Greenburg, 1967).

The 1967 study of Greenburg *et al.* focused on influenza, which was a major factor in January–February 1963, a time when SO_2 was also above normal (up to 0.8 ppm, 24-hour average). The presence of flu viruses constitutes an additional public health stress agent, which may act synergistically with stress due to air pollution. Thus, it is difficult to assign the causes of excess mortality (about 21%) that occurred then solely to air pollution.

Hodgson (1970) was the first to analyze a continuous period of monitoring (November 1962–May 1965) using time-series methods. His model analyzed the sum of heart disease and respiratory deaths separately from all other causes, by three broad age groups, and he employed smokeshade, SO_2, and the absolute value of daily degree-days as predictors. The last variable, a temperature correction, considers the separate effects of heat in the summer and cold in the winter, and thus serves as a seasonal correction. This model was used with both monthly data (31 consecutive months) and daily data, including various lag effects and nonlinear specifications. The results showed smokeshade to be more important than SO_2, and the mean effect (about 12%) on mortality was about the same as the worst years in London (Figure 6-3), despite the less polluted air in New York City and the use of a single monitoring station there to characterize the exposure of the entire city. Hodgson also showed that the monthly models showed larger effects on mortality than the daily models, and that including the previous month's pollution and degree-days (the average of month t and month $t - 1$ were used to predict mortality in month t) resulted in a slight increase in the regression coefficient (but with a slight loss in explanatory power). This result suggests that the short-term effects of pollution exposure may persist considerably more than a few days. However, when lags were explored with the daily models, only the inclusion of the preceding day made a significant improvement and SO_2 became statistically significant. Although for Hodgson's monthly results, the pollution effects were larger than the temperature effects, the two were about equally important for the daily models. Respiratory mortality had about the same effects in monthly and daily models, but the effects on heart disease were stronger for the daily model. It is important to note that the period studied included the 1963 flu epidemic and that no special means were taken to account for this effect. Since respiratory and heart disease deaths were combined, all of Hodgson's results may be confounded by this factor, which may partially explain the significance of monthly deaths.

Hodgson (1970) derived the following relationships for total mortality (based on $\mu g/m^3$ as the units of measure for the pollutants and assuming that 1 COH = $100\,\mu g/m^3$):

For monthly mortality (no lag):

$$\begin{aligned} &\%\text{ excess mortality} = 0.0548\text{ smoke} + 0.0022\ SO_2. \\ &R^2 = 0.73 \qquad\qquad (0.0017) \qquad\qquad (0.0046) \end{aligned} \tag{6-7}$$

For daily mortality (including previous day):

$$\begin{aligned} &\%\text{ excess mortality} = 0.0316\text{ smoke} + 0.0032\ SO_2. \\ &R^2 = 0.37 \qquad\qquad (0.0040) \qquad\qquad (0.0012) \end{aligned} \tag{6-8}$$

On the basis of mean values of $430\,\mu g/m^3$ for SO_2 and $215\,\mu g/m^3$ for smoke, model (6–7) yields an estimate of about 8.2% for the effect of daily variations in air pollution on New York City mortality, from 1962–65.

Hodgson also compared linear and log-log models using the monthly data for respiratory deaths and concluded that the linear form was slightly superior on the basis of equivalent R^2's. However, the smokeshade variable was more significant in the log-log model ($t = 2.8$ versus $t = 2.1$). the elasticities from the log model were 0.23 for smoke and 0.05 for SO_2 (not significant).

Glasser and Greenburg (1971) studied the winters of 1960–64 (October–March), including hourly SO_2, bi-hourly smokeshade, daily mean temperature, sky cover, wind speed, and rainfall as potential predictor variables. The pollution data were from the 121st Street station and were averaged to provide daily means. Mortality data were treated as deviations from 15-day moving averages and as deviations from expected values based on the 5-year record. Day of week effects were also accounted for. However, in their regression analysis, Glasser and Greenburg treated SO_2 as a surrogate for all air pollution; the result was about 2.8% of mortality associated with SO_2. Their dose-response curve for SO_2 (as the sole pollutant) had an overall slope of about 0.005% excess mortality per $\mu g/m^3$, with the suggestion of a threshold at about 400 $\mu g/m^3$ (24-hour average). With smoke as the sole pollutant, the coefficient was about 0.015 per $\mu g/m^3$, with the suggestion of a threshold at about 3.0 COH. Like Hodgson, they ignored the contributions of the 1963 flu epidemic, so that these findings may well be overestimates. The two-way contingency table they presented giving the deviations in daily mortality for fixed levels of SO_2 and smoke allows some independent estimates to be made of the dose-response relationships. Since each table entry (31 in all) represents the aggregate of a different number of days, different weights will be given to the various pollution levels from in the original data. When all the table entries were regressed on the midpoints of the SO_2 and smokeshade ranges, the following model resulted:

$$\begin{aligned} &\%\ \text{excess mortality} = \underset{(0.0034)}{0.0036}\ \text{smoke} + \underset{(0.0013)}{0.0071}\ SO_2. \\ &R^2 = 0.57 \end{aligned} \qquad [6\text{-}9]$$

Based on the means of the smokeshade and SO_2 values, this relationship yields a total mortality effect of 5%. This figure is lower than Hodgson's result, and the roles of SO_2 and smokeshade as principal contributors have been reversed.

Studies by Buechley *et al.*

This same period (1962–66) was also studied by Buechley *et al.* (1973), who used data on mortality time trends for 422 urban places in the United States to develop a normalization curve to serve as a daily mortality trend predictor. For the New York area, they used mortality data from a much larger region, the tri-state air quality control region, which had a population of almost 14 million in 1960. Exposures to SO_2 and smokeshade were based on the single monitoring station at 121st Street in Manhattan. The model accounted for days of the week, holidays, temperature fluctuations, and flu epidemics (important for 1963). Again, the results suggested that SO_2 played a more important role than smoke; the regression coefficient for SO_2 was 0.0045% excess mortality per $\mu g/m^3$. However, the mean value of SO_2 used by Buechley *et al.* (270 $\mu g/m^3$) appears to be about half the value expected from other data sources for New York City (Glasser and Greenburg, 1971). Using the higher SO_2 value would reduce the coefficient to 0.0022% excess mortality per $\mu g/m^3$, while the total effect (regression coefficient $\times$ mean value) on metropolitan area mortality would remain at 1.2%, which is still much lower than the results found by either Hodgson (1970) or Glasser and Greenburg (1971) or by the various London studies. Buechley *et al.* presented a plot of percent mortality deviations according to geometric increments of increasing SO_2, which suggested a semilog model. According to this model, the SO_2 elasticity would be 0.0093, which is reasonably close to the value found with a linear model (0.012).

Buechley (1975) subsequently gave more details of his methods and supplied additional regression results. For example, he described calibration problems with the SO_2 instrument in 1964–65, apparently resulting in high readings. An adjustment procedure was made to set the minimum daily values to zero; this may explain the fact that his mean SO_2 value for 1962–66 was substantially lower than what others have cited. Because of the skewed distribution of the SO_2 variable, Buechley also investigated a transformed "Z-score" SO_2 variable that was detrended to remove the effect of the declining mean values from 1967–72 (factor of 5 decrease). He also included a "particulate" variable, assumed to be COH, and a carbon monoxide variable. He stated that CO was missing from 1962–65: it is not clear whether the regression results given for 1962–66 including CO are, in reality, only for 1966 or if some procedure was used to supply data for the missing 4 years. Buechley (1975) gave results for 27 different stepwise regressions in the appendix to his report, including results for 1967–72, 1962–66, and 1962–72. The temperature, seasonal, and other correcting variables vary somewhat according to the stepwise selection procedure, since typically only about half the possible variables were selected for the optimum step. Buechley pointed out that since particulates have such a strong seasonal cycle, it is important to use an adequate representation for the seasonal mortality trend in the regressions, and that a simple sine function is not adequate. His regressions include one or the other SO_2 variables and usually particulates, but CO was included only if the significance criteria for selection were met. The detrended SO_2 variable (SO_2Z) should be regarded as an index variable for all pollutants with the same daily variability as SO_2. For example, SO_2Z was more highly correlated with particulates and CO than is the untransformed SO_2 variable.

Table 6-1 summarizes Buechley's results for mortality in the New York metropolitan area, and shows that SO_2Z is always more significant than the untransformed variable (however, the mortality effect cannot be computed since the mean Z-score is zero), but this substitution reduces the effects assigned to the other pollutants. The coefficients for SO_2 behave as expected, decreasing with time as the mean SO_2 value dropped. The particulate coefficient is almost constant for all the regressions. The CO data are significant only for the entire period, and since this includes a large block of missing data,

TABLE 6-1 Regression Results for Mortality in the New York Metropolitan Area

	Coefficients (t values)				Mortality Effect (%)			
Period	SO_2	SO_2Z	Particulates	CO	SO	Particulates	CO	Total
1962–66	3.5 (5.4)	—	—	—	0.95	—	—	0.95
	2.7 (4.1)	1.05 (6.3)	—	—	—	—	—	—
	—	—	6.2 (2.8)	—	0.73	1.33	—	2.05
		0.85 (5)	5.5 (2.5)	—	—	—	—	—
1967–72	2.9 (2.1)	—	8.2 (3.3)	0.2 (1.9)	0.16	1.75	0.66	2.57
	—	0.85 (5)	6.4 (2.7)	—	—	—	—	—
	0.6 (0.4)	—	9.0 (3.7)	0.2 (1.6)	0.03	1.93	0.57	2.53
	—	0.53 (3)	7.5 (3.1)	—	—	—	—	—
1962–72	1.7 (2.9)	—	5.7 (3.8)	0.3 (3.9)	0.26	1.23	1.14	2.63
	—	0.53 (4.4)	4.6 (2.9)	0.3 (3)	—	—	—	—

Source: Buechley (1975).

these results should be interpreted cautiously. The total effect of air pollution on mortality in the metropolitan area is estimated to be about 2.6%. If valid CO data had been available for the earlier period, the total effect then may have been around 3%. One should keep in mind that the Buechley studies included mortality for the entire New York metropolitan area, but pollution data came from only one station in Manhattan. The resulting errors in exposure (which would likely have been substantial) would have been expected to bias the regression coefficients downward.

Studies of New York City Mortality by Schimmel and Colleagues

Schimmel and Greenburg (1972) studied the period 1963–68, using SO_2 and smokeshade as predictors and a correction procedure to account for trend and weather effects. Regressions were performed for several different causes of death and for lags up to 7 days. A subset of the city immediately surrounding the monitoring station, comprising about 17% of the population, was also analyzed. Schimmel *et al.* (1974) and Schimmel and Murawski (1976) extended these basic methods to 1972. Schimmel (1978) then extended the work to 1976, using slightly different methods, which were intended to supersede those in the previous papers.

The basis for Schimmel and Greenburg's technique was to "correct" the pollution variables for the effects of interfering (collinear) variables, leaving the mortality variable as measured. While Buechley also stated that the dependent variable should remain "uncorrected," he (and Hodgson) used several additional variables in their regressions to deal with various cyclical factors. Schimmel and Greenburg stated that "correcting" the pollution variables for temperature gave almost the same result as adding a temperature variable to the regression. They presented results for uncorrected variables and two methods of correction, which varied by about a factor of 2 to 3 in terms of the number of excess deaths attributed to air pollution. They recommended the intermediate results as the best estimates, which were (coefficients expressed in percent excess mortality per $\mu g/m^3$):

For total mortality, citywide:

$$\text{\% excess deaths} = 0.029 \text{ smoke} + 0.0033 \text{ SO}_2 \text{ (same day).} \qquad [6\text{-}10]$$

$$\text{\% excess deaths} = 0.046 \text{ smoke} + 0.0047 \text{ SO}_2 \text{ (accumulated over 7 days).} \qquad [6\text{-}11]$$

For the smaller "special district":

$$\text{\% excess deaths} = 0.026 \text{ smoke} + 0.0066 \text{ SO}_2 \text{ (same day).} \qquad [6\text{-}12]$$

$$\text{\% excess deaths} = 0.049 \text{ smoke} + 0.0112 \text{ SO}_2 \text{ (accumulated over 7 days).} \qquad [6\text{-}13]$$

The regression coefficients for smoke appeared to be more sensitive to the various correction methods than those for SO_2. Excess deaths due to heart disease and total mortality were both split, 20% associated with SO_2 and 80% with smoke, while respiratory deaths were only associated with smoke. Thus, although no special attention was given to influenza epidemics (1963), there is no evidence of interaction between respiratory deaths and SO_2. For 1963–68 (mean SO_2 = 450 $\mu g/m^3$, mean smoke = 2.1 COH), the total pollution effects on mortality were estimated to be:

	Citywide	*Special District*
Same day	7.6%	8.5%
Accumulated over 7 days	11.6%	15.3%

The values for the whole city are similar to those obtained by Hodgson (1970) but much higher than Buechley's (1975). The higher results for the special district may reflect errors in measurement of the "true" population exposure for the whole city when a single monitoring station is used (as mentioned in connection with Buechley's results). The lag effects are substantial.

Schimmel *et al.* (1974) presented results, using both uncorrected and "temperature corrected" SO_2 and smokeshade for 1963–68, 1969–72, 1967–69, 1970–72, and 1963–72. Seasonal regressions are also presented, as well as a table of the total pollution effect by year. During this time span (1963–72), SO_2 dropped by about a factor of seven, while smoke remained almost constant. The percentage of excess deaths attributed to air pollution for each year varied from 5.7 to 12.1, independently of the changes in SO_2 concentration from year to year. The year-to-year variations in the regression coefficients appeared to exceed the year-to-year variations in mean pollution levels. By season (bimonthly period), the total mortality effects were similar in all periods except July–August, when they were negligible. The results for SO_2 suggested behavior as a surrogate, in that as the mean SO_2 value dropped, the regression coefficients increased, giving the same or even an increased effect on mortality. This was the case only after 1970; the SO_2 contribution for 1967–69 was markedly lower than for 1963–68. The results for 1963–68 may have been influenced by the 1963 flu epidemic.

The next publication in this series, by Schimmel and Murawski (1976), worked with the same data base but used deviations from seasonal trend lines based on 15-day moving averages for temperature, SO_2, and smokeshade. Results were presented for five different mortality measures (by cause). Corrections for day of week and holidays were neglected, but a special effort was made to account for heat wave mortality. The effect of daily temperature on mortality was positive for all causes of death and all seasons; it had been negative in London. The absolute values of percentages of mortality assigned to pollution decreased relative to those in previous publications, but the relative features stayed the same: smoke was more important than SO_2 and there was no decrease in the mortality effect corresponding to the large reduction in ambient SO_2 that had been accomplished.

In the final paper, Schimmel (1978) essentially disavowed the early findings that recommended use of the intermediate levels of excess mortality attributed to air pollution, in favor of the lowest estimates. He gave a useful table (his Table 1) explaining the differences in the various regression methods and their results. He also extended the data base to 1976, and results were presented for various lags, by cause of death, sex, race, and age group (in appendices). The following relationships summarize these findings, based on the percentage of (total) same-day mortality attributed to either SO_2 or smokeshade. (The figures in parentheses are the estimated standard errors of the coefficients.)

1. 1963–76, deviation mortality versus deviation temperature and deviation pollution (total effect = 1.59%):

$$\% \text{ excess mortality} = \underset{(0.001)}{0.00074\, SO_2} + \underset{(0.0021)}{0.0071 \text{ smoke}}. \quad [6\text{-}14]$$

2. 1963–76, deviation mortality versus deviation temperature and deviation pollution, all deviations corrected for temperature (total effect = 0.59%):

$$\% \text{ excess mortality} = \underset{(0.001)}{-0.0013\, SO_2} + \underset{(0.0021)}{0.0048 \text{ smoke}}. \quad [6\text{-}15]$$

3. 1963–69, deviation mortality versus deviation temperature and deviation pollution, all deviations corrected for temperature (total effect = 1.00%):

$$\% \text{ excess mortality} = \underset{(0.001)}{-0.00054\, SO_2} + \underset{(0.0028)}{0.0057 \text{ smoke}}. \quad [6\text{-}16]$$

4. 1970–76, deviation mortality versus deviation temperature and deviation pollution, all deviations corrected for temperature (total effect = 0.26%):

$$\% \text{ excess mortality} = \underset{(0.0023)}{-0.0039\, SO_2} + \underset{(0.003)}{0.0039 \text{ smoke}}. \quad [6\text{-}17]$$

These relationships are based on joint regressions of SO_2 and smokeshade; when separate regressions were performed, SO_2 remained negative but the smokeshade coefficients decreased slightly. The smoke coefficient for 1970–76 was not significant, which suggests a threshold somewhere between 1.77 and 2.15 COHs (mean values for the two periods).

The lag analysis for total mortality showed no changes in significance for SO_2, but the effect of smoke increased with lag time up to a lag of 7 days (the cumulative effect of pollution for the week before death exceeded that for the day of death alone). The effect was in the range 3 to 4% excess mortality for all three periods. For longer lags, there were negative effects, part of which could be compensation for the "harvesting" of the terminally ill. When the sum of pertinent causes of death was used instead of total mortality, the smoke effects increased by 50 to 100%, for lags from zero to about 4 days. For longer lags, the mortality effects for specific causes decreased relative to total mortality, which places further doubt on the reality of the increased effect for lags beyond about 4 days.

Taken in its entirety, Schimmel *et al.*'s analysis of New York City mortality is a mathematical tour de force. However, it is difficult to accept Schimmel's final (1978) recommendations on statistical methods, for several reasons:

1. Measures of the overall goodness of fit were often not given.
2. Statistical comparisons of the various models were not made.
3. Criteria for "satisfactory" or "superior" results were never stated.
4. Treating SO_2 as an index variable by removing the long-term trend confounds its effects with those of other pollutants.
5. Larger effects on mortality were found in the "special district," but this methodology was dropped in the later analyses.
6. There was no consideration of the quality of the air pollution measures and large outliers were truncated, regardless of possible causes.
7. The effect of temperature perturbations on mortality was positive, whereas it had been mostly negative in London. Since breathing cold air is known to have adverse respiratory effects, it is easier to accept the British finding. The air pollution stagnations in New York (and Donora) were accompanied by warm temperatures, which may be a manifestation of the urban heat island effect and the reductions in convective heat removal that accompany stagnations. Thus, one could argue that correcting mortality for daily temperature *in*creases is inappropriate (apart from heat waves); allowing the variables to compete with one another in a multiple regression seems preferable to the *a priori* "correction" procedure used by Schimmel, in the absence of a physiological hypothesis for an independent year-round effect of temperature.

Other Studies of New York City Mortality

Özkaynak *et al.* (1990), Özkaynak and Spengler (1985), and Özkaynak *et al.* (1986) reanalyzed Schimmel's 1963–76 data set and added data on atmospheric visibility, obtained from the three airports in the metropolitan area. Visibility can be a surrogate for the presence of fine particles in the atmosphere, different from the particles that create the dark stain used in the smokeshade measurement. They used a simple sinusoidal model to correct the data for seasonal trend and restricted their analysis to days for which the visibility records were similar for the three airports. They did not account for days of the week, holidays, or flu epidemics. This regression assigned 1.5% excess deaths to smokeshade, 1.3% to SO_2, and 1.0% to visibility (all statistically significant). They made several other conclusions:

1. Excess mortality attributed to air pollution varied from 2 to 4%, depending on the methods used for preprocessing the data.
2. Based on analysis of quarterly data, they estimated 2% excess deaths associated with SO_2 and smoke for 1963–69, and 1.4% for 1970–74.
3. The association between mortality and airport visibility was sensitive to the selection of the airport.

Bloomfield *et al.* (1980) studied the effects on time-series mortality regressions of using multiple air quality stations and of disaggregating the metropolitan area mortality counts according to monitoring stations. In the first instance, for COH in New York City from 1971 to 1975 (a period of relatively low pollution), they found that the average of two monitoring stations performed better (gave increased statistical significance) than the average of seven and about the same as the best of the seven stations. Thus, there was no great advantage to either averaging the stations or considering them individually. The effects of SO_2 were negative and nonsignificant; for smokeshade, the effect was in the range 1 to 2% excess mortality.

In the second part of Bloomfield *et al.*'s study, three areas of Pittsburgh were considered individually and jointly (see below for additional discussion of Pittsburgh mortality). Statistical significance was only found for one area, which provided useful information. This study appears to have confirmed that attempts at disaggregation will inevitably suffer from the data "noise" introduced by considering smaller populations. Bloomfield *et al.* defined "representative" air monitoring stations on the basis of their statistical relationships to daily mortality, rather than on some *a priori* criterion with respect to population locations, a point that detracts substantially from the validity of their analysis. On this basis, they concluded that the 121st Street Laboratory station, which was used for most of the New York City studies, was not "representative" for the 1971–75 period. If true, this would imply underestimates of the true mortality effect for those studies which used these pollution data (which includes most of them).

Summary and Conclusions from the New York Studies

While most of the studies reviewed here suffered from some defects in data or analysis, it seems safe to conclude that they confirm the finding from the London studies that smoke is a more important indicator for acute mortality than SO_2. This is an important result, considering the different chemical nature of smoke in New York as compared to London, and the much lower concentrations in New York. Table 6-2 compares selected findings from these studies, based on same-day mortality and the inclusion of no other pollutants in the calculation of "total effect." The regression results fall into three groups: group

TABLE 6-2 Results of Time-Series Studies of New York City Mortality (same-day effects)

Author	Period	Regression Coefficients (% mortality per μg/m³) Smokeshade	SO_2	Total Effect (%)
A. Regressions for the entire city				
Hodgson (1970)	1962–65	0.032–0.055	0.0022–0.0032	8.2
Glasser & Greenburg (1967)	1960–64	—	0.005	2.8
Buechley *et al.* (1973)*	1962–66	—	0.0045	1.2–2.4
Buechley (1975)*	1962–66	0.0062	0.0027	2.1
	1967–72	0.009	0.0006	2.0
Schimmel & Greenburg (1972)	1963–68	0.029	0.0033	7.6
Schimmel (1978)	1963–69	0.0057	−0.0005	1.0
	1970–76	0.0039	−0.0039	0.3
Özkaynak & Spengler (1985)	1963–76	0.0077	0.0048	2.8
B. Comparisons by geographic area (1962–68)				
Buechley (1975)	Tri-state area	0.0062	0.0027	2.1
Schimmel & Greenburg (1972)	NY City	0.029	0.0033	7.6
Schimmel & Greenburg (1972)	Special district	0.026	0.0066	8.5

*Tri-state metropolitan area.

1, Hodgson and Schimmel and Greenburg (1972), found large coefficients for smoke and hence large total effects. Neither of these studies used deviations from 15-day averages nor corrected for days of the week, flu epidemics, or holidays. Schimmel and Murawaski (1976) accounted for these factors in some of their results, and the smoke coefficients were greatly reduced. Group 3 is the latest work by Schimmel (1978) with lower coefficients for smoke, negative values for SO_2, and a total net effect of air pollution on mortality of 1% or less. All the other studies constitute group 2, with total effects on mortality of 2 to 3%. Note that including lags would likely increase these values to around 4 to 5%.

The fact that the total mortality effect did not change much as SO_2 was reduced by a factor of 5 tends to support the conclusion that SO_2 is less important than smoke, although we have no information on the actual citywide change in SO_2 exposure during this period.

Pickles (1980) analyzed the effect of errors in estimating pollution exposure. Based on Goldstein's correlation analysis of the 40-station network (1976), he concluded that the use of a single station in New York could bias the results low by about a factor of 2, relative to seven stations used in London. While section B of Table 6-2 does not suggest differences that large, it does imply that larger effects will be found for smaller geographic areas, presumably because of better characterization of exposure. This finding also reinforces the causal nature of the air pollution results, since surrogate effects would not be expected to show this trend.

Other aspects of the New York studies suggested that other pollutants, such as CO and fine particles, may also affect mortality and that lag effects tend to increase the values of the coefficients and their effects. Most studies showed a lag effect of only a few days.

MORTALITY STUDIES IN CALIFORNIA

Los Angeles

The pollutant species of interest in Southern California has traditionally been ozone or total oxidants. Mills (1960) performed a time-series analysis of mortality due to cardiac or respiratory causes in Los Angeles County for 1956–58. Daily maximum oxidant readings from up to 20 stations were used to provide a countywide average. Daily maximum temperature was used as a partial control: days with maximum readings over 96°F were omitted from the trend analysis, which was based on deviations from monthly averages, on two bases: all available days (881 days) grouped according to average maximum oxidant reading, and the same groupings for the months of January through May, only. Both methods provided statistically significant slopes through the grouped data corresponding to about 0.23% per pphm (0.012% per $\mu g/m^3$) (Figure 6-13). There is no suggestion of a threshold effect, even at levels corresponding to the current ambient air quality standard (12 pphm), but there may be a slight suggestion of saturation above about 32 pphm. Using the mean of Mills's tabulated oxidant data, the average mortality effect in Los Angeles County was about 3.2% year round and 2.8% in the winter–spring

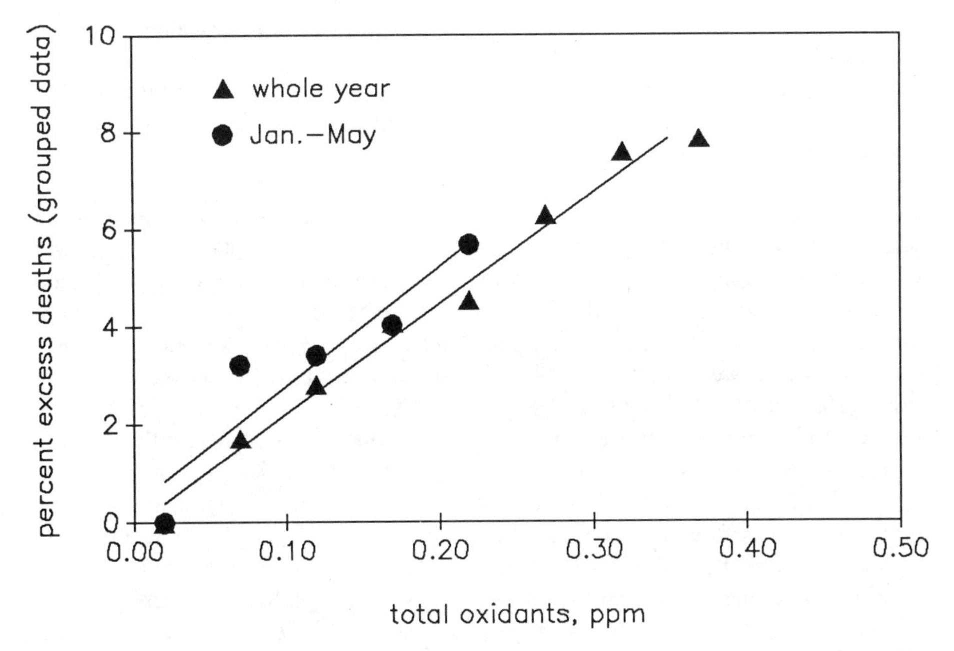

FIGURE 6-13. Regression results for Los Angeles mortality, 1956–58, versus oxidants. *Source: Mills (1960).*

period. These percentages would probably be somewhat lower if expressed on a total mortality basis. Hexter and Goldsmith (1971) analyzed the same data set by adjusting all the variables for seasonality and considering the effects of lags. Although the cross-correlation coefficients were positive, they failed to reach significance on the basis of individual days, and they concluded that the overall effect of pollution on daily mortality was not significant. This may not have been the case for summed lags.

The California Department of Public Health had previously reported (1957) that no correlation could be found between daily nursing home deaths and air quality, but this study probably suffered from lack of power. The population at risk was small (3734 beds) and the period of record, although not specifically stated, appeared to be only 2 to 3 years.

Since Mills tabulated the entire data set, it is possible to get a feeling for the interrelationships by inspecting the data. Heat-wave mortality was apparent, and days with repeated high maximum oxidant readings appeared to be especially lethal. However, there were also high-mortality days that did not coincide with either temperature or oxidant peaks. Joint regression of the monthly average deaths on monthly average maximum oxidants and temperature yielded a nonsignificant result for oxidants (in contrast to Hodgson's [1970] findings for SO_2 in New York), suggesting that the oxidant episodes had minimal lasting effects. Occasionally, there were dips in mortality following the oxidant or temperature peaks, suggesting a "harvesting" effect.

Hexter and Goldsmith (1971) studied total mortality and mortality from heart disease in Los Angeles County for 1962–65 with respect to "basin averages" for CO and oxidants (number of stations not given), in conjunction with cyclic trend and temperature variables. Although the logarithm of CO was statistically significant, the degree of fit for this model was only marginally better than for a model with all other terms except log CO. The total mortality effect (elasticity) was 13.5%, substantially higher than what Wyzga (1977) found in Philadelphia from a single monitoring station with about the same mean CO value, and also substantially higher than Mills's result for oxidants in Los Angeles. If Hexter and Goldsmith's CO coefficient were converted to linear form (as an approximation), the result would be 0.0004% per $\mu g/m^3$, which is not greatly different from Wyzga's value for Philadelphia. It is difficult to compare their results with Buechley's (1975) result for CO from joint regressions in New York because of the uncertainty regarding Buechley's units of measure. However, the mortality effect attributed to CO in New York is small (1%), probably in part because of the use of a single monitoring station. Hexter and Goldsmith did not report their results for oxidants, only that the regression errors were larger when this pollutant was used.

Los Angeles County was analyzed by Shumway *et al.* (1988), using the type of spectral analysis techniques that had been used in London. Total, CV, and respiratory mortality from 1970–79 were analyzed, considering temperature and humidity data averaged over five locations and six air pollutants, averaged over six monitoring stations. The six pollutants were CO, SO_2, NO_2, hydrocarbons (HC), O_3, and a measure of particulates (KM) believed to be similar to COH. The regressions were based on the spatial averages of the daily maxima at each station. Missing data were filled in by interpolation. Spectral analysis indicated that 1 year was the most important frequency (the data were not deseasonalized) and that periods longer than 1 week seemed to be most important. The data were then filtered and smoothed to reduce the original 3652 observations to 508 weekly averages, for all variables. Note that this process will reduce the noise inherent in using daily maximum air quality measures but precludes looking at daily lag effects.

Linear correlations showed that the relationships between mortality and air pollution were almost as strong as those between mortality and temperature; respiratory mortality was slightly weaker. Ozone was nearly collinear with temperature, and most of the other pollutants were highly intercorrelated. The best linear regression model (ordinary least squares—not accounting for serial correlation) was:

$$\text{Total daily deaths} = 1.986\,\text{CO (ppm)} - 0.486\,\text{temp} \quad (R^2 = 0.40) \text{ (standard errors were not given).} \qquad [6\text{-}18]$$

Based on an estimated mean (daily max) CO value of 7.5 ppm and 172 daily deaths, the corresponding elasticity would be about 0.087. Note the large negative effect of temperature, which represents the seasonal component.

Shumway *et al.* (1988) also investigated nonlinear dose-response relationships and used autoregressive modeling to account for the serial correlation among residuals. Smoothed-surface plotting showed that the temperature-mortality relationship was U-shaped but that the pollution-mortality relationships were monotonic, with "some curvature" (apparently concave downward). These findings suggested use of logarithmic transforms for the pollution variables (one at a time) and a quadratic relationship for temperature. Accounting for the serial correlation reduced all of the coefficients by 40 to 50%; the best estimates of the elasticities on total mortality (at the mean) were:

CO: 0.062 HC: 0.064 Particulates: 0.052.

Shumway *et al.* (1988) reported that relative humidity, SO_2, NO_2, and O_3 were "nonsignificant contributors to mortality"; given the high correlation between ozone and temperature (0.85) and the use of daily maximum measures, it is difficult to accept that this has indeed been established for ozone. Although the original contour plots suggested some interaction between pollution and temperature (steeper slopes at lower temperatures), Shumway *et al.* did not investigate this possibility.

Özkaynak *et al.* (1990) and Kinney and Özkaynak (1991) described a regression analysis of 1970–79 total mortality in Los Angeles County, which included lagged O_3, COH, SO_2, NO_2, CO, visibility, and meteorological variables. Filtering was used with all variables to remove the long-term components of variability. Air quality measurements were obtained from up to eight monitoring stations (six for CO). The significant variables included O_3, NO_2, CO, particulates and temperature (+), although NO_2, CO, and particulates were too highly intercorrelated to be separated from one another. Each was evaluated along with ozone and temperature. The overall elasticity was about 0.04, with slightly less than half of the effect attributed to ozone (see Figure 6-14a, below). The slope of the ozone variable was very close to that obtained by Mills: 0.18% per pphm, or 0.009% excess deaths per $\mu g/m^3$ ($p < 0.0001$). The NO_2 coefficient was slightly larger: 0.018% excess deaths per $\mu g/m^3$. The authors reported that cardiovascular deaths behaved similarly to total mortality but that temperature was the only significant variable for respiratory deaths, perhaps because they were few in number (about 8% of the total). The difficulty of this task may be appreciated from the fact that these three variables explained only about 4% of the mortality variation over a period of 10 years. Kinney and Özkaynak also investigated day-of-week effects, serial correlation, and various types of filters to remove seasonality and concluded that their results were robust. Neither SO_2 nor visibility (associated with sulfates) were significant.

An idea of the variability of individual years in a given location may be obtained from Figure 6-14, for Los Angeles County in this case (Kinney and Özkaynak, 1991). Figure 6-14a shows how the apportionment of the relative shares due to O_3 and NO_2 varies from year to year. In any single year, the coefficients range over an order of magnitude; however, the means for the 10 years and their standard deviations are quite close to the values obtained from the overall regression. Figure 6-14b is intended to place the pollution results in context with the effects of temperature. The additional deaths associated with increasing the daily temperature from its annual mean to 90°F are contrasted with the deaths associated with the mean values of the pollutants; the temperature effects are much larger. Wyzga (1978) found a similar relationship for Philadelphia summers. Figure 6-14b also suggests that the effects of temperature have been decreasing over the years, perhaps because of increased use of air conditioning. However, the pollution effects show no such trend, in spite of declines in mean concentrations of SO_2 and CO. The overall conclusions from

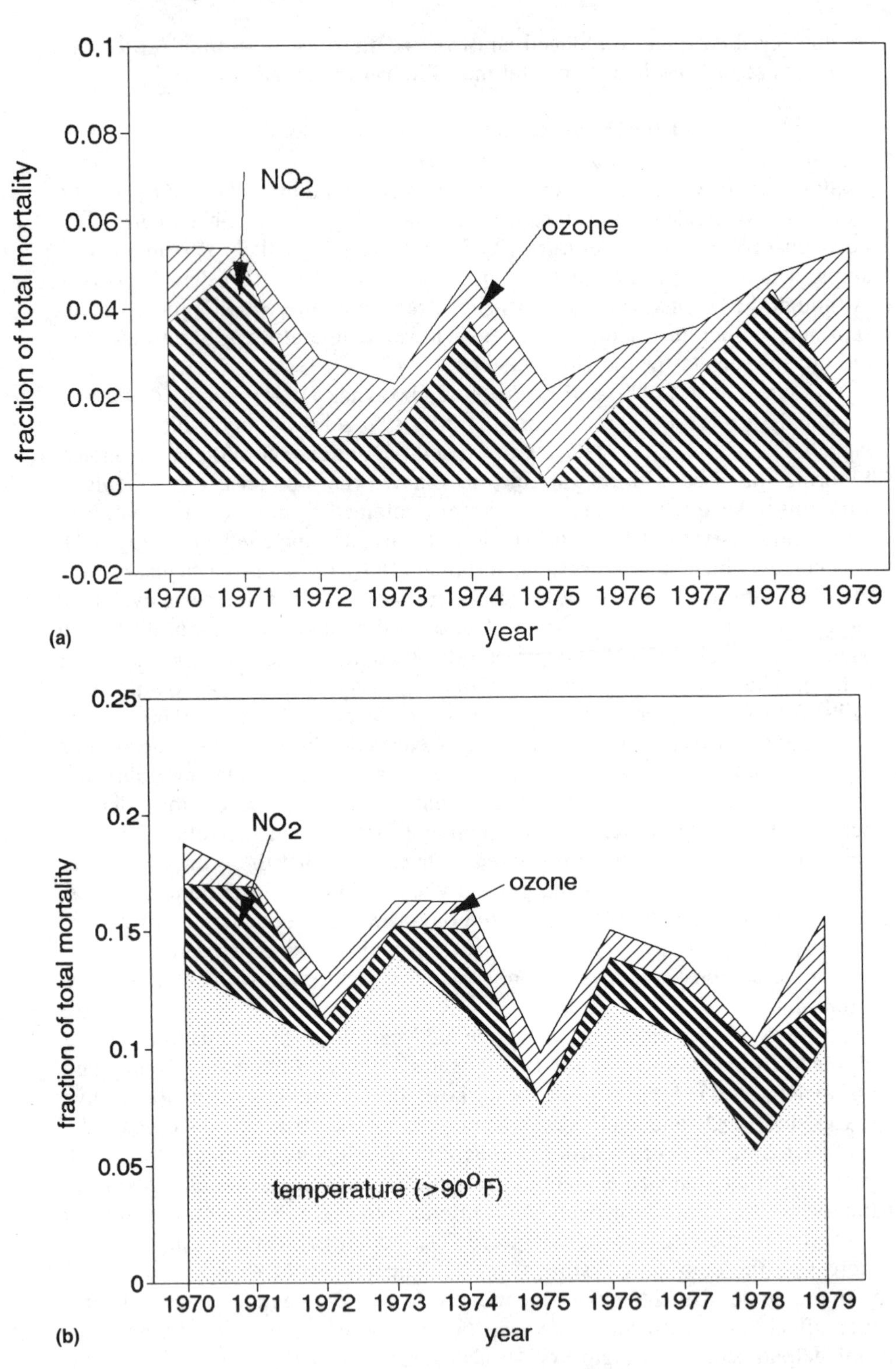

FIGURE 6-14. Regression results for Los Angeles County mortality, 1970–79. (a) Comparison of mean contributions from O_3 and NO_2. (b) Comparison of pollution contributions with the effect of a 90°F day, relative to mean temperature. *Source: Kinney and Özkaynak (1991).*

Figure 6-14 are that daily mortality analyses must be conducted over several years to obtain stable results, even for large cities, and that there are other sources of variation to consider besides air pollution.

Santa Clara County (San Jose)

One of the most interesting recent time-series studies is that of Fairley (1990), since it was performed in an area not generally known for high levels of air pollution. Coefficient of haze measured in downtown San Jose was used to represent the entire county (3300 km^2). (The unit of analysis used by Fairley was the daily *sum* of twelve 2-hour readings, multiplied by 10.) The other regression variables were temperature and relative humidity. The maximum COH value over the period of study (November–January of 1980–86) was about 3.0 (as measured), which corresponds roughly to about 300 $\mu g/m^3$ as TSP (220 $\mu g/m^3$ as PM_{10}). There are no major industrial pollution sources involved, but the use of wood stoves could be a factor. Some of the monitor readings were reported to have been influenced by nearby construction. In a personal communication, Fairley reported that no effects could be found when CO concentrations were used instead of COHs. The dependent variable was daily deaths from nonaccidental causes, which averaged about 19 per day. Poisson models were used in addition to ordinary least squares and gave about the same results.

Fairley tried a number of different regression modeling approaches, beginning with a bivariate regression and accounting successively for temporal trend, weather, seasonal trend, deviations from a 15-day moving average and lags. Adjusted R^2 values ranged from 0.003 to 0.13; regression coefficients varied from 1.2 to 6% per COH unit, on a same-day basis. The lagged values were important; the sum of lags 0, 1, and 2 was 42% higher than the same-day value alone (elasticity = 0.077). Since the highest single coefficient value was for lag 2, it is possible that the true cumulative effect could have been larger. It is also possible or even likely that COH is acting in part as a surrogate for the entire urban-suburban air pollution mix, but, because of the absence of sulfur sources, acidic aerosol particles are *not* a likely constituent (contrary to Fairley's suggestion).

Fairley also investigated deaths by age and by cause. The highest significance was achieved for respiratory deaths, followed by circulatory causes. Accidents, cancers, and "other" deaths were not significant. The regressions by age suggested that persons 70 and over were more at risk.

TIME-SERIES MORTALITY STUDIES IN PHILADELPHIA

The 1957–1966 Period

Wyzga (1977, 1978) studied a 10-year record of data from the city of Philadelphia, classified by 6-month summer and winter periods for 1957–60, 1961–63, and 1964–66. He felt that separate analysis by major season was required to properly handle the weather corrections. For the two earlier periods, only particulate measurements were available (TSP and COHs). Data were available from either two or three sampling stations, depending on the period, which were averaged for use in the mortality regressions. Missing data from individual stations were estimated and added to the record. For the last period (1964–66), data were also available from one station for SO_2, CO, NO_2, NO, hydrocarbons, and oxidants. Wyzga's model accounted for serial correlation, seasonality, heat waves, epidemics, and daily variations in temperature. Day-of-week effects were not mentioned. Lags were investigated for a limited set of data, which showed that only the day preceding death made a significant additional contribution to the same-day regressions. However, use of a geometrically decreasing lag parameter added from 38 to 68% to the same-day regression coefficients.

Wyzga found that COH was more significant than TSP, and thus used COH for studies of the six different periods, as shown on Table 6-3 (coefficients in percent excess mortality per $\mu g/m^3$ for same-day regressions).

When lags were added, the winter 1964–66 coefficient increased by about 50%. These data suggest that a threshold of no effect may exist around a mean COH value of about 1.0 (ca. 100 $\mu g/m^3$), and that air pollution in summer may be just as lethal as in winter (at the same concentration level).

Table 6-4 shows results obtained for the winters of 1964–66 for other pollutants (means in $\mu g/m^3$, coefficients in percent excess mortality per $\mu g/m^3$).

The last column in the table gives the correlation between each species and

TABLE 6-3 Mortality Regression Results from Philadelphia

Period	Mean COH	No. of Sites	Coefficient*	t	Effect (%)
Winters					
1957–60	1.89	2	0.0146	2.00	2.8
1961–63	1.61	3	0.0188	2.02	3.0
1964–66	1.31	2	0.0350	2.85	4.6
Summers					
1957–60	1.22	2	0.0488	3.53	6.0
1961–63	0.92	3	0.0338	1.82	3.1
1964–66	0.87	2	0.0086	0.41	0.8

Source: Wyzga (1978).

* % excess deaths per $\mu g/m^3$.

TABLE 6-4 Effects of Different Pollutants on Philadelphia Mortality

Species	Mean*	Coefficient†	t	Effect (%)	r_{COH}
COH	1.31	0.035	2.85	4.6	1.00
TSP	162	0.017	1.85	2.8	0.80
NO	78	0.028	3.11	2.2	0.81
NO_2	64	0.020	1.28	1.3	0.60
SO_2	251	0.0035	0.94	0.9	0.67
CO	8640	0.00024	1.35	2.1	0.33
HO	1483	0.0031	2.17	4.6	0.69
Oxidants	40	0.016	0.53	0.6	0.40

* Mean values in $\mu g/m^3$ except for COH.

† % excess deaths/$\mu g/m^3$.

COH for 1964–66, which may be used to speculate whether the effects shown for each species relate to the health effects of that species or just to a statistical artifact arising because of correlated measurements (not to be confused with correlated *exposures*). A cross-plot of the regression results against this correlation coefficient suggested that CO, HC, and possibly NO may have had effects over and above those gained by association with COH. Of the pollutants above, only COH, NO_2, and possibly SO_2 may be assumed to be widely distributed throughout a metropolitan area. CO, NO, and hydrocarbons (HC) are primary pollutants from traffic and would not be expected to persist in nontraffic areas. Oxidant levels (primarily O_3) tend to be depressed in traffic areas because of chemical reactions with NO, and thus levels may be much higher in the suburbs. Also, one would not expect oxidants to be important in winter. The result for NO is more difficult to interpret, since there are other sources of nitric oxide emissions besides traffic (all stationary combustion sources) and this species was more statistically significant than COH. NO could have been acting as a surrogate for CO since the mean effects were about the same.

The 1973–1980 Period

A later period of data on daily mortality and air pollution was studied by Schwartz and Dockery (1992a). (Detailed discussion of similar studies follow.) Air pollutants were TSP and SO_2 averaged over all the available monitors; mortality data were for the city of Philadelphia from 1973 to 1980, for non-accidental deaths and for several groups of specific causes. The approach was to fit the mortality data against weather and season first, and then to regress the residuals against air pollution. Poisson models were used in addition to an autoregressive procedure to account for serial correlation. The significant weather variables were hot days (mean > 80°F), previous day's temperature, dew point, and winter temperature.

With TSP, both lags 0 and 1 were significant with about the same coefficient (0.066% excess deaths per $\mu g/m^3$, on the basis of 2-day average pollution). Sulfer dioxide gave similar but slightly less significant results (0.050% excess deaths per $\mu g/m^3$). When both variables were entered into the regression, only TSP remained significant. The total air pollution effect on mortality was about 5%. Regressions on mortality by age showed the over-65 group to be more susceptible to the effects of air pollution. By cause, the largest percentage effect was for chronic obstructive lung disease (e = 0.14). Pneumonia and CV deaths also had stronger relationships (e = 0.079 and 0.071) than total (non-accidental) mortality (e = 0.051), while cancer was slightly weaker (e = 0.028). Accounting for day-of-week effects increased the elasticity for nonaccidental mortality to 0.056.

The later Philadelphia study shows some similarities and some differences with Wyzga's (1978) findings; the earlier study involved air pollution levels more than twice as high. The similarities include the lag structure and the minor role for SO_2 (even when taken as the sole pollutant at much higher concentration levels). The major difference is in the relative magnitudes of the coefficients for TSP, which are much larger in the later study, so that the reduction in average TSP that occurred between the two periods did not reduce the numbers of deaths ascribed to TSP (it *in*creased slightly). Possible explanations include the use of only one monitoring station by Wyzga and the added effect of lags. The average of the elasticities shown in Table 6-3 checks well with the later results if multiplied by 1.5 to account for lag effects. However, Wyzga's average regression coefficient is still substantially lower (0.0004). The only major differences in the way the two studies account for confounding variables are in Wyzga's use of variables for short-term temperature fluctuations and to account for the presence of epidemics, and his separate analyses of summers and winters. However, other studies (London and New York) have also shown increasing coefficients with decreasing mean values of pollution, and it would be of interest to see both Philadelphia data sets analyzed according to a common protocol.

TIME-SERIES MORTALITY STUDIES IN OTHER LOCATIONS

Boston

An extensive tabulation of data concerning the effects of the 1966 Thanksgiving episode in the Boston area was presented by Heimann (1970). The mortality analysis was part of a larger attempt to document a variety of adverse health effects; emergency room visits are discussed in Chapter 9, for example. One of the advantages of this data set is the extensive air monitoring network that was in place, providing daily values for COH, SO_2, and TSP (with some missing days) from 10 to 20 stations. The maximum 24-hour values reached during this episode (and a similar one the preceding October) were 2.2, 0.124 ppm, and 226 $\mu g/m^3$, respectively, averaged over the network. Heimann used 5-day averages to check for excess mortality during the two episodes and concluded that none was apparent, in part because of the high day-to-day variability. However, a scan of the tabulated data showed that a large part of the variability was due to day-of-week effects and that some degree of lag was apparent. There was also an overall downward trend during this 41-day period. I averaged all of the data by day-of-week and calculated daily deviations from the averages. I found that 2-day averages with a 2-day lag seemed to fit the best. After accounting for the temporal trend, the total mortality data showed a weak relationship with SO_2 that did not quite reach statistical significance ($p = 0.07$). The slope was about 0.034% excess deaths per $\mu g/m^3$; the elasticity, about 0.05.

Pittsburgh

Mazumdar and Sussman (1983) continued the most recent methodology developed by the Schimmel group, for both mortality and morbidity in Allegheny County, PA (Pittsburgh metropolitan area), for 1972–77. They used deviations from 15-day moving averages, corrected for temperature by a two-stage regression procedure, with smoke and SO_2 measured at three different stations. Dummy variables were used to account for day-of-week effects. Data from each pollution measurement station were entered separately in the regression; the correlations between stations were all less than 0.6. In general, only the most polluted station showed statistical significance, for smoke in a joint regression with SO_2 (which was negative). None of the pollutants was significant in separate regressions. The smoke coefficient for total mortality was about 0.008% per $\mu g/m^3$, for the station with a mean COH of 1.27 (SO_2 = 0.046 ppm). When mortality for the local area around the monitoring station was regressed against the local smoke values (Bloomfield *et al.*, 1980), the smoke coefficient increased to about 0.012% per $\mu g/m^3$, which is consistent with many of the previous studies. The finding of nonsignificant SO_2 effects is not surprising, given the modest concentration levels present (mean values from 65 to 123 $\mu g/m^3$). Note that the period studied includes the 1975 "episode" in Pittsburgh, during which TSP reached 770 $\mu g/m^3$ and SO_2, 200 $\mu g/m^3$. Regressions for heart disease mortality provided higher coefficients, but they were no more statistically significant than for total mortality.

Chicago

Namekata *et al.* (1979) performed a two-phase study of air pollution in the city of Chicago from 1971 to 1975. The first phase was cross-sectional and studied differences among 76 neighborhoods (community areas) for several causes of death (age-adjusted), averaged over the 5 years; this portion of the work is reviewed in Chapter 7. The second phase examined daily variations averaged over the 76 areas for total nonaccidental deaths and heart disease. The pollutants of interest were TSP and SO_2. Socioeconomic variations were accounted for through the use of three index variables that were developed by an outside agency. These variables included income measures, housing status, and education; the racial composition of the population was not specifically accounted for, but the authors reported that it was highly correlated with these three socioeconomic factors and thus was accounted for indirectly. Smoking habits were not accounted for. This study affords an opportunity to compare the results of cross-sectional and time-series regressions.

The daily mortality study used no filtering or seasonal corrections but employed average temperature in the regressions along with dummy variables for the days of the week. Air pollution was not measured every day, so that the analysis was limited to those dates with at least 20 TSP or 18 SO_2 measurement sites. These numbers of sites represent a substantial improvement over the estimates of daily exposure made for most other time-series studies. As discussed above, a model of this sort is likely to predict more excess mortality for air pollution than one that employs filtering or seasonal adjustment, since a single temperature variable cannot account for both heat waves and the normal seasonal cycle in mortality. Either TSP or SO_2 (separate regressions) accounted for about 7% of Chicago daily mortality. When the product of $SO_2 \times$ TSP was entered (alone), the result was less significant and accounted for about 3% of mortality. Lags of 3 and 6 days gave lower results, as did regressions on heart disease deaths.

Osaka

Watanabe and Kaneko (1970) briefly reported the results of a 5-year (1962–67) correlation study in Osaka, Japan, for SO_2, suspended matter, and temperature. Although they presented no multiple regression results, they concluded that SO_2 had the greater effect, at cold temperatures. Sulfur dioxide levels occasionally exceeded 0.2 ppm and suspended matter, 1000 $\mu g/m^3$. No data were given on the magnitude of the effect nor on regression coefficients.

Dublin

Kevany *et al.* (1975) used a partial correlation analysis on mortality and hospital admissions data (by cause) in Dublin for the winters of 1970–73. An eight-station monitoring network for smoke and SO_2 was used; two other stations had been rejected because of the influences of local sources. Maximum daily temperature was used to control for weather effects; the study was limited to October–March. The statistical technique was that of partial correlations of selected variables, controlling for the others (regression coefficients were not given). Cardiovascular disease mortality was associated with SO_2 only on the same day; smoke was not so associated. The primary contributing cause of death was ischemic heart disease. To examine threshold effects, the data set was successively truncated below increasing pollutant concentration levels and the analyses repeated. The partial correlation coefficients increased with threshold level; for SO_2, in excess of about 75 $\mu g/m^3$. At longer lag levels (7 days), smoke levels above 100 $\mu g/m^3$ showed a strong partial correlation. The authors concluded that "SO_2 has immediate effects on cardiovascular mortality which are compensated for by reductions on the following days. However, smoke showed a sustained increase in mortality over a longer time span." The mortality results were supported by hospital admissions correlations for both smoke and SO_2. The correlations for respiratory mortality were significant only for SO_2 over 75 $\mu g/m^3$ for acute bronchitis (no lag) and chronic bronchitis (7-day lag). No respiratory causes were significant for smoke.

The Netherlands

The relationships between mortality and ozone and heat waves were explored in Rotterdam by Biersteker and Evendijk (1976) for the summers of 1974 and 1975. They stratified the data by temperature and ozone, using 100 $\mu g/m^3$ (0.05 ppm) as a threshold (1-hour maximum). They found significantly higher mortality on the high-ozone days in 1975 but not in 1974, and thus concluded that the 1975 effect was really due to a concurrent heat wave (maximum temperature reached 35°C). As supporting evidence, they noted that mortality among the elderly increased during this period in small towns as well as in large cities, implying that temperature was thus a more likely cause than air pollution.

Since we now know that ozone often peaks downwind of large cities, I thought it might be useful to reexamine the data of Biersteker and Evendijk, which were tabulated for both years in their report. My multiple regression found that same-day mortality for 1975 was due to temperature alone but that, as lag increased up to 3 days, the emphasis shifted to ozone. The correlation between temperature and ozone was about 0.7. Figure 6-15a presents the dose-response function, using a 2-day average and accounting for day-of-week effects. The elasticity of this relationship was about 0.19. For 1974, no such effect was apparent, but ozone levels were lower and did not occur in conjunc-

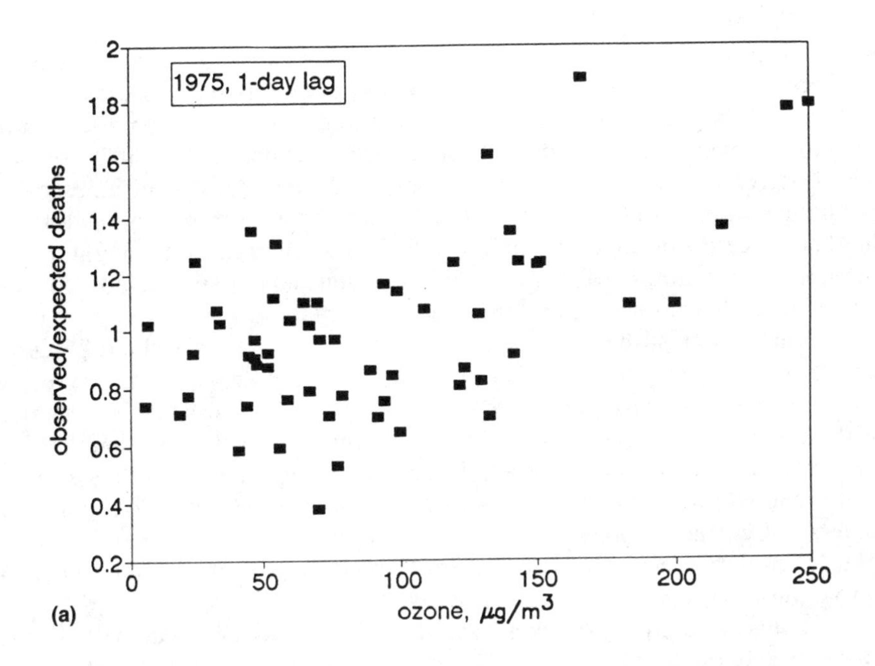

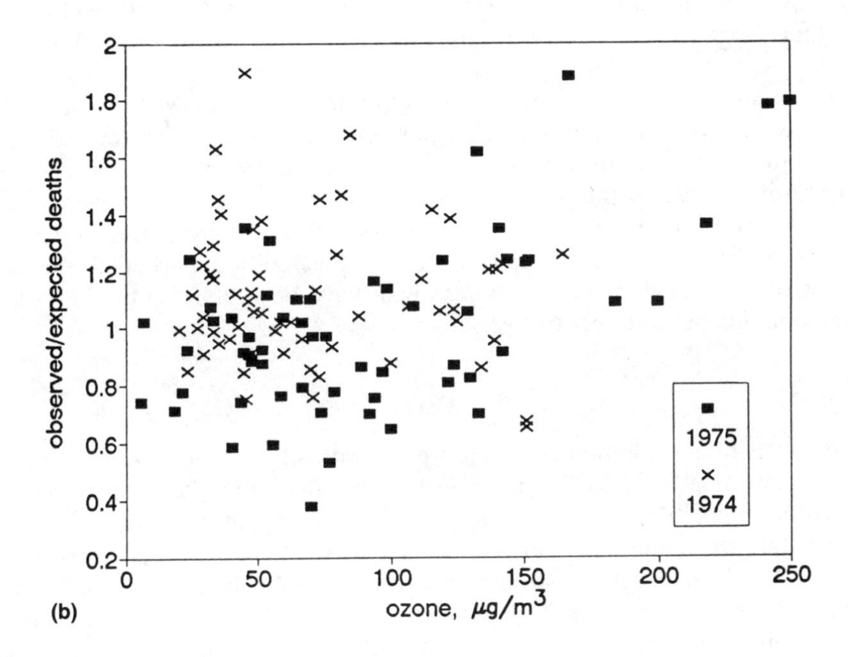

FIGURE 6-15. Rotterdam mortality versus ozone. (a) Summer 1975. (b) Summers of 1974 and 1975. *Source: Biersteker and Evendijk (1976).*

tion with a heat wave. In 1975, there were 4 straight days with maximum ozone over 200 $\mu g/m^3$, and maximum temperature over 33°C. When the 2 years were pooled (Figure 6-15b), the ozone relationship remained significant with a slightly lower elasticity (about 0.16). This example illustrates how additional knowledge of the behavior of air pollutants may shed new light on old data.

More recent time-series studies of Dutch mortality have focused on the role of SO_2 and its interactions with temperature variations, based on a six-station monitoring network intended to cover the whole country, from 1979 to 1987. The first report (Kunst *et al.*, 1991) concluded that SO_2 (as an index of air pollution) was associated with total, CV, and respiratory causes of death in log-linear regression models that controlled for temperature, temperature difference, precipitation, wind speed, influenza incidence, and included a non-linear dependence on temperature. The association with SO_2 was stronger in the warmer months; the overall average level of SO_2 was only about 26 $\mu g/m^3$, and it accounted for about 0.33% of all deaths on an annual basis (regression coefficient = 0.000126 per $\mu g/m^3$). This coefficient is reasonably consistent with other studies of SO_2 and daily mortality (in New York, for example). In a personal communication in 1992, Kunst *et al.* reported that they did not observe anything special associated with the January 1985 episode reported by others in Central Europe (see Chapter 5). No other pollutants were analyzed, but the authors speculated that summer "smog components" might be playing a role during heat waves.

In a follow-up study, Mackenbach *et al.* (1992) added a number of other weather and timing variables to the model and showed that the SO_2 effect became nonsignificant. The largest interaction was with a variable for cold weather, lagged 1 to 5 days. Information on the results for warm months with the new variables was not reported. There are several other interesting aspects to these studies. The plot of (crude) mortality ratio versus SO_2 showed a concave downward relationship (i.e., the SO_2 mortality "effect" leveled off at about 100 $\mu g/m^3$ and higher). However, they also reported that use of log transforms or lags for the SO_2 variable did not affect the results. Several other authors have found that a logarithmic transform improved the fit of SO_2 (Shumway *et al.*, 1983, for example); one thus wonders whether the outcomes of some of these other studies might have been affected by the lack of additional temperature control variables.

Paris

The effects of air pollution and meteorology on daily mortality were studied in Paris from 1969 to 1976 by Loewenstein *et al.* (1983). Pollutants included smoke and SO_2 by the "strong acid" method (hydrogen peroxide), measured at 30 stations in Paris and environs. The correlation between the two pollutants was 0.89, making separation practically impossible. Meteorological data included temperature, humidity, wind speed, atmospheric pressure, insolation, precipitation, and synoptic classification. The seasonal variations were as expected, with maximum mortality in December and minimum in August. Heat waves and epidemics were also noted. Seasonal variability was controlled in the statistical analysis by taking deviations from "normal" mortality. None of the meteorological parameters was correlated with total mortality; the correlations for SO_2 and smoke were 0.11 and 0.12, respectively. Smoke and SO_2 were negatively correlated with wind speed, suggesting that area sources were important for both; however, SO_2 was more strongly (negatively) correlated with temperature, suggesting that there were other sources of smoke besides space heating.

As a means of displaying the interactions between total mortality, SO_2, and smoke, Loewenstein *et al.* tabulated the data in octiles (following the example of Mazumdar *et al.*, 1980), giving the numbers of deaths and the average percentage excess in each of 64 combinations of SO_2 and smoke classification "bins." Because of the strong correlation between smoke and SO_2, most of the data were clustered along the diagonal, and 22 of these bins were empty. I used these data to explore dose-response relationships on two different bases, ordinary least squares (OLS) and weighted least squares, using the square root of the numbers of days in each bin for weighting factors. Only smoke was significant; weighting reduced the elasticity from 0.057 to 0.025; the slope of the lower value was 0.022% per $\mu g/m^3$. Elasticities were slightly higher when the sum of SO_2 and smoke was used. I limited the graphical analysis to bins with at least 5 days' data. The relationship with smoke alone was clearly more systematic than with SO_2 alone; when plotted versus the sum of SO_2 and smoke, the suggestion of a smoke threshold around $100\,\mu g/m^3$ vanished. The importance of the higher levels of SO_2 as an additional factor besides smoke was seen from plots that showed mortality increasing at constant smoke levels as SO_2 rose from 300 to 400 to more than $400\,\mu g/m^3$.

Athens

Hatzakis *et al.* (1986) analyzed daily mortality in the Greater Athens area for 1975–82. The dependent variable was the deviation from the expected daily mortality count, that is, "adjusted" mortality. Expected mortality was based on fitting a sinusoidal curve to the monthly mortality data for 1956–58. The independent variables were the 24-hour averages of SO_2 (mean = $86\,\mu g/m^3$) and British smoke (mean = $57\,\mu g/m^3$), calculated as the geometric means of data from five measuring stations. During the 8 years studied, air quality (maximum monthly average) improved by about a factor of 5. Controls (dummy variables) for day-of-week, month, season, year, holidays, and temperature were included in the analysis. The authors reported that adjusted daily mortality was significantly associated with SO_2 ($p = 0.05$), but not with smoke. If a threshold were present for SO_2, it lay slightly below $150\,\mu g/m^3$. No regression results were reported for smoke or for SO_2 and smoke taken jointly, although one of the authors stated in a personal communication (Katsouyanni, 1988) that the p-value for smoke alone was 0.67 and that they felt that this lack of significance resulted from the greater spatial variability of smoke relative to SO_2. A subsequent paper (Katsouyanni *et al.*, 1990b) showed a spatial coefficient of variation for SO_2 of 9.6% (five stations), while the same statistic for smoke was 45%.

Hatzakis *et al.* presented a table of partial regression coefficients for SO_2 showing how the value decreased as additional controls for confounding variables were added to the model. The bivariate coefficient was 0.12% excess deaths per $\mu g/m^3$; accounting for long-term and seasonal trends reduced it to

0.043%; accounting for holidays and days of the week and their seasonal interactions reduced it to about 0.029%, and their final value was 0.020% excess deaths per μg/m^3 (elasticity = 0.017). One might argue that the penultimate value was more appropriate, since the last step involved dummy variables for each month, only two of which were significant.

This study is important, since it represents a finding of daily mortality effects at or below the current U.S. primary standard for ambient SO_2. The analysis may have been compromised by evaluating only one pollutant at a time, which the authors felt to be necessary because of the high intercorrelation (0.73). The regression coefficient reported is larger than any reported for either SO_2 or smoke in several analyses of daily mortality in New York and London, although the SO_2 result just met the significance criterion ($p = 0.05$). A stepwise regression approach with both pollutants initially in the regression would be particularly interesting, since so many potential confounding variables were investigated. There are also possible contributions from other (unmeasured) air pollutants; a mean of 158 μg/m^3 was reported for NO_2, measured from 8:00 A.M. to 2:00 P.M. on 834 days (Katsouyanni, personal communication, 1988), which is high by U.S. standards.

Additional information on Athens was provided by Katsouyanni *et al.* (1990a), who adopted a case-control approach toward analyzing deaths by age and cause. They defined 199 "index" days as those for which SO_2 exceeded 150 μg/m^3 and matched them with 398 "comparison" days that were alike in all respects except that SO_2 was under 150 μg/m^3. Some of the index days were 2- to 3-day episodes. Excess mortality was reported on index days for the total of all causes, for cardiac, respiratory, and other causes on both an underlying and immediate cause-of-death basis, but the cardiac causes were only significant at the 0.10 level. Deaths were stratified at age 75 to examine the effects of age, and it was shown that only respiratory deaths were more significant for the older age group. Overall, the excess mortality on index days was 0.82% for all ages and 0.26% for those aged 75 and over. This compares with a mean effect of 1.7% from the regression study of Hatzakis *et al.* (1986), which implies that either a substantial portion of the pollution effect occurred at SO_2 levels below 150 μg/m^3 or that the procedure of matching days provided better "control" of the nonpollution effects than the regression methods.

The final paper in this series (Katsouyanni *et al.*, 1990a) extended the Athens data set to 1988 and studied the contributions of smoke, SO_2, O_3, and CO, based on a reorganized monitoring network. Both smoke (140 μg/m^3) and NO_2 (162 μg/m^3 in the central area) were high by U.S. standards. In winter, smoke was the most important pollutant ($p = 0.003$); I estimated the regression coefficient at about 0.02% per μg/m^3. In summer, an interaction was shown between NO_2 and temperature; for days below 29°C, NO_2 was not significant, but it was highly significant ($p < 0.0001$) for hotter days. Because of this ambiguity in results for Athens, the conclusion of Katsouyanni *et al.* (1990b) seems apt: "SO_2 is only an index of air pollution."

Vienna

Neuberger *et al.* (1987) studied the period 1972–83 in Vienna, with emphasis on influenza morbidity and mortality among the elderly (age 70+). Sulfur dioxide was the only pollutant considered, with temperature as a covariate. Both the presence of influenza and SO_2 (probably as an index of all air pollution) influenced total, cardiovascular, and respiratory mortality. For total mortality, the SO_2 coefficient for the over-70 age group was about 0.036 per $\mu g/m^3$ and did not vary substantially between men and woman or between normal periods and periods of influenza (the overall mortality rates were higher during influenza periods). This finding implies that there is no interaction between air pollution and influenza and that neglecting influenza in regression analysis should not create a major error, at least for this age group and level of air pollution (SO_2 up to 500 $\mu g/m^3$).

Lyons and Marseilles

Air monitoring data from the PAARC study (see Chapter 12) were used to study daily mortality variations in Lyons and Marseilles by Derriennic *et al.* (1989), from 1964 to 1976. The pollutants available included smoke, TSP, and SO_2 by "strong acid" and by a more specific method, measured at four locations in each city. Causes of death examined included respiratory, circulatory, and others. Statistical analyses were limited to the data on TSP and the specific method for SO_2. Seasonal adjustment was performed on all data by dividing by the 31-day moving average over the 3-year period; the regression coefficients are thus in the form of elasticities. Cross-correlations were used to examine the lag structure; total mortality was significantly correlated on the same day in Lyons, where respiratory mortality lagged SO_2 by 6 to 7 days; circulatory deaths were found to lag SO_2 by 5 to 7 days in Marseilles but were not significant in Lyons; TSP showed no causal relationships in either city for respiratory or circulatory mortality; results for total mortality were not discussed. The regression analysis accounted for autocorrelation and showed elasticities for SO_2 and respiratory deaths of 0.21 to 0.23 in both cities and 0.07 for circulatory deaths in Marseilles. The temperature effects (after deseasonalization) had conflicting signs in the two cities. It is unfortunate that results were not also presented for smoke, in order to determine the robustness of the findings for SO_2.

RECENT TIME-SERIES MORTALITY STUDIES BY SCHWARTZ AND COLLEAGUES

The basic methodology used by Schwartz and Dockery in Philadelphia has been replicated in several other cities. This protocol has the following general features:

1. Deaths from accidents and those resident deaths that occurred outside the study area are excluded.
2. Poisson models are used, in which

$$\ln(M) = \text{sum } \beta_i + \text{error}.$$

3. The mortality data (M) are fit against weather and seasonal variables first and the residuals are then regressed against air pollution. The variables examined in this phase of the analysis include a continuous time variable, seasonal dummy variables, 24-hour mean temperature and dew point, lagged temperature and dew point, dummy variables for hot, cold, humid, hot and humid days, including lags. Interaction terms were also considered. Weather and season terms were retained in the model when $t > 1$.
4. Each pollutant is evaluated singly, and sometimes in combination. Twenty-four-hour averages are used for all variables.
5. The generalized estimating equations of Liang and Zeger (1986) are used to account for possible "overdispersion" and serial correlation. This procedure is thought to give conservative estimates of the significance of the pollution variables.
6. The linearity of the implied dose-response function is assessed by fitting the residuals against a set of dummy variables corresponding to n-tiles (where $n = 4$ or 5) of the pollution variable. The regression coefficients for these dummy variables represent the average excess mortality risk in that pollution n-tile, independent of the other n-tiles. This approach does not require the assumption of the shape of a particular dose-response function *a priori*. Significance tests are not applied to these n-tile coefficients, since the range of the variable has been truncated so severely by this subdivision.

The studies that follow all use this protocol; the Philadelphia study (Schwartz and Dockery, 1992a) was discussed previously.

Detroit

Based on the average number of daily deaths given, the study area in the Schwartz (1991b) study appears to be Wayne County, MI, rather than the city of Detroit. Since daily TSP measurements were not available in sufficient number, airport visibility was used to predict a TSP value for each day of the study. The predictive model was "calibrated" against concurrent data for about 500 days. An average of about 14 TSP measuring sites was used to develop this model. Twenty-four-hour averages for SO_2 and both peak and average O_3 were also investigated for association with weather-corrected daily mortality. The TSP levels on the day before death had the most predictive power; the average of the prior 2 days was less significant. Sulfur dioxide was significant as the sole pollutant, but with a significantly lower elasticity (0.010 versus 0.0475). When TSP and SO_2 were regressed jointly, the effect of SO_2 was greatly reduced, leading Schwartz to conclude that SO_2 was acting only through its collinearity with TSP, rather than as an independent agent. Ozone was never significant. The linearity analysis showed that the average risk was higher than that of the lowest pollution quintile (about 65 $\mu g/m^3$, on average) in each of the other four quintiles. However, the overall slope of the quintile plot was lower than the regression coefficient (4% excess risk per 100 $\mu g/m^3$ as opposed to 6% per 100 $\mu g/m^3$).

Steubenville, OH

In Steubenville (Schwartz and Dockery, 1992b), mortality was affected only by the combination of heat (daily mean >70°F) and humidity (dew point >65°F) days, not by temperature *per se*. Both TSP and SO_2 were found to be significant predictors of residual mortality, but only TSP survived in joint regressions. The lag structure was not specifically discussed, but little autocorrelation was reported, and the timing of the TSP data ended at 8:00 A.M., corresponding to a lag of 16 hours. The TSP coefficient was smaller than the values for Philadelphia and Detroit, in spite of the higher mean TSP level; the elasticity was 0.042. With both pollutants in the regression, the total effect was slightly larger, but the SO_2 contribution was not statistically significant. There appears to be a typographical error in the SO_2 regression coefficients, as published; the correct values should be a factor of 10 smaller. The linearity plot showed that the highest three quartriles all carried higher average risks than the lowest quartile (mean of about 40 $\mu g/m^3$). It is also interesting to note that, since the linear time trend variable was not significant in this study, the entire 9% improvement in mortality that occurred during the 10 years would appear to be due to reductions in air pollution. One wonders if studies over periods as long as 10 years should not work with mortality *rates* rather than counts.

This study was remarkable in its ability to find small effects with so few daily deaths* and with moderate air pollution levels. However, the two-stage regression procedure carries some risk of assigning a portion of the air pollution effect to weather, and there is the possibility that the pollution coefficients have been underestimated because of the use of a single monitoring station for an entire SMSA.

St. Louis

This (and the companion study in Eastern Tennessee discussed in the next section) represent the first attempts at time-series studies of particles classified by size and chemistry (Dockery *et al.*, 1992). Unfortunately, they were somewhat compromised by the short periods of record used and the large geographic areas that were required to obtain sufficient daily death counts. The St. Louis SMSA comprises eight counties, plus the city, for a total of almost 13,000 km^2.

A single monitoring site was used, which recorded daily values for PM_{10}, $PM_{2.5}$, SO_4^{-2}, H^+, SO_2, O_3, and NO_2. Elemental composition data were also obtained from the particulate catch. This site was located on the south side of the city in a residential area (Carondelet). The TSP data collected for use in a 1980 cross-sectional study (Lipfert, 1992) offer a means of comparing mean particulate values across the SMSA. In 1980, the SMSA-wide mean of 27

*There may be some uncertainty as to the exact geographic area that was analyzed in this study, as judged from the reported number of daily deaths (3.07 per day or 1120 per year). Based on the average of 1975 and 1985 published data, the SMSA should have recorded 1558 resident nonaccidental deaths, and Jefferson County (OH), 954. If we take these figures at face value, we would have to conclude that almost 30% of the SMSAs' residents died outside the area during this period, which seems excessive. The SMSA population decreased by 1.5% between censuses.

stations was 84.2 μg/m^3, with a standard deviation of 17.2. The difference between the means for the city and for the SMSA was statistically significant. Based on TSP data recorded at Carondelet in 1980 (80 μg/m^3 [Dockery *et al.*, 1989]), this site did not seem out of line with the rest of the SMSA on the basis of mean values; however, the question of interest here is the temporal correlation, which was not available.

Compared to the studies discussed previously, significance levels for St. Louis were considerably lower; this was probably a result of the short period of record (1 year). The only pollutant of the seven species to achieve statistical significance was PM_{10}, although $PM_{2.5}$ was close, as was the coarse particle fraction (difference between PM_{10} and $PM_{2.5}$). Thus this study did not indicate a major difference in mortality association by particle size. One- and 2-day lags were about equally effective; results for the 2-day average were not reported. Neither SO_4^{-2} nor H^+ were even close to significance, nor were any of the gaseous species. Note that SO_4^{-2} is a major component of the fine particle fraction; elements normally associated with the coarse fraction were reported to have positive associations with mortality. The elasticities for PM_{10} and $PM_{2.5}$ were 0.041 and 0.030, respectively. A similar value was found for SO_4^{-2} (0.049), even though it was not significant. The linearity plot showed that the first three PM_{10} quintiles all had about the same risk (1.0 – 1.01), with mean values from about 12 to 27 μg/m^3. The highest two quintiles both had an average risk of about 1.03 (note that these results could also be interpreted as an approximate kind of linearity, at a lower slope than that derived by the regression). The authors pointed out that the highest PM_{10} value in this data set was only 97 μg/m^3, well below the 24-hour NAAQS level of 150 μg/m^3. However, we have no assurance that levels are this low throughout the SMSA, although the point seems well taken that the association of mortality and air pollution does in fact extend to levels well below the NAAQS. My estimate of the average TSP level for St. Louis for this time period was about 64 μg/m^3.

Eastern Tennessee

This portion of the study (Dockery *et al.*, 1992) was structured around a monitoring data set for Harriman, TN, which was similar to the St. Louis data. In order to accumulate sufficient daily deaths for a 1-year analysis, the geographic area was extended to include 11 counties (total area of about 12,000 km^2). There was more topographical relief in the Tennessee study area, and almost half of the population lived in one county (Knox), which was not represented by air quality monitoring. None of the air pollutants was even close to statistical significance, although the regression coefficients were generally similar to those found in St. Louis. The elasticities were 0.048 for both PM_{10} and $PM_{2.5}$ and 0.07 for SO_4^{-2}. There were more significant weather variables in Eastern Tennessee than in the other cities. The linearity plot showed that the second highest PM_{10} quintile had a risk of about 1.0, relative to the lowest quintile, while the remaining quintiles formed an approximate linear relationship. My estimate of the average TSP level for Knox County (the major population center of the area under study) for this time period was about 55 μg/m^3.

Minneapolis-St. Paul

Only minimal information was available about this study at press time (Schwartz, in press). Analyses were based on days for which a TSP reading was available for both cities; it is not clear how missing data were handled, but it appears that gaps were left in the record. The mean TSP value for the ensemble was $78\,\mu g/m^3$; the mean SO_2 value, $31\,\mu g/m^3$. However, no regression results were given for SO_2. The TSP coefficient was 0.000525, with a *t*-value of 3.5. This corresponds to an elasticity of 0.041. No information on lags or significant weather variables was given in this summary account.

Utah County, Utah

Much of the methodology developed by Schwartz and Dockery was extended by Pope *et al.* (1992) to Utah County, which includes the cities of Provo and Orem in the Utah Valley. The meteorological model consisted of dummy variables for various ranges of temperature (always included) and for humidity, season, and temperature/PM_{10} interaction (included only when significant). Use of dummy variables rather than a continuous variable for temperature allows for a nonlinear temperature response, which was in fact observed. Same-day pollution and pollution averaged over various lag periods were considered; a 5-day lag period was reported to fit best, with an elasticity of 0.069 at the mean PM_{10} value of $47\,\mu g/m^3$.

The regression coefficient for PM_{10} on mortality compares well with the other PM_{10} estimates discussed above. In this regard, the lack of significant levels of other pollutants in the Utah Valley should be noted. For example, aerosol acidity was monitored daily during one winter and never exceeded $0.5\,\mu g/m^3$ (as H_2SO_4). The Utah Valley may be one of the few areas that have been studied where most of the health effects of air pollution appear to be due to one type of pollutant (particulate matter). Although sulfates and nitrates were not measured during this period, Caka *et al.* (1992) reported data from 1990 that provide some idea of what the levels might have been. The maximum SO_4^{-2} level reported was about $11\,\mu g/m^3$; NO_3^-, $29\,\mu g/m^3$ (compared to the maximum PM_{10} level of $365\,\mu g/m^3$). Sulfates and nitrates thus do not appear to be important components of the aerosol at Provo. My estimate of the average TSP level in Utah Valley was $87\,\mu g/m^3$.

Comparing the average daily deaths on the same day in the highest pollution category (2.98 deaths/day, or about 14% above the baseline) to those in the highest 5-day average pollution category (3.24 deaths/day, or 24% above the baseline) shows that increasing the dose by a factor of 4.6 increased the incremental mortality response by less than doubling. This comparison suggests a nonlinear dose-response in terms of averaging time, whereas the response to concentration changes *per se* appeared to be linear. It also strengthens the causal hypothesis, since the comparison shows that mortality responds at least partially to the dose of pollution and not just to the conditions that result in elevated concentrations.

Pope *et al.* also derived regression models by cause of death, for respiratory, cardiovascular, and "all other" causes. The mean effects were 17%, 8.4%, and

2.3%, respectively; these findings are quite similar to those of Schwartz and Dockery (1992a) for Philadelphia. Pope and colleagues have also studied the responses of the Utah Valley population with respect to other health endpoints; these findings are discussed in Chapters 9 and 11. Finally, an interesting aspect of the Utah Valley studies is the finding that health responses tracked air pollution levels in the long term as well as daily, as evidenced by improvements on average during a 13-month shutdown of the major pollution source, followed by worsening when it resumed operations. The mortality aspects of this particular finding are discussed further in Chapter 8.

Birmingham, AL

Schwartz's analysis (1993) of daily mortality in Birmingham featured comparisons of several different regression approaches, but the analysis was limited to a single pollutant, PM_{10}. Daily deaths from 1985 through 1988 were analyzed, and, although the geographic area was reported to be the Birmingham SMSA (4 counties), the daily death counts corresponded more closely to the central county of the SMSA (Jefferson). An additional PM_{10} monitor was in operation for about half of the period. Three different regression models were used: (1) the Poisson model described above, together with

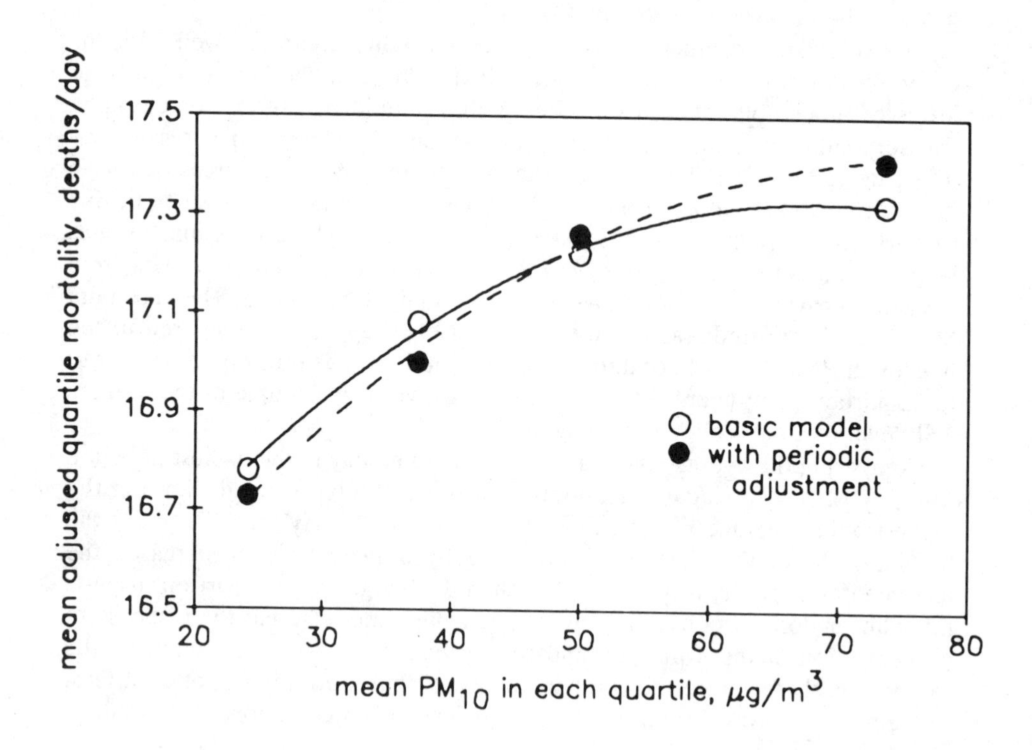

FIGURE 6-16. Dose-response relationships for Birmingham, AL. *Data from Schwartz (in press).*

the Liang and Zeger (1986) method of accounting for over-dispersion and serial correlation; (2) a generalized additive model that accounted for non-linearities in the weather effects on mortality; and (3) use of trigonometric filters to remove the long-term (seasonal) components of variation in daily mortality.

Essentially the same results were achieved with all three models. Mortality (all causes and cardiovascular) was significantly associated with PM_{10}, averaged over the preceding 3 days. The regression coefficient was 0.11% per $\mu g/m^3$, yielding an elasticity of 0.053. The dose-response function based on quartiles is shown in Figure 6-16. The response is seen to level off for both of the regression models shown, although Schwartz reports that the departure from linearity was not statistically significant when the generalized additive model was used.

SUMMARY AND RECONCILIATION OF THE RECENT TIME-SERIES STUDIES

In his synthesis of these studies, Schwartz (in press [b]) emphasizes the similarity of the regression coefficients that have been derived for the particulate measures, and the fact that the best results are not specific to size or composition. He concludes that neither O_3 nor SO_2 plays an important role in daily mortality. He compares the PM_{10} results to the TSP results by multiplying the PM_{10} coefficients by the estimated average ratio of PM_{10} to TSP (0.5); based on the national averages given in his Table 2-9, this factor may be on the low side.

There are two important questions to consider in comparing results from this group of very similar studies:

1. How should results for different pollutants be compared?
2. Should we expect to find the same time-series regression coefficients in all cities?

We have used the concept of elasticity to compare pollutants and studies elsewhere in this book, primarily because this measure is unit-free and lends itself readily to conceptualization of the relative magnitude of effects. However, in comparing measurements that differ only in particle size, one might argue that the regression coefficient *per se* is more relevant. The concept behind changing the U.S. NAAQS for particulate matter was that the smaller particles were the ones that affected health, by virtue of their ability to penetrate into the deep lung. This being the case, we would expect the same regression coefficient for both PM_{10} and TSP, since only the inhalable portion of TSP should be physiologically active. According to this hypothesis, the large particle fraction of TSP should be inert and thus have no effect on the regressions.

The regression coefficients from this group of studies are plotted against mean concentration in Figure 6-17; we see that all of them except the one from Utah fall on or very near a line of constant elasticity, instead of constant coefficient. The hypothesis of smaller particles being more lethal is thus apparently rejected. This implies that the two particulate measures are virtually

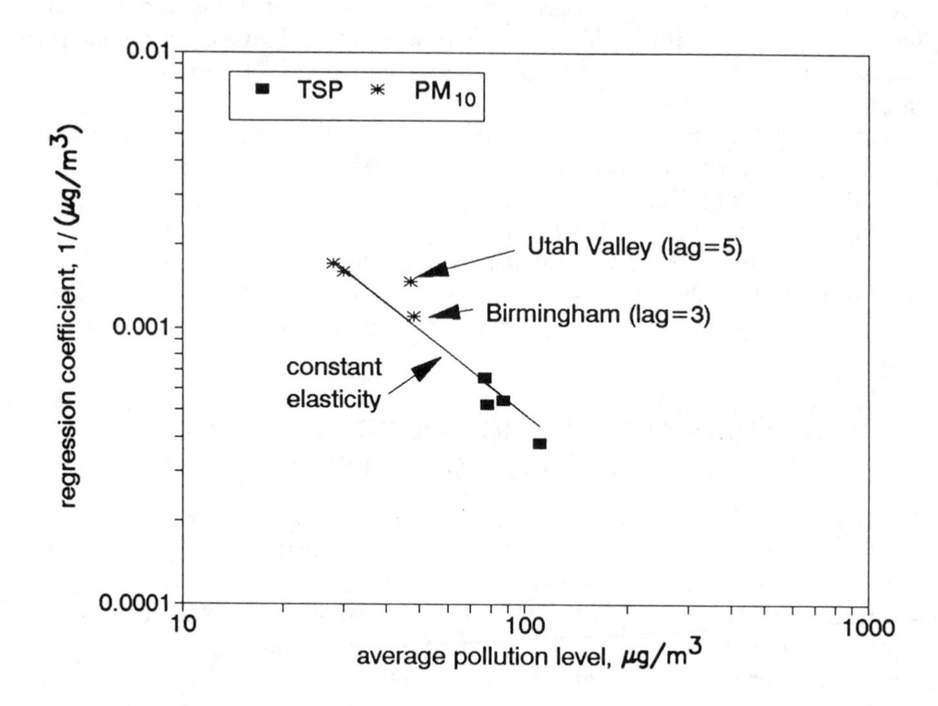

FIGURE 6-17. Comparison of particulate regression coefficients from recent time-series studies by Schwartz and colleagues.

interchangeable and that a constant percentage of deaths is associated with air pollution, regardless of the measure used.

Figure 6-17 also implies that elasticity is the preferred metric for comparison among studies (as well as for comparison among pollutants for a given study). If we use my estimates of average TSP levels in the 4 areas for which PM_{10} studies were made (Lipfert, 1992), we derive TSP regression coefficients of 0.00077, 0.00087, 0.00079, and 0.00070, for St. Louis, Eastern Tennessee, Utah Valley, and Birmingham, respectively. Together with the 4 studies actually based on TSP, the coefficient of variation of the group is 25%. On the basis of elasticities, however, the coefficient of variation of the 7 locations is only 17% and, if Utah Valley is separated, only about 8%. Thus, use of elasticity to compare locations "explains" a good part of the variation. Keeping in mind the lessons of Appendix 3B, we might expect that the regression coefficients of Schwartz and colleagues, based on bivariate regressions, may represent the sum total effect of all the correlated pollutants in each city, rather than a mechanistic response to certain kinds of particles, *per se*.

Figure 6-18 plots the particulate elasticity values from this group of studies against the annual average mortality rates. We see a slight, generally increasing relationship, except for Utah Valley and Birmingham, which were based on longer lag periods. The increase in elasticity with annual mortality rate is consistent with the findings by age for Philadelphia—that older people are

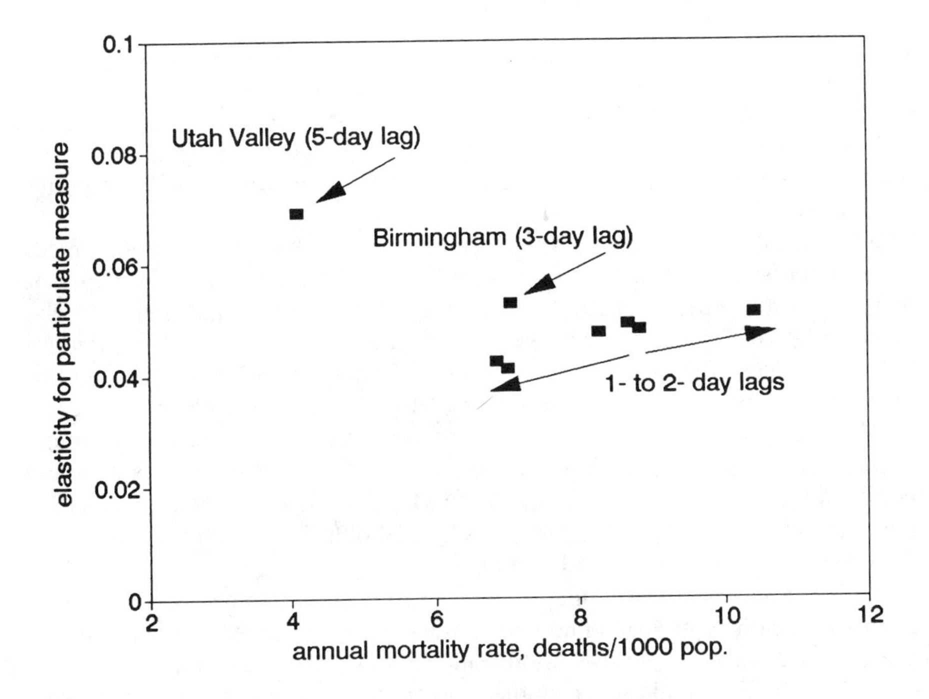

FIGURE 6-18. Comparison of particulate elasticities from recent time-series studies by Schwartz and colleagues.

more susceptible to the acute effects of air pollution. A city with a higher fraction of older people will have a higher annual mortality rate and should show a stronger response to acute air pollution. The increase with lag period, based on averaging over the lagged days, allows inclusion of additional cases which may have had delayed reactions.

One of the most remarkable findings from this set of studies is the apparent exclusion of all other pollutants except suspended particulates and the lack of sensitivity to particle size and composition. The eight locations studied by Schwartz and his colleagues vary considerably in terms of climate, population, and the presence of other pollutants. We thus tend to support the claim that some measure of particulate matter is the active agent, even as we maintain that the case cannot be proven statistically.

A note of caution should be sounded against putting too much stock in the single-year findings from St. Louis and Tennessee, on the lack of significance of sulfates and fine particles. Kinney and Özkaynak (1991) showed very clearly how variable individual years can be, and it is entirely possible that a longer study in those locations could have produced different results. This possibility is also shown by the wide range of the simulation results in Appendix 3B.

OVERALL CONCLUSIONS FROM THE TIME-SERIES STUDIES

The works reviewed above present a range of methodologies and results. They confirm that air pollution has a significant effect on mortality, even at concentration levels much lower than those that occurred during the episodes and lower than present air quality standards, in some cases. In some of the studies reviewed, it is difficult to identify the most important pollutants with certainty.

While reliable thresholds for these pollutant effects have not been identified, none of the studies reviewed were specifically structured to find such thresholds. Several studies suggested thresholds for both SO_2 and smoke in the range of 100 to 200 $\mu g/m^3$. However, logarithmic models (which do not admit thresholds) were found to be appropriate in London and Los Angeles and to be roughly equivalent to linear models in New York (since these are the cities with the largest ranges in daily pollutant concentrations, they are the most likely candidates for logarithmic models). In addition, since studies at lower pollution levels have also found significant effects of pollution on mortality, the bulk of evidence points to the absence of thresholds, at least in the concentration ranges that have been studied thus far.

Table 6-5 summarizes the major studies reviewed in this chapter and shows their statistically significant findings. Fifteen different cities are represented over a wide range of air pollution (smoke levels from around 10 $\mu g/m^3$ to nearly 2000 $\mu g/m^3$). Only seven studies had sufficient air monitoring stations to yield (subjectively) acceptable estimates of exposure. The studies employing only a single station tended to yield lower estimates of the effect on mortality, although there are several other design factors to consider when comparing studies, such as lags. The tendency toward constant elasticity (rather than constant linear regression coefficients) over the range of pollutant concentrations involved in the various studies also suggests a logarithmic relationship.

For mortality from all causes and for cardiovascular causes, elasticities were in the general range 0.01 to 0.08; for respiratory mortality, 0.15 to 0.23. Separate analyses (not shown in Table 6-5) showed that older people were affected more. Consideration of lags increased the effect on mortality. All of these findings are consistent with a causal relationship.

The methods of accounting for confounding variables were also found to be important. The seasonal cycle is the most crucial of these and probably represents biological cycles rather than climatic effects *per se*, since it is important in all types of climates. Based on available information, it is not possible to identify any one method of seasonal adjustment as "superior" to any other. The effects of daily temperature fluctuations were seen to be variable, aside from heat waves. Such effects were shown in London, but their magnitudes were substantially less than those of air pollution; in Los Angeles and Philadelphia, the temperature effects were larger than the air pollution effects. Aside from heat waves, temperature effects in New York were smaller than those of air pollution, according to Hodgson's (1970) analysis.

Although most studies that compared SO_2 effects to smoke effects at the higher concentration levels concluded that smoke was the greater hazard, SO_2 cannot be completely exonerated in those cases where the two pollutants are

TABLE 6-5 Comparison of Major Time-Series Mortality Studies

Location	Reference	Time Period	Seasonal Control	Other Variables	Pollutants (no. of sites)	Expos Acc'y	Coeff. for Significant Pollutants-Diseases*	Elasticity (lag days)
London	Mazumdar, *et al.* (1982)	1958–72 winters	15-day avg. temp. corr.	Day-of-week, year, episodes	Smoke (7)	Fair	Total: smoke, 0.025†	0.042 [1]
					SO_2 (7)	Fair	SO_2, 0.0012†	0.003 [1]
	Ostro (1984)	1958–72 winters	15-day avg.	Temp., RH	Smoke (7)	Fair	Total: smoke > 150, 0.0141	0.043 [1]
					SO_2 (7)	Fair	Total: smoke < 150, 0.050	0.041 [1]
	Schwartz & Marcus (1986)	1958–72 winters	15-day avg.	Serial corr., temp., RH	Smoke (7)	Fair	Total: smoke, 0.024–0.027	0.04–0.045 [1]
					SO_2 (7)	Fair		
	Shumway *et al.* (1983)	1958–72 winters	Filter	Temp., lags	Smoke (7)	Fair	Total: SO_2††	0.076 [1]
					SO_2 (7)	Fair	Resp.: SO_2††	0.15 [4]
							CV: SO_2††	0.071 [1]
New York	Hodgson (1970)	11/62–5/65	Degree-days		COH (1)	Poor	Total, COH, 0.032	0.082
					SO_2 (1)	Poor		
	Glasser & Greenburg (1971)	1960–64 winters	15-day avg.	Weather	SO_2 (1)	Poor	Total: SO_2, 0.005	0.028
					COH (1)	Poor	Total: smoke, 0.015	
	Buechley *et al.* (1973)	1962–66	National trend	Day-of-week, holidays, epidemics, heat wave	SO_2 (1)	Poor	Total: SO_2, 0.0035	0.012–0.024
					COH (1)	Poor		
	Buechley (1975)	1967–72	National trend	Day-of-week, holidays, epidemics, heat wave	COH (1)	Poor	Total: COH, 0.0057†	0.026
					SO_2 (1)	Poor	SO_2, 0.0017†	
					CO (1)	Poor	CO, 0.0003†	
	Schimmel & Greenburg (1972)	1963–68	Temperature correction		COH (1)	Poor	Total: COH, 0.046†	0.116 [7]
					SO_2 (1)	Poor	SO_2, 0.0047†	
	Schimmel *et al.* (1974)	1963–72	Temperature correction		COH (1)	Poor	Total: SO_2 + COH†	0.087
					SO_2(1)	Poor	Resp: SO_2 + COH†	0.0221
							Heart: SO_2 + COH†	0.088
	Schimmel & Murawski (1976)	1963–72	15-day avg.		COH (1)	Poor	Total: SO_2 + COH†	0.033 [5]
					SO_2 (1)	Poor	Resp.: SO_2 + COH†	0.12 [5]
							Heart: SO_2 + COH†	0.025 [5]
	Schimmel (1978)	1963–76	Temperature correction		COH (1)	Poor	Total: COH, 0.0071	0.016
					SO_2 (1)	Poor	SO_2, 0.0007	0.03–0.04 [7]
	Özkaynak & Spengler (1985)	1963–76	Sinusoidal trend line		COH (1)	Poor	Total: COH + SO_2 + vis†	0.038
					SO_2 (1)	Poor		
					Visibility	Good		

TABLE 6-5 *Continued*

Location	Reference	Time Period	Seasonal Control	Other Variables	Pollutants (no. of sites)	Expos Acc'y	Coeff. for Significant Pollutants-Diseases*	Elasticity (lag days)
Los Angeles	Mills (1960)	1956–58	Monthly avg.	Temperature	Oxid. (20)	Good	Cardiac, resp.	0.032
	Hexter & Goldsmith (1971)	1962–65	Cyclic trend	Temperature	CO	Fair	Total: CO[††]	0.16
					Oxidants (basin avg.)	Good		
	Shumway *et al.* (1988)	1970–79	Quadratic temperature	(Weekly avgs.)	CO (6)	Fair	Total: CO[††]	0.062
					HC (6)		Total: HC[††]	0.064
					COH (6)		Total: COH[††]	0.052
	Kinney & Özkaynak (1991)	1970–79	Filter	Serial corr., day-of-week, temperature, rel. humidity	Ozone (8)	Good	Total: Ozone[†]	0.015 [1]
					CO (6)	Fair	NO_2[†]	0.025
					SO_2 (8)	Fair	Total: ozone[†]	0.019 [1]
					COH (8)	Good	CO[†]	0.021
					NO_2 (8)	Good	Total: ozone[†]	0.018 [1]
					Visibility	Good	COH[†]	0.021
Philadelphia	Wyzga (1977)	1957–66	30-day temp.	Serial corr., heat waves, temp., epidemics	COH (2–3)	Poor	Total, 0.009–0.049	0.008–0.06 [var.]
					TSP (1)		Total, 0.017	0.028
					NO (1)		Total, 0.028	0.022
					HC (1)		Total, 0.0031	0.046
	Schwartz & Dockery (1992a)	1973–80	Weather (2-stage)	Serial corr.	TSP	Fair	Total[§], 0.066	0.05 [1]
					SO_2		Total[§], 0.050	
Pittsburgh	Mazumdar & Sussman (1983)	1972–77	15-day avg.		COH (3)	Poor	Total, 0.008	0.01
					SO_2 (3)			
San Jose, CA	Fairley (1990)	1980–86	15-day-avg.		COH (1)	Poor	Total[§], 0.0077	0.077 [2]
							Circ., 0.003–0.004	0.03–0.04
							Resp., 0.0026–0.003	0.14–0.17
Steubenville, OH	Schwartz & Dockery (1992b)	1974–84	Weather (2-stage)	Serial corr., temp., RH	TSP (1)	Poor	Total[§], 0.038	0.042 [1]
					SO_2 (1)		Total[§], 0.038	0.028 [1?]
Detroit	Schwartz (1991b)	1973–82	Weather (2-stage)	Serial corr., temp., RH	TSP[‖]	Good	Total[††], 0.055	0.048 [1]
					SO_2 (12)	Good	Total[††], 0.033	0.010 [1?]
					Ozone	?		

City	Study	Years	Seasonal control	Other covariates	Pollutants (lag)	Quality	Mortality, coefficient	Coefficient*
Minneapolis-St. Paul	Schwartz (in press [b])	1973–82	Weather (2-stage)	Serial corr., temp., RH	TSP (2)	Poor	Total§, 0.052	0.041 [1]
St. Louis	Dockery *et al.* (1992)	1985–86	Weather (2-stage)	Serial corr., temp., RH	PM_{10} (1) $PM_{2.5}$ (1) SO_4 (1) Ozone (1) SO_2 (1) H^+ (1)	Poor	Total§, 0.171	0.049 [2]
Provo, UT	Pope *et al.* (1992)	1985–89	Weather (2-stage)	Serial corr., temp., RH	PM_{10} (1)	Poor	Total§, 0.147	0.069 [5]
Paris	Loewenstein *et al.* (1983)	1969–76	Deviations from normal	Weather	Smoke (30) SO_2 (30)	Good	Total, 0.022	0.025
Athens	Hatzakis *et al.* (1986)	1975–82	Sinusoidal curve	Weather, holidays, day-of-week, month	SO_2 (5)	Fair	Total, 0.020	0.017
Lyons	Derriennic *et al.* (1989)	1964–76	31-day avg.	Temperature	SO_2 (4) TSP (4)	Fair	Resp.	0.23 [6–7]
Marseilles	Derriennic *et al.* (1989)	1964–76	31-day avg.	Temperature	SO_2 (4) TSP (4)	Fair	Resp. Circ.	0.21 [2–9] 0.07 [5–7]

* Regression coefficients in % excess deaths per $\mu g/m^3$.
† Joint regression.
†† Log model used, no linear coefficient available.
§ All causes less accidents, homicides, suicides.
‖ TSP estimated from airport visibility.

highly correlated. Several studies found that SO_2 had a significant effect while particulates did not. Other species found to be significant were CO, oxidants, hydrocarbons, and NO. No studies performed statistical tests to determine the confidence in the indicated superiority of alternative models or pollutants.

It is also noteworthy that few studies have been performed in cities where air conditioning is heavily used. For example, according to the 1980 census, 60% of Los Angeles housing units had no air conditioning. In St. Louis, only 19% were without air conditioning and 53% had central air. Air conditioning tends to reduce the penetration of outdoor air pollution and thus may be protective against acute responses (it can also intensify the effects of indoor air pollutants). It would also be interesting to evaluate the hypothesis that particle size is unimportant by studying a location heavily influenced by wind-blown dust, such as in the Western United States.

REFERENCES

Ball, D.J., and Hume, R. (1977), The Relative Importance of Vehicular and Domestic Emissions of Dark Smoke in Greater London in the Mid-1970s, *Atm. Env.* 11: 1065–73.

Bates, D.V., Macklem, P.T., and Christie, R.V. (1971), in *Respiratory Function in Disease*, 2nd ed., Saunders, Philadelphia. chap. 17.

Biersteker, K., and Evendijk, J.E. (1976), Ozone, Temperature, and Mortality in Rotterdam in the Summers of 1974 and 1975, *Env. Res.* 12:214–17.

Bloomfield, P., Mazumdar, S., and Schimmel, H. (1980), *Regional Analysis of the Impact of Air Pollution on Human Mortality*, Technical report No. 165, Series 2, Department of Statistics, Princeton University.

Buechley, R.W. (1975), SO_2 Levels, 1967–72 and Perturbations in Mortality, Report to the National Institutes of Health (NIEHS), Contract NO1-ES-5-2101.

Buechley, R.W., Riggan, W.B., Hasselblad, V., and Van Bruggen, J.B. (1973), SO_2 Levels and Perturbations in Mortality, *Arch. Env. Health* 27:134–37.

Bull, G.M. (1973), Meteorological Correlates with Myocardial and Cerebral Infarction and Respiratory Disease, *Br. J. Prev. Soc. Med.* 27:108–13.

Bull, G.M., and Morton, J. (1975), Seasonal and Short-Term Relationships of Temperature with Deaths from Myocardial and Cerebral Infarction, *Age and Aging* 4:19–31.

California Department of Public Health (1955–1957), *Clean Air for California*, Reports I–III, Berkeley, CA.

Commins, B.T. (1963), Determination of Particulate Acid in Town Air, *Analyst* 88: 364–67.

Commins, B.T., and Waller, R.E. (1967), Observations from a Ten-Year Study of Pollution at a Site in the City of London, *Atm. Env.* 1:49–68.

Derriennic, F., Richardson, S., Mollie, A., and Lellouch, J. (1989), Short-Term Effects of Sulphur Dioxide Pollution on Mortality in Two French Cities, *Int. J. Epidemiology* 18:186–97.

Derwent, R.G., and Stewart, H.N.M. (1973), Elevated Ozone Levels in the Air of Central London, *Nature* 241:342.

Dockery, D.W., Schwartz, J., and Spengler, J.D. (1992), Air Pollution and Daily Mortality: Associations with Particulates and Acid Aerosols, *Envir. Res.* 59:362–73.

Dockery, D.W., Speizer, F.E., Stram, D.O., Ware, J.H., Spengler, J.D., and Frris, B.G., Jr. (1989), Effects of Inhalable Particles on Respiratory Health of Children, *Am. Rev. Resp. Dis.* 139:587–94.

Ellison, J. McK., and Waller, T.E. (1978), *Env. Res.* 16:302.

Fairley, D. (1990), The Relationship of Daily Mortality to Suspended Particulates in Santa Clara County, 1980–1986, *Env. Health Persp.* 89:159–68.

Fleisher, J.M., and Nayeri, K. (1991), Mortality and Air Pollution in London: A Time-Series Analysis (letter), *Am. J. Epidemiology* 133:631–32.

Glasser, M., Greenburg, L., and Field, F. (1967), Mortality and Morbidity during a Period of High Levels of Air Pollution, *Arch. Env. Health* 15:684–94.

Glasser, M., and Greenburg, L. (1971), Air Pollution and Weather, *Arch. Env. Health* 22:334–43.

Goldstein, I.F. (1976), *Use of Aerometric Data to Monitor Health Effects*, APCA Paper 76–32.6, presented at the 69th Annual Meeting, APCA, Portland, OR.

Goldstein, I.F., Landau, E., and Van Ryzin, J. (1983), Letter to the Editor, *Arch. Env. Health* 38:122.

Gore, A.T., and Shaddick, C.W. (1958), Atmospheric Pollution and Mortality in the County of London, *Br. J. Prev. Soc. Med.* 12:104.

Greenburg, L., Jacobs, M.B., Drolette, B.M., Field, F., and Braverman, M.M. (1962), *Report of an Air Pollution Incident in New York City*, November 1953, Public Health Reports 77:7.

Greenburg, L., Field, F., Ehrardt, C.L., Glasser, M., and Reed, J.I. (1967), Air Pollution, Influenza, and Mortality in New York City, *Arch. Env. Health.* 15:430.

Hammerstrom, T., Silvers, A., Roth, N., and Lipfert, F. (1991), Statistical Issues in the Analysis of Risks of Acute Air Pollution Exposure, *Risk Analysis*, submitted.

Hatzakis, A., Katsouyanni, K., Kalandidi, A., Day, N., and Trichopoulos, D. (1986), Short-Term Effects of Air Pollution on Mortality in Athens, *Int. J. Epidemiology* 15:73–81.

Heimann, H. (1970), Episodic Air Pollution in Metropolitan Boston, *Arch. Env. Health* 20:239–51.

Hexter, A.C., and Goldsmith, J.R. (1971), Carbon Monoxide: Association of Community Air Pollution with Mortality, *Science* 172:265–67.

Hodge, W.T., and Nicodemus, M.L. (1980), Human Biometeorology: A Selected Bibliography, NOAA Technical Memorandum EDIS NCC-4.

Hodgson, T.A., Jr. (1970), Short-Term Effects of Air Pollution on Mortality in New York City, *Env. Sci. Tech.* 4:589.

Holland, W.W., *et al.* (1979), *Am. J. Epidemiology* 111:525.

Ito, K., and Thurston, G.D. (1989), Characterization and Reconstruction of Historical London, England, Acidic Aerosol Concentrations, *Env. Health Persp.* 79:35–42.

Ito, K., Thurston, G.D., and Lippmann, M. (1991), Association of Daily Mortality with Ambient Exposure to an Air Pollutant Mixture: Particulate Matter, Sulfur Dioxide, and Acidic Aerosols, Paper 91-137.1, presented at the 84th Annual Meeting of the Air and Waste Management Association, Vancouver.

Junge, C., and Scheich, G. (1971), Determination of the Acid Content of Aerosol Particles, *Atm. Env.* 5:165–75.

Katsouyanni, K., Hatzakis, A., Kalandidi, A., Trichopolous, D. (1990a), Short-Term Effects of Atmospheric Pollution on Mortality in Athens, *Arch. Hellen. Med.* 7: 126–32 (in Greek).

Katsouyanni, K., *et al.* (1990b), Air Pollution and Cause Specific Mortality in Athens, *J. Epid. Comm. Health* 44:321–24.

Kevany, J., Rooney, M., and Kennedy, J. (1975), Health Effects of Air Pollution in Dublin, *Irish J. Med. Sci.* 144:102.

Kinney, P.L., and Özkaynak, H. (1991), Associations of Daily Mortality and Air Pollution in Los Angeles County, *Env. Res.* 54:99–120.

Kunst, A.E., Loman, C.W.N., and Mackenbach, J.P. (1991), Determinanten van binnenjaarlijkse fluctaties in der sterfte (Determinants of Within-Year Fluctuations in Mortality), *T. Soc. Gezondheidsz* 69:123–31.

Lebowitz, M.D. (1973), A Comparative Analysis of the Stimulus-Response Relationship between Mortality and Air Pollution-Weather, *Env. Res.* 6:110–18.

Lebowitz, M.D., Toyama, T., and McCarroll, J. (1973), The Relationship between Air Pollution and Weather as Stimuli and Daily Mortality as Responses in Tokyo, Japan, with Comparisons with Other Cities, *Env. Res.* 6:327–33.

Liang, K.Y., and Zeger, S.L. (1986), Longitudinal Data Analysis Using Generalized Linear Models, *Biometrika* 73:13–22.

Lipfert, F.W. (1977), The Association of Air Pollution and Human Mortality: A Review of Previous Studies, APCA Paper 44.1, presented at the 70th APCA Annual Meeting, Toronto.

Lipfert, F.W. (1978), *The Association of Human Mortality with Air Pollution: Statistical Analyses by Region, by Age, and by Cause of Death*, Eureka Publications, Mantua, NJ.

Lipfert, F.W. (1988), *Exposure to Acidic Sulfates in the Atmosphere: A Review and Assessment*, Electric Power Research Institute Report EA-6150.

Lipfert F.W., and Hammerstrom, T. (1991), Temporal Patterns in Hospital Admissions and Air Pollution, submitted.

Lipfert, F.W. (1992), *Community Air Pollution and Mortality: Analysis of 1980 Data from U.S. Metropolitan Areas*, report prepared for U.S. Department of Energy, Brookhaven National Laboratory.

Lipfert, F.W., Malone, R.G., Daum, M.L., Mendell, N.R., and Yang, C-C. (1988), *A Statistical Study of the Macroepidemiology of Air Pollution and Total Mortality*, BNL Report 52122, U.S. Dept. of Energy.

Loewenstein, J.-C., Bourdel, M.-C., and Bertin, M. (1983), Influence de la pollution atmosphérique (SO_2 poussières) et des conditions météorologiques sur la mortalité à Paris entre 1969 et 1976, *Rev. Epidém. et Santé Publique* 31:163.

Logan, W.P.D. (Jan. 8, 1949), Fog and Mortality, *Lancet* 78.

McCarroll J., and Bradley, W. (1966), Excess Mortality as an Indicator of Health Effects of Air Pollution, *Amer. J. Public Health* 56:1933.

Macfarlane, A. (1977), Daily Mortality and Environment in English Conurbations. 1. Air Pollution, Low Temperature, and Influenza in Greater London, *Brit. J. Prev. Soc. Med.* 31:54–61.

Macfarlane A., and White, G. (1977), Deaths: The Weekly Cycle, *Population trends* 7:7–78.

Mackenbach, J.P., Looman, C.W.N., and Kunst, A.E. (1992), Sulfur Dioxide Air Pollution, Lagged Effects of temperature, and Mortality: The Netherlands, 1979–1987, *J. Epid. Comm. Health*, submitted.

Martin, A.E. (1964), Mortality and Morbidity Statistics and Air Pollution, *Proc. Roy. Soc. Med.* 57:969–75.

Martin, A.E., and Bradley, W.H. (1960), Mortality, Fog, and Atmospheric Pollution—An Investigation during the Winter of 1958–59, *Mon. Bull. Minist. Health Service Lab. Serv.* 19:56–73.

Mazumdar, S., and Sussman, N. (1983), Relationships of air pollution to health, *Arch. Env. Health* 38:17–24.

Mazumdar, S., Schimmel, H., and Higgins, I. (1980), *Relation of Air Pollution to Mortality, An Exploration Using Daily Data for 14 London Winters*, 1958–72, report prepared for the Electric Power Research Institute, Palo Alto, CA.

Mazumdar, S., Schimmel, H., and Higgins, I. (1982), Relation of Daily Mortality to Air Pollution: An Analysis of 14 London Winters, 1958/59–1971/72, *Arch. Env. Health* 37:213.

Mazumdar, S., Schimmel, H., and Higgins, I. (1983), Response to Letter to the Editor, *Arch. Env. Health* 38:123.

Mills, C.A. (1960), Respiratory and Cardiac Deaths in Los Angeles Smogs during 1956, 1957, and 1958, *Arc. Med. Sul.* p. 9981, 307–15.

Neuberger, M., Rutkowski, A., Friza, H., and Haider, M. (1987), Grippe, Luftverunreinigung und Mortalität in Wien, *Forum-Staedt-Hygiene* 38:7 (in German).

Ostro, B. (1984), A Search for a Threshold in the Relationship of Air Pollution to Mortality: A Reanalysis of Data on London Winters, *Env. Health persp.* 58:397–99.

Özkaynak, H., and Spengler, J.D. (1985), Analysis of Health Effects Resulting from Population Exposures to Acid Precipitation Precursors, *Env. Health Perspectives* 64:45.

Özkaynak, H., Spengler, J.D., Garsd, A., and Thurston, G.D. (1986), Assessment of Population Health Risks Resulting from Exposures to Airborne Particles, in *Aerosols: Research, Risk Assessment, and Control Strategies*, S.D. Lee, T. Schneider, L.D. Grant, and P.J. Verkerk, eds., Lewis Publishers, Chelsea, Michigan, pp. 1067–80.

Özkaynak, H., Kinney, P.L., and Burbank, B. (1990), Recent Epidemiological Findings on Morbidity and Mortality Effects of Ozone, Paper #90-150.6, presented at the Annual Meeting Air & Waste Mgmt. Assoc.

Pickles, J.H. (1980), *The Use of Regression Models in Epidemiological Analyses of Daily Mortality Data*, Laboratory Note RD/L/N 213/79, Central Electricity Research Laboratories, Leatherhead, UK.

Pope, C.A., III, Schwartz, J., and Ransome, M.R. (1992), Daily Mortality and PM_{10} Pollution in Utah Valley, *Arch. Env. Health* 47:211–17.

Rogot, E., and Blackwelder, W.C. (1970), *Public Health Rep.* 85:25.

Roth Associates, Inc. (1986), *Assessment of the London Mortality Data*, draft report for the Electric Power Research Institute, Palo Alto, CA.

Roth, H.D., Wyzga, R.E., and Hayter, A.J. (1986), Methods and Problems in Estimating Health Risks from Particulates, in *Aerosols: Research, Risk Assessment, and Control Strategies*, S.D. Lee, T. Schneider, L.D. Grant, and P.J. Verkerk, eds., Lewis Publishers, Chelsea, Michigan, pp. 1047–66.

Russell, W.T. (1924), The Influence of Fog on the Mortality from Respiratory Disease, *Lancet* 2:335–39.

Russell, W.T. (1926), The Relative Influence of Fog and Low Temperature on the Mortality from Respiratory Disease, *Lancet* 2:1128–30.

Schimmel, H. (1978), Evidence for Possible Acute Health Effects of Ambient Air Pollution from Time Series Analysis: Methodological Questions and Some New Results Based on New York City Daily Mortality, 1963–1976, *Bull. NY Acad. Med.* 54:1052–1109.

Schimmel H., and Greenburg, L. (1972), A Study of the Relation of Pollution to Mortality, New York City, 1963–1968, *J. APCA* 22:607–16.

Schimmel, H., and Murawski, T.J. (1976), The Relation of Air Pollution to Mortality, *J. Occ. Med.* 18:316–33.

Schimmel, H., Murawski, T.J., and Gutfield, N. (1974), Relation of Pollution to Mortality, APCA Paper 74-220, presented at the 67th Annual Mtg. APCA.

Schwartz, J. (1991a), Response to Letter by Fleisher and Nayeri, *Am. J. Epidemiology* 133:632.

Schwartz, J. (1991b), Particulate Air Pollution and Daily Mortality in Detroit, *Env. Res.* 56:204–13.

Schwartz, J. (1993), Air Pollution and Daily Mortality in Birmingham, AL, *Am. J. Epidemiol.* 137:1136–47.

Schwartz, J. (in press [a]), What Are People Dying of on High Pollution Days? *Env. Res.*

Schwartz, J. (in press [b]), Particulate Air Pollution and Daily Mortality: A Synthesis, *Public Health Reviews*.

Schwartz, J., and Dockery, D.W. (1992a), Increased Mortality in Philadelphia Associated with Daily Air Pollution Concentrations, *Am. Rev. Resp. Dis.* 145:600–4.
Schwartz, J., and Dockery, D.W. (1992b), Particulte Air Pollution and Daily Mortality in Steubenville, Ohio, *Am. J. Epidemiology* 135:12–25.
Schwartz, J., and Marcus, A.H. (1986), Statistical Reanalyses of Data Relating Mortality to Air Pollution during London Winters 1958–1972, U.S. Environmental Protection Agency, Washington, DC. Also see *Am. J. Epidemiology* 131:185–94 (1990) and 133:632–33 (1991).
Shils, M.E., and Skolnick, J.N., eds. (1978), Proc. Symposium on Environmental Effects of Sulfur Oxides and Related Particulates—1978, *Bull. NY Acad. Med.* 54:983–1278.
Shumway, R.H., Azari, A.S., and Pawitan, Y. (1988), Modeling Mortality Fluctuations in Los Angeles as Functions of Pollution and Weather Effects, *Env. Res.* 45:224–41.
Shumway, R.H., Tai, R.Y., Tai, L.P., and Pawitan, Y. (1983), Statistical Analysis of Daily London Mortality and Associated Weather and Pollution Effects, California Air Resources Board, Sacramento, CA.
Simon, C. (1978), Discussion of paper by Merill Eisenbud, *Bull. NY Acad. Med.* 54:1012–24.
Thurston, G.D., Ito, K., Lippmann, M., and Hayes, C. (1989), Reexamination of London, England, Mortality in Relation to Exposure to Acidic Aerosols during 1963–72 Winters, *Env. Health Persp.* 79:73–82.
Watanabe, H., and Kaneko, F. (1970), Excess Death Study of Air Pollution, *Proc. 2nd. Int. Clean Air Congress*, pp. 199–201.
Wolman, A. (1965), The Metabolism of Cities, *Scientific American* 213:186.
Wyzga, R.E. (1977), Urban Air Pollution and Mortality: Ten Years of Philadelphia Data, *Proc. Social Stat. Section*, American Statistical Association, Washington, DC, pp. 660–65.
Wyzga, R.E. (1978), The Effect of Air Pollution upon Mortality: A Consideration of Distributed Lag Models, *J. Amer. Stat. Assoc.* 73:463–72.

APPENDIX 6A THE STIMULUS-RESPONSE METHOD OF ANALYSIS

Lebowitz *et al.* (1973) devised a methodology for examining the concordance between daily mortality and the daily environment without recourse to formal time-series methods. This methodology is interesting to the present inquiry, because it treats simultaneous peaks in one or more pollutants and weather variables as useful (with respect to the power of the analysis), rather than as something to be avoided. Variables were converted to standard scores, and the analysis was done separately for summers and winters. Events were identified for which the standard scores exceeded 1.0. The average of several air pollutants and temperature or humidity scores constituted the file of stimuli, which could be any number of consecutive days. The magnitude of the stimulus was the sum of its daily average scores (i.e., the total dose). In this sense, the analysis follows the treatment of episodes as discussed in Chapter 5. The responses were the mortality scores, and pairs were identified in which a response occurred either during a stimulus or from 1 to 3 days later.

Lebowitz (1973) examined data from New York City (1962–65), Philadelphia (1963–64), and Los Angeles (1962–65). Lebowitz, Toyama, and McCarroll analyzed data from Tokyo, 1966–69. In all cases, there were more stimulus-response pairs than would have occurred randomly, and the magnitudes of the stimuli with responses were significantly greater than for the stimuli without reponses. In addition, the magnitudes of the responses with stimuli were greater than those of the responses that occurred without stimuli. Lebowitz thus concluded that "the relationship of abnormal environment and high (excess) mortality is very significant."

REFERENCE

Lebowitz, M.D., Toyama, T., and McCarroll, J. (1973), The Relationship between Air Pollution and Weather as Stimuli and Daily Mortality as Responses in Tokyo, Japan, with Comparisons with Other Cities, *Env. Res.* 6:327–333.

7

Cross-Sectional Studies of Long-Term Effects on Mortality

There are lies, damned lies, and statistics.

Mark Twain

As stated by Lazar (1981), "An observed variance is at the source of any epidemiological knowledge." In the short-term studies discussed in Chapters 5 and 6, these variances occurred over time. Since populations, life-styles, and medical practices tend to change over time in the long term, it has usually been more convenient to use variations with respect to place (location) to examine the long-term effects of air pollution.

The study of long-term or chronic health effects of air pollution has been fraught with difficulty and controversy, more so than the short-term studies (Smith, 1975; Lipfert, 1980; Ware *et al.*, 1981; Ricci and Wyzga, 1983; Evans *et al.*, 1984a). In this chapter, the primary method of analysis considered involves comparing the health statistics of populations of places that have had different environments over the long term. However, the comparisons are often complicated or even compromised by other differences that may be related to the sources of air pollution, such as industrialization. Long-term studies often use data from only one specific year, which may or may not be truly representative of the long term.

Spatial patterns of U.S. mortality rates show some well-defined trends, as depicted in Figures 7-1 and 7-2, which were taken from the Health Care Financing Administration (HFCA, 1990) report on 1986 Medicare hospitalization and mortality, for persons 65 and over. In general, heart disease is higher east of the Mississippi (Figure 7-1a); ischemic heart disease shows even sharper gradients and peaks in the Northeast (Figure 7-1b). Cancer death rates tend to be higher in the Northeast, but not exclusively in industrialized states (note Vermont, New Hampshire, and Maine, as well as Alaska) (Figure 7-2a). Gorham *et al.* (1990) argue that breast cancer is higher in northern latitudes because of reduced intake of vitamin D; such patterns are seen in the Soviet Union, for example. Lung cancer deaths are more evenly distributed (Figure 7-2b) and the hot spots include some of the states with high tobacco use

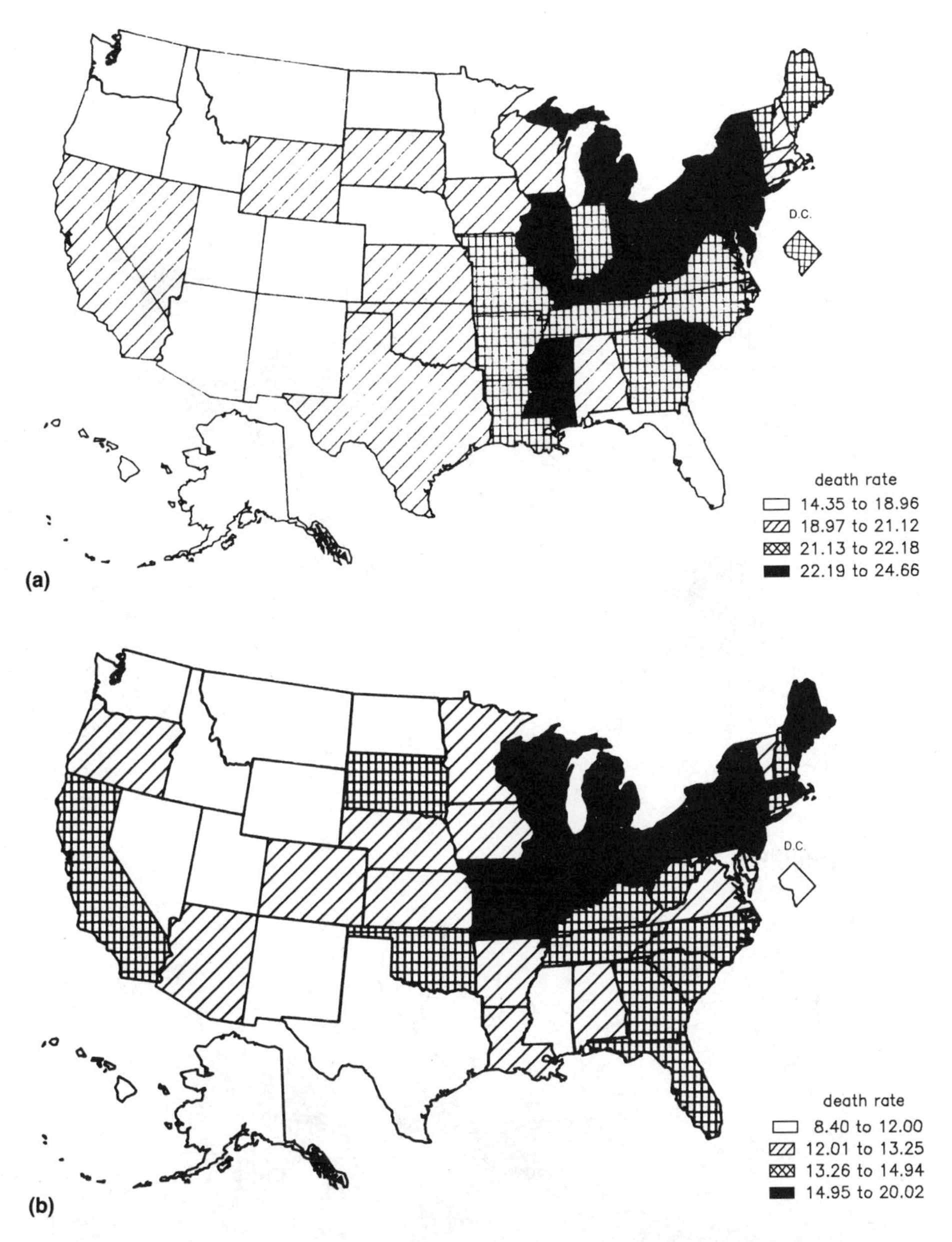

FIGURE 7-1. Maps of 1986 heart disease mortality rates for the population aged 65 and over. (a) All heart disease. (b) Ischemic heart disease. *Source: Health Care Financing Corporation (1990).*

(Nevada, Kentucky, Virginia, but not North Carolina). Pneumonia and influenza deaths are distributed across the country but tend to be higher north of about the 36th parallel. These patterns may suggest associations with certain air pollutants as well as with other variables; the challenge to the epidemiologist is to establish whether the associations are causal or merely circumstantial.

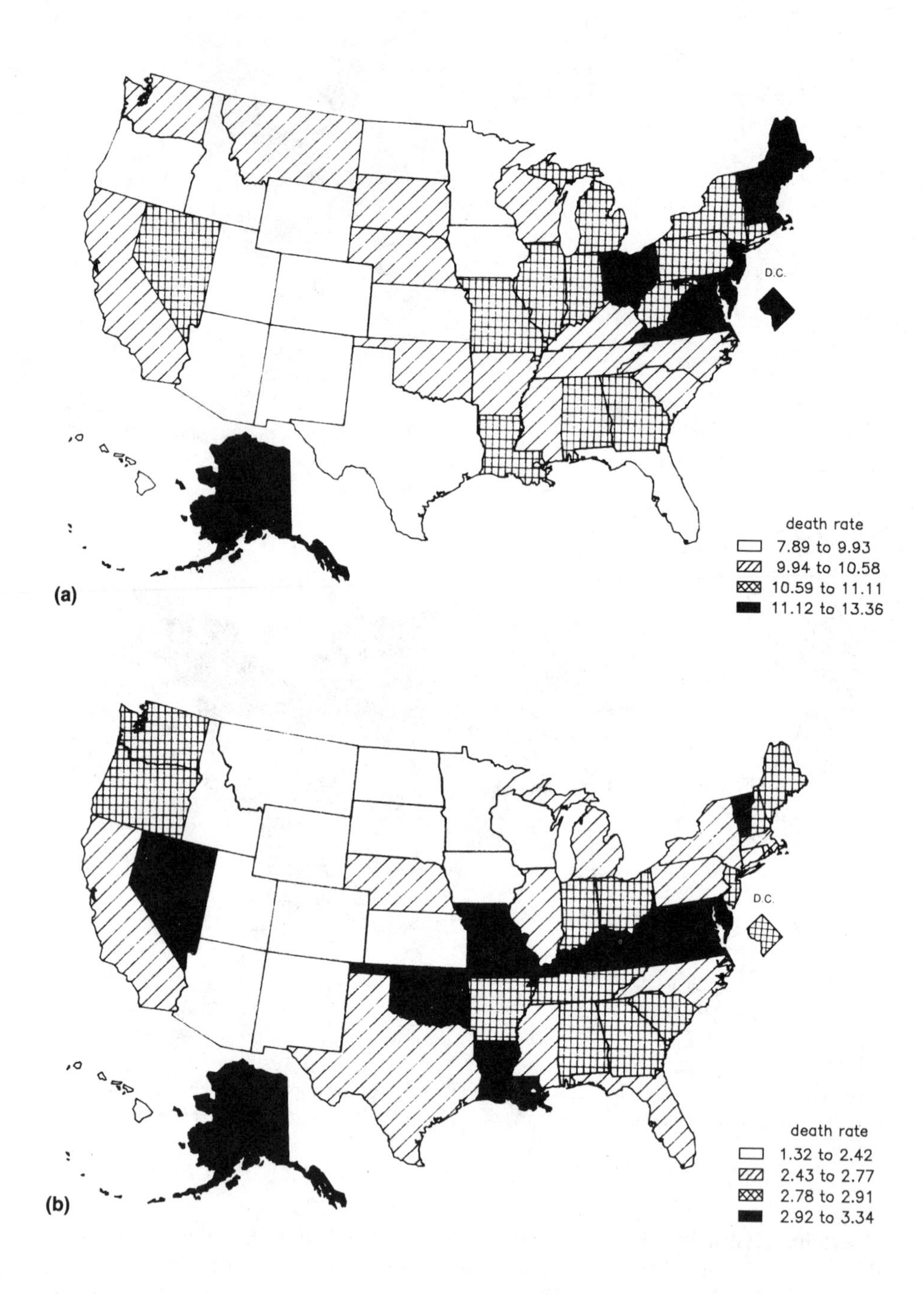

FIGURE 7-2. Maps of 1986 cancer mortality rates for the population aged 65 and over. (a) All cancers. (b) Respiratory system cancer. *Source: Health Care Financing Corporation (1990).*

METHODOLOGICAL ISSUES AND INDEPENDENT VARIABLES FOR CROSS-SECTIONAL STUDIES

A general discussion of statistical and methodological issues was presented in Chapter 3. Cross-sectional studies are intended to reveal associations between air pollution and health by virtue of similarities in spatial patterns, averaged over some suitably long term. In order to be able to interpret any resulting statistical associations as potentially causal, one must account for all other spatial factors that covary with mortality rates or air pollution. One must also be aware of any spatially varying temporal factors, such as flu epidemics, heat waves, and so on, that might influence health in certain locations. This section briefly reviews some of the common sources of spatial variation and discusses some of the regression variables that various authors have used as controls for these effects.

Demography. Mortality rates vary by age, race, and sex, both within a given population and between communities, as a reflection of differences in population averages. Some cross-sectional studies have been able to work with specific population groups, such as white females, aged 45–64, for example. However, the gain in specificity resulting from such subdivision may be offset by Poissonian noise arising from the smaller numbers of deaths. Increasing the size of the geographic units of analysis to reduce random noise creates errors in the estimates of exposure to air pollution.

The choices for the dependent variables used in a cross-sectional regression may then include age-sex-race *specific* mortality rates, age-sex-race *adjusted* rates, or *crude* (unadjusted) rates. A variant in age-sex-race adjustment involves computing the observed-expected mortality ratio, based on the demography of the community in question. Unless otherwise specified, in this chapter *mortality rate* refers to the *crude* rate.

When crude mortality rates are used, independent variables are required that describe the demography of each community in the analysis. These may include the percentages of the population found in certain age or race strata. "Percentage of population 65 years of age or more" is the most common population age descriptor variable. Statistics on median age add little new information where it is most needed: the age distribution within the 65-and-older group. In 1985, 70% of U.S. deaths occurred in this age group (74% when accidents, suicides, and homicides are subtracted).

Population variables describing racial and ethnic distributions include percentage of Black, other nonwhite (Asian, American Indian, etc.), and Hispanic persons. Each of these groups tends to have mortality rates different from whites and thus cities with higher than average percentages of these groups would be expected to have correspondingly different mortality rates for the total population. In the 1980 U.S. Census, race was self-defined, which leads to a certain amount of confusion, mainly in heavily Hispanic cities in the Southwest. Many of the older cross-sectional studies used the combined percentage of all nonwhites, which is no longer an adequate racial descriptor in the United States.

Socioeconomic Factors. It has long been known that the upper socioeconomic strata of a society tend to enjoy better health (Rogot *et al.*, 1992).

This may result from better access to medical care or from better personal health habits that may come from better education. Per capita income, poverty status, and educational attainment are often used as general indicators of personal socioeconomic status. Education may be a better socioeconomic variable than income, since some persons may have a low income because they have poor health, not vice versa. Educational attainment as a socioeconomic indicator will not be changed by subsequent illness; Liberatos *et al.* (1988) found that education was the indicator of social class most frequently used in 76 epidemiological studies they reviewed. Poverty status may be a better health-related variable than, say, median income, since, if there is an effect of income on mortality, it would be expected to be most obvious at the low end of the scale. Very wealthy people do not necessarily enjoy better health. Similarly, attainment of literacy may be a better health-related education variable than attendance at college. Some of these trends are displayed in Appendix 7A.

The socioeconomic "health" of a community may be reflected by its rate of population growth (in- versus outmigration). Ill health can be a factor in the decision to migrate; one hypothesis is that healthy and more economically advantaged people are more likely to retire to advantageous climates (Bultena, 1969), leaving behind those less able to cope. It has also been shown that a (smaller) reverse migration stream of retirees can take place after the loss of a spouse or the onset of ill health, to take advantage of family and other support services (Patrick, 1980). Migration thus can have both direct and indirect effects on cross-sectional analyses, since long-term exposure to air pollution is also affected by migration. However, Cohen (1992) recently reported results from a telephone survey that estimated that people spend over 70% of their lives within 25 miles of their location of death, as a national average. These percentages are higher in the Northeast (up to 90%) and lower in Florida, California, and Arizona (ca. 50% in these high-migration states). At least one of the cross-sectional studies reviewed included a variable on the percentage of residents living in their state of birth.

Life-Style Variables. Most of the improvement in life expectancy that was attained in the United States following the 1970s has been attributed to better personal habits: more exercise, better diet, less tobacco consumption. These variables are difficult to measure across an entire community. Smoking is the only one for which quantification attempts have been made on a national scale (Lipfert, 1978; Lipfert *et al.*, 1988), although survey data on prevalence of certain other risk factors are now becoming available for most of the states.

Earlier analyses of smoking patterns typically found large urban-rural differences, and it has long been assumed that city people smoke more. In a study of 1980 smoking data (Lipfert *et al.*, 1988), SMSA tobacco sales data from the 1977 Census of Retail Trade were compared to statewide sales data from the same source, and a consistent urban-rural relationship was found, amounting to an annual urban-statewide difference of about five packs per year per person (out of an average of 185 per year). This small (but still statistically significant) difference suggests that regional smoking patterns are now probably more important than urban-rural differences within regions, which supports the use of state-level data in the analysis of mortality effects.

Finally, a comparison was made of estimates of cigarette consumption with independent state-level survey data on the percentage of people who smoke (smoking prevalence). The correlation coefficient relating these two measures was only about 0.5 (explaining 25% of the variance) for the 29 states that had conducted surveys. Possible explanations for this rather poor result include variations in amount consumed per smoker and underreporting by those responding to the survey. I prefer to use consumption data rather than data on prevalence of the habit, because heavy smokers have a much higher relative mortality risk than light smokers, and since consumption data may reflect the possible effects of passive (involuntary) smoking. Cigarette consumption rates are analogous to air pollution emission rates in this sense.

For an analysis of chronic health effects, it is not clear whether current cigarette smoking rates or some integral over time is the appropriate metric (this question exists for air quality data as well).

Occupational Stresses. Some occupations involve direct and indirect health risks, including exposure to toxic agents (for manual workers) and lack of exercise and job-related stress (for office workers). Also, some occupations and industries are characterized by important differences in personal habits (Brackbill *et al.*, 1988).

In previous studies, iron (Fe) and manganese (Mn) have been found to be significant predictors of spatial variations in mortality (Lipfert, 1978; Lipfert, 1984; Lipfert *et al.*, 1988). However, these species are also markers for ferrous metal manufacturing activities, which may have other associations with health, either directly because of occupational hazards or indirectly because of life-style differences. For example, Brackbill *et al.* (1988) found that the metal industries were among the highest in terms of percentages of workers who smoke. Lipfert (1984) found that Mn was only significant for males (65+), which suggests long-term occupational effects rather than community air pollution *per se*.

Regional versus Local Characteristics. Although the United States has tended to become more homogeneous in recent years, there are still regional differences in population ethnicity, culture, diet, and the nature of the industrial and commercial base. Such gradients exist in other developed countries as well. Some of these factors contribute to regional patterns in health status. To the extent that there are also regional gradients in the types of air pollution present, notably sulfur oxides, the opportunity for confounding may exist. However, air pollution patterns are inevitably characterized by local perturbations about the regional patterns, and the analyst should try to determine whether statistical associations result primarily from the regional patterns or from the local perturbations. This is analogous to separating daily perturbations from seasonal trends in a time-series analysis.

Climatic Factors. Beginning with Hippocrates, both climate and weather have been shown to influence longevity and health (see Chapter 8). Few studies have attempted to capture these effects, however. For example, average annual heating degree days is essentially a climate variable reflecting long-term rather than current weather conditions. A more relevant variable is probably the severity of any heat waves that occurred during the period of study. Heat wave mortality has been shown to respond to the extent and duration of departures of daily maximum temperatures above normal, especially above about 33°C (Kalkstein, 1991). A necessary auxiliary variable in this regard is the prevalence of air conditioning, which not only protects against heat but also against outdoor air pollution.

Other Environmental Factors. As discussed in Chapter 2, before the conquest of infectious diseases, population density was an important determinant of mortality (Farr, 1885). When applied to county or larger units, this statistic is now of limited use because of the heterogeneity of land use typically found within larger areas, and it is unlikely to capture the local density at which people actually live. However, even average population density can distinguish the 100% urban SMSAs (such as Jersey City) from most of the others, which are usually mixed urban and suburban. When measured properly, population density may reflect the combined effects of air pollution from urban area sources, exposure to infectious disease agents, and access to medical care.

Indoor air pollution sources (other than smoking) should also be considered in a cross-sectional study to the extent that they may differ from place to place. Census data on the types of space heating devices and fuels in use may be helpful in this regard.

Other possible environmental factors influencing health may include the quality of drinking water supplies; previous studies (Lacey, 1981; Lipfert, 1984) have implicated soft water as a contributing factor in heart disease, primarily for males.

Temporal Interactions. The cross-sectional analyses reviewed in this chapter are intended to define associations between premature mortality and air pollution by analyzing spatial patterns. Typically, annual mortality rates are computed for a set of observational units (political subdivisions), appropriate demographic control variables are defined and data obtained from census sources, and air pollution exposures are estimated from extant monitoring data. Rumford (1961) pointed out the necessity of considering the "total socioeconomic complex" if valid estimates of the pollution effects are to be obtained. Ideally, these data will all be for the same period, unless it is desired to test for a long-term lag between air pollution exposure and subsequent mortality (when pollution levels have varied over time). Multiple regression analysis is then used to develop a "model" for mortality rates. These studies are usually described as seeking chronic health effects, since no account is taken of daily variations, except as reflected in the annual averages or totals. However, it is generally not possible to distinguish between long-term (chronic) effects and the annual sum of acute effects when only 1 year is analyzed. To estimate long-term effects where the situation has changed over time, historic values of the independent variables (especially air pollution exposures, i.e., cumulative dose) should be used.

Certain years may be unique in this regard. For example, 1980 (a census year and thus important for cross-sectional analyses) was marked by a major volcanic explosion (Mt. St. Helens) and by a severe heat wave and drought. Events of this nature may have both direct and indirect effects on any mortality–air pollution relationships.

Accuracy of Mortality Rates. Death registration in the United States is now complete; however, there may be uncertainties as to the cause of death listed on the death certificate. The studies reviewed here rely on the primary cause of death; however, contributing causes may be important with respect to air pollution etiology, and these have rarely been studied. There is an extensive literature on the validation of causes of death (Gittelsohn and Royston, 1982); these studies indicate that the quality of information on the death certificate "varies greatly by cause of death and characteristics of the decedent." It thus follows that regional biases could exist as well and could be a source of error in cross-sectional studies. For this reason, it may be difficult to interpret changes in statistical significance as one considers particular rather than general causes of death.

Mortality rates can also be affected by geographic coding and by uncertainties in population figures. Census undercounts are a problem, and occasional discrepancies in published figures have been discovered (Lipfert, 1978).

ORGANIZATION OF THE CHAPTER

Because of the importance of the size of the unit used for geographic analysis in ecological regressions (see Chapter 3), the reviewed studies are grouped in increasing order of the size of these units. Section 1 deals with analyses within cities, for example, based on census tracts. Section 2 reviews studies of whole cities, and Section 3, studies of counties and SMSAs. We assume that English "county boroughs" are analogous to U.S. cities, since each comprises less than a whole county. Finally, Section 4 reviews studies based on even larger geographic units, such as groups of counties or entire states. Within each of the sections, we begin with the earliest studies; foreign studies are included as appropriate within these groupings. Studies that examined more than one type of geographic unit are grouped with the unit for which most of the analysis was done. As was the case with time-series studies in Chapter 6, some of the early studies are given more attention than their age might seem to warrant; I find many of these studies interesting because of the typically higher levels of air pollution and sharper geographic gradients, in addition to the historical perspective they provide.

SECTION 1: INTRAURBAN CROSS-SECTIONAL STUDIES

Early Studies Based on Dustfall Measurements

Perhaps the first paper on air pollution and health in the United States was presented by White and Marcy (1913) at the 15th International Congress on Hygiene and Demography, held in Washington in 1912. They presented data on smoke density, as measured by monthly average sootfall, population density, and pneumonia and tuberculosis (crude) death rates for the 27 Wards of Pittsburgh, as reproduced in Figure 7-3. White and Marcy cited prior work in Germany by Ascher, who concluded, on the basis of animal experiments and epidemiological observations, that there was a relationship between smoke and deaths from acute lung diseases. However, although White and Marcy noted the strong relationship between air pollution and pneumonia (and a weaker relationship with tuberculosis) within Pittsburgh, they were unable to confirm such a relationship by comparing various U.S. cities. There was little influence of population density (intended to be an index of poverty) on these rates; the apparent elasticities were about 1.0 for pneumonia (R = 0.93) and 0.35 for TB

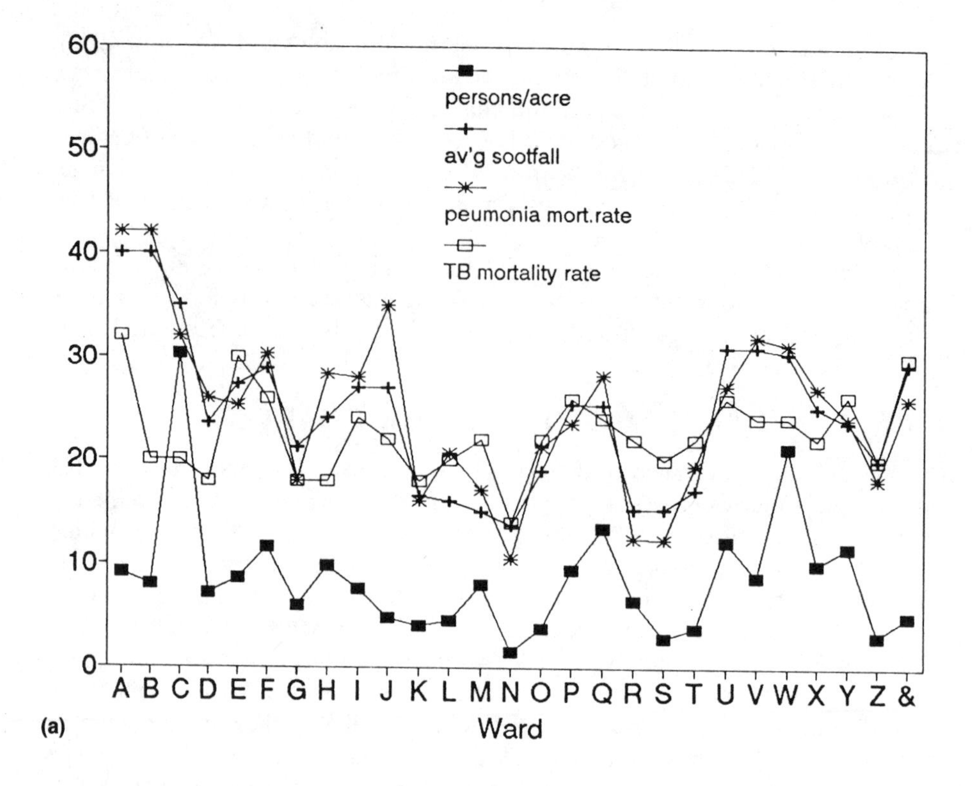

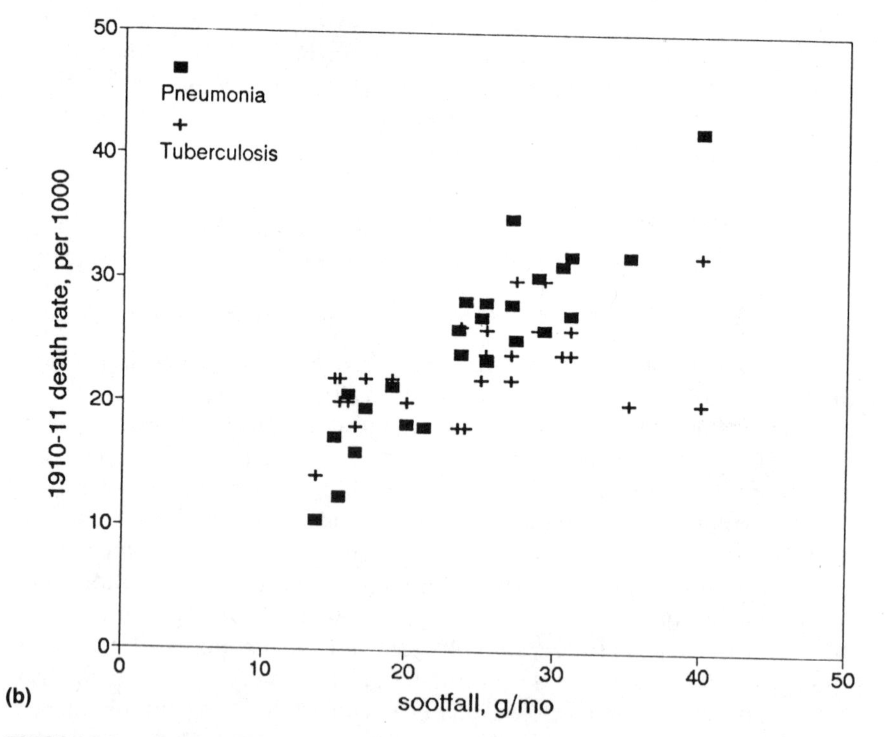

FIGURE 7-3. Pneumonia and air pollution in Pittsburgh wards, 1912. (a) Data by ward. (b) Mortality versus sootfall. *Data from White and Marcy (1913).*

(R = 0.54). These estimates could have been confounded by lack of age adjustment or by other socioeconomic factors. Note that the units of measure reported in this study appear questionable; sootfall is not referred to a collection area, and death rates appear to be per 10,000 population, not per 1000 population.

Mills (1943) continued this type of analysis, using dustfall data from districts in Cincinnati and Pittsburgh, and later from Chicago, Detroit, Pittsburgh, Cincinnati, Nashville and Atlanta (Mills and Mills-Porter, 1948). The earlier data from Cincinnati are shown in Figure 7-4; note that the pneumonia rates in Cincinnati are about one-third of the Pittsburgh values (assuming that White and Marcy's data were too high by a factor of 10; Mills's pneumonia data for Pittsburgh were in the range of 25 to 95 per 10,000. The U.S. average rate was 7 per 10,000). The elasticity in Cincinnati was about 0.28 (R = 0.79).

The mortality indices used in the 1948 paper were deaths for white males and females for various years from 1929 to 1946 due to pneumonia, tuberculosis, and respiratory cancer (Chicago and Detroit only). The statistic used to establish a relationship between mortality and air pollution was the chi-square test for differences between the five dirtiest and the five cleanest areas within a city. These differences were statistically significant for males in all cases, but the effects for females were often substantially smaller. This result could indicate confounding either by smoking (which the authors recognized as a possible synergistic effect) or by occupational exposures. Also, no mention was made of age adjustment, a possible further source of confounding. Comparison

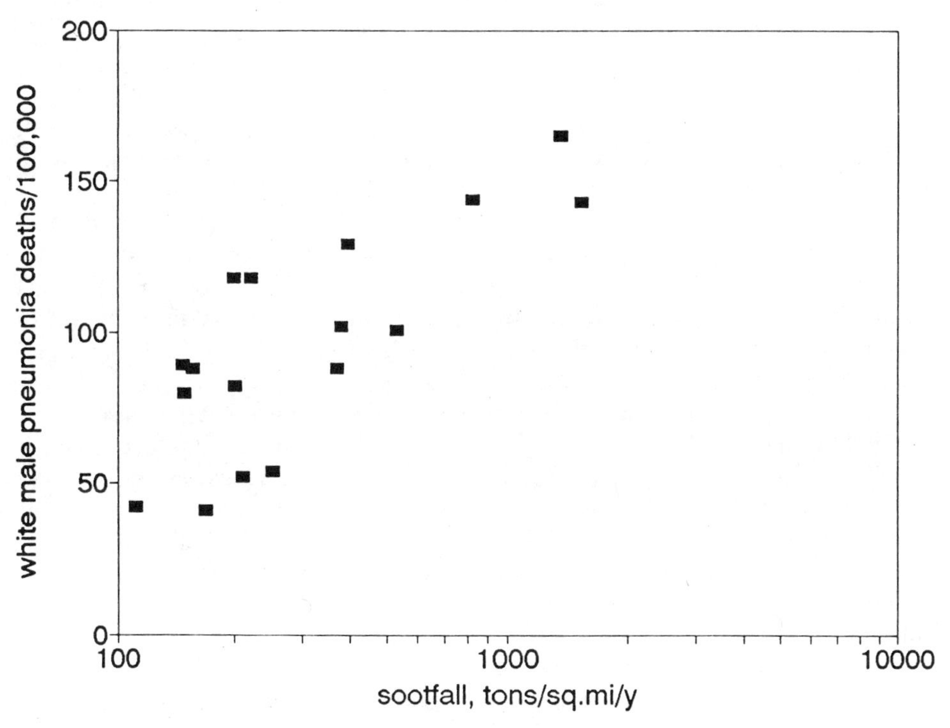

FIGURE 7-4. Pneumonia death rates vs. sootfall for Cincinnati districts, 1929–31. *Data from Mills (1943).*

of the mortality rates for the "cleanest" areas of each city for the same time periods gave no suggestion of regional effects or effects of city size; the "clean" rates in Nashville were substantially higher than in the other cities.

In a subsequent report, Mills (1952) extended the Chicago analysis to include age-specific mortality (no changes in conclusions) and discussed the parallel air quality trends for SO_2, confirming that the dustfall statistics used in the first paper should be regarded as a general index of air pollution.

Rumford (1961) described three early rudimentary cross-sectional studies, one of which was based on dustfall measurements in Philadelphia. That study found the strongest air pollution effects (in relation to four different socio-economic indices) for chronic rheumatic heart disease, followed by arterio-sclerotic heart disease and pneumonia. The other studies used residential distance from major air pollution sources as a surrogate for exposure and a case-control approach. Socioeconomically matched census tracts showed a 30% increase in white female mortality for the tracts within 1 mile of the source.

The Nashville Studies

Zeidberg and colleagues conducted an epidemiological study in Nashville, TN, examining data from 1949 to 1960 on mortality and respiratory disease. Mortality studies comprised respiratory diseases (Zeidberg *et al.*, 1967a), cardiovascular diseases (Zeidberg *et al.*, 1967b), cancer (Hagstrom *et al.*, 1967) and infant/fetal mortality (Sprague and Hagstrom, 1969). This study employed 123 air monitoring stations across the SMSA for sulfation rate and dustfall and 36 stations for SO_2 (gas) and soiling (COH), for 1 year: 1958–59. Mortality data by census tract were aggregated for the years 1949–60. Socioeconomic factors included median tract family income, education, and the average number of persons per household room. These data were combined into three socioeconomic classifications when analyzing deaths by pollution level. Similarly, three classifications of air pollution were used to examine deaths by economic class. The highest sulfur oxide level was about $35\,\mu g/m^3$; the highest soiling, 1.1 COH (about $110\,\mu g/m^3$ equivalent TSP), and the highest dustfall level corresponded to about $80\,\mu g/m^3$ TSP. Thus, although this study characterized exposure well for the year of study, compared to other cross-sectional studies, it may have been limited by the ranges of variability in air pollution. In addition, smoking habits were not considered, which probably limits the useful results to females.

Total respiratory deaths were higher for both males and females in the middle economic class for the highest air pollution group, regardless of which pollutant defined the group (Zeidberg *et al.*, 1967a). However, deaths due to lung cancer and bronchitis or emphysema showed the opposite trend, suggesting that smoking may have been negatively correlated with air pollution. If so, the true effect of air pollution on total respiratory deaths may have been even larger. The mortality differences by economic class were large (about a factor of 2).

Mortality from cardiovascular disease (Zeidberg *et al.*, 1967b) showed more modest differences by economic class (a factor of 1.3) but inconsistent patterns according to air pollution (for the middle class). Total cardiovascular deaths were highest for the highest pollution level for all four pollutants, but only "soiling" (an index of small particles) showed a linear trend. Deaths from arteriosclerotic heart disease (the largest subgroup, about half the total), showed no air pollution trends. Hypertensive and other myocardial degenerative deaths showed strong air pollution trends. Females showed mortality trends consistent with air pollution trends (soiling) for all causes of death, suggesting that the trends for males may have been confounded by the implied opposing trends for smoking.

In contrast, cancer mortality (all sites) was more uniformly distributed by economic class (Hagstrom *et al.*, 1967). Rates for the middle class were slightly lower than for the other two classes. This was also true when rates for the middle class were grouped by air pollution level. However, stomach cancer was significantly elevated by dustfall level (factor of 6). By sex and soiling level, often only the highest level of pollution had a significantly higher mortality rate.

The analysis of fetal and infant mortality (Sprague and Hagstrom, 1969) concluded there was an association between postneonatal mortality of white infants and sulfur oxides but that the effect could not be completely separated from socioeconomic factors.

The Nashville study could have examined interactions between economic status and air pollution, as did the Erie County, NY, studies (see next section). However, the authors presented only selected "slices" of the possible two-way contingency tables. The study design illustrates one of the fundamental difficulties with cross-sectional studies; small geographic units (census tracts) are required to characterize exposure to locally distributed air pollutants. However, a long period is then needed to stabilize the mortality rates; for example, 1 year's air monitoring record may be inadequate to estimate the cumulative exposure. Also, there may have been substantial residential relocation during the 12-year period of record for mortality.

The Erie County Air Pollution–Respiratory Function Study

Erie County (Buffalo, NY, and environs), was studied by Winkelstein and colleagues during the 1960s (Winkelstein *et al.*, 1967, 1968; Winkelstein and Kantor, 1969a,b; Winkelstein and Gay, 1970, 1971; Brown *et al.*, 1975). Pollutants were measured during 1961–63 at 21 sampling stations and included TSP, dustfall, and sulfation rate (SO_x). The method of statistical analysis was the two-way contingency table for census tracts grouped by location. Only one pollutant was considered at a time, in conjunction with "economic level" (median family income for the group of census tracts). Data were also given on years of school completed, percentage of laborers in the labor force, and percentage of sound housing for the groups of census tracts in the contingency tables. Thus, the Erie County study was among the first to recognize the need to consider socioeconomic factors in detail in conjunction with cross-sectional analysis. The study populations were white males and females, age 50 to 69, and 70 and over.

Published Findings
In the first two papers (Winkelstein *et al.*, 1967, 1968) the authors examined 1959–61* mortality from all causes, chronic respiratory disease, respiratory cancer, and those death certificates with any mention of asthma, bronchitis, or emphysema; stomach cancer was examined in the third paper (Winkelstein and Kantor, 1969a), and deaths related to failures of the circulatory system were analyzed in the fourth (Winkelstein and Gay, 1970). While the results for males may have been confounded by smoking patterns, there was no consistent relationship between the average level of air pollution and lung cancer, which suggests that such confounding was not very serious. The contingency tables show a consistent increase in mortality with TSP for all the causes of death analyzed; the trends persisted when mortality ratios were standardized for economic levels and plotted against air pollution level. On this basis, the effect of TSP appeared stronger for females than for males (all causes of death and heart disease), which could imply minimal confounding by smoking or by occupational exposure. In contrast, the results for sulfur oxides appeared less consistent (as published), and when both pollutants were considered together, the effect seemed to be entirely due to particulates. However, the correlation between the two species was very high (0.98), so that it seems futile to attempt to separate the effects (see Appendix 3B).

The analysis of arteriosclerotic heart disease (ASHD) and cerebrovascular (stroke) mortality (Winkelstein and Gay, 1970) considered only TSP and showed strong effects for both males and females, only in the 50–69 age group. For ASHD, the effects of economic grouping were strongest for the lowest economic group, and the air pollution effect was ambivalent within the highest economic group.

The stomach cancer study (Winkelstein and Kantor, 1969a) was conducted to investigate the possible role of ethnicity in these cross-sectional studies, because certain ethnic groups were thought to be more susceptible to stomach cancer, presumably because of their diet. The authors concluded that stomach cancer was positively associated with TSP for both males and females, aged 50–69, independent of economic level and ethnic origin. They cited support for this finding from previous studies by Stocks (1960) in Britain; Manos and Fisher (1959) also identified an association between indices of air pollution and stomach cancer in the United States. Presumably, the causal mechanism involves swallowing carcinogenic particles that have been cleared from the respiratory tract.

The study of prostatic cancer mortality was intended to replicate findings from the Nashville study, which it did (Winkelstein and Kantor, 1969b). The authors also pointed out the independence of prostatic cancer from economic status. The study of mortality from cirrhosis of the liver (Winkelstein and Gay, 1971) showed an interaction with economic level (stronger air pollution effects at the lower economic levels); the air pollution relationship was present with or without mention of alcoholism on the death certificate. These studies could be interpreted as showing the nonspecificity of health effects of particulate air

---

*Since the time periods did not coincide, the authors assumed stationary patterns in mortality and air pollution.

pollution; since many different compounds can be present in TSP, some of them toxic, this finding is not surprising.

Reanalysis

One of the other important findings of the Winkelstein studies is the interaction between economic level and air pollution with respect to residence; in Buffalo, the contingency tables implied that persons of higher economic status tended to avoid the more polluted residential areas. Based on the tables, this was more the case for TSP than for sulfur oxides (SO_x); however, the correlation matrix in Table 7-1 shows the same effect for both pollutants. This correlation between independent variables can lead to errors in the contingency table analysis (as discussed in the original paper), thus, for the purposes of this chapter, the Erie County data were reanalyzed using multiple regression methods.

The purpose of this reanalysis was to confirm the authors' conclusions, which had been based on analysis of the two-way contingency tables and standardized mortality ratios. Some of the results are shown in Figures 7-5 to 7-8, which are plotted against TSP, as a matter of convenience. Since TSP and SO_2 are so highly correlated in this data set, either pollutant could have been used for the graphs. Figure 7-5a shows the all-cause mortality data for white males, age 50–69. The interaction between economic level and TSP is readily apparent, since the individual regression slopes against TSP for each economic group become weaker as economic level rises. Thus, a unit of TSP appears to have a larger effect on a male of lower economic status. This effect, if real, highlights the importance of obtaining detailed data on air pollution exposure in conjunction with socioeconomic factors, rather than using citywide averages (in a multicity study).

Figure 7-5b presents the all-cause mortality data for males aged 70 and over. The TSP effect is not statistically significant, and the economic effect seems confined to the lowest economic group; these results may be due to the lower range of variation among census tracts for this age group. Combining the two age groups (Figure 7-5c) provides a closer comparison to the crude (total) mortality used in many other studies; the economic effect again appears confined to the lowest group, the overall TSP effect is significant ($p = 0.05$), and the interaction is weaker than in Figure 7-5a. For comparison, the 1960 U.S. annual average mortality rates for white males (all causes) are 22.5 per

TABLE 7-1 Correlation Matrix for 1959–1961 Erie County, NY, Data

					Mortality Rates (ages 50–69)		
	TSP	Income	Education	Laborers (%)	Male	Female	Average
SO_x	0.98	−0.39	−0.52	0.69	0.71	0.77	0.76
TSP	1.00	−0.39	−0.54	0.67	0.72	0.74	0.75
Income		1.00	0.91	−0.88	−0.86	−0.67	−0.78
Education			1.00	−0.79	−0.80	−0.68	−0.76
Laborers (%)				1.00	0.96	0.84	0.92
Male mortality					1.00	0.90	0.97
Female mortality						1.00	0.98

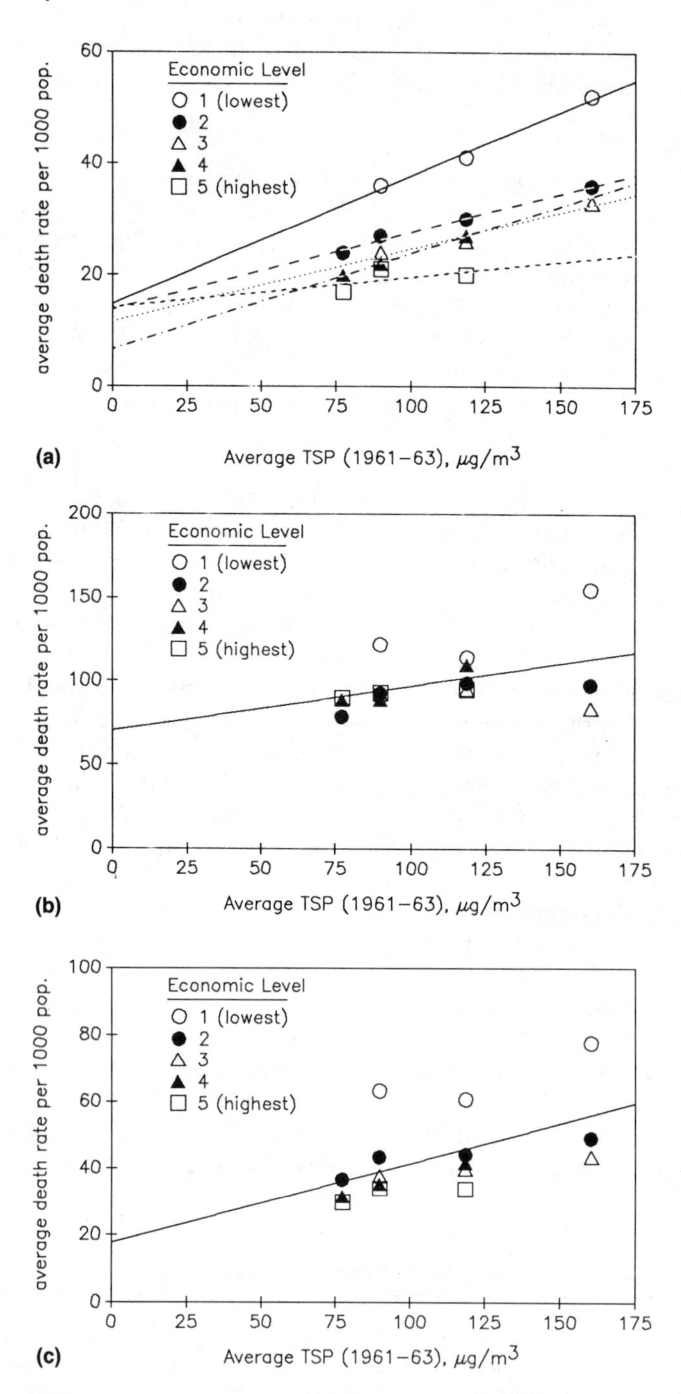

FIGURE 7-5. All-cause mortality for white males in Erie County, NY, 1959–61, by average TSP level. (a) Ages 50–69 (regression lines for each economic level). (b) Ages 70 and over (regression line for all data). (c) Ages 50 and over (regression line for all data). *Data from Winkelstein* et al. *(1967).*

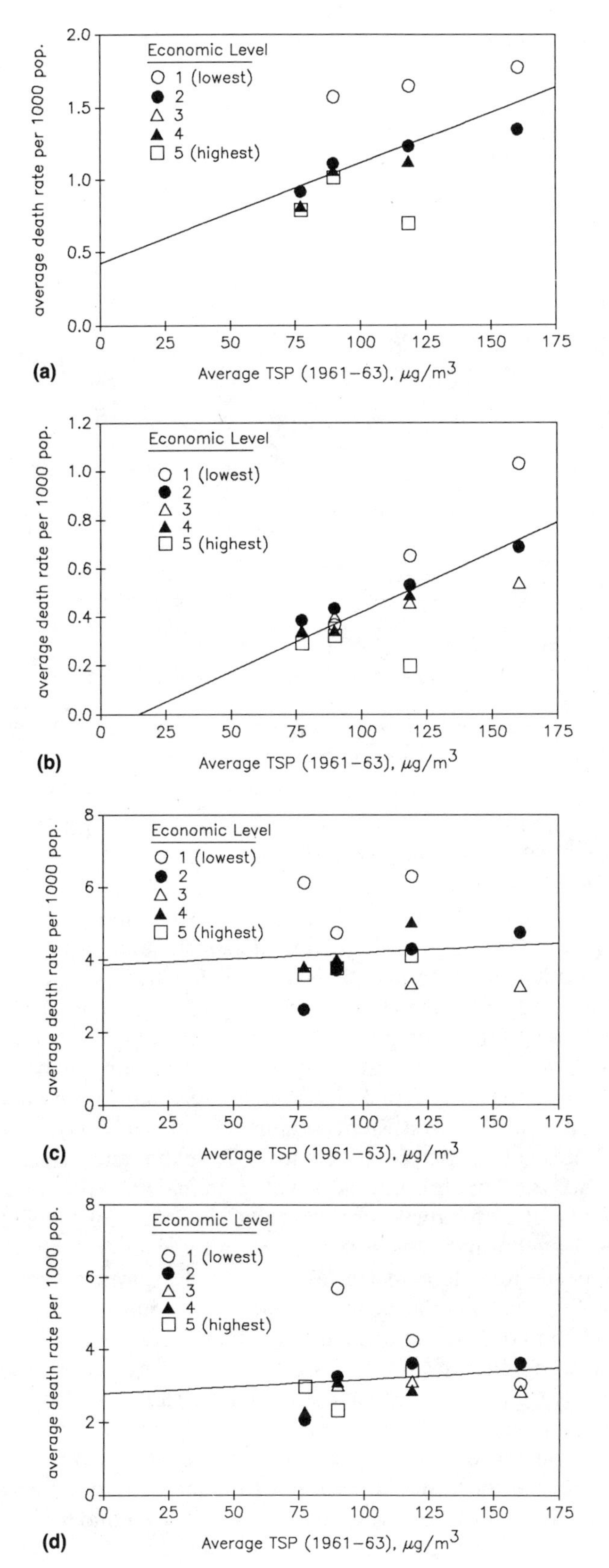

FIGURE 7-6. Arteriosclerotic heart disease mortality in Erie County, NY, 1959–61, by average TSP level. (a) Males, ages 50–69. (b) Females ages 50–69. (c) Males ages 70 and over. (d) Females ages 70 and over. *Data from Winkelstein and Gay (1970).*

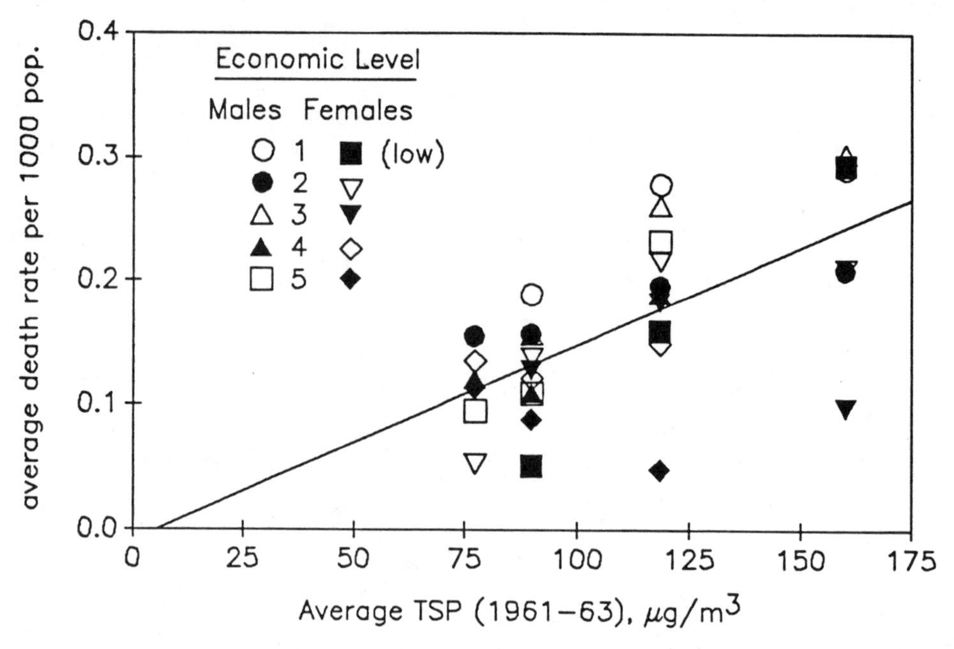

FIGURE 7-7. Cerebrovascular mortality for ages 50–69 in Erie County, NY, 1959–61, by average TSP level. *Data from Winkelstein and Gay (1970).*

thousand population (age 50–69), 38.6 (age 50+), and 77.8 (age 70+) (Grove and Hetzel, 1968). These figures imply an effective TSP intercept or threshold of around 50–75 μg/m^3.

Figures 7-6 and 7-7 present circulatory system mortality for males and females. Figure 7-6 shows the strong dependence of arteriosclerotic heart disease mortality on TSP for both males and females in the 50–69 age group (Figures 7-6a and 7-6b); the slopes of the overall regression lines are about the same, even though the average mortality rate for females is about half that for males. Note that the air pollution effect is inconsistent for both sexes in the highest economic level group. For the 70+ age group, both air pollution and economic effects are inconsistent (Figures 7-6c and 7-6d). Cerebrovascular mortality rates are displayed in Figure 7-7; males and females are plotted together since the gender differences were minimal. The economic effects tend to be less consistent for this cause of death, especially for females.

Figures 7-8a through 7-8c present mortality data for various respiratory causes of death for white males, age 50–69. Respiratory cancer deaths are plotted in Figure 7-8a; the economic effect is evident, but the overall TSP effect, although positive, is not significant. Figure 7-8b presents similar data for deaths due to chronic respiratory disease (underlying cause); the overall TSP effect is highly significant, although not all of the slopes for individual economic strata are consistent (not shown). When mortality with any mention of respiratory disease on the death certificate was analyzed (Figure 7-8c), the results became somewhat more stable because of the increase in the absolute number of deaths analyzed. The zeros plotted on Figure 7-8b and 7-8c may be due to the small populations of these cells.

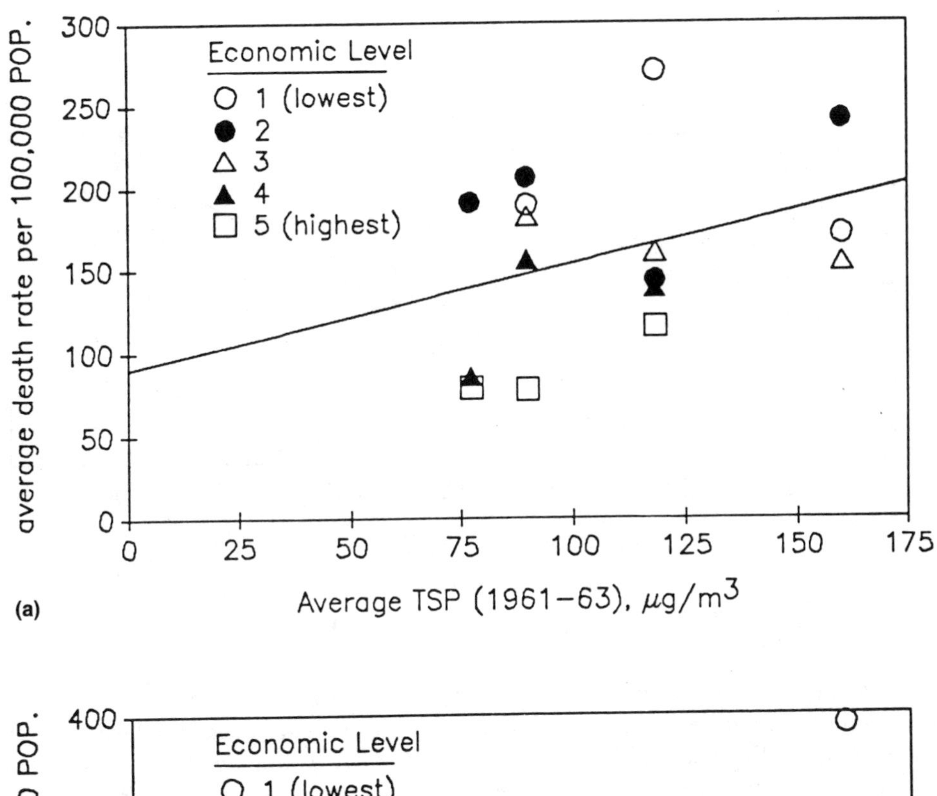

(a)

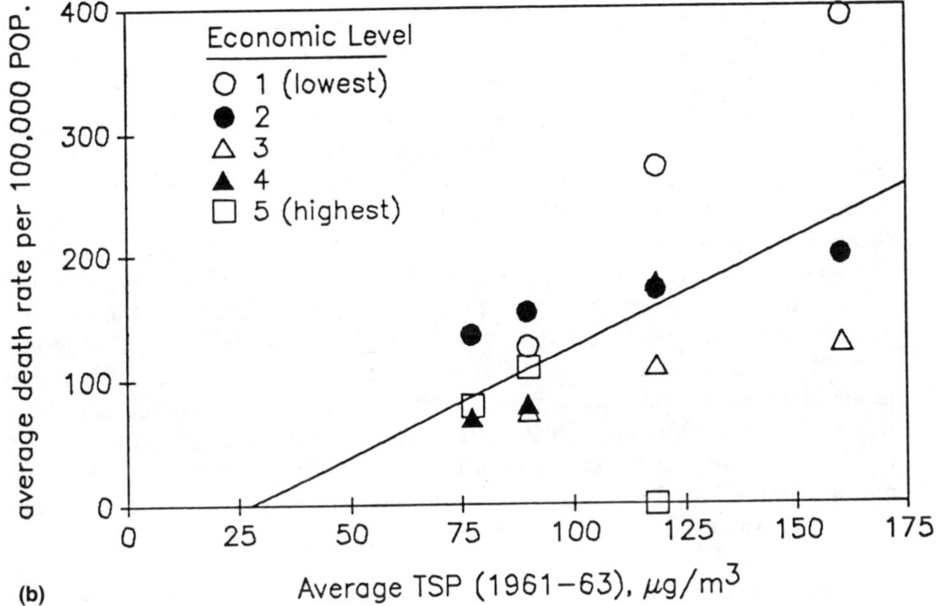

(b)

FIGURE 7-8. Respiratory mortality for white males ages 50–69 in Erie County, NY, 1959–61, by average TSP level. (a) Respiratory cancer. (b) Chronic respiratory disease. (c) Any mention of respiratory disease. *Data from Winkelstein* et al. *(1967).*

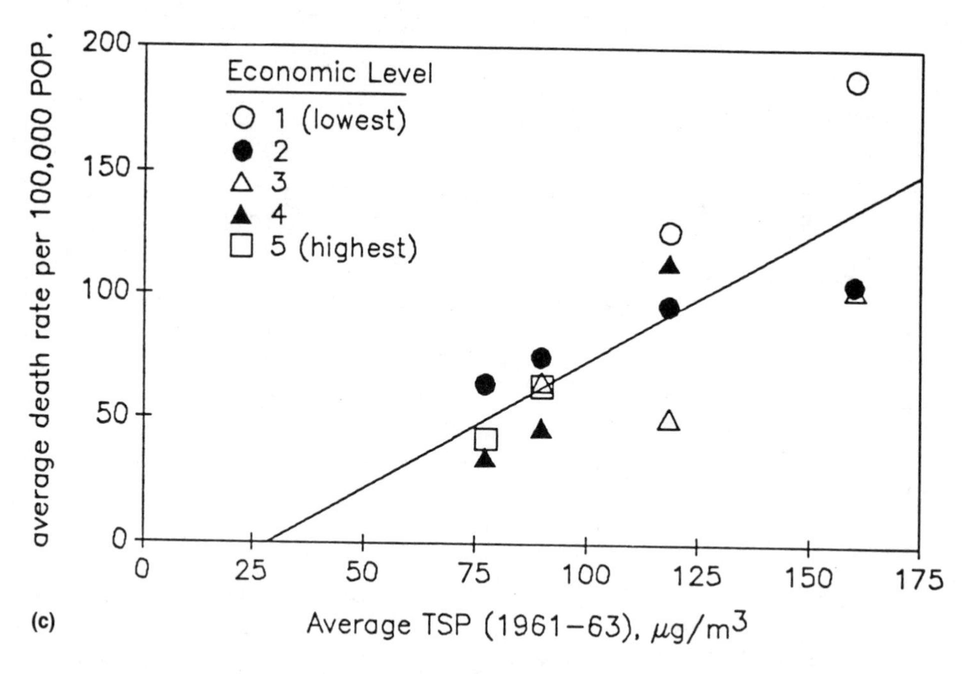

FIGURE 7-8. *Continued*

I performed some multiple regressions of these data to compare the results when the actual socioeconomic data are used instead of "economic levels," and to compare the TSP and SO_x effects when they are estimated in different ways. In the original paper (Winkelstein *et al.*, 1968), the census tracts were redistributed to form three groups differing in level of sulfation rate during the heating season. This procedure increased the contrast between groups; however, outdoor exposure to air pollution is generally lower during the heating season, so that this grouping may sacrifice realism in exposure. For this reanalysis, I calculated the annual sulfation rates coincident with the four TSP groups (Winkelstein *et al.*, 1967, 1968) and used the annual sulfation rates as the independent variable for the three sulfation-rate groups defined by the authors. The average TSP and SO_x (converted to SO_2) are about 110 and 102 $\mu g/m^3$, respectively; maximum values are 206 and 385 $\mu g/m^3$, respectively, and occurred in the same census tract. In addition, Finch and Morris (1979) noted that there was potential confounding due to differences among the census tracts in age distribution within the broad age group 50–69, and they presented standardized mortality ratios based on indirect age adjustment, based on the age distributions of each census tract. The correlation matrix for these variables (n = 16) is given in Table 7-1.

These correlation coefficients indicate a great deal of collinearity in the data set and suggest that a multiple regression model should use only one pollution and one socioeconomic variable. The finding of similar correlations for male and female mortality suggests the influence of environmental as opposed to personal or genetic factors.

Multiple regressions using various combinations of these variables illustrate some of the pitfalls with collinear data. If "% laborers" is selected as the socioeconomic variable, R^2 is maximized, and the pollutant variables lose significance (for TSP, $p = 0.2$; for SO_x, $p = 0.07$). The regression coefficients are of the order of 0.15 to 0.4% excess deaths/μg/m^3. If "Income" or "Eduction" is selected, R^2 is slightly reduced, and the pollutants become highly significant ($p < 0.001$). We can only conclude that it is highly likely that either TSP or SO_x (or both) is significantly associated with both male and female mortality. Thus, although sulfation effects on mortality were ruled out by the original analysis using three sulfation levels (Winkelstein *et al.*, 1968), sulfur oxides may have contributed to the mortality effects shown in the first paper by Winkelstein *et al.* (1967). Considering that sulfation plates do not always provide reliable measures of air concentrations of SO_2 makes it more probable that sulfur oxides could play an important role in explaining the Erie County mortality gradients. Changing the sulfation data from heating season to annual average had no effect on the conclusions of the second paper (Winkelstein *et al.*, 1968).

Table 7-2 compares the TSP regression coefficients. There are five different ways to estimate the change in mortality with respect to TSP:

- Calculating the slope of the standardized mortality plots in the original paper, using the actual average TSP values for each air pollution level.
- Calculating the slope from the marginal totals in the two-way contingency tables.
- Averaging the slopes within the contingency tables.
- Regressing mortality on TSP (bivariate).
- Making multiple regressions of mortality on TSP and various socioeconomic variables.

Table 7-2 shows that there is generally good agreement between the standardized mortality results (left-hand column) and the multiple regressions (right-hand column), but that most of the other estimation methods overestimate the TSP effect, in comparison, as might be expected. I also investigated whether a logarithmic transform for TSP improved the fit and found only minimal differences between linear and logarithmic models.

The hypothesis of interaction between TSP and economic levels for white males, age 50–69, was tested formally by adding the variable (TSP × economic

TABLE 7-2 Comparison of TSP Regression Coefficients for White Males (% excess deaths/μg/m^3)

Cause of Death	Standard Mortality Ratio	Contingency Table	Average Economic Level	Bivarariate Regression	Multiple Regression
All (50–69)	0.36	0.87	0.51	0.80	0.073*–0.25
All (70+)	0.18	0.36	0.25	0.27	0.18*
All (50+)	–	–	0.33	0.53	0.38–0.42
Respiratory cancer	0.03*	0.61	0.20	0.40	0.09*
CRD (50–69)	1.20	1.30	1.60	1.20	1.10
Asthma, (any mention)	0.40	1.30	0.75	1.15	0.67

Source: Winkelstein *et al.* (1967).

* Not statistically significant ($p < 0.05$).

level) to the multiple regression model. The interaction variable was significant, while the variables for TSP and economic level (main effects) lost significance.

The reanalysis of the stomach cancer findings (ages 50–69) showed that TSP was significantly ($p < 0.05$) associated with stomach cancer mortality for both males and females in multiple regressions and that the socioeconomic variables (Income, Education, Laborers) were not. The air pollution effect appeared quite strong, with an elasticity of about 1.7 for both sexes. The correlation between sexes was only moderate ($R = 0.51$, $p = 0.05$), suggesting that factors common to a household (such as diet) may have not been overwhelming.

Follow-Up Studies

In follow-up studies in Erie County using data from around 1970, Fleissner *et al.* (1974) concluded that the mortality patterns for white males were largely unchanged, even though TSP levels had declined about 30%. However, her data showed that all-cause mortality in this age group had declined about 13%, which is consistent with this magnitude of air pollution improvement. Finch *et al.* (n.d.) looked at changes in both male and female mortality rates and concluded that the longitudinal changes were consistent with the previous cross-sectional gradients. In considering the time history of air pollution in Buffalo and its effect on mortality, it should be noted that TSP levels in the mid-1950s were about a factor of 2 higher than during the 1959–61 period studied by Winkelstein *et al.* (1967). Longitudinal studies should thus consider a longer period of exposure than just the past few years, if truly chronic effects are to be assessed. However, since the middle-age group (50–69) seemed to be most sensitive to the effects of air pollution, there is no suggestion of a cumulative dose effect. Deaths in this age group may reflect the presence of a susceptible subgroup, whose numbers have diminished by the time the cohort reaches age 70.

Other Intraurban Studies

Pittsburgh

Gregor (1977) studied intraurban variations in mortality, using the census tracts of Allegheny County, PA (Pittsburgh). His study used 5-year average death rates, stratified by age, race, sex, and broad categories of causes (total, "pollution-related," and "non pollution-related"). Although a large number of independent variables was initially proposed, including an estimator for smoking based on age and income, the final model employed only education, population density, two climate variables, SO_2, and TSP. The pollution data were interpolated to the census tract level, using a computerized mapping program. Thus, the numbers of degrees of freedom appropriate for testing the significance of these regression coefficients is actually much smaller than the number of census tracts (assumed to be about 200; the figure was not given in the paper). There were from 4 to 14 TSP stations, depending on the year, and 42 to 47 SO_2 stations (sulfation plates). The interpolation procedure was also used for the climate data. Since climate variability across an urban area is small, the climate variables in this study should thus be regarded as "nuisance variables."

Gregor reported statistical significance for TSP but not for SO_2. His coefficients and elasticities for TSP were an order of magnitude higher than has typically been found from interurban studies but similar to the values reported by Winkelstein *et al.* (1967) for Erie County, NY. His values for SO_2, although not statistically significant, were similar to those found from some of the time-series analyses (0.012% per $\mu g/m^3$). Gregor's TSP results were strongest for the pollution-related causes of death and for the age group 45–64, and were slightly stronger for females. Although these results are similar to those of Winkelstein *et al.* (1967), both studies may be confounded by socioeconomic factors related to the geographic distribution of TSP sources. An alternative explanation could involve the fact that TSP concentrations 20 to 30 years earlier were much higher (probably by as much as a factor of 2) (Lipfert, 1978). It is also possible that the TSP regression coefficients from the interurban regression studies were depressed because of errors in exposure arising from the use of only one or several measurement stations per city. In addition, I find it difficult to accept that the largest "effects" on mortality in this study were from the number of rainy days, a variable that would not be expected to vary much across a metropolitan area and that has no conceivable physiological connection with mortality.

Chicago

Namekata *et al.* (1979) performed a two-phase study of air pollution in the city of Chicago from 1971 to 1975. The time-series analysis phase was discussed in Chapter 6. The cross-sectional phase studied differences among 76 neighborhoods (community areas) for several causes of death (age adjusted), averaged over the 5 years. The pollutants of interest were TSP and SO_2. Socioeconomic variations were accounted for through the use of three index variables that were developed by an outside agency. These variables included income measures, housing status, and education; the racial composition of the population was not specifically accounted for, but the authors reported that it was highly correlated with these three socioeconomic factors and thus was accounted for indirectly. Smoking habits were not accounted for. This study affords an opportunity to compare the results of cross-sectional and time-series regressions.

The cross-sectional regressions found about 25% excess deaths due to TSP, mainly in heart disease, which is similar to the findings of the Erie County study. No independent effects were found for SO_2; when both pollutants were entered, the SO_2 coefficient was negative. Although the magnitude of the TSP effect seems quite large considering the mean value of 83 $\mu g/m^3$, it is possible that the effects of historical exposures were still being felt (previous TSP levels were 2 to 3 times this figure). As a rough check on the plausibility of an effect of this magnitude, I noted that the total nonaccidental age-adjusted mortality for Chicago was about 33% above the national average after taking into account the racial composition of the population. It is therefore possible (though unlikely) that most of this difference is due to air pollution. Heart disease deaths accounted for most of Chicago's excess mortality with respect to the national average.

The indicated pollution effects in this study were probably confounded with either racial differences, with smoking habits, or with both. Examining the results for diseases that may be associated with these factors can facilitate interpretation. With regard to confounding by smoking, we note that emphysema deaths were strongly associated with TSP, but that lung cancer was not. This comparison must therefore be regarded as inconclusive. "Marker" causes for blacks may include diabetes (weakly associated with TSP), cerebrovascular disease (not associated with TSP), cirrhosis of the liver (strongly associated with TSP), pneumonia (not associated with TSP), and cancer of the genitourinary organs (not associated with TSP). This comparison is also equivocal but certainly does not suggest strong confounding.

After comparing with the time-series portion of this study (discussed in Chapter 6), I concluded that the cross-sectional or long-term mortality effects appear to substantially exceed the daily (short-term) effects in Chicago, the shortcomings of both studies notwithstanding.

Cleveland County, England

Dean *et al.* (1978) studied local variations in mortality and air pollution in Cleveland County (North-East England), from 1963 to 1972. Smoking effects were also emphasized; data on smoking habits of individual decedents were obtained by interviews with relatives, as were qualitative descriptions of occupational exposures. Air pollution exposure was assessed by defining three different residential areas having the following 1972 winter average smoke and SO_2 levels ($\mu g/m^3$): High − 175, 129; Medium − 66, 85; Low − 35, 57. From 1963 to 1972, both smoke and SO_2 levels declined in the "High" area, but only smoke declined in the other areas. For all causes of death examined (lung cancer, chronic bronchitis, coronary heart disease, and cerebrovascular disease for males; lung cancer and chronic bronchitis for females), only the "High" area showed an elevated risk (about 60%, $p < 0.001$), after standardizing for age and smoking. Thus one pollutant cannot be designated as "responsible"; on the basis of smoke, the effect would be about 0.48% per $\mu g/m^3$; on the basis of both smoke and SO_2, about 0.33% per $\mu g/m^3$. The magnitudes of these effects are comparable to those of the U.S. intraurban studies. Dean *et al.* (1978) considered several other risk factors, singly. For example, with the exception of lung cancer, the effect of social class was comparable to that of air pollution, which raises the possibility of confounding. With the exception of female bronchitis, the effect of air pollution was smaller than that of the lowest smoking category (1–12 cigarettes/day). Some speculative insights may be gained from the effects of occupational exposures (males): Exposure to dust had significant effects on lung cancer and bronchitis, but not on heart or cerebrovascular disease, suggesting SO_2 effects for the latter causes.

In Part 2 of Dean *et al.*'s report, similar findings were shown for three other residential subdivisions, for lung cancer and bronchitis. In addition, the residential distribution of social class was shown and was discounted as a confounding factor.

Cracow

Mortality rates from random samples of two areas in Cracow were compared after 10 years of followup (1968–78) by Krzyżanowski and Wojtyniak (1982). This is one of the few nonecological cross-sectional mortality studies, since 4355 individuals were studied and their individual characteristics were identified (including smoking habits) and used in the mortality models. The two residential areas were characterized as: high pollution, smoke = 180 μg/m^3, SO_2 = 114 μg/m^3; lower pollution, smoke = 109 μg/m^3 and SO_2 = 53 μg/m^3. Both of these levels would be considered high in comparison to current conditions in the United States. Fifteen percent of the study sample lived in the high-pollution area (center of the city). Because of the relatively small numbers of deaths involved, several different models were evaluated, rather than one comprehensive model.

The authors found a marginally statistically significant (positive) effect of air pollution on male mortality (after age adjustment and consideration of confounding variables), but a *negative* effect on female mortality. There was also an interaction between air pollution and smokers for males. The magnitude of the air pollution effect (both smoke and sulfur) was approximately consistent with the findings for other intraurban studies: Buffalo (Winkelstein *et al.*, 1967) and Pittsburgh (Gregor, 1977). The counterintuitive (negative) effect for females may have resulted from the small number of deaths (21) in the polluted area, lack of interaction with smoking, or differences in exposure due to spending more time indoors. Also, the finding of air pollution effects for men may be an artifact, and there may have been no significant pollution effect for either sex.

The validity of the study is supported by the findings of strong effects for smoking (men only), occupational hazards (both sexes), and education (women only). A rural place of birth was beneficial for both men and women. This study should be viewed as providing qualified support for the hypothesis that long-term exposure to air pollution is associated with increased mortality, although it provides no information as to the harmful pollutants. In that sense, the results are similar to those of Morris *et al.* (1976), who compared two nearby towns in Pennsylvania and showed an additive (positive) effect of air pollution and smoking for males (with a tendency toward negative effects for females).

Munich

Gottinger (1983) regressed 1974–78 male and female mortality rates for 41 census tracts in Munich, using both weighted linear and logarithmic models. Air pollutants were TSP, SO_2, and CO, and values for each census tract were obtained by interpolation. The average TSP level was 230 μg/m^3, and the average SO_2 level was given as "38 ppt/24h," which I interpreted as 0.038 ppm or about 100 μg/m^3 on a 24-hour average basis. Gottinger used two climate variables but did not state why climate should be a variable within a metropolitan area or how many stations were used. In this respect (and in several others, including similar text passages, variable names, and table layout) his study closely resembles Gregor's 1977 study of the Pittsburgh area, which Gottinger did not cite. Gottinger considered using a smoking index variable but decided against it since "inclusion of the index would reduce the air pollution coefficients throughout and in some cases destroy their statistical significance." For this reason, the author advises caution in interpreting his results. The socioeconomic variables used were education and population density. Five-year average mortality rates were computed for three broad age groups (<45, 45–64, >64).

Using weighted linear models, Gottinger reported statistically significant effects for TSP but not for SO_2 or CO: for both genders, for the upper two age groups, and for all causes, and for both pollution-related causes and for nonpollution-related causes. The magnitudes of the TSP coefficients were remarkably similar to Gregor's findings and reasonably consistent with the findings from other intraurban studies. Gottinger also reported the estimated elasticities, but I could not reconcile his figures with the regression coefficients and the corresponding mean values. For example, he reports an elasticity value of 0.027 for male mortality from all causes, age 45–64, whereas I estimated about 0.26, based on the linear model coefficient. For comparison, Gregor (1977) reports an elasticity value of 0.40. Like Gregor's results from Pittsburgh, the largest "effects" on mortality in this study were from the number of rainy days, which is remarkable given the differences in geography. I also noted several errors in the tabulated data and thus decided to use Gottinger's results with "caution," as he recommended.

Cross-Sectional Mortality Analysis of London Neighborhoods, 1954–1956

Gore and Shaddick (1958) presented data on 10 London communities, including standardized mortality ratios (SMRs) by sex for all causes, bronchitis, other respiratory diseases (including tuberculosis), lung cancer, other cancer, and cardiovascular diseases, for the years ending September 30, 1956. Independent variables included average levels of smoke and SO_2 (from the seven-station network), social class, and the percentage of community population born in London. Since Gore and Shaddick presented only bivariate correlations of these various factors, I attempted a multiple regression analysis as a check on the intraurban relationships I described, using lung cancer SMRs as a surrogate for smoking information, the lack of which was a major concern in previous

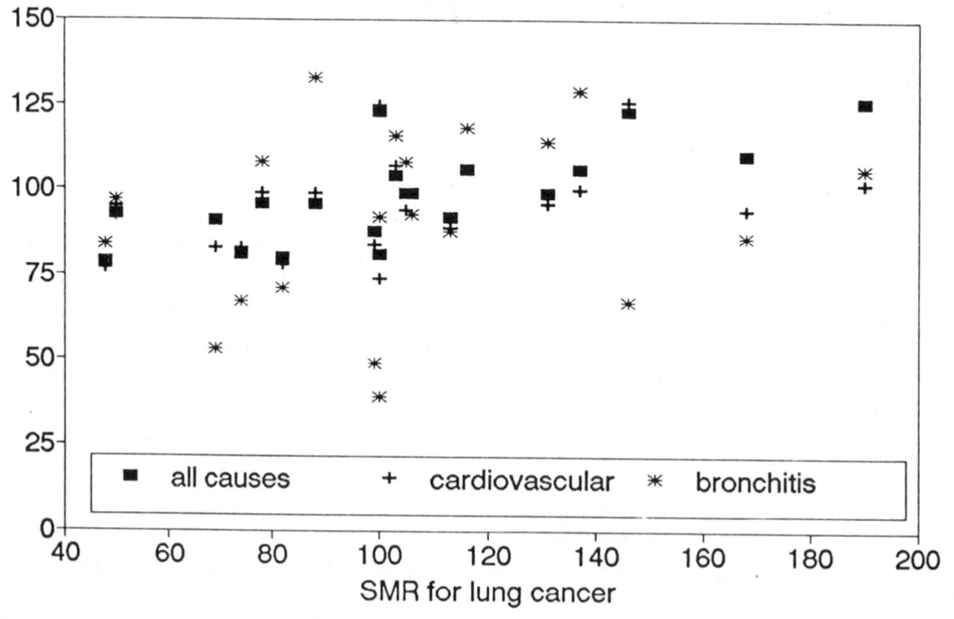

FIGURE 7-9. Comparison of standardized mortality ratios for various causes in London neighborhoods, 1954–56. *Data from Gore and Shaddick (1958).*

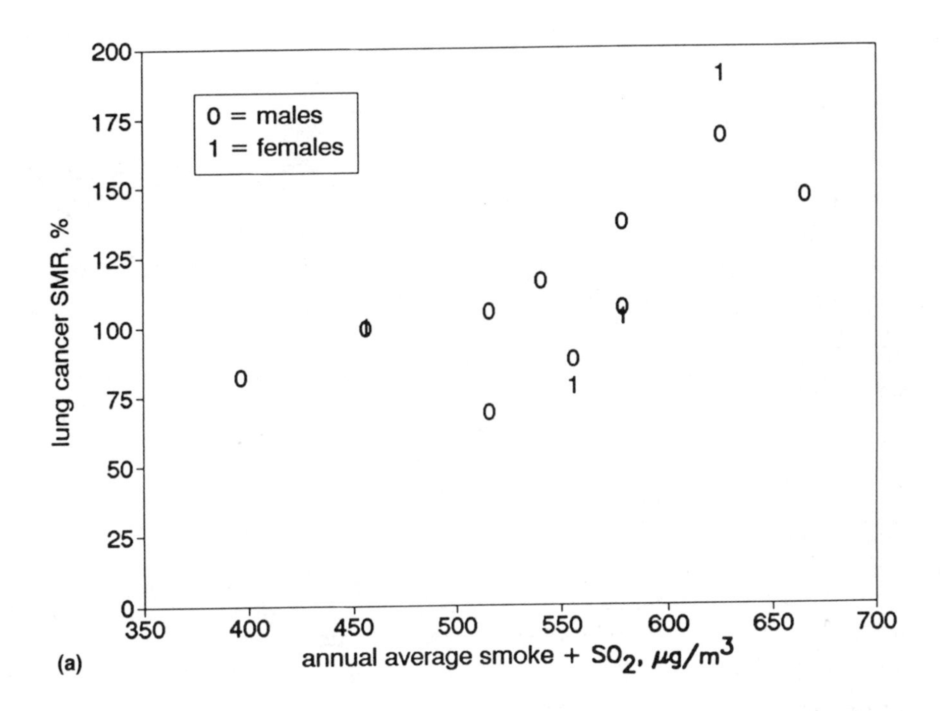

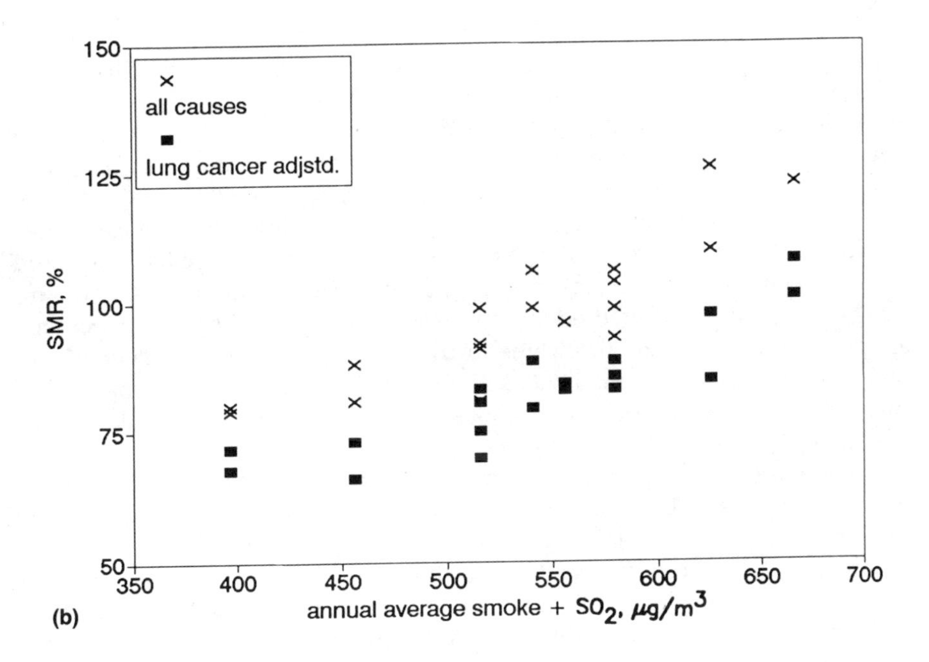

FIGURE 7-10. Standardized mortality ratios in London neighborhoods, 1954–56. (a) Lung cancer vs. SO_2 + smoke. (b) All-cause mortality vs. SO_2 + smoke, with and without adjustment for lung cancer. (c) Bronchitis, adjusted for lung cancer. *Data from Gore and Shaddick (1958).*

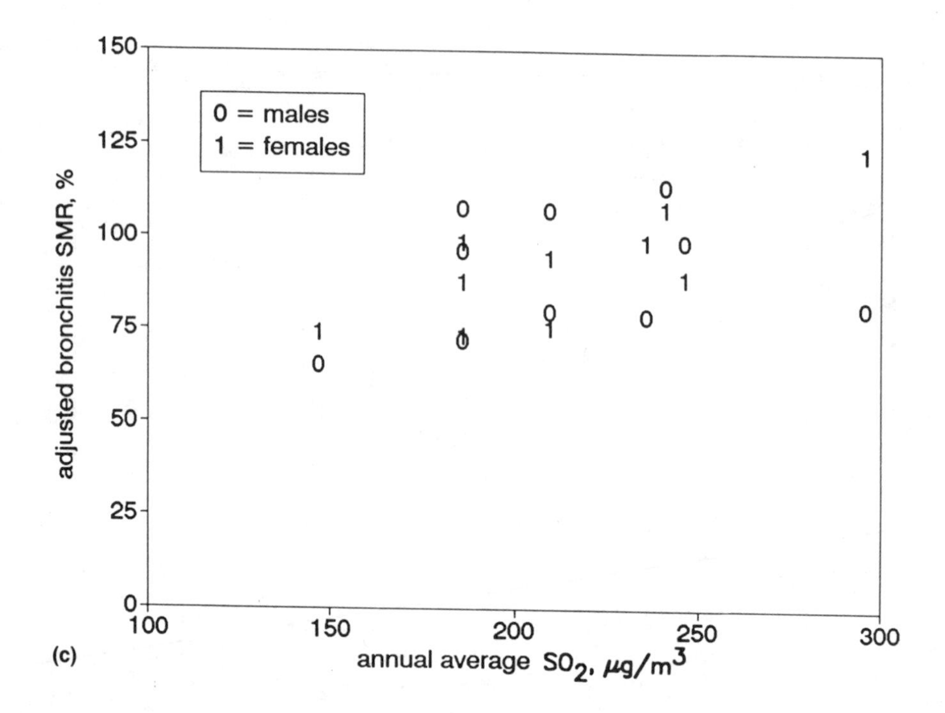

FIGURE 7-10. *Continued*

intraurban studies. (I was unable to corroborate all of Gore and Shaddick's published correlation coefficients, which raises the possibility of typographical errors in their tabulated data.) I pooled male and female SMRs to increase statistical power, with a dummy variable to distinguish any differences in their SMRs. The results of this analysis are presented in Figures 7-9 and 7-10. Figure 7-9 shows the interdependencies of some of the mortality rates. All-cause and cardiovascular SMRs are similar, as expected, since cardiovascular diagnoses normally comprise such a large part of the total. Their relationships with lung cancer could be interpreted as a mutual dependence on smoking (or perhaps air pollution). However, bronchitis SMRs appear to be independent of the other causes. For such a small data set, it seemed fruitless to try to separate the two air pollutants; for simplicity, I used their sum as in the episode studies (Chapter 5).

Lung cancer was significantly associated with air pollution (Figure 7-10a); whether this was a direct relationship or one mediated through smoking patterns could not be determined without data on individuals. This association required that lung cancer SMRs be included as a cofactor in regressions of other diagnoses. Figure 7-10b shows the relationship between all-cause SMRs and air pollution, with and without consideration of lung cancer as a surrogate for smoking. The slope of the regression drops by almost 50%, from about 0.17% per $\mu g/m^3$ to about 0.09% per $\mu g/m^3$. However, the data adjusted for lung cancer suggest a threshold at about 500 $\mu g/m^3$, which was also suggested by some of the episode data in Chapter 5 (Figure 5-2e). The slope of these cross-sectional data is considerably higher than that obtained from temporal data. The socioeconomic variables were not important for all-cause or lung cancer mortality, nor was the dummy variable for gender. This raises the possibility that some other socioeconomic variables should have been included. For bronchitis SMRs, which had been shown to be strongly associated with air pollution in interurban studies (discussed below), only SO_2 appeared to have a weak relationship ($p = 0.06$), as shown in Figure 7-10c, which was not improved by using a logarithmic transform. Being born in London was an important risk factor for bronchitis mortality. This analysis provided corroboration of the existence of intraurban mortality gradients associated with air pollution, but it also pointed out the need to consider potential confounding factors.

SECTION 2: CROSS-SECTIONAL STUDIES OF CITIES

Analysis of Mortality in British Boroughs and Towns

Early Cross-Sectional Studies

Early studies of long-term mortality and air pollution in Britain focused on bronchitis and cancer, in relation to the degree of urbanization and various measures of air pollution, some of which were collected over a 10-year period ending in 1954. Pemberton and Goldberg (1954) regressed bronchitis mortality rates in 35 county boroughs alternatively against SO_2 (from lead peroxide sulfation candles) and "total solids" (not defined, but assumed to be settleable particulates) and found the better correlations mostly with SO_2. The long-term SO_2 levels were in the approximate range of 50 to 120 $\mu g/m^3$. Most of the correlations were significant, although for females they just missed the 0.05 level. Pemberton and Goldberg also noted a negative (not significant) relationship between SO_2 and two socioeconomic variables and thus concluded that confounding of the bronchitis-pollution relationship was not likely. For men, the relationship was stronger in the 45–64 age group; for women, the opposite was true.

Gorham (1958) performed a simple regression analysis of bronchitis mortality rates for 53 county and metropolitan boroughs for 1950–54, in relation to sulfate deposits and precipitation pH. He found both to be statistically significant (after logarithmic transformations), the latter with $p < 0.001$. In his discussion, Gorham felt that the responsible agents might actually be acidic aerosols, a pollutant that has been receiving attention in the United States more recently.

Daly (1954) also studied correlations between bronchitis mortality and air pollution for 1946–50 but was dissatisfied with the available ambient measurement data and devised an index based on the density of consumption of domestic coal (space heating). This procedure produced higher correlations for both males and females (age 45–64), which were not substantially diminished by controlling for socioeconomic conditions. However, none of these early studies considered possible confounding by tobacco smoking. Daly also pointed out that only 18% of the coal consumed in Britain was used for domestic purposes, yet this was the index that was associated with bronchitis mortality.

He felt that this was due to poor combustion conditions in domestic fires, which produced much more smoke per unit of mass consumed than industrial combustors (but only proportional amounts of SO_2 or less). For this reason, he felt "it would be premature to attribute the apparently bad effects of air pollution primarily to sulphur dioxide." In addition, he performed a rudimentary analysis of the impact of sulfur pollution from power stations and concluded that there was no effect on bronchitis mortality.

Stocks (1959) looked at the relationships between bronchitis and various cancers (lung, stomach, intestine, breast) and both smoke and "deposit" (settleable particulates) in four different sets of cross-sectional data, from 58 county boroughs in England and Wales. The correlation between smoke and "deposit" was just significant; the long-term average smoke levels ranged from 60 to 490 $\mu g/m^3$ (mean = 260 $\mu g/m^3$). He found significant relationships between bronchitis, lung cancer, and stomach cancer and both smoke and deposit for both sexes after controlling for population density. Stocks hypothesized that the stomach cancer finding could relate to "exposure of food to dirty air" and pointed out that stomach cancer rates had declined in the United States where the wrapping of food had long been practiced, whereas such wrapping was just beginning in Britain.

Stocks (1960) continued his cross-sectional analyses by adding various trace metals and polycyclic organic compounds as independent (air pollution) variables. Most data were from the 1950s, and the overall annual average benzopyrene level was 35 ng/m^3 (an order of magnitude higher than comparable data from the U.S.). Socioeconomic variables were population density and an index of social class. Although cross-sectional data on smoking habits were not available, Stocks reported from a survey of hospital patients that lung cancer among heavy smokers was not affected by urban residence, but that among light smokers there was a rural-urban mortality ratio of about 2.5:1. For 26 locations in northern England and Wales, he found strong correlations between smoke or benzopyrene and mortality due to lung cancer (males), bronchitis, pneumonia, and a weaker correlation with stomach cancer (both sexes). Strong correlations were also found for many of the trace metals. Similar findings were reported when settleable particulates were regressed against mortality for two other cross-sectional data sets, but the relationships were weaker for subareas of London.

In a more detailed analysis, Gardner *et al.* (1969) used measurements of British smoke, SO_2, and Daly's domestic fuel consumption index (ca. 1951) to examine relationships with 1948–54 and 1958–64 deaths from all causes, cardiovascular disease, bronchitis, lung cancer, and stomach cancer. They devoted considerable effort to dealing with possible confounding variables, including drinking water hardness (calcium content), "social conditions," latitude, and rainfall. It is not clear how the latter two factors might be causally linked to mortality, except that the north of Britain tends to be more industrialized. Including these factors in a multiple regression analysis may thus partially account for regional factors, leaving only local variations to be explained by air pollution and "social conditions." The authors noted that these last two factors were highly correlated (R = 0.8) in their data set. Daly's air pollution index was better correlated with all-cause mortality than either smoke or SO_2 for both sexes and thus was used in the multiple regressions. These regressions showed the following, by cause of death:

All causes: air pollution was highly significant for males aged 45–64 for both periods and for age 65–74 in the earlier period. Social factors were more important for females. Rainfall, water calcium, and latitude were also important.

Cardiovascular disease: air pollution was just significant for males ages 45–64. Rainfall, water calcium, and latitude were important for all subgroups.

Bronchitis: Air pollution was significant for six of the eight subgroups. The other factors were mostly not significant.

Lung cancer: Air pollution was significant for seven of the eight subgroups. The other factors were mostly not significant.

Stomach cancer: Air pollution was significant for three of the four male subgroups. The other factors were mostly not significant.

The authors commented on the possible role of cigarette smoking, which they were unable to account for. However, their results do not suggest major confounding, because female lung cancer was also associated with air pollution (although females then smoked less), and heart disease (known to be associated with smoking) was not associated with air pollution, for the most part. It is unfortunate that they did not also present multiple regression results for the individual measured air pollutants. We are unable to examine the association between SO_2 and heart disease, for example.

Stratified Cross-Sectional Analysis of Towns in England and Wales

Buck and Brown (1964) emphasized towns with air monitoring in residential areas in their analysis of bronchitis and lung cancer mortality. However, the unavailability of sufficiently complete monitoring records required some unfortunate compromises in their study. They used data on smoke and SO_2 from only one month (March 1962) to represent long-term exposure. The mortality data were for males and females separately, but averaged over 1955–59, and age-adjusted based on the 1951 age distributions. Socioeconomic variables in this study were population density and an index of social class. A separate study was done on a subset of locations for which smoking survey data were available. The 219 locations found were subdivided into five subsets for separate regression analyses: London boroughs, county boroughs, boroughs, urban districts, and rural districts. Average smoke levels for the London boroughs were lower than all other areas except rural districts, reflecting the imposition of smoke controls under the Clean Air Act of 1956.

Age-adjusted lung cancer rates in England were rising at this time, yet they were much higher then than U.S. rates are at present and showed a positive dependency on urban density. About 75% of the male population smoked, regardless of residence. However, there was no spatial relationship between air pollution and lung cancer, and the relationship with smoking was weak. Both of these findings could have been affected by the long latency period of lung cancer (about 20 years) and, thus, by the inappropriate timing of the independent data.

Bronchitis mortality was found to be highly correlated with smoke and SO_2, as well as with social index. Regression coefficients were not given, but I made a crude estimate based on the mean values of each of the five groups. This estimate indicated about 50% of male and female bronchitis mortality to be associated with SO_2 (mean value = 200 $\mu g/m^3$). If bronchitis has a long induction period like lung cancer, it is possible that the high bronchitis mortality in London (female SMR = 172) derives from smoke levels that existed before controls were imposed. Also, it is likely that other socioeconomic or climate factors not included in this study play a role in bronchitis aetiology. For example, Fairbairn and Reid (1958) found significant correlations with an index of atmospheric visibility they called a "fog index." For the age group 45–64, this index was correlated with male and female bronchitis mortality and male pneumonia mortality, but not lung cancer, influenza, or pulmonary tuberculosis (deaths from 1948–64).

Comparison of the Spatial Distributions of Bronchitis and Lung Cancer in England and Wales

Ashley (1967) investigated the hypothesis that bronchitis might be protective of lung cancer, since a deficit in lung cancer had been seen in bronchitic coal miners. He assembled a data set of SMRs for male bronchitis and lung cancer (1958–63), population density, and smoke and SO_2 averages for March 1963 (following the example of Buck and Brown, 1964), for 53 county boroughs and urban areas. I performed multiple regressions to confirm Ashley's findings; they confirmed that deaths from both diagnoses were positively correlated with population density, but that only bronchitis was (positively) associated with air pollution (Figure 7-11a through 7-11c). The data fit against smoke slightly better (e = 0.20, R = 0.61) than against SO_2 or the sum of smoke and SO_2. The fact that lung cancer showed negative associations with smoke (p = 0.03) and with SO_2 (p = 0.01) provided some suggestive support for Ashley's hypothesis of antagonistic effects. However, if both SMRs were adjusted for population density (the regression coefficients were nearly identical), there was still a positive relationship between the two, suggesting a positive common factor that could be smoking. As a crude means of "adjusting" the bronchitis data for smoking, I divided the two, as suggested by Ashley. This new parameter was more strongly associated with air pollution (Figure 7-12a through 7-12c), with an elasticity of 0.31 for smoke (R = 0.68) and 0.38 for the sum of smoke and SO_2 (R = 0.68). Without data on smoking habits, it is difficult to find support for Ashley's original hypothesis in these data.

Follow-Up Study in Britain

Chinn *et al.* (1981) analyzed mortality in 104 counties and boroughs in England and Wales from 1969 to 1973. The dependent variables were mortality for infants (aged 0–1); males and females aged 45–54, 55–64, and 65–74; for all causes, stomach cancer, lung cancer, breast cancer, influenza, pneumonia, bronchitis, hypertension, ischemic heart disease, cerebrovascular disease, "other" heart disease, and suicide. The independent variables were (then) current SO_2 and particulates (British smoke), various socioeconomic variables, latitude, temperature, and rainfall. Limited analysis was done using data for

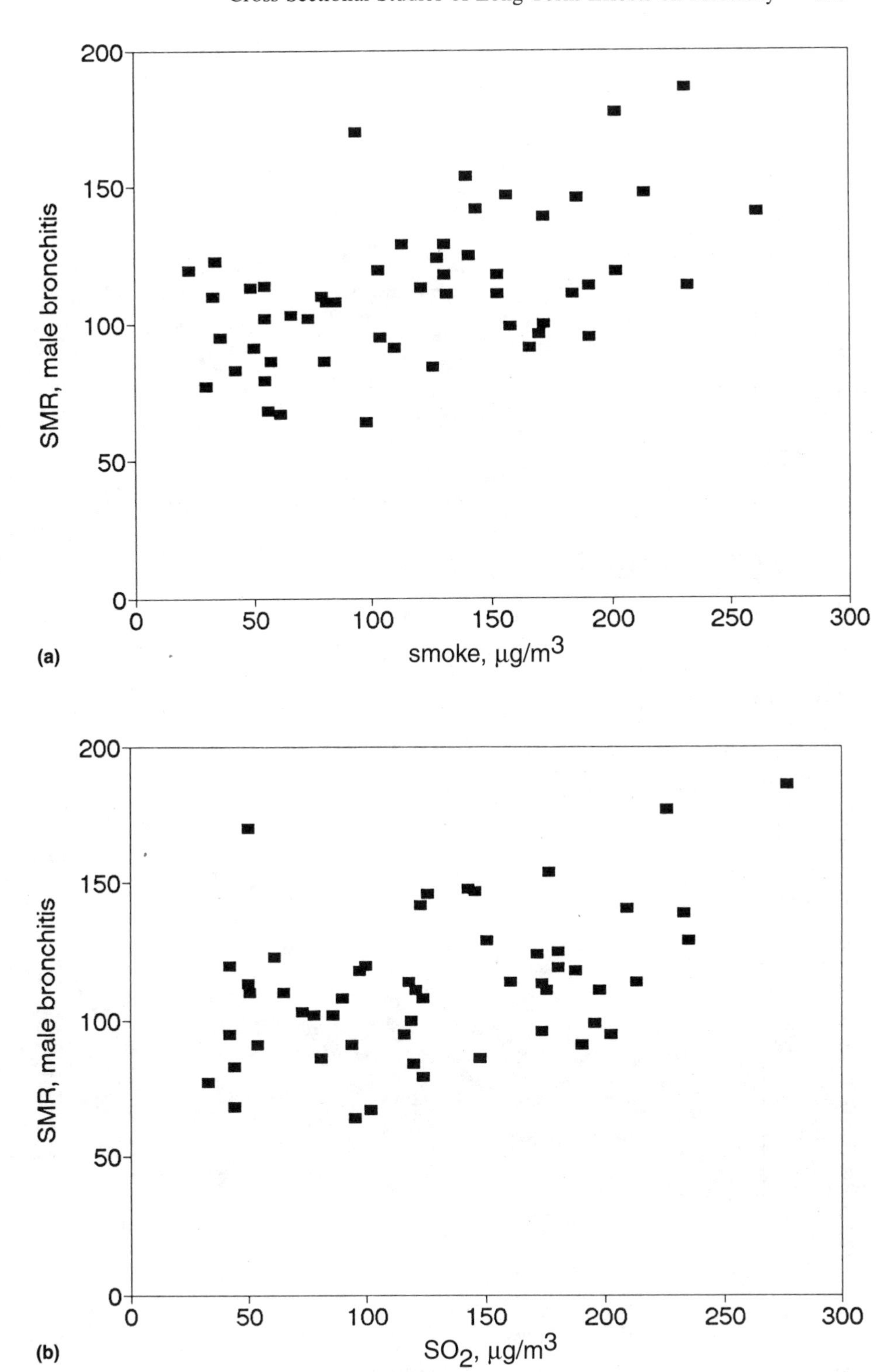

FIGURE 7-11. Standardized mortality ratios for male bronchitis for British county boroughs and urban areas, 1958–63. (a) By smoke. (b) By SO_2. (c) By population density. *Data from Ashley (1967).*

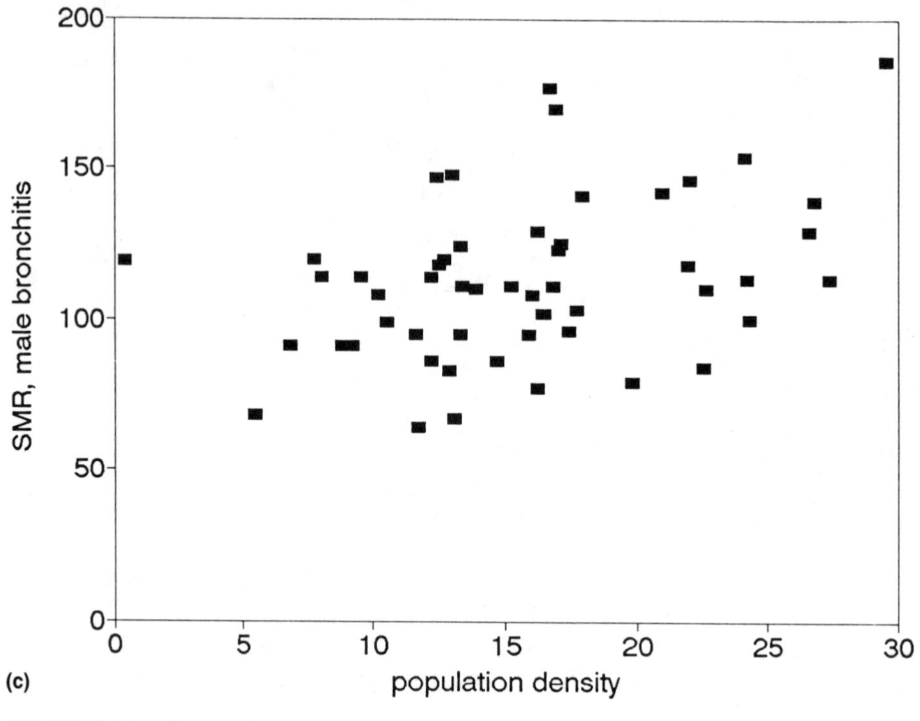

FIGURE 7-11. *Continued*

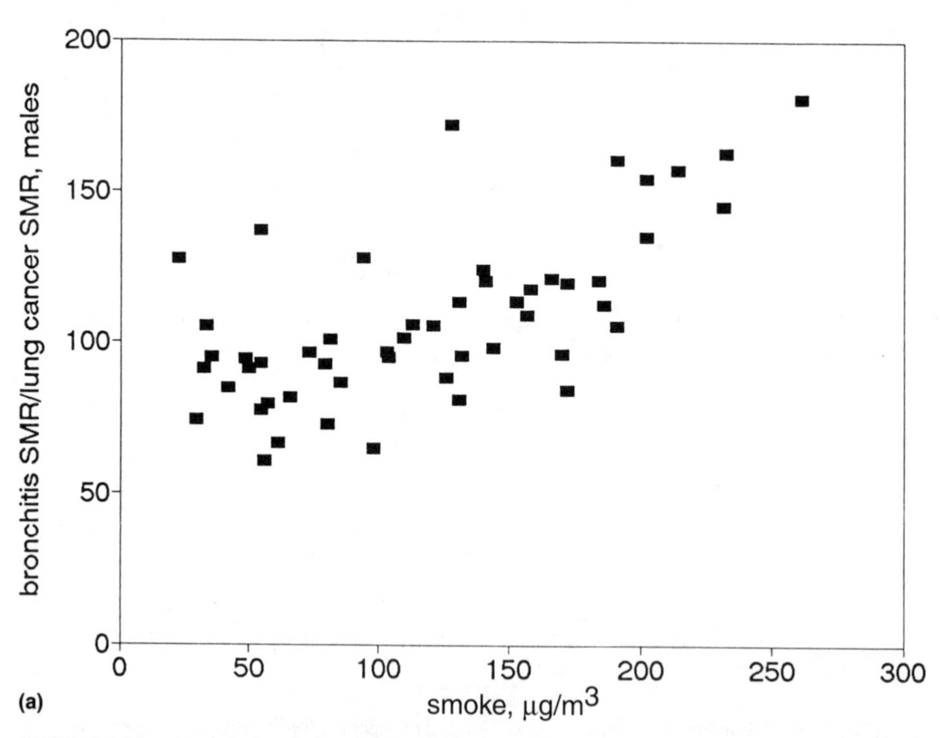

FIGURE 7-12. Ratios of male bronchitis and lung cancer for British county boroughs and urban areas, 1958–63. (a) By smoke. (b) By SO_2. (c) By population density. *Data from Ashley (1967).*

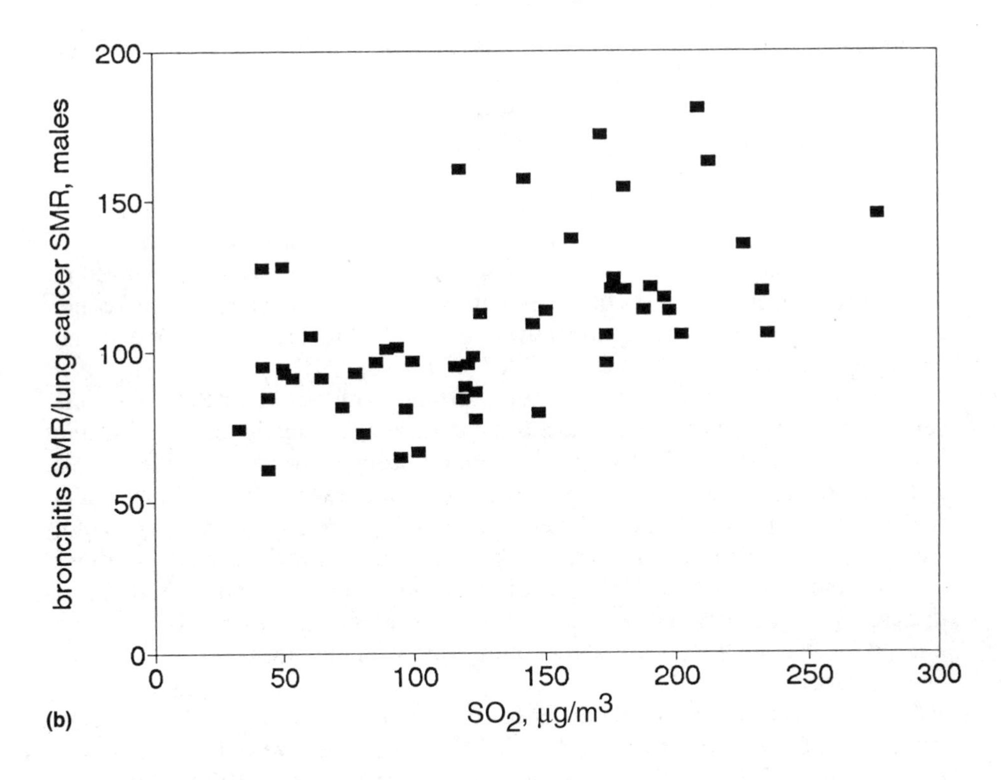

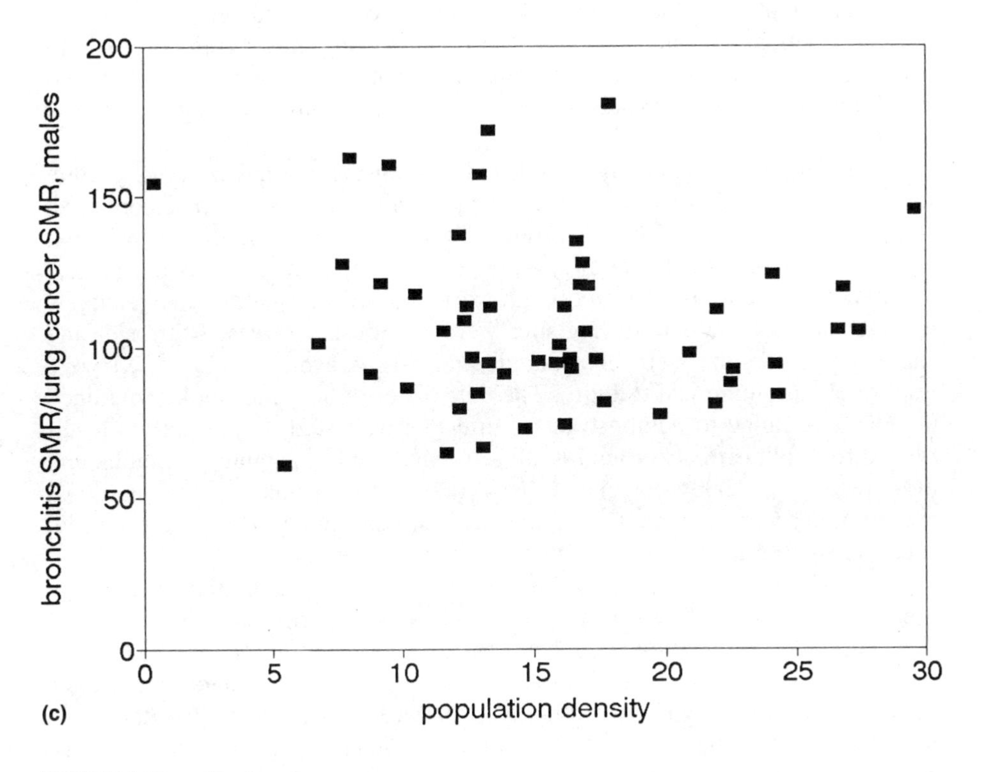

FIGURE 7-12. *Continued*

water hardness and smoking (limited data) and Daly's (1954) index of domestic fuel consumption.

The authors concluded that there was "no consistent relation of smoke or sulphur dioxide with mortality from all causes or with mortality from specified causes postulated *a priori* to be related to pollution. In particular there was no significant association between smoke and mortality rates for respiratory illness. Comparison with results from similar analyses of data from two previous decades suggested that a decline in the strength of associations had occurred in parallel with declining levels of the pollution." The mean values of smoke and SO_2 were 55 and 119 $\mu g/m^3$, respectively (compared to 1963 values of 119 and 130 $\mu g/m^3$, respectively, in Ashley's 1967 study).

They found no associations of infant mortality with smoke or SO_2 and some sensitivity in the results to the selection of data sets, especially for all-cause and heart disease mortality. Deleting the London boroughs, which were the highest in SO_2, strengthened some of the SO_2 correlations, especially for females and for deaths from heart disease. Several of these became significant ($p < 0.05$). Including either water hardness or smoking variables weakened the associations of male mortality with SO_2. Even in 1948–50, female mortality from heart disease was not associated with air pollution (as an index variable) (Gardner *et al.*, 1969). Associations were shown more often for the 45–64 group than for ages 65–74.

The importance of this study is its potential to provide an independent validation of the findings of similar U.S. studies. I believe it to be the most comprehensive cross-sectional study of general mortality in Britain based on actual air quality monitoring data. Replications of U.S. studies *per se* cannot provide independent validation because of the strong autocorrelation of both pollution levels and mortality rates (all years tend to resemble one another). The Chinn *et al.* study provides several interesting insights, although it also has faults.

The strengths of this study included: the geographic areas used were quite small by U.S. standards (maximum 70 square miles) and on average contained 4 monitoring sites each. Thus, exposure to the specific air pollutants considered was characterized much better than in most comparable U.S. studies. The authors report that the intraborough variation in air quality was relatively modest and that few monitoring sites were in industrial areas. Both age- and cause-specific deaths were considered, and 5-year averages (1969–73) were used to obtain numerical stability. The data for both SO_2 and smoke contained an adequate range to demonstrate an effect, if it existed; concentration levels ranged from almost background to almost twice the U.S. standard. Smoke and SO_2 were only weakly correlated ($R = 0.25$). Both drinking-water hardness and smoking were included, although the smoking analysis was quite crude, since consumption data were available only on a regional basis.

There are some weaknesses in this study: Only winter pollution levels were used, since summer levels were "uniformly low." If this were strictly true, the degrees of association would not be affected but the (nonstandardized) regression coefficients would be too low. Using year-round data would have provided a more straightforward comparison with other studies. No data were included on frequency of fog, sulfate particles, oxidants, or acid aerosols. No attempt was made to construct long-term average pollution levels from the available historical data. Sulfur dioxide was reported to be correlated with male smoking, making it difficult to separate the effects. Latitude was included as an independent variable in the regressions, even though there is no *a priori* reason to suspect that latitude *per se* affects health. Since latitude was moderately correlated with air pollution ($R = 0.5$), its inclusion could have resulted in the underestimation of pollution effects. The data of Craxford *et al.* (1967) clearly show an air pollution gradient increasing from north to south in the mid-1960s, due in part to differences in the sulfur content of local coal. Latitude may thus be a surrogate for chronic exposure, in addition to competing with the current smoke and SO_2 variables.

Although Chinn *et al.* presented regressions of current mortality rates on indices of past pollution levels (back to ca. 1950), no indications are given as to what typical concentrations might have been at that time or whether the results were consistent with an expected dose-response relationship. The publications give only sketchy details of all the results that must have been available. In particular, only standardized regression coefficients (SRCs) are given (SRC = $\beta[\sigma_x/\sigma_y]$); the means and standard deviations of the variables are not given. I calculated the mean values for smoke and SO_2 from the tabulated data and assumed that the standard deviations (σ) could be estimated by (max − min)/6, to provide a basis for estimating elasticities and ordinary regression coefficients from the information given in the paper.

The findings of this study appear to contradict many of the U.S. cross-sectional studies in that it finds almost no effects of small particles (British smoke). The study confirms the importance of considering tobacco smoking and drinking-water hardness but provides no confirmation of the hypothesis that the very young and very old are the most susceptible to the effects of air pollution.

The authors downplayed the findings of significant effects of SO_2 on mortality, giving the following reasons:

- The number of significant SO_2 findings was not greatly different from what might be expected due to chance.
- The authors stated that for SO_2 to be harmful to health in the absence of simultaneous high particle loadings "seems biologically implausible."
- When the London boroughs were deleted from the analysis, SO_2 increased in significance, even though most of the higher SO_2 readings were found there.
- Since inclusion of a crude index of smoking reduced the effect of SO_2 on males, it was felt that use of a better measure of smoking would reduce it still further.

However, the effects of SO_2 on female mortality remained, unaffected by the inclusion of smoking or water hardness, for all causes of death, hypertension, and bronchitis. In addition, no (positive) air pollution effects were seen for the "null hypothesis" causes of death: breast cancer and suicide. The lack of an effect in London could be due to increased mobility of the population there or possibly to adaptation to previous levels there, which had been much higher. It is curious that ischemic heart disease failed to show any air pollution effects, although it was the most important cause of death for many of the age-sex groups and the bivariate correlations with air pollution were about the same as for all-cause mortality. The report mentioned ambiguities in diagnosis, which may be an explanation. I estimated the elasticities for female all-cause mortality with respect to SO_2 at about 0.04 to 0.05. A similar value was obtained for all-cause male mortality, without considering smoking. There was insufficient information to make a similar estimate with smoking included. These elasticities are virtually identical to values obtained for all-cause mortality in the United States (see next section) and Canada (Plagionnakos and Parker, 1988 [discussed in Chapter 8]), and thus I concluded that the Chinn *et al.* study supports the hypothesis that air pollution has an effect on mortality at present concentration levels (in contrast to the authors' conclusions).

Cross-Sectional Studies of U.S. Cities

Preliminary Studies of 1969 U.S. City Mortality

I analyzed total mortality from 1969 in up to 136 U.S. cities (Lipfert, 1977b); in addition, the findings for 60 SMSAs were compared to their central city counterparts. The independent variables in these regressions included: percentage over 65, percentage of nonwhite, percentage living in poverty, population density, birth rate, percentage of housing built before 1950, mean and maximum SO_2, mean and minimum SO_4^{-2}, TSP, iron (Fe), manganese (Mn), and benzo(alpha)pyrene (BaP). Stepwise regression optimizations were used to define the "best" models. The main objective of this study was to point out some of the pitfalls in this type of analysis; I reached the following conclusions:

- Statistical significance of SO_4^{-2} depends on which other variables are included in the model.
- The significance of the pollutant variables was sensitive to the use of cities versus SMSAs as the observational unit.

This preliminary study did not include a variable for cigarette smoking and used only 1 year's mortality data, which could be unstable in the case of smaller cities (see Lipfert, 1978; Lipfert, 1984). The use of housing age as a surrogate for housing adequacy was questionable and may serve mainly to illustrate the collinearity between sulfate and other socioeconomic indicator variables for the Northeastern United States. The instability of the socioeconomic coefficients between cities and SMSAs suggested that important variables were omitted in both specifications.

1959–1971 City Mortality Studies

The main body of my analysis of city mortality for 1959–71 has been published in several formats (Lipfert, 1977a, b; Lipfert, 1978; Lipfert, 1980a, b, c). It consists of analysis of 1969–71 mortality in up to 201 U.S. cities and a limited analysis of 1959 mortality in a subset of these cities. I analyzed total mortality for both periods; for 1969–71, I also examined mortality by 10-year age groups and by broad cause of death groups.

The independent variables included percentage over 65, percentage of nonwhite, percentage of poor, amount of old housing, birth rate, smoking, education, population density, population change, region of residence, mean concentrations of SO_4^{-2}, TSP, Fe, Mn, and benzo(alpha)pyrene. The 1969–71 mean values of sulfate and TSP were 9.1 and 86 $\mu g/m^3$, respectively; in 1959,

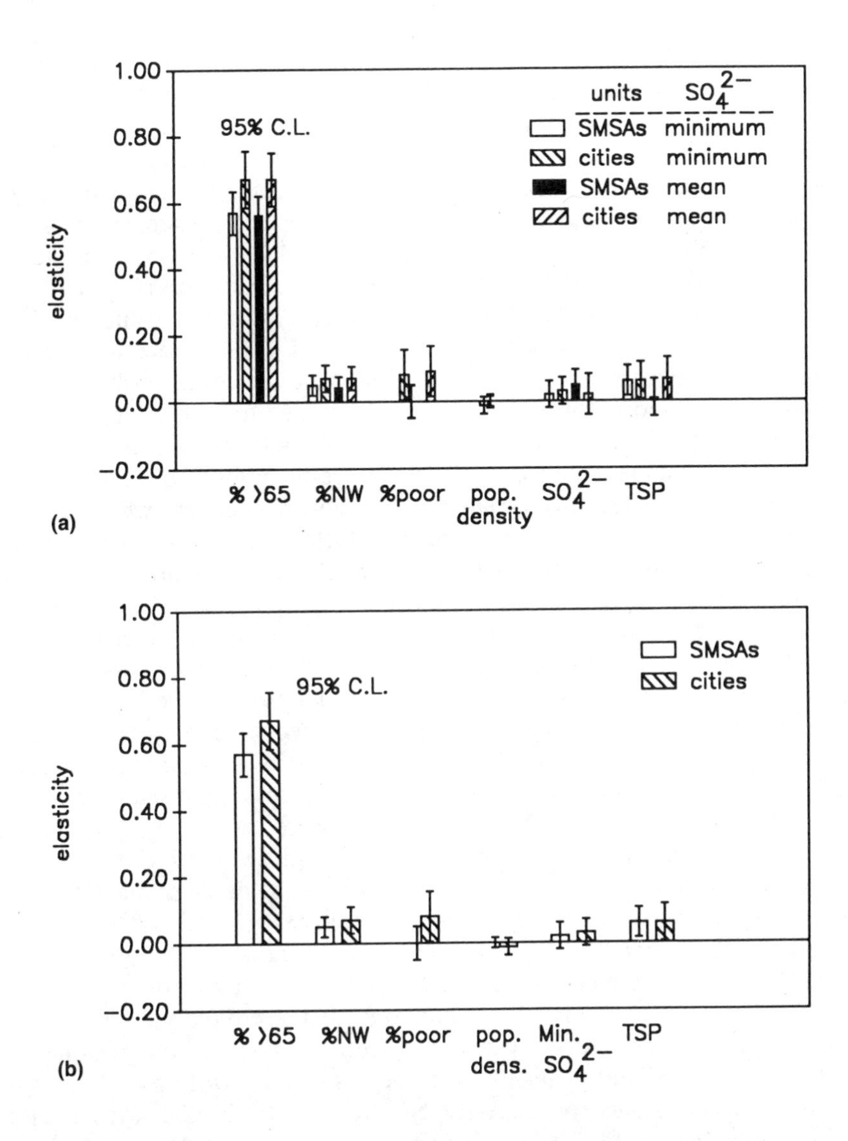

FIGURE 7-13. Multiple regression results for U.S. SMSAs and cities, 1969. (a) Using different sulfate measures with the model of Lave and Seskin (1977). (b) Using the model of Lipfert (1978).

they were 11.1 and 113 μg/m^3; the average TSP for the years 1953–57 was 149 μg/m^3. The "1959" data set was constructed from pollution data for any of a number of years from 1958–62, in part because air pollution was not monitored each year in all cities during this period.

This analysis began with a comparison with the SMSA results of Lave and Seskin (1978), with emphasis on the choice of geographic units and of pollution variables. Lave and Seskin used SMSAs and emphasized the *minimum* value of sulfate aerosol recorded in each location (see the detailed review of their work in Section 3 of this chapter). For a data set of 60 locations across the United

States for which (1969) air quality data were reasonably complete, I compared regression results using mortality and demographic data for the central city (where the air monitor was located) with the SMSA surrounding it, using both the minimum sulfate data and the mean SO_4^{-2} values. The same pollution data were used for both cities and SMSAs. Figure 7-13a compares these four regressions, using the model of Lave and Seskin (1978). Minimum sulfate was not significant for this data set; mean sulfate was significant, and TSP was not significant with SMSA data, but the situation was reversed when city data were used instead. When the model was expanded to include data on birth rate and housing, TSP was more significant for SMSAs than for cities (Figure 7-13b), even though the accuracy of estimated pollutant exposure would surely be much poorer for the SMSAs. The behavior of the birth rate and housing variables in the two regressions is also of interest. In cities, birth rate has a positive effect on mortality; this could be either the effect of infant mortality, or (more likely) it could reflect errors in the population base, due to census undercount, for example, which would affect death and birth rates by the same percentage. In SMSAs, this variable had a negative effect, which was not significant, indicating that the more fertile SMSAs had lower mortality. The housing variable (% of dwellings built before 1950) was positively associated with mortality in both cases, but was not significant for cities. For SMSAs, this variable helps delineate those SMSAs that are dominated by older cities from those with more new construction. Evidently, this distinction is less important in a data set of central cities, perhaps since this data set is more homogeneous. This portion of the analysis showed that cross-sectional regressions can be sensitive to the selections of locations, geographic units, and variables included.

The next consideration was that of accounting for regional differences in smoking habits. The only reliable data available were tax receipts by state (and by city for those cities having separate taxes). I analyzed these data separately to determine whether gradients in taxes across state lines seemed to affect sales, as might be expected due to interstate transfers (Lipfert, 1978). I found such an effect, of about 4% lower sales for each cent of higher average tax in neighboring states. Incorporating smoking data into the cross-sectional air pollution regressions would only be important if these data had patterns correlated with the air pollution patterns; Figure 7-14 shows the regional distributions for 3 different years. In all three cases, the low-smoking states are also low-pollution states (in the West North Central Division, for example). For the larger city data sets, smoking was positively correlated with sulfate ($p < 0.05$), and dropping this variable increased the t value for sulfate by about 0.5. Figure 7-14 also shows how cigarette consumption was getting more uniform across the nation with time. The measure of smoking used was the annual number of packs per person 18 years old or more, adjusted for the tax gradient. I feel that this measure is better than "percentage of current smokers," for example, since it does not suffer from inaccurate self-reporting, accounts for amount smoked, and is analogous to a pollutant emission rate. In that sense, this measure may also account for any passive smoking effects.

Because of the latency period of the health effects of smoking, it is not clear which year should be used for the smoking data; Figure 7-15 provides some comparisons of the estimated elasticity of 1969 total mortality on cigarette sales. It was about 0.20 when based on 1969 sales and about 0.10–0.11 when

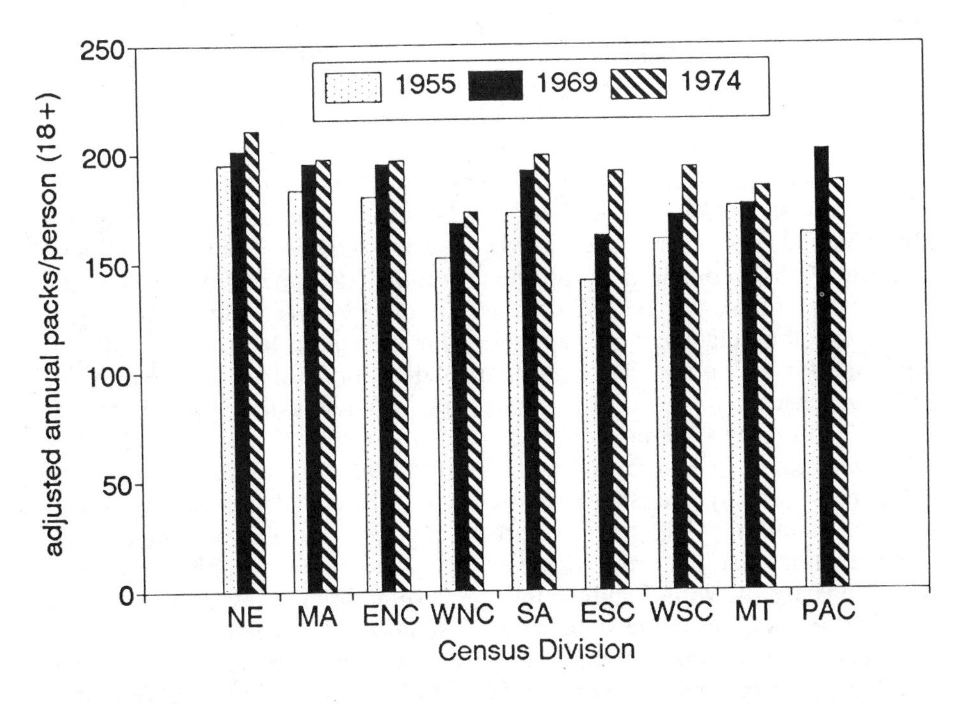

FIGURE 7-14. Spatial and temporal comparisons of cigarette consumption. *Note: NE = New England Census Division, MA = Middle Atlantic, SA = South Atlantic, ENC = East North Central, ESC = East South Central, WNC = West North Central, WSC = West South Central, MT = Mountain, PAC = Pacific. Source: Lipfert (1978).*

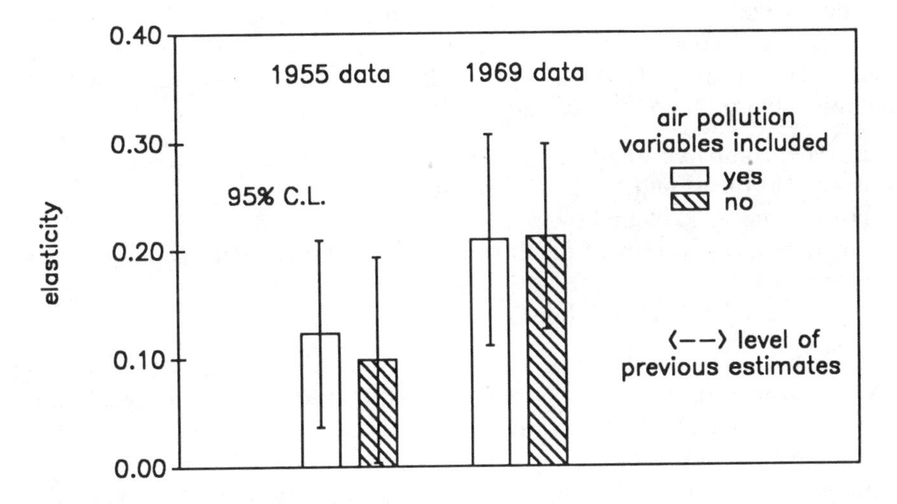

FIGURE 7-15. Multiple regression results for the effect of cigarette consumption on total mortality for U.S. cities, 1969–71. *Data from Lipfert (1978).*

based on 1955 sales. The surgeon general had estimated the elasticity on total mortality at about 0.15; previous regression estimates by others (Morris *et al.*, 1976; Schwing and McDonald, 1976) were around 0.10, perhaps because of their neglect of interstate transfers. It made little difference to the smoking coefficients whether the air pollution variables were included in the regressions. The appropriate year for the smoking data was probably between 1955 and 1969; the 1969 data were used in the analysis of air pollution effects.

The only significant pollutant for the 1959 total mortality regressions was Mn, which may well have been a surrogate for activities from ferrous metal industries. This is somewhat surprising, given the higher pollution levels in the earlier years. However, the TSP data used in this regression were from various years (and thus did not represent the *acute* effects of 1959), and TSP was significant for the pooled data set with a coefficient of 0.00046 per $\mu g/m^3$ (p = 0.02), which checks well with the time-series studies reviewed in Chapter 6. Figure 7-16 presents the air pollution regression results for the 1969 city data sets. On the left in Figure 7-16a, the results from regressions entering each pollutant separately are shown, with their 95% confidence limits. Sulfate was not significant; TSP had the largest effect, about 6%. Manganese and benzo(alpha)pyrene had the tightest confidence limits. Multiple pollutant regressions are shown in Figure 7-16b, for several different assumptions about smoking. Without using the smoking data, sulfate had a negative effect on city mortality; dropping this variable reduced the magnitude of the TSP indicated effect. The regressions with either 1955 or 1969 smoking data were about the same and showed that the nearly 6% effect of air pollution on mortality could be apportioned as about half due to TSP and one-quarter each to B(a)P and Mn. There was no indication that sulfate had an effect.

To compare with intraurban studies, a weighted regression on 1969 total mortality for 158 cities was performed with TSP as the sole pollutant and yielded a regression coefficient of 0.093% per $\mu g/m^3$. This value is much smaller than found in most of the intraurban studies (which also suffered from inadequate socioeconomic controls) but compares well with my brief study of London neighborhoods, which used lung cancer as a surrogate for smoking.

The overall conclusion from the 1959–71 city mortality analysis was that there was a small (~5%) but statistically significant association between air pollution and urban mortality that was stronger in certain cases:

- In 1969, compared to 1959.
- In the Northeast and North Central States.
- For persons aged 65 and over.
- For nonspecific causes of death (in this case, this category is dominated by heart disease).
- For Mn and TSP air pollution.

Sulfate was significant (positive) only for older age groups (75+) and then only when regressed as the sole pollutant with an *a priori* model specification (as opposed to stepwise regression). However, the analysis of the health effects of sulfate may have been compromised by multicollinearity with both other pollutants and socioeconomic variables. No thresholds were suggested for SO_4^{-2} or Mn; a threshold of 85 to 130 $\mu g/m^3$ was suggested for TSP (1969–71). Regressions with lagged air pollution data were also compromised by being

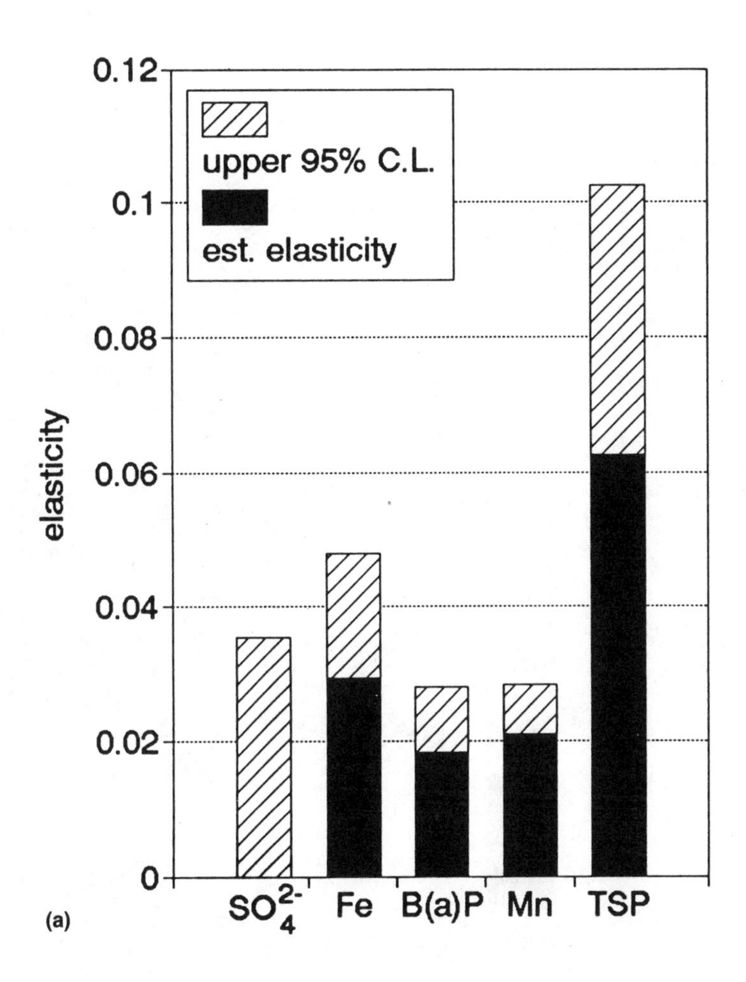

FIGURE 7-16. Multiple regression results for the effects of air pollution on U.S. city mortality, 1969–71. (a) Single pollutant regressions using 1969 smoking data. (b) Multiple pollutant regressions for various treatments of smoking. *Data from Lipfert (1978).*

based on fewer observations, but they suggested that the lagged variables were often slightly superior.

These studies were limited by the failure to study adequately long-term exposures and by omitting other potentially confounding variables such as drinking-water hardness and diet. Other criticisms include:

- The use of the age of housing as a surrogate variable for the adequacy of housing was questionable; however, this variable demonstrated the sensitivity of the sulfate variable to model specification, especially for total mortality.
- Drinking-water hardness, ozone, and migration should have been included as independent variables.

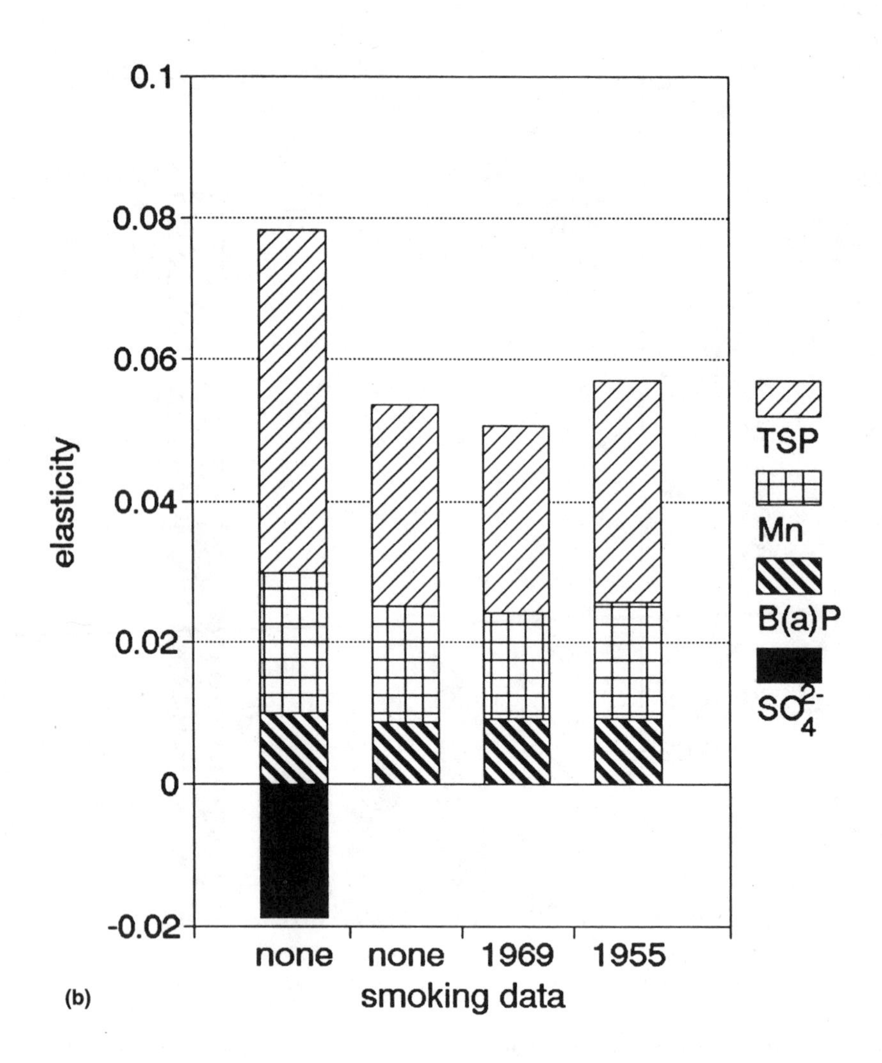

FIGURE 7-16. *Continued*

- The spatial autocorrelation of regression residuals should have been examined.
- The results for Mn may have been overinterpreted and most likely are a surrogate for occupational exposure. Analysis by sex would be helpful to verify this hypothesis. A physiological hypothesis is lacking for the effects of this pollutant at low concentrations.
- Given the availability of air quality data for some species back to 1953, an attempt should have been made to construct a lifetime exposure index to examine very long-term effects. Analysis of mortality differentials in a given year cannot distinguish between chronic effects and the sum of acute effects (Evans *et al.*, 1984b).

- The data for all-cause mortality were not corrected for external causes of death (accidents, etc.), in most cases.

This analysis showed that for age-specific mortality, in most cases neither SO_4^{-2} nor TSP was significant. If the effect of Mn were a surrogate for occupational effects, then there would be essentially no indicated effects of pollution on mortality, a point that has been overlooked by many who have cited this work. By implication, the need for better age adjustment is implied when total mortality is analyzed.

Other Studies of 1970 Mortality in U.S. Cities

Crocker *et al.* (1979) studied total mortality and certain diseases in 60 U.S. cities for 1970. Their analysis featured a simultaneous equation approach to estimating the effects of medical care on mortality, a factor that had not been considered by most previous authors. Their dependent variables were 1970 mortality (all ages) for all causes, for pneumonia and influenza, and for early infant diseases. Independent variables included age, income, education, smoking, diet, medical care, climate, and mean concentrations of TSP, NO_2, and SO_2. Sulfate was not included in this study.

The authors found no pollutants to be statistically significant for total mortality, TSP to be significant for pneumonia and influenza, and SO_2 to be significant for early infant diseases. A cost-benefit analysis was performed with these results.

Unfortunately, there were substantial flaws in the air quality data used that prevented this study from reaching meaningful conclusions (in addition, given the number of independent variables, 60 observations [locations] are insufficient for reliable results). The only measure of sulfur oxides was SO_2, from bubblers, which tend to give low readings. For example, the SO_2 value used for New York City was too low by a factor of 4, compared to data reported by the city. The data reported for many cities were below the detection limit for this monitoring method. Measurement noise of this type will bias the regression coefficients downward. The NO_2 data used in this study were based on the Jacobs-Hochheiser method, which is no longer considered valid because of substantial interferences (Chapter 2). The data base used mixed mean and median (50th percentiles) measures in the same data set, which amounts to biasing part of the data set downward. The pollution data cited show anomalous values for several cities.

The dietary data were from 1955, not from 1965 as reported. Thus they are too outdated to be meaningful as a 1970 health predictor. The simultaneous equations method used to study the effect of medical care was not very successful in predicting mortality ($R^2 = 0.39$) and nearly canceled the effect of smoking, since this variable appeared in the complete specification twice, with opposite signs.

Gerking and Schulze (1981) continued the analysis of 1970 total mortality in 60 U.S. cities using ordinary and two-stage least-squares procedures (apparently the same data set used by Crocker *et al.*, 1979), emphasizing the simultaneous equations approach. They concluded that the estimated effects of air pollution are highly sensitive to model specification and that, as a result, relatively little is known about the effects of long-term low-level air pollution exposures on human mortality.

While the above conclusions may be true, the points would have been more convincing had a larger data set been used, comprised of better quality data; results are usually conflicting when faulty data are used.

1980 U.S. City Mortality

The most recent analysis of city mortality (Lipfert *et al.*, 1988) comprised a statistical analysis of spatial patterns of 1980 U.S. urban total mortality (all causes), evaluating demographic, socioeconomic, and air pollution factors as predictors. Specific mortality predictors included

Nonpollutant terms
Percentage 65 and over
Racial and ethnic distribution (% Black, % Asian, % Hispanic)
Percentage below poverty level
Percentage of college graduates
Percentage of population change since 1970
Ratio of males to females*
Birth rate
Cigarette smoking
Hardness of drinking water
Source of drinking water* (surface versus ground)
Heating fuel use and degree days*
Population density
Log of city population*
Percentage born in state of current residence

Air pollutants (annual averages for 1978–82)
Carbon monoxide*
Ozone*
Sulfate aerosol (from hi-vol samplers)
TSP (exploratory use only)
Particulate lead*
Particulate iron
Particulate cadmium*
Particulate manganese
Particulate vanadium*
Inhalable particles* (IP) (diameter $<15\,\mu m$)
Fine particles* (diameter $<2.5\,\mu m$)
Sulfate aerosol (from IP samplers)
Modeled SO_2
Modeled SO_4^{-2}
Modeled NO_x

The ozone data were the annual means of maximum hourly concentrations. The data on inhalable particles (IP) were obtained from dichotomous samplers accepting particles, 50% of which were less than about $15\,\mu m$ in diameter, with a further 50% size cut at about $2.5\,\mu m$. The modeled concentrations were obtained from a long-range transport diffusion model (Shannon, 1981); data from these computations are essentially a transfer function between emissions and air quality, and present the advantage of being available for all cities (see Figure 7-23). Because the number of cities with valid data on air quality and water hardness varied considerably by pollutant, we considered several different data sets ranging from 48 to 952 cities. The mean computed sulfate concentration was about $5\,\mu g/m^3$; the mean inhalable particle mass was about $40\,\mu g/m^3$. This measure of particle mass is probably more comparable to British smoke than it is to TSP. These mean values confirm that average air quality in the United States had improved considerably over the situations previously studied. However, no effort was made in this study to model air quality specific to the year 1980; all air quality measurements were averaged over the period 1978–82, depending on the availability of data.

*Denotes variables that were not significant in the regressions or that entered with the (apparent) wrong sign.

TABLE 7-3 Stepwise Regression Results (each pollutant entered separately)*

	Full Data Set				Water Hardness Data Set			
Pollutant (x_i)	n	Adjusted R^2	b_i (standard error)	e_i	n	Adjusted R^2	b_i (standard error)	e_i
Old SO_4^{-2} (computed)	908	0.814	0.085 (0.016)	0.045	454	0.853	0.098 (0.016)	0.052
New SO_4^{-2} (computed)	908	0.817	0.103 (0.016)	0.066	454	0.854	0.123 (0.017)	0.081
SO_2 (computed)	908	0.816	0.035 (0.005)	0.061	454	0.854	0.037 (0.005)	0.063
NO_x (computed)	908	0.817	0.035 (0.005)	0.065	454	0.848	0.032 (0.006)	0.050
SAROAD SO_4^{-2}	187	0.831	0.100 (0.024)	0.087	185	0.858	0.082 (0.022)	0.071
IP SO_4^{-2}	78	0.813	0.195 (0.052)	0.083	68	0.883	0.177 (0.053)	0.077
Fine particles	80	0.806	NS	[0.059]	68	0.867	NS	[0.059]
Total (inhalable) mass	80	0.806	NS	[0.047]	68	0.867	NS	[0.027]
Fe	172	0.851	0.401 (0.115)	0.041	172	0.851	0.401 (0.115)	0.041
Mn	172	0.849	9.60 (2.54)	0.033	172	0.849	9.61 (2.54)	0.033
$SO_4^{-2} \times O_3$	211	0.806	NS	[0.018]	201	0.849	NS	[0.007]

Source: Lipfert *et al.* (1988).

Key: ( ) = standard error of regression coefficient; E_i = elasticity = $b_i(x_i/y)$ (evaluated at the mean); NS = not significant (pollutant did not enter stepwise regression); [ ] = elasticity if pollutant had entered.

The regression models and data were tested for outliers, influential observations, and behavior of residuals. The models that passed these tests showed sulfate aerosol, iron particles, and (to a lesser extent) total particle mass to be associated with mortality, although, depending on the data set considered and the choice of variables or factors to account for demographic effects on mortality, there were variations in the specific pollutants that showed statistical significance.

The initial exploration of these new data sets was by means of stepwise regressions, taking one pollutant at a time. These calculations were made for both the largest possible data sets and for the data sets limited by the subset of cities having water hardness data. The results are given in Table 7-3 and show virtually identical results in terms of adjusted R^2 values for all of the computed pollutants. There are differences in statistical significance and in elasticity, however.

Table 7-3 also shows very similar elasticities for all three sulfate variables, indicating that they are essentially interchangeable. Note, however, that the regression coefficients differ, in roughly inverse proportion to the mean values of the three variables. The three SO_4^{-2} variables differ in the following ways:

- The computed value reflects only combustion products and smelter emissions and does not account for local primary sulfate emissions or natural compounds (such as calcium sulfate, $CaSO_4$).
- The values measured from high-volume samplers are thought to reflect artifacts formed on the filters from SO_2 in the air being sampled. The amount of artifact will vary with local conditions.
- The SO_4^{-2} values obtained from the IP samplers are thought to represent the truest measurements of the local airborne sulfate.

The only attribute these three variables would appear to have in common is thus the regional component of sulfate, rather than local concentrations. If this were in fact the case, the possibility of confounding by some (unspecified) regional characteristic variable must be considered. The water hardness variable usually had a negative coefficient (as expected) but never achieved statistical significance. Thus the chief difference between the two data sets was in the selection of cities, not the effects of the water hardness variable *per se*. The "water hardness cities" tended to be larger in population.

Although the overall regressions for computed SO_4^{-2} and NO_x provided virtually identical fits, there were some interesting differences in the models defined by the final steps and in the stepwise regression sequences. Both models had the same ($p < 0.05$) regression coefficients for the following variables:

Percentage over 65
Birth rate
Population change
Smoking
Percentage of (prior) residents

The coefficients for the two stepwise models were statistically different for the percentages of poor and of Hispanics (both larger in absolute magnitude for

TABLE 7-4 Sequences of Stepwise Regressions (separate models)

	Partial Regression Coefficients	
Regression Step	SO_4^{-2}	NO_x
2	0.119	0.057
3	**0.212***	0.087
4	**0.160**	0.059
5	**0.140**	**0.059**
6	**0.134**	**0.067**
7	**0.127**	**0.058**
8	**0.125**	**0.085**
9	**0.133**	**0.110**
10	**0.117**	**0.117**

* Bold type indicates that the variable has entered the model.

the NO_x model), and the percentages of Blacks and Asians failed to enter in the NO_x model. Table 7-4 gives the results of the stepwise regression sequences. Sulfate enters on the third step with a substantial partial regression coefficient, which then decays about 50% as additional variables enter the model. On the other hand, NO_x entered its model on the 7th step and its value *improved* with each succeeding step, ending with a partial regression coefficient identical to that of sulfate. Thus, if the analysis had been terminated on the 8th step or earlier, the conclusions regarding the comparison of the two variables might have been quite different (SO_4^{-2} appearing to be "better" than NO_x).

Based on the stepwise regression models for the three computed pollutants (SO_4^{-2}, SO_2, NO_x), a fixed demographic model was selected to further examine pollutant effects. From the list of all demographic variables entering for each of three pollutants, six were chosen for the fixed model: percentage of population and 65 and over; birth rate; percentage of population change since 1970; percentage of population of Hispanic origin; average of 1969 and 1980 cigarette smoking estimates (packs per person per year), and percentage of population classified as below the poverty line. Three other variables were not included in the fixed model: percentage of population classified as Black or Asian, which entered only with one or two of the computed pollutants and were not significant when they entered; the percentage of population residing in their state of birth, which was significant only with NO_x. Results of the fixed-model regressions are shown in Table 7-5. The pollutant regression coefficients are not significantly different from those given in Table 7-3; adjusted R^2 values are slightly lower for the fixed-model regressions, as expected, since they are not "optimal" models.

The analysis of the computed sulfate variable presented in Chapter 2 showed that the comparisons between computed and measured sulfate varied considerably by geographic region. It is thus possible that use of the computed sulfate variable introduced bias, even though it reduced measurement noise. Accordingly, we performed regression analyses by region, using the regional boundaries shown in Figure 7-17, which differ somewhat from the conventional Census Regions (the western boundary has been moved east, for example).

TABLE 7-5 Results from Fixed Demographic Model Regressions*

Model #	4–1	4–2	4–3	4–4	4–5	4–6
≥65 (%)	0.546††	0.547††	0.551††	0.670††	0.670††	0.672††
	(0.010)	(0.010)	(0.010)	(0.015)	(0.015)	(0.016)
Birth Rate	0.187††	0.182††	0.180††	0.219††	0.209††	0.203††
	(0.011)	(0.011)	(0.011)	(0.014)	(0.014)	(0.014)
Population	−0.013††	−0.013††	−0.014††	−0.010†	−0.009†	−0.014††
	(0.001)	(0.001)	(0.001)	(0.004)	(0.004)	(0.004)
Spanish (%)	−0.025††	−0.031††	−0.042††	−0.031††	−0.035††	−0.042†
	(0.004)	(0.004)	(0.005)	(0.005)	(0.005)	(0.005)
Smoking (78)	0.009††	0.008††	0.009††	0.011††	0.011††	0.013†
	(0.002)	(0.002)	(0.002)	(0.003)	(0.003)	(0.003)
Poor (%)	0.041††	0.053††	0.064††	0.067††	0.076††	0.086†
	(0.008)	(0.008)	(0.008)	(0.009)	(0.009)	(0.009)
Water Hardness	X	X	X	−0.0006	−0.001	0.0007
				(0.0005)	(0.0005)	(0.0005)
Computed SO_4^{-2}	0.112††	X	X	0.109††	X	X
	(0.015)			(0.016)		
Computed SO_2	X	0.038†	X	X	0.038††	X
		(0.005)			(0.005)	
Computed NO_x	X	X	(0.038)††	X	X	0.032††
			(0.005)			(0.006)
Adjusted R^2	0.813	0.814	0.813	0.850	0.851	0.845
n	916	916	916	459	459	459

* A value in parentheses denotes standard error of coefficient; X denotes a variable not entered.
† $p < 0.05$.
†† $p < 0.001$.

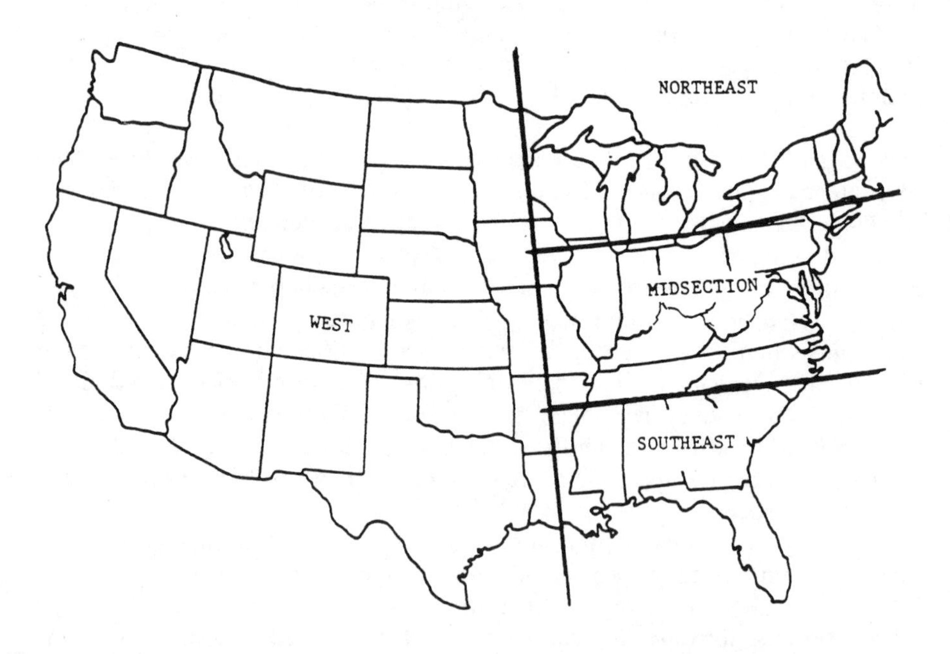

FIGURE 7-17. Map showing regional subdivisions used for 1980 city mortality analysis. *Source: Lipfert* et al. *(1988).*

TABLE 7-6 Pollution Regression Coefficients by Region (stepwise regressions with one pollutant at a time)*

		Region			
		Northeast	Mid-Section	Southeast	West
Full data set	n	137	260	128	383
	SO_4^{-2}	0.162	0.154	0.168†	
		(0.043)	(0.034)	(0.092)	
	SO_2	0.089	0.040		0.025
		(0.012)	(0.010)		(0.009)
	NO_x	0.108	0.030		0.021
		(0.015)	(0.010)		(0.005)
Cities having water hardness data	n	88	175	59	132
	SO_4^{-2}	0.165	0.146	0.220	
		(0.044)	(0.039)	(0.073)	
	SO_2	0.059	0.037		
		(0.014)	(0.011)		
	NO_x	0.068	0.027		
		(0.018)	(0.011)		

*A value in parentheses denotes standard error of regression coefficient; blank indicates that the variable did not enter stepwise regression.
†This coefficient is not statistically significant but is provided for reference.

These subdivisions are less arbitrary than state boundaries in many cases and were also chosen with a view toward equalizing the numbers of cities in the three eastern subdivisions. The results are given in Table 7-6.

Of the pollutant variables, only the computed species showed statistical significance for these subsets; it was not possible to determine whether the lack of significance of Fe, Mn, and measured SO_4^{-2} was due to the reduced number of observations in each subset or due to collinearity between other regional factors and pollutant levels. It should also be recognized that the use of statewide average smoking data may have compromised this regional analysis, since there may not be adequate range for the smoking variable in all regions, since local city differences are not accounted for. Table 7-6 shows that the pollution regression coefficients are reasonably consistent across regions, except for a high value for NO_x in the Northeast. All four regions showed at least some mortality response to pollution, but the Southeast showed the least response; it also had the lowest average pollutant levels. The computed pollutants were significant in 15 out of 24 regressions. SO_4^{-2} was never significant in the West; SO_2 and NO_x were never significant in the Southeast. Ozone generally showed negative regression coefficients within regions, including the West.

Because the analysis of the regional subsets suggested that different pollutants could be having their primary effects in different parts of the country, some multiple pollutant regression models were evaluated. For the combination of computed SO_4^{-2} and NO_x, O_3, and Fe ($n = 119$), O_3 did not enter, SO_4^{-2} and NO_x showed reduced significance ($p = 0.038$ and 0.065, respectively), and Fe showed increased significance. However, the combined elasticity, 0.135 ±

0.06, was not significantly greater than that given by SO_4^{-2} alone. This result suggested that Fe might be the most important pollutant for this set of cities. When Mn was substituted for Fe, SO_4^{-2} became less significant and NO_x, more so. Iron and SO_2 showed equal significance with NO_x losing significance. A similar calculation for 172 cities, not using ozone, resulted in similar findings. It was difficult to conclude that any one pollutant was more important than another; however, Fe and Mn consistently tended to show higher statistical significance when either was entered in combination with the computed pollutants.

These results were compared with previous results and data for 1970. We found that 1970 pollution was not a better predictor for 1980 mortality than was 1980 pollution. This finding could be interpreted to support the hypothesis that the cross-sectional regressions are measuring the annual sum of acute responses, occurring in the same year. The only coefficients that underwent a statistically significant change since 1970 were those for poverty and the percentage of population 65 years of age or older (borderline significance for the latter). Since sulfate was not significant in 1970, the large standard error found for that data set does not preclude the 1980 coefficient for that period.

Because of the high degree of similarity of alternative regression models employing different pollutants as predictors, three different bases were used to compare regression models. With these varying criteria, SO_2, sulfate aerosol, and Mn particles all appeared to provide better regression fits than the other pollutants. A two-stage procedure was used employing mortality "adjusted" for the nonpollution variables (demographics, smoking, poverty). The adjusted mortality rates were computed as the residuals from a first-stage regression, including all the nonpollutant variables. This procedure was carried out twice, with and without drinking-water hardness in the model. These residuals were then regressed against each pollutant variable in turn as a bivariate procedure, using only the cities for which the pollutant data were available. This protocol assured that all the different data sets defined by the availability of pollutant data were treated alike with respect to nonpollutant variables but carried the risk that some portion of the pollutant effects were inadvertently included in the "adjustment" procedure due to collinearity between demographic and pollutant variables. The differences in fit among the various sulfate measures, SO_2, NO_x, Fe, Mn, fine particles and total particles, were minimal, as seen in Figure 7-18. The correlation coefficients and their (positive) 95% confidence limits are shown in Figure 7-18a; the metal particles (Fe and Mn) consistently showed the highest values for both data sets. The elasticities are compared in Figure 7-18b; most of the values are between 0.03 and 0.05, with the exception of measured sulfates for the all-city data set (without water hardness).

Our inability to identify a single pollutant most closely associated with mortality rates may be interpreted in two ways. Since this was a total mortality analysis, it is possible that all the pollutants are associated in some way with either a component of the population or with some particular cause of death. It is also possible that none of them is causally related to mortality and that they all reflect the effects of some additional, unmeasured variable. Thus, we concluded that statistical criteria alone were not sufficient to define air pollution–mortality relationships.

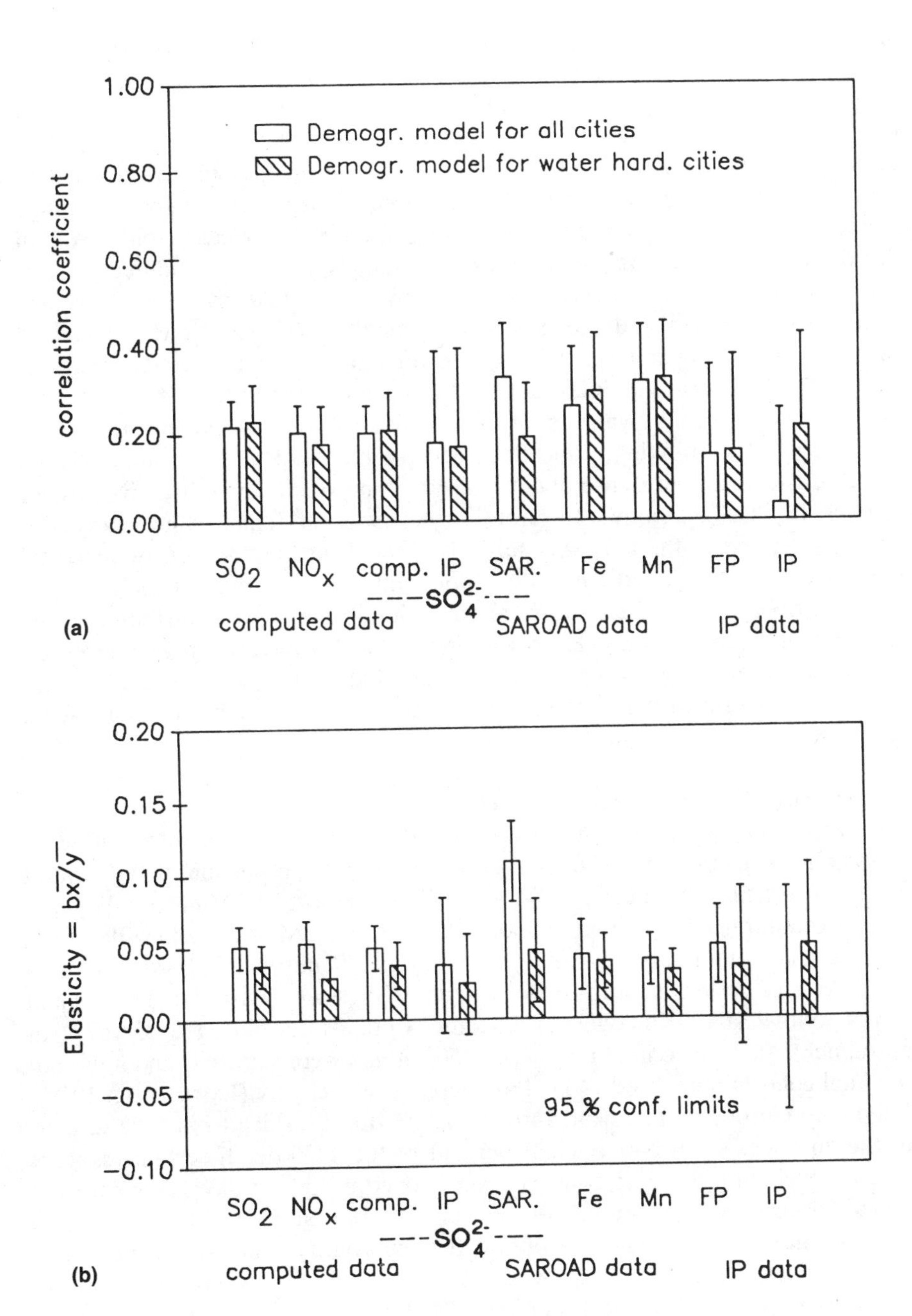

FIGURE 7-18. Results from two-stage regression analysis of 1980 U.S. city mortality. (a) Correlation coefficients. (b) Elasticities. *Source: Lipfert* et al. *(1988).*

Prospective Studies on Individuals

California Seventh-Day Adventists

Abbey *et al.* (1991) described a prospective study of 6000 nonsmoking long-term California residents who were followed for 10 years, beginning in 1976. Monthly ambient air quality data dating back to 1966 were used, and the study was restricted to those who had lived within 5 miles of their current residence for at least 10 years. All the air quality monitors in the state were used to create individual exposure profiles (duration of exposure to specific cutoff levels) for each participant, by interpolating to their zip code centroid based on the three nearest monitoring stations. Pollutant species were limited to TSP and ozone; oxidants were used in the early part of the monitoring record. Endpoints evaluated and the numbers of deaths included: Newly diagnosed cancers (any site for males, 115; any site for females, 175; respiratory cancer, 17), definite myocardial infarction (62), mortality from any external cause (845), and respiratory symptoms (272). Individual risk factors considered included age, past smoking, education, occupational exposures, history of high blood pressure, and presence of definite symptoms of airway obstructive disease in 1977. Of these endpoints, respiratory symptoms and female cancer (any site) were associated with TSP exposure. Neither heart attacks or nonexternal mortality were associated with either pollutant. The authors felt that possible errors in their estimated exposures to air pollution may have contributed to the lack of significant findings. However, it would also have been of interest to repeat the regressions using pollution data linked to the month of death, rather than averaged over the entire period, as a test of chronic versus acute responses.

Prospective Cohort Study in 6 U.S. Cities

Pope *et al.* (1993) analyzed survival probabilities among 8111 adults who were first recruited in the mid-1970s in 6 cities in the eastern portion of the United States. The cities are: Portage, WI, a small town north of Madison; Topeka, KS; a geographically-defined section of St. Louis, MO; Steubenville, OH, an industrial community near the West Virginia–Pennsylvania border; Watertown, MA, a western suburb of Boston; and Kingston–Harriman, TN, 2 small towns southwest of Knoxville. The adults were white and aged 25 to 74 at enrollment. In each community, about 2500 adults were selected randomly, but the final cohorts numbered 1400–1800 persons in each city (Ferris *et al.*, 1979). Follow-up periods ranged from 14 to 16 years, during which from 13% to 27% of the enrollees died. Ninety-eight percent of the 1430 death certificates were located, including those persons who had moved away and died elsewhere.

This cohort has been studied extensively for respiratory health (see Chapter 11). Air monitoring data were obtained from routine sampling stations and from special instruments set up by the research team. This study differs importantly from purely ecological studies in that individual characteristics of the decedents are known, including smoking habits, occupational exposure indices, body mass index, and education. However, the details of exposure to community air pollution are not known for each subject; this information would require long-term personal monitoring, which is clearly impractical for over 8000 subjects. Thus, this prospective study suffers from the same kinds of uncertainties in the exposure variables as do the ecological studies. For example, the 6 cities differ substantially in the types of air pollution sources present and in the types of heating and air conditioning systems used, which can influence indoor air quality.

Pope *et al.* (1993) used the individual data on personal characteristics to develop risk factors for smoking, education, obesity, and occupational exposures to dusts and fumes, using the Cox proportional hazards model. The adjusted relative risks were then compared across the 6 cities, using cross-plots to examine possible dose-response relationships against any of several air pollutants (TSP, PM_{15}, PM_{10}, $PM_{2.5}$, SO_2, SO_4^{-2}, H^+, and O_3). Thus, the study is cross-sectional in design, with 6 observations. Most of the air quality measures were averaged over the period of study, in an attempt to study long-term (chronic) responses. Pope reported (personal communication, 1993) that using time-sequenced air quality data made little difference in the results.

There were large differences in survival probabilities among the six locations. The long-term average mortality rate in Steubenville was 16.2 deaths per 1000 person-years; in Topeka, it was 9.7, yielding a range in average (crude) relative risk of 67% among the 6 cities. After adjustment for age, smoking status, education, and body-mass index, the range in average relative risk was reduced to 26%. By way of comparison, the range in adjusted lung function for adult nonsmokers among the 6 cities was only 12% (Dockery *et al.*, 1985), but the data of Speizer *et al.* (1989) show that FEV_1 was a powerful predictor of mortality (see Chapter 12). The elasticity of mortality on FEV_1 was about −2.4, so that the 12% and 26% figures appear to be consistent. However, the data of Dockery *et al.* (1985) show that the most polluted city (Steubenville) did not exhibit the poorest average lung function for nonsmokers.

Occupational exposure to dusts or fumes was not a significant risk factor. Pope *et al.* (1993) report that the strongest associations between adjusted relative risks and air pollution were for PM_{15} or PM_{10}, $PM_{2.5}$, and SO_4^{-2}. Since no relationship was shown for aerosol acidity (H^+), one would conclude that the sizes and mass of inhaled particles are more important than their chemistry. Figure 7-19 is a plot of relative all-cause mortality risk against PM_{15} (data obtained from Speizer, 1989), for example. When the populations were stratified by smoking habits and occupational exposure to dusts and fumes, these relationships persisted; they were slightly higher for current and former smokers and for those with occupational exposure. In comparing the most and least polluted cities, they also report elevated risks for cardiopulmonary causes (1.37) and lung cancer (1.37, not significant). When Pope *et al.* considered the 6 cities individually, only Steubenville showed a statistically significant ($p < 0.05$) elevated risk with respect to the least polluted city (Topeka).

This study is of particular interest, not only for the findings with respect to air pollution, but as a means of comparing ecological and prospective studies. I created a parallel data set for this purpose by extracting appropriate data for the county surrounding each of the 6 cities, from the *County and City Data Book, 1988* (U.S. Bureau of the Census, 1988). These data were used in an ecological analysis of the same locations, using the multiple regression methods of previous studies (Lipfert *et al.*, 1988; Lipfert, 1992). To account for the long-term trends, I averaged county mortality rates for 1984, 1980, and 1975 (the latter two were obtained from the previous two editions of the *County and City*

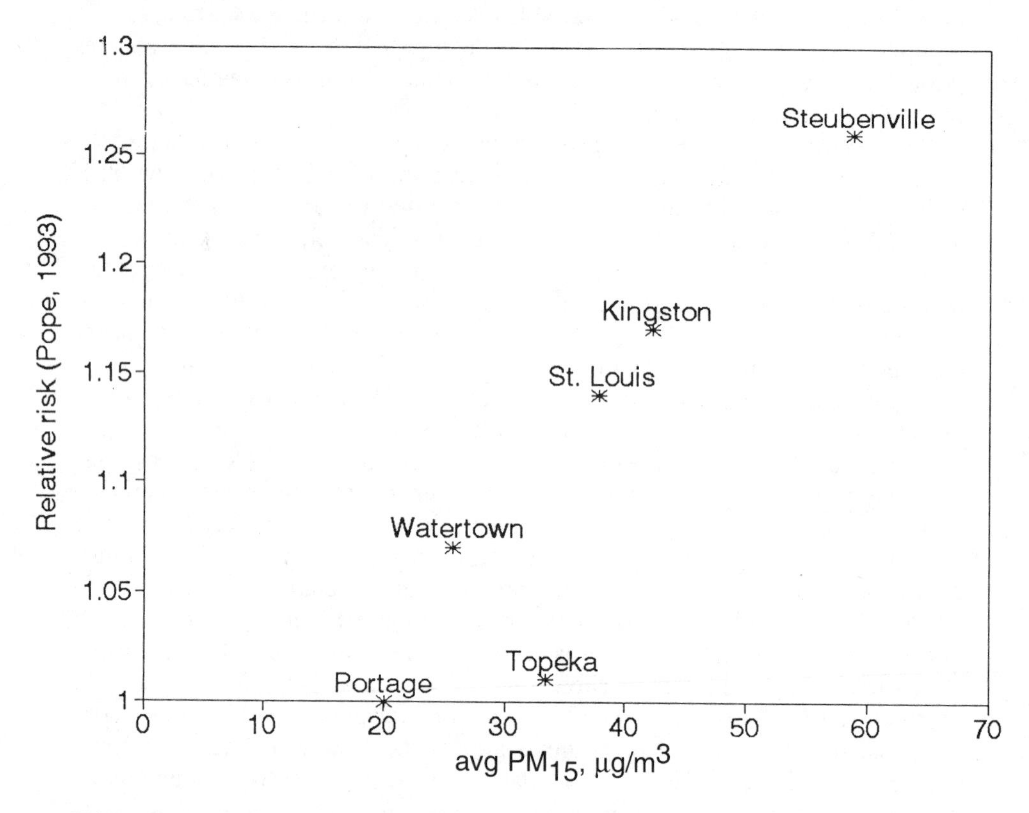

FIGURE 7-19. Relative mortality risk based on individual survival rates in 6 cities (Pope *et al.*, 1993) vs. average PM_{15} levels (Speizer, 1989).

Data Book.) The average crude county mortality rate was 10.3 per thousand population. The 1980 census data extracted included 1980 percentages of the county populations that were over 65, nonwhite, in families with incomes below poverty level, or with less than a high school education. State-level data on smoking were obtained from Lipfert *et al.* (1988). These data were used as independent variables in multiple regressions to adjust the crude mortality rates to correspond with the relative risk factors developed by Pope *et al.* (1993). The most successful regression included variables for age, race, and smoking. The education variable was not significant; its effects were probably captured by the nonwhite and smoking variables, whose coefficients tended to be somewhat larger than found in previous studies.

In spite of the few degrees of freedom, the regression coefficients developed in this way were reasonably robust and in agreement with previous findings from ecological studies. Next, we compare the findings for smoking and education with the results of Pope *et al.* (1993), who report a doubling in individual risk for persons who currently smoke at the rate of 25 pack-years (a pack-year is a 1 pack per day habit persisting for 1 year). For the (average) 15 years of follow-up, this corresponds to an annual cigarette consumption rate of 608 packs per person. The corresponding regression coefficient would thus be $1 \times 10.3/608$ or 0.0169. The value obtained by multiple regression was 0.037 ±

0.017, with smoking as the sole socioeconomic variable, and slightly lower when additional variables were included, which provides a reasonable check. The regression coefficients for cigarette consumption found by Lipfert *et al.* (1988) for national data sets were in the range 0.007 to 0.019, which also checks. Pope *et al.* (1993) estimated a relative risk of 1.19 for those individuals not completing a high-school education; the corresponding regression coefficient would thus be $0.19 \times 10.3/100 = 0.0196$. The actual value obtained by multiple regression was 0.024; again, a reasonable check. These comparisons, although limited, suggest that ecological methods may be used to develop appropriate risk factors, when data on individuals are lacking.

Two sets of adjusted county mortality rates, expressed as fractional increments from the average, were developed to compare with the results of Pope *et al.* (1993). First, residuals from a multiple regression on average county mortality rate against age, race, and smoking were used. Then, because the resulting smoking coefficient seemed on the high side and the smoking variable was admittedly crude, as an alternative, the mortality rates were pre-adjusted for smoking using a coefficient of 0.0167, and these adjusted rates were regressed against age, race, and education. On this basis, the education coefficient (0.016), although not significant, agreed well with the results of Pope *et al.* (1993), based on individuals.

Figure 7-20 compares both sets of county level mortality residuals (fractional differences from an average mortality rate of about 10.3 per 1000 population) with the relative risks developed by Pope *et al.* (1993); the main points of agreement are that Portage is a low mortality location and Steubenville is high. The range in relative risk developed by Pope *et al.* (26%) is substantially larger than the corresponding range in county-level regression residuals (8%). Even when the county level rates were only adjusted for age and race, the range from highest to lowest mortality rates was only about 12.5%. Part of the difference in the two approaches is the treatment of age: Pope's relative risks pertain to adults, while the county level mortality rates are for the whole population. However, since the mortality measures are in percentage form in both cases, age differences are unlikely to explain the entire disparity shown in Figure 7-20.

County-level data were not available on body-mass index, but this was the smallest of the individual risks analyzed by Pope *et al.* (1993) (relative risk of 1.08 for a 30–40 pound weight increment), so that it seems unlikely that its omission from the county-level regressions could explain the large differences seen. In addition, the relationship between weight and mortality has been shown to be U-shaped, with the lowest mortality rates found for those close to average weights (Lew and Garfinkel, 1979).

In other words, the members of the cohorts in the most polluted cities seem to be dying at much higher rates than the corresponding county populations as a whole, after adjustment for age, race, smoking and education. Given the random selection procedures used to develop the cohorts that were reported by the authors, this is a perplexing finding and no explanation is immediately forthcoming. It is always possible in cross-sectional regressions that some unknown risk factor has been omitted; candidates here (which could influence both types of regressions) may include diet, exercise, use of air conditioning, or migration to locations with differing air quality.

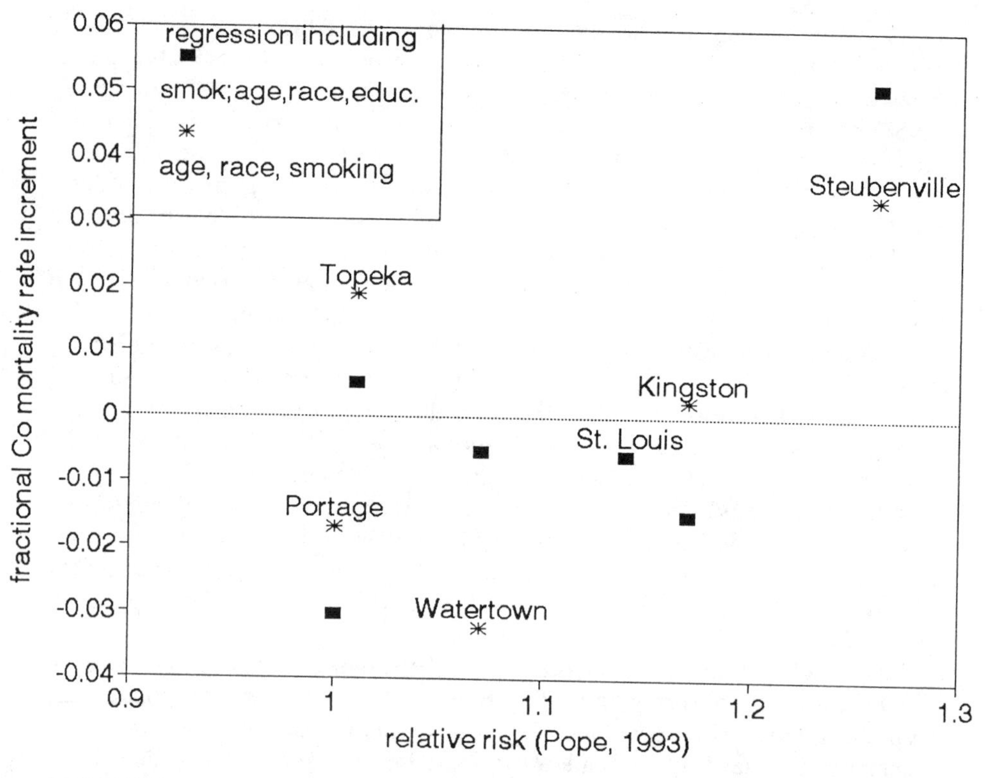

FIGURE 7-20. Residuals from county-level mortality analysis versus individual relative risks.

Since the dependent variables developed with ecological methods qualitatively track the individual relative risk factors developed by Pope *et al.* (1993), it follows that the pollution relationships should also be similar, and this is indeed the case. Figures 7-21 plots fractional residual mortality rates against PM_{15}, and Figure 7-22, against $PM_{2.5}$. While the relationships are much more variable than those based on relative risk (Figure 7-19), the uncertainties in air pollution exposure also tend to be much larger in an ecological study. The elasticities of the apparent dose-response relationships of Figures 7-21 and 7-22 are in the range of 0.04 to 0.06; PM_{15} had the better fits, which were statistically significant ($p < 0.05$). These results conform well with other ecological studies, both time-series and cross-sectional; the corresponding elasticity range for the relative risk data of Pope *et al.* (1993) is 0.22 to 0.24. Plots of the residuals against TSP, SO_2, NO_2, and average ozone did not suggest dose-response relationships; however, the range in average ozone was quite small, 0.04 ppm to 0.051 ppm. One should also keep in mind that regression coefficients based on one pollutant at a time necessarily capture the effects of all other pollutants with which it is correlated. Thus, Steubenville is not only high in particulates, but also in SO_2, NO_2, and metals such as Mn (Lipfert, 1988). Indeed, estimates of Mn levels from either around 1970 or around 1980 fit the 6 city mortality data about as well as particulates and also yield coefficients in line with those from previous studies (Lipfert, 1978; Lipfert *et al.*, 1988). Finally, one must recognize that Steubenville constitutes an "influential observation" in this data

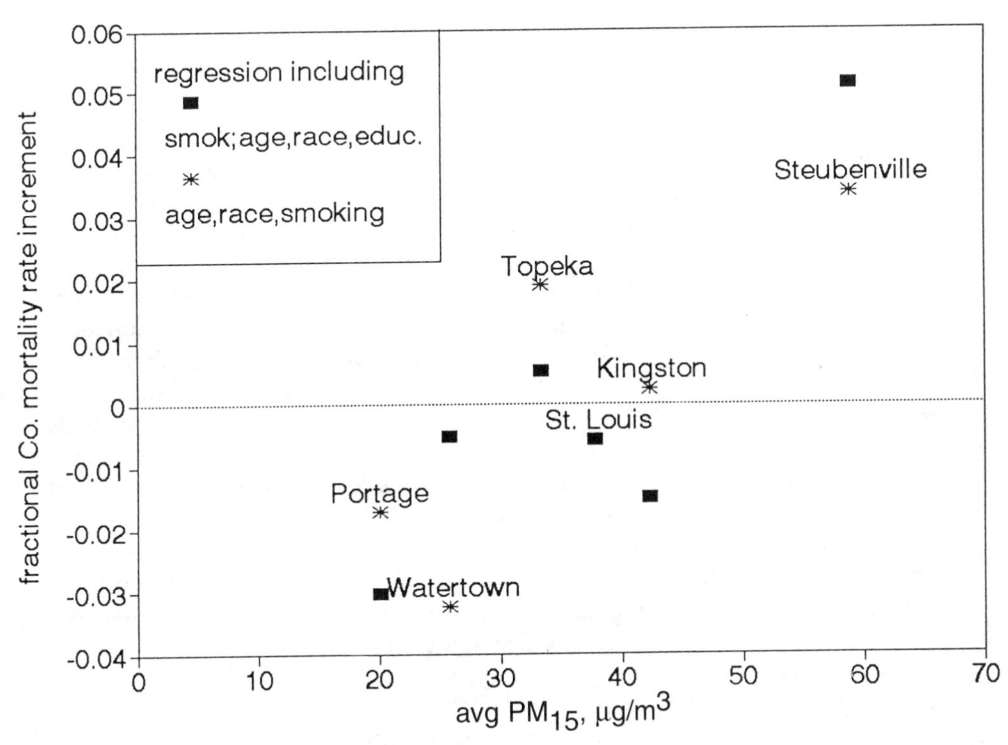

FIGURE 7-21. County-level mortality residuals versus average PM_{15}.

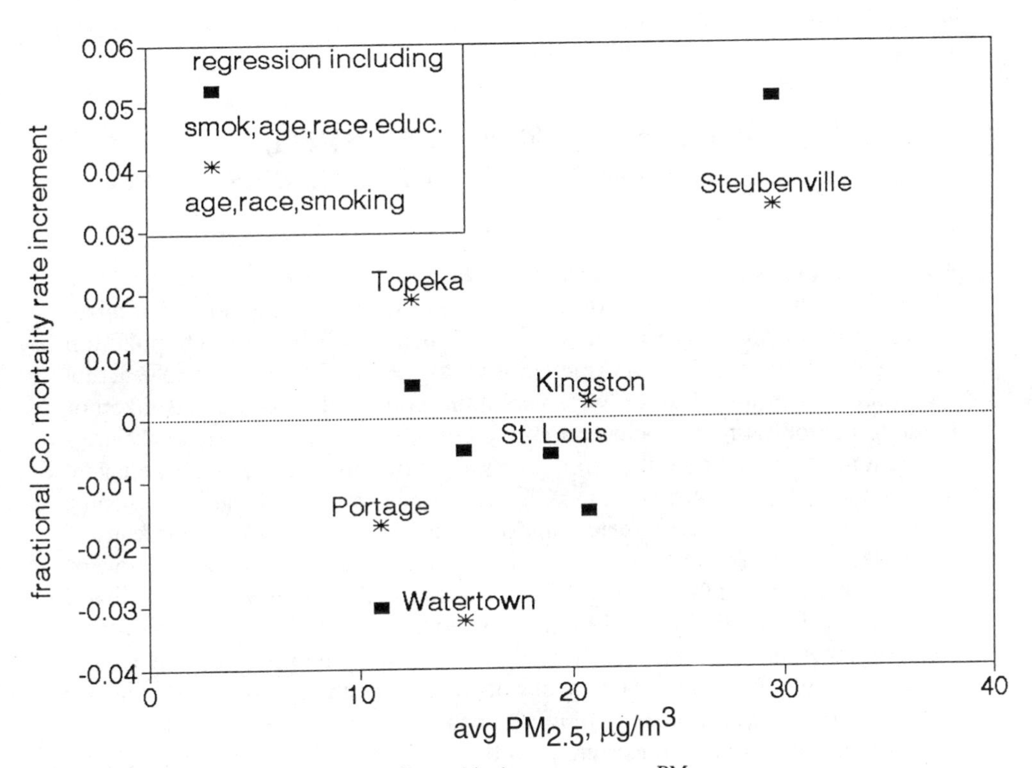

FIGURE 7-22. County-level mortality residuals versus average $PM_{2.5}$.

set and, thus, that it is possible that some other attribute may have contributed to the relatively low survival rates there.

In summary, the prospective cohort study reported by Pope *et al.* (1993) is consistent with and strengthens the findings from ecological studies in general. The risks due to smoking and the benefits of education are confirmed quantitatively, but the apparent risks of air pollution exposure appeared to be much larger in the study of individuals, compared to either time-series studies of acute mortality or cross-sectional studies across many locations for individual years. However, the elasticities based on individuals agree with findings from the few intraurban studies that have been reported (for example, Winkelstein *et al.*, 1967; Gregor, 1977), and with the only other study of individuals (Krzyżanowski and Wojtyniak, 1982). Intraurban studies may be a better paradigm for the prospective study because their population groups are smaller and air pollution exposures may be better characterized. By implication, a corollary finding from this comparison would be that cohort exposures to air pollution within counties or SMSAs may differ substantially (this hypothesis would favor PM_{15} over fine particle measures, because the latter tend to be spatially distributed more uniformly). If so, aggregate national estimates of the effects of air pollution (such as excess mortality) should not be based on the product of regression coefficients obtained from cohort studies and exposures based on extension of local monitoring data to whole counties or SMSAs.

If we choose to accept these higher response rates, then we must also conclude that air pollution has both chronic and acute effects on human mortality. Additional studies should be designed to test this hypothesis in a larger data set, paying particular attention to estimating actual cohort exposures to air pollution, including long-term temporal changes.

SECTION 3: CROSS-SECTIONAL STUDIES OF COUNTIES AND SMSAS

Early Studies

Correlation Analysis of U.S. SMSAS. 1949–1951

Manos and Fisher (1959) performed an exploratory investigation of interrelationships between several indices of industrialization and air pollution potential and standardized mortality ratios for 50 causes of death. No actual measurements of air pollution were used, and no specific account was taken of possible confounding by socioeconomic factors or smoking. The seven causes of death most frequently linked to the various pollution indices were esophageal and stomach cancer, coronary heart disease, lung cancer, endocarditis, chronic rheumatic heart disease, diabetes, and oral cancer. For the remaining causes, there was a noticeable shift in the distribution of correlation coefficients toward nonsignificant or negative values. Since most of these diseases are also linked to smoking (U.S. DHEW, 1979), there is strong suggestive evidence of confounding of these results by differences in smoking habits, stomach cancer being a notable exception. The causes of death potentially associated with air pollution (or smoking) on physiological grounds or by other studies that were *not* well-correlated with these air pollution indices included pneumonia, hypertension, bronchitis, tuberculosis, influenza, emphysema, and asthma. This technique of exploratory analysis was supported by some researchers (Schiffman and Landau, 1961) and criticized by others (Prindle, 1959). The latter paper discusses possible confounding by smoking habits and presents data on mortality by sex and also on the ratios of mortality in central city counties to noncentral city counties, by cause and sex. It also speculates that the role of air pollution in heart disease might be to exacerbate the symptoms of those with already impaired circulatory systems.

Analysis of Geographic Differentials in 1960 SMSA Mortality

Kitagawa and Hauser (1973) included some multivariate regressions in their extensive study of socioeconomic factors influencing mortality. They considered the following variables: the logarithm of population, percentage living in central cities, central city population density, percentage employed in manufacturing, percentage of substandard housing, percentage of crowded housing, median family income, median year of schooling completed, percentage who completed one or more year of college, geometric mean TSP concentration, number of days in 1960 with maximum temperature above 90°F, and percentage of nonwhite population. Regressions were performed separately by sex and race, using age-adjusted mortality rates. Not all variables were included in all regressions, but consistent β-coefficients were reported for TSP for all races (both sexes) and for whites. The coefficients reported for nonwhites were more variable, depending on the covariates used. Significance levels were not reported. In order to estimate elasticities from these β-coefficients, data on the coefficients of variation for age-adjusted mortality and for TSP are needed. These could be obtained from the appendix data of Lave and Seskin (1978), as 78/1015 = 0.077 for mortality, and 41/118 = 0.35 for TSP. An estimate of the elasticity is thus obtained by multiplying the average β-coefficient, 0.16, by 0.077/0.35 = 0.035. This value checks quite well with elasticities obtained from time-series studies and other cross-sectional studies.

Association of Cardiovascular Deaths with Particulate Cadmium

Carroll (1966) examined the hypothesis (which was based on laboratory tests) that cadmium (Cd) may be associated with hypertension and could thus be linked to mortality from heart disease. He identified 28 cities for which the Cd content of TSP had been determined in 1960–61, and performed a bivariate rank correlation between these data and county-level 1959–61 age-adjusted death rates for "diseases of the heart except rheumatic diseases" ($R = 0.76$). Carroll also identified a group of five Southwestern cities that correlated with Cd along a parallel trend with about 20% lower heart disease mortality rates. This suggests corroboration of more recent findings that Hispanics may be at lower risk of heart disease (Rosenwaike, 1987; Mitchell *et al.*, 1990, and total mortality Rogot *et al.*, 1992).

While this study was appropriately designed for testing a specific hypothesis, it creates doubts as to the possible roles of other variables that might have similar spatial distributions. I used data collected for my dissertation (Lipfert, 1978) to augment Carroll's data set and performed some multiple regressions. First, I tagged six Southwestern cities (I added Los Angeles to the five identified by Carroll) with a dummy variable, which could represent the presence of sizable Hispanic populations. This regression found both Cd and "Spanish" to be significant, with $R = 0.81$. The elasticity of Cd was about 0.06. Next, I added data on smoking (state level), poverty, and concentrations of SO_4^{-2}, TSP, and Mn. This required the data set to be reduced to 24 cities because of missing data. Of these new variables, only SO_4^{-2} was significant, with an elasticity of 0.32 as the sole variable and 0.16 when entered with Cd and "Spanish." The elasticity of Cd was reduced to about 0.04 (the correlation between Cd and SO_4^{-2} was 0.46). These findings illustrate some of the difficulties encountered in cross-sectional regressions. SO_4^{-2} is regionally distributed and thus acts as an indicator variable for the Northeast. When the dummy variable for Hispanics was added, the mortality associated with SO_4^{-2} dropped by about 50%, for example. Cadmium was less affected by collinearity, and since county mortality was regressed against city pollution levels, ecological bias problems may be present.

Cadmium was included in our 1988 city mortality study of total mortality but was not significant, perhaps since ambient concentrations were much lower than in the data used by Carroll.

Mortality Studies by Lave and Colleagues

Lave and Seskin

Lave and Seskin published a national cross-sectional regression analysis (Lave and Seskin, 1978) that concluded that about 9% of annual U.S. metropolitan mortality (ca. 1960) was associated with air pollution. The analysis was based on multiple linear regression analysis of annual mortality rates in the major SMSAs in relation to annual air quality levels (as measured at city centers) and a few other explanatory variables. This study was the first to attempt to characterize the air pollution exposure of an entire SMSA using (often fragmentary) data from a single monitoring station.

Lave and Seskin's work began with a journal article (Lave and Seskin, 1970) that couched the findings in the general terms of "air pollution." The later publications in 1977 and 1982 abandoned this caution and called for emission controls for sulfur oxides on the basis of their findings with respect to air concentrations of sulfate particles. This extension of research results into the policy arena may have been a factor with respect to the ensuing critical reviews (Landau, 1978; Cooper and Hamilton, 1979; Thibodeau *et al.*, 1980; Lipfert, 1980). While the Lave and Seskin work broke new ground for cross-sectional studies and included many innovations in methods, because of its many shortcomings and internal inconsistencies, its results should be used with caution. This also applies to similar cross-sectional studies that followed during the 1970s and early 1980s.

Lave and Seskin's handling of the aerometric data generated much controversy. They entered the yearly mean, maximum and minimum 24-hour average values for TSP and sulfates into their regression models, thus defining six pollutant variables. Their rationale for this practice was given as one of ignorance about the "true" specification. However, these six variables are highly correlated with one another, and some may relate to different types of effects (acute versus chronic). To test one type of effect versus another, regressions should have been performed with each pollutant measure entered separately. Moreover, the "1960" sulfate data were a mixed set, consisting partly of 26 biweekly 24-hour samples per monitoring station and partly of quarterly composite averages (Lipfert, 1980). Over a year's time, the minimum and maximum values of a set of 26 readings are quite different than those of a set of four readings; mixing these data sets thus introduced a bias for a large number of locations. Further, the sulfate data were not all from the same year but were taken from 1957 to 1961. The limitations of the sulfate data in 1960–61 are important, since the bulk of the exploratory regression analysis was done with this data set, and since the "minimum" sulfate variable was selected as the most important measure of sulfur oxides. It is this measure that was most affected by the confusion between biweekly and quarterly samples. However, this was not the case for the 1969 sulfate data set, which was used in a more limited set of analyses (Lave and Seskin, 1978). Methods of measurement were inadequate for all of the other pollutants studied by Lave and Seskin, especially NO_3^- and NO_2. The average TSP and sulfate concentrations for the 1960 data set were about 118 and 10 $\mu g/m^3$, respectively. In 1969, these values were about 95 and 11 $\mu g/m^3$, respectively.

Lave and Seskin studied 1960 and 1961 mortality rates in 117 SMSAs, 1969 mortality in 112 SMSAs, and mortality rates in 81 SMSAs common to both data sets. Their dependent variables were 1960 and 1961 total mortality rates (all ages and causes); 1960 mortality rates for four age groups; 1960 and 1961 mortality rates for 15 causes of death; and 1969 total mortality (crude and age-sex-race–adjusted rates). In addition to the six air pollution variables, their independent variables included population density, population (log of counts), the percentage of population aged 65 and over, the percentage of population classified as nonwhite, and the percentage of population classified as poor. Variables for population migration, census region, home- and water-heating fuels, occupations, and climate variables were used at various times. A limited analysis was included of suspended nitrates, NO_2 and SO_2 (mean, minimum, and maximum yearly values in all cases).

Because each metropolitan area was characterized by a single air monitoring station, the Lave and Seskin study should be regarded as examining the regional structure of the air pollution–health effects question, as opposed to the local structure. Of the two pollutants used in the study, sulfate would be characterized as a regional pollutant, since the typically small sulfate particles tend to travel far from their original sources. On the other hand, TSP is usually a local pollutant, and high ambient concentrations tend to be localized around the monitoring site. Thus, if TSP has a significant association with mortality across an entire SMSA based on a single monitoring site, it is more likely that the source of the particles (such as heavy industry) is the causal factor through occupational and socioeconomic factors (which could be diffused throughout the area) than exposure to the heavy particles *per se* (which will be localized).

Lave and Seskin found both sulfates and TSP to be statistically significantly associated with total SMSA mortality in 1960 and in 1969. When identical locations were compared across the two years, there were no significant differences, although the TSP effect tended to be stronger in 1969 (even though the overall mean TSP declined from 118 to 95 $\mu g/m^3$). The TSP coefficient values in % excess deaths per $\mu g/m^3$ were 0.039 and 0.066, for 1960 and 1969,

FIGURE 7-23. Isopleths of computed sulfate aerosol annual average concentrations (μg/m^3). *Data from ASTRAP (Shannon, 1981). Graph from Lipfert* et al. *(1988).*

respectively, which are roughly equivalent to the time-series results (Chapter 6) but lower than Winkelstein *et al.*'s (1967) results for Erie County, NY (Figure 7-5). If taken at face value, these results suggest that chronic exposure to TSP has effects well beyond the annual sum of short-term effects (as suggested, for example, by Evans *et al.*, 1984b) and that the substantial difference between inter- and intracity findings results from the poor characterization of exposure in SMSAs (compared to groups of census tracts).

Sulfur oxides tend to have the most definitive geographic pattern (Figure 7-23), being widely distributed throughout the North Central and Northeastern regions of the United States. When Lave and Seskin tested for regional influences by adding a dummy variable for each of the nine Census Divisions (regression 3.5-9 in Lave and Seskin, 1978), the sulfate coefficient (minimum values) dropped by 40% and lost statistical significance; the TSP coefficient (mean values) increased by 24%. Large decreases in the sulfate coefficient also occurred when population migration was taken into account. However, Lave and Seskin described these effects as "pollution coefficients . . . not substantially affected."

When nonlinear pollution specifications were examined (Figures 3.2, 3.3, 7.1, and 7.2 in Lave and Seskin, 1978), the authors discounted the nonlinear shapes of the resulting dose-response curves (which suggested a threshold for TSP at 110 μg/m^3 in 1960 and at 3.5 to 5.5 μg/m^3 for minimum sulfate in 1969), emphasizing instead statistical tests on the superiority of the overall fit of the model. This approach established the linear, no-threshold model as the specification of choice, unless it could be shown that a more complicated model was statistically superior. The use of a threshold is an important distinction with respect to air quality standards and requires more sophisticated model comparison tests (see p. 83). For total mortality, log-log models were also investigated, which are of interest since such models were found to fit some of the time-series data better (Chapter 6). For 1960, the log-log model was slightly inferior to the linear model, although these results are suspect since the log-log coefficients, which should be numerically equal to the elasticity (see Chapter 3), were an order of magnitude larger than expected. For 1969, the log-log model had a higher R^2 than the linear model and appropriate values for the coefficients (equivalent to the linear model elasticities). Since the pollution coefficients had about the same t-values as with a linear model, one can only conclude that the log-log model was no worse.

By cause of death, the most significant air pollution effects were for cardiovascular disease (both sulfate and TSP) and total cancers (sulfate). Stomach cancer was not associated with TSP, and respiratory cancer was not associated with either pollutant. Deaths from respiratory disease were not associated with sulfate but were weakly associated with minimum TSP. This latter association was stronger for tuberculosis and asthma; influenza, pneumonia, and bronchitis were not associated with either pollutant. Lave and Seskin also examined some variables not plausibly associated with air pollution (suicide, venereal disease, and crime rates) to check for spurious (noncausal) associations; none were found. However, these noncausal findings were assigned a noncausal interpretation, whereas findings of nonsignificant air pollution effects with similar R^2 values for respiratory diseases were ascribed to "sampling variability." Sampling variability could have been a factor with suicide or venereal disease as well.

By age, sex, and race, air pollution effects were stronger for nonwhites and for females, for whites aged 45–64, and for nonwhites 65 and over. For most of these subgroups, TSP was more important than sulfates. It was also important for nonwhites younger than 45, for which nondisease deaths tend to dominate. Since age, race, and disease groups were investigated only separately, additional caution is advised when considering causality.

The authors concluded from these regressions that "levels of certain pollutants in the air (as they prevailed in certain cities during the period 1960–69) **caused** increases in mortality, and presumably in morbidity as well" (my emphasis). However, the design of these studies and the data they used make even the published statistical associations suspect.*

There are several problems with the Lave and Seskin studies in addition to the flaws in the air quality data:

- The regression model used was derived on an *ad hoc* basis, without a well-specified *a priori* hypothesis, and without optimization or consideration of alternative socioeconomic variables.
- Important demographic variables and personal risk factors were omitted from the analysis, including smoking habits, diet, and education. Subsequent analysis by others (see below) showed all of these factors to be important and to be correlated with pollution variables.
- The authors' conclusions are inconsistent with their own analysis. In nearly every case, adding explanatory variables to the data set decreased the significance of the pollution variables, especially for sulfate. Pollution was not

* Ellis (1972) also chided Lave and Seskin (1970) for neglecting the simultaneous effects of heat and air pollution in their "otherwise comprehensive review." Ellis felt that Lave and Seskin's claim of a "10–15% reduction in the mortality and morbidity rates for heart disease" in response to a "substantial abatement of air pollution" was a "naive, extravagant, and optimistic claim."

significant for respiratory causes of death, in contrast to most of the previous studies reviewed above. The authors ignored these indications of noncausal associations.

- The age groups used to analyze age-specific mortality were too broad (45–64, 65 and over). Age adjustment is still needed within such groups.
- For some pollutants, the SMSA is too large a political subdivision for estimating representative air pollution exposures from monitoring data. Most SMSAs are groups of whole counties and may have only one monitoring station, often in the central business district. Thus, no information is provided on exposures in the suburbs, where often the bulk of the SMSA population lives. Where multiple stations were available for an SMSA, they were apparently not used in this study.
- The study included a cross-lagged analysis of all combinations of 1960 and 1969 mortality, air pollution, and socioeconomic variables and found that 1969 mortality was better predicted by 1960 air pollution than by data from the same year. However, since the 1960 sulfate data were so badly flawed, this finding is meaningless.

The Lave and Seskin studies were among the first to apply econometric methods to the health effects of air pollution. As a result, their emphasis was on regression analysis, which imposes a linearized structure on the data and makes the analysis of interactions more difficult. They made no attempts to explain their results from a physiological perspective, yet they called for revisions to the health-based standards as a result of their findings. In their summary, they stated that their study was focused on "how human mortality is affected by a *reduction* (my emphasis) in air pollution." However, a cross-sectional study only examines variations by place, and thus changes over time are not part of the model development. To apply the spatial differentials and regression coefficients to changes over time requires the establishment of a causal model, including physiological explanations. Such a model should be verified by application to actual case histories of pollution abatement (see Chapter 8).

The 1982 Follow-Up Study

Partly as a result of the criticisms of the earlier work and the publication of contrary findings (Crocker *et al.*, 1979; Lipfert, 1978), Lave and his colleagues (Chappie and Lave, 1982) performed a similar cross-sectional analysis using more recent data and attempted to address some of the criticisms. The 1974 mortality rates for about 100 SMSAs were analyzed, and the regression coefficients were compared to Lave and Seskin's results for 1960 and 1969.

The dependent variables were 1974 total mortality and mortality from disease (total less external causes) for SMSAs. There was also a limited analysis of corresponding data for counties and cities. The independent variables followed the previous study (population density, population [log of counts], percentage over 65, percentage of nonwhite and of poor, sulfate particles [mean, minimum, and maximum], TSP [mean, minimum, and maximum]); income, percentage of college graduates, tobacco and alcohol expenditure per capita, occupation, nutrition variables, physicians per capita, and birth rate were used in addition. The 1974 mean air pollution values for sulfate and TSP were 9.6 and 75 $\mu g/m^3$, respectively.

Chappie and Lave found that in 1974 the sulfate effect became much stronger (accounting for about 12% of total mortality) and the TSP effect dropped to almost zero. Eliminating nondisease causes from the dependent variable slightly strengthened the sulfate effect; adding an education variable weakened it, and adding variables for smoking, alcohol use, and occupation had little effect. When regressions for SMSAs, counties, and cities were compared (96 locations), there were slight reductions in the sulfate effects and small nonsignificant increases in the TSP effects, but the smoking variable became nonsignificant and negative in some cases. The authors concluded: "A strong, consistent and statistically significant association between sulfates and mortality persists. . . . These results can be used to support stringent abatement of sulfur-oxides air pollution."

It is unfortunate that the 1982 study was structured as a defense of the earlier Lave and Seskin work rather than as an independent replication. It might have been more interesting to derive an independent regression model from the later data, and then to compare the air pollution effects. Since geographic patterns in U.S. mortality have been stable over the years (Crocker *et al.*, 1979), one would expect consistent regression results over time. However, most air pollution levels have decreased since the 1960s, and it would be interesting to determine to what extent these changes are reflected in the regression results. It is particularly important that the use of mean, minimum, and maximum measures of each pollutant concentration were retained, since this algorithm was apparently derived from the flawed 1960 sulfate data originally used by Lave and Seskin (1978). No statistical or physiological justification has ever been advanced for this procedure, and most authors who have reanalyzed these data elected to use only the mean pollution values (Evans *et al.*, 1984a; Lipfert, 1984).

Chappie and Lave showed that the 1974 regression coefficients were not statistically different from the 1960 and 1969 coefficients; however, their model allowed each year to have its own intercept, which suggests that important factors are missing from the model. Year-to-year consistency in the data was shown by this analysis, which was expected. In most cases, adding new variables to the specification and using only one measure rather than three for each pollutant reduced the importance of the sulfate variable. Eliminating external (nondisease) causes of death had the opposite effect, since it is the spatial pattern of heart-disease mortality that appears to be associated with sulfate concentration patterns.

The tobacco and alcohol consumption variables ($/capita) used by Chappie and Lave are poor surrogates for actual consumption, since they do not account for price variations or sales to nonresidents (due to differences in state/local taxes, for example) and are based on self-reported data on retail sales that suffer from varying degrees of underreporting. This criticism is borne out by the weak effects shown for these variables in this study, in contrast to previous findings (Lipfert, 1978). Thus, including these quasi-surrogate variables does not constitite "controlling" for smoking or drinking, as the authors claimed.

The effects of smaller aggregation units (counties and cities vs. SMSAs) were also examined with all six pollution variables and with total mortality rates. Sulfate was relatively stable, but the TSP coefficients increased with the smaller areas, as expected (but not enough to reach statistical significance).

This procedure should have been extended to the cases with only two pollution variables and with nonexternal causes of death (external causes are likely to be higher in industrial [i.e., polluted] cities, but the effect would be lost in an SMSA analysis when suburbs are included). R^2 values were not as high as in some other studies (Lipfert, 1978), and some variables had coefficients of the "wrong" sign, all of which indicate problems with the specification (i.e., model) or perhaps input data. Part of the advantage of using cities rather than SMSAs is to increase the number of observations and to evaluate robustness. Chappie and Lave examined only one set of cities, the central cities corresponding to the SMSAs. However, many SMSAs have more than one city; in the 1980 Census, there were about 300 SMSAs and over 900 cities (with a population larger than 25,000).

The results of Chappie and Lave are presented for a *priori* specific models. A stepwise exploration procedure would give insight into the covariance structure and the extent of the collinearity problems. Birth rate was used only as a possible surrogate variable explaining the demand for medical care. However, a better use of birth rate data is as a predictor for mortality to represent possible undercounting of the population, which can inflate both mortality and birth rates (by the same percentage). This is especially important when a mid-census data year is used (1974), since population errors may be considerable, especially in poorer (and often more polluted) areas. The diet (nutrition) data used were for only four regions of the country and were out-of-date (1955). Finally, Chappie and Lave presented no analysis of residual patterns or of possible spatial autocorrelation.

Comparing the Lave and Seskin with the Chappie and Lave pollution coefficients across years shows that the TSP coefficients dropped significantly, which may be consistent with the substantial drop in the ambient TSP concentrations over this period. However, the sulfate coefficients *increased* substantially (more than doubled), in spite of a *drop* in average ambient air concentration. The net total pollution effect on mortality appears to have remained approximately constant over the years, and the proportions for the two pollutants have shifted. Since no reasons for such a change have been advanced based on statistics, air quality measurements, or physiology, a possible conclusion is that the indicated pollution effects on mortality are in fact surrogates for other phenomena.

The Chappie and Lave analysis sheds no light on whether the effects claimed are long-term (chronic) or just the sum of acute exposures over a year. This might have been resolved with a cross-validation analysis, which could provide insight into the physiological situation. Although the Chappie and Lave study is more complete than the Lave and Seskin work in some respects, no other air pollutants were explored (such as CO or O_3, both of which might be expected to be more physiologically important for heart disease than sulfates); drinking-water hardness was not considered, nor was population migration.

Other SMSA Mortality Studies

Studies on 1969–1970 SMSAs

I reanalyzed Lave and Seskin's 1969 total mortality data set for 112 SMSAs, using corrected data and many new independent variables (Lipfert, 1984). My objective was to provide a more rigorous analysis of the portion of the Lave and Seskin data that had valid sulfate data and to compare it with the original analysis by Lave and Seskin (1977).

The dependent variables were 1969 total SMSA mortality from Lave and Seskin and total mortality, mortality by sex, and mortality for two broad age groups, for 1970 (to correspond with the census). The independent variables from Lave and Seskin were: percentage over 65, percentage of nonwhites and of poor, population density, log of population, mean TSP, mean and minimum SO_4^{-2}. Additional new variables included percentage of Blacks, of other non-whites, net migration, diet, smoking, percentage of coal heat times degree-days, percentage of wood heat times degree-days, drinking-water variables, O_3, Fe, and Mn.

The main conclusion that I drew from this analysis of 1970 SMSA mortality was that pollution effects on mortality were sensitive to both model specification and the data set used. Specifically:

- Minimum SO_4^{-2} had no advantage over mean SO_4^{-2} and was less significant, other things being equal. When a complete model specification was used, sulfate was rarely significant (as either measure); these cases occurred more often for females. TSP was more important for deaths of those under 65 years of age.
- Drinking-water quality, ozone, migration, and racial variables all had important effects on the regressions. Only when I used a complete model specification did the coefficients for age, race, poverty, and smoking assume values consistent with exogenous estimates; this result implies that these additional variables were necessary.
- Since Mn was only important for males (65 and over), this variable may have been acting as a surrogate for long-term occupational effects. This suggests that the city mortality studies should be replicated by sex of decedent.
- The analysis was incapable of distinguishing between linear and threshold models and thus could not rule out the applicability of a threshold.

Because data for all the new variables were not available for all 112 SMSAs, this study had problems with missing data, depending on the model specification. Thus, I could not determine whether sulfate lost significance because of the addition of new variables to the data set or because of the deletion of the observations lacking data on the new variables. The sensitivity of the results to age grouping implies that finer age groups should be explored.

Regardless of the missing data problems, the study showed that sulfate effects are not robust and suggested that the Mn effect was probably a surrogate effect rather than a *bona fide* effect of air pollution on mortality.

Studies on 1980 SMSAs

Özkaynak and Thurston (1987) analyzed 1980 total mortality in 98 SMSAs, using additional data from the EPA inhalable particle monitoring network for 38 of these locations. The independent variables were: percentage over 65, median age, percentage of nonwhites, log of population density, education, poverty, TSP, suspended sulfates, inhalable particles (IP, diameter $< 15\,\mu m$), and fine particles (FP, diameter $< 2.5\,\mu m$). Mean values were reported as ($\mu g/m^3$): sulfate, 11; FP, 23; total IP, 48; and TSP, 78. They also estimated some of the IP and FP concentrations based on measured SO_4^{-2} and TSP, to fill in missing data. The sulfate measurements that Özkaynak and Thurston used may have been affected by artifacts from the high-volume sampler filters (Lipfert *et al.*, 1988); this is also suggested by the fact that their mean value exceeds those of previous years and of other studies. Lipfert (1992) compared average sulfate concentrations to national SO_2 emissions for the years 1978–82 and concluded that the concentrations for 1980 and 1981 were higher than expected. A partial explanation may lie with the low rainfall amounts experienced in those years (U.S. EPA, 1986).

Özkaynak and Thurston concluded that the results were "suggestive" of an effect of particles on mortality decreasing with particle size; SO_4^{-2} and FP were significant, IP and TSP were not. However, the elasticities for SO_4^{-2}, $PM_{2.5}$, and PM_{15} were not significantly different from one another, and the TSP data they used were not all representative of their respective SMSAs (Lipfert, 1992). Thus, alternative interpretations of these findings are certainly possible, especially since statistical significance may not be a reliable indicator of physiological importance. In addition, as discussed above, when SMSAs or other large geographic areas are chosen for the observational unit, the finding of increased statistical significance for a small particle variable does not necessarily have health implications, since the effects of large particles will be confined to an area closer to the measuring site and thus may not be detectable by a study using large areas.

In addition, lack of a complete model specification sheds doubt on the validity of the results, since smoking, diet, water hardness, and migration were not accounted for. Contrary to the authors' statement, the confounding effect of smoking cannot be assessed solely by the simple correlation between a smoking index and some air pollution variable. Because of the multicollinearity present, the entire covariance structure must be examined. No other pollutants were evaluated, such as ozone or trace metals. Thus, Özkaynak and Thurston's findings may not be very specific.

As discussed above, the use of SMSAs as the geographic unit of observation may be responsible for the lack of significance of TSP, since central city measurements will not be good estimates of total SMSA exposure. Estimation, rather than measurement, of IP and FP may contribute enough error to make them less significant than sulfate, which is expected to be less variable over the whole SMSA. The authors reported investigating spatial autocorrelation and regional influences but did not present sufficient results to permit an independent evaluation.

General Analysis of 1980 SMSA Mortality Rates

Data were compiled on mortality, demographic, socioeconomic, and environmental factors for U.S. SMSAs, around 1980 (Lipfert, 1992), following the general model that Lipfert *et al.* (1988) used for 1980 city data. Selection of locations was largely determined by availability of air quality data. Death rates were analyzed for all causes, major cardiovascular causes, and chronic obstructive pulmonary disease, in addition to all nonexternal causes.

This analysis differed from previous published cross-sectional studies in two important ways. First, great care was taken with the air quality data to obtain values as representative as possible for each SMSA and to include as many locations as possible. A wide range of pollutants was examined as well. Second, the analysis featured the use of log-linear models, in the same manner as used in many of the recent time-series mortality studies (dicussed in Chapter 6).

Locations Studied

The study employed SMSAs as the geographic unit of analysis. The U.S. Bureau of the Census defines an SMSA as a group of counties (except in New England) that have a total population of at least 100,000 with an urbanized area population of at least 50,000. Two counties in Montana, with populations of 77,000 (Missoula) and 34,000 (Silver Bow County, which includes the city of Butte), which do not qualify as SMSAs, were also added to the data set in order to take advantage of their air quality data. Consolidated Metropolitan Areas, which combine several SMSAs, such as Los Angeles, New York, or Chicago, were not used in this analysis.

The 112 SMSAs first studied by Lave and Seskin (1970) and later by Evans *et al.* (1984a) and others comprised the primary list of locations. These were originally selected on the basis of the availability of air monitoring data around 1960, but the actual geographic definitions in terms of the counties included have changed somewhat over the years, as defined in each decennial census. In general, SMSAs are comprised of whole counties but include independent cities in Virginia and portions of counties in New England (CT, MA, ME, NH, RI, VT). For comparability of mortality and socioeconomic data, New England County Metropolitan Areas (NECMAs, comprising whole counties) were used in these six states.

The Air Quality Data Base

As discussed, cross-sectional studies have usually been intended to study long-term differences among locations. For this reason, it has not generally been regarded as particularly important to use environmental data taken exclusively during the nominal year of study (1980, in this case), although clearly this would be desirable from the standpoint of uniformity and in order to deal with specific attributes of that year. For example, a heat wave occurred in 1980 in the eastern portion of the nation (Bair, 1992), and a major volcanic eruption occurred in the West. Missing or incomplete air quality data are a common problem with observational epidemiological studies; for example, Mendelsohn and Orcutt (1979) used 1974 air quality data in their study of 1970 mortality patterns, arguing that the geographic patterns were stable in time and that the later measurements were more complete. Others have averaged over several years in order to obtain more reliable long-term averages (Lipfert, 1978; Lipfert *et al.*, 1988).

Sulfate Aerosol Data

1980 was an especially problematic year for particulate pollution measurements. Size-classified measurements were being explored then, but the PM_{10} network had not yet been established; PM_{15} data were being acquired on a research basis (Watson *et al.*, 1981). The glass-fiber filters used in the routinely operated high-volume samplers for TSP and their chemical constituents (SO_4^{-2}, NO_3^-, etc.) were found to be unusually alkaline for the years 1979–81 (U.S. EPA,

1984). One of the well-known characteristics of such filters is their tendency to convert SO_2 (gas) in the ambient air being sampled to SO_4^{-2} particles on the filter (Stevens, 1981); this problem was thought to be especially acute during 1979–81. The outcome would be reported values for TSP and SO_4^{-2} that are biased high on locations with appreciable ambient SO_2 levels.

For this study, all the sites assigned to a given SMSA, as defined by the 1980 Census, were combined to provide SMSA-wide estimates. These data were retrieved from the EPA Aerometric Information Retrieved System (AIRS) data base (T. Link, personal communication, 1991). Annual median SO_4^{-2} values, which tend to run 10 to 20% lower than annual mean values, were used because of the typically skewed frequency distributions and the relatively sparse frequency of SO_4^{-2} measurement found in most locations. Lipfert analyzed SO_4^{-2} data by year from 1978 to 1982 and concluded that the 1980 and 1981 data were about 10% higher than the other years, on the basis of the reported national SO_2 emissions (U.S. EPA, 1986). These 2 years were also marked by lower than expected rainfall, which could be part of the explanation for the anomalously high SO_4^{-2} readings.

The sulfate data used for multiple regressions for the 149 locations studied by Lipfert (1992) were then obtained by averaging all the observations available for the period 1978–82. Data for 13 missing locations were estimated either from nearby locations or from alternate time periods.

Other sources of sulfate air quality data include measurements from the PM_{15} filters (also referred to as IP) and estimates made with air emission and transport computer models (Shannon, 1981). The IP data were obtained with unreactive (Teflon) filters and are thought to be more reliable than data obtained with high-volume samplers using glass-fiber filters; the correlation between the two measures was 0.63 (Lipfert *et al.*, 1988). The two SO_4^{-2} measures were related by

$$\text{AIRS } SO_4^{-2} = 3.5 + (1.18 \pm 0.23) \times \text{IP-}SO_4^{-2} \text{ (two } \sigma \text{ confidence limits).}$$

Thus, the slope was not significantly different from unity.

The overall levels of the IP SO_4^{-2} values were in better agreement with SO_4^{-2} values obtained from various air quality research efforts carried out during this period than the AIRS values. In most cases, there was only one IP monitor per city.

Total Suspended Particulate Data

Lipfert *et al.* (1988) made only a few cursory regressions employing TSP. That data base consisted of 1978, 1982, or their average values for each city, with no attempt to derive citywide averages. A similar approach was used by Özkaynak and Thurston (1987), in that a single monitoring site was used to represent each SMSA.

In an effort to improve the estimation of actual exposures to particulates within each SMSA, data from the EPA AIRS data base were used to construct spatial averages for 1980. All TSP monitors with at least 11 observations for 1980 were used; the annual means were averaged (without weighting) to provide an SMSA-wide estimate. There were a few cases of source-oriented networks in the data base (Granite City, IL [St. Louis SMSA] and networks surrounding some the Tennessee Valley Authority (TVA) power plants). These subsets were averaged separately and then entered into the data set for the SMSA in question as a single observation, in order to preclude undue weighting because of the large number of monitors representing a limited geographic area. The standard deviation of 112 SMSA averages was 14.9 μg/m^3; this compares with the average within-SMSA standard deviation of 13.9 μg/m^3, which suggests that there is typically almost as much variation within SMSAs as between SMSAs. If the within-SMSA variance constitutes an "error" in measurement for those cross-sectional studies that relied on a single TSP monitor per city, a serious downward bias could have resulted for the TSP regression coefficients and their significance levels.

The overall mean for 149 SMSAs was 68.4 μg/m^3. A total of 1581 monitoring stations was used in this effort. The maximum annual mean value for an individual monitor was 280 μg/m^3 (East Chicago, IN, near a car wash); the minimum value was 22 μg/m^3 (near Portland, OR). The maximum SMSA average was in Spokane, WA (142 μg/m^3); the minimum was in Atlantic City, NJ (41 μg/m^3).

Year-by-year TSP comparisons were made on the basis of the maximum annual means recorded in each SMSA for the years 1978–82. The averages for 112 SMSAs decreased from 90 μg/m^3 in 1978 to 69 μg/m^3 in 1982. However, when compared to the national estimates of particulate emissions (U.S. EPA, 1986), it appears that the ambient TSP data for 1980 and 1981 were about 5% higher than expected. As was the case with sulfates, this could have resulted from either sulfate artifacts on the filters or from the low rainfall that occurred in those years (U.S. EPA, 1986).

Other Pollutants

Ozone (*ppm*). Two separate sources of ozone data were utilized in this analysis. Peak 1-hour values were available for 1980 for 72 SMSAs (U.S. EPA, 1984), which were generally the largest SMSAs in the nation; these data represented the highest readings for each SMSA, and are not necessarily representative of average exposure across the entire SMSA. Seasonal average values were available for the entire data set, as obtained from a smooth isopleth map around 1978 (T. McCurdy, personal communition, 1991). Most of the analysis was conducted for the seasonal average data set, because of its completeness and the likelihood of better representing spatial averages across each SMSA.

Manganese (μg/m^3). Mn data were based on analysis of high-volume sampler filters; data were estimated from previous years for several SMSAs.

Data from the dichotomous sampler network (1979–1983) (μg/m^3) Total mass and fine particle mass, total sulfate, and fine particle lead (Pb). Samples were taken every 3 or 6 days. The IP data for SMSAs with more than one IP monitoring site were averaged over all the sites in that SMSA. The size-fractionated particulate (IP) data, described by Watson *et al.* (1981), were based on Teflon filters and show systematically lower sulfate values; these values are generally regarded as the "true" sulfate measures. The differences between the two sulfate measures are not consistent and presumably depend on a number of site-specific environmental factors. These measurements were replaced by PM_{10}, which began with a few sites in 1983, too late to be used with the 1980 Census and mortality data.

Mortality Data (Dependent Variables)

Mortality counts were taken from *Vital Statistics, 1980, Part B* (Table 8-6), for which the SMSA boundaries were based on the 1981 definitions. These definitions are consistent with the *State and Metropolitan Area Data Book* (SMADB), from which population data were taken. In New England, death counts are given only for NECMAs, which are comprised of whole counties. The demographic data for New England were therefore also based on NECMAs.

Four different groupings of causes of death were analyzed. Rates were computed by dividing the numbers of deaths in each group for the calendar year 1980 (all ages, races, both sexes) by the population estimated by the U.S. Census as of April 1, 1980. Thus, small errors would be entailed by any population changes that took place during the year; an independent variable for percentage population change was included in the regressions, in part for this reason. Deaths were assigned to locations on the basis of usual residence rather than on the basis of the location at which the death actually occurred. In this report, the term "mortality" should be interpreted as the crude (unadjusted) figure, unless otherwise specified.

The causes of death analyzed and their mean values and standard deviations (deaths per 1000 population) are listed below, based on 149 SMSAs. The ICD9 codes refer to the Ninth Revisions of the International Classification of Diseases. These selections were made to eliminate causes of death unlikely to have resulted from air pollution (external causes) and to specifically examine those major causes that have previously been linked with air pollution (heart and lung disease). No distinctions were made by age, race, or sex.

NonExternal causes. All causes less accidents, homicides, and suicides (ICD9 1–800). Mean = 7.82 (deaths per 1000 population), standard deviation = 1.48.

Major cardiovascular diseases. Includes acute heart attacks, chronic heart disease, hypertension, and stroke (ICD9 390–448). Mean = 4.19, standard deviation = 0.95.

Chronic obstructive pulmonary disease. Includes bronchitis, emphysema, and chronic airways obstruction, but not acute respiratory disease, pneumonia, influenza, or occupational pneumoconiosis (ICD9 490–496). Mean = 0.251, standard deviation = 0.075.

All causes. Included primarily to facilitate comparison with other studies that did not remove external causes of death. Mean = 8.50, standard deviation = 1.48.

In terms of the coefficients of variation, chronic obstructive pulmonary disease was the most variable grouping and all-cause mortality was the least variable.

Demographic and Socioeconomic Variables

Population descriptive data were obtained for 149 U.S. SMSAs from the 1982 SMADB, as follows. A brief rationale for each variable is also given. The variable names used are given in bold in parentheses.

Percentage of population 65 years of age or more (**65+**). Above about age 35, mortality rates are exponentially related to age, but this variable is the only useful age statistic available from SMADB. Data on the age distribution within the 65-and-older group would have been desirable, for example.

Racial and ethnic distribution. Includes the percentage of Blacks (**BLACK**), of other nonwhites (**OTHERNW**: Asian, American Indian, etc.), and the percentage of Hispanics (**HISP**). Each of these groups tends to have mortality rates different from whites, and thus cities with higher than average percentages of these groups would be expected to have correspondingly different mortality rates for the total population. Race was self-defined in the 1980 Census, which leads to a certain amount of confusion, mainly in heavily Hispanic cities in the Southwest. We consider three groupings: whites, Blacks, and others; they sum to 100%. "Hispanics" are a separate grouping not defined by race. In most cities, the fraction of "other" is small, but in El Paso, TX, it is about 38%, apparently because many Latinos do not consider themselves white. However, the classification of deaths by race uses a different criterion; as an example, the 1980 deaths for El Paso were listed as 97% white. This means that death rates cannot be computed accurately by race for these locations.

Percentage of individuals below the poverty level (***POOR***). We feel this is a better income variable than, say, median income, because if there is an effect of income on mortality, it would be expected to be most obvious at the low end of the scale.

Percentage with 4 or more years of college (***COLLEGE***). Education may be a better socioeconomic variable than income, because some persons may have low income because they have poor health, not vice versa. Educational attainment, as a socioeconomic indicator, will not be changed by subsequent illness.

Percentage of population change since 1970 (***CHNG70***). This variable is intended to capture population stability and migration, which can be important, since ill health can be a factor in the decision to migrate, and because data on long-term exposure to air pollution would be affected by migration.

Average annual heating degree-days (***HDD***). This is essentially a climate variable reflecting long-term rather than current weather conditions.

Population density (*logarithm,* ***LPD***). Before the conquest of infectious diseases, population density was an important determinant of mortality (Farr, 1885). When applied to county units or larger, this statistic is now of limited use because of the heterogeneity of land use typically found in larger areas, and it tends not to capture the average density at which people actually live. However, it is capable of distinguishing the 100% urban SMSAs (such as Jersey City) from most of the others, which are usually mixed urban and suburban.

Drinking-Water Quality Data

Previous studies (Lacey, 1981; Lipfert, 1984) have implicated soft water as a contributing factor in heart disease, primarily for males. Data on drinking water hardness (HARDNESS) were obtained from a data base compiled by the National Institutes of Health (Feinleib *et al.*, 1979). These data were for the 1970 time period and earlier, but it was felt that drinking-water supply data would be reasonably stable over time. The NIH data base was for cities rather than SMSAs; the value for the main city of each SMSA was selected; no attempt was made to average over all the component cities of an SMSA. Data were available for 144 SMSAs out of the set of 149 we described; the other five values were obtained by telephone from the respective water supply authorities.

Data on Smoking Habits

Cigarette consumption data have been estimated from state sales tax data for three time periods: 1955, 1969, and 1980 (Lipfert, 1978; Lipfert *et al.*, 1988). The estimates are based on regression analysis on state-level sales data (per capita for the population aged 18 and over), using various economic and demographic variables as predictors. The presence of lower sales taxes in adjoining states was found to be an important factor in explaining cigarette sales differences. These regression results were then used to predict cigarette consumption in each state. It was not possible to derive cigarette consumption data at finer geographic resolution, and thus we are forced to assume uniform consumption throughout the state with discontinuities at the borders. These errors are likely to lead to an underprediction of the effect of smoking on mortality, particularly for interstate SMSAs.

Earlier analyses of smoking patterns typically revealed large urban-rural differences, and it has long been assumed that city people smoke more. In the study of 1980 smoking data, SMSA tobacco sales data from the 1977 Census of Retail Trade were compared to statewide sales data from the same source, and a consistent relationship was found, amounting to an annual urban-statewide difference of about 5 packs per year per person (out of an average of 185 packs per year per person). This small (but still statistically significant) difference suggests that regional smoking patterns are now probably more important than urban-rural differences within regions, which supports the use of state-level data in the analysis of mortality effects.

Finally, a comparison was made of our estimates of cigarette consumption with independent state-level survey data on the percentage of people who smoke (smoking prevalence). The correlation coefficient relating these two measures was only about 0.5 (explaining 25% of the variance) for the 29 states that had conducted surveys. Possible explanations for this rather poor result include variations in amount consumed per smoker and underreporting by those responding to the survey. We prefer to use consumption data rather than prevalence since heavy smokers have a much higher relative mortality risk than light smokers, and since consumption may reflect the possible effects of passive (involuntary) smoking. Cigarette consumption rates are analogous to air pollution emission rates.

For an analysis of chronic health effects, it is not clear whether current cigarette smoking rates or some time integral is the appropriate metric (the same question exists for air quality as well). For this reason, we considered two possible smoking variables: the 1980 data, as described above, and 1969 data. Because of collinearity between the two (R = 0.48), regression models are limited to one or the other (or alternatively, the average (SMOKING78), which was used in the regression runs reported below).

Regression Results

Both linear and log-linear models were evaluated with these data; slightly better fits were obtained with the log-linear models, and they seemed to be less sensitive to outliers. Only the log-linear results are reported here.

The results reported here are based on two-stage regressions, in which the log of mortality rate was first fit to a suite of nonpollution variables. The residuals from this regression were then fitted to the available data for a variety of air pollutants, some of which were missing for some of the locations. This approach has the advantage of providing a common basis for comparing the associations with different pollutants; it has the disadvantage that, if any air pollutant is correlated with a demographic or socioeconomic variable, some portion of its association with mortality may be incorrectly assigned to the other variable. In the cases reported here, the two-stage pollution-mortality coefficients were very similar to results from single-stage regressions. Furthermore, in the case of nonexternal mortality, the residuals were highly correlated ($R^2 = 0.988$) with "adjusted" mortality rates computed by subtracting all the nonpollution effects, based on a regression in which TSP, O_3, and SO_4^{-2} were included. Thus it appears that the two-stage approach may not be unduly conservative.

The only pollutants that were significantly associated with the residuals from the model for nonexternal mortality were TSP and O_3. However, Table 7-7 shows that the elasticities estimated on this basis for fine particles and for PM_{15} were only slightly lower, and it is possible that a larger data set might have achieved significance (experiments with successive truncation of the 149-city data set suggested that a minimum of 100 cities might be needed to achieve significance [Lipfert, 1992]). None of the sulfur-related pollutants were significant or even close; however, subtracting SO_4^{-2} from TSP (after accounting for the approximate molecular weight of the sulfate aerosol) resulted in a decreased association, thus suggesting that the sulfur portion of TSP is at least as harmful as the balance. Figures 7-24 and 7-25 present scatter plots of the

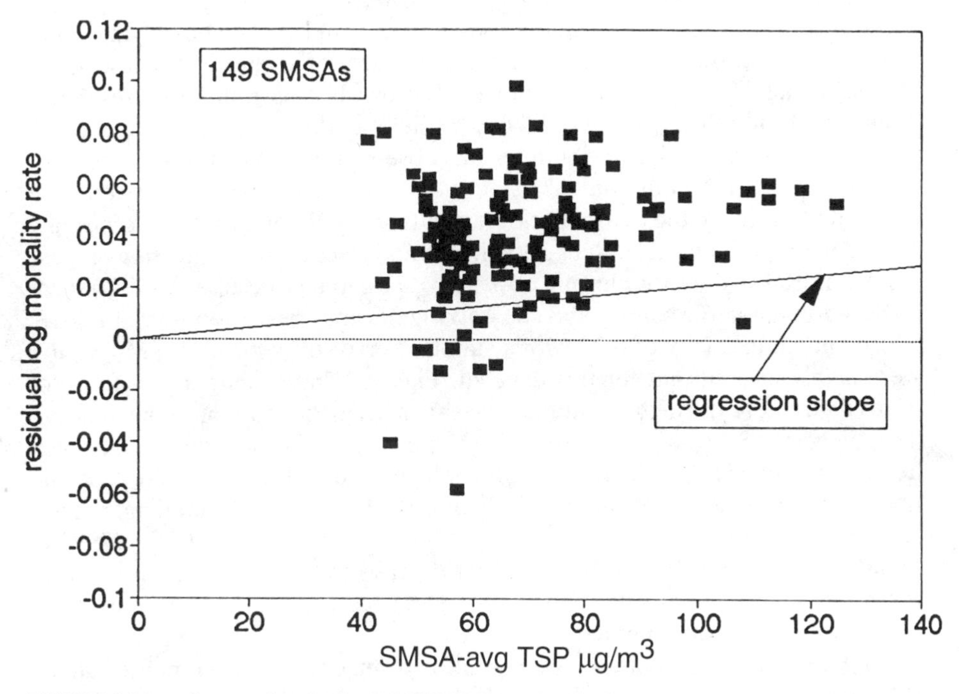

FIGURE 7-24. Scatter plot of adjusted 1980 SMSA mortality from all non-external causes vs. SMSA averaged TSP. *Source: Lipfert (1992).*

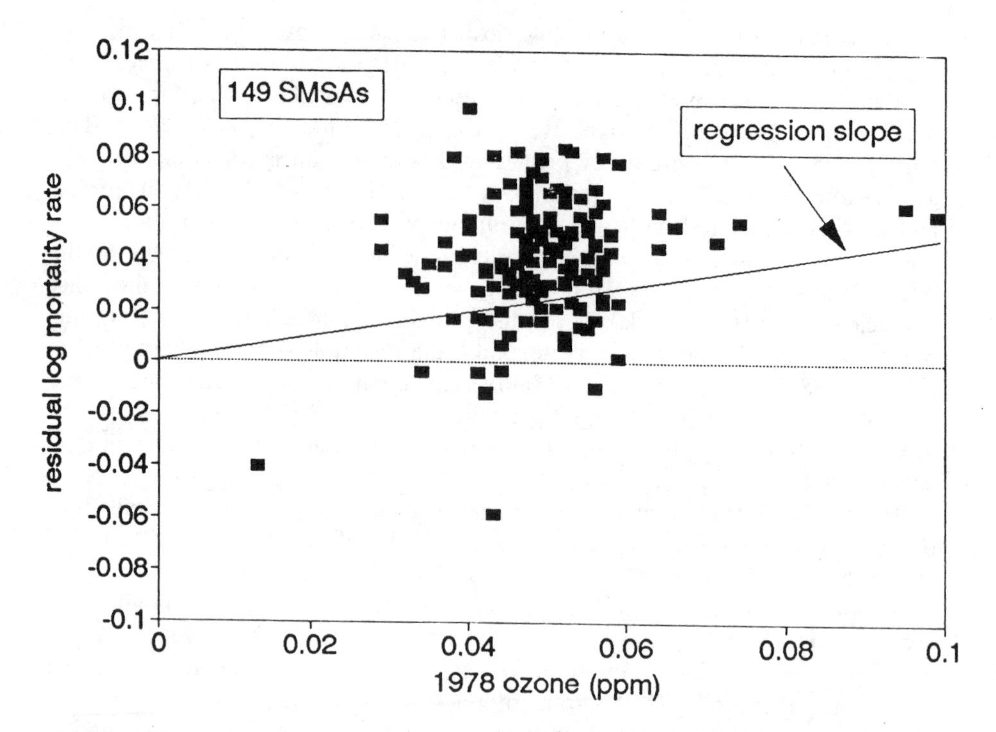

FIGURE 7-25. Scatter plot of adjusted 1980 SMSA mortality from major cardiovascular causes vs. estimated 8-hour average ozone level. *Source: Lipfert (1992).*

residuals, against TSP (Figure 7-24) and against ozone (Figure 7-25); they allow an examination for influential observations and the shapes of the dose-response functions. Both plots display a great deal of scatter at the lower and middle ranges of the pollution variables. The two low mortality rate residuals, Tampa and Honolulu (which are both outliers in the residual frequency distribution), are seen to contribute to the dose-response functions, since they both have low air pollution. Thus it is possible that these regressions are dominated more by the low-pollution cities than by the high values, especially since deleting the highest values had almost no effect on the regression slopes.

The mortality data for chronic obstructive pulmonary disease (COPD) were analyzed in much the same way. The only significant association was with TSP, which had an elasticity of 0.23 with a standard error of 0.062. The PM_{15} values were highly insignificant for this data set. Figure 7-26 presents the scatter plot of residuals; here the high pollution values seem to make more of a contribution. The highest TSP value represented Spokane, WA, which may have been affected by the Mt. St. Helens eruption, which occurred in May 1980. It would not appear possible to draw any conclusions about possible nonlinear dose-response functions from Figures 7-24 to 7-26. Although the COPD–TSP relationship appeared to be robust, COPD deaths tend to be a small fraction of the totals, and it is possible that a systematic regional bias in cause-of-death coding could have confounded the results.

Ambient air quality values based on a long-range transport computer model (Shannon, 1981) were used extensively in the study of mortality and air pol-

TABLE 7-7 Two-Stage Regression Results for Nonexternal Mortality (1980 U.S. SMSAs, one pollutant at a time)

Pollutant	No. of Locations	Correlation Coefficient	Elasticity	Standard Error
Measured pollutants				
SO_4^{-2}	149	0.036	0.006	0.013
TSP	149	0.157	0.033*	0.017
$TSP - SO_4^{-2}$	149	0.142	0.024	0.014
Average O_3	149	0.220	0.061*	0.022
Manganese	138	0.089	0.0006	0.0006
Fine particles	63	0.125	0.020	0.020
PM_{15}	63	0.098	0.0165	0.021
SO_4^{-2} (from PM_{15})	63	0.053	0.005	0.012
Lead	63	0.030	0.0	0.00001
Computed pollutants				
SO_4^{-2}	149	0.106	0.009	0.007
SO_2	149	0.112	0.009	0.006
NO_x	149	0.132	0.012	0.0075

$*p \leq 0.05$.

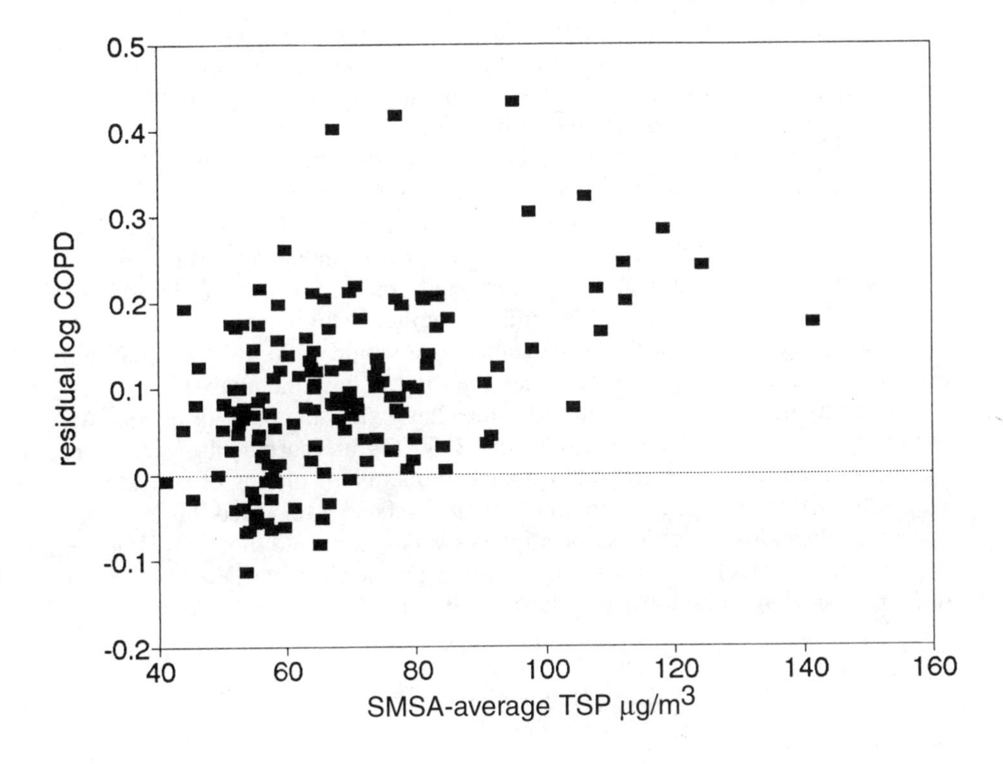

FIGURE 7-26. Scatter plot of adjusted 1980 SMSA COPD mortality vs. SMSA averaged TSP. *Source: Lipfert (1992).*

lution in U.S. cities performed by Lipfert *et al.* (1988); SO_2, SO_4^{-2}, and NO_x were all found to be important predictors of excess mortality. However, subsequent evaluations of these air quality estimates shed some doubt on their validity. Scatter plots against the IP data and against data from the Sulfate Regional Experiment (SURE) (Mueller and Hidy, 1983) show good agreement in a national context and within some regions but differences between regions. The correlation between computed SO_4^{-2} and IP-SO_4^{-2} was 0.70. Since the long-range transport model is essentially a transfer function between source emissions and ambient air, averaged over grid cells of about 120 km on each side, a correlation between health and computed air quality may also represent a correlation with industrial activity and the various accompanying socioeconomic factors. The computed SO_4^{-2} values are based only on combustion and smelter emissions and thus do not include sulfates from natural sources or particles such as $CaSO_4$. The grid-averaged values are incapable of reflecting local phenomena that might affect SO_2 oxidation rates or local primary emissions of SO_4^{-2}.

Conclusions

This study supported the previous findings of associations between TSP and premature mortality and also the hypothesis that improving the accuracy of pollutant exposure tends to increase statistical significance. Similarly, the lack of significance for SO_4^{-2} may partly relate to the flawed measurement methods used at the time. The finding of associations of premature mortality with ozone was unexpected and should be verified independently. The ambiguity between linear and log-linear models should also be investigated further.

After the preparation of the final report on this study (Lipfert, 1992), I acquired a national set of ozone measurements for the year 1980, based on monthly averages from 9:00 A.M. to 4:00 P.M. (A.S. Lefohn, personal communication, 1993). I used these data to create annual and warm season (April–October) averages and then repeated some of the ozone regression analyses. The objective was to determine whether using ozone measurements that were contemporary with the mortality data would yield a more significant association than the long-term averages used in the baseline analysis.

This comparison was compromised somewhat by missing data for 62 SMSAs. However, for the available SMSAs, the 1980 O_3 measurements were more significant than the long-term average measurements for the log of nonexternal mortality, with an elasticity of about 0.021 ($p = 0.066$). I regarded this as confirmation of the existence of an association between ozone and mortality. Further exploration of this relationship would require expansion of the original 149 SMSA data set to include additional locations covered by the 1980 measurements.

Ridge Regression Methods

McDonald and Schwing

McDonald and Schwing (1973) and Schwing and McDonald (1976) applied ridge regression methods to this problem, as a way to deal with the multicollinearity (nonorthogonality) typically found among the independent variables required for a complete mortality model. Ridge regression is a technique that relaxes the least-squares constraint by introducing an arbitrary parameter, k, to the diagonal entries of the correlation matrix, which biases the estimators. When $k = 0$, the least-squares solution obtains. Typically, multiple runs are made for varying values of k, and the regression coefficients are plotted versus k. A judgment must then be made as to the value of k for which regression estimates have "stabilized." With severe collinearity, the coefficients may change sign as k is increased from zero. Confidence limits are not possible for these biased coefficients.

McDonald and Schwing's studies had several other interesting features; in the first paper, air pollution "potential" was used in lieu of concentration measurements. The potential was based on emission density, SMSA dimensions, and dispersion characteristics and thus freed the data from biases due to locations of monitors and from measurement problems. However, this approach cannot handle the transport of pollutants between SMSAs, which may be important for sulfur compounds (long-range transport models account for this phenomenon). The second paper included variables representing cigarette smoking and background ionizing radiation. Both studies used age-adjusted or age-stratified mortality rates.

McDonald and Schwing (1973) used a data base of 60 SMSAs and total age-adjusted mortality rates for 1960. They found substantial instability when 15 independent variables were used (which is not surprising given only 60 observations), but the regressions using six variables were more stable. In both cases, SO_2 potential was an important variable, with a standardized coefficient of about 0.25, contributing about 1.5% of the mean mortality. The R^2's for these models were around 0.7, substantially higher than achieved by Lave and Seskin (1978) with 1960 age-adjusted data. No measures of particle concentrations were included in this initial study.

Schwing and McDonald (1976) expanded the variable list to include estimates of cigarette smoking, some measured air pollutants (1965 or 1969 values), and background ionizing radiation. However, they had data on only 46 SMSAs. They studied 1959–61 mortality rates for specific age groups and causes of death for white males and presented results for least-squares, ridge, and sign-constrained (minimum residual sum of squares) regressions. Although the sulfate and nitrate contents of TSP were included as independent variables, no general measures of particle loading were included. The results of this study were highly variable for air pollution, in part because several different measures of the same pollutant were often used simultaneously, and because the number of observations was small in comparison to the number of independent variables. The authors reported that the total elasticity for all sulfur compounds on all-disease mortality was about 0.02 to 0.03 for ridge regression and 0.03 to 0.04 for sign-constrained least-squares. Corresponding elasticities for smoking were in the range of 0.07 to 0.11. When compared to the corresponding results from Lave and Seskin (1978), the regression coefficients (elasticities) of Schwing and McDonald were generally substantially lower.

Thomas

Thomas (1975) and Thomas and Malone (1975) also found air pollution elasticities in the range of 0.02 for 1960 mortality, using ridge regression and measured values of both sulfates and TSP. Their analysis was performed for SMSAs using 24 independent variables.

Lipfert

I also tried (Lipfert, 1984) ridge regression in conjunction with my reanalysis of the Lave and Seskin 1969 SMSA data set. I found the sulfate coefficient to be stable when their model was used but that the TSP coefficient increased with increasing departure from least squares. On the basis of these various studies, I concluded that the ridge regression technique offered no particular advantage.

SECTION 4: CROSS-SECTIONAL REGRESSION ANALYSES OF LARGER UNITS (COUNTY GROUPS AND STATES)

Entire United States, Aggregated into County Groups

Mendelsohn and Orcutt

Mendelsohn and Orcutt (1979) studied mortality for the entire United States for 1970 by defining 404 groups of contiguous counties. Air quality was based on a limited set of measurements and interpolated estimates, for 1974. Death certificates were merged with the Census Public Use Sample to supply more details on individual versus group demographic characteristics, which was a major contribution of this study.

The dependent variables were 1970 mortality rates grouped by sex, race, and broad classifications for age and causes of death. Independent variables included (1970) age, income, education, and marital status, by sex, race, and broad age groups (24 data cells); county average age, education, income, number of children, marital status, employment status, housing status, migration, central city residence, climate, and regional residence. The pollutants included estimated annual means for SO_4^{-2}, NO_3^-, TSP, CO, SO_2, NO_2, and O_3. The authors concluded that sulfate was "closely associated with many deaths" and that O_3 and NO_2 had no apparent effect on life expectancy.

Mendelsohn and Orcutt also performed regressions allowing the pollutants to assume a quadratic (nonlinear) relationship; the squared terms were not significant. As a means of exploring the role of daily (acute) variations, they also added variables describing the "dispersion" of each pollutant about its mean at a given measuring station. These were also nonsignificant, which the authors interpreted as evidence that the effects being measured were in fact chronic. However, as discussed in Chapter 2, one cannot expect single measuring stations to accurately reflect short-term fluctuations over the multicounty areas used by Mendelsohn and Orcutt; thus, this finding should be viewed as inconclusive.

The chief methodological contribution of this study was the matching of data on individuals with group-mean data, a partial response to the "ecologic fallacy" criticism. The study thus compared the effects of certain demographic variables on mortality averaged across all residents of a county with the effects averaged for age-race-sex specific population cells. The results showed that the countywide variables were still important in the regressions, even when entered along with the variables that pertain better to individuals (Orcutt *et al.*, 1977). This counterintuitive finding may indicate that these individual variables were surrogates for some other effects on mortality rather than just aggregate averages of individual characteristics. If so, the arguments against literal interpretation of the results of "ecological" analyses would be strengthened, a point that these authors overlooked.

Among the main criticisms of the Mendelsohn and Orcutt study is the use of 1974 air quality data to "explain" 1970 mortality data. Concentrations of most primary pollutants declined substantially in many parts of the United States from 1970 to 1974 (1974 mean sulfate and TSP were 10.5 and 70 $\mu g/m^3$, respectively), and the preferred air quality measures would be averages for say 5 to 10 years *before* the date of death, not 4 years *after*. In addition, it is practically impossible to estimate exposure for the entire United States on the basis of existing monitoring networks, and thus this study may be affected by spatial autocorrelation, especially because of the process used to estimate air quality data where measurements were lacking. The one possible exception to these criticisms might be for sulfates, which tend to be distributed more uniformly spatially and which have not changed as much over time as have most primary air pollutants. However, the statistical significance of the sulfate variable could be enhanced relative to other pollutants just by virtue of having less measurement (exposure) error. In addition, the omission of smoking and drinking-water quality (hardness) may have overstated the effects of sulfate, given the regional distributions of these variables and subsequent findings of their significance (Lipfert, 1984).

This study showed that there was no particular benefit from a more rigorous approach to estimating the effects of socioeconomic variables on mortality, and thus it tends to validate the use of cruder methods, at least on a relative basis. The findings on air pollution effects on mortality must be viewed with caution because of the questionable procedures used to estimate air quality exposures. In addition, many of the R^2 values given appear to have been seriously overstated (Lipfert, 1983).

Selvin et al.

Selvin *et al.* (1984) analyzed total (all-cause) mortality by county for 1968–72, for both 3082 individual counties and 410 county groups according to the 1970 Census Public Use Sample (see Crocker *et al.*, 1979). The dependent variables were mortality rates for all causes for white males and white females, ages 45–54. The independent variables included: land area, population, net migration percentage of change, percentage urban, percentage black, percentage foreign stock, divorce rate, percentage of persons under 5 years of age, percentage over 65 years of age, percentage of persons with income under $3000, with income over $15,000, percentage of people with 4 years of college, occupational variables, housting variables, climate variables, elevation, median 1974–1976 TSP, median 1974–1976 SO_2, and median 1974–1976 NO_2.

The authors concluded that they found "no persuasive evidence of a link between air quality and general mortality levels" and that causal relationships cannot be inferred from this type of analysis.

The Selvin *et al.* study is a variant of that by Mendelsohn and Orcutt (1979) but retained many of its problems:

- The air quality data postdate the mortality data and were based on interpolated values.
- The study erroneously assumed that it is sufficient to analyze only one age

group (since the average rates for other groups are exponentially related to one another), and that it is necessary for all the variables to have normal distributions.

- Residuals were examined for heteroscedasticity but not for spatial autocorrelation.
- Air quality data were pooled without regard to measurement method, which could give a bias for SO_2 and NO_2, assuming some stations used bubblers and others used continuous electronic instruments.
- In their discussion, the authors ignored the findings of statistically significant SO_2 coefficients for the whole country and the South. They presented no detailed comparisons of their results with those of Mendelsohn and Orcutt, in spite of the strong parallels between the two studies. The elasticities for SO_x appear to be quite similar for the two studies.
- Suspended sulfates and ozone should have been included in the list of variables, in part to check previous findings, and in part because their spatial patterns tend to be smooth and thus missing values would be easier to estimate.

Regression Analyses of State Mortality

Winkelstein

Winkelstein (1985) performed a semiquantitative study of U.S. lung cancer and ischemic heart disease mortality for 25 states, from 1968 to 1978. The dependent variables were age-specific lung cancer and ischemic heart disease mortality rates for white males and females. The independent variable was the prevalence of cigarette smoking, taken from surveys of behavioral risk factors.

The author concluded that smoking appeared to be the common factor linking spatial patterns of lung cancer and heart disease and that the associations revealed in ecological analyses should receive serious consideration. Although air pollution and other environmental factors may be important issues "susceptible to epidemiological investigation," he concluded that they are unlikely to have an important effect on lung cancer or ischemic heart disease mortality.

The analyses in this paper should be considered exploratory rather than definitive, as was the author's apparent intent. They suggest that smoking could account for even more of the ischemic heart disease variance than shown directly (25%), since lung cancer "accounted" for 74% of it, and smoking is believed to be responsible for 80 to 90% of lung cancers (based on studies of individuals).

Winkelstein felt that the 25-state smoking prevalence data were accurate, because their total came within 12% of the Tobacco Institute's estimate of national consumption. However, since lung cancer rates were shown to be a better predictor of ischemic heart disease rates than the indicated smoking prevalence rates, it is possible that these prevalence rates do not give an accurate picture of the actual distribution of cigarette consumption. Reasons could include inaccurate (self-) reporting, lags between use of tobacco and resulting illness, varying consumption per smoker, and bias from the omitted states, which include both very low-consumption states (Utah) and high-consumption states (Nevada). Also, lung cancer may have a longer lag period than ischemic heart disease; if smoking time trends vary spatially, this could confound the interrelationships.

Although the supporting analyses appear to be relatively crude, Winkelstein's main findings appear sound. They suggest that ecological studies of mortality patterns must include smoking as a possible explanatory factor. The results also provide support for the use of lung cancer mortality rates as a "marker" for cigarette smoking, that is, an independent variable that is highly correlated with lung cancer is likely to also be correlated with smoking habits.

Summary of Studies of Long-Term Effects of Air Pollution on Mortality

Previous Summaries of Cross-Sectional Studies

There have been many previous critiques and summaries of air pollution—mortality studies including those by Lave and Seskin, 1970; Lipfert, 1977, 1978, 1980, 1985; Landau, 1978; Cooper and Hamilton, 1979; Thibodeau *et al.*, 1980; IERE, 1981; Ware *et al.*, 1981; Ricci and Wyzga, 1983; Evans *et al.*, 1984a. However, few of these adequately considered the effects of errors in estimating air pollution exposures, and there are now some new results that were not available to previous reviewers. For example, Thibodeau *et al.* (1980) reanalyzed the Lave and Seskin data set (1978) using the same basic data (including the flawed SO_4^{-2} data) and variables, but correcting for errors in the data and adding some statistical techniques. This type of reanalysis provides confidence in the results of the original authors with respect to the absence of numerical errors but cannot provide insight into robustness with respect to choices of variables and observations, which can be a more important issue.

With respect to cross-sectional studies, Ware *et al.* concluded that "The model can only be approximately correct, the surrogate explanatory variables can never lead to an adequate adjusted analysis, and it is impossible to separate associations of mortality rate with pollutant and confounding variables. This group of studies, in our opinion, provides no reliable evidence for assessing the health effects of sulfur dioxide and particulates." However, they did not support this conclusion by attempting to explain the differences among study results.

Ricci and Wyzga (1983) emphasized statistical issues and proposed a list of issues for further research but arrived at no overall conclusion regarding the validity of long-term effects of air pollution on mortality.

Evans *et al.* (1984a) critically reviewed some of the published cross-sectional mortality studies, including a quantitative comparison of regression coefficients and a reanalysis of the Lave and Seskin 1960 data with new variables added. For this reanalysis, the dependent variable was total 1960 mortality for 66 to 98 SMSAs. The independent variables included percentage over 65, median age, percentage of nonwhites, log of population density, percentage of college graduates, smoking index, percentage of poverty, and mean sulfate concentration. Other variables explored included occupational, home heating, climate, and housing variables; mean values for TSP, Fe, Mn, and B(a)P. In general, the sulfate variable lost statistical significance in this reanalysis (in part because the mean, rather than minimum concentrations, were used; see the discussion of Lave and Seskin, 1978; the elasticity with respect to total mortality was about 0.03). None of the other pollutants provided any better results.

Evans *et al.* concluded that "important questions related to the interpretation of cross-sectional studies remain unanswered. For example, it is unclear which specific pollutants are responsible for any observed effect; the shape of the exposure-response relationship is unresolved; and it is possible that the observed effects are due, in part, to confounding or systematic misclassification." Nevertheless, the authors were of the opinion that "the cross-sectional studies reflect a causal relationship between exposure to airborne particles and premature mortality."

My 1985 review covered both short- and long-term studies and discussed the interpretation of numerical differences in coefficients. I was reasonably confident that the short-term studies reflected causal relationships but held that it was not possible to determine quantitative relationships with confidence. In our analysis of 1980 mortality rates in U.S. cities (Lipfert *et al.*, 1988), we tried to approach the causality issue by comparing the ecological regression coefficients for nonpollutant variables (such as smoking, poverty, and demography) to previous estimates based on individuals. The degree of success of such comparisons would then provide support for a causal interpretation of the pollution effects.

In the United States, health professionals seldom mention environmental pollution factors in discussing current mortality patterns and rates of change. For example, for heart disease, smoking, blood pressure, and levels of serum cholesterol are thought to be the primary risk factors. However, Kleinman *et al.* (1981) attempted to reconcile the observed differentials in regional and urban-suburban rates of coronary heart disease mortality in terms of the corresponding differentials in these risk factors. They concluded that "differentials in these risk factors cannot account for the lower coronary heart disease death rates observed in the West as compared to other regions or among suburban as compared to urban residents." Thus, their finding implicitly supports an air pollution risk factor, in addition to those listed.

Summary of the Major Studies Reviewed

Table 7-8 summarizes the major long-term mortality studies reviewed in this chapter. The table attempts to explain differences in findings in terms of differences in the design and characteristics of the studies. The study characteristics to consider include the degree of age-sex-race control, the adequacy of control for socioeconomic effects (including life-style variables such as smoking), the range of variability in air pollution, and the adequacy of estimation of exposure, in addition to the times and places studied. Many of these factors are somewhat subjective, and thus I have not attempted to use them in a formal quantitative meta-analysis.

Some conclusions arise from considering Table 7-8 (some of which we have already discussed in more detail):

- The indicated effects of air pollution on mortality are sensitive to the data set analyzed and to the socioeconomic variables used to explain nonpollution effects.
- Respiratory deaths tend to be associated with particulate air pollution, especially in the pre-1970 period when particulate concentrations were higher.

TABLE 7-8 Comparison of Cross-Sectional Studies

Authors	Time period	Locations	Age-Sex Control*	Socioeconomic Variables[†]	Pollutant Ranges[††]	Exposure Accuracy[§]	Significant Pollutants[‖]	Diseases
A. Intraurban Studies								
Mills & Mills-Porter (1948)	1929–46	5 cities	Sex	None	TSP, 100–700	Poor	Dust	Resp. dis.
Mills (1952)	1946	Chicago	Age-sex	None	TSP, 100–500	Poor	Dust } SO_2 }	Resp. dis.
Zeidberg *et al.* (1967a, b)	1949–60	Nashville	a-s-r	Income, educ., housing	COH, <0.35–>1. SO_2, <13–>34	Poor Poor	TSP } Soiling } SO_2 }	All CV Stomach cancer Total resp.
Winkelstein *et al.* (1967, 68)	1959–61	Buffalo	a-s-r	Income, educ., occupations	TSP, 75–206 SO_2, 61–385	Good ?	TSP } SO_2 }	All causes Arteriosclerotic heart disease Stroke, age 50–69 Cardiorespiratory disease M + F Stomach cancer Prostate cancer Liver cirrhosis
Namekata *et al.* (1979)	1971–75	Chicago (76 areas)	Age-adj.	Income, housing, education	TSP, 83 SO_2, 41	Good Good	TSP TSP TSP	Heart dis. Liver cirh. Emphys.
Gregor (1977)	1968–72	Allegheny Co., PA (cens. tracts)	a-s-r	Education, pop. dens., climate	TSP, 122 SO_2, 100	Good Good	TSP TSP	All causes poll. rel. causes
Dean *et al.* (1978)	1963–72	Cleveland Co., UK (3 areas)	Sex age-adj.	Smoking	Smoke, 35–175 SO_2, 57–129	Good Good	Smoke } SO_2 }	{ LC Bronchitis Coronary heart disease males Stroke }
Krzyżanowski & Wojtyniak (1982)	1968–78	Cracow, Poland (2 areas)	Sex age-adj.	Smoking, occupational exposure, birth place education	Smoke, 109–180 SO_2, 53–114	Good Good	Smoke } SO_2 }	All causes

TABLE 7-8 *Continued*

Authors	Time per Locations		Age-Sex Control*	Socioeconomic Variables[†]	Pollutant Ranges[††]	Exposure Accuracy[§]	Significant Pollutants[‖]	Diseases
B. Studies of Cities and Boroughs								
Stocks (1959, 60)	1950–54	England & Wales	Sex age-adj.	Pop. density, social class	Smoke, 15–562	OK	Smoke	LC
					B(a)P, 0.001–0.1	?	Smoke	Resp. dis.
					(other H/C, tr. metals), settl. dust	?	Smoke	Stomach cancer
							Dust	LC
							Dust	Bronchitis
							Dust	Stomach cancer
Lipfert (1978)	1959	U.S. cities	Age#, race#	Pop. dens., poverty, housing, smoking, education	TSP, 40–224	OK	Mn	All causes
					SO_4, 3–28	OK		
					Fe, 0.5–13	Poor		
					Mn, 0.01–2.2	?		
Lipfert (1978)	1969–71	U.S. cities	Age, race#	Pop. dens., poverty, housing, smoking, educ., birth rate	TSP, 29–188	OK	TSP + Mn	All causes
					SO_4, 2–19	OK	Mn	All cancers
					Fe, 0.2–9.5	Poor	Mn	Resp. Cancers
					Mn, 0.01–2.0	?	Fe	Resp. Dis.
							TSP + Mn	Other, incl. cardiovascular
Crocker *et al.* (1979)	1970	U.S. cities	Age#, race#	Income, education, smoking, diet, medical care, climate	TSP, 102	OK	(None)	All causes
					SO_2, 27	Poor	TSP	Pneum., flu
					NO_2, 143	Poor	SO_2	infant
Chinn *et al.* (1981)	1969–73	U.K. boroughs	Age-sex	Pop. dens., housing, social class, climate, smoking, water hard.	Smoke, 15–225	Good	SO_2	All causes
					SO_2, 20–320	Good	SO_2	Hypertension
							SO_2	Bronchitis
Lipfert *et al.* (1988)	1980	U.S. cities	Age#, race#	Pop. chng., poverty, smoking, migration, water hard., birth rate	SO_4, 0.1–15	Good	SO_4^{-2}	All causes
					SO_2, 0.3–46	Good	SO_2	
					NO_x, 0.7–51	Good	NO_x	
					Mn, 0.02–0.2	?	Mn	
					Fe, 0.2–4.3	Fe		
					O_3, 37–200	Poor		
					CO	Poor		
					IP	OK		
					FP	Good		
					TSP	OK		

C. Studies of Counties, SMSAs, and Larger Units								
Lave & Seskin (1978)	1960	U.S. SMSAs	a-s-r	Pop. dens., poverty, pop; others	TSP, 40–224 SO_4, 3–28	Poor Poor	TSP** + SO_4^{-2} TSP + SO_4** TSP** SO_4 SO_4	All causes All cardiovascular dis. Resp. dis. All cancer Stomach cancer
Lave & Seskin (1978)	1969	U.S. SMSAs	a-s-r	Pop. dens., poverty, population	TSP, 29–188 SO_4, 2–19	Poor OK	TSP + SO_4^{-2}	All causes
Chappie & Lave (1982)	1974	U.S. SMSAs	Age#, race#	Pop. dens., income, pop., education, smoking, alcohol use, occupation mix	TSP, 75 SO_4, 9.6	Poor OK	SO_4	All dis. causes
Lipfert (1984)	1970	U.S. SMSAs	Age#, race#	Poverty, migration, pop. dens., pop., diet, smoking, water hard., home heat fuel use	TSP, 29–188 SO_4, 2–19 Fe, 0.2–9.5 Mn, 0.01–2.0 O_3, 30	OK Poor Poor Poor Poor	TSP TSP + Mn	{ All causes Males, <65 } { Females, <65 Males, >65 }
Özkaynak & Thurston (1987)	1980	U.S. SMSAs	Age#, race#	Pop. dens., poverty, education	TSP, 78 SO_4, 11 IP, 48 FP, 22	Poor Poor Poor OK	FP SO_4^{-2}	All causes All causes
Lipfert (1992)	1980	U.S. SMSAs	Age#, race#	Smoking, education, migration, ethnicity, water hardness	TSP, 41–142 SO_4^{-2}, 2–17 O_3, 0.013–0.099	Good Good ?	TSP O_3 TSP O_3	Non-ext. causes COPD CV
Mendelsohn & Orcutt (1979)	1970	U.S. County Groups	a-s-r	Income, education, mar. status, migration, pop. dens., housing, region	TSP, 70 SO_4, 10.5 NO_3, 3.0 SO_2, 21 NO_2, 50 O_3, 28 CO, 4600	Poor OK Poor Poor Poor Poor Poor	SO_4^{-2}	All diseases

* age-sex indicates separate regressions by age and sex; a-s-r = age-sex-race; age# (race#) indicates that a population age (race) descriptor variable was used as an independent variable in the regressions; age-adj. = age adjusted.

† Independent variables listed are consistent with the regression results shown; many authors ran regressions for various deletions and combinations from the list.

†† Mean values are given when ranges were not available. Units are mg/m^3.

§ Subjective judgment involving numbers of pollution monitors and temporal coincidence of pollution and mortality.

‖ TSP + SO_4 indicates that both pollutants were significant in a joint regression. Brackets indicate it was not possible to distinguish among the pollutants in the group.

** Significant with additional variables.

Cardiovascular mortality may be associated with sulfur oxides air pollution and, more recently, with ozone.

- During the mid-1970s in the United States, the association between sulfur oxides and mortality became substantially stronger, for unknown reasons.
- The association between (all-cause) mortality and particulate air pollution has become somewhat less significant in both the United States and Britain. This could be the result of reductions in ambient concentrations resulting from air pollution control efforts.
- The association between air pollution and mortality tends to be substantially stronger within urban areas than between urban areas. It is not clear how much of this results from better characterization of exposure to air pollution and how much from sharper gradients in personal life-style characteristics (which may not be fully accounted for in the analyses).
- As better account is taken of confounding variables, associations between air pollution and those causes of death without physiological links with air pollution (such as "all cancers") tend to lose significance.
- There is ambiguity as to which pollutants are associated with all-cause mortality. It is possible that this reflects varying combinations of different relationships between specific pollutants and specific diseases.

These considerations relate primarily to the *existence* of associations between long-term exposure to air pollution and mortality. Table 7-8 provides no guidance as to the magnitude of effects or whether they are related to current or previous exposures. It should also be noted that cross-sectional analyses are limited to those pollutants for which suitable data are available, which excludes carbon monoxide, for example. Ozone may present a different problem; since the distribution of ozone tends to be uniform, on average, over much of the eastern half of the country, there may be insufficient range to detect health effects with cross-sectional analyses. Also, it is not clear which measure of ozone is most appropriate, daily peaks or long-term averages. I used 1975 annual averages in my reanalysis of 1969 SMSA mortality (Lipfert, 1984), which was only marginally successful (perhaps because of the mismatch in time periods). The EPA no longer routinely compiles annual averages for ozone, which makes this type of analysis more difficult. Seasonal daytime averages of ozone were used in the cross-sectional studies of 1980 mortality.

The available studies of long-term lag effects (Lave and Seskin, 1978; Lipfert, 1978; Lipfert *et al.*, 1988) were all inconclusive and related primarily to the effects of sulfur oxides. The evidence on long-term lags is less clear with respect to particulates. Neither data from Britain (Chinn *et al.*, 1981) or the Winkelstein *et al.* (1967) studies in Buffalo, NY, support an extensive lag period for smoke pollution. However, particulate effects from U.S. inter-urban studies may be badly compromised by insufficient numbers of monitoring stations and the resulting poor characterization of population exposures.

The magnitude of air pollution effects on total mortality was estimated at about 5% in three separate studies, all of which controlled for smoking: U.S. cities (Lipfert, 1978; Lipfert *et al.*, 1988), and British boroughs (females only) (Chinn *et al.*, 1981). This agreement among independent data sets provides the best evidence that such associations are not likely to be due to data artifacts. Several independent studies found substantially stronger effects within cities or urban areas (as much as an order of magnitude higher): Buffalo, NY; Pittsburgh; Cleveland County (U.K.), and Cracow. Also, the Cracow study dealt with individuals and thus was not "ecological"; however, it also found negative associations between female mortality and air pollution. While the various flaws in these studies, especially the intraurban ones, may have inflated their coefficients in some cases, this unanimity in the **direction** of the pollution effects makes it increasingly difficult to deny their existence. Also, the existence of large intraurban pollution effects on mortality due to differences in exposure is inconsistent with a major role for regionally distributed pollutants such as sulfate.

The prospective cohort study reported by Pope *et al.* (1993) is consistent with and strengthens the findings from ecological studies in general, notwithstanding the failure of Abbey *et al.* (1991) to find such effects in a similar prospective study in California (failure to find an effect does not rule out its existence). The risks due to smoking and the benefits of education estimated by ecological methods were confirmed quantitatively, but the apparent risks of air pollution exposure appeared to be much larger in Pope *et al.*'s study of individuals, compared to either time-series studies of acute mortality or cross-sectional studies across many locations for individual years. However, the elasticities based on individuals agree with findings from the few intraurban studies that have been reported (for example, Winkelstein *et al.*, 1967; Gregor, 1977), and with the only other study of individuals (Krzyżanowski and Wojtyniak, 1982). Intraurban studies may be a better paradigm for the prospective study because their population groups are smaller and air pollution exposures may be better characterized.

If we choose to accept these higher response rates, then we must also conclude that air pollution has both chronic and acute effects on human mortality. Additional studies should be designed to test this hypothesis in a larger data set, paying particular attention to estimating actual long-term exposures to air pollution, including indoor air pollution effects.

REFERENCES

Abbey, D.E., Mills, P.K., Petersen, F.F., and Beeson, W.L. (1991), Long-Term Ambient Concentrations of Total Suspended Particulates and Oxidants As Related to Incidence of Chronic Disease in California Seventh-Day Adventists, *Env. Health Perspect.* 94:43–50.

Ashley, D.J.B. (1967), The Distribution of Lung Cancer and Bronchitis in England and Wales, *Br. J. Cancer* 21:243–59.

Bair, F.E., editor (1992), *The Weather Almanac*, Gale Research, Inc., Detroit.

Brackbill, R., Frazier, T., and Shilling, S. (1988), Smoking Characteristics of U.S. Workers, 1978–80, *Am. J. Ind. Med.* 13:5–42.

Brown, S.M., Selvin, S., and Winkelstein, W., Jr. (1975), The Association of Economic Status with the Occurrence of Lung Cancer, *Cancer* 36:1903–11.

Buck, S.F., and Brown, D.A. (1964), *Mortality from Lung Cancer and Bronchitis in Relation to Smoke and Sulphur Dioxide Concentration*, Population and Social Index, Research Paper No. 7, Tobacco Research Council, London.

Bultena, G.L. (1969), Heath Patterns of Aged Migrant Retirees, *J. Amer. Ger. Soc.* 17:1127–31.

Carroll, R.E. (1966), The Relationship of Cadmium in the Air to Cardiovascular Disease Death Rates, JAMA 198:267–69.

Chappie, M., and Lave, L.B. (1982), The Health Effects of Air Pollution: A Reanalysis, *J. Urban Econ.* 12:346–76.

Chinn, S., Florey, C. du V., Baldwin, I.G., and Gorgol, M.J. (1981), The Relation of Mortality in England and Wales, 1969–73, to Measurements of Air Pollution, *J. Epidem. Comm. Health* 35:174–79. Also see Florey, C. du V., Chinn, S., Baldwin, I.G., and Gorgol, M. J., Final Report on Mortality and Air Pollution in the County Boroughs of England and Wales, 1969–1973, Department of Community Medicine, St. Thomas' Hospital, London. July 1980.

Cohen, B.L. (1992), Percentage of Lifetime Spent in Area of Residence at Time of Death, *Envir. Res.* 57:208–11.

Cooper, D.E., and Hamilton, W.C. (Jan. 1979), Mortality and Sulfates—the Phantom Connection? *Amer. Mining Congress J.*

County and City Data Book, 1988, U.S. Bureau of the Census, U.S. Government Printing Office, Washington, DC, 1988.

Craxford, S.R., Clifton, M., and Weatherley, M.-L.P.M. (1967), Smoke and Sulphur Dioxide in Great Britain: Distribution and Changes, *Proc. 1966 Int. Clean Air Congress*, London, pp. 213–16.

Crocker, T.D., Schulze, W., Ben-David, S., and Kneese, A. (1979), *Methods Development for Assessing Air Pollution Control Benefits*. Vol. I, Economics of Air Pollution Epidemiology, EPA-600/5-79-001a, U.S. Environmental Protection Agency, Washington, DC.

Daly, C. (1954), Air Pollution and Causes of Death, *Br. Med. J.* ii:687–88.

Dean, G., Lee, P.N., Todd, G.F., and Wicken, A.J. (1978), *Report on a Second Retrospective Study in North-East England*, Tobacco Research Council Research Paper 14, Parts I and II, London.

Dockery, D.W., Ware, J.H., Ferris, B.G., Jr., Glickberg, D.S., Fay, M.E., Spiro, A., III, and Speizer, F.E. (1985), Distribution of Forced Expiratory Volume in One Second and Forced Vital Capacity in Healthy, White, Adult Never-Smokers in Six U.S. Cities, *Am. Rev. Respir. Dis.* 131:511–20.

Ellis, F.P. (1972), Mortality from Heat Illness and Heat-Aggravated Illness in the United States, *Env. Res.* 5:1–58.

Evans, J.S., Tosteson, T., and Kinney, P.L. (1984a), Cross-Sectional Mortality Studies and Air Pollution Risk Assessment, *Env. Int.* 10:53–83.

Evans, J.S., Kinney, P.L., Koehler, J.L., and Cooper, D.W. (1984b), Comparison of Cross-Sectional and Time-Series Mortality Regressions, *J. APCA* 34:551–53.

Fairbairn, A.S., and Reid, D.D. (1958), Air Pollution and Other Local Factors in Respiratory Disease, *Brit. J. Prev. Soc. Med.* 12:94–103.

Farr, W. (1885), *Vital Statistics: A Memorial Volume of Selections from the Reports and Writings of William Farr*, Humphreys, N., ed., Sanitary Institute of Great Britain, London, p. 164.

Feinleib, M., Fabsitz, R., and Sharrett, A.R. (1979), *Mortality from Cardiovascular and Non-Cardiovascular Disease for U.S. Cities*, U.S. Dept. of HEW, PHS, Nat'l. Inst. of Health, DHEW Pub. No. (NIH) 79–1453, Washington, DC.

Ferris, B.G., Jr., Speizer, F.E., Spengler, J.D., Dockery, D., Bishop, Y., Wolfson, M., and Humble, C. (1979), Effects of Sulfur Oxides and Respirable Particles on Human Health, *Am. Rev. Resp. Dis.* 120:767–79.

Finch, S.J., and Morris, S.C. (1979), Consistency of Reported Health Effects of Air Pollution, in *Advances in Environmental Science and Engineering*, Pfaffin, J.R., and Ziegler, E.N. eds., Gordon and Breach Science Publishers, London, pp. 106–17. Also available as BNL Report 21808-R (1977).

Finch, S.J., Novak, K., Ouyang, S.P., and Schaik, J.V., Longitudinal Study of Association Between Air Pollution Reduction and Changes in Mortality Ratios in Buffalo, unpublished BNL memo.

Fleissner, M.L., Gregory, A., and Mosher, W.E. (1974), The Contribution of Air Pollution to the Risk of Chronic Respiratory Disease Mortality, presented at the Annual Meeting of the Amer. Public Health Assn., New Orleans.

Gardner, M.J., Crawford, M.D., and Norris, J.N. (1969), Patterns of Mortality in Middle and Early Old Age in the County Boroughs of England and Wales, *Brit. J. Prev. Soc. Med.* 23:133.

Gerking, S., and Schulze, W. (1981), What Do We Know about Benefits of Reduced Mortality from Air Pollution Control? *AEA Papers and Proceedings* 71:228–34.

Gittelson, A., and Royston, P.N. (1982), Annotated Biobliography of Cause-of-Death Validation Studies, 1958–80, *Vital and Health Statistics*, Series 2, No. 89, DDHS Pub. No. (PHS) 82-1363, Public Health Service, Wasington, DC, U.S. Gov't Printing Office.

Gore, A.T., and Shaddick, C.W. (1958), Atmospheric Pollution and Mortality in the County of London, *Br. J. Prev. Soc. Med.* 12:104.

Gorham, E. (1958), Bronchitis and the Acidity of Urban Precipitation, *Lancet*, ii, 691.

Gorham, E.D., Garland, F.C., and Garland, C.F. (1990), Sunlight and Breast Cancer Incidence in the USSR, *Int. J. Epidemiology* 19:820–24.

Göttinger, H.W. (1983), Air Pollution Health Effects in the Munich Metropolitan Area: Preliminary Results Based on a Statistical Theory, *Env. Int.* 9:207–20.

Gregor, J.J. (1977), *Mortality and Air Quality, the 1968–72 Allegheny County Experience*, Report #30, Center for the Study of Environmental Policy, Pennsylvania State University, University Park, Pennsylvania.

Grove, R.D., and Hetzel, A.M. (1968), *Vital Statistics Rates of the United States, 1940–1960*, U.S. Dept. of H.E.W., PHS Publ. 1677, U.S. Gov't. Printing Office, Washington, DC.

Hagstrom, R.M., Sprague, H.A., and Landau, E. (1967), The Nashville Air Pollution Study. VII. Mortality from Cancer in Relation to Air Pollution, *Arch. Env. Health* 15:237.

Health Care Financing Corporation (HFCA) (1990), *Hospital Data by Geographic Area for Aged Medicare Beneficiaries: Selected Diagnostic Groups, 1986*. U.S. Dept. of Health and Human Services, Washington, DC. HFCA Pub. 03300.

Helsing, K.J., Sandler, D.P., Comstock, G.W., and Chee, E. (1988), Heart Disease Mortality in Nonsmokers Living with Smokers, *Am. J. Epidem.* 127:915–22.

International Electric Research Exchange (IERE), *Effects of SO_2 and Its Derivatives on Health and Ecology, Vol. 1. Human Health*, Nov. 1981., available from EPRI Research Reports Center, P.O. Box 50490, Palo Alto, CA 94303.

Kalkstein, L.S. (1991), A New Approach to Evaluate the Impact of Climate on Human Mortality, *Envir. Hlth Persp.* 96:145–150.

Kitagawa, E.J., and Hauser, P.M. (1973), *Differential Mortality in the United States: A Study in Socioeconomic Epidemiology*, Harvard University Press, Cambridge, MA.

Kleinman, J.C., DeGruttola, V.G., Cohen, B.B., and Madans, J.H. (1981), Regional and Urban–Suburban Differentials in Coronary Heart Disease Prevalence and Risk Factor Prevalence, *J. Chron. Dis.* 34:11–19.

Krzyżanowski, M., and Wojtyniak, B. (1982), Ten-Year Mortality in a Sample of an Adult Population in Relation to Air Pollution, *J. Epid. Comm. Health* 36: 262–68.

Lacey, R.F. (1981), Changes in Water Hardness and Cardiovascular Death-Rates, Technical Report TR 171, Water Research Center, Medmenham, England.

Landau, E., *The Nation's Health*, March 1978.

Lave, L.B., and Seskin, E.P. (1978), *Air Pollution and Human Health*, Johns Hopkins University Press.

Lave, L.B., and Seskin, E.P. (1970), Air Pollution and Human Health, *Science* 169:723–33.

Lazar, P. (1981), Geographical Correlations between Disease and Environmental Exposures, in *Perspectives in Medical Statistics: Proceedings of the Rome Symposium on Medical Statistics 1980*, Bithell, J.F., and Coppi, R., eds., pp. 21–38, London, Academic Press.

Lew, E.A., and Garfinkel, L. (1979), Variations in Mortality by Weight Among 750,000 Men and Women, *J. Chron. Dis.* 32:563–76.

Liberatos, P., Link, B.G., and Kelsey, J.L. (1988), The Measurement of Social Class in Epidemiology, *Epidemiologic Reviews* 10:87–121.

Link, T. personal communication, 1991.

Lipfert, F.W. (1977a), "The Association of Air Pollution and Human Mortality: A Review of Previous Studies," APCA Paper 44.1, presented at the 70th APCA Annual Meeting, Toronto.

Lipfert, F.W. (1977b), The Association of Air Pollution and Human Mortality: Regression Analysis for 136 U.S. Cities, 1969, APCA Paper 18.7, presented at the 70th APCA Annual Meeting, Toronto.

Lipfert, F.W. (1978), *The Association of Human Mortality with Air Pollution: Statistical Analyses by Region, by Age, and by Cause of Death*, Ph.D. Dissertation, Union Graduate School, Cincinnati, Ohio. Available from University Microfilms.

Lipfert, F.W. (1980), Differential Mortality and the Environment: The Challenge of Multicollinearity in Cross-sectional Studies, *Energy Systems and Policy* 3:103–22.

Lipfert, F.W. (1985), Mortality and Air Pollution: Is There a Meaningful Connection? *Env. Sci. Tech.* 19:764–70.

Lipfert, F.W. (1988), *Estimates of Historic Urban Air Quality Trends and Precipitation Acidity in Selected U.S. Cities (1880–1980)*, Brookhaven National Laboratory Report to the National Park Service.

Lipfert, F.W. (1984), Air Pollution and Mortality: Specification Searches Using SMSA-Based Data, *J. Env. Econ. & Mgmt.* 11:208–243.

Lipfert, F.W. (1983), Comment to Robert Mendelsohn and Guy Orcutt Regarding "An Empirical Analysis of Air Pollution Dose-Response Curves," *J. Envir. Econ. & Mgmt.* 10:184–6.

Lipfert, F.W. (1980), Statistical Studies of Mortality and Air Pollution, Multiple Regression Analysis Stratified by Age Group, *Sci. Total Env.* 15:103–22.

Lipfert, F.W. (1980), Statistical Studies of Mortality and Air Pollution, Multiple Regression Analysis by Cause of Death, *Sci. Total Env.* 16:165–83.

Lipfert, F.W. (1980), Sulfur Oxides, Particulates, and Human Mortality: Synopses of Statistical Correlations, *J. APCA* 30:336–71.

Lipfert, F.W., Malone, R.G., Daum, M.L., Mendell, N.R., and Yang, C.-C. (1988), *A Statistical Study of the Macroepidemiology of Air Pollution and Total Mortality*, BNL Report 52122, to U.S. Dept. of Energy, Brookhaven National Laboratory, Upton, NY.

Lipfert, F.W. (1992), *Community Air Pollution and Mortality: Analysis of 1980 Data from U.S. Metropolitan Areas*, report prepared for U.S. Department of Energy, Brookhaven National Laboratory, Upton, NY.

Manos, N.E., and Fisher, G.F. (1959), An Index of Air Pollution and Its Relation to Health, *J. APCA* 9:5–11.

McCurdy, T., personal communication, 1991.

McDonald, G.C., and Schwing, R.C. (1973), Instabilities of Regression Estimates Relating Air Pollution to Mortality, *Technometrics* 15:463–81.

Mendelsohn, R., and Orcutt, G. (1979), An Empirical Analysis of Air Pollution Dose-Response Curves, *J. Envir. Econ. & Mgmt.* 6:85–106.

Mills, C.A. (1943), Urban Air Pollution and Respiratory Diseases, *Am. J. Hygiene* 37:131–41.

Mills, C.A. (1952), Air Pollution and Community Health, *Am. J. Med. Sci.* 224: 403–407.

Mills, C.A., and Mills-Porter, M. (1948), Health Costs of Urban Air Pollution, *Occup. Med.* 5:614–33.

Mitchell, B.D., Stern, M.P., Haffner, S.M., Hazuda, H.P., Patterson, J.K., and Stern, M.P. (1990), Risk Factors for Cardiovascular Mortality in Mexican-Americans and Non-Hispanic Whites, *Am. J. Epidem.* 131:423–33.

Mitchell, B.D., Hazuda, H.P., Haffner, S.M., Patterson, J.K., and Stern, M.P. (1991), Myocardial Infarction in Mexican-Americans and Non-Hispanic Whites, *Circulation* 83:45.

Morris, S.C., Shapiro, M.A., and Waller, J.H. (1976), Adult Mortality in Two Communities with Widely Different Air Pollution Levels, *Arch. Env. Health* 31:248–54.

Mueller, P.K., and Hidy, G.M. (1983), *The Sulfate Regional Experiment: Report of Findings, Vols. 1–3, EPRI EA-1901*, Electric Power Research Institute, Palo Alto, CA.

Namekata, T., Carnow, B.M., Reda, D.J., O'Farrell, E.B., and Marselle, J.R. (1979), *Model for Measuring the Health Impact from Changing Levels of Ambient Air Pollution*, Report to U.S. Environmental Protection Agency, EPA-600/1-79-034.

Orcutt, G., Franklin, S.D., Mendelsohn, R., and Smith, J.D. (1977), Does Your Probability of Death Depend on Your Microenvironment? A Microanalytic Study, *Amer. Econ. Rev.* 67:260–64.

Özkaynak, H., and Thurston, G. (1987), Associations Between 1980 U.S. Mortality Rates and Alternative Measures of Airborne Particle Concentration, *Risk Analysis* 7:449–62.

Patrick, C.H. (1980), Health and Migration of the Elderly, *Research on Aging* 2:233–41.

Pemberton, J., and Goldberg, C. (1954), Air Pollution and Bronchitis, *Br. Med. J.* ii:567–70.

Plagiannakos, T., and Parker, J. (1988), *An Assessment of Air Pollution Effects on Human Health in Ontario, Report No. 706.01 (#260)*, Energy Economics Section, Economics and Forecast Division, Ontario Hydro, Toronto.

Pope, C.A., Dockery, D.W., Xu, X., Speizer, F.E., Spengler, J.D., and Ferris, B.G. (1993), Mortality Risks of Air Pollution: A Prospective Cohort Study, presented at *Aerosols in Medicine*, 9th ISAM Congress, Garmisch–Partenkirchen, Germany, March 30–April 4, 1993.

Prindle, R.A. (1959), Some Considerations in the Interpretation of Air Pollution Health Effects Data, *J. APCA* 9:12–19.

Ricci, P.F., and Wyzga, R.E. (1983), An Overview of Cross-Sectional Studies of Mortality and Air Pollution and Related Statistical Issues, *Env. Int.* 9:177.

Rogot, E., Sorlie, P.D., Johnson, N.J., and Schmitt, C. (1992), *A Mortality Study of 1.3 Million Persons*, NIH Publication No. 92-3297, National Heart Lung and Blood Institute, Bethesda, MD.

Rosenwaike, I. (1987), Mortality Differentials Among Persons Born in Cuba, Mexico, and Puerto Rico Residing in the United States, 1979–1981. *Am. J. Public Health* 77:603–606.

Rumford, J. (1961), Mortality Studies in Relation to Air Pollution, *Am. J. Public Health* 51:165–73.

Samet, J.M., Speizer, F.E., Bishop, Y., Spengler, J.D., and Ferris, B.G., Jr. (1981), The Relationship between Air Pollution and Emergency Room Visits in an Industrial Community, *J. APCA* 31:236–40.

Schiffman, R., and Landau, E. (1961), Use of Indexes of Air Pollution Potential in Mortality Studies, *J. APCA* 11:384–6.

Schwing, R.C., and McDonald, G.C. (1976), Measures of Association of Some Air Pollutants, Natural Ionizing Radiation, and Cigarette Smoking with Mortality Rates, *Sci. Tot. Envir.* 5:139–69.

Selvin, S., Merrill, D., Wong, L., and Sacks, S.T. (1984), Ecologic Regression Analysis and the Study of the Influence of Air Quality on Mortality, *Env. Health Perspectives* 54:333–40.

Shannon, J.D. (1981), A Model of Regional Long-term Average Sulfur Atmospheric Pollution, Surface Removal, and Wet Horizontal Flux, *Atm. Env.* 15:689–701.

Smith, V.K. (1975), Mortality–Air Pollution Relationships: A Comment, *J. ASA* 70:341–43.

Speizer, F.E. (1989), Studies of Acid Aerosols in Six Cities and in a New Multi-city Investigation: Design Issues, *Env. Health Perspect.* 79:61–68.

Sprague, H.A., and Hagstrom, R.M. (1969), The Nashville Air Pollution Study: Mortality Multiple Regression, *Arch. Env. Health* 18:503.

Stevens, R.D.S. (1981), *The Sampling and Analysis of Airborne Sulphates and Nitrates: A Review of Published Work and Synthesis of Available Information*, prepared for Environment Canada, EPS 5-AP-82-14.

Stocks, P. (1959), Cancer and Bronchitis Mortality in Relation to Atmospheric Deposit and Smoke, *Br. Med. J.* i:74–79.

Stocks, P. (1960), On the Relation Between Atmospheric Pollution In Urban and Rural Localities and Mortality from Cancer, Bronchitis, and Pneumonia, with Particular Reference to 3:4 Benzopyrene, Beryllium, Molybdenum, and Arsenic, *Brit. J. Cancer* 14:397–418.

Thibodeau, L., Reed, R.B., Bishop, Y.M.M., and Kammerman, L.A. (1980), Air Pollution and Human Health: A Review and Reanalysis, *Env. Hlth. Persp.* 34:165.

Thomas, T.J. (1975), *An Investigation into Excess Mortality and Morbidity Due to Air Pollution*, Ph.D Dissertation, Purdue University.

Thomas, T.J., and Malone, D.W. (1975), The Effects of Particulate Air Pollution on Expectation of Life, *Proc. Int. Symp. on Recent Advances in the Assessment of the Health Effects of Environmental Pollution*, pp. 1569–73, Commission of European Communities, Luxembourg.

U.S. Dept. of Health, Education and Welfare, *Smoking and Health*, U.S. Gov't Printing Office, 1979. DHEW Publ. No. (PHS) 79-50066.

U.S. Dept. of Health, Education, and Welfare, Mortality Trends for Leading Causes of Death, United States, 1950–69, DHEW Publication No. (HRA) 74-1853, Rockville, MD (1974).

U.S. EPA (1984), *National Air Quality and Emissions Trends Report, 1982*. Report EPA-450/4-84-002, March 1984. U.S. Environmental Protection Agency, Research Triangle Park, NC.

U.S. EPA (1986), *National Air Quality and Emissions Trends Report, 1984*. Report EPA-450/4-86-001, April 1986. U.S. Environmental Protection Agency, Research Triangle Park, NC.

Ware, J., Thibodeau, L.A., Speizer, F.E., Colome, S., and Ferris, B.G., Jr. (1981), Assessment of the Health Effects of Atmospheric Sulfur Oxides and Particulate Matter: Evidence from Observational Studies, *Env. Health Persp.* 41:255–76.

Watson, J.G., Chow, J.G., and Shah, J.J., Analysis of Inhalable and Fine Particulate Matter Measurements, EPA-450/4-81-035, December 1981, U.S. Environmental Protection Agency, Research Triangle Park, NC 27711.

White, W.C., and Marcy, C.H. (1913), A Study of the Influence of Varying Densities of City Smoke on the Mortality from Pneumonia and Tuberculosis, in *Trans. Fifteenth International Congress on Hygiene and Demography, Vol. III*, pp. 1020–27, U.S. Gov't Printing Office, Washington, DC.

Winkelstein, W., Jr. (1985), Some Ecological Studies of Lung Cancer and Ischaemic Heart Disease Mortality in the United States, *Int. J. Epidemiol.* 14:39–47.

Winkelstein, W., Jr., Kantor, S., Davis, E.W., Maneri, C.S., and Mosher, W.E. (1967), The Relationship of Air Pollution and Economic Status to Total Mortality and Selected Respiratory System Mortality in Men I. Suspended Particulates, *Arch. Env. Health* 14:162–71; II. (1968) Sulfur Oxides, *Arch. Env. Health* 16:401–5.

Winkelstein, W., Jr., and Kantor, S. (1969), Stomach Cancer, Positive Association with Suspended Particulate Air Pollution, *Arch. Env. Health* 18:544–7.

Winkelstein, W., Jr., and S. Kantor (1969), Prostatic Cancer: Relation to Suspended Particulate Air Pollution, *Am. J. Public Health* 59:1134–38.

Winkelstein, W., Jr., and Gay, M.L. (1970), Arteriosclerotic Heart Disease and Cerebrovascular Disease. Further Observations on the Relationship of Suspended Particulate Air Pollution and Mortality in the Erie County Air Pollution Study, in *Proceedings Inst. Env. Sciences*, 16th Annual Meeting, Boston, pp. 441–47.

Winkelstein, W., Jr., and Gay, M.L. (1971), Suspended Particulate Air Pollution, Relationship to Mortality from Cirrhosis of the Liver, *Arch. Environ. Health* 22:174–77.

Zeidberg, L.D., Horton, R.J.M., and Landau, E. (1967a), The Nashville Air Pollution Study V. Mortality from Disease of the Respiratory System in Relation to Air Pollution, *Arch. Env. Health* 15:214–25.

Zeidberg, L.D., Horton, R.J.M., and Landau, E. (1967b), The Nashville Air Pollution Study V. Cardiovascular Disease Mortality in Relation to Air Pollution, *Arch. Env. Health* 15:225–36.

APPENDIX 7A THE U.S. NATIONAL LONGITUDINAL MORTALITY STUDY

The U.S. federal government is conducting a long-term prospective study of about 1.3 million persons (Rogot *et al.*, 1992). Census records are matched with death certificates to provide socioeconomic and demographic data on an individual basis. There are 12 cohorts in the study, which vary by size and by year of entry into the study. The longest period of followup available thus far is 7 years. The method of analysis is to compute standardized mortality ratios (SMRs) based on the numbers of deaths expected from all cohorts and then to examine the variation of these SMRs according to the various population descriptors available. These include race, geographic location (state and the 50 largest SMSAs), SMSA status, nativity (U.S. versus foreign-born), Hispanic status, education, income, household size, marital status, employment status, major occupation, and major industry. Tabular data and cross-plots are presented by Rogot *et al.* for white males, white females, black males, and black females. No data are given for multiple factors, such as income and location.

The results of these cross-tabulations confirm most of the expected trends:

- Mortality in the age 25–64 group is inversely related to education. However, the relationship is not monotonic: the highest SMRs (averaging 119) are for those who failed to complete high school. Additional education beyond high school continues to improve life expectancy for males, but only until 4 years of college for females. The average SMR for all cohorts with postgraduate education was 64.

- Family income has the largest effect at the lowest levels for ages 25–64. However, additional income continues to improve life expectancy, even at the highest family income level ($50,000+). The curves for whites and blacks are qualitatively similar, suggesting that income has a more direct effect than, say, education.
- Household size for ages 25–64 is only important for single person households and group quarters (institutions). For age 65+, household size has no effect for men but tends to increase mortality for females.
- Married people (ages 25–64) have the highest life expectancies, for both races and sexes.
- Employment increases life expectancy (ages 25–64); being employed is preferable to doing housework, for both sexes and races.

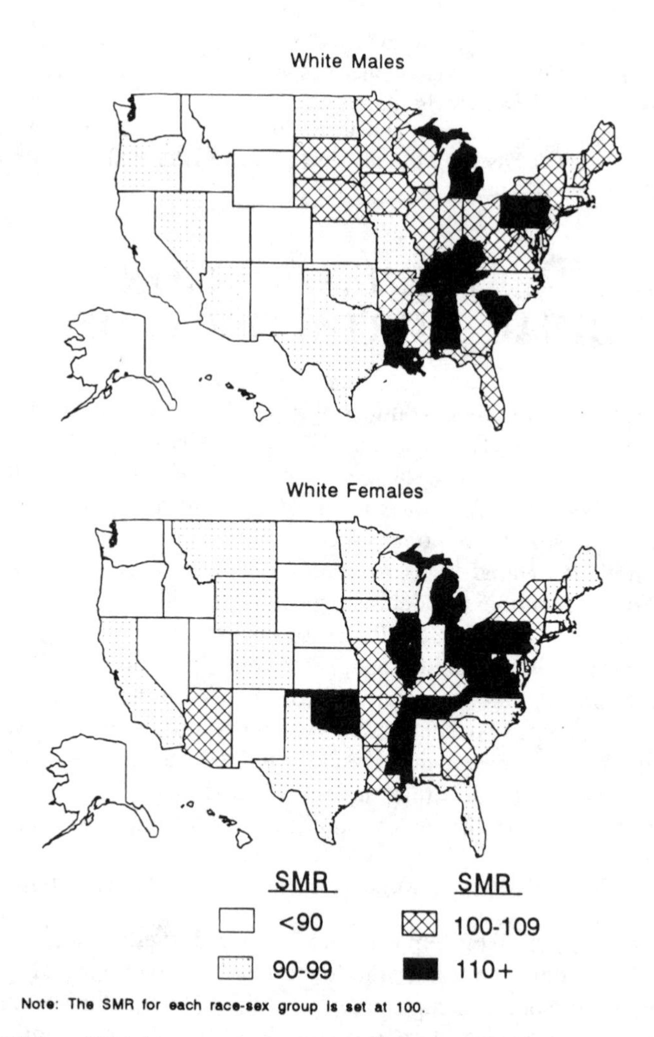

FIGURE 7A-1. SMRs for acute myocardial infarction for white males, 1979–85, all ages, by state. *Source: Rogot* et al. *(1992)*.

- There is a slight advantage in living in the suburbs of an SMSA (ca. 2%), in terms of mortality from all causes, and a slight disadvantage (ca. 2%) in living in the central city of an SMSA. However, these findings should be adjusted to account for higher homicide rates in central cities and higher accident rates outside SMSAs.
- Foreign-born members of the cohorts died at lower rates than those born in the U.S.
- Hispanics, especially Cubans, had longer life expectancies than non-Hispanics (SMR = 0.84). This was reflected in lower rates for cancers and for cardiovascular causes.
- For white males, professional/technical workers had the lowest mortality, followed closely by farmers. Service workers, laborers, and transport workers had the highest mortality. Social and recreation workers had high SMRs for females but not for males. By industry, the highest mortality occurred in wood products, laundering, advertising, rubber, some transportation categories, miscellaneous manufacturing, and mining. Postal service employment had a high SMR for females and a low value for males.
- By location, white all-cause mortality was highest in Appalachia and Nevada (male SMRs over 110) and lowest in the upper Midwest and Great Plains states. Hawaii was very low (SMR less than 90) for both sexes, but Alaska was high for males. If these trends were adjusted for mortality due to external causes, SMRs for the Western states would be reduced by 5–10% relative to the East.
- Cardiovascular mortality (white male and female combined) was highest in Appalachia and the Southeast (SMRs over 110) and lowest in the Northwest

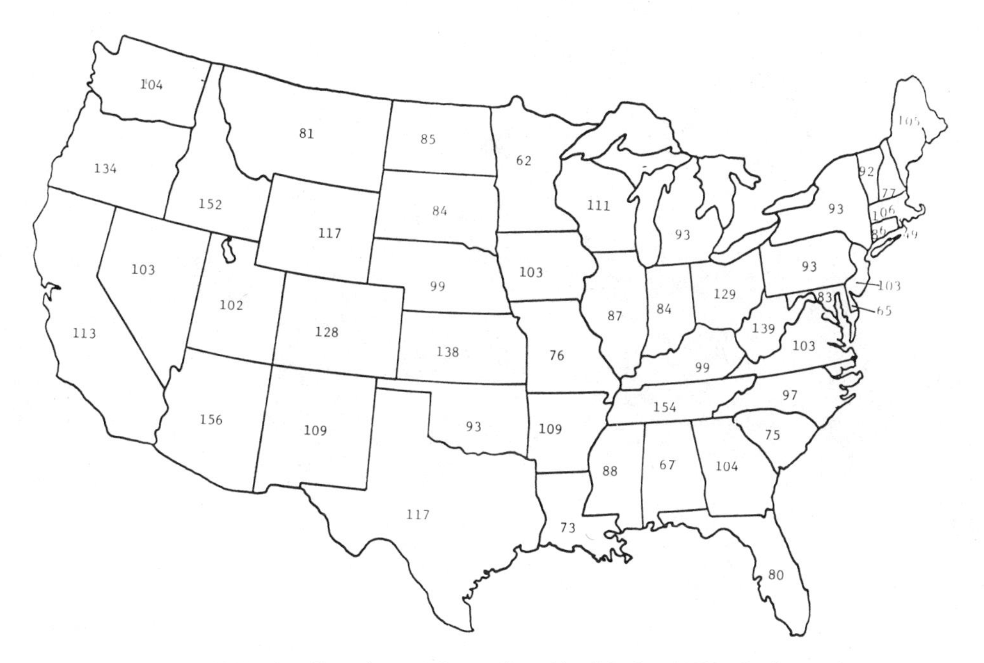

FIGURE 7A-2. SMRs for all respiratory disease for white females, 1979–85, all ages, by state. *Data source: Rogot* et al. *(1992).*

(SMRs less than 90). Heart attack mortality (Figure 7A-1) for white males followed somewhat similar patterns but tended to be higher in the Southeast and also in the upper Midwest. Heart attacks were quite low in the West.
- Respiratory disease mortality (ICD 460–519 [excluding lung cancer]) was high for white females in portions of Appalachia, and in portions of the West (Figure 7A-2).
- High mortality SMSAs for males (all causes) included New Orleans, Cincinnati, Birmingham, and Pittsburgh. For females, high SMRs were seen in Columbus (OH), Toledo, Dayton, St. Louis, and Albany.

REFERENCE

Rogot, E., Sorlie, P.D., Johnson, N.J., and Schmitt, C. (1992), *A Mortality Study of 1.3 Million Persons*, NIH Publication No. 92-3297, National Heart Lung and Blood Institute, Bethesda, MD.

8

Long-Term Temporal Studies of Mortality

Chapters 5 and 6 examined studies that demonstrated the existence of the adverse effects of air pollution on human mortality and provided estimates of their magnitude on a short-term or event basis. Chapter 7 discussed the evidence for long-term effects and the difficult methodological problems that cross-sectional studies pose. Quantitative comparison of these two different types of studies is important for two reasons:

- Time-series studies provide no indication of the degree of prematurity of deaths hastened by air pollution. Although the studies of lag structure seem to indicate that these increases in death rates are not accompanied by corresponding decreases within a few days or weeks, they lack the statistical power to explore longer delays in response. To do this, one must examine whether the annual rates are affected similarly by acute air pollution events, which generally requires a cross-sectional approach.
- As pointed out by Lipfert (1978) and Evans *et al.* (1984), if short-term mortality increases are not canceled by subsequent decreases, the cross-sectional analysis of annual mortality rates should show at least this amount of increase. Only incremental death rates beyond the annual sum of short-term effects should be considered "chronic." This is an important distinction from an etiological perspective and in terms of our initial hypothesis (Chapter 1), that community air pollution tends to exacerbate existing illness rather than to create disease *de novo*.

This chapter examines analyses that feature aspects of both cross-sectional and time-series methods, as a way of "bridging the gap" between studies of chronic and of acute effects. One of the common uses of the quantitative relationships between air pollution and health being examined here is in estimating the benefits of air pollution controls or of the potential adverse effects of new industrial development. It is thus crucial that such relationships be transferable beyond their original contexts. One way to improve the reliability of this process is to require consistency with actual case studies of long-term changes.

In addition, seasonal trends in both air pollution and mortality are examined in this chapter, since one of the major factors in improving urban air quality has been improvements in winter due to use of cleaner fuels for space heating. As a result, peak periods for some pollutants have shifted to summer. The chapter tries to deduce whether there are corresponding shifts in the temporal patterns of mortality.

MORTALITY STUDIES INVOLVING LONG-TERM TEMPORAL CHANGES

Studies Involving Changes in Pollution Emission Levels

Dublin

Leonard *et al.* (1950) examined weekly records of air pollution and deaths from "respiratory" diseases (not otherwise specified) in Dublin during a period of changing fuel use. During World War II, there was a shift from coal to peat as the main space heating fuel in the city. Smoke and SO_2 levels dropped as a result, but the seasonal cycle in respiratory deaths was essentially unchanged. After the war, coal burning resumed and pollution levels increased again, but to lower levels than those experienced before the war. Leonard *et al.* point out the lack of concordance between winter peaks of pollution and respiratory mortality as evidence of the lack of a relationship. However, there are some complicating factors that may bear on this conclusion. First, there is an overall downward trend in both respiratory deaths and air pollution, as shown most clearly by the minimum (summer) levels. Second, the absolute numbers of deaths were small, especially in summer. The winter peaks undoubtedly reflect mainly epidemic periods; it should surprise no one that there are important time-dependent factors in respiratory disease *other* than air pollution. It is unfortunate that all-cause or heart disease mortality data were not given, since the larger numbers involved might have permitted the extraction and separation of the three explanatory factors of interest: seasonal cycle, secular trend, and air pollution. No definitive conclusions are thus possible from this study.

Charleston, SC

Jacobs and Langdoc (1972) studied changes in long-term mortality in parts of Charleston, SC, as a function of both time and place. They identified a portion of the county that was heavily industrialized and that underwent an extensive cleanup from 1968 to 1970. Annual TSP decreased from 228 to 120 $\mu g/m^3$ during this period; there were undoubtedly changes in other air pollutants as well, but no data on other species were reported. The authors contrasted the mortality experience of this area (about 38% of the total) to the whole county and to nine adjacent census tracts (about 20% of the total), for which a TSP monitor about four miles away from the industrial area was deemed representative. The mean TSP values were 74 and 55 μg/m^3 for 1968 and 1970, respectively. Significant changes were noted in deaths from all causes, for heart disease and stroke (all ages and sexes) and for heart disease for both males and females, aged 45–64. The authors reported that there were no significant population shifts during this 2-year period. However, the spatial comparison neglected possible confounding effects due to racial and socioeconomic differences and possible differences in smoking habits. Since these confounders are unlikely to have changed significantly in the 2-year period, I felt that a better comparison would be one taken over time within each detailed area, using the change over time in the entire county as a reference. On this basis, there were substantial decreases in the mortality ratios for both male and female heart disease deaths (age 45–64) and in total deaths from all causes in the polluted area, but not in the comparison area. The slopes of these changes with respect to TSP were similar to the values obtained by Winkelstein *et al.* (1967) and Gregor (1977). The fact that these changes represent long-term increments over time rather than over space lends important support to the causal hypothesis and suggests that pollution affects mortality rather quickly. It is possible that the mortality effects observed in Charleston are acute rather than chronic, and thus it would be interesting to perform a time-series analysis. However, such an analysis of daily or weekly deaths might be problematic because of the small population involved.

San Francisco Bay Area

Effects on mortality due to reductions in motor vehicle use were inferred by Brown *et al.* (1975), based on a sudden drop in rates in two counties in the San Francisco Bay area in the first quarter of 1974. The decreases were 8 to 13% in total mortality (after subtracting automobile accidents), 11 to 17% in cardiovascular causes, and 33 to 38% in respiratory causes. Decreases of this magnitude had not occurred during the previous 4 years. They reported that the changes were greatest in the under-45 age group, which would be consistent with "differential exposure." The corresponding reduction in retail gasoline sales was 9.5%, but the authors were of the opinion that emissions could have been reduced even further because of changes in usage patterns.

Their conclusions were challenged by Waller and Lawther (1976), who urged caution before making conclusions on the basis of a single quarter's data and in the absence of data on the actual changes in ambient air concentrations. They noted no change during this period in London, when there were similar limitations on the uses of fuel and a 3-day workweek was imposed. However, they also provided data showing a marked drop in mortality in the first quarter of 1957, during which fuel rationing had been imposed in London as a result of the Suez Canal crisis. Reductions in CO levels were also noted at that time. They felt that changes in weather and in influenza mortality were more likely factors than changes in motor vehicle emissions. Brown *et al.* responded (1976) that other social disruptions in London in 1974 (associated with 3-day workweek, for example) may have masked the effect there and that weather or influenza factors did not appear to be important in Californiaa in 1974. I noted that 1975 mortality rates were also lower than 1973 (higher than 1974 in one county and lower in the other), so that the 1974 changes did not appear to be a singularity. Thus, I concluded that the hypothesis of Brown *et al.* remains an interesting speculation, pending further detailed analysis.

Hokkaichi, Japan

Imai *et al.* (1986) combined cross-sectional and time-series analysis of mortality in the area of Hokkaichi, Japan. Trends in age- and cause-specific mortality in a "polluted" area were compared to those in a "clean" area, from 1963 to 1983. The dependent variables were mortality by 20-year age groups for bronchitis and bronchial asthma; the independent variable was SO_2 (measured by sulfation plates). The authors concluded that mortality from bronchitis and bronchial asthma responded to improvements in air quality and also showed a corresponding spatial gradient.

This study has only limited application to community air pollution. First, the "polluted" area was also known for the presence of sulfuric acid mist (Kitagawa, 1984), which is characterized by large particles and thus differs from H_2SO_4 formed in the atmosphere. This industrial pollutant was probably more irritating than the SO_2 present and was not specifically considered by Imai *et al.* Second, the specific causes of death involved are not very important in the United States (chronic bronchitis deaths = 0.17% of total; bronchial asthma deaths are not listed as such). Sulfur dioxide was measured by sulfation plates, which would also be sensitive to sulfuric acid mist, as well as to other environmental factors, including wind speed. According to Kitagawa (1984), H_2SO_4 was present in the area from 1965 and at least until 1969, even though pollution control equipment was reportedly installed in 1967. A ratio of 0.48 between SO_3 and SO_2 was reported, which would mean that there may have been as much as 40 to 60 $\mu g/m^3$ acid in the polluted area (annual average). Such high acid levels do not exist in community air in the United States, where the H_2SO_4/SO_2 ratio is closer to 0.05 to 0.10 (Lipfert, 1988).

The strongest mortality effects for bronchial asthma were for children, given as the 0–19 age group. Infant mortality (<1 year) usually dominates this age group and also has a strong socioeconomic component. Since one might expect a lower socioeconomic group to reside in the polluted area, the mortality gradient shown may be only coincidentally associated with the pollution gradient. In addition, the numbers of deaths in each category were quite small; a change in diagnosis of one or two deaths per year could have changed the outcome of the analysis.

Since a court case had been settled in favor of exposed people in this area (1972), there was probably a high degree of public awareness of respiratory health issues that could have influenced cause-of-death diagnoses at that time. Verification of cause of death by autopsy was not mentioned. Death rates from bronchitis in the polluted area peaked in 1975 and had returned to about the same as 1965 levels by 1977. There was also a strong increase in reported bronchitis deaths in the "clean" area from 1972 to 1975, possibly also because of increased public awareness.

Age adjustment was done on the basis of the 1935 distribution of the Japanese population, which may have been substantially different from that during the period of study, given the impact of World War II. Smoking habits were not taken into account in this analysis, nor were deaths given by gender. Thus, it is not possible to judge whether either temporal changes or spatial gradients in smoking habits might have influenced the results. Bronchitis mortality increased dramatically in the "clean" area, especially after 1974 (by a factor of 3 for the over-60 age group), implying the presence of other factors not accounted for in the analysis.

In spite of what may be an effect on respiratory health in the area, including morbidity effects as reported by Kitagawa, Imai (personal communication; May 31, 1986) reported no effects on heart disease. Thus, their findings do not support those U.S. studies that associate sulfates with increased heart disease mortality (discussed in Chapter 7).

Utah

Archer (1990) examined the spatial differences in long-term trends in lung cancer and nonmalignant respiratory disease in three Utah counties: one containing a steel mill that began operating in the early 1940s (Utah County), a nearby urban county (Salt Lake County), and a nearby "background" county (Cache County). Mortality rates were age adjusted, and estimates of county-wide smoking rates were obtained by survey. Air pollution was considered only indirectly, by contrasting the time history of disease in Utah County with the other two counties. Archer concluded that much of the 50 to 90% mortality increase in the polluted county, which began about 20 years after the mill began operating, could be attributed to increased air pollution. Salt Lake County experienced even higher increases in lung cancer, but this county was more urban and had a higher percentage of smokers. For nonmalignant respiratory disease, the largest increase over the rural control (Cache County) was seen in the county with the new steel mill. It is interesting to note that in 1986 for persons 65 and over, the Provo-Orem SMSA (which comprises Utah County) had the lowest lung cancer mortality rate of all SMSAs (HFCA, 1990).

Daily mortality variations in Utah County from 1985 to 1989 were studied by Pope *et al.* (1992), a period that encompassed a 13-month shutdown by the major source of particulates in the area. (This study was also discussed in Chapter 6.) The regression model used accounted for daily variations in PM_{10}, temperature, and humidity and included a linear time variable, which was not statistically significant. Thus there was no independent secular mortality trend in these data. The model predicted an average mortality improvement of 2.2% $\pm$ 0.9% due to the shutdown of the source, by virtue of a drop in PM_{10} of 15 $\mu g/m^3$. The actual improvement was 3.2%, which provides an excellent check and confirmation that the excess daily deaths during periods of high pollution were not subsequently canceled by unnoticed mortality reductions due to premature "harvesting." Although sulfates and nitrates were not measured during this period, Caka *et al.* (1992) reported data from 1990 that provide some idea of what the levels might have been. The maximum SO_4^{-2} level reported was about 11 $\mu g/m^3$; NO_3^-, 29 $\mu g/m^3$ (compared to the maximum PM_{10} level of 365 $\mu g/m^3$). Sulfates and nitrates thus do not appear to be important components of the aerosol at Provo.

Analysis of Long-Term Trends in New York City Mortality

In the early to mid-1960s, New York City was one of the most polluted major cities in the United States. By mandating higher quality fuels for space heating

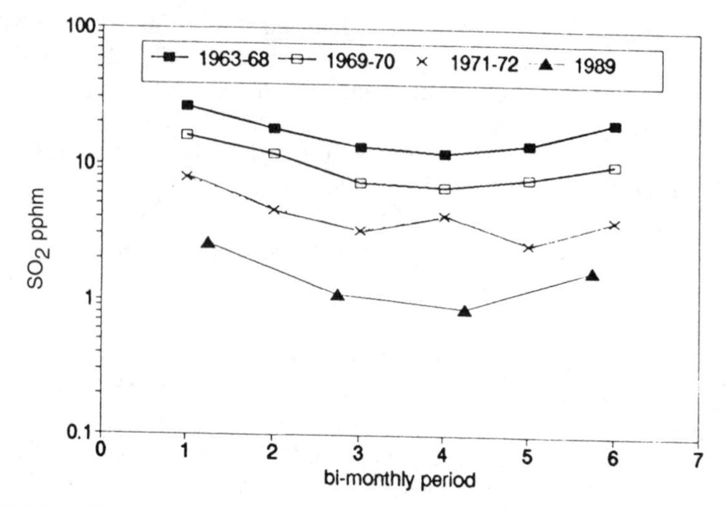

FIGURE 8-1. Changes in seasonal average SO_2 concentrations in New York City, 1963–89. *Data from Schimmel* et al. *(1974) and the New York State Department of Environmental Conservation.*

and industry, the city reduced sulfur dioxide (SO_2) air concentrations by a factor of 7 by 1972 and another 50% since then, as shown in Figure 8-1. Particulate air pollution (measured as smokeshade = coefficient of haze [COH]) was initially reduced more slowly (Figure 8-2) and is more difficult to compare with recent years because of changes in measurement methods. Figure 8-3 shows the seasonal cycles in New York City mortality for various time periods. It appears that the substantial improvements in air quality that have been realized have not materially altered the amplitude of the seasonal mortality cycle, at least not for all causes of death combined.

I used the data tabulated by Schimmel *et al.* (1974) for New York City to explore these relationships as deduced from long-term temporal variations, in this case, 60 two-month periods. During a 2-month period, minor daily temperature variations would be expected to cancel out; the net temperature effect constitutes the seasonal variation. The data show that during this 10-year period, the climate was quite uniform but that 1990 appears to have had a warmer winter. One could also argue that air pollution effects might be expected to cancel out in the long term if the indicated excess deaths were premature by only a few days or weeks. Lag studies of daily mortality (Chapter 6) have ruled out compensation within a few days (Schimmel, 1978; Wyzga, 1978) but lack the statistical power to look further out in time. If the air pollution effects were indeed "excess," they should show in longer term analyses as well.

However, problems with temporal confounding can arise with long-term temporal studies, as discussed in Chapters 3 and 6. In addition to seasonal cycles, over 10 years the population can change, improvements in medical care are made, and epidemics and weather anomalies can occur. In order to confound, an exogenous factor must affect the outcome (mortality) *and* be correlated with the independent variable of interest (air pollution). By and large, epidemics and weather anomalies do not meet the latter requirement. In

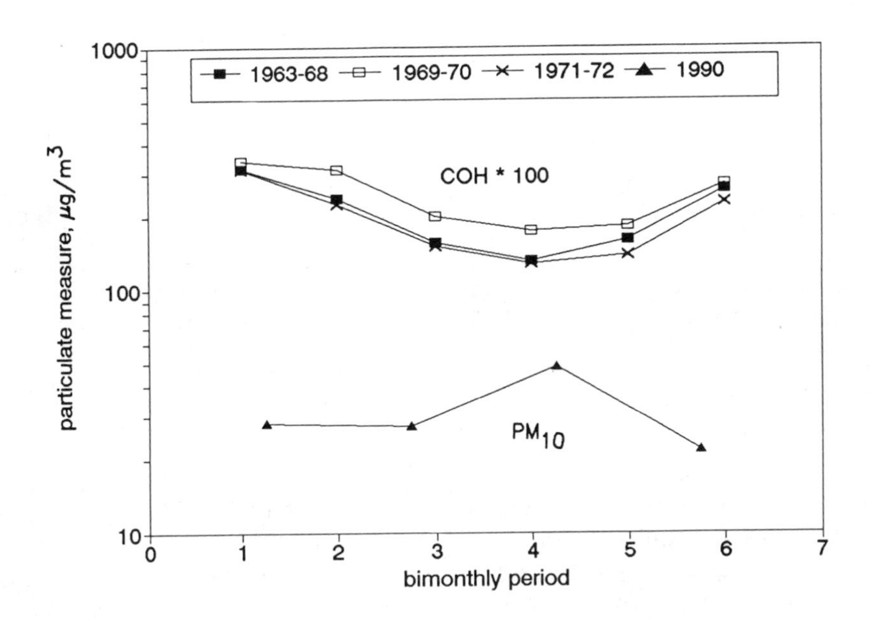

FIGURE 8-2. Changes in seasonal average particulate concentrations in New York City, 1963–89. *Data from Schimmel* et al. *(1974) and the New York State Department of Environmental Conservation.*

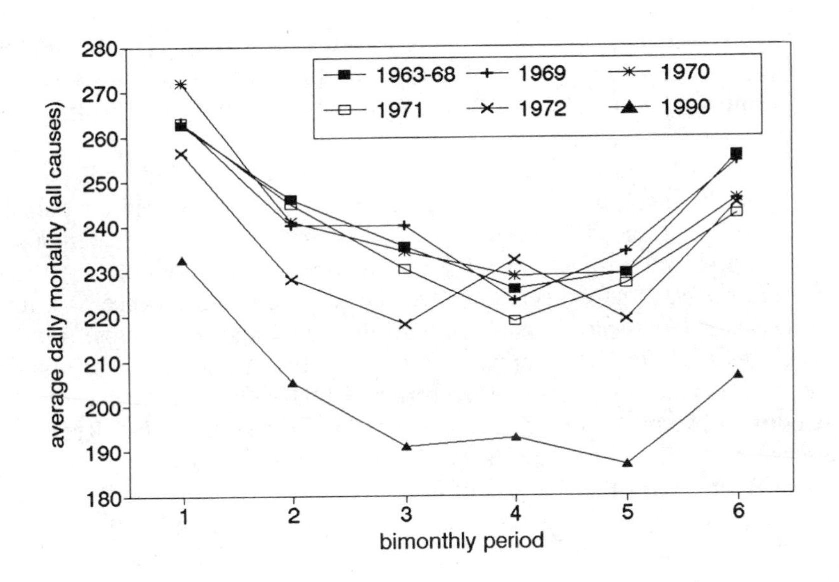

FIGURE 8-3. Changes in total mortality in New York City, 1963–89. *Data from Schimmel* et al. *(1974) and Mortality and Morbidity Weekly Report (data for 1990).*

TABLE 8-1 Seasonal Regressions of New York City Mortality, 1963–1972 (60 bimonthly periods)

	Regression Coefficients (as elasticities)			
Variable	Temperature	Year	SO_2	COH
All causes ($R^2 = 0.73$)	−0.10*	−0.002	0.017	0.065**
Respiratory ($R^2 = 0.61$)	−0.24	0.87	0.039	0.28**
Heart disease ($R^2 = 0.78$)	−0.17*	−0.06	0.014	0.060
Other circulatory disease ($R^2 = 0.80$)	−0.11	−1.45*	0.069*	0.089
Other causes ($R^2 = 0.40$)	−0.02	0.33**	0.003	0.032
Resp./other ($R^2 = 0.60$)	−0.21	0.56	0.035	0.25*
Heart/other ($R^2 = 0.76$)	−0.15*	−0.40**	0.010	0.027
Circ./other ($R^2 = 0.82$)	−0.09	−1.80*	0.10*	0.034

*$p < 0.01$.
**$p < 0.05$.

my regression analysis, I used the mean temperature (averaged over 2 months) as the seasonal indicator and a linear time variable ("year") to represent gradual changes in population and medical care. In addition, as an alternative analysis, I normalized the average daily death counts for respiratory causes, heart disease, and "other circulatory" causes by dividing by the counts for all other causes. All of these data were unadjusted for age. The results are given in Table 8-1.

Table 8-1 provides some interesting information not usually available from a daily perturbation analysis:

- All-cause mortality was significantly affected only by temperature (season) and by particulates, to about the same extent as predicted by the first Schimmel *et al.* study (1974). During the 1963–72 period, there was no significant temporal trend in all-cause mortality beyond that resulting from improvements in air quality.
- Respiratory mortality was increasing during this time period (it is also currently on the increase nationwide) after the effects of reduced air pollution were accounted for. As expected, this was the most strongly affected diagnosis category, and particulates were the most important pollutant. Although temperature had the expected sign, it did not reach significance as a linear term. Since a crossplot versus temperature suggested a U-shaped relationship (opposite effects in summer and winter), the relationship was refit with linear and quadratic temperature terms; COH remained significant (after deletion of two outliers that were presumed to be flu epidemics). Respiratory disease mortality had an elasticity about four times that of total mortality, which is similar to the findings of Winkelstein *et al.* (1967) in Buffalo on a cross-sectional basis, in contrast to the findings of Lave and Seskin (1978).
- Heart disease mortality was not significantly associated with air pollution, using this linear model. However, additional regressions using the log of SO_2 or the product $SO_2 \times$ COH did reach significance. Temperature was very important for heart disease mortality, but there was no significant temporal trend.
- The "other circulatory disease" category, which includes stroke and hyper-

tensive disease, showed a significant temporal decrease. However, part of this may be due to a change in the International Classification of Diseases (ICD) coding in 1969, which could also affect the correlation with SO_2. Nevertheless, addition of a dummy variable delineating a step change at 1969 did not change the significance of the SO_2 variable, and the results of the bimonthly analysis agreed with those of the daily analysis (Schimmel *et al.*, 1974).

- The "other" category was not associated with air pollution, as expected. It also showed a significant increase with time, probably because of demographic changes to the population over the 10-year period.
- Using the "other" category to account for the expected long-term temporal changes did not change any of the pollution associations. It did show a temporal decrease for heart disease, which was expected from national trends (Stamler, 1985).

Examples of some of the resulting "dose-response" functions are given in Figures 8-4 to 8-6. These plots show reasonably well-behaved trends that are not dominated by a few outliers; they were made by adjusting the dependent variables to 1970 and to 50°F, using the regression coefficients from the multiple regressions. Although use of the logarithmic transform for SO_2 improved the fits for heart disease and circulatory causes, this was not the case for respiratory causes. Respiratory mortality was adversely affected by cold in winter and by heat in summer. Given the experience with confounding of SO_2

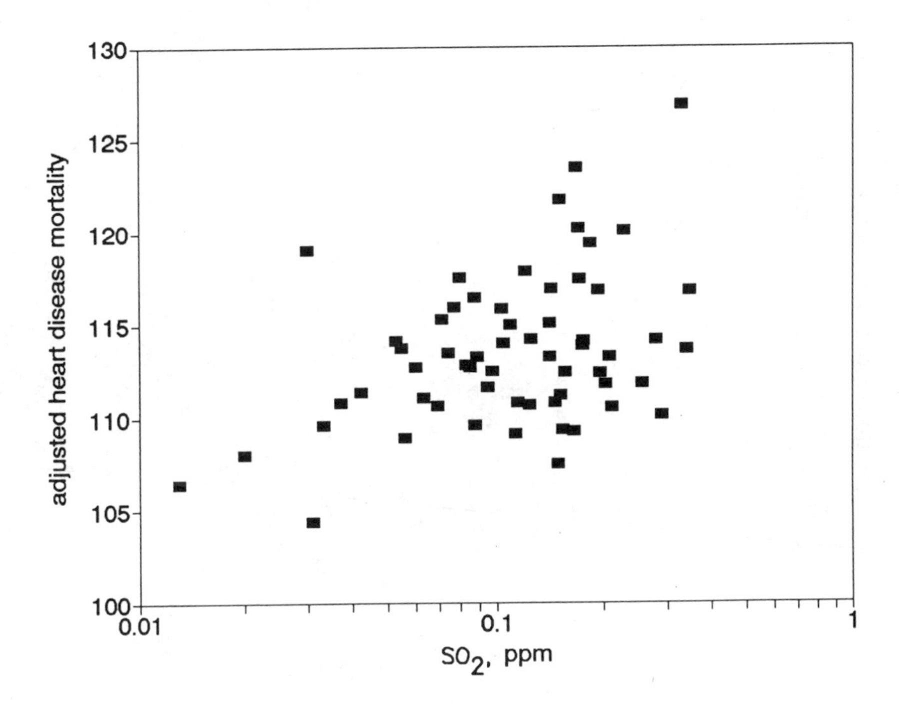

FIGURE 8-4. Adjusted heart disease mortality in New York City versus SO_2, 1963–72. *Data from Schimmel* et al. *(1974).*

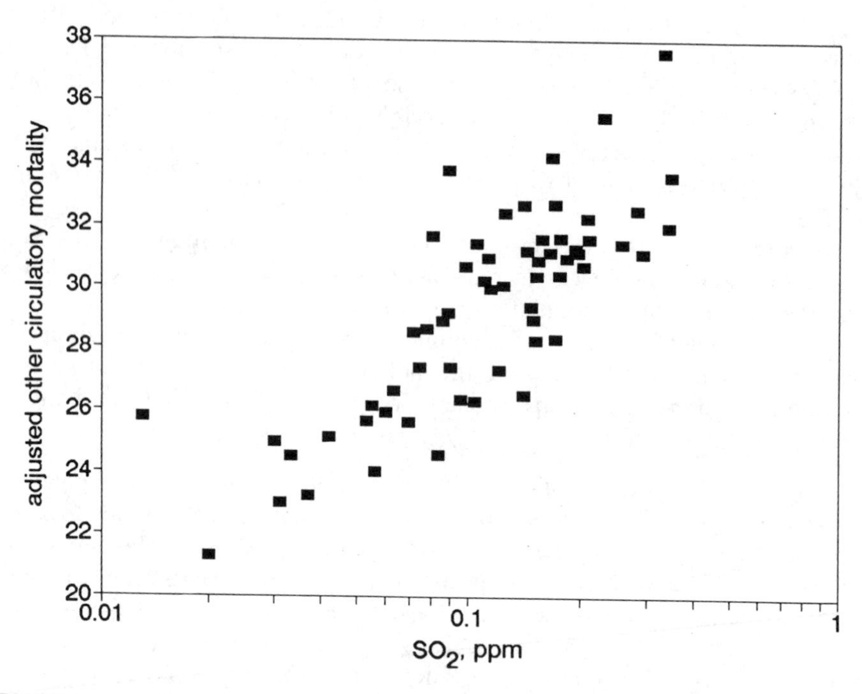

FIGURE 8-5. Adjusted other circulatory mortality in New York City versus SO_2, 1963–72. *Data from Schimmel* et al. *(1974).*

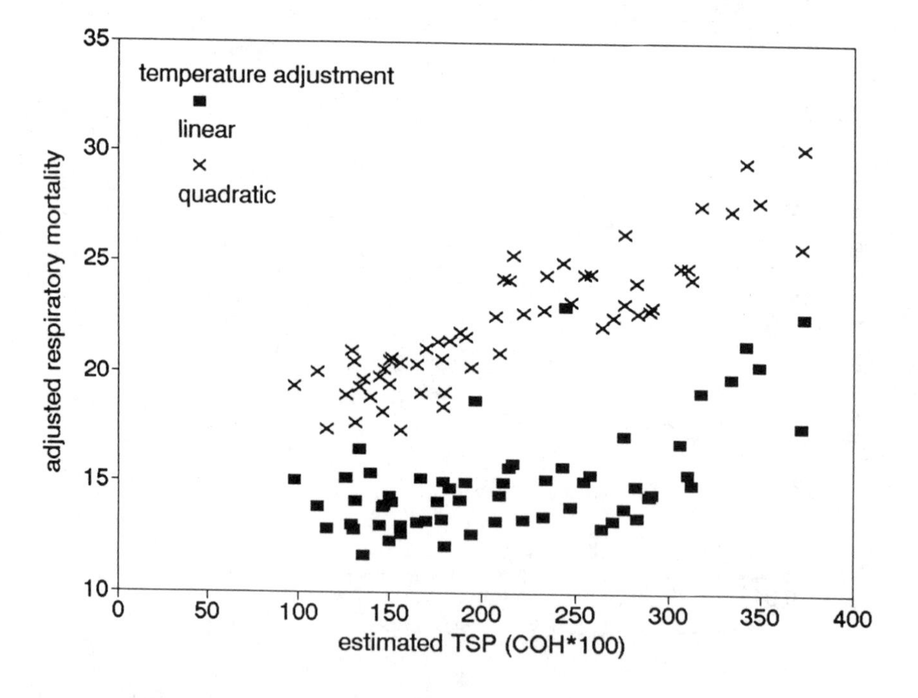

FIGURE 8-6. Adjusted respiratory system mortality versus COH, for two different temperature corrections. *Data from Schimmel* et al. *(1974).*

by temperature shown in the Dutch studies discussed in Chapter 6, these results should probably be regarded as suggestions for further detailed study. However, this reanalysis suggests that the effects of daily air pollution changes may also be seen in the longer term, implying that air pollution previously shortened lives in New York City by more than a few weeks or months.

Air pollution in New York has now been reduced by about 90% since the peak level of the 1960s, which should correspond to a reduction in the daily death rate of 5.85% or 13.5 deaths per day in summer. Figure 8-2 shows that this theoretical amount of reduction is well within the actual reduction that has been experienced. If this level of reduction has been in place for 10 years or more, an estimated 50,000 lives would have been prolonged by at least a few months.

Combined Cross-Sectional Time-Series Studies

Lave and Seskin

In their 1978 compendium of regression analyses, Lave and Seskin recognized the need to examine temporal as well as spatial variability. They identified 26 U.S. SMSAs having data from 1960 to 1969 and pooled their annual data on mortality and air pollution into one large data set. Socioeconomic factors were obtained by interpolating linearly between the census years. They explored three ways to deal with long-term temporal variability not otherwise represented (such as improvements in medical care): first, the U.S. national mortality rate was used as a baseline for temporal changes. Second, a continuous (linear) time variable was used. In the third method, dummy variables for each year were added to the regression. This approach is the most general, because it allows for anomalous years that may not fit the linear pattern. However, none of these approaches allows for spatial interactions in temporal nonuniformity (all SMSAs may not behave the same way). They used the same basic linear model and pollutant definitions that had been used in their other cross-sectional analyses (discussed in Chapter 7).

Lave and Seskin's pooled results confirmed the significance of the air pollution effect on mortality but also showed that their "minimum" sulfate variable was more sensitive to model specification changes (i.e., the three ways of accounting for temporal changes) than was the TSP variable. The sulfate elasticity was cut in half relative to the previous cross-sectional regressions for 1960 and 1969 (which were based on approximately 100 locations), while the TSP elasticity increased slightly. Christainsen and Degen (1980) examined the validity of the statistical procedures involved in Lave and Seskin's pooled analysis by performing regressions for each year that had complete data and by making estimates of coefficients, using more general procedures not requiring least-squares assumptions. The year-by-year results showed that 1960 was an outlier for "minimum sulfate" (probably because of the flawed data used for that year) and that 1962 was an outlier for TSP. As a result, Christainsen and Degen concluded that the hypothesis of a pollution-mortality relationship that is constant over time must be rejected, although they agreed that such relationships were indeed significant. Constancy over time is one of the requirements of a pooled analysis, if valid tests of significance are to be made.

I plotted the ordinary least-squares pollution coefficients from Christainsen

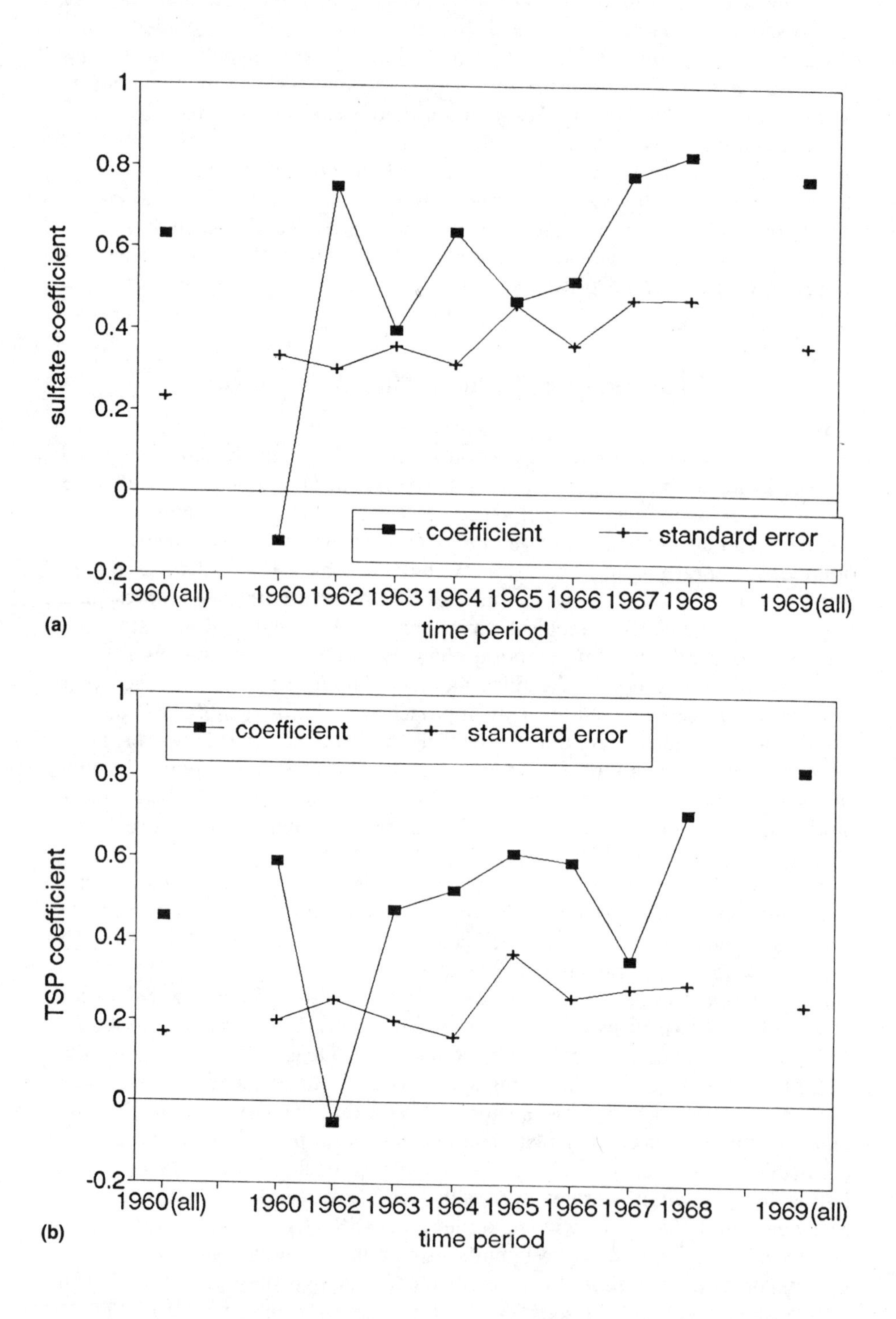

FIGURE 8-7. SMSA mortality-air pollution regression coefficients for various time periods. (a) Sulfate aerosol. (b) TSP. *Data from Christainsen and Degen (1980).*

and Degen's analysis for eight individual years along with the Lave and Seskin results for 1960 and 1969 (Figure 8-7). Figure 8-7a shows that the standard errors of the sulfate coefficients are more or less in line with the results from the two larger data sets. However, the regression coefficients vary greatly, probably reflecting the vagaries of year-to-year sampling in such a small data set. Figure 8-7b shows similar data for TSP; the coefficients are seen to be somewhat more regular. The sulfate coefficient and its standard error obtained from the original pooled analysis (0.33 and 0.1, respectively) were both much lower than the values shown on Figure 8-7a, which averaged 0.53 ± 0.10 (standard error of the mean of eight values). The pooled TSP coefficient was about the same (0.51) as the values given on Figure 8-7b, and its standard error was about 0.07, compared to 0.47 ± 0.08. This seems to imply that the pooled analysis is measuring something different from the regressions for individual years for sulfate, but not for TSP.

Neither Lave and Seskin nor Christainsen and Degen seemed to realize that the use of "minimum" sulfate measure may have been responsible for this instability. After 1961, each year's sulfate record was based on only 26 samples; the smallest value of such a data set is not a robust statistic. Use of the mean SO_4^{-2} values, as was done for TSP, may have been more stable. In addition, it would have been interesting to examine the temporal effects *per se*, as "minimum" sulfate dropped from 4.7 to 3.4 $\mu g/m^3$ and mean TSP dropped from 118 to 96 $\mu g/m^3$, in order to partition temporal changes from spatial differences. (This was probably not a real change, since the mean value of sulfate increased slightly from 1960 to 1969. More likely, the apparent change in minimum SO_4^{-2} reflects the differences in the sampling schedules used in 1960 and discussed in Chapter 7.)

Ontario

In a combined time-series cross-sectional study of nine Ontario counties, Plagiannakos and Parker (1988) studied changes in mortality rates from 1976 to 1982. Their model included SO_2, TSP, and sulfate aerosol, as well as variables for smoking, alcohol consumption, medical care, income, and education. Their approach pooled cross-sectional and time-series data and employed a dummy variable for the temporal trend (it was significant and negative). The counties were selected on the basis of the availability of air monitoring data and included metropolitan areas and the industrial areas of Algoma and Sudbury. The time steps were annual. The authors reported also using a dummy variable for each county, but since these variables were not significant, they were dropped from the analysis. They analyzed total mortality (excluding injury and poisoning), respiratory disease mortality, and hospital admissions for all causes and for respiratory causes. They used a logarithmic model, for which the regression coefficients are numerically equal to elasticities (Chapter 3) and the effects are multiplicative rather than additive. Thus air pollution would have a larger absolute effect in the presence of other stress factors such as poverty, smoking, older populations, and so on.

Air pollution data were obtained from routine monitoring stations, ranging from two to twelve per county for SO_2, two to 27 for TSP, and one to eleven for sulfate. Air quality improved significantly during the period studied. In 1976, annual average SO_2 ranged from 25 to 65 $\mu g/m^3$ and 8 to 30 $\mu g/m^3$ in

TABLE 8-2 Regression Results for Southern Ontario

	Mortality		Hospital Admissions	
	All Causes	Respiratory	All Causes	Respiratory
SO_2	0.04–0.06	0.12–0.14	0.06–0.08	0.15–0.18
SO_4^{-2}	0.05	0.10*	0.05	0.18
TSP	—	0.11–0.18	—	—

*$p < 0.10$; for all others, $p < 0.05$; – indicates a nonsignificant result.

1982. Sulfates were almost constant over this period, ranging from about 8 to 14 μg/m³ by location. The TSP data fell into two clearly delineated groupings; the urban areas were in the range of 80 to 95 μg/m³ in 1976, while the other counties (including Algoma and Sudbury) ranged from about 45 to 70 μg/m³. By 1982, the two groups had decreased to 65 to 80 and 40 to 60 μg/m³, respectively. As a result, there was less collinearity in this study than in some previous ones. Annual average SO_2 and sulfate were not strongly correlated with socioeconomic variables, and the correlations with TSP were only 0.57 and 0.46, respectively. TSP showed a number of significant correlations with socioeconomic variables. The correlation over the region between maximum 24-hour and annual average SO_2 was 0.45, but the 24-hour measure was more highly correlated with socioeconomic variables (the correlation with alcohol consumption was 0.68), which might have made it difficult to deduce any acute effects from SO_2. The maximum 24-hour SO_2 reported was about 800 μg/m³. Since the largest SO_2 sources involved in this study are smelters, it is possible that metal particulates may have contributed to any health effects attributed to high values of SO_2.

All four dependent variables were significantly associated with air pollution, with the following elasticities, which are alternatives, not to be summed (see Table 8-2). The 24-hour average SO_2 coefficient had about the same significance and elasticity as annual average SO_2; given the likelihood of poorer exposure estimates for the 24-hour measure, one cannot rule out that the actual effects were acute rather than chronic.

In spite of the limited data set used, these findings appear to be self-consistent with no obvious contradictions or omissions; the results for total mortality are also consistent with findings for 1980 urban mortality in the United States (Lipfert *et al.*, 1988; Lipfert, 1992, discussed in Chapter 7).

Long-Term Temporal Regression Studies

Steubenville, Ohio

Respiratory health in Steubenville has been studied for a number of years (Ferris *et al.*, 1979), and more recently a time-series analysis of daily mortality (Schwartz and Dockery, 1992) was performed for the Steubenville SMSA for 1974–86 (see Chapter 6).* We obtained monthly data from the Ohio Department of Health on mortality occurring from 1965 to 1990 in the county

---

*This material originally appeared in Lipfert and Wyzga, 1992.

containing Steubenville (Jefferson County) for various causes of death and corresponding air quality data from the U.S. EPA. Time series were constructed on monthly, annual, and seasonal bases, with winter defined as December–March, and summer as June–September. The pollutant data consisted of TSP measured at two sites over the entire period, usually at one site or the other, and SO_2 measured at two sites from 1974 to 1975 to 1990. The bulk of the analysis was performed for TSP.

During this period, Jefferson County lost about 20% of its population, and its average age increased. The mortality expected on the basis of the U.S. population first decreased and then increased. It is important to account for these demographic changes when analyzing time series covering a decade or more. Whereas most analyses of daily deaths deal with *counts*, when the population changes, it is important to use *rates*. Also, the improvements that have occurred throughout the nation in longevity, medical care, and life-styles must be taken into account.

In our analysis of annual and seasonal long-term trends, we divided the death rates observed in Jefferson County by the rate expected if each of its 10-year age groups experienced the U.S. average age-specific all-cause mortality. Cause-specific U.S. trends were not used. Linear interpolation was performed between censuses to estimate the age distribution for each year. This degree of detail was not practical for the monthly analysis; these mortality rates were based on the interpolated population and were regressed against a linear time variable and pollution variables. A dummy variable was also used to account for changes in the TSP measuring station location. Since TSP generally decreased with time, collinearity was present among these three variables (TSP,

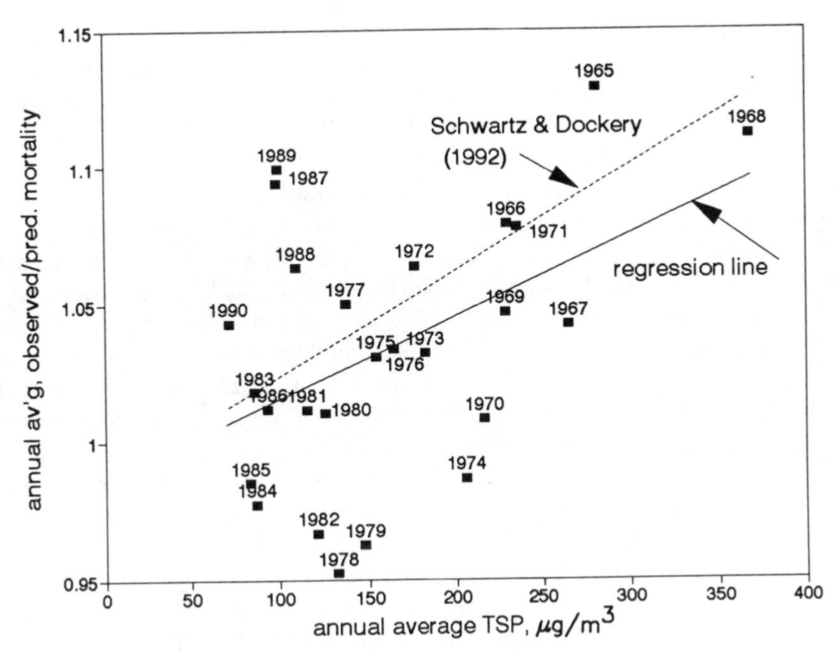

FIGURE 8-8. Observed/expected annual mortality for Steubenville, OH, versus TSP, 1965–1990. Regression line is from Schwartz and Dockery, 1992.

time, and measuring station). Seasonality was accounted for in the monthly analysis by using a sinusoidal function that peaked in February (minimum in August); this phasing was found to perform the best with major cardiovascular deaths and was retained throughout the monthly analysis.

With a log-linear model for total mortality (less external causes), TSP just achieved statistical significance ($p < 0.05$), with a coefficient about half of the value obtained by Schwartz and Dockery (1992) (Figure 8-8). Although the single high TSP value seen in this plot would appear to be an influential observation, removing it had no effect. The other variables in this model included temporal trend and season.

With linear models, only major cardiovascular (MCV = heart disease and stroke) death rates were found to show an association with TSP in the monthly analysis (Figure 8-9). The elasticities and *p*-values ranged from 3.3% ($p < 0.15$) to 8.4% ($p < 0.0001$), depending on how many of the collinear variables (mentioned above) were included in the regressions. Overall R^2's were of the order of 0.17 ($n = 312$). As a means of partitioning the effect of monthly variability against the general long-term trend, the TSP series was shifted one month forward and backward relative to mortality; the in-phase case performed best ($e = 4.9\%$, $t = 2.49$), compared to $t = 1.23$ for a lag and $t = 0.52$ for the noncausal case (negative lag). The long-term time trend variables gained strength for both out-of-phase cases, suggesting that the relationship between TSP and mortality was specific to the coincident month and that inclusion of additional long-term trend variables in the model tends to obscure this relationship.

As discussed in Chapter 6, Schwartz and Dockery (1992) found that a linear time trend variable was not significant in their analysis of daily mortality, implying that the entire 9% improvement in mortality that occurred in Steubenville from 1974 to 1984 resulted from reductions in air pollution.

Analysis by season has the advantage of reducing the need for seasonal corrections, at the expense of fewer degrees of freedom ($n = 26$). Separate bivariate correlations were calculated for TSP in winter (D, J, F, M) and summer (J, J, A, S), using ratios of observed MCV rates to expected annual all-cause rates as the dependent variable. The residuals from these regressions were checked for time dependence. The data from 1968 appeared to be an outlier; perhaps the high TSP values measured that year were influenced by local dust sources during this period. The elasticities and R^2's were found to be 6.6% and 0.36 (winter) and 6.8% and 0.35 (summer). On an annual basis, these values were 7.6% and 0.42. Thus, the findings from this preliminary analysis are consistent for three different time scales (monthly, seasonal, annual).

Taking into account that we used a potentially more sensitive end point than did Schwartz and Dockery (1992), these findings appear to be reasonably consistent with their daily analysis. However, since the U.S. downward time trend for MCV deaths is steeper than the trend for all causes, which we used as a basis, we may have underestimated the degree of interference between the decrease due to improved TSP levels and that due to better medical care and life-styles. The analysis should be confirmed with the appropriate normalization data and with attention to flu epidemics, holiday periods, and heat waves, and alternative model specifications should be explored and tested against one another.

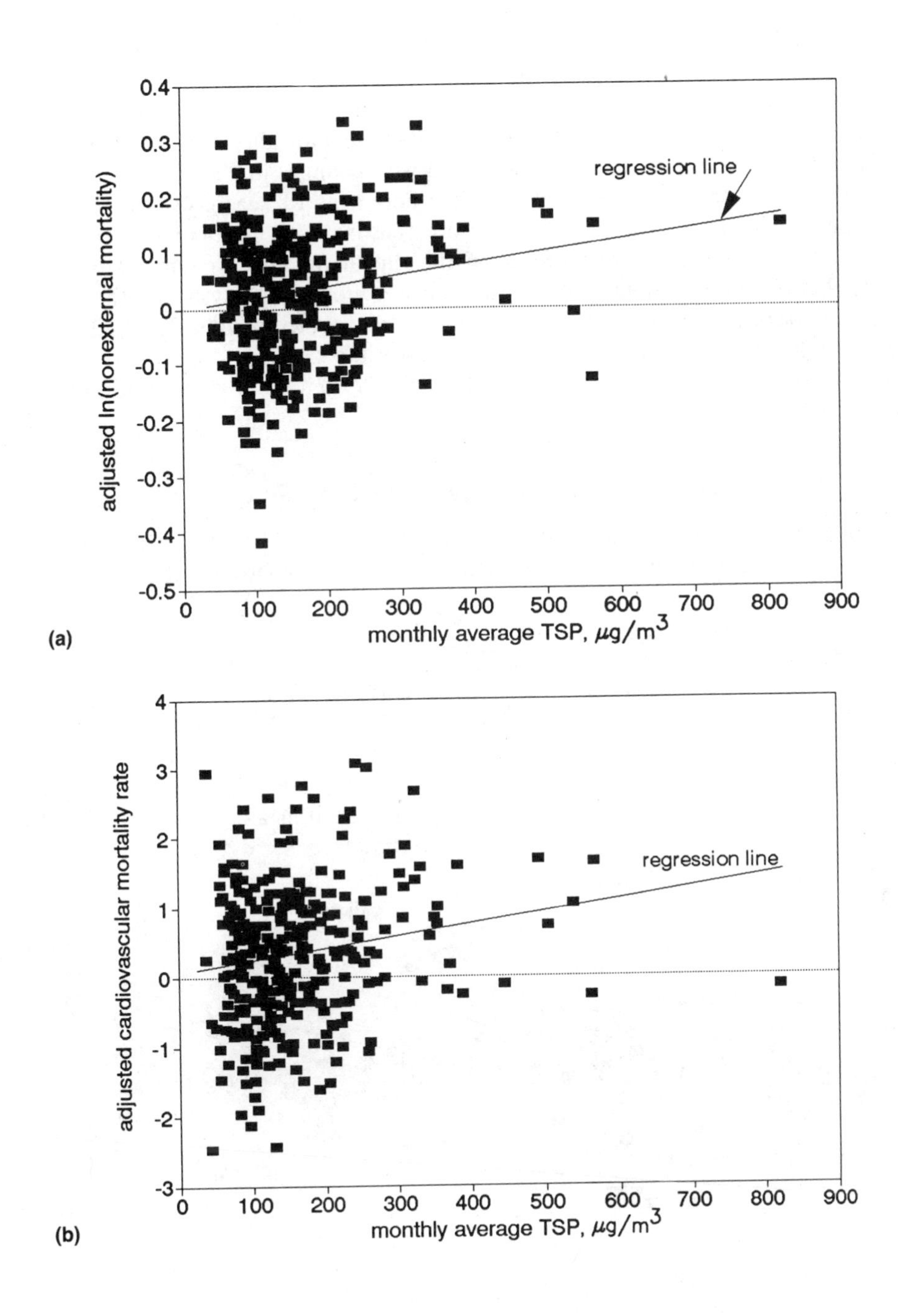

FIGURE 8-9. Monthly mortality in Steubenville, OH, for major cardiovascular causes versus TSP, 1965–1990, adjusted for seasonality and time trend. (a) Log of mortality rate. (b) Mortality rate. Regression lines are the best fits through the data.

Estimates for Los Angeles

In a similar vein, we obtained monthly mortality data from 1960 to 1990 for Los Angeles County from the California Department of Health, together with corresponding center-city TSP values from the U.S. EPA. The population age distribution was not readily available, however, so in this instance we used data for the balance of the State of California as a temporal "control." The hypothesis at issue is whether the change in all-cause mortality in Los Angeles relative to the rest of the state is associated with changes in TSP levels. Using separate bivariate regressions for summer and winter "seasons," we found statistically significant relationships in both seasons with elasticities near the high end of the range found in Steubenville (linear models). It remains to be seen whether these findings are robust against a monthly analysis and the use of actual mortality rates and ratios of observed to expected deaths. The findings for a log-linear model are shown in Figure 8-10; the slopes and elasticities are substantially higher than Kinney and Özkaynak (1991) found for daily mortality in Los Angeles.

Seasonal Patterns of Health and Mortality

Seasonal patterns of illness are well-known to residents of the temperate zones, where seasonal changes create gradients in various environmental factors,

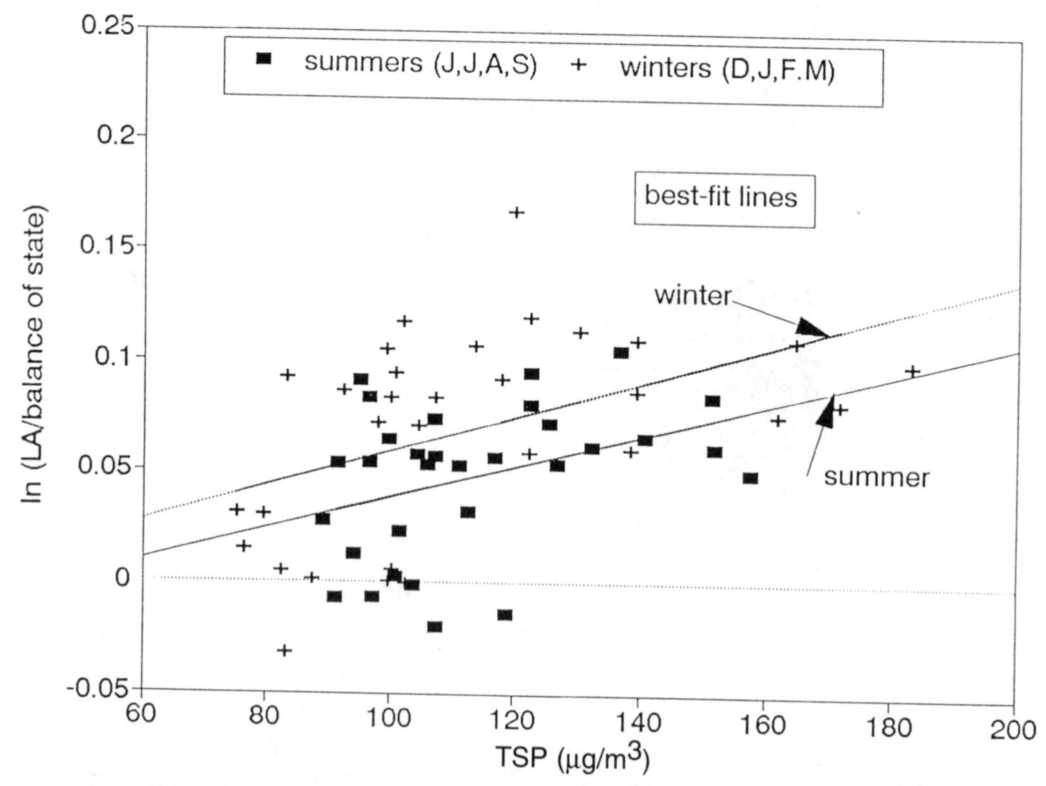

FIGURE 8-10. Los Angeles all-cause mortality, relative to the rest of the state, versus TSP, 1960–1990, by season.

including temperature, precipitation, viability of infectious organisms, air pollution, and aeroallergens. Interactions among these factors also occur; for example, cold temperatures can create physiological stresses, increase the demand for space heating, which produces air pollution, and reduce atmospheric ventilation, which increases ambient concentrations of all the pollutants present. Warmer temperatures may promote the spread of infectious organisms. Near large urban centers, heat waves may be accompanied by higher concentrations of O_3 and particulate matter. This section of the chapter examines the extent to which air pollution effects on health might be inferred from seasonally resolved data.

Background

Seasons were suspected of influencing health long before air pollution was suspected to also play a role. According to Macfarlane (1977), Hippocrates wrote, "if the winter be southerly, showery, and mild but the spring northerly, dry and of a wintry character, . . . the aged. . . have catarrhs from their flabbiness and melting of the veins, so that some of them die suddenly and some become paralytic on the right side or the left." Farr (1985) pointed out over a century ago the necessity of examining day-to-day changes in weather factors, rather than mean values. Most time-series analyses of air pollution and health have found it necessary to account for seasonal factors in some way. One of the most interesting questions is whether the observed seasonal mortality patterns are extrinsic (resulting from environmental factors) or intrinsic (resulting from innate biological rhythms). There is an extensive literature on biometeorological effects on health (Hodge and Nicodemus, 1980). However, Driscoll (1991) discounts most of these studies as insufficiently rooted in physiology.

Changes in Seasonal Patterns

Seasonal patterns of air pollution have changed over the years in the United States and in other industrialized countries. As discussed in Chapter 2, substituting natural gas and distillate fuel oils for solid fuels for space heating has reduced winter emissions, mainly SO_2 and particulates; increases in electric utility emissions and transportation fuel use has increased summer levels of O_3 and sulfate aerosols. One hypothesis of interest is thus: have changes in air pollution from space heating affected the seasonal mortality cycle?

Seasonal patterns in mortality in England and Wales were compared with the United States by Lewis-Faning (1940) for the early 1930s, a period when solid fuels were used extensively for space heating in both countries. He noted that the seasonal mortality cycle was more severe in England but that Germany (large cities) and the United States were about the same after 1935. He selected New England as a subset of the United States that might be comparable to England in climate and demography. Both the English and American data included both urban and rural areas. Age standardization was used to compare across areas. Figures 8-11a through 8-11c present these data. Figure 8-11a compares death rates due to all causes; New England (NE) and England and Wales (E-W) are similar in summer, but E-W are higher in winter. The entire U.S. is quite similar to NE in winter but higher in summer, perhaps because of the effects of infectious diseases in summer. The comparison for heart disease (Figure 8-11b) shows NE to be parallel to E-W but higher than

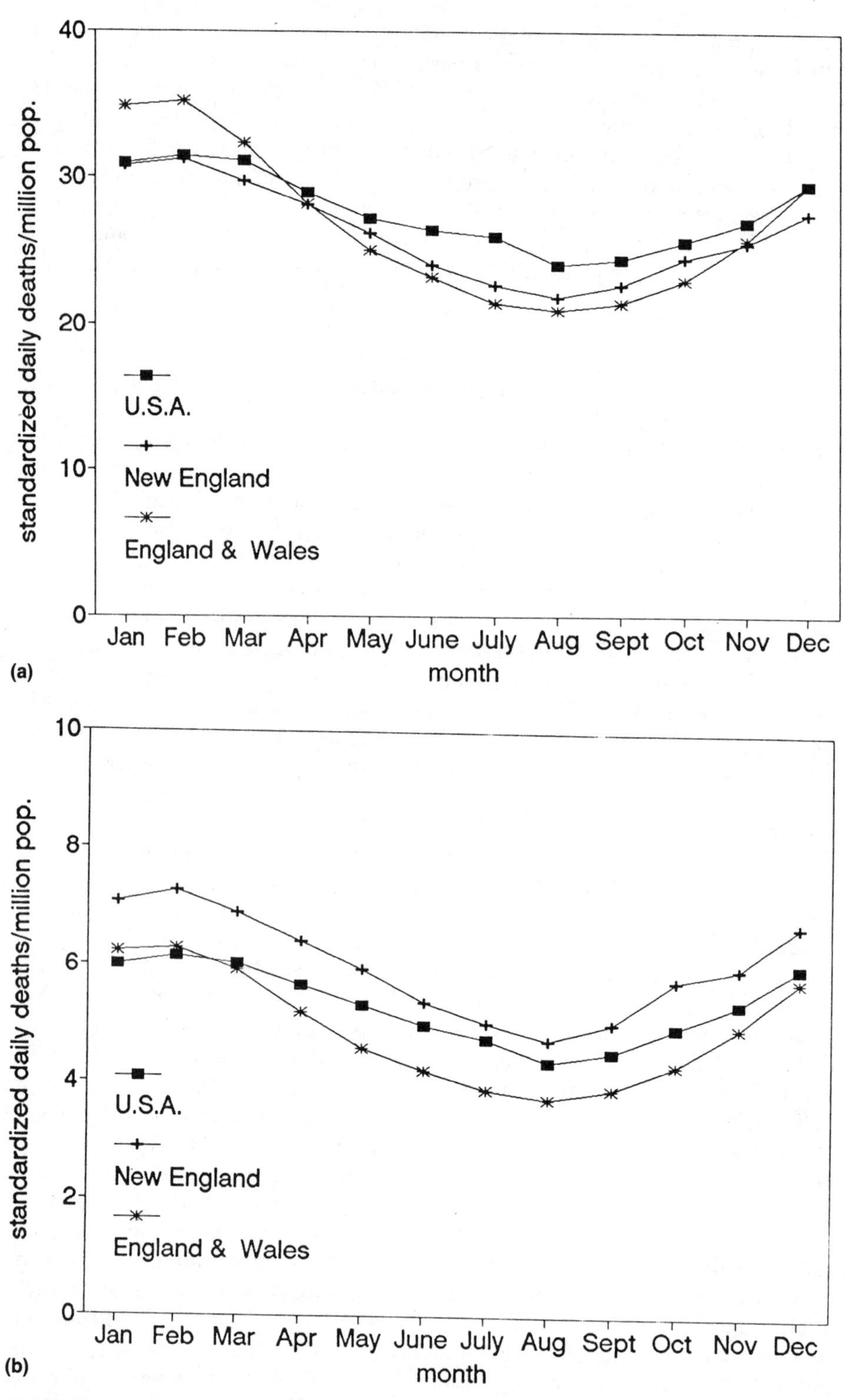

FIGURE 8-11. Comparison of seasonal mortality trends in the early 1930s. (a) All causes. (b) Heart disease. (c) Bronchitis and pneumonia. *Data from Lewis-Faning (1940).*

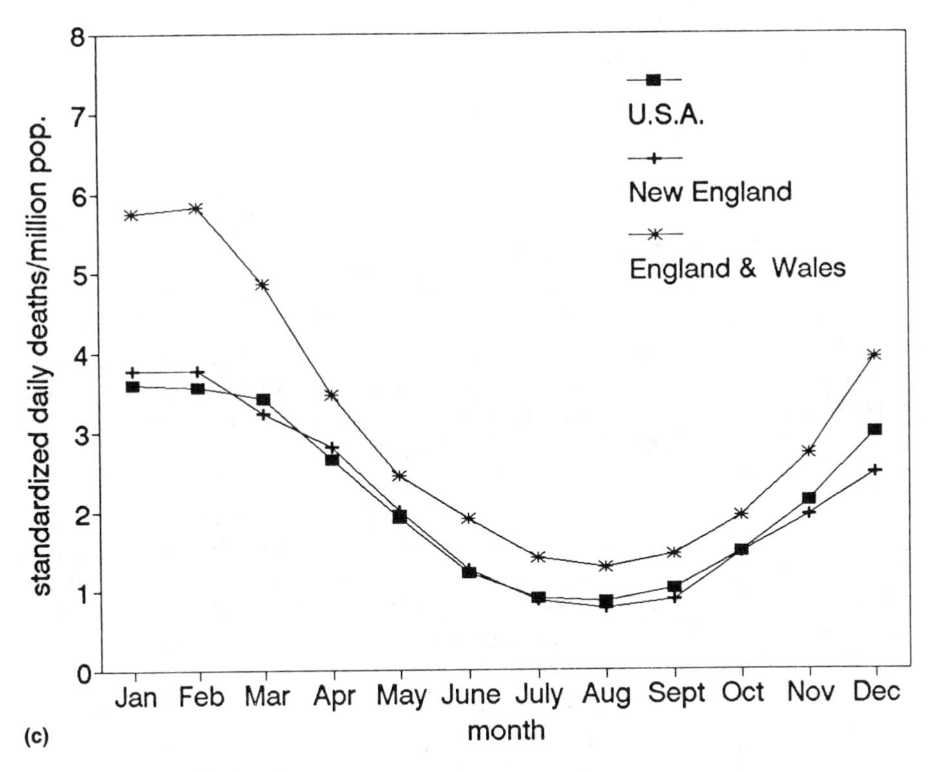

FIGURE 8-11. *Continued*

US in winter. For the sum of bronchitis and pneumonia (Figure 8-11c), the two American sets are almost identical, while E-W show a substantial excess in winter, which accounts for much of the all-cause excess. If we hypothesize that winter air pollution follows the order E-W > NE > US, we find that support for a discernible effect of air pollution on these seasonal differences seems to be limited to respiratory diseases.

Perhaps the most thorough work on seasonal patterns in mortality is the book by Sakamoto-Momiyama (1977). Her analysis covers many different countries and extends back to the beginning of the century, with particular emphasis on changes in seasonal patterns over the decades. Sakamoto-Momiyama defines a "seasonal" mortality pattern as one whose monthly coefficient of variation (CV) exceeds 12%, and a "deseasonalized" pattern as one whose CV is less than 8%. The United States has been in the latter category since about the 1920s. The seasonality of Canadian mortality is very similar to that of the United States (Bako *et al.*, 1988).

Statewide and Regional Trends in Seasonality

The U.S. Department of Health and Human Services (DHHS) publishes annual reports on vital statistics that contain summary tables of deaths by month, for states, race-sex groups, and causes of death. I compiled monthly total mortality rates from these reports for the nine Census Divisions, for the years 1940, 1950, 1960, 1970, and 1985. Monthly mortality trends for the United States as a whole were also examined by race and by major cause of death in hopes that these factors could help explain the observed spatial and temporal gradients in total mortality. The seasonality of mortality was judged from scatter plots, from the coefficient of variation of the 12 months (standard deviation/mean) (Sakamoto-Momiyama, 1977), or from the ratio of the maximum-minimum monthly mortality rates during the year.

The first three decades (1940–60) represent a period of greatly changing patterns in air pollution, especially for home heating sources (Table 2-5, p. 23). At that time, space heating was one of the largest contributors to seasonal differences in air pollution. Most of these changes resulted from the availability of utility natural gas from pipelines constructed in the late 1940s and 1950s and from the advent of less expensive electricity from hydropower in the 1960s. Replacement of coal and wood by these fuels reduced winter levels of SO_2 and particulates. Changes in heating fuels since then have been modest, involving a minor resurgence in wood fuel in the mid 1970s and some conversions from oil to gas in the East.

In 1985, differences in the seasonality of mortality by race were minor, following the order whites > Blacks > others. Based on the parameters of best-fit quadratic curves through each data set, whites and Blacks were not significantly different, but "others" were significantly different from both Blacks and whites. This suggests that geographic differences in the seasonality of total mortality are not likely to be due to geographic racial differences, *per se*.

By cause of death, respiratory mortality showed the strongest seasonal patterns in all 5 years. Figure 8–12 plots the rates by year and month; in summer, only 1940 stands out from the other years, undoubtedly because of the improvements in drug therapies that came into use after 1945. It is difficult to interpret the changes in winter pneumonia and influenza deaths after 1940 because individual years may be influenced by epidemics (such as the peaks in 1960 and 1950), but 1985 represents the year with the lowest urban levels of particulate air pollution. Longitudinal trends are discussed in the next section.

Cardiovascular deaths (Figure 8-13) also showed a definite seasonality that has lessened only slightly over time. The seasonality in 1985 was different from that in 1940 and 1960, but not 1950. External causes of death (accidents, homicides, suicides) showed the opposite pattern in 1985, with higher rates in summer (Figure 8-14). Cancer deaths showed only modest seasonal fluctuations, with no particular patterns (Figure 8-15).

Spatial patterns in the seasonality of 1985 U.S. mortality, as measured by the coefficient of variation (CV), were also examined by state for 1985. Few regional trends were apparent; the highest CVs appeared in Mississippi, the District of Columbia, California, Delaware, South Dakota, and Rhode Island, which are states with no immediately obvious common factors. The map suggested that climate had an inverse role in seasonality, if any; states with severe winters such as Vermont, Michigan, and Idaho, had less seasonality of mortality than those with milder winters (Florida, California, Texas).

Data availability limited the examination of historical trends by region to the 9 census divisions. Scatter plots of monthly trends normalized with respect to the annual average were prepared for each division for the years 1940, 1950, 1960, 1970, and 1985, including best-fit curves for each divisions, i.e., polynomials of the same order through each year's data (Lipfert, 1991). In 1940, there was substantial diversity among divisions, but a pronounced

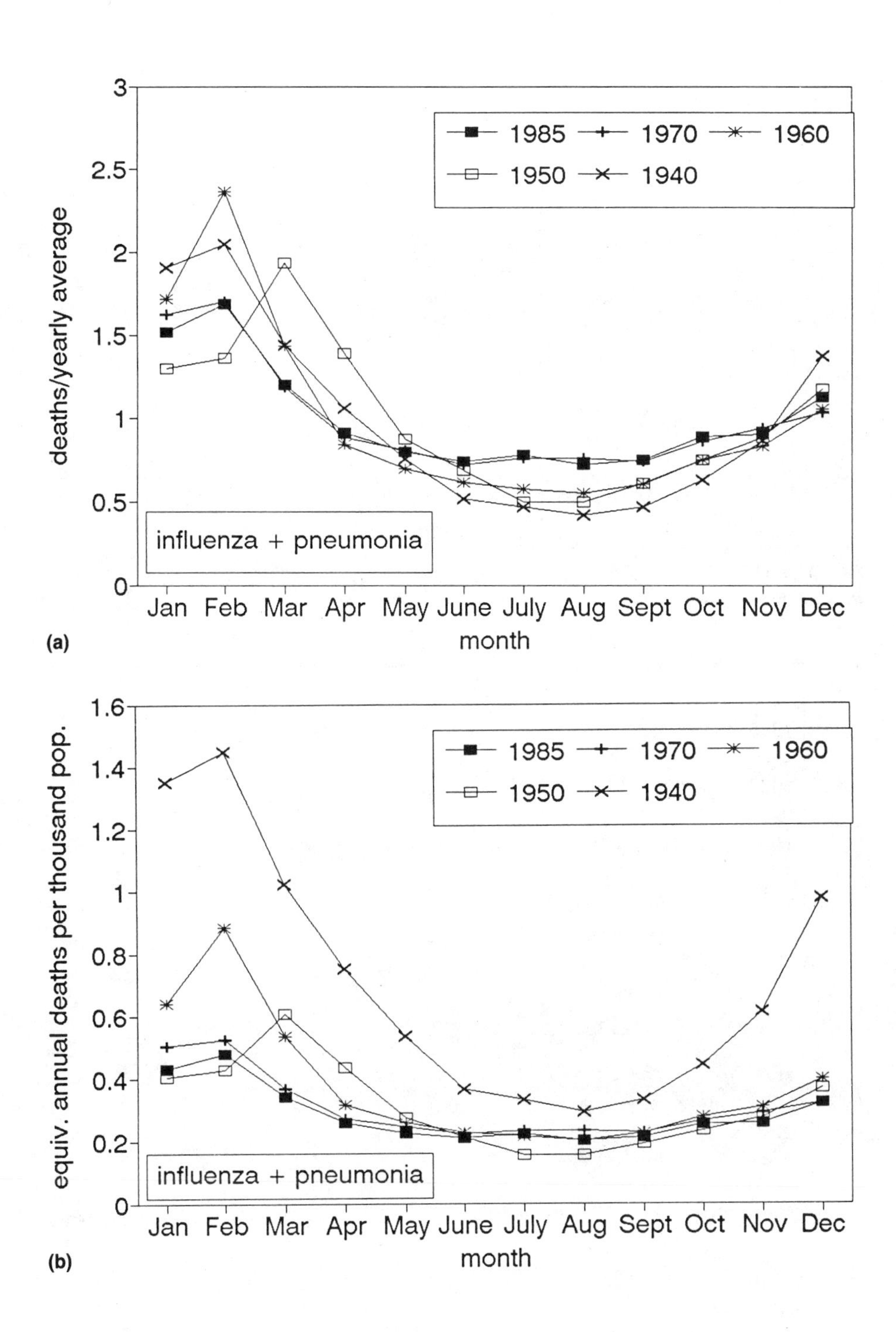

FIGURE 8-12. Changes in seasonal trends in mortality due to inflenza and pneumonia, United States 1940–85. (a) Monthly deaths/annual average. (b) Equivalent annual death rate by month. *Data from U.S. HEW.*

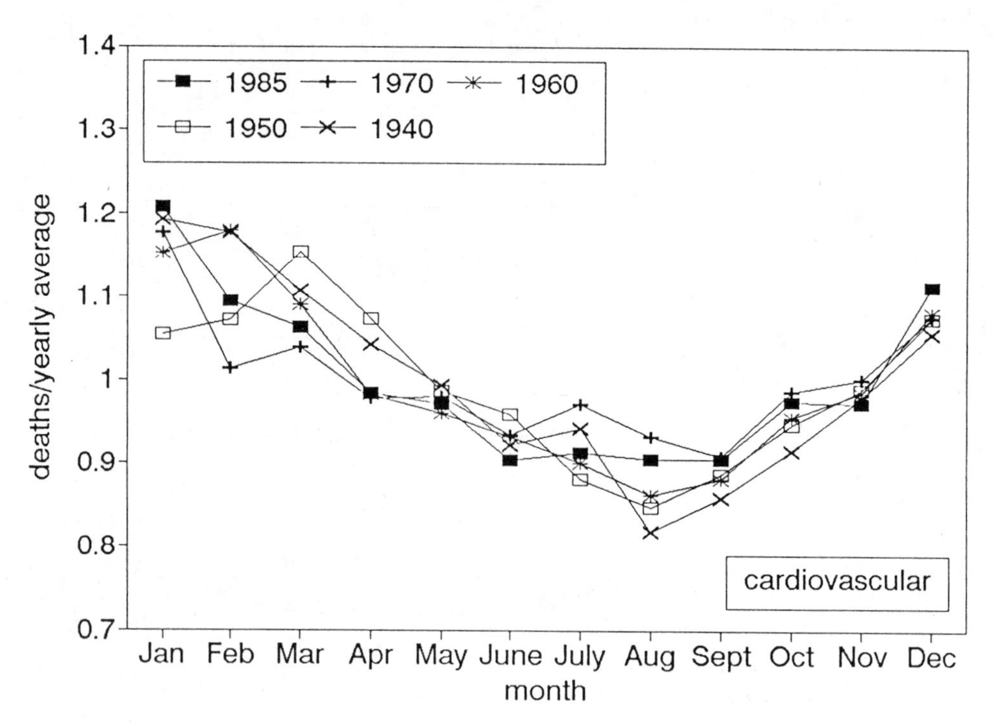

FIGURE 8-13. Changes in seasonal trends in cardiovascular mortality, United States 1940–85. *Data from U.S. HEW.*

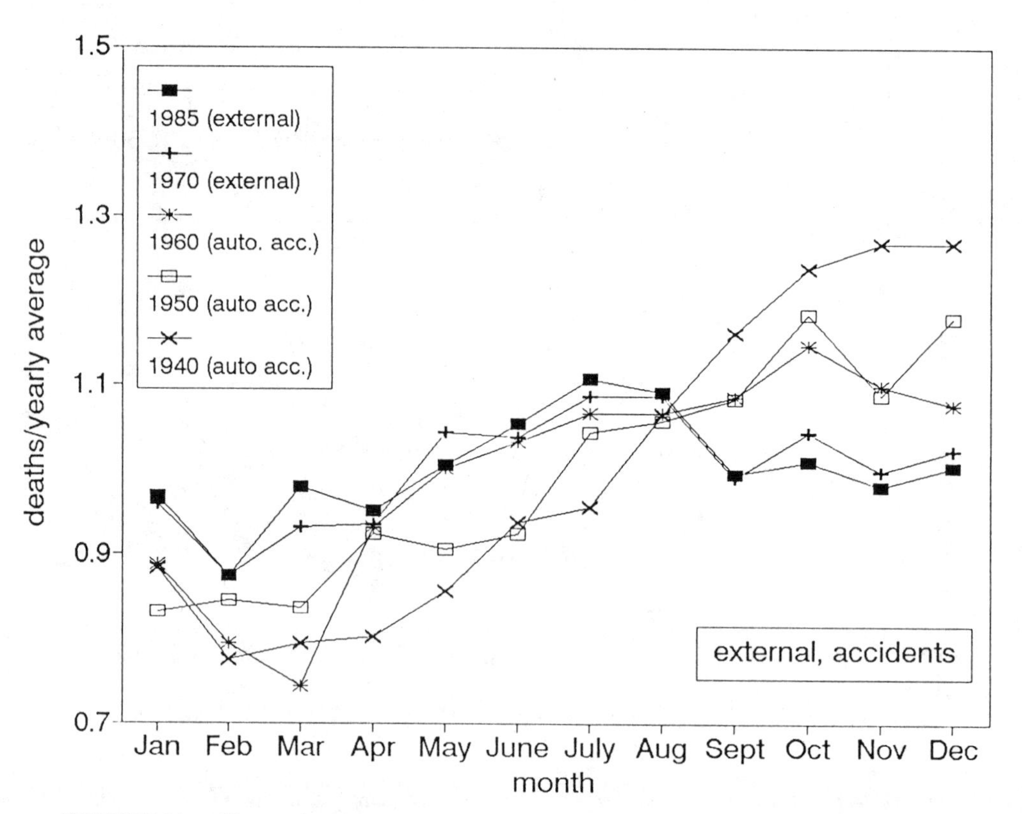

FIGURE 8-14. Changes in seasonal trends in mortality due to accidents and external causes, United States 1940–85. *Data from U.S. HEW.*

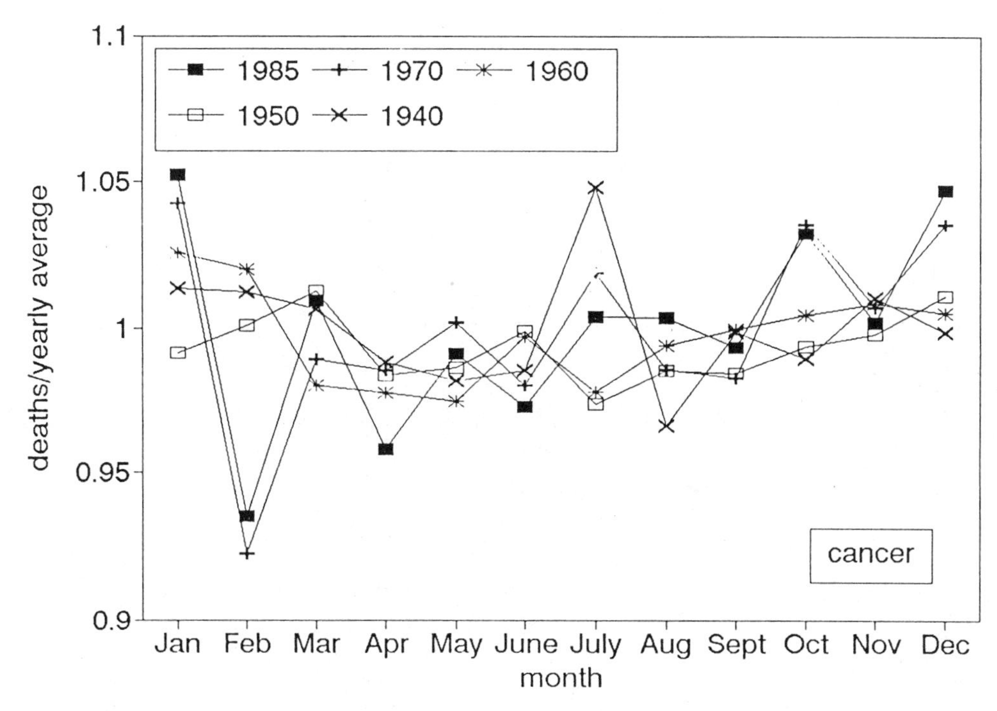

FIGURE 8-15. Changes in seasonal trends in cancer mortality, United States 1940–85. *Data from U.S. HEW.*

summer peak in mortality was present in all, a departure from the normal annual periodic trend. Gastrointestinal causes of death were important in summer at that time. The diversity among divisions gradually decreased over time, such that, by 1985, the 9 division plots virtually coincided. Most census divisions still showed a minor summer peak in 1985, presumably due to heat waves. Comparisons were made across regions for each year and across time for selected regions. In general, the differences across regions over time were modest and required close examination. Table 8-3 presents the coefficients of variation and max/min statistics, which confirm that the seasonality of mortality has decreased since 1940. However, seasonal mortality varied more in 1960 than in 1950 or 1985.

In 1940 there were marked geographic differences in the seasonality of mortality (Table 8-3). The Mountain and Pacific Divisions showed the least seasonality, with peaks occurring in December and the minimum levels in May–June. The three southern divisions showed the most seasonality, with symmetrical curves and minimum levels in July–August. The remaining four divisions, which differed greatly in terms of demography and population density, were virtually indistinguishable, in phase with the southern divisions but with less seasonality.

All nine divisions were approximately in phase in 1950, with the peak mortality occurring in March. However, the two western divisions had less seasonality. The East South Central Divisions had the most seasonal mortality pattern. In 1960, mortality in all the divisions peaked in February except the Pacific Division (January peak). There again appeared to be three groups: the

TABLE 8-3 Trends in Monthly Mortality Variability

Region	Coefficient of Variation					Maximum/Minimum Month				
	1940	1950	1960	1970	1985	1940	1950	1960	1970	1985
New England	0.0838	0.0643	0.0984	0.0634	0.0728	1.33	1.23	1.38	1.26	1.25
Middle Atlantic	0.0952	0.0815	0.0753	0.0663	0.0739	1.39	1.34	1.24	1.30	1.28
E.N. Center	0.0909	0.0616	0.0785	0.0506	0.0692	1.35	1.22	1.29	1.22	1.26
W.N. Center	0.0951	0.0784	0.0953	0.0515	0.0739	1.32	1.32	1.41	1.22	1.27
S. Atlantic	0.1328	0.0800	0.1101	0.0669	0.0666	1.48	1.29	1.38	1.27	1.20
E.S. Atlantic	0.1403	0.0883	0.1101	0.0666	0.0756	1.48	1.32	1.43	1.25	1.25
W.S. Central	0.1278	0.0748	0.0994	0.0613	0.0667	1.52	1.25	1.38	1.25	1.24
Mountain	0.0637	0.0556	0.0666	0.0302	0.0583	1.27	1.22	1.25	1.10	1.23
Pacific	0.0953	0.0472	0.0785	0.0442	0.0795	1.38	1.20	1.34	1.15	1.32

western divisions were less seasonal; the South Atlantic and East South Central Divisions had the most seasonal mortality patterns. The data for 1970 showed a remarkable degree of national homogeneity. Summer rises in mortality were seen in the North Central and Mid-Atlantic Divisions; the Mountain and Pacific Divisions were noticeably flatter. In 1985, the seasonality of mortality was virtually identical in all census divisions; the only significant difference was between the Mountain and Pacific Divisions. Interestingly, Macpherson *et al.* (1967) showed a similar mortality cycle in Sydney, Australia, over a much smaller temperature range (50–90°F).

While the patterns of seasonal mortality from all causes have become substantially uniform across the country, the overall measures of seasonality have changed only slightly since 1950. Because the differences between winter and summer urban air pollution have decreased substantially since the 1950s and 1960s, we conclude that air pollution played only a minor role in affecting the timing of deaths on the spatial and temporal scales examined. The regions most likely to have been affected by air pollution showed among the smallest changes in seasonality as air quality improved over the years. Similar plots of deaths due to respiratory causes alone would likely be required to demonstrate any effects of air pollution, on this scale.

Seasonal Patterns of Urban Mortality in the United States, 1990

I used the 1990 weekly tabulations of "Deaths in 121 U.S. cities" as presented in *Morbidity and Mortality Weekly Report* (*MMWR*) to create several time series of mortality for cities within different climatic and environmental characteristics. To determine the presence of seasonality, I ranked these data in order, beginning with the highest mortality week of the winter, and regressed the deaths in each city against the week number and its square. Statistical significance of the quadratic term in these regressions was taken as evidence of the presence of seasonality. By this definition, seasonality also depends on the regularity of the data; data anomalies (such as noise or spikes, both of which were present for some cities) could mask the patterns. Also, *MMWR* data are preliminary and subject to missing weekly data in the tabulations.

For the entire United States in 1990, "seasonality" was shown by this definition for total mortality, for pneumonia and influenza, and for ages 65 and over (data were not available at this level for other causes). Since random variations would likely mask seasonal variations for the younger age groups in individual cities, I emphasized the 65-and-over age group. Seasonal patterns in mortality for this age group were found in Boston, New York City, Columbus (OH), Detroit, Minneapolis, St. Louis, Atlanta, Memphis, Denver, and Phoenix. There seemed to be no relationship between climate and the degree of seasonality; Phoenix had a more pronounced pattern of elevated winter mortality than Denver or Minneapolis, for example. However, Honolulu, Los Angeles, San Diego, and Seattle showed no seasonality. Honolulu showed a surprisingly high fraction of mortality due to pneumonia and influenza compared to other cities with mild climates; this may have been due to differences in diagnostic practices. Miami showed seasonal mortality patterns for all ages and for ages 45–64, but not for ages 65 and over.

Of the cities analyzed, New York had the most definitive seasonal patterns, in part because its weekly counts are larger and thus would be less affected by Poissonian noise. Seasonal patterns were found in most sections of the country; the idiosyncrasies of individual cities may have been responsible for the several contrary findings.

U.S. National Mortality Trends

The final trend study of this chapter concerns the use of contrasts between "summer" and "winter" mortality rates as a tool for analysis of long-term trends. For this purpose, "summer" was defined as June through September, which will always include the lowest mortality months. "Winter" is defined as January through March, plus the preceding December; this combination seems to always include the highest mortality month. In order to account for changes in population age distribution over the years using readily available data, I referred the total numbers of deaths to the population 65 and over, even though about one-third of the deaths occurred to younger persons.

This analysis follows a similar study by Rose (1966), who contrasted British deaths in December and June for ischemic heart disease (IHD) and noted that the June data were much more regular. He tried to separate the effects of air pollution from those of temperature by plotting (and correlating) the December/June mortality ratio against either the December/June temperature ratio or the December/June smoke pollution ratio. Both of these independent variables were measured near London, while the mortality data were apparently for all of England and Wales. Not surprisingly, he found a much stronger relationship for temperature, in part because weather is more regionally distributed than air pollution. However, Rose's data can be used to estimate the elasticity of the air pollution effect, however crudely, which was in the range 0.02 to 0.06.

Figure 8-16 plots the trends in deaths due to all causes (Figure 8-16a, linear scale) and due to pneumonia and influenza (Figure 8-16b, log scale). The increase in summer influenza and pneumonia mortality since 1980 may represent a previously undetected trend. The relatively steep decreases after 1970 are especially intriguing since they correspond to improvements in urban air quality following implementation of the 1970 Clean Air Act (and perhaps the effects of the energy crisis); no new notable medical intervention practices are known to have occurred at this time. According to the national emissions trends presented in Chapter 2, particulates seem to be the only species whose temporal trends correspond to these mortality trends.

Figure 8-17 plots the influenza and pneumonia rates since 1950 (in order to avoid the changes in medical care known to have been implemented from 1940 to 1950) against estimated national particulate emissions, a crude surrogate for ambient air quality. Unfortunately, summer pollution levels cannot be distinguished from winter levels before the mid-1950s. The plot for winter suggests a quasi-linear dose-response relationship, although a portion of this slope is likely due to other factors, such as improved medical care.

Summer mortality, as defined, was used as a baseline for the comparison of trends, assuming that it represents a measure of the lowest possible rates at a given point in time and thus the least population stress. Figure 8-18 shows how the winter and summer influenza and pneumonia "excess" explains the winter-

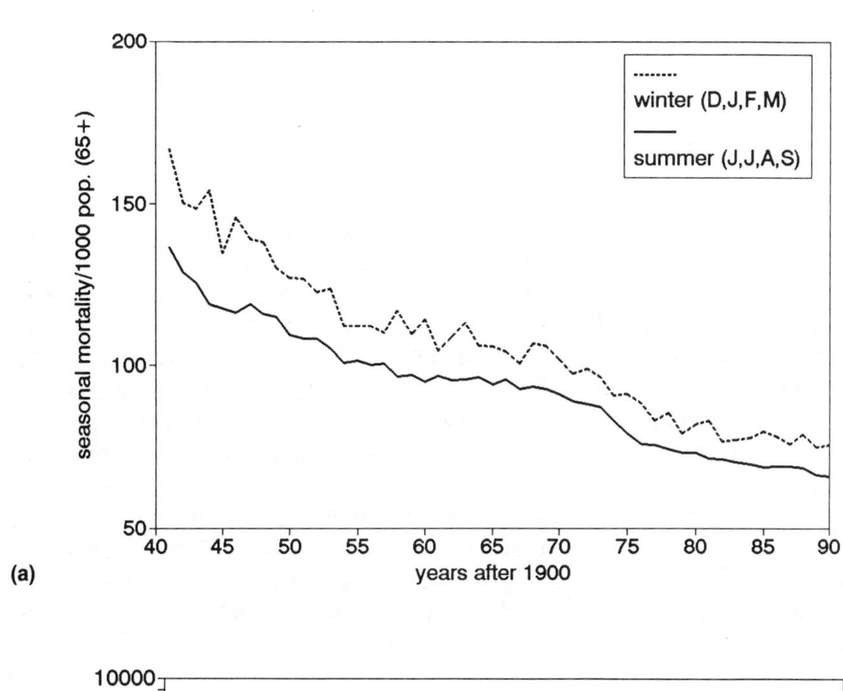

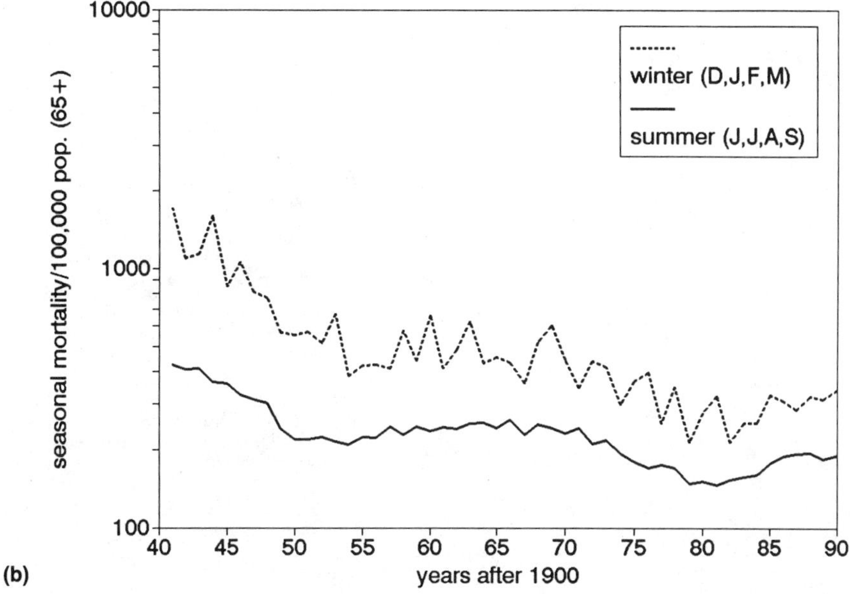

FIGURE 8-16. Trends in U.S. total seasonal deaths, referred to the population 65 and over. (a) All causes. (b) Influenza and pneumonia. *Data from U.S. HEW.*

summer ratio for all causes. Since the regression line does not pass through the point (1, 1), contributions of other causes of death on the seasonality of total mortality are indicated. However, the scatter plot implies that these causes must be highly correlated with influenza and pneumonia deaths and thus potentially of common etiology.

Winter average temperatures were explored as a potential explanatory factor for the variation in excess influenza and pneumonia mortality; there was a generally positive relationship, but a great deal of scatter. Particulate emissions were similarly explored, but the effect of particulates could not be separated from the effect of time *per se*.

Figure 8-19a plots the winter-summer ratio for all-cause mortality versus time, and Figure 8-19b, the corresponding data for influenza and pneumonia mortality. These plots show that the lowest levels of the ratios have been essentially constant since the late 1950s for all causes, but only since around 1970 for influenza and pneumonia. The highest ratios on both plots constitute the envelope of flu epidemics. The severity of these events has been constant for all causes but decreasing for influenza and pneumonia. The slight upswing in the ratio in the last few years may be a normal fluctuation but bears watching.

Separating summer and winter mortality trends offers the advantage of reducing the random "noise" created by winter epidemics. Because advances in medical care and changes in population would be expected to operate year round, this procedure offers a simple way to normalize the data and examine

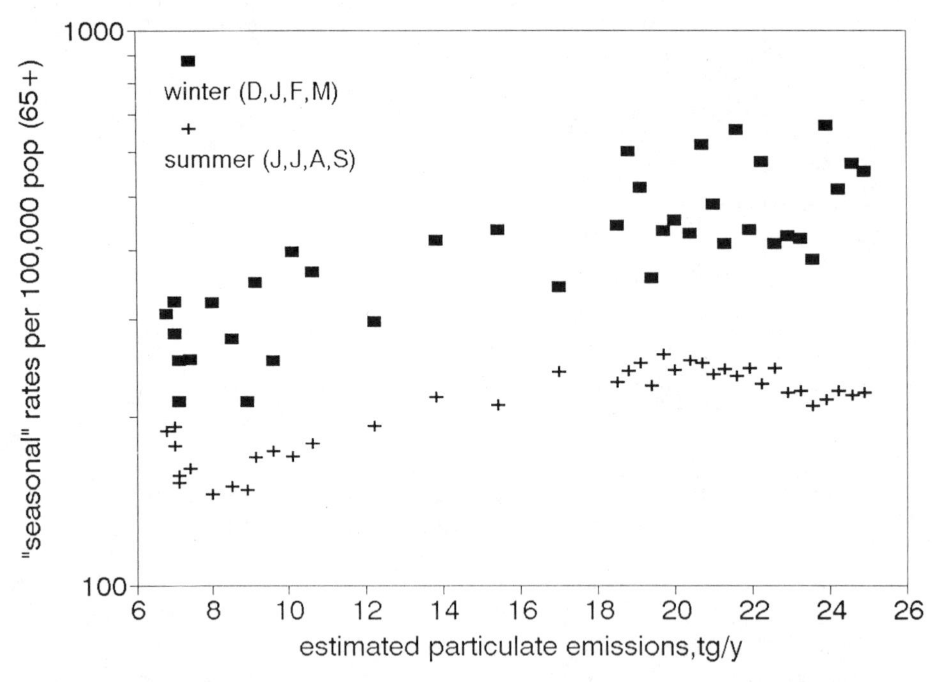

FIGURE 8-17. Seasonal U.S. mortality due to influenza and pneumonia vs. annual particulate emissions. *Data from U.S.HEW and U.S. EPA.*

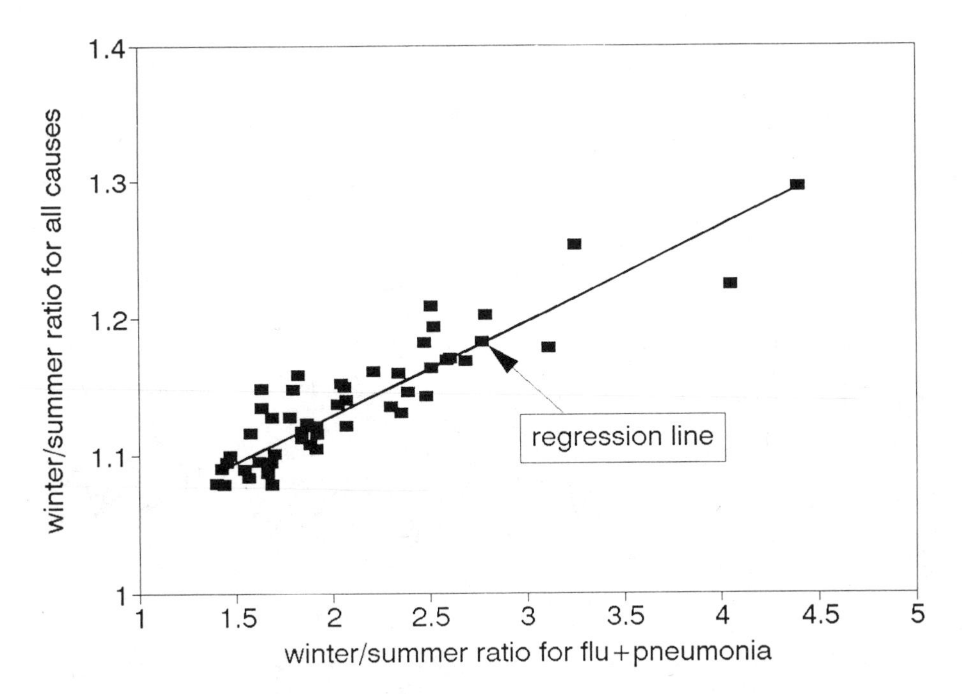

FIGURE 8-18. Relationship between winter/summer ratios for all-cause mortality and influenza–pneumonia mortality. *Data from U.S. HEW.*

trends. Some of the changes observed may relate to changes in environmental quality, but a spatially resolved analysis would be required to generate specific hypotheses toward this end.

CONCLUDING DISCUSSION

Like all the ecological studies discussed in this book, these analyses can only hypothesize, rather than *prove*, their assertions. However, the evidence that air pollution affects mortality became more compelling by the finding that temporal effects extend to bimonthly periods in New York City, monthly periods in Steubenville and Los Angeles, and to case studies of air pollution abatement in several locations. The pooled time-series cross-sectional studies imply that stable regression coefficients can be derived from small data sets.

The magnitudes of seasonal cycles in U.S. mortality have changed little in 40 years, but the nation has become more homogeneous in this regard. There have been large improvements in winter urban air quality, and the peak periods have shifted from winter to summer for several pollutants. For this reason, detailed regression analysis is required to deduce air pollution-mortality relationships based on seasonal data. On the basis of the data presented here, it appears that only a small part of winter excess mortality around 1960 and before was due to air pollution, since the seasonal air pollution curves have flattened, while the mortality curves have not.

The seasonality of mortality does not seem to be related to climate *per se*; the conclusion thus follows that the primary drivers must be biological, but

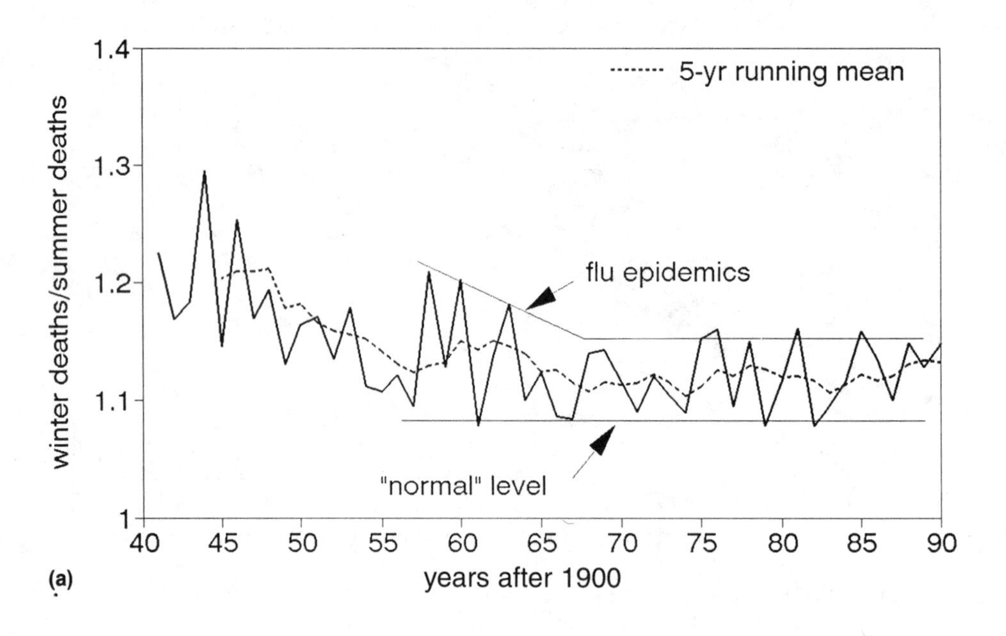

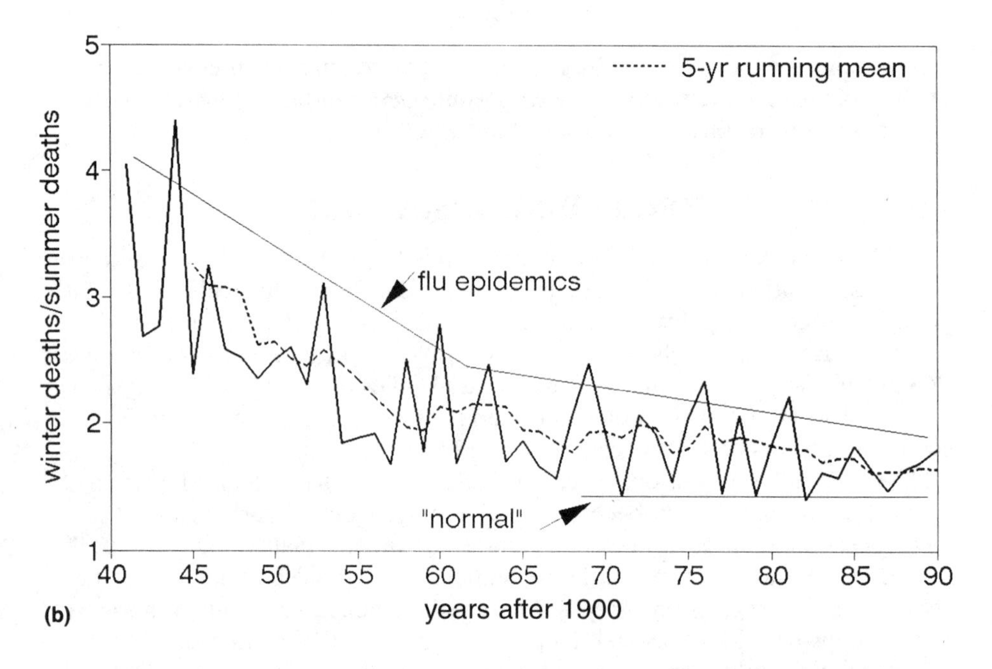

FIGURE 8-19. Trends in winter/summer mortality ratios. (a) All causes. (b) Influenza and pneumonia. *Data from U.S. HEW.*

there may be other important factors in certain cities (including air pollution). Rogot *et al.* (1992) used summer-winter mortality ratios to deduce that the presence of central air-conditioning reduced hot-weather mortality by 42%.

Winter-summer mortality ratios appear to be less sensitive to exogenous temporal changes and thus may be useful in examining long-term trends, due to both air pollution and climatic changes. There is suggestive evidence that part of the drop in mortality rates that occurred during the 1970s may have been due to the concomitant improvements in air quality. This hypothesis is strengthened by the facts that most of this decrease was in cardiovascular deaths, and that some investigators have reported that most of the improvement came in out-of-hospital deaths (Goldman *et al.*, 1982; Burke *et al.*, 1989). Out-of-hospital deaths are more likely to have been influenced by ambient (outdoor) air pollution.

REFERENCES

Archer, V.E. (1990), Air Pollution and Fatal Lung Disease in Three Utah Counties, *Arch. Env. Health* 45:325–34.

Bako, G., Ferenczi, L., Hill, G.B., and Lindsay, J. (1988), Seasonality of Mortality from Various Diseases in Canada, 1979–83, *Can. J. Public Health* 79:388–89.

Brown, S.M., Marmot, M.G., Sacks, S.T., and Kwok, L.W. (1975, 1976), Effect on mortality of the 1974 fuel crisis, *Nature* 257:306–07 and 259:560–61.

Caka, F.M., Lewis, E.A., and Eatough, D.J. (1992), Sulfate and Nitrate Formation in Utah Valley during Winter Inversions, AWMA Paper 92-62.05, presented at the 85th Annual Meeting of the Air & Waste Management Association, Kansas City, MO.

Christainsen, G.B., and Degen, C.G. (1980), Air Pollution and Mortality Rates: A Note on Lave and Seskin's Pooling of Cross-Sectional and Time-Series Data, *J. Env. Economics & Mgmt.* 7:149–55.

Driscoll, D.M. (1991), Alternatives to the Theory of Meteorotropicity, *Proc. 10th Conf. on Biometeorology and Aerobiology*, Salt Lake City, American Meteorology Society, pp. 1–2.

Evans, J.S., Kinney, P.L., Koehler, J.L., and Cooper, D.W. (1984), Comparison of Cross-Sectional and Time-Series Mortality Regressions, *J. APCA* 34:551–53.

Farr, W. (1885), *Vital Statistics: A Memorial Volume of Selections from the Reports and Writings of William Farr*, Humphreys, N., ed., Sanitary Institute of Great Britain, London, p. 415.

Ferris, B.G., *et al.* (1979), Effects of Sulfur Oxides and Respirable Particles on Human Health, *Am. Rev. Resp. Dis.* 120:767–79.

Goldman, L., Cook, F., Hashimoto, B., Stone, P., Muller, J., and Loscalzo, A. (1982), Evidence that Hospital Care for Acute Myocardial Infarction Has Not Contributed to the Decline in Coronary Mortality Between 1973–1974 and 1978–1979, *Circulation* 65:936–942.

Gregor, J.J. (1977), *Mortality and Air Quality, The 1968–72 Allegheny County Experience*, Report #30, Center for the Study of Environmental Policy, Pennsylvania State University, University Park.

Health Care Financing Corporation (HFCA) (1990), *Hospital Data by Geographic Area for Aged Medicare Beneficiaries: Selected Diagnostic Groups, 1986*. U.S. Dept. Health and Human Services, Washington, DC. HFCA Pub. 03300.

Hodge, W.T., and Nicodemus, M.L. Human Biometeorology: A Selected Bibliography, NOAA Technical Memorandum EDIS NCC-4, Sept. 1980.

Imai, M., Yoshida, K., and Kitabatake, M. (1986), Mortality from Asthma and Chronic Bronchitis Associated with Changes in Sulfur Oxides Pollution, *Arch. Env. Health* 41:29–35.

Jacobs, C.F., and Langdoc, B.A. (1972), Cardiovascular Deaths and Air Pollution in Charleston, S.C., *Health Services Reports* 87:623–32.

Kinney, P.L., and Ozkaynak, H. (1991), Associations of Daily Mortality and Air Pollution in Los Angeles County, *Environ. Res.* 54:99–120.

Kitagawa, T. (1984), Cause Analysis of the Hokkaichi Asthma Episode in Japan, *J. APCA* 34:743–46.

Lave, L.B., and Seskin, E.P. (1978), *Air Pollution and Human Health*, Johns Hopkins University Press.

Leonard, A.G., Crowley, D., and Belton, J. (1950), Atmospheric Pollution in Dublin during the Years 1944–1950, *Royal Dublin Soc. Sci. Proc.* 25:166.

Lewis-Faning, E. (1940). *A Comparative Study of the Seasonal Incidence of Mortality in England and Wales and in the United States of America*, Med. Res. Council Special Report 239, HMSO, London.

Lipfert, F.W. (1978), *The Association of Human Mortality with Air Pollution: Statistical Analyses by Region, by Age, and by Cause of Death*, Ph.D. Dissertation, Union Graduate School, Cincinnati, Ohio. Available from University Microfilms.

Lipfert, F.W. (1988), *Exposure to Acidic Sulfates in the Atmosphere: A Review and Assessment*, Electric Power Research Institute Report EA-6150.

Lipfert, F.W. (1991), Changes in the Seasonality of Mortality During the Past Half-Century, *Proc. 10th Conference on Biometeorology and Aerobiology, Am. Meteor. Soc.*, Boston, pp. 3–10.

Lipfert, F.W., and Wyzga, R.E. (1992), Observational Studies of Air Pollution Health Effects: Are the Temporal Patterns Consistent? Paper IU-21A.09, presented at the 9th World Clean Air Conference, Montreal, Canada.

Lipfert, F.W., Malone, R.G., Daum, M.L., Mendell, N.R., and Yang, C-C. (1988), *A Statistical Study of the Macroepidemiology of Air Pollution and Total Mortality*, BNL Report 52122, to U.S. Dept. of Energy.

Macfarlane, A. (1977), Daily Mortality and Environment in English Conurbations. 1. Air Pollution, Low Temperature, and Influenza in Greater London, *Brit. J. Prev. Soc. Med.* 31:54–61.

Macpherson, R.K., Ofner, F., and Welch, J.A. (1967), Effect of the Prevailing Air Temperature on Mortality, *Br. J. Prev. Soc. Med.* 21:17–21.

Plagiannakos, T., and Parker, J. (1988), *An Assessment of Air Pollution Effects on Human Health in Ontario*, Report No. 706.01 (#260), Energy Economics Section, Economics and Forecast Division, Ontario Hydro, Toronto.

Pope, C.A., III, Schwartz, J., and Ransome, M.R. (1992), Daily Mortality and PM_{10} Pollution in Utah Valley, *Arch. Env. Health*. 47:211–17.

Rogot, E., Sorlie, P.D., and Backlund, E. (1992), Air-conditioning and Mortality in Hot Weather, *Am. J. Epidemiology* 136:106–16.

Rose, G. (1966), Cold Weather and Ischemic Heart Disease, *Brit. J. Prev. Med.* 20:97–100.

Sakamoto-Momiyama, M. (1977), *Seasonality in Human Mortality*, University of Tokyo Press.

Schimmel, H. (1978), Evidence for Possible Health Effects of Ambient Air Pollution from Time Series Analysis, *Bul. NY Acad. of Med.* 54:1052–109.

Schimmel, H., Murawski, T.J., and Gutfield, N. (1974), Relation of Pollution to Mortality, Paper 74-220, presented at the 67th Annual Mtg. APCA.

Schwartz, J., and Dockery, D.W. (1992), Particulate Air Pollution and Daily Mortality in Steubenville, Ohio, *Am. J. Epidemiology* 135:12–19.

Stamler, J. (1985), The Marked Decline in Coronary Heart Disease Mortality Rates in the United States, 1968–1981; Summary of Findings and Possible Explanations, *Cardiology* 72:11–22.

Waller, R.E., and Lawther, P.J. (1976), Mortality and the 1974 fuel crisis, *Nature* 259:559–60.

Winkelstein, W., Jr., Kantor, S., Davis, E.W., Maneri, C.S., and Mosher, W.E. (1967), The Relationship of Air Pollution and Economic Status to Total Mortality and Selected Respiratory System Mortality in Men. I. Suspended Particulates, *Arch. Env. Health* 14:162–71.

Wyzga, R.E. (1978), The Effect of Air Pollution upon Mortality: A Consideration of Distributed Lag Models, *J. Amer. Stat. Assoc.* 73:463–72.

PART III

Studies of Selected Morbidity Effects of Air Pollution

All disease of Christians are to be ascribed to demons.

St. Augustine, 354–430 A.D.

Community health effects encompass much more than premature mortality and comprise a vast literature. The mortality findings of Section II would gain credibility if supported by similar morbidity findings. Two types of morbidity responses are reviewed in this section for this purpose: hospital admissions and emergency room visits, which are discussed in Chapters 9 and 10, and effects on respiratory function, which are discussed in Chapter 11. These endpoints were selected primarily because their measurements are objective. There is also an extensive literature on respiratory symptoms and absence from work or school; use of these endpoints involves a compromise between increased prevalence and reduced objectivity. Perhaps a future edition of this book can incorporate these studies.

Increased rates of hospitalization were most obvious during the air pollution disasters discussed in Chapter 5. Time-series analysis techniques similar to those used for mortality (Chapter 6) have been used by several authors to try to deduce quantitative morbidity relationships, for example, for rates of hospital admission and emergency room visits; these studies are reviewed in Chapter 9. A less common study design has attempted to relate long-term measures of hospital use (length of stay or admission or discharge rate per capita) and air quality on a spatial or cross-sectional basis; these studies are discussed in Chapter 10.

Respiratory health in the United States can be measured by several different statistical measures, none of which is likely to represent the true "underlying morbidity" of the population. Annual mortality due to respiratory conditions totals about 0.7 per 1000 population; annual hospital admissions with respiratory diagnoses, about 13 per 1000; chronic respiratory conditions (bronchitis, emphysema, asthma) total about 70 per 1000 population (of which about 95% have one or more physician visit per year); acute respiratory conditions occur at the rate of about 1100 per 1000 population. As an example, Ayres *et al.*

(1989) reported a winter weekly respiratory case load of 5 to 8 per 1000 population for his general practice in Britain, which might sum to an annual rate of about 200 per 1000 when the normal seasonal variation is taken into account. A further example of the range in morbidity statistics may be seen from the data of Richards *et al.* (1981) for children's asthma in Los Angeles; about 14% of emergency room cases were admitted. The range in these various per capita measures of health response is thus over three orders of magnitude; hospital admissions are in the lower end of this range.

However, it is important to understand that mortality and hospitalization studies may involve different segments of the population. For example, children's hospital admissions are an important part of total respiratory admissions but constitute only a small part of mortality. Also, studies of lung function variability are usually based on a general population, most of which is healthy. These differences in the populations involved must be considered when their responses to air pollution are compared.

9

Temporal Studies of Air Pollution and Hospital Use

STUDY METHODOLOGIES

Studies of rates of hospitalization in relation to air pollution attempt to infer dose-response relationships on the basis of concordant temporal (Chapter 9) or spatial (Chapter 10) patterns. With severe air pollution episodes, the responses may be obvious in relation to "normal" patterns of hospital use, but when longer time periods are considered, a number of methodological issues may arise as discussed in conjunction with mortality studies. Some of these issues relate to the "ecological" nature of such observational studies, that is, the attempt to infer effects on individuals on the basis of observations of the behavior of groups. Other statistical problems arising with time-series studies include lag effects, serial correlation, confounding by exogenous long-term trends, and collinearity among independent variables (Chapter 3).

Validity of the Dependent Variable Measures. Among the various types of air pollution health effects that have been studied, there are certain advantages to using hospitalization statistics. Studies of mortality deal with an inevitable endpoint and are thus concerned with determining the degree of "prematurity" in addition to cause and effect relationships. In contrast, a visit to a hospital is a voluntary response by the affected individual, based on perceived needs and often after the advice of a physician, and thus all hospitalizations could be considered as "excess." (In this chapter, "excess" hospital usage is that portion associated with air pollution.) In addition, since hospital admissions are more than an order of magnitude more numerous than deaths (for a given population), using this endpoint facilitates the study of more specific diagnoses.

Most of the studies reviewed in this section employed one of two dependent variables. Some studies dealt with emergency room (ER) visits, for which the action taken can be assumed to have been determined solely by the affected individual. Hospital admissions, on the other hand, require the patient to be examined by a physician, which could remove an element of subjectivity from the data (Bates and Sizto, 1983). Admissions are less frequent than ER visits by a factor of 2 to 5; in 1981, the average number of yearly visits to a hospital ER or outpatient department for all diagnoses varied from 0.5 to 0.7 per person, depending on age; hospital admissions varied from 0.07 to 0.37 per person. Both of these endpoints are likely to be less subjective than diary studies of symptoms, for example. Also, it may be unreasonable to expect a person affected by air pollution to be hospitalized on the day of exposure, except in the most severe cases; a lag may result from either the unavailability of a hospital bed or the appropriate physician. Moreover, it is likely that there will be a range of such lags in a heterogeneous population.

Most of the studies examined short-term variation in hospital admissions using time-series methods. However, any significant short-term excess admissions should also be reflected in the long-term statistics (either admissions or discharges) and possibly as differences in lengths of hospital stays. Cross-sectional methods are used to examine differences in long-term averages and must account for a host of possible confounding factors, such as those influencing the supply and delivery of medical care. It also follows that if decisions to hospitalize are influenced by factors other than the patient's immediate needs, short-term cause-and-effect relationships could be obscured as well.

Reliability of Air Pollution Exposure Estimates. None of the studies in the literature used personal air pollution monitors or activity patterns to derive detailed estimates of air pollution exposure. It was necessary to assume that the time-series of community air pollution monitor readings was an adequate proxy for total integrated exposure (with the possible exception of the Portland, OR, study [Jaksch and Stoevener, 1974]). This constitutes the "ecological" nature of this group of studies.

Two types of exposure errors are possible: confounding, in which an unmeasured agent affecting hospitalization is highly correlated with a measured pollutant; and random errors due to unmeasured indoor pollutants or to randomly located unmeasured outdoor pollutants. To deal with confounding by collinear variables, it is useful to compare similar types of studies in different geographical settings where the confounding variables are likely to differ, which is one of the objectives of this review. Random errors in independent variables, on the other hand, will usually bias any derived correlations or regression coefficients downward. Thus, one could expect that the dose-response relationships derived by studies employing imprecise pollution measures will be underestimates of the true effects, *ceteris paribus*. As discussed, selection of the appropriate averaging time for the air quality data can influence the precision of exposure estimates.

Use of Control Diagnoses. Controls are intended to test for "false positives," that is, statistically derived associations between health measures and air pollution that are physiologically implausible. In studies of air pollution episodes, the population serves as its own control in that it is expected that its health status will return to normal values after the episode has abated. For time-series studies, it may be useful to select as controls those diagnoses that are unlikely to repond to air pollution or populations that are known to be unexposed to the pollutants in question.

Confounding Variables. All time-series studies of air pollution effects must take into account the "natural" temporal patterns in air quality, which can confound statistical analyses if the response variables have similar patterns that could conceivably be due to other factors. These include both long- and short-term temporal cycles. On a long-term basis, seasonal trends include higher ozone levels in summer and higher levels of pollutants associated with space heating in winter. Such seasonal cycles may be confounded with seasonal weather cycles, which can also affect health responses. Respiratory illness shows a seasonal pattern, higher in winter, in part because of periodic outbreaks of viral infections. There may also be longer term air quality trends associated with community growth or with air pollution abatement efforts. Weekly cycles may be expected in traffic-related air pollutants or in those pollutants related to local industrial sources that operate with reduced emissions on weekends. Hospital usage also tends to exhibit weekly patterns, perhaps because of the reduced availability of physicians on weekends. On weekends, ER usage is frequently higher, while admissions are usually reduced.

One of the traditional methods of avoiding such temporal confounding is the use of "deviation" variables, that is, the difference between the observed value and the value expected for that day of week, season, year, and so on. This practice is essential for the dependent variables; its use for the air pollution variables is tantamount to assuming linear responses, which are often assumed by the use of linear regression models in any event. A metric in less common use for studying specific diagnoses, such as repiratory disease, is the percentage of all diagnoses represented by the selected category. This implicitly assumes that the confounding variables such as day-of-week are randomly distributed among all diagnoses, which may not always be a valid assumption.

STUDIES OF AIR POLLUTION DISASTERS

Air pollution disasters or "episodes" lasting only a few days allow more detailed examinations of the time course of events than any other type of study. However, the rarity of these events and the accompanying high concentrations of pollutants may make extension to the more general case somewhat problematic.

Donora, PA

The Donora disaster in October 1948 affected a high percentage of residents and was studied extensively, albeit retrospectively (Shrenk *et al.*, 1949). No concurrent air monitoring was performed. Of the 50 persons hospitalized (compared to 20 deaths), records were available for 32, and two of these were workers on riverboats that were passing by Donora. This suggests that prior chronic exposure, which might be assumed for residents, was not a prerequisite for susceptibility to this incident. Based on the available hospital records, Shrenk *et al.* found no diagnostic information that related directly to air pollution; all of the findings were attributed to other disease conditions present. For example, "all patients appeared to have in their clinical histories a component of cardiorespiratory disease."

London

The data collected during the fogs of December 1952 and December 1957 lend themselves to detailed analysis of mortality and hospitalization. The most useful measure of hospitalization rate is "requests for emergency bed service" for Greater London (Abercrombie, 1953; Ministry of Health [MOH], 1954), since this measure accounts for possible supply constraints due to unavailability of beds. These figures account for about 25% of daily admissions under normal conditions.

The episode of December 1952 in London appears to be the first for which both air quality and health response data were collected. Fog persisted for five days and smoke and SO_2 levels exceeded 4000 μg/m^3 locally in central London; Meetham (1981) estimated that sulfuric acid and CO levels would also have been elevated, although no measurements were made of these species. The winter average of emergency bed requests for acute (noninfectious) cases was about 150 to 180 per day (Abercrombie, 1956). When the episode began on December 5, weekly admissions were already elevated above normal levels (>200 per day) for reasons that may relate either to weather or to an excess in the numbers of pneumonia cases (MOH, 1954). The peak daily emergency bed request was reported to be 492 on December 9 (an excess of about 150%), on the day after the peak pollution was recorded. These figures compare to a daily average of 750 acute respiratory hospital admissions for the week preceding the fog and an increase to 1110 on December 9, of which 460 were for respiratory disease. Weekly admissions did not return to the pre-episode level until December 25. Most of the excess applications were for respiratory diagnoses, which increased by a factor of 4. Cardiac disorders tripled, but from a much lower base level. By age, most of the increase was reported for the less-than-5-year and over-45-year age groups.

The time histories of mortality, emergency hospital bed applications, and the sum of British smoke and SO_2 are plotted in Figure 9-1. Air quality data are based on the 24 hours ending at noon on the day in question. Since they tend to be highly correlated during London winters, values of smoke and SO_2 have been summed to provide a crude index of the combined pollution effect during the episode, for the purpose of illustration. The increases in smoke and SO_2 were reasonably coincident; the period of heavy fog lagged air pollution by about two days (perhaps because the polluted air contributed to fog formation). Mortality lagged air pollution by about one day (actually by one and a half days). Hospital applications increased more slowly and stayed elevated after the fog (note the periodic dips in hospital applications on Sundays; I used an additive day-of-week correction of 100 applications for the Sunday values for the pupose of estimating dose-response curves). These pollution data are based on the average of 11 stations for smoke and 10 for SO_2; the average values are considerably lower than the figures that are usually used to characterize this episode (4000+ μg/m^3).

Cross-plots of these data are given in Figure 9-2a and 9-2b for various assumed lags between pollution and response; the arrows indicate the direction of increasing time. This model assumes an average response over a period of k days after the pollution measurement. Note that the peak pollution levels occurred on the weekend during this episode, during which only 48-hour

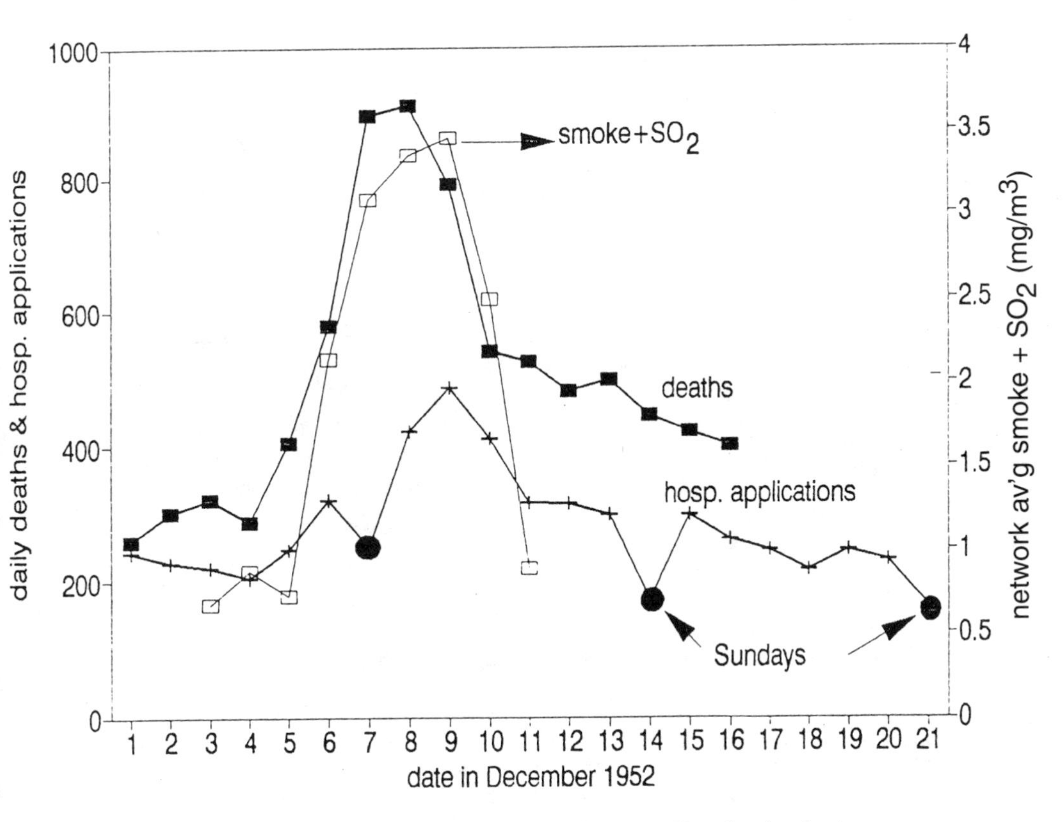

FIGURE 9-1. Time line for the 1952 air pollution episode in London, showing deaths, emergency applications for hospital beds, smoke, and SO_2 (network averages). *Data source: British Ministry of Health (1954).*

samples were taken at most of the air monitoring stations. Thus, the dose response curves should actually pass through the average response (indicated on the graphs) for the two peak pollution days. Figure 9-2a plots mortality and morbidity responses for assumed lags of 2 days. The mortality responses form a reasonable dose-response curve except for the last day, which shows that excess mortality persisted beyond the assumed 2-day lag. Hospital applications did not form a linear dose-response relationship for this lag. Figure 9-2b presents these same data for a lag of 3 days for mortality and 5 days for hospital applications, which results in more conventional-looking dose-response curves. The nondimensional slopes or elasticities of these dose-response relationships are about 0.70 for mortality and 0.50 for morbidity. It is noteworthy that, in spite of the severity of this episode, "excess" deaths exceeded "excess" hospitalizations (as measured by emergency requests), which suggests that the primary mechanism may have been exacerbation of existing disease.

These dose-response relationships are seen to resemble either of two types of patterns, depicted schematically in Figure 9-3. When the assumed lag period is too short (top), the response persists after the pollution has abated. When it is too long, there is an induction period at low pollution levels during which the response seems to increase precipitously with only minimal changes in

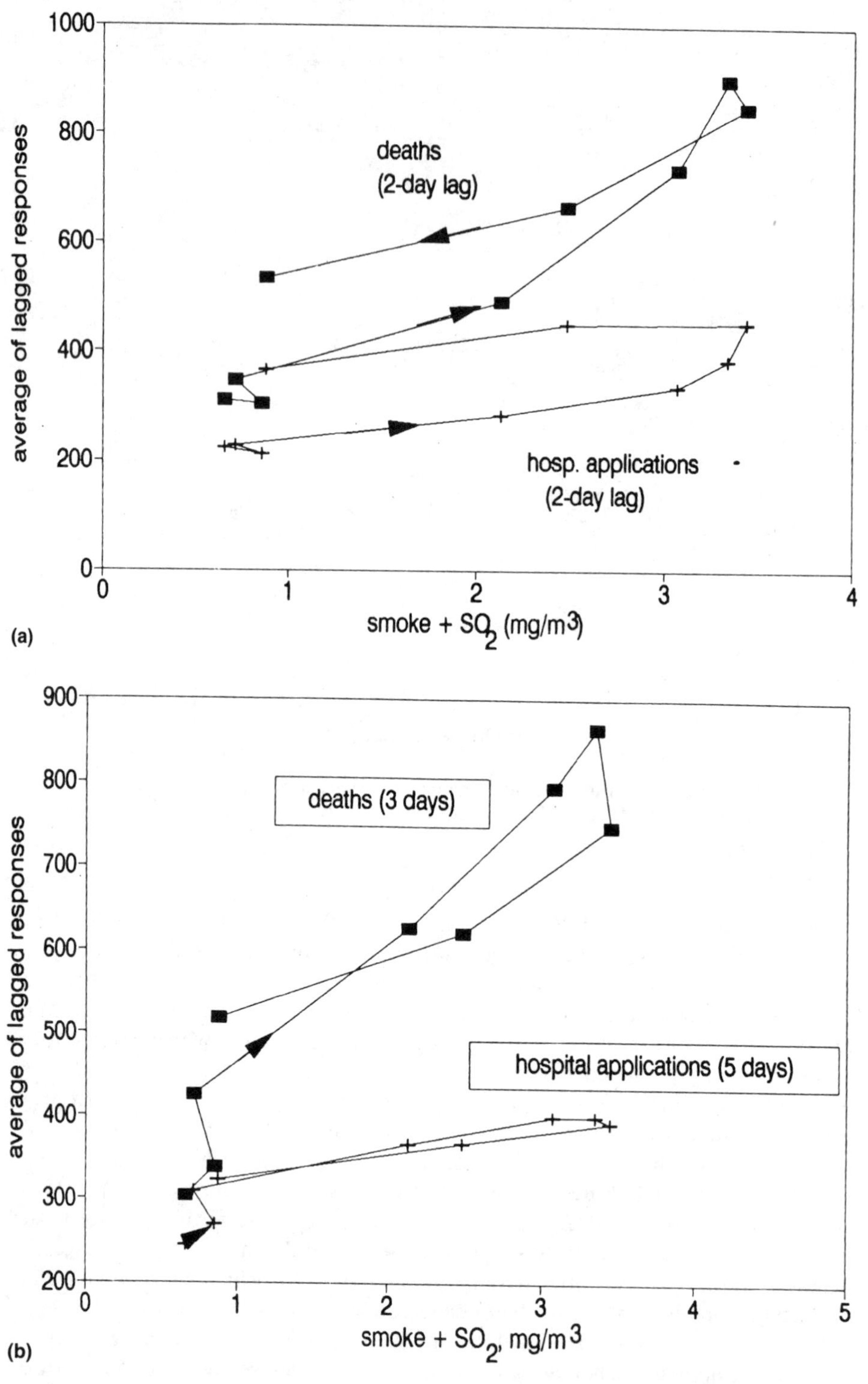

FIGURE 9-2. Dose-response functions for the 1952 London episode, for mortality and hospital applications. (a) Based on the sum of smoke and SO_2, lagged 2 days. (b) Based on the sum of smoke and SO_2 with a 3-day lag for mortality and a 5-day lag for hospital applications.
Data source: British Ministry of Health (1954).

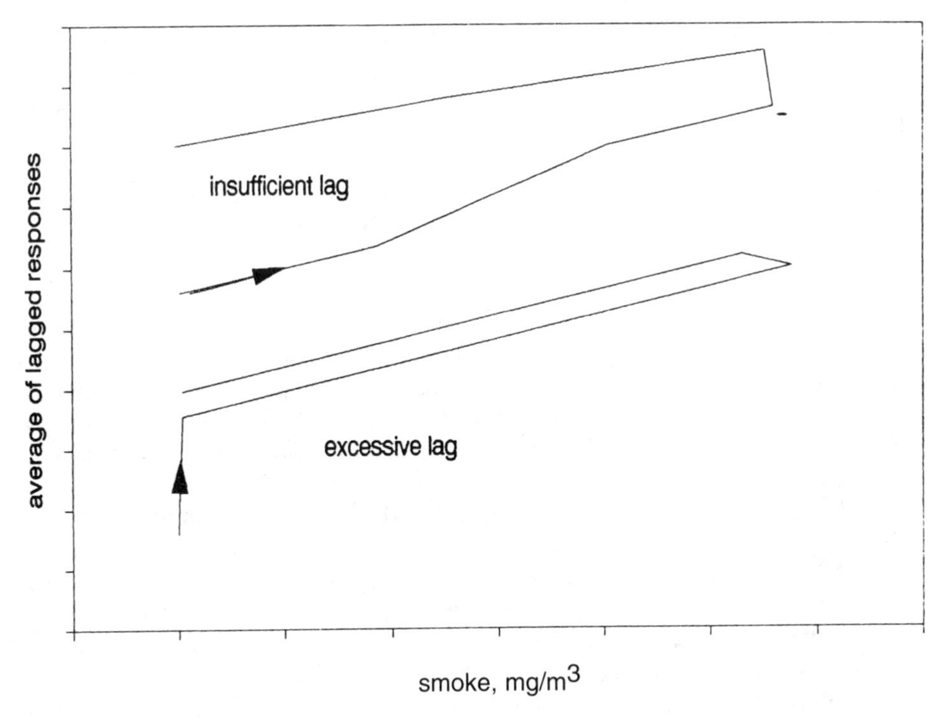

FIGURE 9-3. Examples of generic dose-response functions for various lags.

pollution. The patterns of Figure 9-2 thus seem to indicate that during a severe episode, the lag increases during the course of the episode, probably because the most susceptible individuals respond first. If an attempt is made to define such an episode by means of a linear slope, it appears that use of insufficient lag will underestimate the slope (with a reduced correlation coefficient), and excessive lag may either overestimate the slope or indicate a quadratic relationship with leveling off at higher pollution levels.

Fry (1953) reported the impact of the 1952 fog on his general medical practice (4500 patients) in the southeastern outskirts of London for the week of the episode; upper respiratory tract complaints increased by about 150%, while "respiratory disorders" (all conditions with symptoms referable to the lungs and where abnormal signs were detected) increased by about 400%. The increased patient load continued for at least another week, having begun on the third or fourth day of the fog (corresponding to maximum pollutant concentrations). Fry identified 43 patients with lower respiratory tract involvement, 37 of which had a history of previous chest trouble. These were concentrated in the age group 60–80 and consisted of 29 men and 14 women. He considered 16 of these to be severely ill; of the remainder, "quite a number . . . were able to remain up and about in their homes and some even went out to work." Two of the group died within 8 days and one was admitted to hospital. Fry reports that none of his pediatric asthmatic patients was affected by the fog episode. The MOH report (1954) shows an increase in upper respiratory disorders of about 100% lasting at least 2 weeks after the incident. Fry also reported that six patients were seen with sudden attacks of vomiting at the peak of the fog, not accompanied by other gastric symptoms. He concluded that the cause was related to "swallowing fog." This was also the case at Donora, where some instances of gastric distress were reported (Roueché, 1984). If Fry's somewhat anecdotal report could be taken as representative, one would conclude that many more people suffered symptoms from the episode than either died or were admitted to hospital (but nevertheless, only a small fraction of the total population was seriously affected).

The January 1956 fog in London (and elsewhere in England) was somewhat less severe, since only one 6-hour period of visibility less than 100 m was recorded (Bradley *et al.*, 1958) and there were far fewer casualties (Logan, 1956). The reported morbidity data (Logan, 1956) included the increase in claims for "incapacity benefits" for 2-week periods, which was about 70% for Greater London and the South-Eastern Region and 41% for the remainder of England and Wales. The difference between these two figures may represent the local morbidity effect of the fog.

In December 1957, London was again struck by a 4-day polluted fog episode (Bradley *et al.*, 1958). Average pollution levels were about 25 to 30% lower than 1952 when based on a common set of measuring stations, but the fog was not as thick, and visibility was severely restricted for a smaller fraction of time. Also, calm winds were reported much less frequently in 1957. Excess mortality was only about 20% of that experienced in 1952. Figure 9-4 presents the time histories of the various parameters; these trends are qualitatively similar to those from 1952 (Figure 9-1) in that both deaths and hospital applications lag pollution by several days and hospital admissions are low on Sundays.

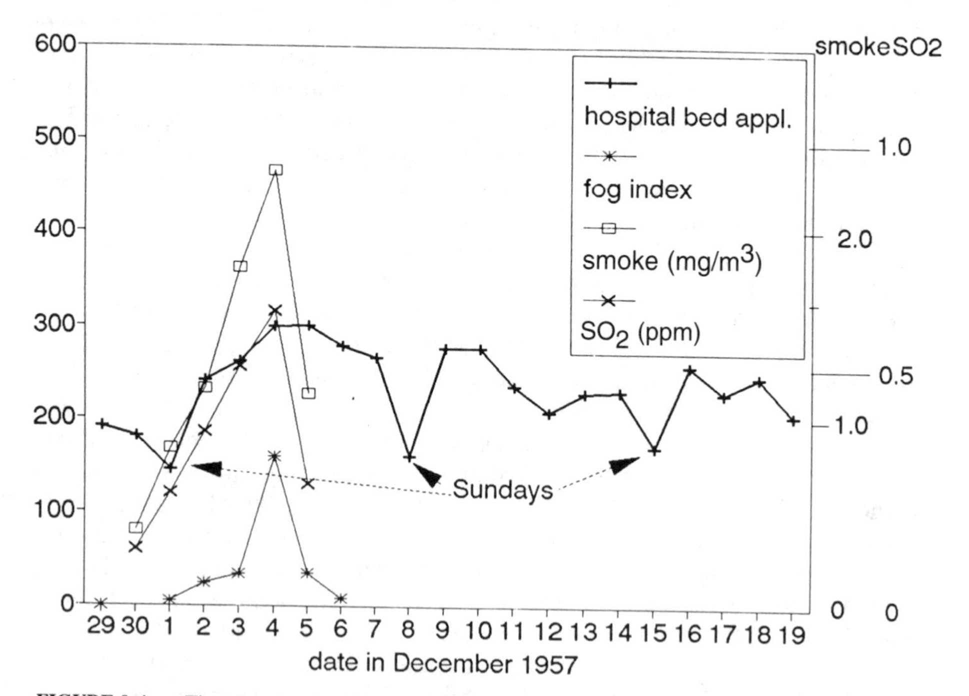

FIGURE 9-4. Time line for the 1957 air pollution episode in London, showing deaths, emergency applications for hospital beds, air pollution (network averages of smoke + SO_2). *Data from Bradley* et al. *(1958).*

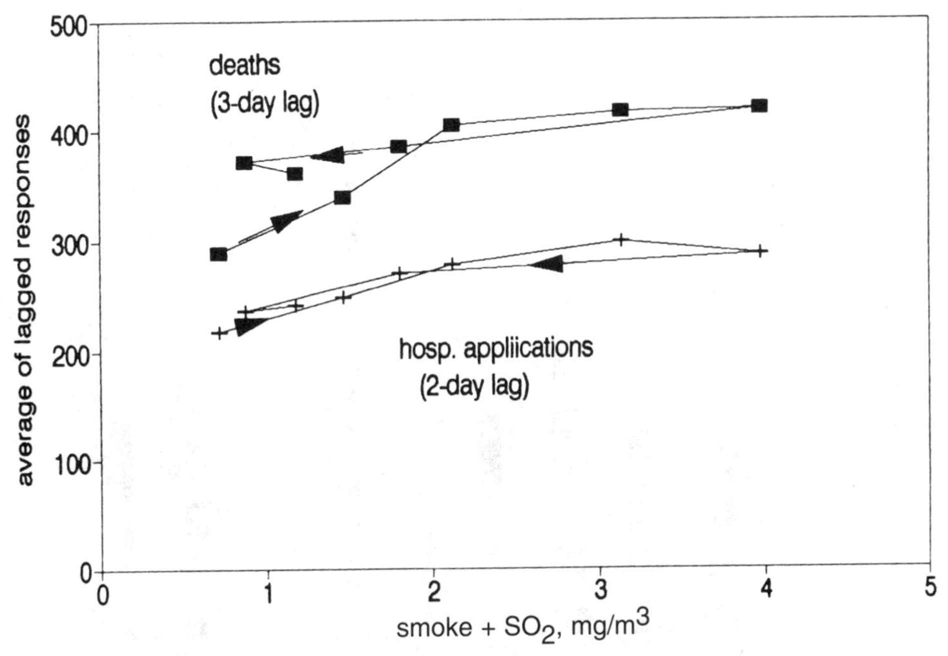

FIGURE 9-5. Dose-response functions for the 1957 London episode for mortality and hospital applications, with a 3-day lag for mortality and a 2-day lag for hospital applications. *Data from Bradley* et al. *(1958)*.

I attempted to compensate for this by adding 50 applications to each Sunday's total as a day-of-week correction for the purpose of plotting dose-response curves (Figure 9-5). The most consistent dose-response curve for hospital applications was based on the sum of smoke and SO_2 with a 2-day lag; the elasticities for both deaths and hospital applications were about 0.20.

The 1952 and 1957 episodes differed greatly in the severity of their impacts, 1952 being much more lethal. However, similar maximum air concentrations were reached in both, but at different locations *vis-à-vis* central London. Figure 9-6 plots smoke and SO_2 against radius from the central railway station (Charing Cross). In 1952 maximum concentrations occurred downtown; in 1957 they were 9 miles away. Perhaps the population density was less at the more distant location, and thus fewer people may have been exposed. This example underscores the need to have more than one monitoring station in a large city.

Few morbidity data were reported from the 1962 episode in London, during which SO_2 levels exceeded those reported in 1952, but smoke levels were considerably lower. Marsh (1963) reported a 50% increase in new sickness benefit claims. Waller *et al.* (1969) showed a sharp (1-day) peak of about 100% in emergency bed applications coincident with the day of maximum air pollution; the mortality peak was only about 40% but lasted for 3 days. This paper also shows a less severe event in January 1963; the emergency bed application response to both of these peaks was about 3% per 100 $\mu g/m^3$ smoke. When referred to SO_2, the responses were about 2.6% and 1% per 100 $\mu g/m^3$, respectively. These figures are comparable to the slopes of Figure 9-2.

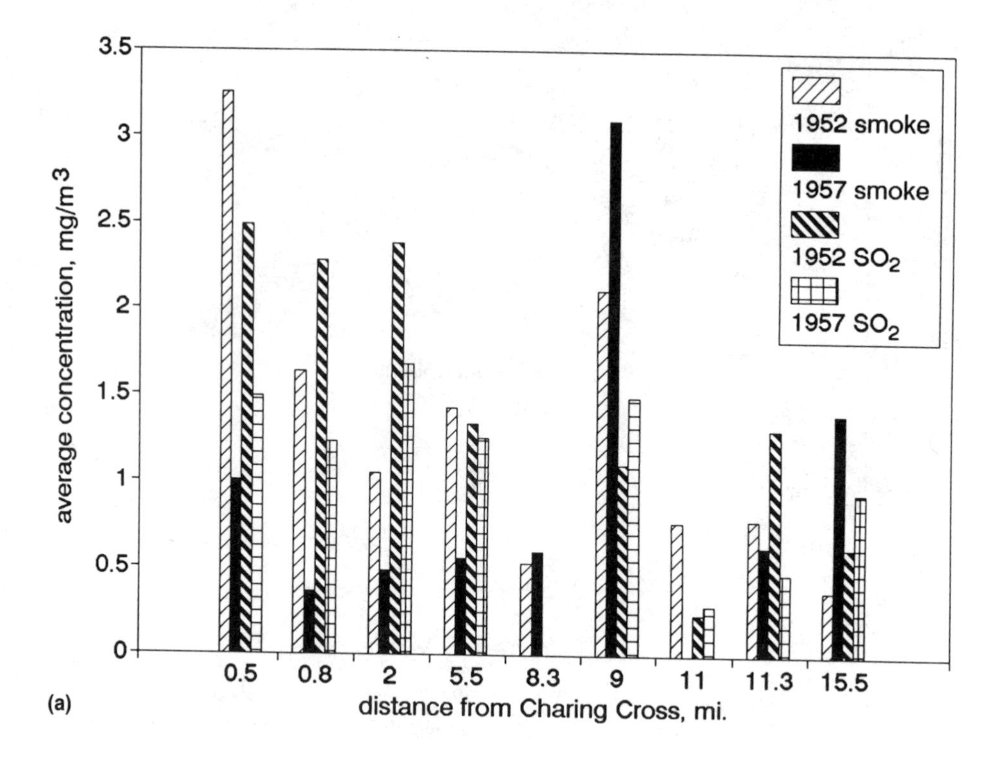

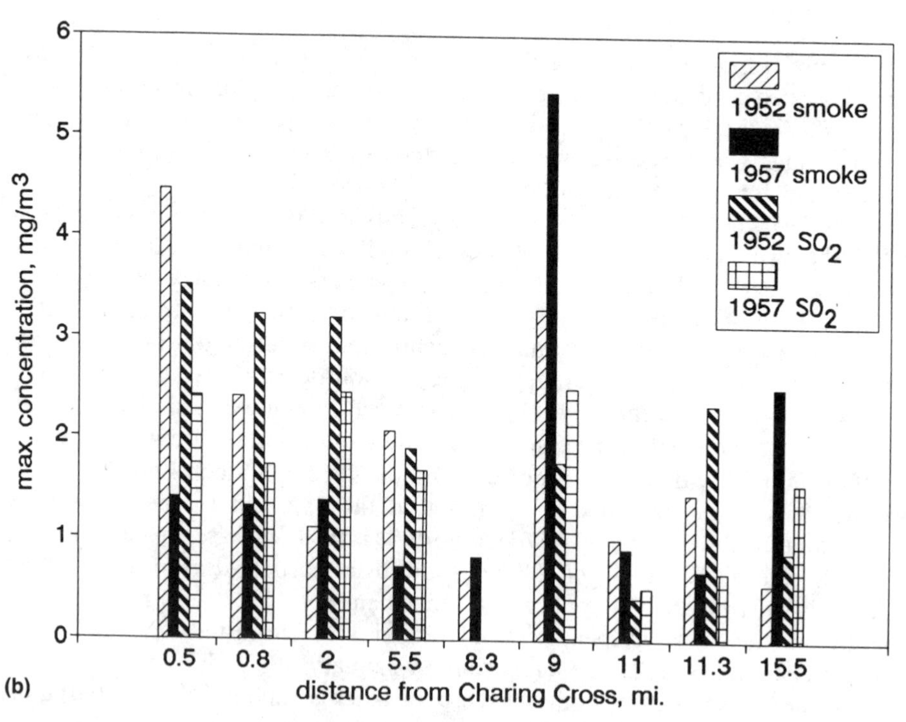

FIGURE 9-6. Relationships between air pollution and distance from central London, for the episodes of 1952 and 1957. (a) Concentrations averaged over the episode. (b) Maximum 24-hr concentrations during the episode.

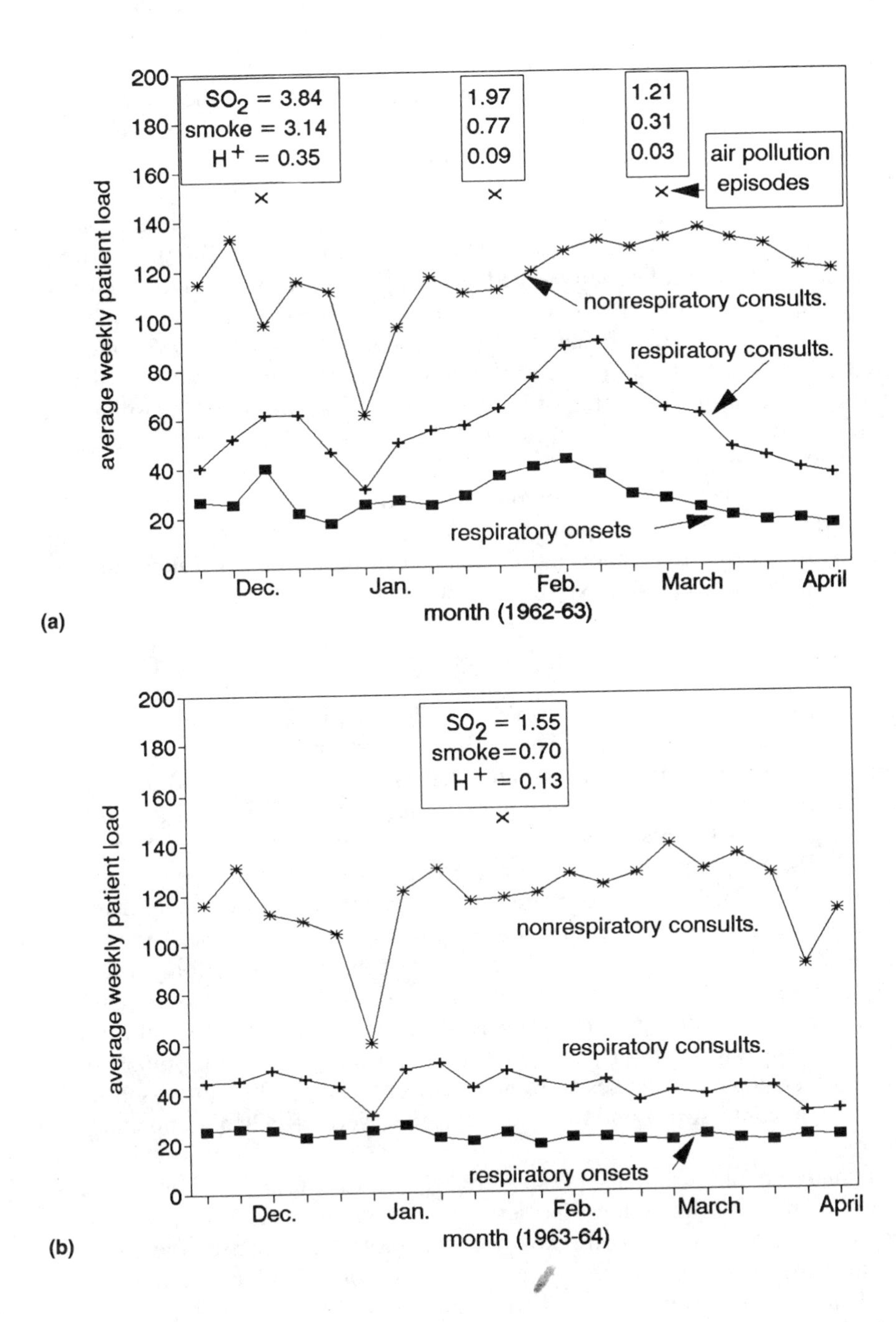

FIGURE 9-7. Time lines of general practice morbidity indices during the 1962 and 1963 London episodes (indicated by the air pollution levels at the top of each plot). (a) Winter of 1962–63. (b) Winter of 1963–64. *Data from Carne (1967).*

Effects of the 1962 fog on general medical practitioners were reported by Carne (1967), as part of an analysis of pollution and weather effects for the winter seasons of 1962–64. Corresponding air quality data for episodic events were reported by Waller and Commins (1967). These two data sets have been combined in Figure 9-7, which plots weekly time lines for new cases of respiratory disease ("onsets"), and physicians' consultations for either respiratory disease or other causes. The pollution data are the maximum daily values for the event; the "H^+" data represent the total acidity of the aerosol reported in equivalent units of H_2SO_4, in $\mu g/m^3$. The December 1962 fog is seen clearly in terms of a sharp peak in onsets and a broader peak for respiratory consultations (Figure 9-7a). There is a sharp coincident drop in nonrespiratory consultations; this could be interpreted as a supply constraint or as the fact that perhaps people did not wish to venture out during these adverse atmospheric conditions. Such avertive behavior was mentioned by Marsh (1963) and by Scott (1963). The sharp drop in consultations at Christmas is notable, as is the broad increase at the beginning of February. No peaks are seen corresponding to the other two air pollution "episodes" of that season (although the first one could have been obscured by the general rise in respiratory effects at that time); their pollution levels were considerably lower. Corresponding data for the winter of 1963–64 are plotted in Figure 9-7b. The effects of the one "episode" indicated by Waller and Commins can barely be discerned, but the pollution levels were modest (in terms of normal London levels).

New York City

The first reported air pollution episode in New York occurred in November 1953; the lack of atmospheric ventilation caused smoke levels to reach a coefficient of haze (COH) value greater than 8.0, and SO_2 reached 0.85 ppm; fog was reported sporadically (Greenburg *et al.*, 1962a). The total suspended particulate (TSP) sampler routinely operated by the National Air Sampling Network (NASN) recorded a 24-hour average of 642 $\mu g/m^3$ during this event. Widespread complaints of eye irritation were reported, and data on visits to emergency clinics at four of the city's hospitals were analyzed (Greenburg *et at.*, 1962b). The numbers of emergency clinic visits for each of four hospitals before, during, and after the episode were calculated, and it was concluded that statistically significant increases were seen for upper respiratory infections at three of the four hospitals, and for cardiac diagnoses at two hospitals. The New York stagnation episodes were generally characterized by increased temperatures, in contrast to London, where episodes were usually accompanied by colder weather.

I reanalyzed this event by pooling the visits to all four hospitals and analyzing them as a single daily time series with varying lags. Day of week was accounted for, but temperature was not; the short time span of the analysis (29 days) makes it unlikely that seasonal trends could have interfered. Since day-of-week effects were marginal, the analyses were performed as bivariate regressions, using either smoke or SO_2 as the dependent variable. The SO_2 monitoring record was fragmentary (data available about 2 hours per day, morning and afternoon, for 10 days), which made it impossible to discriminate between the two pollutants on the basis of these data alone. The results are presented in Figure 9-8a through 9-8d. Respiratory clinic visits appeared to be associated with either SO_2 or smokeshade, with no lag (Figure 9-8a and 9-8b). Cardiac visits fit better with a 3-day lag (Figure 9-8c and 9-8d). The largest percentage response is seen for cardiac causes, although the absolute numbers are small (13 visits per day for the four clinics). The three smokeshade relationships are statistically significant ($p < 0.01$). Regressions using the data grouped for three periods in November 1953 and the averages of the Novem-

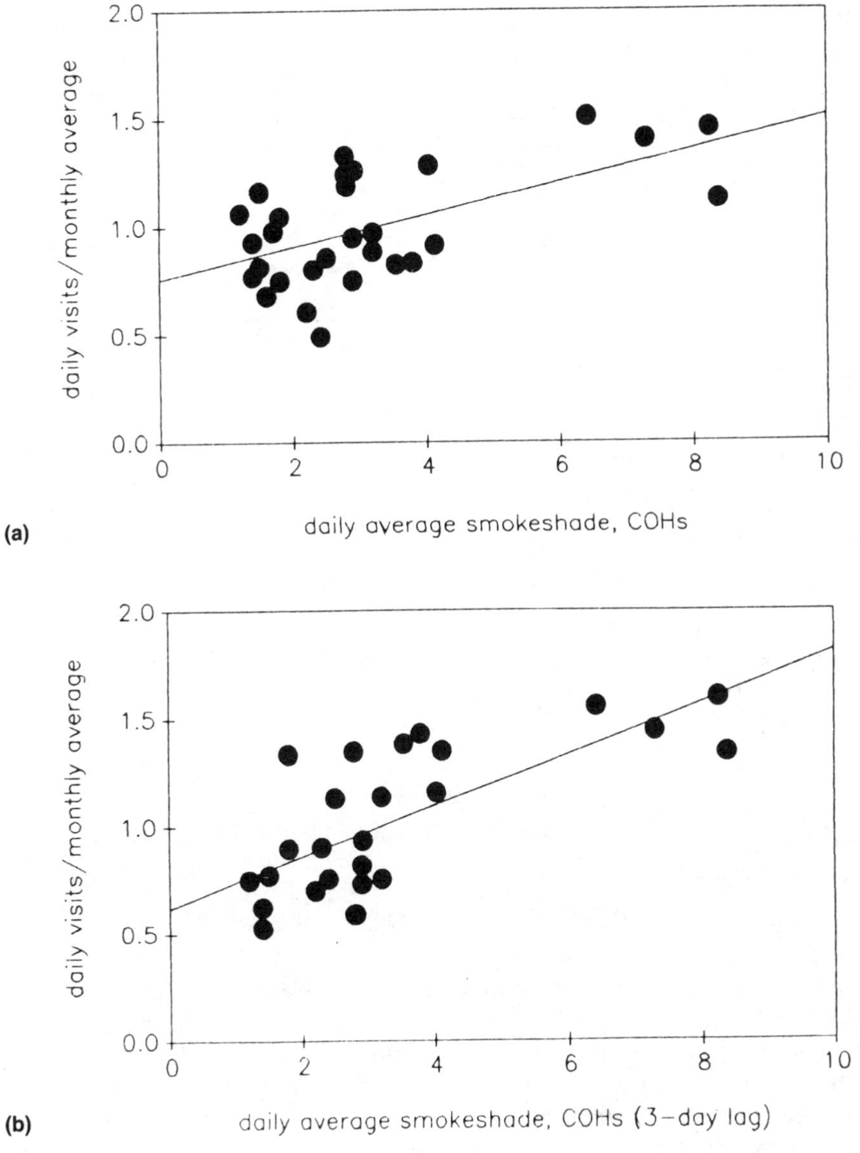

FIGURE 9-8. Dose-response functions for emergency clinic visits during the 1953 air pollution episode in New York City. (a) Respiratory visits, based on smokeshade. (b) Respiratory visits, based on SO_2. (c) Cardiac visits, based on smokeshade. (d) Cardiac visits, based on SO_2. *Data from Greenburg* et al. *(1962a)*.

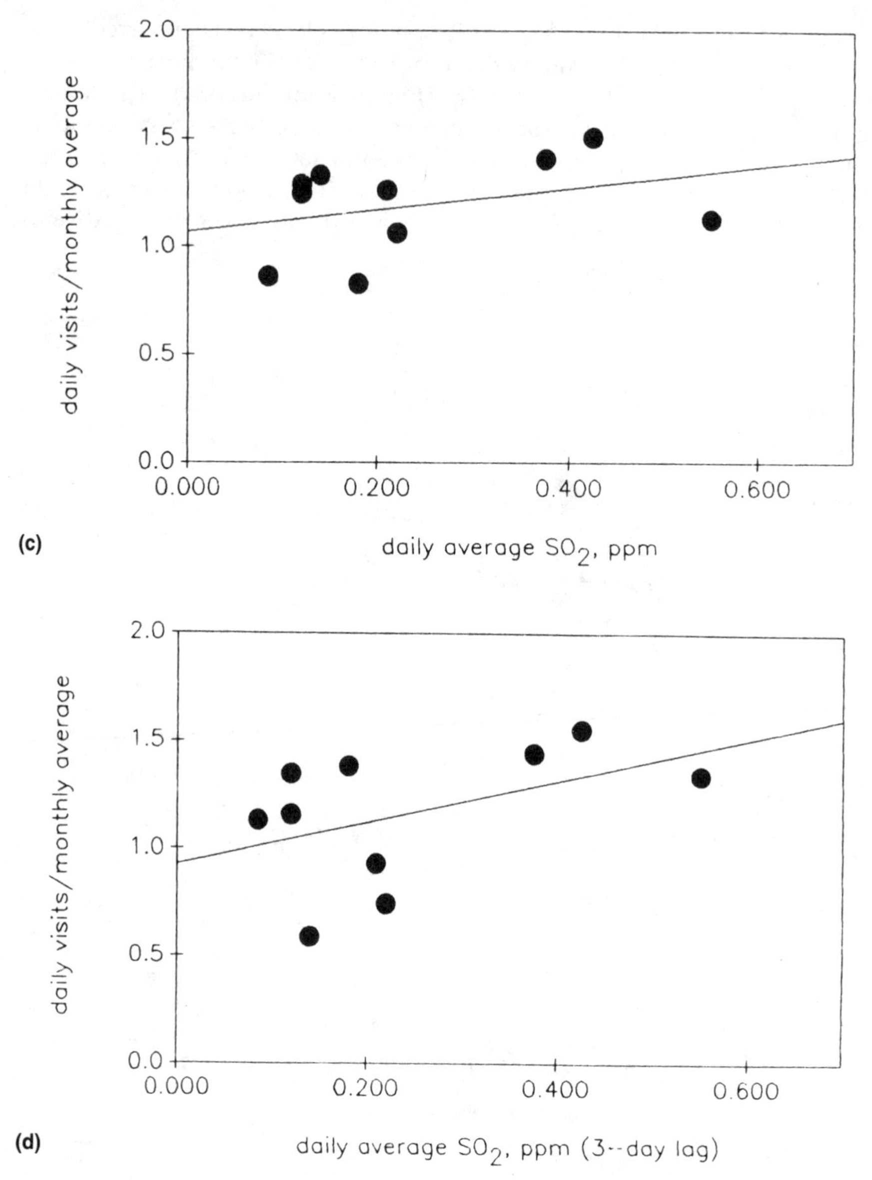

FIGURE 9-8. *Continued*

bers of 1950–52 and 1954–56 gave very similar results. Assuming that the many other hospitals in New York had similar responses, the morbidity effect would appear to have been larger than the mortality effect during this episode (about 200 excess deaths or about 9%) (Greenburg *et al.*, 1962b).

In December 1962, similar air pollution levels were recorded in New York, but for intermittent periods over 8 days (Greenburg *et al.*, 1963). Each day, SO_2 reached 1.4 ppm for several hours but dropped to normal levels; smokeshade was elevated on a more persistent basis, reaching about 8 COHs. Mortality counts and visits to six clinics were analyzed, plus data from Blue Shield (apparently insurance claims) and four old-age homes. Again, the statistical technique was to compare grouped time periods; the authors report significant increases during the episode only for the old-age homes (it was not clear how the tests of significance were performed). Since the authors did not report daily morbidity data, it was not possible to perform a reanalysis. It appeared that the sum of morbidity responses across all institutions was elevated during the episode period. A reanalysis of the mortality data showed a nearly significant association with smokeshade with about the same slope as in the 1953 episode.

Data from the 3-day Thanksgiving 1966 episode in New York (Glasser *et al.*, 1967), during which smoke reached about 600 μg/m^3 and SO_2 reached 1300 μg/m^3, were confounded by the closing of the outpatient clinics on the holiday (Thursday). This may have created an increased patient load on the ERs, which showed excess use only on Friday, at the end of the episode. Thus the excess could have resulted from the perturbation caused by the holiday. The numbers of ER visits were small, however; the "excess" respiratory visits totaled about 40%, based on an average of the holiday and the day after.

Los Angeles

Periods of high air pollution were experienced in the Los Angeles basin in the fall of 1954 (Air Pollution Study Project, 1955). The initial report focused on the period of October 15–20, during which oxidants increased by about 15% over the control period; no consistent relationships with mortality or morbidity could be found. However, examination of the published raw data shows two spikes in particulates, October 8–10 and November 4 (from "normal" levels of about 200 μg/m^3 to 500–650 μg/m^3 at Pasadena). The periods of highest hospital admissions for all diagnoses and for asthma were October 16 and November 6 (asthma admissions were approximately doubled). In general, there were no anomalous temperatures during this period, except for an increase of about 10 to 15°F on November 5, to 85°F. Other morbidity measures also peaked during these two weeks. This response is consistent with a lag of a few days, which had apparently not been considered by the original authors.

Pittsburgh

Air stagnation occurred near Pittsburgh in August 1973, with 24-hour TSP levels around 400 μg/m^3 (Riggan *et al.*, 1976). Sulfur dioxide reached about 340 μg/m^3 (R. Savukas, Allegheny County Bureau of Air Pollution Control, personal communication, May 1989). Using Medicare records, Shiffer and Parsons (1980) deduced an increase in respiratory hospitalizations during this 13-day period compared to the 2 previous years (35 admissions vs. 13 and 21, respectively). They also noted four deaths following hospital admission with a respiratory diagnosis, which was more than the previous years but too few for reliable statistics.

Mount St. Helens Volcanic Eruptions

In May and June 1980 volcanic eruptions blew ash into the stratosphere and increased atmospheric suspended particulate levels over a wide area, with ash deposition levels of up to 2 inches in Spokane County, Washington, and elsewhere in the region. Most of the ash was of respirable size, although the fine particle fraction mass was only around 100 $\mu g/m^3$ (Fruchter *et al.*, 1980). The TSP levels were much higher, up to 30,000 $\mu g/m^3$ in the plume (Baxter *et al.*, 1981). There was little sulfur or chlorine in the ash, and the pH after solution was essentially neutral (Fruchter *et al.*, 1980).

At least 18 people died in the impact zone due to asphyxia from ash inhalation (Baxter *et al.*, 1981); the population at risk was not reported, but the 150 square-mile area was described as "largely uninhabited." Kraemer and McCarthy (1985) reported that Spokane residents complained of eye and throat irritation and that local hospitals experienced increased pediatric asthma admissions. Comparing 1980 with 1981 hospital admissions showed an excess of 34% for the year, but not all of this excess was associated with the eruption. The 1980 admissions rate for May was double that of 1981, but this was also the case for the month of April (before the eruption). A regression of monthly percentage of excess asthma admissions against the monthly average difference in TSP was not statistically significant, although the regression coefficient was similar in magnitude to that estimated from other episodes and studies. In a time-series analysis without reference to a "normal" year for standardization, TSP would undoubtedly be a significant predictor. Biweekly pediatric asthma admissions responded to TSP spikes of about 1000 and 10,000 $\mu g/m^3$ with approximate doubling in admission rates (in both cases).

Following the first eruption, Baxter *et al.* (1981) reported increased ER visits and admissions in eastern Washington by about a factor of 3 for respiratory diseases in the first week and persisting for 4 weeks. After the May 25 eruption, the local increase was about 100%. They also reported that some patients who initially had breathing difficulties were later diagnosed with cardiac problems. The largest category of respiratory ER visits was for asthma or wheezing.

Tests of children's lung function (see Chapter 11) found essentially no effects due to deposited ash (Buist *et al.*, 1983); this evaluation was based on tests conducted at a summer camp after the eruption at which average respirable dust levels were around 170 $\mu g/m^3$. This may indicate that only certain susceptible individuals were affected by the ash.

Baxter *et al.* (1983) studied ER utilization over a 4-week period associated with this episode in Yakima, Washington, and noted a doubling for asthma and bronchitis visits. The rate tripled for the first week of the episode. A case-control study of these patients showed them to be characterized by previous history of asthma or bronchitis, rather than by additional exposure to particulates. Note that because the Mount St. Helens incident did not involve excess SO_2, these data imply that SO_2 is not an essential ingredient of the community air pollution mix with respect to excess respiratory hospitalizations.

West Germany

Wichmann *et al.* (1989) reported that about 19% excess hospital admissions were recorded during a 5-day episode in central Europe in January 1985. Peak smoke levels reached 850 $\mu g/m^3$ (3-hour average) and SO_2, 2170 $\mu g/m^3$ (30-minute average). Maximum 24-hour average values of NO_2 and CO were

230 μg/m^3 and 7 ppm, respectively. The excess hospital admissions were all due to cardiovascular causes; no significant increase was seen in respiratory admissions. In a multiple regression model for the sum of respiratory and cardiovascular admissions, which included the effects of temperature, results were much more significant when a 2-day lag was used, for smoke, SO_2, NO_2, and CO (daily averages).

This data set also allows a comparison of the absolute numbers of deaths, hospital admissions, and consultations in physicians' offices. The numbers of hospital admissions for cardiovascular and respiratory causes were about the same order as the numbers of deaths (more for respiratory; fewer for cardiovascular). The number of physicians' consultations was 100 to 500 times the number of deaths, but no correlation was found between air pollution and physicians' consultations.

Northern California Forest Fires

Dry lightning strikes ignited more than 1500 forest fires in Northern California during a 5-day period in August–September 1987 (Duclos *et al.*, 1990). Local TSP readings reached more than 4000 μg/m^3, and local visibility was reduced by 90% for several days. Data were gathered on ER visits in six counties, and the effects of the fire were assessed by comparing the numbers of respiratory visits with the numbers that would normally be expected at that time of the year (O/E = ratio of observed to expected). Significant excesses were shown for asthma ($O/E = 1.4$), chronic obstructive pulmonary disease ($O/E = 1.3$), laryngitis ($O/E = 1.6$), sinusitis ($O/E = 1.3$), bronchitis ($O/E = 1.2$), and other upper respiratory infections ($O/E = 1.5$). The excess ER visits for pneumonia, tonsillitis, pharyngitis, mental health problems, and bee sting reactions failed to reach significance. Excess ER visits for coronary problems and otitis were nearly significant ($p = 0.1$). Since this is a relatively sparsely populated area with no routine air pollution monitoring, it was not possible to estimate the actual air pollution exposures.

Summary of Episode Hospitalization Studies

These air pollution episodes span a range in levels of air quality and in responses. Respiratory diagnoses were the only category of admissions evaluated in all studies. Figure 9-9 plots these responses against daily maximum smoke concentrations (the Mount St. Helens point is plotted at the log mean of 1000 and 10,000 μg/m^3; two data points are shown for London [1957], corresponding to the maximum daily smoke reading, which occurred in the outskirts of the city and the highest central city reading). The Pittsburgh episode is clearly an outlier on this plot, but since fewer than 20 excess admissions were recorded for this episode, this datum is quite uncertain. Similarly, the New York data point may be an upper limit because the potentially confounding effect of the holiday was not removed. On a logarithmic plot basis, the slope is not significantly different from unity, implying a linear relationship. Plotted in linear coordinates, the slope is about 3 to 4% excess respiratory admissions per 100 μg/m^3 of smoke. The use of smoke (alone) as a correlating parameter is an arbitrary choice; for example, if SO_2 were added to smoke as in Figure 9-2, all

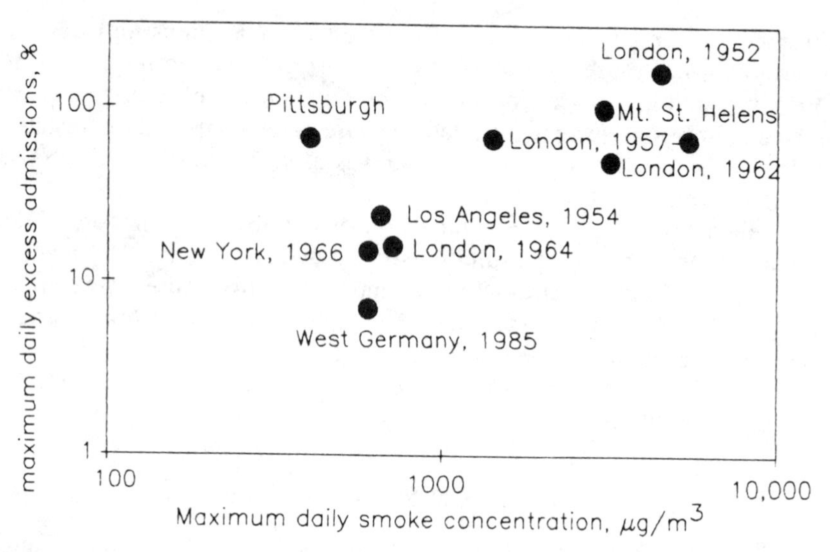

FIGURE 9-9. Comparison of hospitalization responses during various air pollution disasters, based on maximal measures.

of the points would shift to the right, except the Mount St. Helens datum. We also note that data from the Northern California forest fire disaster are consistent with Figure 9-9 (30 to 40% excess visits and particulates probably in excess of several hundred of $\mu g/m^3$).

TIME-SERIES STUDIES

London

In his analysis of winter fogs in London during the late 1950s, Martin (1961, 1964) studied deviations from 15-day moving averages of deaths and of emergency requests for hospital admission. He noted that morbidity was less sensitive than mortality, was unable to separate smoke effects from SO_2 effects, and found bivariate correlation coefficients in the 0.25–0.34 range for respiratory conditions, 0.20–0.28 for cardiovascular conditions, and 0.17–0.32 for all causes. Log transforms were used for the pollution variables in calculating these correlations, but lag periods were not considered (beyond that inherent in the pollution data reporting convention, i.e., 24-hour averages ending at 9:00 A.M. on the day of reporting). Day-of-week effects were removed using a statistical adjustment procedure, which resulted in a 35% increase in Sunday admissions requests, for example (Martin, 1961).

Martin tabulated mortality and morbidity (respiratory and cardiac) data for selected high-pollution days during the winters of 1958–59 and 1959–60. The health response data were on a same-day basis, and days preceded by days of higher pollution were eliminated by Martin to preclude confounding by lag effects. I replotted these data in Figures 9-10; Figure 9-10a plots data for 26 days in which both smoke and SO_2 were high, using their sum as a correlating parameter, as in Figure 9-2. The mortality and morbidity responses are crossplotted in Figure 9-10b; after allowing for the fact that the morbidity measure

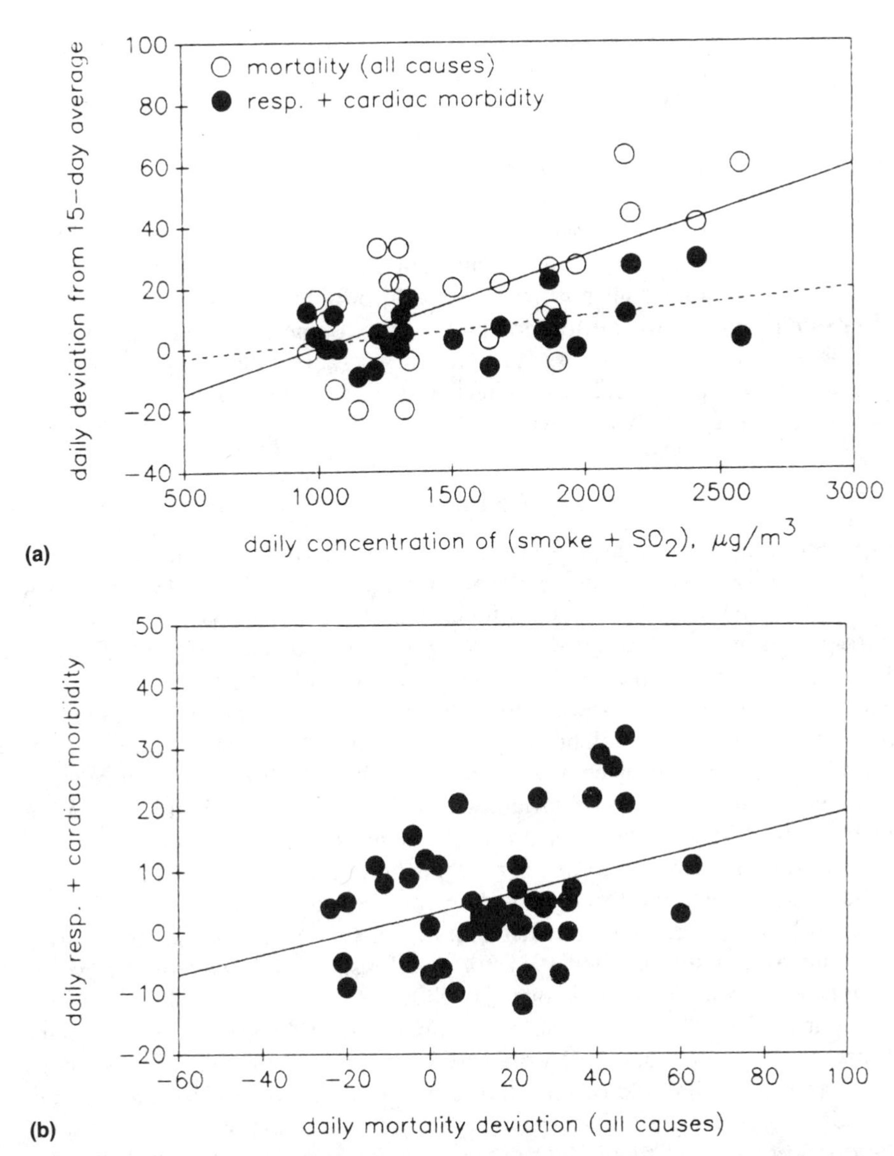

FIGURE 9-10. Dose-response functions for mortality and emergency bed applications in London, for the winters of 1958–59 and 1959–60. (a) Based on the sum of smoke and SO_2. (b) Crossplot of morbidity vs. mortality responses. *Data from Martin (1964).*

used here (emergency bed applications) accounts for only about 20% of admissions, the figure suggests that for this data set, the "excess" mortality and morbidity responses to air pollution (as measured by hospital admissions), are approximately equivalent. Martin did not consider lag effects, and thus probably underestimated the relationships between episode mortality and morbidity and air pollution. Since no absolute values were given, elasticities cannot be calculated with confidence; if the 1952 winter "normal" value of 200 per day is used (based on the Emergency Bed Service), the elasticity values for linear dose-response functions would be 0.057, 0.07, and 0.067 for smoke, SO_2, and their sum, respectively. Elasticities for mortality were higher, especially for smoke, and the dose-response curve was concave-upward, suggesting a log-linear model. Consideration of lag effects would probably have increased these values considerably. Neither Martin's tabulated data nor the scatter plots suggested the presence of pollution thresholds.

Holland *et al.* (1961) also studied London hospital admission data for 1958, in relation to smoke concentrations and certain weather variables, on a monthly average basis. They found significant effects for smoke with respect to respiratory admissions (but not heart disease), only for ages over 15, together with a significant negative effect for mean daily temperature. The same technique was used for a data set for Royal Air Force personnel in various locations; only the temperature effect was replicated (smoke levels were described as "low," but values were not reported).

California

Hospital admission patterns in Southern California have been studied since the 1950s, primarily in conjunction with air pollution by oxidants. The first such study (California Department of Public Health [1955–1957]) examined a 4-month period in 1954 and failed to find significant associations between any of several admissions categories and weekly oxidant levels. Brant and Hill (1964) performed a similar short-term study for the same period and claimed to have found a significant relationship with oxidant levels lagged by 4 weeks. However, their study utilized a very small number of observations in relation to the number of independent variables and cited 24-hour oxidant levels in the range of 1 to 2.8 ppm (about an order of magnitude too high to be credible). For these reasons, I chose to disregard their findings.

The Los Angeles studies by Sterling and colleagues (1966, 1967, 1969) appear to be more credible. In their first study, they examined 223 successive days from March to October 1961 for relationships with day of the week, weather, and air pollution: oxidants, O_3, SO_2, NO_2, CO (often labeled as CO_2 in the paper), NO, NO_x, particulate matter (apparently COH), and "oxidant precursor." Pollution data were obtained from eight monitoring stations about 5 miles apart; hospital admissions were used from all hospitals with more than 100 beds located within 5 miles from an air monitoring station. Apparently, all locations were pooled to provide a single time-series data set, resulting in a total of about 30,000 admissions (an average of about 134 per day).

Since admissions and pollution showed noticeable day-of-week patterns, both were adjusted for day-of-week by taking residuals from the overall average for that day. Admissions were grouped by degrees of "relevance" to potential air pollution effects: "highly relevant" diagnoses included allergic disorders, inflammatory diseases of the eye, acute upper respiratory infections, influenza, and bronchitis (about 5/day, or 3.7% of the total); "relevant" diagnoses included diseases of the heart, rheumatic fever and vascular diseases, other diseases of the respiratory system (about 17 per day); "total relevant" diagnoses were defined as the sum of the above; "irrelevant" diagnoses constituted all others.

Correlations were presented for these deviation variables; no seasonal corrections were made. Statistical significance ($p < 0.01$) was shown for all pollutants versus "highly relevant" diseases (except for SO_2, for which $p < 0.05$). For the "relevant" diseases, CO, NO_2, NO_x, oxidant precursors, and particulates were not significant, while SO_2 and O_3 remained significant. Neither temperature nor humidity was significant. "Irrelevant" causes showed negative relationships. For the grouping "total relevant," O_3, SO_2, and total oxidants had the highest correlations (0.25–0.27). Correlations for total admissions (all causes) were mixed positive and negative. The mean SO_2 level was about 34 $\mu g/m^3$; the mean ozone level about 0.4 ppm (although not stated, this is apparently the mean of hourly maxima). Sulfur dioxide was singled out for a detailed analysis of "excess" admissions, by disease. Possible lag effects were not mentioned, nor was serial correlation.

Unexpected associations were found for some diagnostic categories: Carbon monoxide was not associated with heart disease admissions; the strongest SO_2 effect was for "infectious diseases," although upper respiratory, bronchitis, and heart disease were also elevated on high SO_2 days. I approximated the relative magnitudes of these effects by estimating elasticities: For this study, a 1% change in air pollution would be associated with a change in admissions of about 0.1% (i.e., $e = 0.10$).

In the second report (Sterling *et al.*, 1967), lengths of hospital stays were studied, using the same data set. The paper shows that length of stay is also influenced by the day of week of admission, averaging 1 to 2 days longer for Saturday admissions. Unfortunately, the study emphasized the average pollution levels *during* the hospital stay rather than the levels preceding it. Perhaps the authors were not aware of the large differences that can exist between indoor and outdoor pollution levels (depending upon air exchange rates). Significant (positive) correlations were shown for SO_2, NO_2, and particulates, but in general the values were lower than in the admissions data. Significant negative correlations were found for ozone (but not for oxidants). The authors noted that the analysis did not account for the simultaneous effects of temperature and humidity, which also had significant (negative) effects on lengths of stay. In addition, one might expect substantial random variations in lengths of stay because of differences in medical practice.

Interactions and lag effects were examined in the third report (Sterling *et al.*, 1969), using multiple regressions and the data set of 223 days from Los Angeles. With respect to admissions, some significant lag effects were shown for "relevant" diseases, especially for a 3-day lag. The showing for "irrelevant" diseases was similar, except that 2-day lags were often significant. No information was presented on correlations between pollutants (which could greatly affect the multiple regression results) or on serial correlation effects. In summary, while the Sterling *et al.* studies may have been compromised by neglecting seasonal effects and serial correlation, the facts that temperature was not significant and some of the air pollutants were highly significant ($p < 0.01$) lead to the conclusion that *bona fide* associations were shown.

A specialized population was studied by Durham (1974), who used health center records from 1969 to 1971 from seven California universities to study air pollution–morbidity relationships. This study was notable in several regards:

- Air pollution data included daily means and peaks for eight pollutants, collected from within about 5 miles of the campuses where the students

resided (patients living more than 5 miles from each campus were excluded from the data base). Gravimetrically measured particulates were not included, however.
- Both temporal and spatial gradients were analyzed. Two of the universities were in the San Francisco Bay area and five in the Los Angeles area, which allowed time series to be studied at different overall pollution levels.
- Scheduled and repeat visits were excluded from the data base.
- Data on smoking habits were collected.

Seventy-eight percent of the records showed a physician's diagnosis; gastroenteritis was used as a control diagnosis. In order to control for day-of-week and other noncausal temporal effects, the dependent variables were coded as the ratios of specific diagnoses to total health center visits for that day (note that this procedure will not account for the possibility that certain diagnoses will be more likely than others to show spurious temporal relationships).

Time lags up to 7 days were studied; analyses were based on both the date of the visit and the date on which the symptom was first noticed. Analyses were performed by university and by calendar quarter. In addition to correlation analysis, factor analysis was used to combine the effects of weather and air pollution. Correlation results were aggregated by diagnosis for all significant pollutants and by pollutant for all significant respiratory diagnoses.

According to Durham, the results of this study showed that pharyngitis, bronchitis, tonsillitis, common cold, and sore throat in the Los Angeles schools had the highest associations with air pollution. The most important species were peak oxidants and mean SO_2 and NO_2 (all had about the same degree of association). Essentially no association was found for the control diagnosis or for particulate matter (as measured by COH). The date that the symptom was first noticed produced better results than the date of first visit, and lags of several days were common. Confounding effects were shown due to the interactions of air pollution with weather changes. Males, nonathletes, and smokers had more respiratory complaints, which were more strongly associated with air pollution. Average pollutant levels in Los Angeles were: oxidants, 0.025 ppm; SO_2, 0.013 ppm; NO_2, 0.069 ppm. Since few individual pollutant-disease relationships were reported, it is difficult to judge whether any of the positive findings may have been due to chance. Examples of the lag relationships were shown for total respiratory diagnoses and oxidants and SO_2 for one school; approximate elasticities estimated from these results were in the range of 0.07 to 0.10.

A different analytical approach was used by Goldsmith *et al.* (1983) in analyzing emergency room visits (all diagnoses) at four Los Angeles hospitals in 1974–75. Because of the difficulty in separating the pollutant variables, path analysis was used as a means of developing structural equations prescribing the interrelationships among variables representing weather, air pollution, and ER visits. This technique depends on prior knowledge of the qualitative interrelationships among the variables; regression techniques are then used to estimate the coefficients defining the quantitative relationship. In this case, the stated intent was to derive relationships between concentrations of sulfur oxides and morbidity. On the basis of bivariate correlations, Goldsmith *et al.* found significant (+) associations with ER visits and temperature and oxidants at all four hospitals, NO_2 and haze at Azusa and Riverside, sulfate at Long Beach and Lennox, SO_2 at Long Beach and Riverside, and CO at Riverside. However, using path analysis, the only remaining significant associations were oxidants at Azusa and sulfates at Long Beach and Lennox. Lag effects were not analyzed in this study, which suggests that the associations found might be underestimated somewhat.

Portland, OR

Jaksch and Stoevener (1974) studied the daily outpatient "numbers of contacts" with the medical system and their relative costs, during 1969–70 in the Portland, Oregon, SMSA, using records from the Kaiser-Permanente Medical Care Program. Since Kaiser clinics serve only members, the population at risk was known. The analysis was based on a 5% random sample of the membership, but it was not known whether this sample group was representative of the SMSA as a whole. For example, the yearly admission rate was given as 9%, whereas for the Western United States, the expected rate is over 12% (Lewis, 1974). Use of this sample allowed smoking habits and occupational exposures to be accounted for, and thus the study was an improvement over the usual ecological design. The air pollution data were limited to TSP samples obtained every fourth day from nine stations. The maximum local annual average TSP was about 70 $\mu g/m^3$; the overall SMSA average was 61 $\mu g/m^3$. The TSP data were interpolated in time using spline fits and geographically to provide values for each patient's home and work address for each day of the 2-year period. Meteorological data consisted of temperature-humidity index (THI).

Diseases were grouped as respiratory (upper and lower), respiratory allergies, other allergies, circulatory, digestive eye, genitourinary, and other. The dependent variables were the cost per visit and the frequency of visits. Day-of-week effects were not mentioned, except as an index for estimating air pollution exposure. For the cost of respiratory disease visits, air pollution (TSP) was statistically significant for lags of zero and 1 day, with the best results for the 1-day lag period. Significance levels dropped for 3- and 4-day lags. R^2 was low (about 0.02), but there were about 1600 (individual) observations. The effect on cost of a 20 $\mu g/m^3$ change in TSP was stated to be quite small, about 3 cents per medical system contact, which corresponds to a very low elasticity. (For the \$0.03/contact figure to be roughly consistent with the episode slope of 3–4% per 100 $\mu g/m^3$ mentioned on p. 363, the average contact value would have to be about \$200 to \$300, which seems unduly high.) The results were qualitatively similar for circulatory and respiratory diseases combined, except that air pollution did not reach significance (p = 0.4). Air pollution was not significant with respect to the numbers of visits (nor was any other variable). It should perhaps be noted that Portland has relatively little air pollution and thus it should not be surprising that the air pollution–morbidity effects found were much smaller than for Los Angeles, for example.

Washington, DC

Seskin (1977) performed a similar study in Washington for 1973–74, focusing on a group medical care practice. Dependent variables were numbers of unscheduled visits to the pediatrics, internal medicine, ophthalmology, and emergency clinics. Respiratory diagnoses were not identified. Air pollutants included oxidants, CO, and SO_2, measured at several different stations. Both 1973 and 1974 were analyzed separately, as was each pollutant-station combination. Missing data were estimated by linear interpolation. Meteorological data consisted of temperature, wind speed, and precipitation. Day-of-week effects were handled by dummy variables for Saturdays and Sundays. No attempts were made to estimate pollutant exposure by combining data from more than one station. Of the 48 regressions (2 years × 3 pollutants × 2 measuring stations × 4 diseases), eight (17%) were significant at the 10% level or better (four oxidants, two CO, and two SO_2), and two (4%) at the 1% level (oxidants and CO with respect to ophthalmology visits). Maximum 1-hour values of these pollutants were used in the regression analysis. The magnitudes of the effects shown were in the range of a 0.5 to 4.3% change for a 10% change in air pollution (elasticities from 0.05 to 0.43). No consistent lag effects were found. In view of the poorly characterized exposures, the "true" effects (if any) could have been larger, although this source of underestimation could have been balanced by the failure to consider serial correlation. Weather effects were minimal compared to the day-of-week effect.

Chicago

Emergency room admissions data from the Cook County Hospital from September 1971 to March 1973 were analyzed by Carnow (1975), Namekata and Carnow (1976), and Fishelson and Graves (1978). Namekata *et al.* (1979) also collected and analyzed admissions data from this hospital during a later period.

Carnow (1975) used deviations from moving averages as a means of seasonal adjustment and presented results of stepwise regression analysis for ER admissions on Tuesdays during the two winters of the study period ($n = 30$). Diagnostic groups studied were asthma, acute bronchitis, pneumonia, total respiratory, heart attacks, congestive heart failure, and total cardiac diagnoses. Air pollutants were SO_2, COH, and CO, based on an 8-station monitoring network. Exposure data were computed by weighting the monitoring data to reflect proximity of the patients' home addresses to the various stations. Weather variables were also included in the regression procedure, giving a total of 12 independent variables that could be selected. The results showed significant (+) associations between SO_2 and respiratory admissions and between CO and congestive heart failure admissions. There were no significant associations with COH. The absolute magnitude of the effects appeared to be small; assuming that the regression coefficients are given as daily admissions per ppm, the largest seasonally adjusted SO_2 deviation was associated with an increase in total respiratory admissions of less than 1%.

Fishelson and Graves (1978) considered SO_2 and COH, averaged from five monitoring stations near the hospital, in addition to temperature, relative humidity, and sky cover. Mean values for SO_2 and COH were about 60 and 88 $\mu g/m^3$, respectively. The paper is not explicit but implies that 24-hour averages were used for the air quality data. The data set consisted of ER visits on 81 Tuesdays broken down by age, sex, race, and cause. Dummy variables were used to account for holidays. Both linear and logarithmic models were examined, including lags up to 3 days. Serial correlation was considered by means of the Durbin-Watson statistic. In addition, some regressions incorporated a lagged dependent variable as an explanatory (independent) variable to "capture various effects prevailing in the study area that could charge over time and affect admissions." Using this variable increased the R^2's. In general, the linear regression results associated cardiac admissions with SO_2, with a 1-day lag ($p = 0.005$). For most of the significant regressions, the Durbin-Watson test was either inconclusive or indicated no autocorrelation. Males and females participated about equally in the cardiac–SO_2 relationship, but when grouped by age, only the over-59 group was not important.

By implication, the 5–40 age group must have also been important (results not shown). However, when the regression included the sum of same-day and lagged pollution and the lagged dependent variable, all the age groups over 1 year were significant. Coefficient of haze was significant (positive) only for cardiac admissions in the 40–59 age group. Respiratory admissions were only significant for the over-59 age group, for lagged SO_2.

As shown in Chapter 3, for a regression using logarithmic transforms for dependent and independent variables, the elasticity is numerically equal to the regression coefficient. These elasticity values seemed unduly large in comparison to the studies we reviewed, with many values around unity and higher, for example. Fishelson and Graves did not include a table of mean values in their paper, but the pertinent values may be obtained from Krumm and Graves (1982); the resulting linear model elasticity estimates are an order of magnitude lower than the log models (placing them more in line with the other studies), which suggests large differences between linear and logarithmic specifications. Also, in their calculations of the potential benefit of reducing ambient SO_2 in Chicago, they apparently erred by not considering that the regressions employing the sum of same-day and lagged SO_2 produce regression coefficients one-half the size of those employing single-day figures.

Krumm and Graves (1982) reanalyzed this data set (but did not refer to the first paper), emphasizing total respiratory plus cardiac admissions and including interactions and higher order terms for the pollutants. In their ordinary least-squares regression using logarithms, SO_2 was not significant (negative) and COH was significant and positive. In subsequent regressions, the interaction term, $SO_2 \times$ COH, became significant. This paper emphasized theory and presented few useful new results. This data set yielded very different findings, depending on which of three different methodologies was used and is thus mainly of theoretical interest.

Namekata *et al.* (1979) studied admissions to two Chicago hospitals (including Cook County) for Tuesdays, Wednesdays, and Thursdays, for 1 year beginning April 1977. They included patients age 15 or older for five disease groups: all respiratory diagnoses, allergic conditions and upper respiratory infections, acute bronchial and lower respiratory infections, all cardiac diagnoses, hypertension and vascular heart disease. Weather data (temperature, wind speed, precipitation, relative humidity, hours of possible sunshine, and sky cover) were from Midway Airport. Daily community exposure estimates were made for 76 residential areas for three pollutants (TSP, SO_2, and NO_2), based on an interpolation procedure using routine monitoring network data. Only days with reasonably complete records were retained in the study (20 days for SO_2 and NO_2; 42 for TSP; 131 when a single hospital and monitoring station were used (CO and NO were added); 36 days for ozone in the summer).

Regressions were run with one pollutant at a time. For total respiratory diagnoses, only NO was significant, although SO_2 was close ($p = 0.15$) for the small data set. For allergic conditions and upper respiratory infections, only NO approached significance ($p = 0.10$). For acute bronchial and lower respiratory infections, SO_2 was significant in both data sets, but negative in the larger one. For cardiac diagnoses, SO_2 was significant ($p < 0.05$, $n = 20$); NO and NO_2 were close ($p < 0.10$). For hypertension and vascular heart disease, O_3 was significant (negative) and NO was close (+). The SO_2 elasticities were about 0.50 for total respiratory diagnoses and about 4.0 for cardiac conditions (this extraordinarily high value suggests collinearity problems). Serial correlation was not examined in this study, but since it was not a complete time series (because of missing data), this problem may be less important. Lag effects were not mentioned, presumably for the same reason.

Steubenville, OH

Samet *et al.* (1981) analyzed ER visits in the industrial city of Steubenville, for the months of March, April, October, and November during 1974–77. They used deviations from the expected numbers of admissions based on day of week, season, and year as the dependent variable, for "all respiratory diseases," "all diseases except trauma," and "all diseases." Air pollution variables were 24-hour averages from a nearby station for SO_2 (mean = $90\,\mu g/m^3$), TSP (mean = $156\,\mu g/m^3$), NO_2, CO, and O_3. Deviation variables were not used for the pollutants, since the authors noted no weekly cycles or long-term trends, based on inspection. The correlation between SO_2 and TSP was 0.69; all other pairwise pollutant correlations were lower.

Results were presented in terms of quartile analyses, for two subsets stratified by ambient temperature. Dose-response relationships were not clearly evident from these data. Linear regressions found that either (unlagged) SO_2 or TSP was associated with respiratory disease visits ($p < 0.05$); TSP was also associated with trauma* and all diseases. The elasticities for respiratory visits were about 0.045; for all diseases, about 0.02. The regression coefficients for respiratory diseases were about 3% per $100\,\mu g/m^3$ for smoke and 5% per $100\,\mu g/m^3$ for SO_2, which are consistent. The possibility of serial correlation was mentioned, but no data were given.

Pittsburgh

Mortality and morbidity from 1972–77 were studied in Allegheny County, PA, by Mazumdar and Sussman (1983). The morbidity variables were "daily numbers of emergency and urgent hospital admissions" from all acute care hospitals in the county, for all causes, respiratory diseases, heart disease, and "other circulatory disease." Respiratory diseases constituted about 8% of the total urgent and emergency admissions (rate = 6.7/1000 pop.). The data were

* Based on data provided to me by the senior author.

analyzed as deviations from 15-day moving averages, corrected for temperature variation, and adjusted for day-of-week effects. Results were given in terms of the percentage of that morbidity measure attributable to a given pollutant. Smoke and SO_2, measured at three different locations, were the pollution variables with means of about 94 and 100 μg/m^3, respectively; results were presented for each pollutant separately and for both jointly. Significant results were obtained for smoke separately in 7 out of 12 regressions, with the pollution effects ranging from 0.9% for all causes to 4.4% for heart disease. As a separate pollutant, SO_2 was significant in 3 of the 12 cases, but one of these was negative. Sulfur dioxide was never significant in the joint regressions; the joint results for smoke were not greatly different than in the separate regressions. There was essentially no relationship between the results for morbidity and mortality variables; the morbidity results were larger in magnitude and more statistically significant.

Carbon Monoxide Effects in Denver

Kurt *et al.* (1978, 1979) studied cardiorespiratory complaints (CRC) presented to a Denver hospital ER over a 3-month winter period in 1975–76, in relation to CO concentrations and other air pollutants. Carbon monoxide was emphasized in part because it is thought to be a particular problem in Denver, because of the high altitude and traffic density there. Diagnoses studied included nontraumatic chest pain and shortness of breath, dyspnea, or wheezing. The two presentations of this study (1978 and 1979) differ only in the statistical analysis techniques used. The CO data were obtained from a monitoring station located 300 m from the hospital; data on O_3, NO_2, and SO_2 were also obtained from this station. The 24-hour mean CO levels showed a better correlation with CRC than did daily peaks. The average daily maximum CO concentration was 32 ppm on the 24 highest CO days (Figure 9-11). No cor-

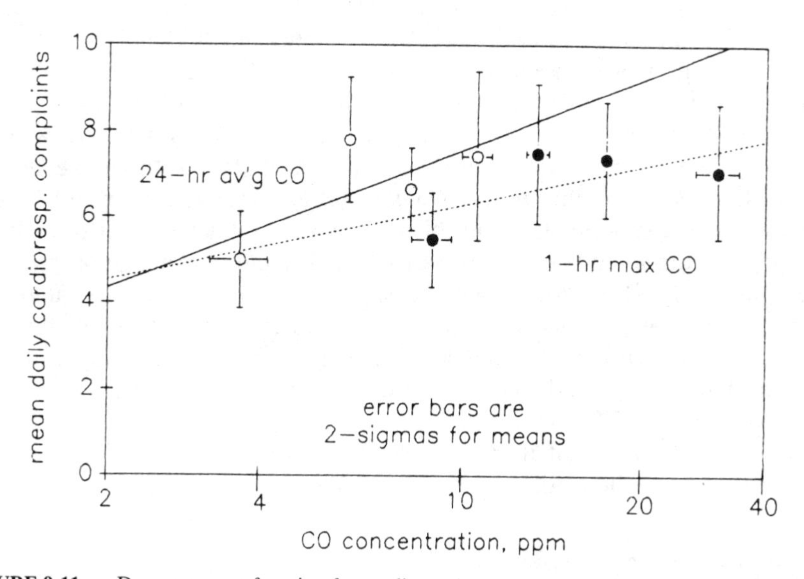

FIGURE 9-11. Dose-response function for cardiorespiratory complaints at a Denver hospital emergency room. *Data from Kurt* et al. *(1979).*

rections were made for weather variables or for seasonal or weekly cycles; the scatter plots shown suggested that weekly cycles may have been present. Carbon monoxide was positively correlated with temperature; no information was given on the effect that this may have had on the CRC–CO relationship. No control analyses were presented, either for other symptoms versus CO or for CRC versus other pollutants, although the authors reported that the other pollutants showed no consistent relationships with trends in cardiorespiratory complaints.

Linear correlations between CRC and CO were not significant on a same-day basis; neither were the grouped mean data shown in Figure 9-11. However, the relationship strengthened when a 1-day lag was considered (Kurt *et al.*, 1978). Significant differences in CRC were found when these 2-day periods were grouped according to "high" and "low" CO concentrations. The grouped mean data (from the 1979 paper) shown in Figure 9-11 show the following tendencies in the data: a nonlinear dose-response relationship (the x axis is logarithmic), a CRC intercept of three to four complaints per day, a more consistent relationship when 24-hour mean CO is used as the dose metric, and elasticities of about 0.4 to 0.5. The elasticities are lower if a linear CO variable is used.

Considering the typical spotty spatial distributions of CO and hence the likely errors in population exposure (only one monitoring station was used, and its proximity to the hospital is not relevant to population exposures *before* admission), this study must be judged as showing an important relationship between air pollution and cardiorespiratory health. It is unfortunate that other pollutants were not given the same level of detailed analysis or that "control" diagnoses were not examined.

Utah

A period of "heavy smog" conditions in the Salt Lake Valley was studied by Lutz (1983), in relation to patient visits to a family practice clinic in Salt Lake City. The analysis was based on 13 weekly totals of patients with "pollution-related" diseases,* as a percentage of the total number of patients for that week. Air pollutants considered included particulates (TSP), CO, O_3; weather variables were percentages of cloud cover and of days with smoke or fog. The author felt that some portion of any sulfur oxide effects would be embodied in the TSP variable. Only bivariate correlations were performed; the most highly correlated variables were particulates and percentage of days with smoke or fog ($p < 0.005$). Ozone was significantly negatively correlated with the percentage of patients with "pollution-related" diseases; this finding was ascribed to collinearity with cloud cover (negative). However, temperature was not considered in this study and could thus be a confounding variable (ozone is highest in summer; TSP, in winter).

In a brief reanalysis of these data, I added daily minimum and maximum temperature data obtained from the *World Weather Disc* (1990) for Salt Lake

*Asthma, cough, dyspnea, acute bronchitis and bronchiolitis, pneumonia, emphysema or chronic lung disease, acute upper respiratory tract infection, laryngitis, sinusitis, conjunctivitis, ischemic heart disease. Only 2% of the patients had heart-related complaints.

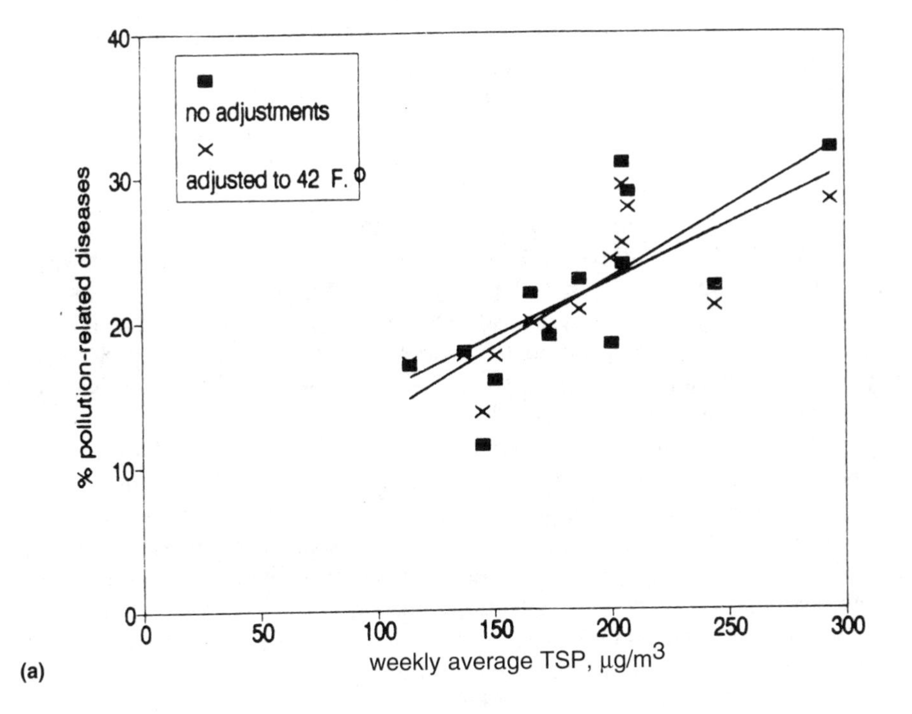

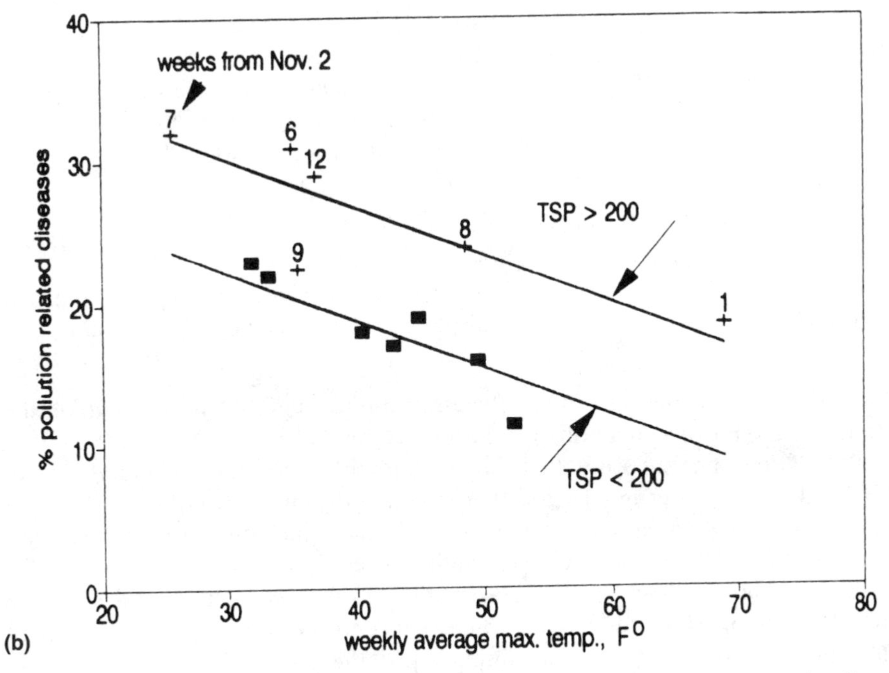

FIGURE 9-12. Relationships between visits to a family practice clinic for "pollution-related" diseases in Salt Lake City and environmental factors. (a) Based on TSP, after adjustment for temperature. (b) Based on temperature, showing weeks of high TSP (labels are sequential week numbers). *Data from L.J. Lutz, Health Effects of Air Pollution Measured by Outpatient Visits.* The Journal of Family Practice *16(2):307–313, copyright 1983. Reprinted by permission of Appleton and Lange, Inc.*

City, averaged to provide weekly data corresponding to the pollution and patient data (daily temperatures from Ogden were quite similar). I converted the pollutants standards index (PSI) data used by Lutz to TSP by setting TSP = 260 μg/m^3 at a PSI value of 100. In a multiple regression, both (maximum) temperature and TSP were significantly associated with the percentage of "pollution-related" outpatient visits. These relationships are displayed here in two different ways. Figure 9-12a (adapted from Lutz (1983)) shows the relationship with TSP, as the sole predictor (squares) and adjusted to a constant temperature of 42°F (+'s), based on the regression coefficient; the overall trend is not greatly affected by the adjustment. However, as seen in Figure 9-12b, if temperature is considered to be the primary predictor, only 5 weeks stand out as different from the rest of the data; these 5 weeks consisted of five of the six highest TSP values. These 5 weeks were not consecutive, which rules out an infectious epidemic as the source of the increased prevalence of "pollution-related" visits. The maximum point was week 7, just before Christmas. Deleting this observation had little effect on the regression slope, which was about 12% per 100 μg/m^3. Although this study was brief and simple (and did not use a "control"), it does not suffer from day-of-week or lag effects, and the use of percentages rather than the absolute numbers of visits helps account for seasonality. The regression slope (the elasticity was about 0.70) was substantially higher than found in previous studies of hospital admissions based on daily data. Explanations could include the use of a less severe measure of morbidity (clinic visits vs. hospital admissions), or the use of weekly data, which effectively sums the lag effects. The multiple regression would predict about 11% pollution-related diseases for a summer week in which TSP averaged 50 μg/m^3 and maximum temperature, 80°F, which does not seem unreasonable (although it is outside the range of the data).

In comparing this study to others, one must keep in mind that the dependent variable is defined differently:

- Upper respiratory tract infections are included.
- Visits are to a family practice clinic, not a hospital.
- The variable is a percentage of all visits; if the clinic was capacity limited, respiratory problems may have displaced other diagnoses at certain peak times.

Nevertheless, the findings are sufficiently interesting that replication of the study design in other locations would be recommended.

Pope (1989) analyzed inpatient hospital admissions in Utah County (Provo, UT, and vicinity) during a period that included the shutdown of a nearby steel mill, which was a major source of air pollution, including particulates and, presumably, sulfur oxides. The mortality aspects of this situation were discussed in Chapters 6 and 8. The observations in the 1989 study consisted of 35 monthly averages of inhalable particle concentrations (PM_{10}) and inpatient hospital admission statistics, beginning April 1985. The plant was shut down from August 1986 to September 1987, during which time period PM_{10} levels were noticeably lower than corresponding previous months, especially in December and January. In winter, this region suffers from air stagnations due to inversions, which tend to increase concentration levels of all air pollutants. For example, the maximum 24-hr PM_{10} and TSP readings reported during the study were 365 and about 600 μg/m^3, which are well above the U.S. ambient standards (given in Chapter 2) (Pope, personal communication).

The categories of hospital admissions considered were diagnoses of bronchitis or asthma (mean = 1.5/1000 population), pneumonia or pleurisy (mean = 1.75/1000), and the sum of these two groupings. The average hospitalization rates in Utah County were substantially lower than comparable values for the United States, on the basis of census regions. The lower rates in Utah may be due to lower rates of smoking and a younger population. Pope stratified the population into adults and children, cut at age 18.

Pope recognized that his analysis was compromised by the coincidence of peak air pollution periods with the normal winter peak in respiratory disease and used ambient temperature (monthly mean of daily lows) as a control variable. Control population groups included all nonrespiratory admissions and admissions for out-of-county residents. Regression results were presented using the current month and the previous month (i.e., a lag of 30 days), for PM_{10} alone and with temperature. Temperature was significant in seven of 10 regressions, among them the control groups. Including temperature reduced the PM_{10} regression coefficients for respiratory admissions, but Pope found PM_{10} to be a significant predictor of all categories of respiratory admissions for children, of bronchitis and asthma admissions for adults, and for the combined category of all respiratory admissions. R^2 values ranged up to 0.83 (the high value of explained variance results in part from the use of monthly rather than daily data). Use of a lagged variable usually increased the coefficient. In general, the particulate regression coefficients were much larger than found by previous authors. For example, Pope's results imply an approximate doubling of children's admissions and a 25% increase in total respiratory admissions in response to an increase of about 90 μg/m^3. On the basis of equivalent TSP, this corresponds to about 16% per 100 μg/m^3. This large response was attributed by Pope to the high percentage of nonsmokers (95%) in the study population; an additional possibility is the use of monthly averages. In addition, as discussed, when both lagged and unlagged variables included in a multiple regression and autocorrelation exists, the "true" effect of the parameter is the sum over all the lags.

Because of these interesting findings, I reanalyzed this data set, using data values read from the graphs presented in the paper. Since temperature data were not presented, I used a seasonal correction factor consisting of periodic functions peaking in January or February with corresponding minima in July or August. As a first check on Pope's findings, I compared the 13-month means for the period of plant shutdown with a similar 13-month mean previous to the shutdown. This gave a 12.5% difference in total admissions, due almost totally to children's admissions. Regression coefficients computed on this basis were consistent with Pope's results, which included the temperature covariates. Next, a two-stage adjustment procedure was used, in which admissions were first adjusted for seasonal effects using the periodic function and then regressed against PM_{10}. This procedure minimizes any pollution effects that have a seasonal pattern similar to the admissions data, since all of the (average) seasonal effect is removed by "adjustment." However, PM_{10} remained significant for both children's and total respiratory admissions. These results are plotted in Figure 9-13. Finally, the high-pollution months of December,

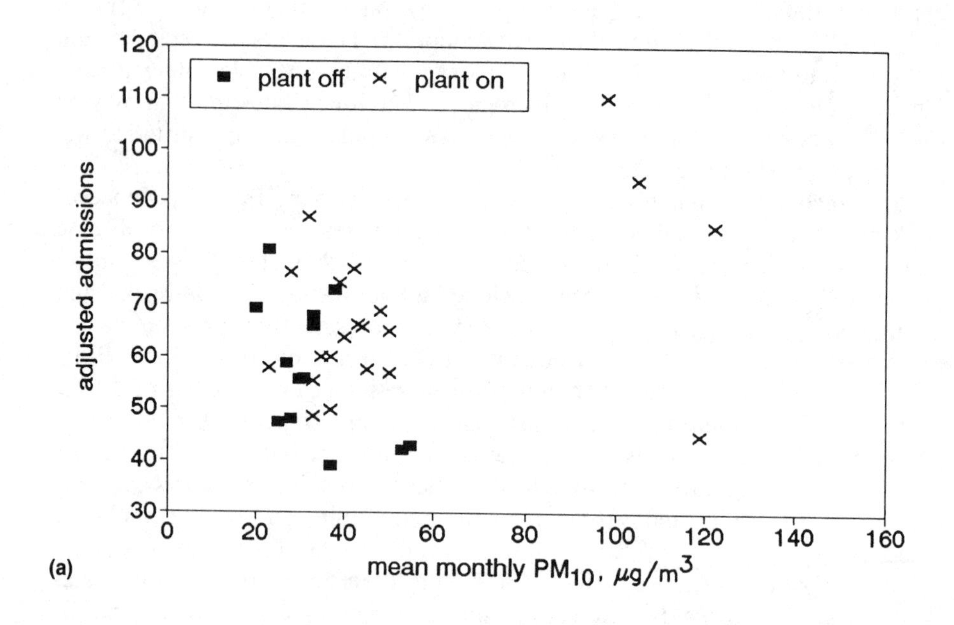

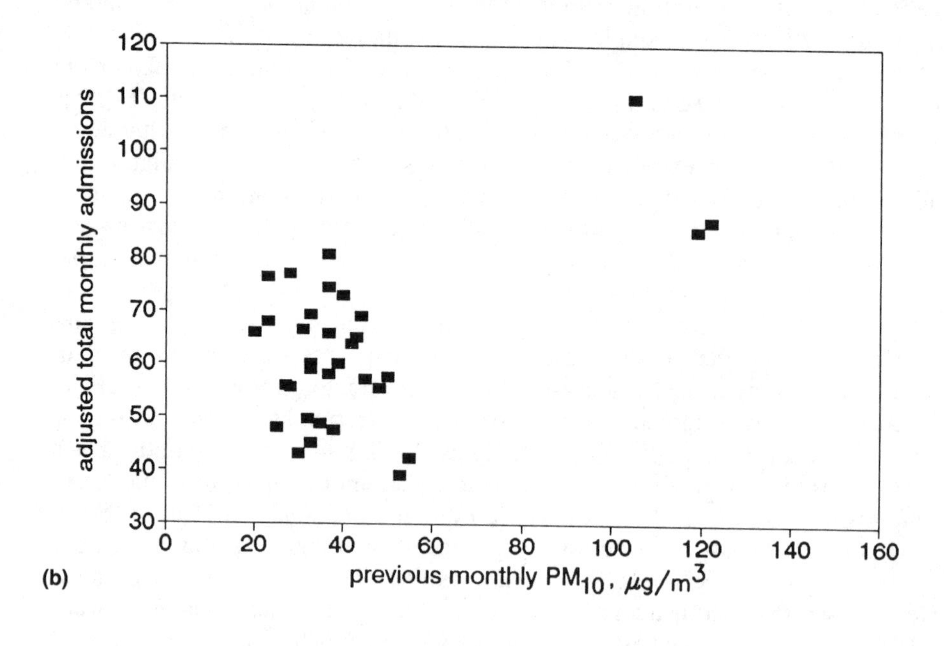

FIGURE 9-13. Dose-response data for hospital admissions in Utah County, UT, after adjustment for seasonality. (a) Based on PM_{10} for the same month. (b) Based on PM_{10} from the previous month. *Data from Pope (1989).*

January, and February were removed from the data set, and admissions were regressed against PM_{10} without seasonal corrections; results were significant for total and adult admissions. Regressions using the periodic function and PM_{10} gave results similar to Pope's. A dummy variable representing the plant's operating condition was never significant when used in conjunction with PM_{10}, implying negligible effects associated with any other emissions from the plant not in phase with PM_{10}.

My reanalysis thus essentially verified Pope's conclusions, but some interesting questions remain that could probably be best explained by an analysis of daily (or perhaps weekly) values: Is the 30-day lag effect real or an artifact of the analysis? Are there effects due to any other pollutants? Are there associations with any other admissions diagnoses?

Pope (1991) extended his analysis to 1989 and included two additional geographic areas. Salt Lake Valley, which includes Salt Lake City and County and the southern portion of Davis County, was intended to provide additional "cases," while Cache County was intended as a "control." Cache County includes the city of Logan and is located more than 100 km north of the other two areas. Populations were demographically similar in all three areas, but Cache County had many fewer people and lower particulate levels (and also less complete air monitoring). The age categories were changed for the second paper, consisting of all ages and preschool-aged children, and ICD-9 (International Classification of Diseases) codes were used instead of Diagnosis-Related Groups (DRGs). In general, the results of the extended analysis were similar to those of the first paper. The coincidence of peaks in hospital admissions and PM_{10} was less pronounced in 1989, so that the overall regression coefficients were reduced somewhat. In Salt Lake Valley, PM_{10} was a significant predictor of hospitalization but not in Cache County, as expected,

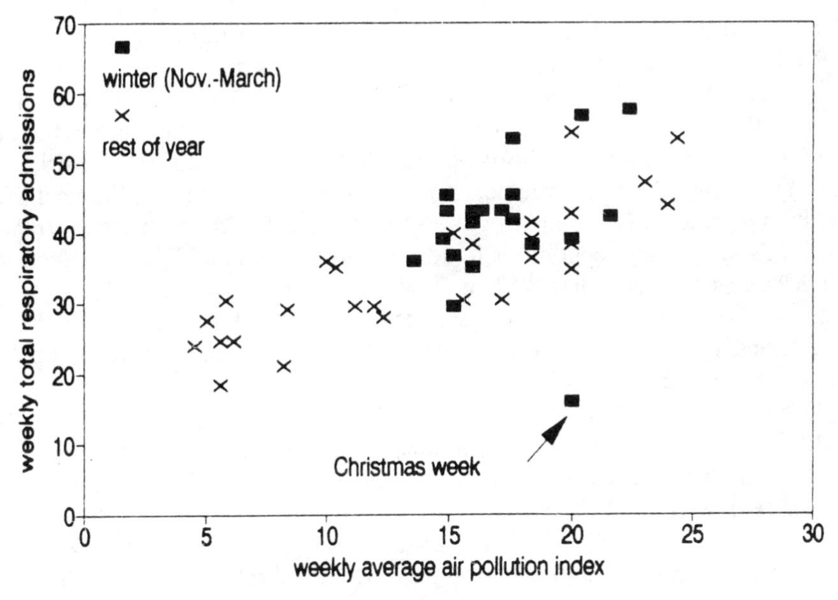

FIGURE 9-14. Dose-response relationship for respiratory admissions in Hamilton, Ontario, 1970–71. *Data from Levy* et al. *(1974)*.

since the PM_{10} data used in the Cache County regressions were measured in Utah County. This verifies that the PM_{10} variable was not serving as some sort of surrogate seasonal indicator, independent of local pollution levels. In addition, Pope reports that "bronchitis and asthma admissions for preschool-age children were approximately twice as frequent in Utah Valley when the steel mill was operating versus when it was not."

The paper by Lamm *et al.* (1991) was intended to address Pope's (1989) earlier finding that particulate air pollution was the major factor in children's hospitalization in Utah County. Lamm's approach introduced data on viral activity as an additional explanatory variable and appears to be the first to consider infectious agents and environmental factors simultaneously. Such an approach is consistent with a model in which air pollution exacerbates rather than causes disease. The infectious agent in question, respiratory syncytial virus (RSV), occurs with great regularity in all climates in winter or early spring, is particularly important in urban areas, and results in hospitalization of young children. It can also affect adults, but the response is usually milder than in young children (Hall, 1981; Levenson and Kantor, 1987).

A variable such as RSV prevalence describes the fundamental cause of illness and would be expected to have a 1:1 diagnostic relationship with the appropriate diseases, just as all patients having the flu exhibit one or more strains of influenza virus. The question of relevance here is, given the presence of an infectious agent, do environmental factors exacerbate its effects? The high degree of temporal collinearity between RSV and PM_{10} in Utah make it difficult to answer this question with confidence.

Lamm *et al.* considered Utah and Salt Lake Counties separately but used a regional surrogate to represent the monthly trends in RSV activity; actual RSV diagnoses from Utah Valley were not used. Respiratory syncytial virus is the major cause of bronchiolitis in young children and was used by Lamm *et al.* to represent RSV activity throughout the area. The paper shows a reasonably close relationship (graphically; no correlation was reported) between the total RSV laboratory analyses for six states and the sum of bronchiolitis admissions in both counties. This admissions variable was then labeled "RSV clinical activity" for use in the regression analysis. It is noteworthy that when Lamm *et al.* regressed bronchiolitis admissions in Utah County against this variable, current and lagged PM_{10} variables, and a lagged temperature variable, PM_{10} remained significant ($p < 0.02$ for the sum of lags), indicating the importance of local environmental conditions in addition to the regional level of infection. Lamm *et al.* concluded that "When RSV activity and temperature have been taken into account, no coherent evidence of statistical associations between PM_{10} levels and hospital admissions has been found for children or adults for any respiratory illness."

However, a careful reading of Lamm *et al.'s* results for daily admissions would take issue with assigning excess hospitalization solely to RSV. They state that only two of 36 regression coefficients other than RSV were significant at the 0.05 level (as would be expected due to chance alone), whereas their tables actually show three such values and a fourth one narrowly misses the mark. The 36 regression possibilities for children include 12 temperature coefficients; because of the strong seasonal dependence, we would expect RSV to primarily displace temperature. Adding a variable with no expected significance to the pool of possibilities always tilts the comparison toward the null hypothesis. If we combine the lagged and same-day PM_{10} variables as discussed above, there are 12 possibilities, three of which are significant at the 0.05 level. Further, there are several individual PM_{10} coefficients that just miss the 0.05 level; six values out of 24 were significant at the 0.15 level, for example. For adults, RSV was not significant, and one of the six summed PM_{10} coefficients reached the 0.05 level. Lamm *et al.* performed no regressions of combined age and diagnosis categories that could be compared with Pope's results (which were significant for PM_{10}). While this statistical performance is certainly not robust, neither does it warrant Lamm *et al.*'s conclusion to totally discount the role of air pollution. The appropriate conclusion would appear to be that air pollution contributes to the exacerbation of respiratory disease that is primarily caused by biological agents. If the significant coefficients reported by Lamm for daily admissions in Utah County are used to estimate the hospital admissions attributable to PM_{10}, the fraction of total pediatric admissions comes to about 0.15, which is reasonably consistent with the elasticity estimates from other studies. However, this may be fortuitous, given the many flaws in the approach used by Lamm *et al.* In addition, we cannot be sure that at least part of the difference between Pope's results and my interpretation of Lamm *et al.'s* results is due to the difference between monthly and daily analyses.

Canadian Studies

Levy *et al.* (1975, 1977) analyzed weekly hospital admissions due to acute respiratory causes (bronchitis, bronchiolitis, emphysema, pneumonia, asthma) in Hamilton, Ontario, from July 1970 to July 1971. The authors reported that only patients with recent histories (<5 days) of acute respiratory complaints were included. Most of the analysis was based on SO_2 and haze and a combined air pollution index (API), using data from a central monitoring station; no significant correlations were found for oxidants or CO. The correlations appeared to be better for API ($p < 0.001$) than for either of its constituents alone. The correlations diminished for hospitals more distant from the monitor, suggesting the influence of local as opposed to regional air pollutants. No seasonal corrections were made (although a seasonal effect was apparent from the scatter plots given), but in a combined regression with temperature, air pollution remained significant ($p < 0.001$). Bivariate correlations suggested that SO_2 may have been slightly more significant than haze, but in the absence of seasonal adjustments, this conclusion would be speculative. Using the weekly data read from a graph (Levy *et al.*, 1975), I reanalyzed the relationship between respiratory admissions and API, including a linear variable for the secular trend and a periodic function to represent the seasonal cycle. Neither of these variables was significant, and including them reduced the API regression coefficient only slightly. The relationship is shown in Figure 9-14 (data are segregated by season) and is one order of magnitude stronger than the typical results from daily time-series analyses. Part of the difference may lie in neglecting weather effects, and part may be due to the use of weekly data, which in effect aggregates all the daily lags. As a further check on these differences, I aggregated the data into monthly periods and repeated the regression analysis. The slope dropped about 30%, but the R^2 remained about the same. One could interpret the difference between the weekly and the monthly slopes as the short-term response; this estimated value agreed well with the results from studies of daily fluctuations in other studies. Whether the long-term component is an artifact of the analysis or a *bona fide* response to air pollution cannot be answered with these data.

Bates and Sizto (1983, 1986, 1987; also see Bates, 1985) studied temporal variations in admissions to a group of acute care hospitals in Southern Ontario. These papers should be considered as a group, since the details and limitations of the methodology become more apparent in the later works. Bates *et al.* (1990) used similar methods to study hospital ER visits in Vancouver.

The Ontario studies involved a population of about 5.7 million people (1977–78 figures) in an area of about 64,500 km^2; the Sudbury area (known for its very large SO_2 source) was not included. Data from 79 hospitals and 15 air monitoring stations were pooled in these analyses, which were limited to the bimonthly periods of January–February and July–August. Weather data consisted of daily mean temperatures, averaged for two stations; air quality data for the first study consisted of daily maximum readings of SO_2, O_3, NO_2, and soiling index (COH); for 0-, 24-, and 48-hour lags. In the later papers (Bates, 1985; Bates and Sizto, 1986, 1987, 1989), sulfate aerosol and relative humidity were added to the data set (and more air sampling stations) and the analysis was extended to 1980, 1982, and 1983, respectively. The data pooling technique used was to sum the admissions over all hospitals and to average the air quality data from all included monitoring stations having a reading for the day in question. In the case of sulfate aerosol, most stations recorded (24-hour averages) only every sixth day on a rotating schedule, so that only a few stations were averaged each day. However, some of the sulfate stations recorded data every day or every third day, a point that seems to have been overlooked by Bates and that could contribute some inadvertent geographic weighting to the averages. Bates selected as his air quality stations those which measured all the pollutants at the same location.

Diagnoses studied included all causes, nine different respiratory diagnoses, and several nonrespiratory (control) causes. Cardiac diagnoses were not studied. Total admissions averaged about 2300 per day, total daily respiratory admissions about 40 in summer and 70 in winter, asthma about 18 per day in winter and 16 per day in summer, and the nonrespiratory causes (in the first paper) were about 20 per day. The respiratory admissions studied did not include most upper respiratory causes or problems with tonsils or adenoids, which accounts for the lower annual average admission rate compared to the U.S. figures for all respiratory causes (3.4 per 1000 population, or about 2.3% of all admissions versus 13 per 1000). The dependent variables used in the time-series analyses were the deviations from the averages for each day of the week and season. In the latter two papers, deviations were also normalized for each year, to remove any spurious correlations that might have resulted from long-term trends in the data. Such "detrending" was not performed within the 2-month seasons, however. Deviation variables were not used for the pollutants; the authors cite the lack of a weekly pattern in ozone as justification.

Over the 9 years studied by Bates and Sizto, there were substantial changes in air quality and in hospital admissions; these long-term trends were specifically excluded by the regression approach they used. Even though long-term

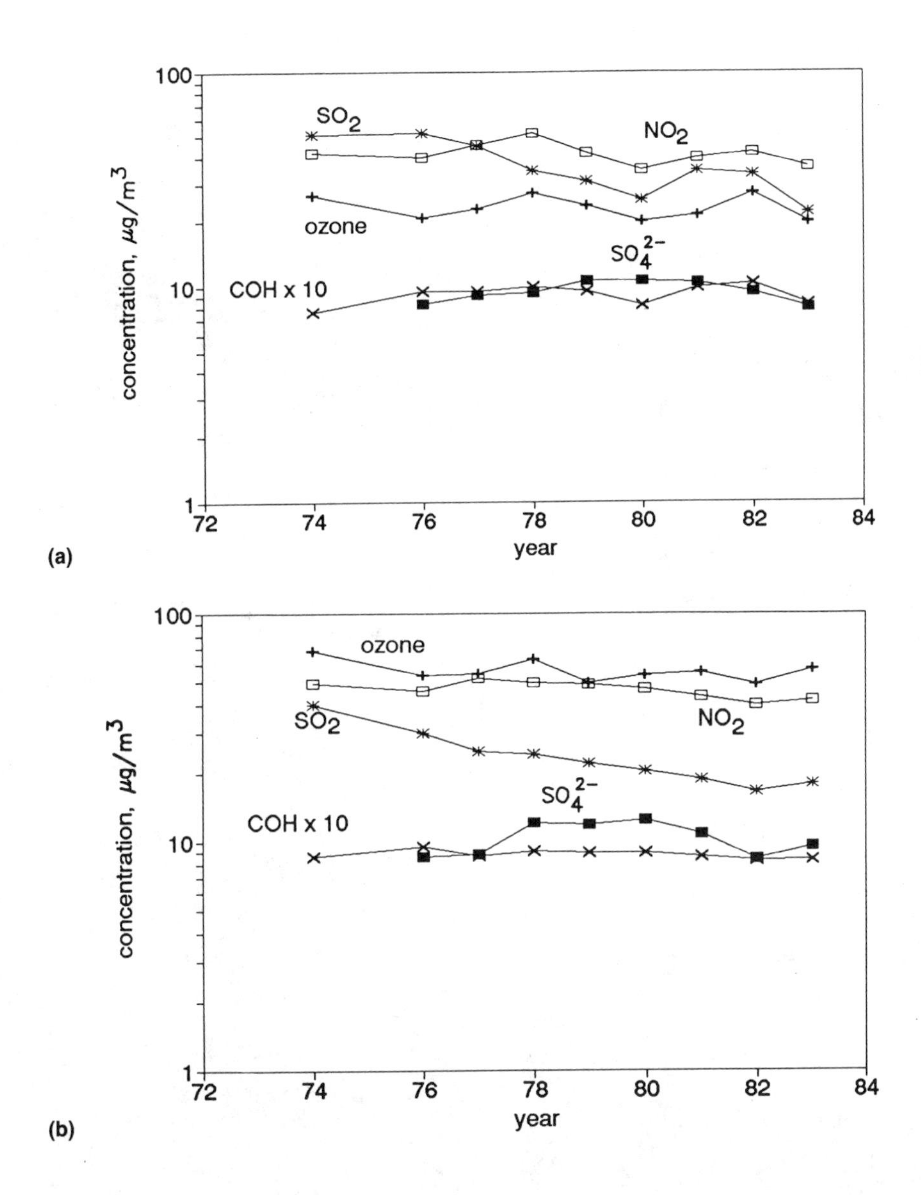

FIGURE 9-15. Long-term trends in air quality in Southern Ontario. (a) January–February. (b) July–August. *Data from Bates and Sizto, (1983, 1987).*

hospitalization trends may have other causes, such as changes in medical or insurance practices, it may be useful to examine these long-term temporal trends for consistency with the daily deviation associations found by the formal time-series analyses. Figure 9-15 shows the air quality trends plotted on a logarithmic scale, so that percentage changes may be compared directly at any concentration level. Sulfur dioxide shows a substantial decline in both summer (July–August) and winter (January–February), sulfate peaks in the middle of

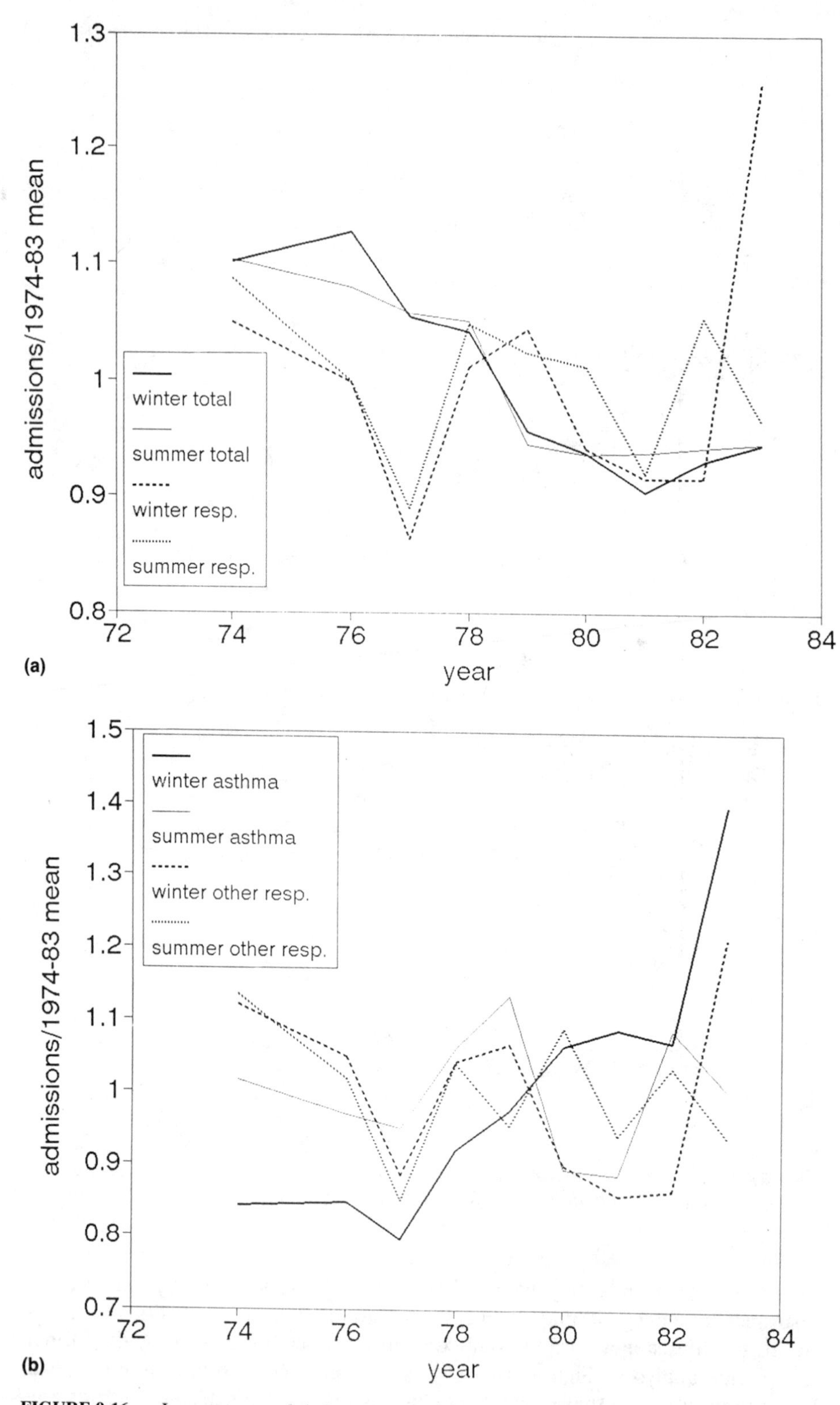

FIGURE 9-16. Long-term trends in hospital admissions in Southern Ontario. (a) All diagnoses and all respiratory diagnoses. (b) Asthma and other respiratory diagnoses. *Data from Bates and Sizto (1983, 1987).*

the period, and the remaining pollutants show no particular trends. Figure 9-16 shows the relative changes in hospital admissions over this period, by dividing the annual admissions by the 9-year mean. Total admissions (all diagnoses, Figure 9-16a) show a more or less continuous decline over time, with the exception of the winter of 1983 (possible flu epidemic?). Respiratory admissions are more variable from year to year. Asthma admissions are compared to the remaining respiratory categories in Figure 9-16b; the increase in asthma admissions is seen to be confined to winter. Note that Bates and Sizto adjusted the asthma admissions figures after 1979 to account for the change in ICD-9 coding.

Bates and Sizto used bivariate correlation coefficients as measures of association, in addition to selected multiple regressions. Table 9-1 presents a comparison of the bivariate correlations from the four papers, for winter and summer periods, respectively. The table shows the maximum (absolute values of r) correlations for any of the three lag periods for the four papers, listed in chronological order. In comparing across studies, one should keep in mind that the long-term relationships have not been removed from the first two papers. These entries confirm the trends shown in Figures 9-15 and 9-16. Bates and Sizto listed all the correlations in the first paper, regardless of significance level; only significant values ($p < 0.01$) were listed in subsequent papers. Thus, blanks in Table 9-1 represent correlations that did not achieve the 1% level.

In the winter period, admissions for all diagnoses were significantly (+) correlated with SO_2 and temperature and negatively with O_3, NO_2, and sulfate. However, these relationships did not survive the detrending process and thus must have been associated mainly with long-term variability. Total admissions were correlated with COH in the latter two (detrended) studies. Admissions for all respiratory diagnoses were significantly correlated with temperature for all four studies (note the reduced values for the detrended studies), as were asthma admissions in three of the four. No pollutant variables were consistently associated with any of the respiratory diagnosis admissions categories in winter. Correlations for the nonrespiratory or "control" admissions were mostly nonsignificant, except for temperature.

There were many more significant correlations in the summer period, except for relative humidity; roughly 40% of the 151 possibilities were significant at the 0.01 level or better. For total (all diagnoses) admissions, COH was significant, as in winter, and had the highest correlations for all the variables in the two detrended studies. Ozone, NO_2, and temperature were also significantly correlated with total admissions. For respiratory admissions, all the variables except COH and relative humidity were correlated; for SO_2 and O_3, detrending resulted in an appreciable diminution of the apparent relationships. Removing the asthma admissions weakened most of these relationships, except for temperature. However, asthma admissions showed only weak correlations, at best, especially for children. Nonrespiratory admissions were highly significantly negatively correlated with temperature; the similar (but weaker) relationships seen for ozone and sulfate may be the result of their collinearity with temperature. Lag effects were more pronounced for respiratory admissions than for all diagnoses; either 24- or 48-hour lags usually produced slightly higher correlations.

Because of the similarity in the apparent relationships between several

TABLE 9-1 Comparison of Hospital Admissions

Pollutant		Diagnostic Category					
		Total	Respiratory	Respiratory-Asthma	Asthma (all ages)	Asthma (0–14)	Nonrespiratory
		January–February Correlation Coefficients					
SO_2	1	0.230	(0.060)	n/a	(−0.090)	(0.030)	−0.160
	2	0.416		0.171	−0.282	−0.248	
	3						
	4						
NO_2	1	−0.310	(−0.060)	n/a	0.130	(0.090)	(0.110)
	2	0.159			−0.215	−0.238	
	3	0.112				−0.117	
	4					−0.117	
COH	1	−0.140	(0.100)	n/a	(0.080)	(0.090)	(0.070)
	2						
	3	0.134					
	4	0.127					
O_3	1	−0.230		n/a	0.120	(0.080)	(0.060)
	2				−0.146	−0.182	
	3						
	4						
SO_4	1	n/a	n/a	n/a	n/a	n/a	n/a
	2	−0.259			0.156	0.174	
	3						
	4						
Temperature	1	0.240	0.280	n/a	0.140	(0.070)	(−0.050)
	2	0.212	0.196	0.149			
	3		0.120		0.186	0.211	0.179
	4		0.120		0.171	0.194	0.150
Relative humidity	1	n/a	n/a	n/a	n/a	n/a	n/a
	2					0.113	
	3						
	4						

		July–August Correlation Coefficients					
SO_2	1	0.160	0.290	n/a	0.170	0.140	(0.100)
	2	0.279	0.207	0.239			
	3		0.147	0.117	0.118		
	4		0.138	0.112	0.106		
NO_2	1	0.050	(0.120)	n/a	(0.110)	(0.090)	(0.080)
	2		0.128				
	3	0.118	0.123				
	4	0.114	0.110			−0.117	
COH	1	(0.110)	(0.080)	n/a	(0.050)	(0.060)	(0.040)
	2						
	3	0.156					
	4	0.159					
O_3	1	0.160	0.280	n/a	0.210	0.140	(0.090)
	2	0.218	0.223	0.217	0.124		−0.236
	3	0.132	0.166	0.154			
	4	0.122	0.147	0.149	0.124		
SO_4	1	n/a	n/a	n/a	n/a	n/a	n/a
	2		0.264	0.186	0.212	0.196	−0.261
	3		0.180	0.129	0.147		
	4		0.171	0.134	0.127		
Temperature	1	0.210	0.230	n/a	(0.120)	(0.070)	(0.080)
	2		0.148	0.151			−0.355
	3	0.129	0.141	0.189			
	4	0.108	0.129	0.168			
Relative humidity	1–4	n/a	n/a	n/a	n/a	n/a	n/a

Sources: Line 1 = Bates & Sizto (1983). Data from 1974, 1976, 1977, 1978. Line 2 = Bates (1985). Data from 1974, 1976–80. Line 3 = Bates & Sizto (1985). Data from 1974, 1976–82, detrended. Line = 4 Bates & Sizto (1987). Data from 1974, 1976–83, detrended.

Note: Largest (absolute) r values for any lag period are shown in table. ( ) denotes $p > 0.05$. All other entries are significant. n/a and blanks = data not provided in original reference.

pollutants and respiratory admissions, Bates and Sizto used multiple regression analysis to try to partition the effects. In the second and third papers, partial correlation coefficients were given rather than regression equations, *per se*. The explained variance decreased markedly when annual detrending was introduced. A regression equation was given in a more recent paper (Bates and Sizto, 1989). However, it featured separate terms for each of the various lags for temperature and air pollution, which can lead to some confusion in interpretation, especially when the signs conflict (using cumulative lags avoids this problem). Assuming that the mean values of lagged and unlagged variables are identical and summing the coefficients for various lag terms yields:

$$\text{\% deviation in respiratory admissions} = -7.1 - 0.172\text{T} + 0.81\ \text{SO}_4^{-2} - 0.036\ \text{O}_3\ (R^2 = 0.056). \qquad [9\text{-}1]$$

Equation [9-1] is problematic since the three environmental terms are highly correlated with one another (positively) and each has a positive (bivariate) relationship with respiratory admissions, yet two of them show a net negative effect on admissions in the multiple regression. This is a symptom of collinearity, which can inflate both the regression coefficients and their significance levels (when collinear terms enter with opposite signs). Further, evaluating Eq. [9-1] at the means of the independent variables yields −4.2% for the mean deviation, rather than the expected value of zero. According to this relationship, it appears that sulfate is the main contributor and that its elasticity is somewhere between 0.04 and 0.08. From contingency tables in an earlier version of the paper provided to me by the senior author, I was able to estimate the elasticity of total respiratory admissions on sulfate as about 0.08. This value is similar to those found in the other studies discussed. Based on 40 respiratory admissions per summer day as an average, Eq. [9-1] predicts that eliminating sulfate entirely as a causal factor would thus save 2 to 3 daily respiratory admissions for this population of about 6 million people. However, the validity of this prediction also depends on concurrent trends in the other collinear pollutants.

Data from southern Ontario were also analyzed by Hammerstrom *et al.* (1991), in an effort that emphasized statistical methodology. Time-series analyses were performed on a 6-year (1979–85) data base of daily hospital admissions, weather data, and concentrations of six different air pollutants. For the gaseous species, 24-hour averages were used. The data were characterized by substantial serial correlation and nonnormal frequency distributions. Analyses were performed for respiratory admissions (causal hypothesis) and for admissions for gastrointestinal illness and accidents (control diagnoses). Statistical analyses progressed from simple bivariate correlations to bootstrapped correlation analysis to multiple regressions with autoregressive error modeling. In general, the numbers of significant correlations decreased as the level of complexity of the analysis increased. It was shown that bivariate correlations are inappropriate when seasonal trends are present in the data and that biased estimators can result when inappropriate models are used. For example, there were many more significant bivariate correlations between admissions for both respiratory and control diagnoses and air pollution than would be expected by chance, but most of these disappeared when multiple regressions were used.

Typically, only 1 to 5% of the variance in admissions was explained by the variables used in this analysis, even when weather effects were included in multiple regressions. The pollutants found to be significant in multiple regressions included various weather variables, but only one pollutant species included O_3 (summer and winter) and SO_2 and SO_4^{-2} (summer only). Although TSP was not investigated in multiple regressions, it had the highest bivariate correlation with respiratory admissions. These results suggest that there are many other factors controlling the timing of hospital admissions; if any of them should be temporally correlated with air pollution, confounding and spurious correlation could easily result.

In a further account of this analysis, Lipfert and Hammerstrom (1992)

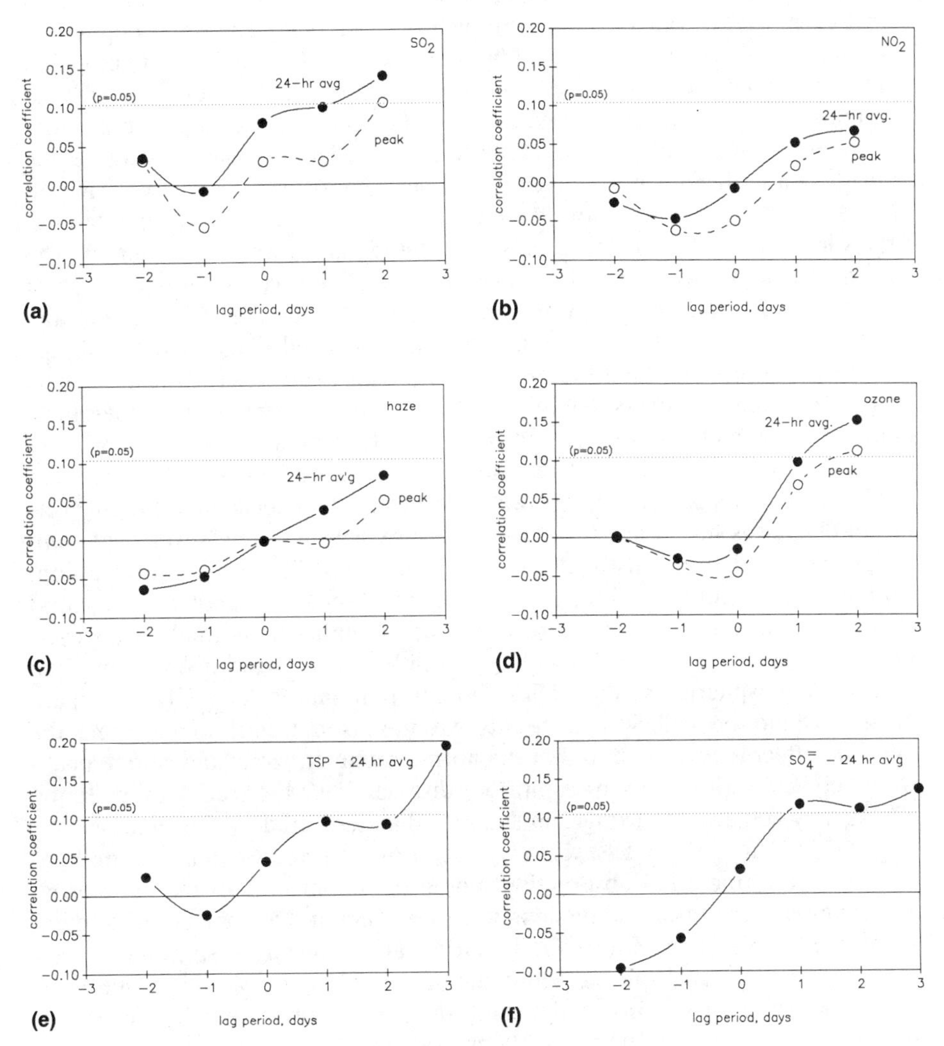

FIGURE 9-17. Bivariate correlations between July–August respiratory admissions in southern Ontario (1979–85) and various air pollutants, comparing 24-hr averages with peak hour data. (a) SO_2. (b) NO_2. (c) COH. (d) Ozone. (e) TSP. (f) SO_4^{-2}. *From Lipfert and Hammerstrom (1992), with permission.*

showed that 24-hour averages gave higher bivariate correlations than peak-hour averages for all the gaseous pollutants (Figure 9-17) and that breaking up the large region into three subregions did not improve the precision of estimates. It appeared that the stability in admission rates achieved by pooling outweighed the loss in precision of air pollution exposure estimates. Based on multiple regressions using cumulative lags up to 3 days, it appeared that the elasticity of air pollution on July–August respiratory admissions was around 0.20, which is considerably higher than estimates by Bates and Sizto based on individual lags. Lipfert and Hammerstrom concluded that, on a bivariate basis, all six pollutants exhibited similar dose-response relationships (Figure 9-17) and that it was not possible to identify the "responsible" pollutants with certainty.

The most recent work by Bates and colleagues (1990) deals with ER visits in Vancouver, B.C., from July 1984 through October 1986. The study population was about 1 million people, served by nine hospitals for which the numbers of daily emergency room visits were pooled. Diagnostic categories were asthma, pneumonia, "all respiratory" (excluding common colds and upper respiratory infections), and all diagnoses. Data were recorded manually by the same individual for all nine hospitals, to provide a consistent data base. Air pollutants included maximum hourly values of SO_2, NO_2, O_3, COH, and 24-hour values for SO_4^{-2} aerosol. Daily maximum temperature was also included. The statistical analysis was limited to bivariate correlation coefficients for lags of zero, 1, and 2 days, by age group and diagnosis. In summer, temperature and ozone were significantly correlated with total visits (all diagnoses) for all age groups; the authors felt that the ozone correlations (which were always weaker) simply reflected the expected relationship between ozone and temperature. Sulfur dioxide and sulfate were correlated with asthma and total respiratory visits, mainly for ages 15–60. In winter, results were similar with somewhat weaker associations; temperature was also positively associated with ER visits for all diagnoses, up to age 61. However, since seasonal trends were not completely removed from this analysis, the results must be viewed with caution. Within each 6-month period, seasonal trends (due to nonpollution factors) could create artifacts in either direction. For example, the normal winter cycle in respiratory disease coincides with cold weather, space heating emissions, and reduced atmospheric mixing. Thus, an air pollutant associated with space heating would be collinear with this within-season trend. In summer, the opposite effect may obtain, in that the normal summer ozone and sulfate peaks would coincide with a dip in respiratory disease. Thus the true daily pollution effects may actually be stronger in summer than indicated by this analysis.

A "spike" in asthma ER visits was seen in all three Septembers in Vancouver, confirming other observations about the importance of the fall season. For example, respiratory admissions were very high in October in the study of Levy *et al.* (1975), in Hamilton, Ontario. The Vancouver spikes were not associated with any of the air pollutants measured in this study; in one year, the rise in asthma visits followed a sharp drop in temperature, which conforms to the hypothesis of Goldstein and Block (1974) (discussed in the next section).

Maarouf and Zwiers (1987) analyzed 5 years of daily ER admissions to 26 hospitals in metropolitan Toronto for asthma, bronchitis, and emphysema, in relation to relative humidity and the nitrate fraction of TSP. The reasons for selecting these independent variables were not given. They chose the mid-September to mid-December period for analysis because of generally higher and more variable admission rates during this period (about 18 per day). Their analysis accounted for day-of-week and long-term trends as well as for occasional outliers with high admissions followed by low admissions on the next day. One of their techniques used a case-control method, where "cases" were defined as the highest 10% of admission days and "controls" as the lowest 10%. The environmental conditions were then compared as a function of lag for both groups. The authors concluded that respiratory admissions were associated with both relative humidity and suspended nitrates, but they performed no multiple regressions.

Studies by Knight *et al.* (1989a, b) in Prince George, BC, offer the opportunity for direct comparisons of admissions with ER visits for a single regional hospital. A 2-year period from 1984 to 1986 was studied, emphasizing respiratory diagnoses in three broad categories: asthma, lower respiratory diseases (bronchitis, emphysema, chronic obstructive lung disease, etc.), and "other" (influenza, pneumonia, upper respiratory diseases, etc.). The study was limited by the sparse air quality data available, consisting of SO_2 (which was not used because of a change in methods partway through the period), total reduced sulfur (TRS), and TSP measured every sixth day (with no coverage on the other 5 days). Because of the presence of local pulp mill sources, TRS, which includes hydrogen sulfide (H_2S), was of interest in this community. The statistical analysis accounted for day-of-week and seasonal trends and used a log-linear Poisson model. The absolute numbers of admissions or ER visits were low, with counts of zero on many days. Respiratory admissions were dominated by children. The ratio of ER visits to admissions was about 2 for asthma, 3 to 4 for lower respiratory diseases, and about 6 for "others." There were no associations between admissions and ER visits other than common temporal patterns, suggesting that the underlying populations and causal factors may be different. The associations with air quality were described as "quite small." For admissions to hospital, TRS had no association, and the TSP association failed to reach significance (with lags of up to 3 days). For respiratory ER visits, TRS was significant with a lag of 2 days ($t = 3$), and TSP was significant for asthma and "other" visits, also with a lag of 2 days. Since log-linear models were used, elasticities depend on the magnitude of the pollution variable; a 10% reduction in TRS was associated with an elasticity of 0.05; a 75% reduction, with an elasticity of 0.09. The elasticities for TSP were about twice as high but less certain. Analysis by season showed the associations to exist only in winter, which is the time of highest air pollution. This is in contrast to the Ontario studies, which found significant results only in summer.

The most recent analysis of daily variations in respiratory admissions to Ontario hospitals is that of Burnett *et al.* (1992, 1993). This study includes a larger portion of the province than the previous Ontario studies; the study area was extended north to include 168 acute care hospitals, for 1983 through 1988. Air monitoring data on ozone (daily 1-hour maximum readings at 22 monitoring stations) and sulfates (24-hour averages at 9 monitoring stations) were included. Taking into account prevailing wind patterns, the province was partitioned into regions, and a single monitoring station was selected to represent each region. Where multiple monitors were available in a region, the station with the highest correlations with its neighbors was selected. The SO_4^{-2} data were obtained from 3 different networks, one of which used glass-fiber filters and two of which employed Teflon filters. Because of the artifact SO_4^{-2} formed on the glass-fiber filters (discussed in Chapter 2), an empirical correction was devised to allow data from the 3 networks to be merged. Monitoring data for particulate matter, SO_2 and NO_2 were judged to be too spotty for use in the analysis; weather data were obtained from 5 different locations in the province.

Hospital admissions were analyzed separately for each hospital, based on urgent and emergency categories of admissions; elective admissions and transfers from other hospitals were not considered. ICD-9 codes 466 (acute bronchitis), 480–486 (pneumonia), 490–1 (bronchitis), 492 (emphysema), 493 (asthma), 494 (bronchiectasis), and 496 (chronic airway obstruction) were included. Separate groups were considered for ages 0 to 1, 2 to 34, 35 to 64, and 65 and over. The statistical analysis used day-of-week adjustments and a linear filter to remove long-term (seasonal) patterns. The time series for each hospital was normalized with respect to the average number of respiratory admissions for that hospital; the data were then pooled by region for the environmental analysis. The methods of Liang and Zeger (1986) were used for estimating the air pollution effects. Burnett *et al.* (1993) report that serial correlation was not present in the filtered data.

For the months of May through August, ozone was found to be a stronger predictor of respiratory admissions in these data than sulfates or temperature. Taken one at a time, the respective elasticities were about 0.06, 0.02, and 0.04. In a multiple regression, the (single-day) effect of ozone was about 4.6% with an additional 1.1% for sulfate. Lags up to 3 days were considered individually; all were statistically significant, and the cumulative effect appeared to be about 3 times the single-day effect. Temperature was not found to be a confounder. The largest percentage effects associated with air pollution were found for ages 0 to 1 for asthma, infection, and all respiratory diagnoses combined. Ages 2 to 34 displayed the largest percentage effect for chronic obstructive pulmonary disease. Summed over all ages, the largest effect was for asthma. Neither O_3 or SO_4^{2-} was associated with respiratory admissions during December through March.

A decile plot of adjusted admissions, summed over all hospitals versus ozone, appeared quite linear; the range of 1-hour maximum ozone levels was from 20 ppb to about 100 ppb. A similar plot for SO_4^{-2} was concave downward, suggesting a slight diminution of response for levels over about 10 $\mu g/m^3$. Burnett *et al.* (1993) also reported nonsignificant associations between air pollution and a nonrespiratory control diagnosis (not specified) and with negative lags for respiratory diagnoses.

Studies of Asthma in Various Cities

Asthma hospitalizations were included in several of the reviewed studies but generally were among the weaker relationships with air pollution. The recent increases in asthma mortality and morbidity make this topic particularly relevant; a number of studies emphasizing the timing of asthma hospitalization are reviewed also. Asthma was included in several of the Canadian studies discussed previously.

Goldstein and coworkers (1974, 1978, 1981a, 1981b, 1986) have investigated temporal and spatial relationships of asthma in New York City, as measured by ER visits. The first paper (1974) used conventional multiple regression methods for relatively small samples (n = 40–104) from the fall seasons of 1970 and 1971, testing the hypothesis that the number of ER visits for asthma was associated with temperature and daily average SO_2. The hypothesis was accepted for Brooklyn but not for Manhattan. Since no account was taken of lags, or day-of-week or seasonal effects, this finding appears problematic. In the subsequent papers, Goldstein *et al.* developed statistical methods based on the frequency distributions of visits, without *a priori* consideration of the pollutant distributions. Goldstein and Dulberg (1981b) and Goldstein and Weinstein (1986) investigated a period of greatly decreasing SO_2 levels (1969–72), using selected data from the 40-station New York City monitoring network; they were unable to find a relationship between ER visits for asthma and either 24-hour averages of SO_2 and COH or hourly SO_2 peaks. (Hourly SO_2 concentrations as high as 0.5 ppm were experienced during this period.) This finding is consistent with Greenburg *et al.*'s (1962b) observation that there was no increase in ER visits for asthma during the November 1953 air pollution episode. Goldstein and Weinstein (1986) also concluded that days with peak hourly values did not necessarily coincide with days of high 24-hour average SO_2. No other pollutants were investigated, in spite of a previous finding (Goldstein and Rausch, 1978) that peak days for asthma tended to coincide throughout the city, suggesting either a weather variable or a regionally distributed pollutant such as SO_4^{-2} or O_3. Given the regular occurrence of asthma episodes in the fall in different cities, weather factors are likely to be important.

Girsh *et al.* (1967) studied children's ER visits for asthma in Philadelphia from July 1963 to May 1965. Visits were found to increase by a factor of 9 on stagnant days with high barometric pressure and "high air pollution." Although no data are given in the paper, mention is made of measurements of NO_x, SO_2, CO, total oxidants, dust, and pollen counts (during the ragweed season). The hospital studied was located in an area of high dust loading. Asthma attacks were not found to be associated with pollen counts or with the peak incidence of upper respiratory infections. The authors reported that no specific pollutant could be "incriminated" and that total oxidants seemed to be least important.

New Orleans's reputation for asthma "epidemics" prompted attempts to try to identify any associated environmental agents. During some episodes, ER admissions increased by one order of magnitude and asthma deaths were reported. The study reported by Lewis *et al.* (1962) ran from June 1960 to June 1962 and included a census of an area thought to have a number of susceptible persons, time-series analysis of hospital ER treatments, and local monitoring of airborne particles. The results reported included a lack of correlation between air pollution and asthma attacks among 67 selected patients in the census area or between attacks in this group and ER treatments for the population at large. A statistically significant relationship was reported between ER visits and particles identified as due to "poor combustion, with silica." These particles were reported to be in the size range over 4 μm; uncontrolled refuse dump fires were the suspected source.

The results of Khan (1977) in Chicago were generally consistent with the findings in other cities. The relationship between daily attacks and ER visits or hospitalization was weak ($R < 0.4$), even for the same subjects. The proportion of asthma variance associated with air pollution was small: 5 to 15%. However, the most significant air pollutant with respect to asthma attack frequency was CO, which seems counterintuitive.

A panel study of 34 asthmatics in Los Angeles from August 1977 to May 1978 (Katz and Frezieres, 1986) found that only 10% were "responders" (subjects with a statistically significant relationship between symptoms or need for medication with air pollution). Sulfate was the most significant pollutant for these three subjects. Thus it appears that only a fraction of even a sensitive subpopulation is likely to respond to daily variations in air pollution. This helps explain the small magnitudes of the effects reported in these various morbidity studies and the need for long time periods or large populations. A later, more comprehensive analysis of Los Angeles asthmatics by Whittemore and Korn (1980) found both oxidants and TSP to be significant predictors of asthma attacks, based on 24-hour averages and a multiple logistic regression model. The magnitude of the effect was small, on average: a 25% excess risk of an attack was associated with either about 0.12 ppm increase in oxidants or 250 $\mu g/m^3$ in particulates. (Note that this dose-response function, 10% per 100 $\mu g/m^3$, is consistent with the episode slope of Figure 9-2, given the higher frequencies of attacks with respect to hospitalization.) The authors pointed out that, because of collinearity, TSP was intended to be a surrogate for all particulates, SO_x, and NO_x. The most important predictor was an asthma attack on the preceding day, which underscores the importance of dealing with serial correlation in this type of study. Also, the most likely day for an attack to be reported was the last day of the weekly reporting period, which raises the possibility of subjective bias. Such a reporting artifact could mask any true relationships but would be less likely in studies of hospitalization.

Richards *et al.* (1981) studied childhood asthma from August 1979 to February 1980 in Los Angeles, using data on ER visits, hospital admissions, and subjective symptoms. Monthly visits and admissions were highest in winter, in spite of a smog episode in September and the findings of other studies of peak asthma attacks in the fall. However, because of the differences in climate, this finding provides support for the hypothesis of Goldstein and Block (1974) that asthma attacks increase with the advent of the first cooler days of the season. On a daily basis, ER visits were positively correlated with COH, NO_x, hydrocarbons, TSP (not significant), days with Santa Ana winds, and allergen counts. Significant negative associations were found with O_3, SO_2, temperature, relative humidity, and sulfate (not significant). These relationships were combined by means of factor analysis; three factors explained 30% of the variance in numbers of asthma ER visits. These ambiguous findings may be due in part to the failure to control for day-of-week or seasonal trends or to consider the simultaneous variations of temperature and air pollution.

Giles (1981) reports an attempt to find meteorological correlates of asthma in Tasmania, including hospital morbidity. Meteorological factors included wind chill and discomfort and the analytical technique was that of seeking coherence among the time series. No such relationships were found.

White *et al.* (1991) conducted a pilot study of the effect of ozone on children's visits to the ER of a large inner-city public hospital in Atlanta for asthma or reactive airway disease, from June through August 1990. Ozone data were available from two monitoring stations. This study did not develop a consistent dose-response relationship between respiratory ER visits and ozone, but did report 38% more visits on 6 days for which peak ozone exceeded 0.11 ppm ($p = 0.04$). The study may have been compromised by lack of controls for day-of-week and seasonal effects, in addition to the short time span covered.

Frequencies of asthma attacks were studied by Ponka (1991), using admissions to Helsinki hospitals from 1987 to 1989. Emergency admissions for asthma were studied separately. The cases analyzed were limited to ICD-9 493 (bronchial asthma); chronic bronchitis was excluded. The air pollution variables and the numbers of monitoring sites ( ) included SO_2 (4), NO and NO_2, (2), CO (2), ozone (1), and TSP (6). Levels were generally low (the average SO_2 was 19.2 $\mu g/m^3$, but the average TSP was 76 $\mu g/m^3$) and 24-hour averages were used in the analysis. Data were also collected for temperature, humidity, and wind speed at one station. Logarithmic transforms were used in addition to linear models. Seasonality was accounted for by standardizing for minimum daily temperature and by multiple regression. The most significant pollution variables, after temperature correction, were NO, NO_2, O_3, and CO; SO_2 and TSP were not significant. Elasticities were not given, but I estimated typical values to be around 0.12 on a same-day basis. Log-transformed variables were reported to fit slightly better. For ozone, the results for 1- or 2-day lags were much stronger than for the same day; lags greater than 2 days were not studied. Results were also presented for three different age groups; 21 of the 36 pollution correlations were significant ($p < 0.05$); the strongest effects were seen in adults (ages 15–64 and 64+). Ponka concluded that traffic-related pollution seemed to be the most important in Helsinki, but no explanation was offered for the lack of response to the fairly high levels of TSP.

Cody *et al.* (1992) analyzed asthma and bronchitis visits (all ages) to ERs of nine northern and central New Jersey hospitals for the summers (May–August) of 1988 and 1989. The method of analysis followed that of Bates and Sizto (1983) in that data from all the hospitals and from five ozone monitoring stations were pooled to create a single regional time series for each year. However, unlike the Bates and Sizto study, it appears that not all the hospitals in the region were included. Emergency visits for finger wounds were used as a control variable. Other environmental variables in the analysis included SO_2 (24-hour average), PM_{10} (every sixth day), daily mean temperature and relative humidity, and atmospheric visibility measured at Newark at noon. Ozone was averaged from 10:00 A.M. to 3:00 P.M. each day. Day-of-week effects were examined by replicating the analysis with weekends excluded. A similar approach was used to remove precipitation effects on visibility, but deleting precipitation days. Regression analyses were performed separately for each year and for the pooled data set ($n = 226$). Lags up to 2 days were considered. Ozone was significantly higher in 1988, but the average daily ER visits for asthma and bronchitis were lower. Since there were fewer than five daily ER visits for asthma or bronchitis on average, a Poisson regression model would have been preferable.

The statistical analysis consisted of bivariate correlations and various multiple regression models. There were no statistically significant relationships for bronchitis visits or for finger wounds. For asthma visits, the only significant relationship was the negative effect of temperature (probably representing seasonal effects); the method chosen to account for this was stepwise multiple regression. Since the correlation between ozone and temperature was much stronger ($r = 0.64$) than between either asthma visits and temperature ($r = -0.23$) or ozone (r values from 0.08 to -0.12, depending on lag), the method of accounting for seasonality in the data could be important.

However, after accounting for the temperature effects, Cody *et al.*'s multiple regression results showed a significant relationship between ozone and ER visits for asthma that was stronger in 1989 than in 1988, in spite of the lower average ozone concentrations in 1989. While there was no significant difference in the temperature regression coefficients between the two years, 1989 showed both same-day and 1-day lagged ozone to be significant, with a combined elasticity of 0.73 ± 0.28 (2-σ limits), while 1988 showed only 1-day lagged ozone to be significant, with an elasticity of 0.41 ± 0.28. The elasticity of asthma ER visits on 1-day lagged ozone for the combined data set was 0.27 ± 0.19 (2-σ limits). For comparison with other studies, this regression coefficient corresponds to about 28% excess visits per $100\,\mu g/m^3$. Accounting for serial correlation had only minor effects on the analysis, as did omitting weekends. No other environmental variables were significant in multiple regressions.

I used the coefficients for temperature and ozone from the pooled data set to estimate the long-term average difference between the two summers. This analysis predicted a decrease of 0.14 visits per day based on environmental factors; there were actually 0.5 *more* visits per day in 1989, which suggests that there are other important factors controlling the frequency of asthma ER visits. It is also possible that the results of Cody *et al.* are sensitive to the method of seasonal adjustment, since it is expected that the indicated effect of temperature on asthma ER visits is at least in part a surrogate for other environmental influences such as the presence of pollen, rather than temperature *per se*.

Schwartz *et al.* (1993) studied emergency room visits for asthma at eight Seattle hospitals for 1 year beginning September 1989, in relation to data on atmospheric visibility and daily PM_{10} data from a residential area (also see Koenig *et al.*, 1992). The analysis controlled for season, weather, day-of-week, hospital, age, serial correlation and time trend, and found a significant association with residential PM_{10}. The elasticity was about 0.12, and the authors reported that the mean of the previous 4 days' PM_{10} was a better predictor than the value from the previous day. No other pollutant showed significant correlations, and the correlation between PM_{10} and gastroenteritis (a control diagnosis) was negative.

In summary, some, but not all, of the studies of the timing of hospitalization for asthma provide convincing evidence of relationships with air pollution, in part because of shortcomings in study designs.

Other Time-Series Studies

Kevany *et al.* (1975) analyzed mortality and hospital admissions in Dublin for the winters of 1970–73, examining partial correlations for respiratory and cardiovascular causes in relation to smoke and SO_2, mainly for the winter of 1972–73. Respiratory admissions showed more significant correlations with smoke than with SO_2. Influenza admissions were not associated with either pollutant. Cardiovascular admissions were highly significant for both smoke and SO_2 for males; results for females were mixed. Kevany *et al.* corrected for temperature effects, but the report was unclear as to how day-of-week effects were handled. Lags varied from zero to 3 days. Collinearity between smoke and SO_2 was not reported, but the authors noted declining long-term trends for smoke and increasing trends for SO_2, which may have helped to separate the pollutants.

Hospitalization studies in Dublin were continued by Sweeney (1982), who studied winters (October–March) from 1975 to 1978 using data from up to 36 monitoring stations for smoke and SO_2. The statistical analysis was based on weekly data and admissions for respiratory or cardiovascular complaints, citywide. Temperature, relative humidity, and wind speed were also included in the analysis. A strong dip in admissions was noted for Christmas weeks in all 3 winters; apparently, this artifact was not removed from the analysis. For 1975–76, both respiratory and cardiovascular admissions were associated with SO_2. For the remaining period, respiratory admissions were associated with smoke. Negative correlations were also shown between admissions and wind speed, but this could simply reflect air stagnations with associated buildup of air pollutants rather than a real effect of wind speed.

Özkaynak *et al.* (1990) presented a brief account of a study of hospital admissions in Boston and three other Massachusetts cities, as part of the development of a health-based air quality–visibility index for the state. Data were pooled for four cities (Boston, Worcester, Springfield, and Fall River/New Bedford) following the methodology of Bates and Sizto (1983). The authors report that many models were developed and that both positive and negative relationships were seen for the various respiratory diagnoses. Examples were shown linking pneumonia and influenza admissions with 24-hour lagged ozone for adults in summer and with 24-hour lagged TSP for children in winter ($p < 0.01$). The earlier report on this project (1985) illustrates some of the pitfalls and general observations seen in many of the other studies we reviewed:

- No single pollutant is identified as most important for respiratory admissions.
- Daily averages are usually more significant than daily peaks.
- Results are unreliable for the smaller populations (including children's admissions).
- Seasonal and temperature effects are important, especially for upper respiratory and pneumonia or influenza admissions.
- Collinearity can result in regression coefficients with opposite signs for correlated pollutants in the same regression.

Autocorrelation was not considered, nor were results given for "control" diagnoses, although since no pollutants were significant for cerebrovascular admissions, this category may have in fact served as a control for respiratory admissions. The pollutant elasticities for respiratory admissions were higher (0.10–0.30) than in most of the studies reviewed, even though day-of-week and seasonal effects were accounted for with dummy variables. Multiple regression R^2's were in the 0.1–0.6 range (including day-of-week and seasonal effects); pollutant partial R^2's were in the 0.02–0.04 range.

Daily ER visits in Barcelona were studied by Sunyer *et al.* (1991a) for 1985 and 1886. Diagnoses were limited to chronic obstructive pulmonary disease (COPD), and data were collected from four large hospitals which accounted for about 90% of all ER visits in Barcelona. The ER data were screened by a clinician to ensure proper diagnosis classification based on a list of symptoms and averaged about 12 visits per day. Air pollution data were obtained from a citywide network for SO_2 and British smoke (24-hour averages from 17 stations); data on SO_2, CO, NO_2, and O_3 (1-hour maximum values) were obtained from the average of two stations. Meteorological data were collected from five different monitoring stations. Pollution levels were within the European Community Guidelines, but WHO guideline values were exceeded on occasion.

The paper presents bivariate correlations between ER visits and air pollution, but these results were influenced by seasonal confounding. When stratified by temperature, the relationships between ER visits for COPD and 24-hour SO_2 and smoke were about the same and were stronger in warm weather. The potentially confounding effects of meteorlogy, season, and day-of-week were handled by "adjusting" the ER visits variable. After adjustment, SO_2 (either 1-hour or 24-hour measures) was significantly associated ($p < 0.01$) with daily ER visits for COPD on the same day and lagged 1 day. The relationship was weaker and lost significance for a 2-day lag; cumulative lag effects were not reported. The smoke and CO variables were also significant ($p < 0.01$), but NO_2 and O_3 were not. Accounting for serial correlation made little difference in these results.

The elasticity for SO_2 on a single-day basis was about 0.095; estimated values for smoke and CO were 0.06 and 0.05, respectively. Rough estimates of the 2-day values would be about double these figures. In an effort to identify a threshold of no effect, Sunyer *et al.* successively truncated the data set to remove SO_2 values above certain cutoff points and then reevaluated the regression slope (after adjustment for potential confounders). The results of this exercise are shown in Figure 9-18 in terms of elasticities (based on 11.9 COPD visits per day in all cases). The slopes were statistically significant for daily SO_2 values in excess of about $70\,\mu g/m^3$, which is well below most air quality standards.

In an invited commentary on this paper, Corn (1991) tried to place these findings in a risk-analysis context and expressed doubt that air quality standards could actually "ensure zero risk to all members of the exposed population." In their rebuttal, Sunyer and Anto (1991b) pointed out that errors in estimating the actual pollution exposure likely biased the true relationship downward, and that subsequent extension of the data set for 3 additional years produced similar findings, in spite of lower pollution levels (Sunyer *et al.*, 1993).

The extended study (Sunyer *et al.*, 1993) also found effects for both smoke and SO_2, which remained significant in a joint regression, using ridge regression because of their collinearity. Nearly linear relationships were found in both summer and winter, although the winter effects were stronger.

Schwartz *et al.* (1991) considered cases of children's croup and bronchitis documented by hospitals and physicians over a 2-year period in Duisburg, Köln, Stuttgart, Tübingen, and Freudenstadt. Not all of these cases resulted in admission to hospital. Pollutants considered were 24-hour averages of SO_2, NO_2, and TSP (measured by a beta-gauge tape sampler). Data from 1 to

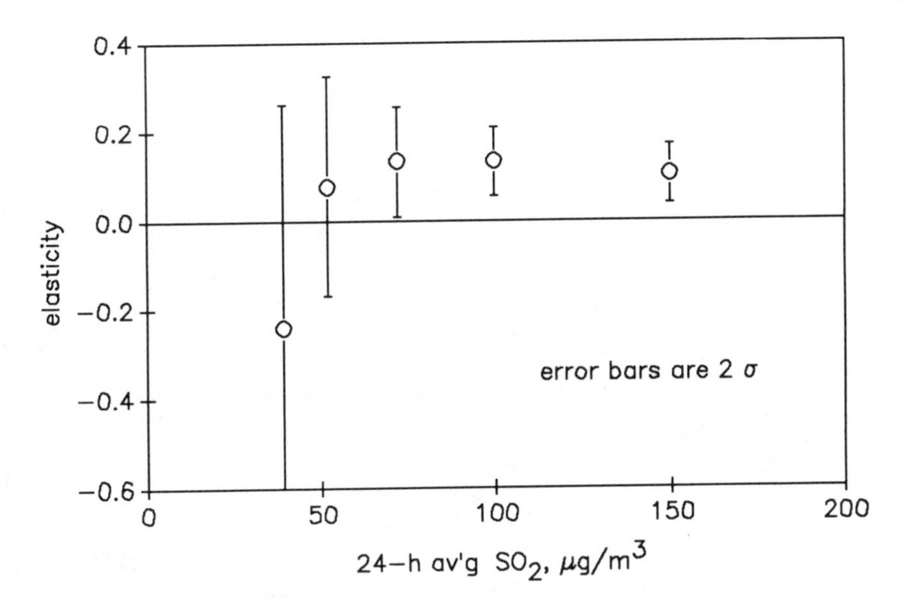

FIGURE 9-18. Elasticity of ER visits for COPD in Barcelona as a function of SO_2 level. Data points are not independent but were created by successively truncating the data set. *Data from Sunyer* et al. *(1991a)*.

4 monitors were averaged for each city. Weather variables included mean temperature and relative humidity. Each city was studied separately and then all five were pooled, taking into account the variance within and between cities. The method of statistical analysis employed Poisson regression (because of the relatively low daily case counts) and a two-stage analysis approach. The case counts were first fit to weather, seasonal, and other temporal variables; air pollution variables were then fit to the residuals for each city, and serial correlation was accounted for by using autoregressive methods. In order to account for the loss of data due to physicians dropping out of the study, variables for time or for the numbers of reporting physicians were included in each model. After seasonal effects were accounted for, weather variables were not significant.

No significant pollution associations were reported for bronchitis. The results for croup showed that all three pollutants were statistically significant ($p <$ 0.05) when the five cities were pooled (the logarithm of TSP was found to fit better than the linear measure). Log TSP was significant in four out of five individual cities, NO_2 was significant in two, while SO_2 was only significant in Köln (where neither TSP nor NO_2 was significant). The analysis was repeated for the months from October to March (the high-pollution months), and the results were quite similar. Serial correlation effects were reported to be small. The average of same-day and 1-day lagged pollution was reported to have the same predictive capability as same-day alone, but it was not reported whether the magnitude of the response increased when lags were included (as would be expected).

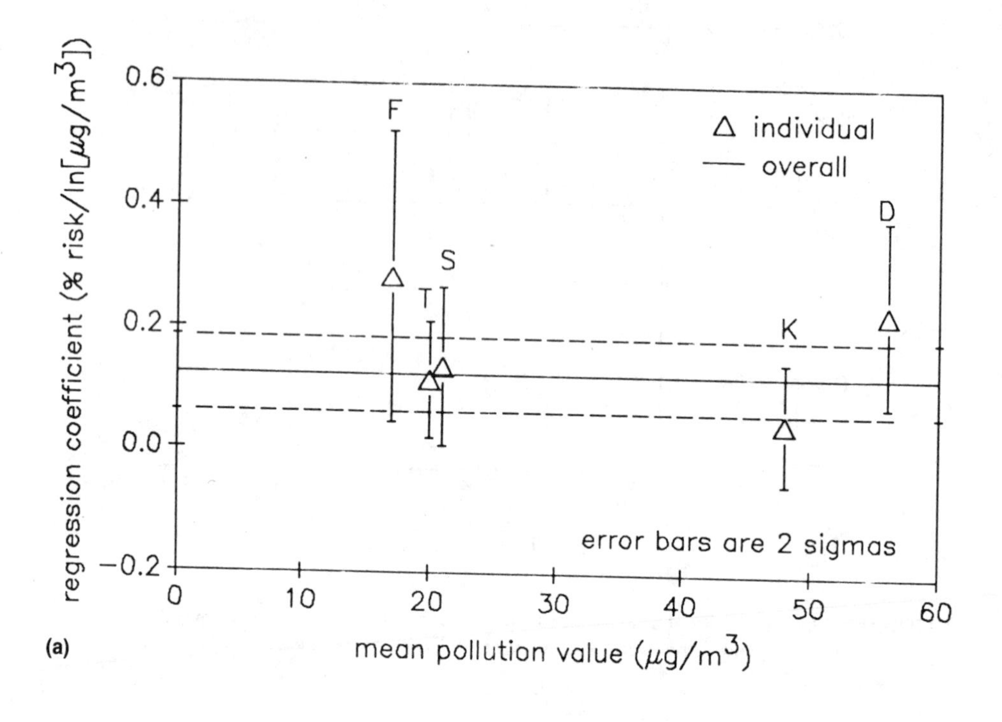

(a)

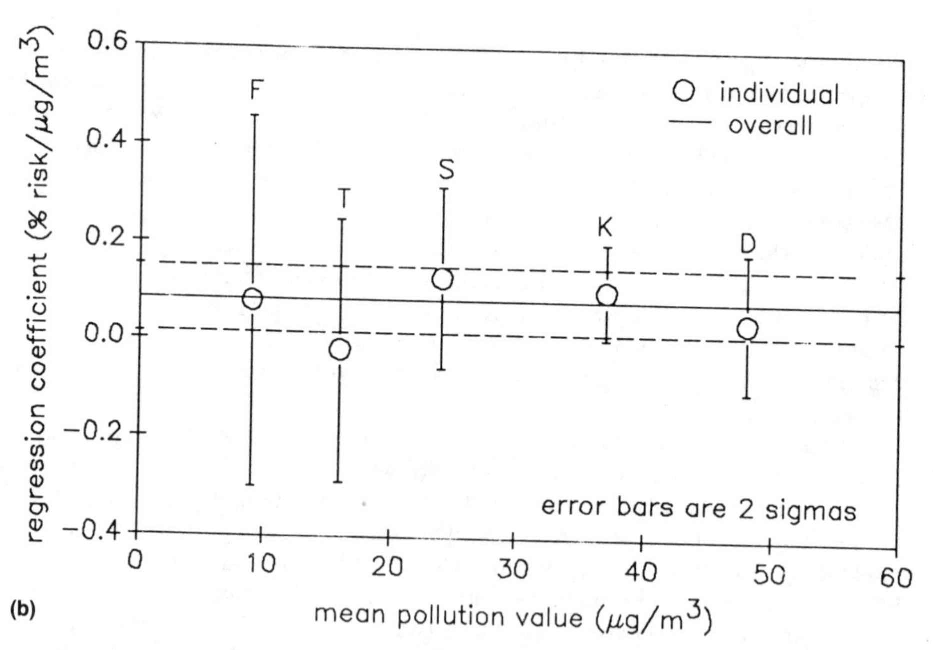

(b)

FIGURE 9-19. Regression coefficients for children's croup and bronchitis in five German cities (D = Duisburg, K = Köln, S = Stuttgart, T = Tubingen, F = Freudenstadt). The horizontal solid line represents the coefficient for all five cities pooled. (a) Log TSP. (b) SO_2. (c) NO_2. *Data from Schwartz* et al. *(1991)*.

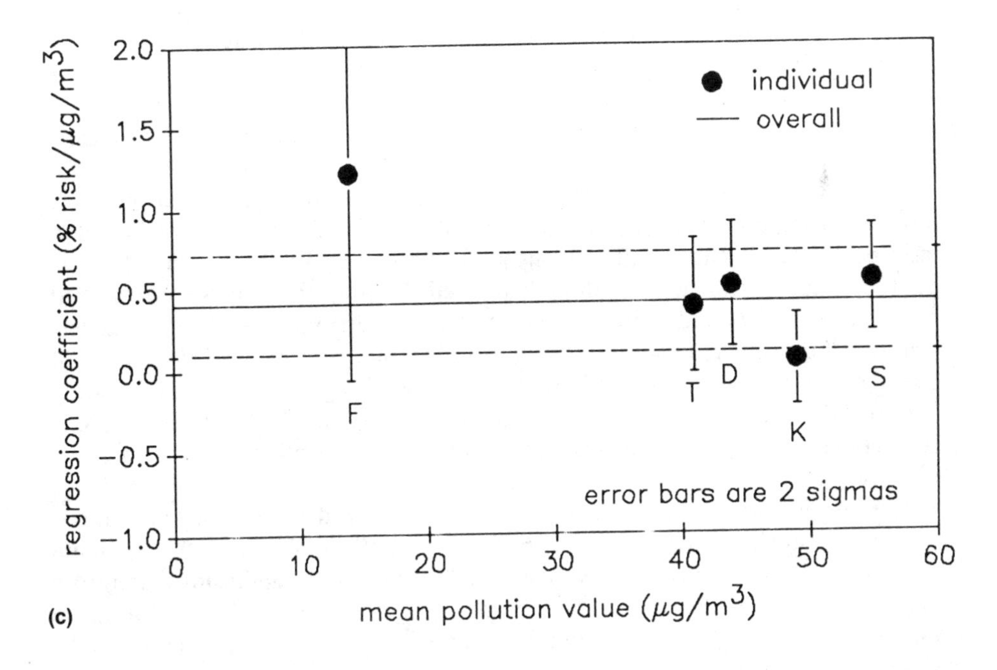

(c)

FIGURE 9-19 *Continued*

The pooled elasticities (at the mean) for log TSP, NO_2, and SO_2 were 0.094, 0.17, and 0.022, respectively; use of the log transform for TSP reduced the elasticity somewhat. Figure 9-19 plots the regression coefficients against mean pollution values for each city, to display dose-response relationships. The plots for SO_2 and NO_2 suggest the possibility of thresholds around 30 to 40 $\mu g/m^3$ (which was also seen in the plot of pollution quintiles for NO_2 presented by Schwartz *et al.*). Note that a trend of regression coefficient increasing with decreasing pollution level suggests a surrogate model. The plot for TSP more closely resembles a linear, no-threshold model, since the regression coefficient tends to remain constant over the whole range of mean values. The effects of NO_2 appeared to be stronger than the other pollutants, which would be further emphasized if a threshold were considered. Schwartz *et al.* noted that NO_2 has been linked with decreased resistance to respiratory infection in clinical experiments at much higher concentration levels. They also noted the weak effects of SO_2 and suggested that they may have been the result of collinearity with the other pollutants.

Miscellaneous Studies

Noble *et al.* (1971) analyzed police ambulance runs and emergency unit visits to three Boston nonprofit hospitals in 1968 in relation to weather variables, days of the week, holidays, and holiday travel patterns. The dependent variables all included trauma events, although one category of visits was defined by substracting the traffic accidents. The only independent variable that might be related to air pollution was "occurrences of smoke or haze," which was not significantly correlated with any of the dependent variables. The patterns that emerged from this analysis were day-of-week and holiday effects, a tendency for more ER visits during warm, sunny weather, and possible adverse weather effects on traffic accidents. Adverse weather, such as precipitation or fog, did not have a negative impact on ER visits, as had been hypothesized by Carnow (1975).

Nagata *et al.* (1979) analyzed daily data on acute respiratory and digestive diseases, using insurance records from July to September 1975 for two Japanese industrial cities on the island of Honshu. To preclude day-of-week effects, only data for Tuesdays through Fridays were used. Lags up to 10 days were studied for maximum SO_2 and oxidant concentrations and several weather variables. The two locations had similar climates and maximum oxidant levels (means of 0.065 and 0.079 ppm), but different maximum SO_2 levels (means of 0.022 and 0.008 ppm). Only bivariate correlations were made, and no seasonal corrections were used (from the scatterplots, seasonal effects appeared to be modest).

For respiratory disease, the high-SO_2 area showed significant correlations with maximum SO_2 for lags of 4, 6, 7, 8, and 10 days, and with maximum temperature for a lag of 6 days. The low-SO_2 area had significant respiratory correlations only for maximum temperature, for lags of 5 and 7 days. For acute digestive diseases (presumably intended to be "controls"), both areas showed a significant correlation with maximum SO_2 (lags of 3 and 5 days, respectively, at the 0.01 level for the low-SO_2 city). There were also scattered significant correlations for digestive disease and weather variables, three at the 0.05 level and one at the 0.01 level. This study may be an example where more restrictive probability levels should be used to allow for the problem of multiple inferences and the large number of correlations presented (Miller, 1981). If the lags are considered to be independent, which depends on the degree of autocorrelation, 264 correlations are presented and one would expect to see 2 to 3 significant at the 0.01 level and about 13 at the 0.05 level. In fact, there were 3 values at the 0.01 level and 12 more at the 0.05 level, so that on this basis, the results could be entirely due to chance. The only compelling arguments to the contrary are the findings of 40 positive correlations for respiratory disease versus temperature and SO_2 and only 4 negative values. There were no significant correlations for oxidants and either disease (25 positive and 19 negative values). If the association of respiratory disease with SO_2 is taken at face value, then an argument can also be made for including lags longer than 1 to 3 days, although the period of time between exposure and taking action on symptoms could also have a cultural component.

Strahilevitz *et al.* (1979) examined the associations between daily use of a psychiatric hospital and air pollution in St. Louis during the summer and fall of 1979. Pollutants included COH, oxidants, hydrocarbons, CO, NO_x, and SO_2. No weather variables or seasonal corrections were included, but a significance level of 0.005 was selected to allow for the problem of multiple inference. Three groups of days were examined: all days, weekends and holidays, and all other days. Carbon monoxide was positively associated with ER visits by all patients for all days ($p < 0.005$); NO_2 was positively associated ($p < 0.01$) with admissions for alcoholics on all days. Since both of these species are associated with vehicle emissions, which are usually lower on weekends as are psychiatric admissions, and significance was only found on all days, it is possible that these findings were influenced by the weekday-weekend patterns.

This theme was also studied by Rotton and Frey (1985), who found an association between family disturbances and assaults and ozone, in addition to effects due to changes in the weather. The study covered 1975 and 1976 in Dayton, Ohio. Ozone was the only pollutant examined.

Hospital records from November to February 1979 in Bombay were examined by Bladen (1983) for the city as a whole and for various subareas. The analysis was limited to "acute respiratory disease related to bronchitis, emphysema, and asthma." Air pollution species included SO_2, particulates, CO, and hydrocarbons. No seasonal or temperature adjustments were described, nor were results given for individual pollutant species. The bivariate correlations reported were in the range 0.50 to 0.75 (monthly averages), and peaks in both air pollution and admissions were reported to occur at times of thermal inversion. No data were reported on concentration levels.

Gross *et al.* (1984) studied respiratory ER visits in the southern part of Israel (Negev), for both adults and children, in relation to TSP and RSP (particles $< 3.5\,\mu m$) from a nearby station. The primary source of particles was reported to be windblown natural dust and chemical industry emissions. Levels of TSP reached almost $1000\,\mu g/m^3$ and RSP, over $100\,\mu g/m^3$, apparently due to dust storms. The authors reported that days with high asthma and shortness of breach admissions (adults) were significantly associated with RSP. This appears to be one of the few studies in the literature that examined the effects of particles from natural sources (note that effluents from the chemical plants are likely to have been well dispersed during dust storms).

Meteorological variables were also emphasized by Diaz-Caneja *et al.* (1991) in their study of air pollution effects on hospitalization in Santander, Spain. They considered admissions to a single hospital for heart failure and chronic obstructive pulmonary disease (COPD) from 1979 to 1982, in response to daily smoke and SO_2 averaged over three monitoring stations. Meteorological variables included barometric pressure, temperature, relative humidity, rainfall, number of calm hours, and wind speed and direction. Since the average number of daily admissions was of the order of one or fewer, a Poisson model should have been used in this study. The stepwise regression models considered these variables raised to various exponents, from 0.25 to 2, lags up to 2 days, and seasonal variations. Same-day effects were found to be the most important. The regression results were difficult to interpret, since significance levels were not given and some of them employed the same variable with different exponents, which entered with opposite signs. However, the authors plotted average admissions for groups of days according to their average pollution levels, which provided a basis for estimating linear dose-response functions. Such estimates will be overestimates, because they do not take into account covariables or seasonal cycles. On this basis, heart failure admissions were not significant, COPD admissions were marginally significant, and their sum was significantly associated with smoke. The elasticities estimated on this crude basis ranged from 0.14 to 0.47, which seem high, but elasticities based on the authors' regression results were even higher. In summary, it appears that there may be a relationship between smoke pollution and hospitalization in

TABLE 9-2 Summary of Major Time-Series Studies of Air Pollution and Hospital Use

Author (Reference)	Area Studied	Time Period	Type of Data	Confounding Effects: Weather	DoW	Seas.	Other	Serial Corr.	Poll. Level $\mu g/m^3$	Adequacy of Pollutant Exposure	Significant Associations: Total	Resp.	Cardiac	Resp.+ Cardiac	Other	Best Lag
Martin (1961, 64)	London	Winters 1958–9 1959–60	Admissions	y		y			smoke = 215	Poor	X@	X	X	X		n/a
									SO_2 = 442	Poor	X@	X	X	X		n/a
Sterling *et al.* (1966)	Los Angeles	Mar–Oct 1961 (n = 223)	Admissions			?	Holiday	?	SO_2 = 34	Good				X	E + R*	1,3
									NO_2 = 83	Good					E + R*	1,3
									CO = 9 ppm	?					E + R*	3,4
									O_3 = 800	Good				X	E + R*	2,3
									part.	Good					E + R*	3,5
Sterling *et al.* (1967)	Los Angeles	Mar–Oct 1961 (n = 223)	Length of Stay			?	Holiday	?	SO_2 = 34	Good			X	X	E + R*	n/a
									NO_2 = 83	Good			X	X	E + R*	n/a
									CO = 9 ppm	?						n/a
									O_3 = 800	Good			X(−)	X(−)		n/a
									part.	Good			X	X		n/a
Durham (1974)	Los Angeles San Francisco (7 universities)	Oct 1970– June 1971	Visits to student health centers			?		?	SO_2 = 13–50	Good		X				
									NO_2 = 60–170	Good		X				1
									CO = 5–7 ppm	Good						
									O_3 = 0.022–0.028 ppm	Good		X				2
									COH = 43–55	Good						
Jaksch & Stoevener (1974)	Portland, OR	1969–70 (n = 1569)	Admissions		X	?	Holiday	?*	TSP = 61	Poor						
		(indiv.)	Cost		X	?	Holiday	?*	TSP = 61	Poor		X				1
Seskin (1977)	Washington, DC	1973,4 (n = 365)	Unsched. Visits			?	Holiday	?	O_3 = 275–350	OK					Eye	
									NO_2 = 85	OK						
									SO_2 = 37–150	Poor						
									CO = 7 ppm	Poor					Eye	
Fishelson & Graves (1978)	Chicago	1971–73 (n = 81)	ER visits			?		No	SO_2 = 60	Good		X**	X			1
									COH = 88	Good			X**			
Namekata & Carnow (1976)	Chicago	1977–78 (n = 20–131)	ER visits			?	Holiday	No?	SO_2 = 25	Poor			X			
									NO_2 = 81	Poor						
									TSP = 76	Poor						

Kurt *et al.* (1978, 79)	Denver	Winter 1975–76	ER visits	y	y	y		?	CO = 18 ppm	Poor			X		1
Samet *et al.* (1981)	Steubenville, Ohio	1974–77 (n = 249)	ER visits				Holiday	?	SO_2 = 90	OK		X			0
									TSP = 156	OK	X	X			0
									NO_2 = 40	OK					
Mazumdar & Sussman (1983)	Pittsburgh	1972–77	Emergency & Urgent Adm.			y		?	SO_2 = 65–120	Poor	X(?)	X(−)			n/a
									COH = 60–130	Poor	X	X(?)	X		
Pope (1989)	Utah Co., UT Salt Lk Co.	1985–89	Monthly Admissions			?		?	PM_{10} = 46	Poor		X		children	1 month
Bates & Sizto (1983)	Southern Ontario	1974–78 (n = 240)	Admissions			?	Trend Holidays	?	***						
									SO_2 = 80	Poor	X( )	X(S)			1,2
									NO_2 = 90	Poor	X(W)(−)				
									COH = 0.9	Poor					
									O_3 = 120	Poor	X(W)	X(S)			1,2
Bates & Sizto (1987)	Southern Ontario	1976–82 (n = 416)	Admissions			?	Holiday	?	***						
									SO_2 = 80	Poor		X(S)			
									NO_2 = 90	Poor	X(S)	X(S)		asthma	1,2
									COH = 0.9	Poor	X (W,S)				
									O_3 = 120	Poor	X(S)	X(S)		asthma	1,2
									SO_4 = 13.3	OK		X(S)		asthma	1,2
Hammerstrom *et al.* (1991) (1991)	Southern Ontario	1979–85	Admissions			?	Holiday	no	SO_2 = 18	Poor		X			2
									NO_2 = 45	Poor					
									O_3 = 40	Poor		X			2
									COH = 0.4	Poor					
									SO_4 = 13	OK		X			3
									TSP = 75	Poor					
Bates *et al.* (1990)	Vancouver, BC	1984–6	ER visits		y	y		?	SO_2 = 38 (max hr)	OK		X		asthma	
									NO_2 = 78			X			1,2
									O_3 = 48		X				
									COH = 0.41						
									SO_4 = 3.35			X		asthma	

* may have been introduced by interpolating for missing TSP data.
** significant for only one age group.
*** (S) summer data (max values).
(W) winter data (max values).
y = confounding effect may have been present.
X = significant association, X(−) = negative significant association, X(?) = marginally significant association.
E + R* = eye and respiratory combined.
@ 1958–59 only.

Santander, but the details of this relationship cannot be deduced from the results presented.

Maarouf (1991) used an episodic analysis approach to examine children's respiratory hospital admissions in Toronto during a 10-day period in October 1982 that included peak values of several air pollutants. He found that admissions peaked 2 days before the highest air pollution values and concluded that the population of susceptible patients may have been depleted by the first part of the "episode." However, the 10-day mean admissions during this period were no different from the monthly mean, so that it is entirely possible that only random fluctuations in admissions were seen during this event (the peak daily admissions were reported to be about two standard deviations above the mean). Maarouf's analysis illustrates the possible pitfalls that may be encountered when looking for weak effects in a small data set.

COMPARISONS OF TIME-SERIES STUDY DESIGNS AND SUMMARY OF RESULTS

Table 9-2 presents summary information on the major time-series studies reviewed. Note that many studies did not explicitly control for seasonal cycles and only Fishelson and Graves controlled for potential effects of holidays on hospital usage. The study by Lamm *et al.* (1991), which linked a specific viral agent to seasonal cycles in pediatric hospitalization, provides ample justification for the need to separate climate-driven air pollution cycles from infection-driven illness cycles. However, in northern climates, use of temperature as an independent covariable will provide a certain amount of seasonal control.

Serial correlation was neglected by most authors; however, Fishelson and Graves (1978) and Burnett *et al.* (1993) did not find it to be a substantial factor. As indicated in the footnotes to the table, interpolating between pollution measurements in time may introduce serial correlation; conversely, leaving gaps in the data record because of missing observations (Namekata *et al.*) may reduce serial correlation.

The Pollutant Exposure column in Table 9-2 is a subjective judgment as to how well the monitoring data used in each study may represent the actual population exposure. Factors considered include the local versus regional nature of the pollutant, the numbers of monitoring stations used, and the way that interpolation was done. Note that poor characterization of exposure will usually lead to *under*estimation of effects. Another variable factor that affects the ability to detect a "true" effect is the number of observations, which varied from 20 for the smallest data set of Namekata and Carnow to 1569 (individual visits) in the Portland study.

By pollutant, statistically significant associations were found for particulate measures (smoke, COH, PM_{10}, or TSP), sulfur measures (SO_2 or SO_4^{-2}), and O_3, in most of the studies, even at relatively low concentration levels. There was little or no correspondence between the adequacy of the exposure estimation and the finding of statistical significance. Nitrogen dioxide was found to be significant in Los Angeles and Ontario, but not in Washington, Chicago, or Steubenville. Carbon monoxide was only studied in two locations

(Los Angeles and Washington) and tends to have poor exposure characterization because of the local nature of emissions. Carbon monoxide was associated with eye irritation in two studies, which seems counterintuitive (eye irritation is not mentioned in the EPA Criteria Document for CO). There is no obvious relationship between average pollutant levels and significance of findings, but this may be due in part to the differences in study designs and execution.

By type of diagnosis, cardiac diagnoses were associated with SO_2, O_3, and fine particles (COH). Respiratory symptoms were associated with all pollutants in Los Angeles and with all pollutants except CO in the various other studies. Respiratory admissions were associated with particulates in about half of the studies reviewed. Note that uncertainties in admission diagnoses may blur any specific disease-pollutant relationships that may have been present.

The lag column (data are given in days) indicates that all daily studies which considered lags found them to vary between 1 and about 3 days, with the exception of the Steubenville study, which found the unlagged variables to fit better. However, only Fishelson and Graves looked at cumulative lag variables, which is probably a better procedure, because it better accommodates individual differences in responses. The finding of a 30-day lag in the monthly data of Pope remains unexplained. For example, Canny *et al.* (1989) analyzed 16 months of data on asthma ER visits in Toronto and found that the average time elapsed between onset of symptoms and arrival at the ER was 41 hours for asthmatic children. Only 3% exceeded 1 week. These accounts of lag time do not include the reaction time required between exposure and onset of symptoms, however. It is also interesting that 75% of the group studied by Canny *et al.* cited a preceding respiratory infection as the cause of their attacks; only 7% implicated allergens.

The wide variety of studies finding significant associations between hospital usage and temporal fluctuations in air pollution makes a persuasive case for a causal interpretation. Such associations have been found in all seasons, in different climates, and for each of the criteria pollutants except lead. As was the case with studies of mortality, it is difficult to assign responsibility to any one species.

REFERENCES

Abercrombie, G.F. (Jan. 31, 1953), December Fog in London and the Emergency Bed Service, *Lancet* 1:234–35.

Abercrombie, G.F. (Nov. 17, 1956), Emergency Admissions to Hospital in Winter, *Lancet* 1:1039–42.

Air Pollution Study Project (1955), *Clean Air for California* (Initial Report), Department of Public Health, State of California.

Ayres, J., Fleming, D., Williams, M., and McInnes, G. (1989), Measurement of Respiratory Morbidity in General Practice in the United Kingdom during the Acid Transport Event of January 1985, *Env. Health Perspect.* 79:83–88.

Bates, D.V. (1985), *Strength and Weaknesses of Evidence Linking Health Effects to Air Pollution*, Carolina Environmental Essay 1985, Institute of Environmental Studies, University of North Carolina, Raleigh.

Bates, D.V., and Sizto, R. (1983), Relationship between Air Pollutant Levels and Hospital Admissions in Southern Ontario, *Can. J. Public Health* 74:117–22.

Bates, D.V., and Sizto, R. (1986), A Study of Hospital Admissions and Air Pollutants in Southern Ontario, in: *Aerosols: Research, Risk Assessment, and Control Strategies*, Lee, S.D., Schneider, T., Grant, L.D., and Verkerk, P.J., eds., Lewis Publishers, Chelsea, MI, pp. 767–77.

Bates, D.V., and Sizto, R. (1987), A Study of Hospital Admissions and Air Pollutants in Southern Ontario, *Env. Res.* 43:317–331.

Bates, D.V., and Sizto, R. (1989), The Ontario Air Pollution Study: Identification of the Causative Agent, *Env. Health Perspectives* 79:69–72.

Bates, D.V., Baker-Anderson, M., and Sizto, R. (1990), Asthma Attack Periodicity: A Study of Hospital Emergency Visits in Vancouver, *Env. Res.* 51:51–70.

Baxter, P.J., Ing, R., Falk, H., French, J., Stein, G.F., Bernstein, R.S., Merchant, J.A., and Allard, J. (1981), Mount St. Helens Eruptions, May 18 to June 12, 1980, An Overview of the Acute Health Impact, *JAMA* 246:2585–9.

Baxter, P.J., Ing, R., Falk, H., and Plikaytis, B. (1983), Mount St. Helens Eruptions: The Acute Respiratory Effects of Volcanic Ash in a North American Community, *Arch. Env. Health* 38:138–43.

Bladen, W.A. (1983), Relationship between Acute Respiratory Illness and Air Pollution in an Indian Industrial City, *J. APCA* 33:226–27.

Bradley, W.H., Logan, W.P.D., and Martin, A.E. (1958), The London Fog of December 2nd-5th, 1957, *Monthly Bulletin of the Ministry of Health* 17:156–65.

Brant, J.W.A., and Hill, S.R.G. (1964), Human Respiratory Disease and Atmospheric Air Pollution in Los Angeles, California, *Int. J. Air Water Poll.* 8:259–77.

Buist, A.S., Johnson, L.R., Vollmer, W.M., Sexton, G.J., and Kanarek, P.H. (1983), Acute Effects of Volcanic Ash from Mount Saint Helens on Lung Function in Children, *Am. Rev. Resp. Dis.* 127:714–19.

Burnett, R.T., Dales, R.E., Raizenne, M.E., Krewski, D., Summers, P.W., Roberts, G.R., and Dann, T.F. (1992), The Relationship Between Hospital Admissions and Ambient Air Pollution in Ontario, Canada: A Preliminary Report, presented at the 85th Annual Meeting of the Air & Waste Management Association, Kansas City, MO. Paper 92-146.05.

Burnett, R.T., Dales, R.E., Raizenne, M.E., Krewski, D., Summers, P.W., Roberts, G.R., Raad-Young, M., Dann, T., and Brooke, J. (1993), Effects of Low Ambient Levels of Ozone and Sulphates on the Frequency of Respiratory Admissions to Ontario Hospitals, *Env. Res.* (submitted).

California Department of Public Health (1955–1957), *Clean Air for California*, First Report, March 1955, Second Report, March 1956, Third Report, March 1957, Berkeley, CA.

Canny, G.J., *et al.* (1989), Acute Asthma: Observations Regarding the Management of a Pediatric Emergency Room, *Pediatrics* 83:507–12.

Carne, S.J. (1967), Study of the Effect of Air Pollution upon Respiratory Diseases in London in the Winters of 1962–63 and 1963–64, *Proc. Int. Clean Air Congress*, pp. 259–61.

Carnow, B.W. (1975), Predictive Models for Estimating the Health Impact of Future Energy Sources, in *Proc. Int. Symp.: Recent Advances in the Assessment of the Health Effects of Environmental Pollution*, Paris, June 1974, published by the Commission of the European Communities, Luxembourg, pp. 313–31.

Cody, R.P., Weisel, C.P., Birnbaum, G., and Lioy, P.J. (1992), The Effect of Ozone Associated with Summertime Photochemical Smog on the Frequency of Asthma Visits to Hospital Emergency Departments, *Envir. Res.* 58:184–94.

Corn, M. (1991), Invited Commentary on "Effects of Urban Air Pollution on Emergency Room Admissions for Chronic Obstructive Pulmonary Disease," *Am. J. Epidemiology* 134:287–88.

Diaz-Caneja, N., Guttierez, I., Martinez, A., Mattoras, P., and Villar, E. (1991), Multivariate Analysis of the Relationship between Meteorological and Pollutant

Variables and the Number of Hospital Admissions Due to Cardiorespiratory Diseases, *Env. Inter.* 17:397–403.

Duclos, P., Sanderson, L.M., and Lippsett, M. (1990), The 1987 Forest Fire Disaster in California: Assessment of Emergency Room Visits, *Arch. Env. Health* 45:53–58.

Durham, W.H. (1974), Air Pollution and Student Health, *Arch. Env. Health* 28:241–54.

Evans, R., *et al.* (1987), National Trends in the Morbidity and Mortality of Asthma in the U.S., *Chest* 91:65s–743.

Fishelson, G., and Graves, P. (1978), Air Pollution and Morbidity: SO_2 Damages, *J. APCA* 28:785–89.

Fruchter, J.S., Robertson, D.E., Evans, J.C., Olsen, K.B., Lepel, E.A., Laul, J.C., Abel, K.H., Sanders, R.W., Jackson, P.O., Wogman, N.S., Perkins, R.W., Van Tuyl, H.H., Beauchamp, R.H., Shade, J.W., Daniel, J.L., Erikson, R.L., Sehmel, G.A., Lee, R.N., Robinson, A.V., Moss, O.R., Briant, J.K., and Cannon, W.C. (1980), Mount St. Helens Ash from the 18 May 1980 Eruption: Chemical, Physical, Mineralogical, and Biological properties, *Science* 209:1116–25.

Fry, J. (Jan. 31, 1953), Effects of a Severe Fog on a General Practice, *Lancet* 1:235–36.

Giles, G.G. (1981), Biometeorological Investigations of Asthma Morbidity in Tasmania Using Co-spectral Analysis of Time Series, *Soc. Sci. Med.* 15D:111–19.

Girsh, L.S., Shubin, E., Dick, C., and Schulaner, F.A. (1967), A Study on the Epidemiology of Asthma in Children in Philadelphia, *J. Allergy* 39:347–57.

Glasser, M., Greenburg, L., and Field, F. (1967), Mortality and Morbidity during a Period of High Levels of Air Pollution, *Arch. Env. Health* 15:684–94.

Goldsmith, J.R., Griffith, H.L., Detels, R., Beeser, S., and Neumenn, L. (1983), Emergency Room Admissions, Meteorological Variables, and Air Pollutants: A Path Analysis, *Am. J. Epidemiology* 118:758–78.

Goldstein, I.F., and Block, G. (1974), Asthma and Air Pollution in Two Inner City Areas in New York City, *J. APCA* 24:665–70.

Goldstein, I.F., and Cuzick, J. (1981a), Application of a Time-Space Clustering Methodology to the Assessment of Acute Environmental Effects on Respiratory Illnesses, *Reviews on Environmental Health* 3:259–75.

Goldstein, I.F., and Dulberg, E. (1981b), Air Pollution and Asthma: Search for a Relationship, *J. APCA* 31:370.

Goldstein, I.F., and Rausch, L.E. (1978), Time Series Analysis of Morbidity Data for Assessment of Acute Environmental Health Effects, *Env. Res.* 17:266–75.

Goldstein, I.F., and Weinstein, A.L. (1986), Air Pollution and Asthma: Effects of Exposures to Short-Term Sulfur Dioxide Peaks, *Env. Res.* 40:332–45.

Greenburg, L., Field, F., Reed, J.I., and Erhardt, C.L. (1962a), Air Pollution and Morbidity in New York City, *JAMA* 182:161–64.

Greenburg, L., Jacobs, M.B., Drolette, B.M., Field, F., and Braverman, M.M. (1962b), Report of an Air Pollution Incident in New York City, November 1953, *Public Health Rep.* 77:7–16.

Greenburg, L., Erhardt, C., Field, F., Reed, J.I., and Seriff, N.S. (1963), Intermittent Air Pollution Episode in New York City, 1962, *Public Health Rep.* 78:1061–64.

Greenburg, L., Field, F., and Erhardt, C.L. (1964), Asthma and Temperature Change, *Arch. Env. Med.* 4:642–47.

Gross, J., Goldsmith, J., Zangwill, J., and Lerman, S. (1984), Monitoring of Hospital Emergency Room Visits as a Method for Detecting Health Effects of Environmental Exposures, *Sci. Total Env.* 32:289– 02.

Hall, C.B. (1981), Respiratory Syncytial Virus, in *Textbook of Pediatric Infectious Diseases*, Feigin, R.D., and Cherry, J.D., eds., Saunders, Philadelphia, pp. 1247–67.

Hammerstrom, T., Silvers, A., Roth, N., and Lipfert, F. (1991), Statistical Issues in the

Analysis of Risks of Acute Air Pollution Exposure, report prepared for the Electric Power Research Institute, Palo Alto, CA.

Heimann, H. (1970), Episodic Air Pollution in Metropolitan Boston, *Arch. Env. Health* 20:239–51.

Holland, W.W., Spicer, C.C., and Wilson, J.M.G. (Aug. 12, 1961), Influence of the Weather on Respiratory and Heart Disease, *Lancet* 2:338–40.

Jaksch, J.A., and Stoevener, H.H. (1974), Outpatient Medical Costs Related to Air Pollution in the Portland, Oregon, Area, EPA-600/5-74-017, U.S. Environmental Protection Agency, Washington, DC.

Katz, R.M., and Frezieres, R.G. (1986), Asthma and Air Pollution in Los Angeles, in *Environmental Epidemiology: Epidemiological Investigation of Community Health Problems*, Goldsmith, J.R., ed., CRC Press, Boca Raton, FL, pp. 117–27.

Kevany, J., Rooney, M., and Kennedy, J. (1975), Health Effects of Air Pollution in Dublin, *Irish J. Med. Sci.* 144:102–15.

Khan, A.U. (1977), The Role of Air Pollution and Weather Changes in Childhood Asthma, *Ann. Allergy* 39:397–400.

Knight, K., Leroux, B., Millar, J., and Petkau, A.J. (1989a), *Air Pollution and Human Health: A Study Based on Hospital Admissions Data from Prince George, British Columbia*, SIMS Technical Report #128, SIMS Publications, New Canaan, CT.

Knight, K., Leroux, B., Millar, J., and Petkau, A.J. (1989b), *Air Pollution and Human Health: A Study Based on Emergency Room Visits Data from Prince George, British Columbia*, SIMS Technical Report #136, SIMS Publications, New Canaan, CT.

Koenig, J.Q., Schwartz, J., Slater, D., Larson, T.V.C., and Pierson, W.E. (1992), Associations between Ambient Particulate Matter and Hospital Emergency Visits for Asthma in Seattle, presented at the American Chemical Society, San Francisco.

Kraemer, M.J., and McCarthy, M.M. (1985), Childhood Asthma Hospitalization Rates in Spokane County, Washington: Impact of Volcanic Ash Air Pollution, *J. Asthma* 22:37–43.

Krumm, R.J., and Graves, P.E. (1982), Morbidity and Pollution: Model Specification Analysis for Time-Series Data on Hospital Admissions, *J. Env. Econ. & Mgmt.* 9:311–27.

Kurt, T.L., Mogielnicki, R.P., and Chandler, J.E. (1978), Association of the Frequency of Acute Cardiorespiratory Complaints with Ambient Levels of Carbon Monoxide, *Chest* 74:10–14.

Kurt, T.L., Mogielnicki, R.P., Chandler, J.E., and Hirst, K. (1979), Ambient Carbon Monoxide Levels and Acute Cardiorespiratory Complaints: An Exploratory Study, *Am. J. Public Health* 69:360–63.

Lamm, S.H., Hall, T.A., Engel, A., White, L.S., and Rueter, F.H. (1991), Assessment of Viral and Environmental Factors as Determinants of Pediatric Lower Respiratory Tract Disease Admissions in Utah County, Utah (1985–1989). Unpublished report from Consultants in Epidemiology and Occupational Health, Inc., Washington, DC. Also see "PM_{10} Particulates: Are They the Major Determinant of Pediatric Respiratory Admissions in Utah County, Utah (1985–1989) in *Proc. 7th Int. Symp. on Inhaled Particles*, Edinburgh.

Levenson, R.M., and Kantor, O.S. (1987), Fatal Pneumonia in an Adult Due to Respiratory Syncytial Virus, *Arch. Int. Med.* 147:791–92.

Levy, D., Gent, M., and Newhouse, M.T. (1975), Relationship between Acute Respiratory Illness and Air Pollution Levels in an Industrial City, in *Proc. Int. Symp. Recent Advances in the Assessment of the Health Effects of Environmental Pollution*, Paris, June 1974, published by the Commission of the European Communities, Luxembourg, pp 1263–76.

Levy, D., Gent, M., and Newhouse, M.T. (1977), Relationship between Acute Respiratory Illness and Air Pollution Levels in an Industrial City, *Am. Rev. Resp. Dis.* 116:167–73.

Lewis, R., Gilkeson, M.M., and McAldin, R.O. (1962), Air Pollution and New Orleans Asthma, *Publ. Health Rep.* 77:947–54.

Lewis, W.F. (1974), Utilization of Short-Stay Hospitals, Summary of Non-Medical Statistics, United States—1971. DHEW Publication No. (HRA)75-1768, National Center for Health Statistics, Rockville, MD.

Liang, K.Y., and Zeger, S.L. (1986), Longitudinal Data Analysis Using Generalized Linear Models, *Biometrika* 73:13–22.

Lipfert, F.W., and Hammerstrom, T. (1992), Temporal Patterns in Hospital Admissions and Air Pollution, *Env. Res.*

Lutz, L.J. (1983), Health Effects of Air Pollution Measured by Outpatient Visits, *J. Fam. Practice* 16:307–13.

Maarouf, A.R. (1991), Possible Health Effects of an Air Pollution Episode in Toronto, Canada, in *Proc. 10th Conf. on Biometeorology and Aerobiology*, American Meteorological Society, Boston, pp. 11–14.

Maarouf, A.R., and Zwiers, F.W. (1987), The Use of Climatological Data in Health-Related Studies, in *Proc. 10th Conf. on Probability and Statistics in the Atmospheric Sciences*, American Meteorological Society, Edmonton, Alberta, pp. 281–84.

Marsh, A. (1963), The December Smog, A First Survey, *J. APCA* 13:384–87.

Martin, A.E. (1961), Epidemiological Studies of Atmospheric Pollution, *Monthly Bulletin of the Ministry of Health* 20:42–49.

Martin, A.E. (1964), Mortality and Morbidity Statistics and Air Pollution, *Proc. R. Soc. Med.* 57:969–75.

Mazumdar, S., and Sussman, N. (1983), Relationships of Air Pollution to Health: Results from the Pittsburgh Study, *Arch. Env. Health* 38:17–24.

Meetham, A.R. (1981), *Atmospheric Pollution*, 4th ed., Pergamon Press, London.

Miller, R.G., Jr. (1981), *Simultaneous Statistical Inference*, Springer-Verlag, New York.

Ministry of Health (1954), Mortality and Morbidity during the London Fog of December 1952, Reports on Public Health and Medical Subjects No. 95. London, HMSO.

Nagata, H., Kadowaki, I., Ishigure, K., Tokuda, M., and Ohe, T. (1979), Meteorological Conditions, Air Pollution, and Daily Morbidity in Summer, *Jap. J. Hygiene* 33:772–77 (in Japanese with English abstract and tables).

Namekata, T., and Carnow, B.W. (1976), Impact of Multiple Pollutants on Emergency Room Admissions, Illinois Institute of Environmental Quality Document No. 77/02.

Namekata, T., Carnow, B.W., Flourno-Gill, Z., O'Farrell, E.B., and Reda, D. (1979), Model for Measuring the Health Impact from Changing Levels of Ambient Air Pollution: Morbidity Study, EPA-600/1-79-024, U.S. Environmental Protection Agency, Research Triangle Park, NC.

Noble, J.H., LaMontagne, M.E., Bellotti, C., and Wechsler, H. (1971), Variations in Visits to Hospital Emergency Care Facilities, *Medical Care* IX:415–27.

Özkaynak, H., Burbank, B., and Spengler, J.D. (1985), Development of a Health-Based Air Quality/Visibility Index for Massachusetts, Phase I, Final Report to Massachusetts Department of Environmental Quality Engineering, Harvard University Energy and Environmental Policy Center, Cambridge, MA.

Özkaynak, H., Kinney, P.L., and Burbank, B. (1990), Recent Epidemiological Findings on Morbidity and Mortality Effects of Ozone, paper 90-150.6, presented at the 83rd Annual Meeting of the Air and Waste Management Association, Pittsburgh.

Ponka, A. (1991), Asthma and Low Level Air Pollution in Helsinki, *Arch. Env. Health* 46:262–70.

Pope, C.A., III (1989), Respiratory Disease Associated with Community Air Pollution and a Steel Mill, Utah Valley, *AJPH* 79:623–28.

Pope, C.A., III (1991), Respiratory Hospital Admissions Associated with PM_{10} Pollution in Utah, Salt Lake, and Cache Valleys, *Arch. Env. Health* 46:90–97.

Richards, W., Azen, S.P., Weiss, J., Stocking, S., and Church, J. (1981), Los Angeles Air Pollution and Asthma in Children, *Ann. Allergy* 47:348–54.

Rotton, J., and Frey, J. (1985), Air Pollution, Weather, and Violent Crimes: Concomitant Time-Series Analysis of Archival Data, *J. Personality and Social Psychology* 49:1207–20.

Roueche, B. (1984), *The Medical Detectives*, vol. II, Dutton, New York.

Samet, J.M., Speizer, F.E., Bishop, Y., Spengler, J.D., and Ferris, B.G., Jr. (1981), The Relationship between Air Pollution and Emergency Room Visits in an Industrial Community. *J. APCA* 31:236–40.

Schwartz, J., Slater, D., Larson, T.V., Pierson, W.E., and Koenig, J.Q. (1993), Particulate Air Pollution and Hospital Emergency Room Visits for Asthma in Seattle, *Am. Rev. Respir. Dis.* 147:826–31.

Schwartz, J., Spix, C., Wichmann, H.E., and Malin, E. (1991), Air Pollution and Acute Respiratory Illness in Five German Communities, *Env. Res.* 56:1–14.

Scott, J.A. (Apr. 26, 1963), The London Fog of December, 1962, *The Medical Officer*, 250–53.

Seskin, E.P. (1977), *Air Pollution and Health in Washington, DC,* EPA-600/5-77-010, U.S. Environmental Protection Agency, Corvallis, OR.

Shiffer, I.J., and Parsons, E.A. (1980), Pilot Study—*Uses of Medicare Morbidity Data in Health Effects Research*, EPA-600/1-80-018, U.S. Environmental Protection Agency, Research Triangle Park, NC.

Shrenk, H.H., Heimann, H., Clayton, G.D., Gafafer, W.M., and Wexler, H. (1949), Air Pollution in Donora, PA. Public Health Bulletin No. 306, Public Health Service, Washington, DC.

Sterling, T.D., Phair, J.J., Pollack, S.V., Schumsky, D.A., and DeGroot, I. (1966b), Urban Morbidity and Air Pollution, *Arch. Env. Health* 13:158–70.

Sterling, T.D., Pollack S.V., and Phair, J.J. (1967), Urban Hospital Morbidity and Air Pollution, *Arch. Env. Health* 15:362–74.

Sterling, T.D., Pollack S.V., and Weinkam, J. (1969), Measuring the Effect of Air Pollution on Urban Morbidity, *Arch. Env. Health* 18:485–94.

Strahilevitz, M., Strahilevitz, A., and Miller, J.E. (1979), Air Pollutants and the Admission Rate of Psychiatric Patients, *Am. J. Psychiatry* 136:205–7.

Sunyer, J., Anto, J.M., Murillo, C., and Saez, M. (1991a), Effects of Urban Air Pollution on Emergency Room Admissions for Chronic Obstructive Pulmonary Disease, *Am. J. Epidemiology* 134:277–86.

Sunyer, J., and Anto, J.M. (1991b), Authors' Response to "Invited Commentary on 'Effects of Urban Air Pollution on Emergency Room Admissions for Chronic Obstructive Pulmonary Disease,'" *Am. J. Epidemiology* 134:289.

Sunyer, J., Saez, M., Murillo, C., Castellsague, J., Martinez, F., and Anto, J.M. (1993), Air Pollution and Emergency Room Admissions for Chronic Obstructive Pulmonary Disease: A 5-year Study, *Am. J. Epidemiology* 137:701–5.

Sweeney, J.C. (1982), Air Pollution and Morbidity in Dublin, *Irish Geogr.* 15:1–10.

U.S. Environmental Protection Agency (1986), *Air Quality Criteria for Ozone and Other Photochemical Oxidants*, EPA/600/8-84/020eF, Environmental Criteria and Assessment Office, Research Triangle Park, NC.

Vollmer, W.M., Osborne, M.L., and Buist, A.S. (1993), Temporal Trends in Hospital-based Episodes of Asthma Care in a Health Maintenance Organization, *Am. Rev. Respir. Dis.* 147:347–53.

Waller, R.E., and Commins, B.T. (1967), Episodes of High Pollution in London, 1952–1966, *Proc. Int. Clean Air Congress*, pp. 228–31.

Waller, R.E., Lawther, P.J., and Martin, A.E. (1969), Clean Air and Health in London, *Proc. Clean Air Conference, Eastbourne*, National Society for Clean Air, London, pp. 71–79.

White, M.C., *et al.* (1991), Childhood Asthma and Ozone Pollution in Atlanta (Pilot Study), presented at the Air & Waste Management Association International Specialty Conference on Tropospheric Ozone and the Environment, Atlanta, GA.

Whittemore, A.S., and Korn, E.L. (1980), Asthma and Air Pollution in the Los Angeles Area, *AJPH* 70:687–96.

Wichmann, H.E., Mueller, W., Allhoff, P., Beckmann, M., Bocter, N., Csicsaky, M.J., Jung, M., Moilk, B., and Schoeneberg, G. (1989), Health Effects during a Smog Episode in West Germany in 1985, *Env. Health Perspectives* 79:89–100.

10

Cross-Sectional Studies of Hospital Use

Summary of Hospitalization Studies

Cross-sectional methods are traditionally used to examine differences in long-term averages and must account for a host of possible confounding factors, including socioeconomic and demographic gradients, differences in life-styles such as smoking habits and alcohol consumption, and those factors influencing the supply and delivery of medical care. The studies examined in Chapter 9 used short-term variations in hospital admissions to deduce dose-response relationships with respect to various air pollutants; the assumption is made that bed availability does not affect short-term variability in admissions. However, any significant short-term excess admissions should also be reflected in the long-term statistics (either admissions or discharges) and possibly as differences in lengths of hospital stays, assuming that short-term excesses were not subsequently canceled by short-term deficits. Thus, one of the purposes of examining cross-sectional studies of longer term relationships is to examine consistency among time scales in the implied relationships between hospitalization and air pollution.

REVIEWS OF CROSS-SECTIONAL STUDIES

Pittsburgh Area

Carpenter *et al.* (1979) performed a cross-sectional analysis of hospitalization costs in Allegheny County (Pittsburgh), PA, for 1972. The observations were drawn from about 38,000 admissions to 28 hospitals, for:

- Respiratory causes (ICD [International Classification of Diseases] 462–515.9, which includes some upper respiratory causes and tonsil or adenoid problems).

- "Suspect" circulatory diseases (ischemic heart diseases, acute cerebrovascular diseases, etc.).
- "Comparison" (i.e., controls) circulatory diseases (rheumatic fever, hypertension, etc.).

The average hospitalization rate for respiratory diseases was 5.8 per 1000 population, ranging from 3.9 to 10.6 per 1000 among the various demographic subgroups defined by air pollution levels. This compares with a value of about 4.0 per thousand from Pope's (1991) study in Utah, although there are some slight differences between the two studies in the ICD codes included.

Air pollution exposures and socioeconomic variables were obtained from the corresponding census tract data; air quality measurements from 49 SO_2 monitors (sulfation plates) and 21 TSP stations were interpolated to census tract centroids. Data on smoking habits were not available. Sixteen demographic groups were identified by race, sex, and age grouping, and the regression analysis used membership in these groups as independent variables explaining hospitalization rates for the entire population (admissions per 1000). Young children were not singled out for special consideration, which is unfortunate given their high rates of respiratory illness and their lack of smoking.

The associations between respiratory disease admission rates and SO_2 and TSP were just significant at the 0.05 level. The effects of TSP and SO_2 were about 0.27 and 0.71 annual admissions per 1000 population, or about 0.06% and 0.12% per $\mu g/m^3$, respectively. These figures are somewhat larger but similar to the results obtained with the time-series analyses we discussed, which suggests that any confounding variables remaining in this data set (such as smoking habits, for example) probably had only modest effects. The corresponding elasticities are 0.05 and 0.12, which should be summed if one wishes to compare with studies that assigned the entire effect to a single pollutant. Neither of the circulatory disease categories was significant, although the "suspect" categories were close for TSP ($p = 0.07$).

A corresponding analysis of lengths of hospital stay also found significance for respiratory diseases and air pollution exposure ($p = 0.03$) and for certain circulatory diseases ($p = 0.009$). This portion of the study was able to account for smoking habits, since it dealt with the population of patients, not with the county population as a whole. Smoking was significant at the $p = 0.0001$ level for respiratory diseases. The pollutant effects on lengths of stay were of the order of one extra day for an additional $20 \mu g/m^3$ of particulates but were not consistent with a conventional dose-response relationship.

Maine Mill Towns

A cross-sectional study of occupational and community exposures to pulp and paper mill effluents (as measured by residential proximity to the sources) was conducted by Deprez *et al.* (1986) for Maine mill towns. Hospitalization diagnoses included respiratory infections and inflammations, respiratory cancer, chronic obstructive pulmonary disease, bronchitis and asthma, respiratory signs and symptoms, all respiratory diagnoses combined, myocardial infarction (controls), and total admissions. Age- and sex-adjusted rates were computed for use in bivariate correlation computations. All but the first two diagnostic groups were significantly correlated with occupational exposure, as indicated by the proportion of a town's work force employed in the production process. No category was significantly correlated with distance to the mill, which was interpreted as lack of a community exposure relationship. The authors stated that potential confounding by smoking and health insurance factors should be examined.

New York City versus Los Angeles

Noting that Northeastern patients tend to stay longer in hospital than West Coast residents, Knickman (1982) studied population samples from New York and Los Angeles obtained from the 1974–76 Health Interview Surveys (see, for example, Ries, 1975). He noted that the admissions rate was higher in Los Angeles but the average length of stay and the number of patient-days were higher in New York. A subgroup of patients with hospital stays longer than 50 days was defined for this study. Knickman studied the components of the geographic difference, including population variables (age, race, sex, income, education, marital status), diagnoses, and the long-stay patient subgroup. He found that this subgroup was responsible for most of the difference between the two samples. After excluding the long-stay patients, he standardized for these variables, which reduced the difference in patient-days between the two locations from 13.3% to 3.9%. This remaining difference was attributed to differences in medical practices. This study illustrates the importance of accounting for the population characteristics when comparing across areas. For example, a portion of the apparent air pollution effects found in the Pittsburgh cross-sectional study by Carpenter *et al.* (1979) might be the result of artifacts due to demographic differences across areas.

Combined Cross-Sectional and Time-Series Study in Ontario

Plagionnakis and Parker (1988) pooled annual data from nine Ontario Counties from 1976 to 1982 in a combined cross-sectional time-series study of mortality and morbidity. These counties included the southern metropolitan areas and the locations around major point sources of SO_2 (Algoma and Sudbury); the latter were not included in the time-series studies of Southern Ontario described previously. Both total rates and respiratory (all subdivisions from ICD 460–519) diagnoses were analyzed. Two morbidity measures were used: annual admissions per 100,000 people, and annual hospital days per person. A linear trend variable was included for time; dummy variables were investigated for each county but were found to be nonsignificant and thus were dropped from the model. Logarithmic transforms were used for all variables. The pollutants investigated were annual average and 24-hour maximum values of TSP, SO_2, and SO_4^{-2}. The socioeconomic variables investigated included age (% 65 and over), education, income, smoking, alcohol consumption, medical staff per capita, and the absolute population of each county (to serve as an index of urbanization).

For annual hospital days, the models included population (−), time (−), the percentage of people 65 and over, alcohol consumption (+), and various pollutants. The pollution regression coefficients were sensitive to the inclusion of these variables. For total hospital days, the "best" models used 24-hour maximum SO_2 (maximum value about 800 μg/m^3) and had an elasticity of about 0.08. For respiratory hospital days, R^2's were lower, but elasticities were higher (up to 0.30 for SO_2). For total admissions, the model included numbers of medical staff per capita, which had a negative coefficient. Maximum 24-hour SO_2 again provided the "best" regression results, with an elasticity of 0.12 to 0.17 for total admissions and of 0.15 to 0.26 for respiratory admissions. Neither TSP variable was significant (maximum 24-hour average about 300 μg/m^3), and annual average SO_4^{-2} was never significant for annual hospital days. Maximum 24-hour average SO_4^{-2} (maximum = 90 μg/m^3) gave slightly better results than annual average SO_4^{-2}, but the overall R^2's were about the same as with SO_2. The finding of better results for maximum 24-hour readings than for annual averages could be interpreted as suggesting that this model might be measuring the annual sum of acute events.

British Children

Douglas and Waller (1966) conducted a survey of the respiratory health of British children up to the age of 5 years. The survey included questions on symptoms and hospitalization; the data were based on 3131 families with consistent residential air pollution exposure histories. The survey data on hospital admissions were corroborated with the hospitals concerned. Air pol-

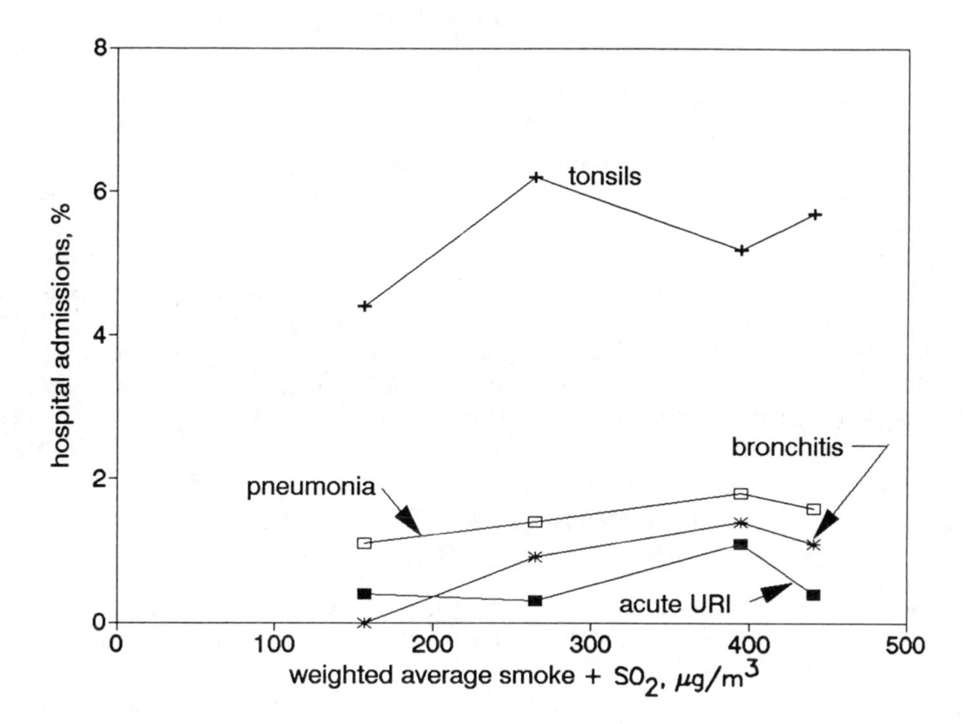

FIGURE 10-1. Percentage of British families with children admitted to hospital for various respiratory diagnoses. *Data from Douglas and Waller (1966).*

lution exposures were estimated by means of classification into one of four categories based on coal consumption. The available SO_2 and smoke monitoring data were then averaged to provide exposure estimates for each category. Differences in social class between the four groups were minimal. I cross-plotted these data in Figure 10-1 to check for consistent dose-response relationships. The rates of hospitalization (cases per 100 children) were lower than comparable U.S. rates and much lower than Canadian rates (Kozak and McCarthy, 1984), in spite of the fact that all of the average pollution levels were above the current U.S. primary standards.

The authors concluded that a relationship had been shown between air pollution and lower respiratory tract infections but not for the upper tract (acute upper respiratory infections [URIs] and tonsillitis). In order to evaluate the data set statistically, I pooled all four diagnoses and regressed against both SO_2 and smoke levels (dummy variables for the different diagnoses were included). Both pollutants were significant (one at a time), and it was not possible to select the better measure. The elasticities were about 0.40 (i.e., 40% of children's respiratory hospital admissions were associated with air pollution). However, the slopes were lower than those Carpenter *et al.* (1979) obtained in Pittsburgh, perhaps because of the errors induced by estimating air pollution exposure. It is unfortunate that Douglas and Waller did not evaluate these data on the basis of individual responses and air pollution exposures, which would have had much greater statistical power.

German Children

Muhling *et al.* (1985) classified children up to age 4 who visited a pediatric clinic from 1979 to 1982 according to their residential air quality levels. Diagnoses considered were "pseudocroup" and obstructive bronchitis; pollutants were SO_2 and dustfall (g/m^2d). Groups of "high" and "low" pollution were defined and tested by chi-squared for significant differences in the numbers of cases during the 4 years. For example, there were about twice as many croup cases when the average SO_2 exceeded 70 $\mu g/m^3$, compared to residential areas where the average SO_2 was less than 70 $\mu g/m^3$. For bronchitis, there were 2.5 times as many cases for areas with average dustfall exceeding 0.35 g/m^2d (about 140 $\mu g/m^3$ as TSP [see Chapter 2]). The authors ruled out confounding by residential density, viral epidemics, or meteorological influences. These findings are qualitatively consistent with those of Douglas and Waller (1966).

REGIONAL COMPARISONS OF HOSPITAL UTILIZATION

Gornick (1982) presents data on hospital use that allow some regional comparisons by diagnosis; these data are potentially of interest since air pollution in the United States also exhibits regional trends (Chapter 2). The population studied was Medicare enrollees aged 65 and over. The trend data (Figure 10-2) show that from 1968 to 1977, for all diagnoses, admission rates increased and the average lengths of stay decreased, nationwide. Patient-days peaked in 1969. Respiratory diagnoses increased slightly less than did all causes. The only

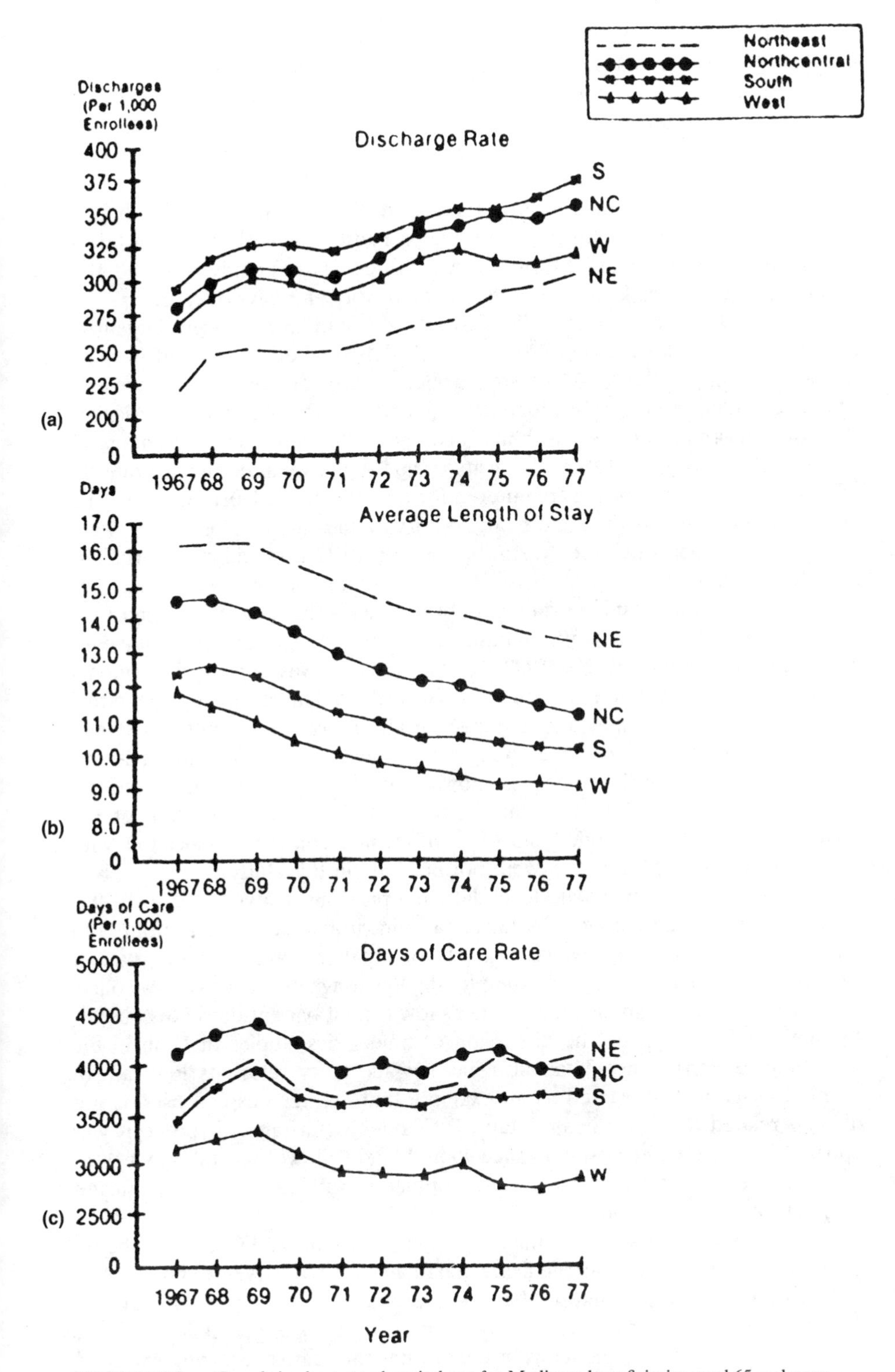

FIGURE 10-2. Trends in short-stay hospital use for Medicare beneficiaries aged 65 and over, by census region. (a) Discharge rate. (b) Length of stay. (c) Days of care. *Reprinted with the permission of Lexington Books, an imprint of Macmillan, Inc., from* Regional Variations in Hospital Use *by David L. Rothberg, editor. Copyright @ 1982 by Lexington Books.*

detailed respiratory diagnosis analyzed was pneumonia, for which the average hospital stay was about 12 days and the median, 9 days; this difference indicates the influence of long-stay patients. By region the time trends were not greatly different, but regional differences persisted for most diagnoses and after adjustment for age, sex, and race. These differences were most pronounced for the long-stay patients and were seen in both surgical and medical cases. The Northeast tended to have the lowest discharge rates but the longest lengths of stays, but the North Central region had the highest values for "days of care per 1000 Medicare enrollees," which decreased slightly over the 10 y period. The West was lowest in both lengths of stay and days of care.

Using 1979 observations for 195 Professional Standards Review Organization areas, Gornick developed regression models for both lengths of hospital stay and discharge rates for Medicare enrollees ages 65 and over (all diagnoses). The independent variables included average enrollee age, sex, race; physicians, short-stay hospital and nursing home beds per 1000 enrollees; percentage of area enrollees living in SMSAs ("density"); and the hospital occupancy rate. In addition, dummy variables were entered for the Northeast, North Central, and South Census Regions. The most important variables for lengths of stay were hospital occupancy rate, the Northeast and North Central dummy variables, and the number of short-stay beds; 83% of the variance was explained. For discharge rates, the most important variables were the number of physicians(−), short-stay beds(+), the South dummy variable(+), and the percentage of enrollees aged 75 and over(+); 60% of the variance was explained. A model was not developed for patient care-days. Gornick explained the negative effect of the supply of physicians as the impact of alternative treatments other than hospitalization. This in turn could lengthen the average stay for those patients who are hospitalized. This analysis showed that both personal and economic variables played a role in hospital utilization and that some portion of the regional differences remained unexplained (which could be associated with environmental variables, which were not included in this study).

Gornick also examined state-level data in a previous analysis (1977). In this work, the high admission and discharge rates in rural states were attributed to "distance to health care," which might favor inpatient rather than outpatient treatment. Personal factors (for example, the low usage of tobacco and alcohol) were cited as being responsible for Utah's low rate of patient-days (40% below the national average; see the discussion of time-series studies in Utah in the preceding section). Since Utah also has low mortality rates, its low rate of hospital usage was also cited as an example that greater hospital use "is not directly related to higher health status." Gornick's (unstated) hypothesis was apparently that greater hospital usage should lead to lower mortality, whereas the presence of excess risk factors in a population will lead to greater rates of hospitalization *and* mortality.

In 1990, the Health Care Financing Administration (HFCA) published a geographic analysis of 1986 Medicare inpatient hospital services and the outcomes of treatment, by major diagnostic group and surgical procedure, including state-level maps of short-stay hospital discharges per 1000 Medicare enrollees aged 65 and over, maps of death rates for deaths occurring within 30 days of discharge, and of all deaths to persons 65 and over. The report also includes tabulations by state and SMSA. All-cause discharges (Figure 10-3a)

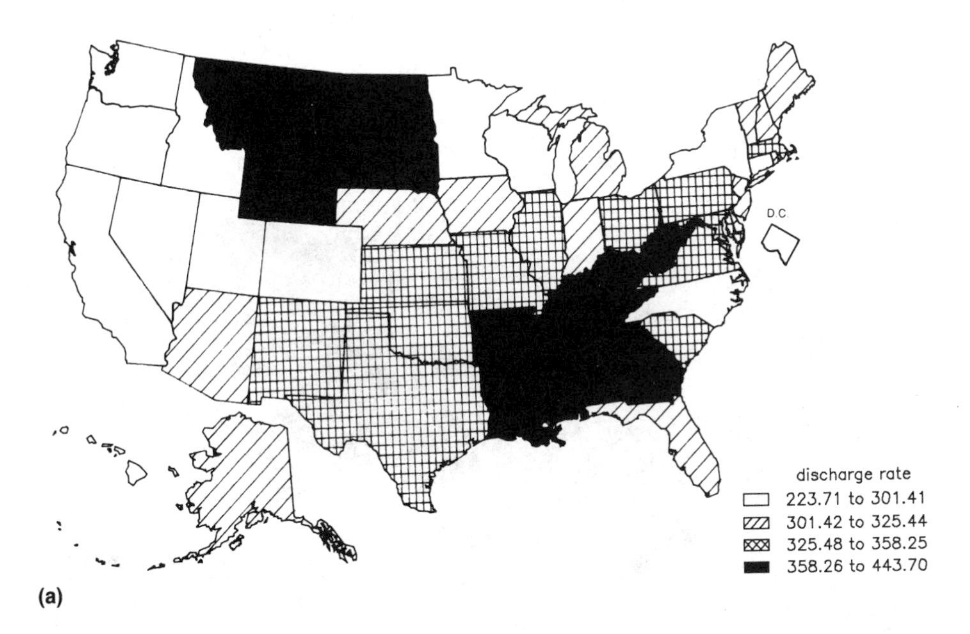

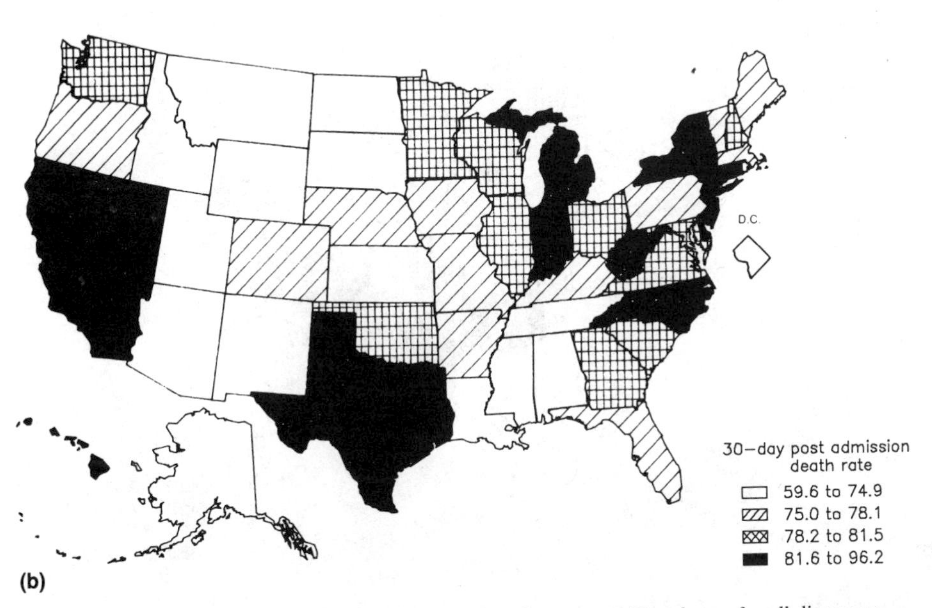

FIGURE 10-3. Maps of 1986 short-stay hospital data for persons 65 and over for all diagnoses. (a) Discharges, per 1000 Medicare enrollees. (b) Deaths within 30 days following discharge, per 1000 discharges. *Source: Health Care Financing Corporation (1990).*

showed several interesting patterns. The highest discharge rates tended to be in rural states in the South Central and West North Central census divisions. Among industrialized states, discharge rates seemed higher in the Midwest. The lowest rates were in the West, including California. Figure 10-3b shows the distribution of deaths within 30 days after discharge, which may be taken as an index of the severity of the cases hospitalized. The anticorrelation between the

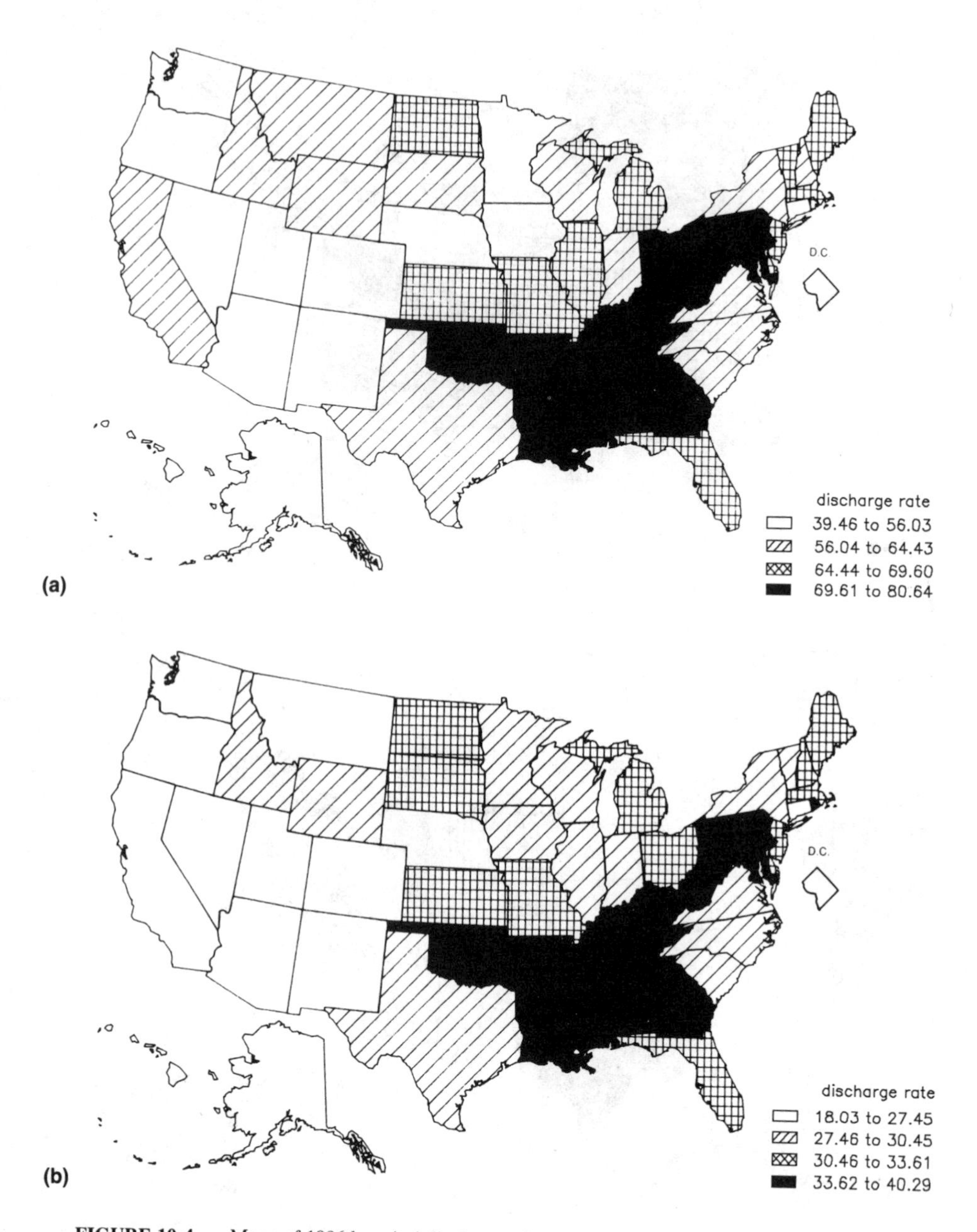

FIGURE 10-4. Maps of 1986 hospital discharges for heart disease per 1000 Medicare enrollees 65 and over. (a) All heart disease. (b) Ischemic heart disease. *Source: Health Care Financing Corporation (1990).*

two maps is evident ($r = -0.42$), suggesting that part of the pattern in discharges (i.e., usage) results from selectively admitting only the worst cases. Texas, West Virginia, Oklahoma, Virginia, Georgia, Ohio, Illinois, and Virginia had high rates for both discharges and deaths, which suggests either that status is poorer there or their hospitals are less effective. Louisiana, Mississippi, Alabama, Montana, Wyoming, Tennessee, and the Dakotas may have higher

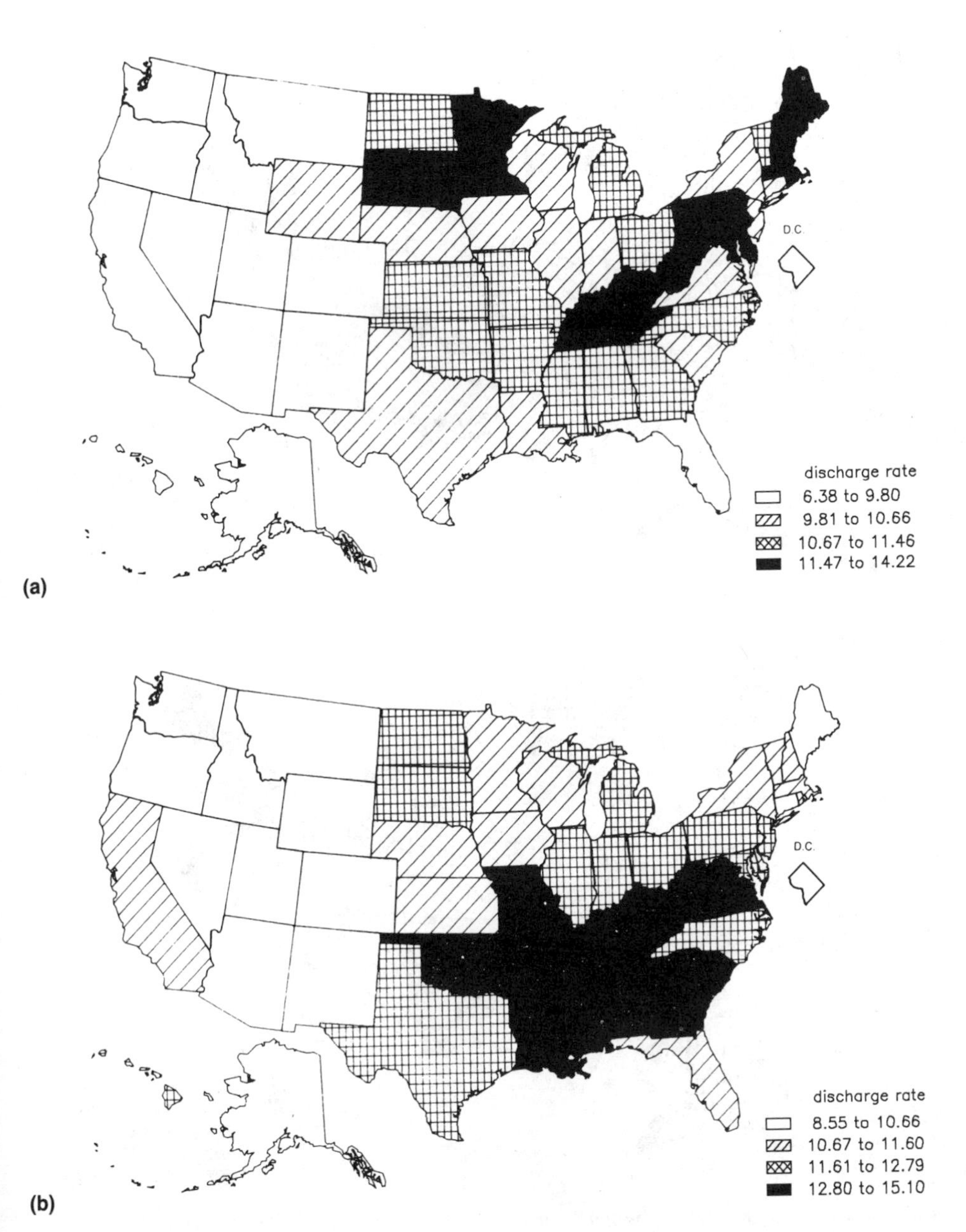

FIGURE 10-5. Maps of 1986 hospital discharges per 1000 Medicare enrollees 65 and over. (a) Heart attacks. (b) Stroke. *Source: Health Care Financing Corporation (1990).*

discharge rates because of the use of hospitals for primary medical care (less severe cases).

The heart disease maps (Figure 10-4a, b) showed regional patterns similar to the mortality pattern (Chapter 7), except that the high-mortality zone extends somewhat to the north of the high-hospitalization zone. Heart attack and stroke patterns (Figure 10-5a, b) are also generally similar, except for the zone of high heart attack hospitalization rates in the upper Midwest.

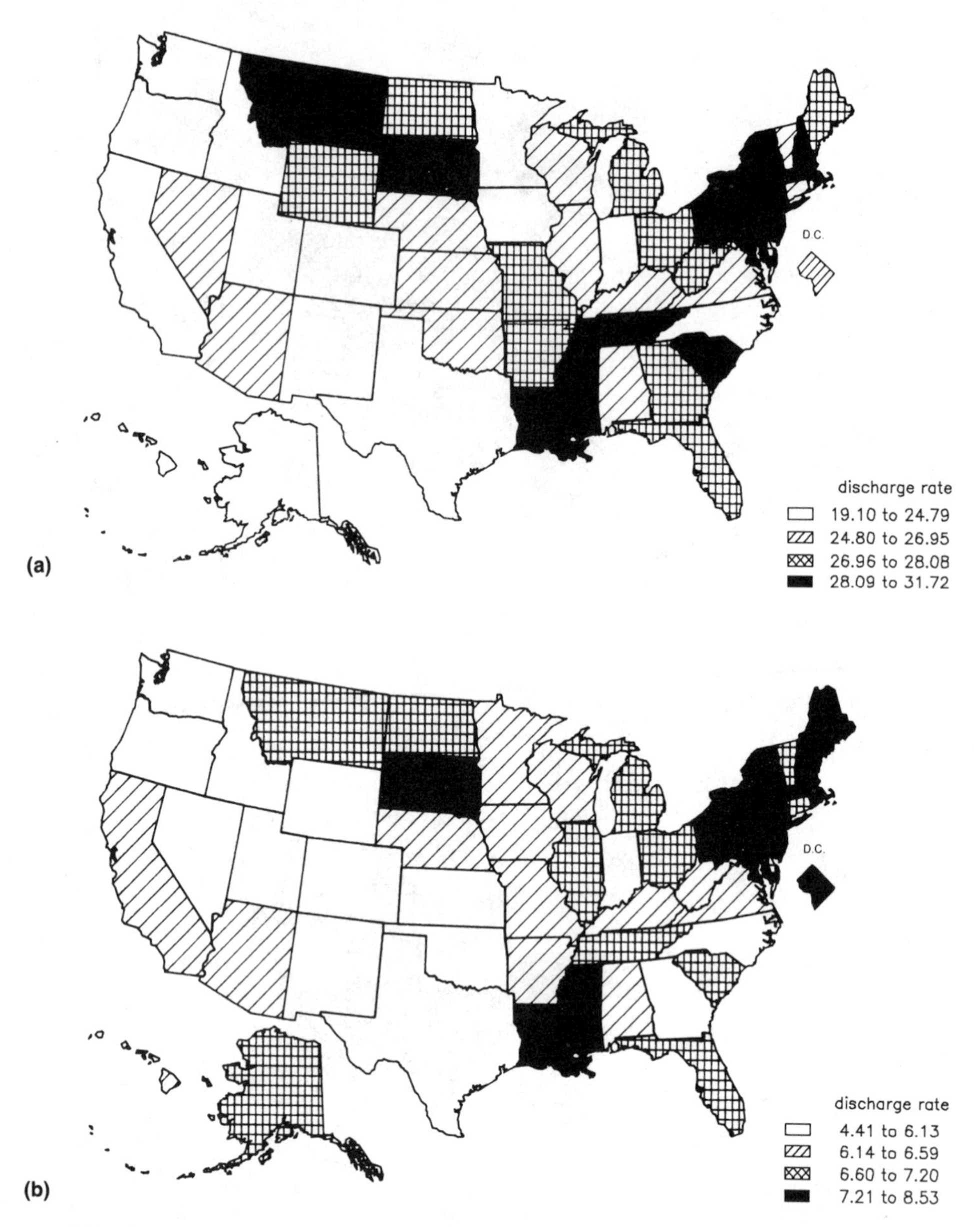

FIGURE 10-6. Maps of 1986 hospital discharges per 1000 Medicare enrollees 65 and over. (a) All cancers. (b) Cancers of the digestive system. *Source: Health Care Financing Corporation (1990).*

Cancer patterns are shown in Figures 10-6a, b and Figure 10-7a. High spots include the Northeast, south Central, and northern Great Plains states. Since many of the states identified as having high discharge rates and low post-discharge mortality rates also have high cancer rates, one should not draw conclusions about their underlying morbidity based only on Figure 10-3; it is also possible that post-discharge mortality occurs after 30 days, which might be the case for some cancers, for example. Figure 10-7 presents the data on

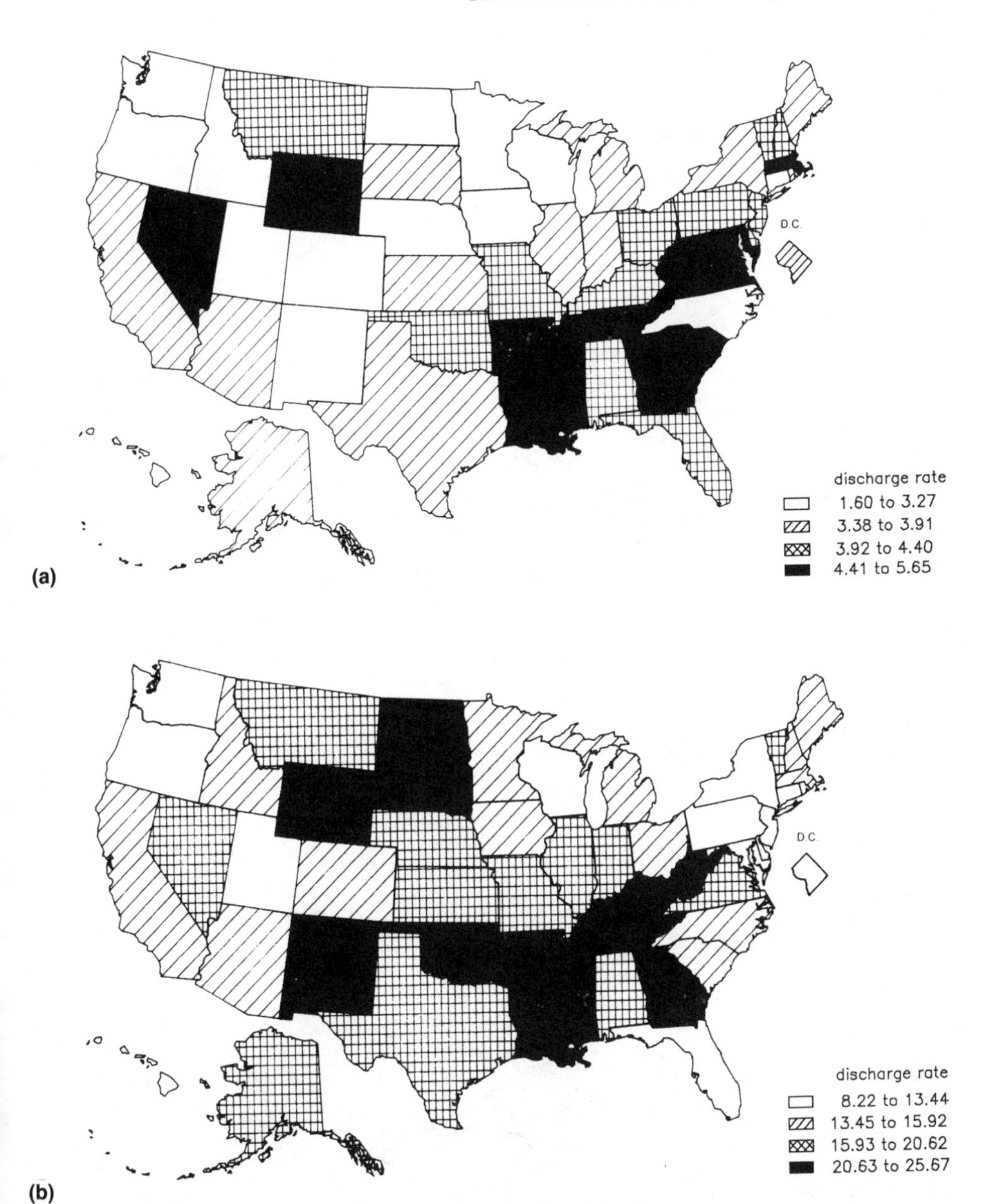

FIGURE 10-7. Maps of 1986 hospital discharges per 1000 Medicare enrollees 65 and over. (a) Respiratory system cancer. (b) Pneumonia and influenza. *Source: Health Care Financing Corporation (1990).*

respiratory diagnoses. Both of these diagnostic groups are scattered across the country; there is also some commonality with Figure 10-3.

Children's hospitalization rates in the United States and Canada were compared by Kozak and McCarthy (1984). They found that overall and respiratory hospitalization rates were higher in Canada and that lengths of stay were longer there also. They attributed this trend in part to the higher availability of hospital beds; these trends are shown for annual respiratory admissions in

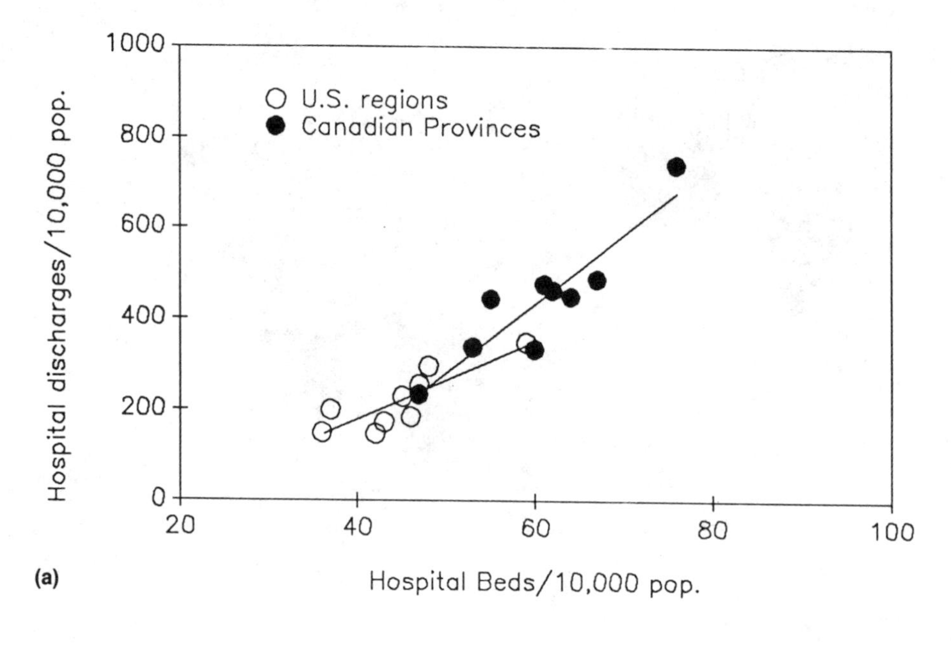

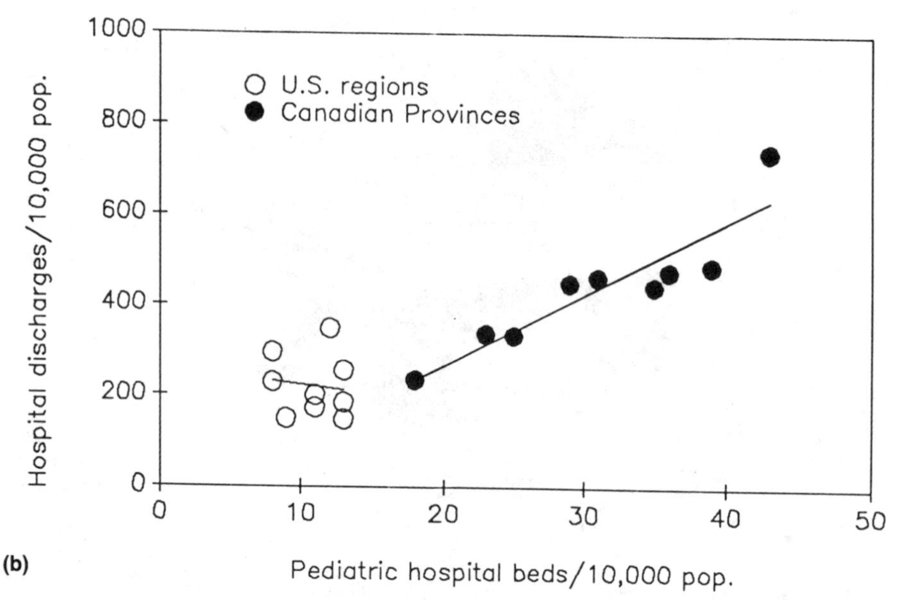

FIGURE 10-8. Relationships between children's hospital discharges for respiratory diagnoses and bed supply, in the United States and Canada, 1978. (a) Based on the total bed supply. (b) Based on the pediatric bed supply. *Data from Kozak and McCarthy (1984).*

Figure 10-8a (total hospital beds) and Figure 10-8b (pediatric beds). This is further evidence of the importance of supply factors in determining rates of hospital use.

As a further example of environmental factors that can affect hospital utilization on a long-term basis, Evans *et al.* (1987) found that asthmatic children from homes containing smokers were more likely to visit the emergency room (ER). The mean annual increase was 63% over homes with nonsmokers, which is a large increase considering that the presence of one smoker raises indoor fine particle loading by only about 20 $\mu g/m^3$ (J.D. Spengler, personal communication). This large response may be more typical of a susceptible subgroup than the population at large. Evans *et al.* controlled for family income, smoking by the children themselves, and allergens in the home, and assumed that all of the families used gas stoves (based on data from the local utility company). However, the excess ER usage was not proportional to estimated passive smoke exposure, so that the possibility of socioeconomic confounding may still exist. It is notable that this is a nonecological study that finds results generally consistent with those of ecological studies.

SYNTHESIS OF HOSPITALIZATION STUDIES

The studies of hospital usage reviewed here and in Chapter 9 showed that only a small fraction of the variance in hospital usage is associated with air pollution. Figure 10-9 depicts the interrelationships among the various factors that can affect hospital admissions; the chart parallels the mortality relationships discussed in Chapter 6, with the addition of the factor of potential constraints imposed by the availability of hospital beds. The effects of bed and physician supplies and medical care practices on long-term hospital usage were shown by the cross-sectional analyses in this chapter. Weather and seasonal effects can also be important in temporal studies. Given the complexity of this situation, it is not surprising that air pollution has such a small effect. Since a visit to a hospital ER is not affected by bed supply, it also follows that ER usage patterns may have stronger relationships with air pollution.

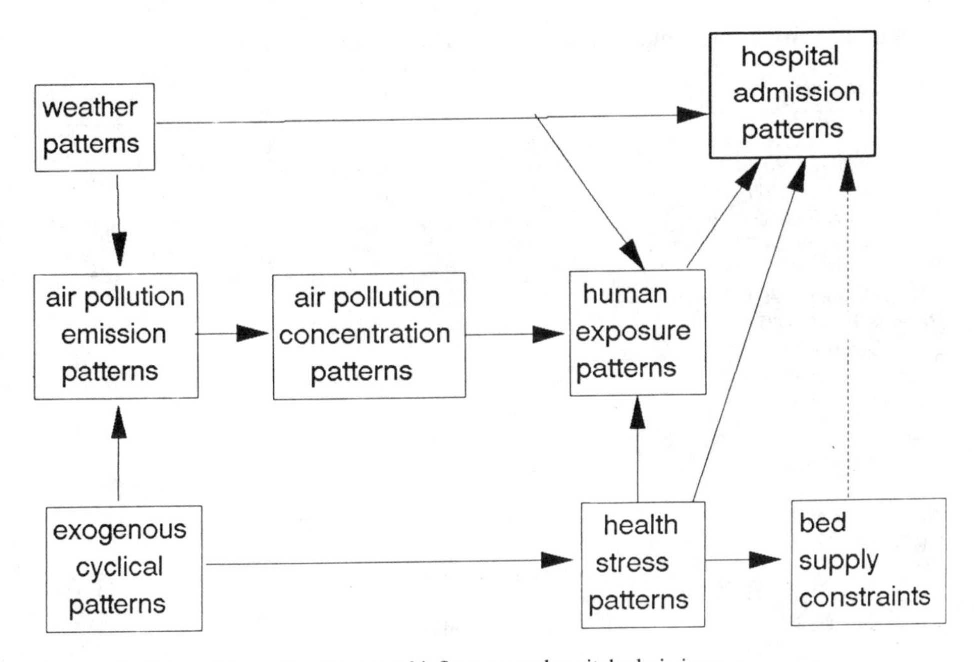

FIGURE 10-9. Schematic of temporal influences on hospital admissions.

Practically all of the studies reviewed in Chapter 9 and in this chapter found statistically significant positive associations between air pollution and hospital use, although the correspondence between specific pollutants and specific diagnoses varied widely. It is possible that negative findings have been excluded because of publication bias, even though the "gray" literature has been included in this review (Lamm *et al.*, 1991, for example). However, there are some important differences among the studies that should be considered.

Selection of Dependent Variables

Diagnostic categories considered varied among the studies. Only four investigators considered cardiac diagnoses, for example, and only two included eye complaints. The definitions of "respiratory" were inconsistent among the various studies. One study (Seskin, 1977), did not specifically investigate respiratory diagnoses.

The fractions of total diagnoses classified as "respiratory" varied considerably among studies (Table 10-1), especially between admissions and ER visits; part of this variation is due to definitional differences, and part may be due to population or environmental differences. Admission to hospital requires concordance by a physician, while a visit to an ER or clinic is based only on the subjective symptoms of the patient. The observation that the respiratory fraction is higher for ER visits reflects the fact that respiratory symptoms may be more obvious than other types of complaints. One could also infer that this percentage increases drastically during severe air pollution episodes, which lends credence to the use of the use of "percentage of admissions (or ER visits) classified as respiratory" for an analysis metric. For comparison with

TABLE 10-1 Definitions of "Respiratory" Cases in Hospitalization Studies

Investigation	Definition	% of total
	Admissions (inpatients)	
Sterling *et al.* (1966) (Los Angeles, 1961)	Includes upper respiratory and eye	3.7
Mazumdar & Sussman (1983) (Pittsburgh, 1972–77)	All respiratory conditions	8.0
Pope (1989) (Utah County, UT 1985–89)	Asthma, bronchitis, pneumonia, pleurisy	4.3
Bates & Sizto (1987) (So. Ontario, 1976–82)	No upper respiratory conditions	2.3
Plagionnakis & Parker (1988) Ontario Counties, 1976–82	All respiratory diagnoses	7–13
	Emergency Room Visits	
Ministry of Health (1954) (London, 1952)	All respiratory	58
Namekata *et al.* (1979) (Chicago, 1977–78)	Upper and lower resp. separated	13
Durham (1974) (California, 1970–71)	All respiratory diagnoses	24
Samet *et al.* (1981) (Steubenville, OH)	All respiratory diagnoses	26

Table 10-1, the U.S. national average percentage of respiratory admissions was 9.35% (Graves, 1987); subtracting acute respiratory infections and chronic tonsil disease leaves 7.4%. Coffey (1983) reported 1977 patient statistics by type of hospital and location inside Standard Metropolitan Statistical Areas (SMSAs); public hospitals had proportionately fewer respiratory admissions inside SMSAs and more outside SMSAs, with an average of about 8%, not counting acute upper respiratory infections or tonsil or adenoid conditions. These figures provide no support for the hypothesis of a positive relationship between urban air quality and hospital admissions.

It is interesting to note that even in Los Angeles, which was a very polluted city in 1961, the sum of upper and lower respiratory admissions (percentages) was no more than corresponding data from Toronto and the Utah Valley. Such comparisons suggest that cross-sectional comparisons of hospital usage will be difficult to interpret. As discussed above, factors relevant to the supply and delivery of medical care must be considered in relation to differences in long-term average rates.

Only two studies considered lengths of hospital stays (Sterling *et al.*, 1976, and Carpenter *et al.*, 1979). Given that admissions decisions may be based in part on factors other than the patients' needs, it would appear that environmental effects on lengths of stay should be given additional emphasis in future studies, as an index of the severity of illness. Another possibility would be to use length of stay as a stratifying variable in conjunction with admissions data.

Quantitative comparison of the relative strengths of the associations found may be made from Table 10-2, which summarizes the various estimates of elasticity of hospital use with respect to air pollution. It is difficult to make such

TABLE 10-2 Comparison of Elasticity Estimates in Hospitalization Studies

Investigator	Location	Significant Pollutants	Elasticity*
	Time-Series Studies		
Martin (1964)	London	SO_2, smoke	0.06–0.07
Sterling *et al.* (1966)	Los Angeles	SO_2, O_3	0.10
Durham (1974)	California	SO_2, NO_2, O_3	0.07–0.10
Samet *et al.* (1981)	Steubenville, OH	SO_2, TSP	0.045
Kurt *et al.* (1978)	Denver	CO	0.32
Pope (1989)	Utah Valley, UT	PM_{10}	0.67
Lamm *et al.* (1991)	Utah Valley, UT	PM_{10}	0.13
Bates & Sizto (1987)	So. Ontario	SO_2, NO_2, O_3, SO_4^{-2}	0.08
Lipfert & Hammerstrom (1992)	So. Ontario	SO_2, TSP, O_3, SO_4^{-2}	0.20
Burnett *et al.* (1993)	So. Ontario	SO_4^{-2}, O_3	0.06
Knight (1989a, b)	Prince George, BC	reduced S, TSP	0.05–0.09
Sunyer *et al.* (1991)	Barcelona	SO_2, CO, smoke	0.09–0.19
Ponka (1991)	Helsinki	NO, NO_2, O_3, CO	0.12
Schwartz *et al.* (1991)	5 German cities	SO_2, NO_2, TSP	0.02–0.17
	Cross-Sectional Studies		
Carpenter *et al.* (1979)	Pittsburgh	SO_2, TSP (joint)	0.17
Douglas & Waller (1966)	England & Wales	SO_2, smoke	0.40
Plagionnakis & Parker (1988)	Ontario	SO_2, SO_4^{-2}	0.13–0.36

* Elasticity values are for individual pollutants, as opposed to joint regressions on combinations (except for Carpenter *et al.*). See appropriate text passages for additional qualifications. Dependent variables include both admissions and ER visits.

estimates for the major episodes, since the responses often did not proceed at a constant rate during the episode; but they appear to be much higher, in the range from 0.20 to 0.70. Most of the elasticities in Table 10-2 are in the range from 0.05 to 0.40. The high value obtained in Utah by Pope (1989) stands out; his study used monthly data, while the analysis by Lamm *et al.* (1991) used (lagged) daily data and included a surrogate variable for the presence of viral activity. The low values found by Martin (1964) may relate to the lack of consideration of lags. The results of Burnett *et al.* (1993) would be increased by considering cumulative lag effects. The elasticity values estimated from cross-sectional studies are reasonably consistent with the time-series findings (when one considers the high levels of air pollution present during the Douglas-Waller [1966] study); this implies that the cross-sectional studies may be mainly measuring the annual sum of acute effects rather than chronic effects *per se*.

Exacerbation versus Toxicological Response

The studies I reviewed did not identify individuals and thus are silent on the question of the basic health status of those persons who sought treatment following exposure to air pollution. As a result, we have no direct information on the applicability of the "exacerbation hypothesis" to hospital usage. Nevertheless, it may be useful to try to apply logical analysis to this question.

Certainly, studies that show effects on asthmatics would comply with the exacerbation model, since it is extremely likely that a person would *become* asthmatic as a result of transient air pollution exposure. Furthermore, O'Halloran and Heaf (1989) analyzed repeat users of emergency treatment for childhood asthma and noted that 37% of the patients were responsible for 63% of the visits, and that the median number of visits for the repeaters was 4.5 (during 1 year). Thus, on a given day, the odds are in favor of a patient having had asthma attacks previously, and there is no apparent reason for this to be any less true on a day with higher than average air pollution. This logic could be extended to the case of cardiac patients responding to elevated CO, since it is well-known that angina patients are more sensitive to CO (Kurt *et al.*, 1979).

For other types of diagnoses, we must resort to examining the nature of the air pollution exposures considered in time-series studies. First, the lag between exposure and the decision to seek treatment is several days, at most, which is insufficient time for a new disease to develop and manifest symptoms. Secondly, in most situations, the types of air pollution exposures experienced are not one-time events but levels that tend to occur repeatedly. As discussed above, repeated exposures may lead to repeat visits for persons with existing impairments. If a new condition were to develop as a result of exposure to air pollution on a specific day, but not on previous occasions with similar exposures, one would question whether the relationship on that particular day were in fact causal or whether other factors might be involved. One needs to compare hospitalization statistics over the long term in order to gain insight into relationships between *prevalence* of disease and the environment. Certainly, none of the maps presented above suggest that high rates of (1986) hospitalization are concentrated in areas of the country that tend to have the worst air pollution (for example, ozone in California, sulfates in the Midwest and Appalachia, or particulates in the mountain and desert states). Earlier data taken when air pollution levels were much higher, for example Wilder (1974), show a slight excess of coronary heart disease in the Northeast and a slight excess of hypertension in the South. At best, we can only conclude that prevalence rates of diseases that have been associated with air pollution deserve more study.*

Since the limited comparison of cross-sectional (long-term) versus time-series (acute) responses suggested the same order of magnitude for both types of responses, the notion is supported that most of the effects are acute. Given this circumstance, it seems likely that the timing of hospital use in response to air pollution results from the juxtaposition of preexisting conditions and excursions in air pollution, as opposed to new cases of disease brought on solely by air pollution exposure. We thus conclude that the exacerbation model has received limited support from studies on hospitalization.

CONCLUDING DISCUSSION

Bennett (1981) presented a brief, mostly qualitative, review of hospital usage studies published through 1979 and illustrated what he considered a number of instances of "misuse" of data on health service utilization as a proxy for the underlying morbidity of the population. He questioned the use of moving averages to eliminate weekly cycles, the assumptions of unchanging relationships over space or time, the effects of individual perceptions as to the seriousness of a symptom, institutional constraints on admissions, adequacy of data reporting, and other extraneous factors that could affect hospital use. However, he did not seem to realize that most of these factors are likely to obscure the nature of any real relationships between the timing of hospital use and air pollution, rather than to create "false positives" as he implied.

Given the qualitative uniformity in findings among the diverse studies reviewed in this section and the likelihood that many of the deficiencies in the studies would act to bias elasticities and their significance downward, it appears that an association between (nonspecific) air pollution and hospital use has been demonstrated. Daily and longer term studies appear to be reasonably consistent in this regard. However, further research is required to identify specific pollutants and concentration thresholds (if any) associated with specific diagnoses, the operative range of lags, and the influence of repeat users of hospital services. One must also note that since medical care supply factors appear to play such an important role in hospital utilization, identification of the role of environmental factors should be geared to elucidating information

*This is not to deny that air pollution has been associated with increased disease prevalence in the past. For example, Lambert and Reid (1970) showed that chronic bronchitis symptoms were consistently more prevalent in areas of Britain with higher air pollution, especially for smokers. However, for nonsmokers, the effect of age on increasing the prevalence differential due to air pollution was weak, even though these data were obtained in 1965, when air pollution was still severe in parts of Britain. One would expect a chronic response to increase with age and thus with cumulative exposure. With respect to the many studies of respiratory symptoms in the literature (which have not been reviewed in this book), one must distinguish between reversible or transient symptoms, which may be triggered by acute responses to irritants, with the prevalence of underlying respiratory or cardiac disease.

about the underlying health of the population (as was urged by Bennett), perhaps from additional data on symptom prevalence. Data on the variability of lengths of hospital stays with respect to environmental factors may also be useful in this regard.

These findings suggest that children's respiratory admissions should be examined separately, especially in winter, when some of the study results seem to be in conflict. For example, Bates and Sizto (1987) found no significant air pollutant associations with children's asthma in Ontario and positive effects due to temperature in winter. Pope (1991) and Lamm *et al.* (1991) showed negative temperature effects, corresponding to the seasonal trend or perhaps due to space heating emissions from wood stoves. Kraemer and McCarthy (1985) examined asthma admissions for ages 0–19 in Spokane, WA, and showed no consistent seasonal pattern or association with particulates (TSP), aside from the Mt. St. Helens perturbation. There was no substantial lag effect apparent from this volcanic eruption episode, while Koenig *et al.* (1992) found a significant lagged association between asthma ER visits and PM_{10} in Seattle. The total seasonal respiratory admissions variation in these studies was much less than in Utah, however.

Additional methodological recommendations from this review include:

- Twenty-four-hour averages should be used for all pollutants.
- Cumulative lags up to at least 3 days should be considered.
- Respiratory diagnoses should be given priority, but cardiac cases should be included as well.
- It is essential to correct for seasonal, day-of-week, and holiday effects.
- Additional studies in other types of settings, such as clinics, should be performed.

Additional information that is needed includes comparisons of long-term trends in hospital usage and air pollution, provided that situations can be found involving both increasing and decreasing trends (such as the Utah studies by Pope [1989, 1991]).

REFERENCES

Bates, D.V., and Sizto, R. (1987), A Study of Hospital Admissions and Air Pollutants in Southern Ontario, *Env. Res.* 43:317–31.

Bennett, A.E. (1981), Limitations of the Use of Hospital Statistics as an Index of Morbidity in Environmental Studies, *J. APCA* 31:1276–78.

Brewer, W.R., and Freedman, M.A. (1982), Causes and Implications of Variations in Hospital Utilization, *J. Public Health Policy* 3:445–54.

Carpenter, B.H., Chromy, J.R., Bach, W.D., LeSourd, D.A., and Gillette, D.G. (1979), Health Costs of Air Pollution: A Study of Hospitalization Costs, *Am. J. Public Health* 69:1232–41.

Coffey, R.M. (1983), Patients in Public General Hospitals, in *Health United States and Prevention Profile, 1983*, U.S. Department of Health and Human Services, Washington, DC.

Deprez, R.D., Oliver, C., and Halteman, W. (1986), Variations in Respiratory Disease Morbidity among Pulp and Paper Mill Town Residents, *J. Occ. Med.* 28:486–91.

Diehr, P. (1984), Small Area Statistics: Large Statistical Problems, *Am. J. Public Health* 74:313–14.

Douglas, J.W.B., and Waller, R.E. (1966), Air Pollution and Respiratory Infection in Children, *Br. J. Prev. Soc. Med.* 20:1–8.

Durham, W.H. (1974), Air Pollution and Student Health, *Arch. Env. Health* 28:241–54.

Evans, D., Levison, M.J., Feldman, C.H., Clark, N.M., Wasilewski, Y., Levin, B., and Mellins, R.B. (1987), The Impact of Smoking on Emergency Room Visits of Urban Children with Asthma, *Am. Rev. Resp. Dis.* 135:567–72.

Gornick, M. (1977), Medicare Patients: Geographic Differences in Hospital Discharge Rates and Multiple Stays, Social Security Bulletin, June:22–41.

Gornick, M. (1982), Trends and Regional Variations in Hospital Use under Medicare, in *Regional Variations in Hospital Use*, D.L. Rothberg, ed., Lexington Books, Lexington, MA, pp. 131–84.

Graves, E.J. (1987), *Diagnosis-Related Groups Using Data From the National Hospital Discharge Survey: United States, 1985*, NCHS Advance Data No. 137, July 2, National Center for Health Statistics, Hyattsville, MD.

Health Care Financing Corporation (HFCA) (1990), Hospital Data by Geographic Area for Aged Medicare Beneficiaries: Selected Diagnostic Groups, 1986, HFCA Pub. 03300, U.S. Dept. Health and Human Services, Washington, DC.

Knickman, J.R. (1982), Variations in Hospital Use across Cities: A Comparison of Utilization Rates in New York and Los Angeles, in *Regional Variations in Hospital Use*, D.L. Rothberg, ed., Lexington Books, Lexington, MA, pp. 23–64.

Knight, K., Leroux, B., Millar, J., and Petkau, A.J. (1989a), *Air Pollution and Human Health: A Study Based on Hospital Admissions Data from Prince George, British Columbia*, SIMS Technical Report #128, February, SIMS Publications, New Canaan, CT.

Knight, K., Leroux, B., Millar, J., and Petkau, A.J. (1989b), *Air Pollution and Human Health: A Study Based on Emergency Room Visits Data from Prince George, British Columbia*, SIMS Technical Report #136, June, SIMS Publications, New Canaan, CT.

Koenig, J.Q., Schwartz, J., Slater, D., Larson, T.V.C., and Pierson, W.E. (1992), Associations between Ambient Particulate Matter and Hospital Emergency Visits for Asthma in Seattle, presented at the American Chemical Society, San Francisco, April.

Kozak, L.J., and McCarthy, E. (1984), Hospital Use by Children in the United States and Canada, *Vital and Health Statistics*, Series 5, No. 1, DHHS Pub. No. (PHS)84-1477, Public Health Service, Washington, DC, GPO.

Kraemer, M.J., and McCarthy, M.M. (1985), Childhood Asthma Hospitalization Rates in Spokane County, Washington: Impact of Volcanic Ash Air Pollution, *J. Asthma* 22:37–43.

Kurt, T.L., Mogielnicki, R.P., and Chandler, J.E. (1978), Association of the Frequency of Acute Cardiorespiratory Complaints with Ambient Levels of Carbon Monoxide, *Chest* 74:10–14.

Kurt, T.L., Mogielnicki, R.P., Chandler, J.E., and Hirst, K. (1979), Ambient Carbon Monoxide Levels and Acute Cardiorespiratory Complaints: An Exploratory Study, *Am. J. Public Health* 69:360–63.

Lambert, P.M., and Reid, D.D. (1970), Smoking, Air Pollution, and Bronchitis in Britain, *Lancet* 1:853–57.

Lamm, S.H., Hall, T.A., Engel, A., White, L.S., and Rueter, F.H. (1991), Assessment of Viral and Environmental Factors as Determinants of Pediatric Lower Respiratory Tract Disease Admissions in Utah County, Utah (1985–1989), unpublished report from Consultants in Epidemiology and Occupational Health, Washington, DC.

Lewis, W.F. (1974), *Utilization of Short-Stay Hospitals, Summary of Nonmedical Statistics, United States—1971*, DHEW Publication No. (HRA)75-1768, National Center for Health Statistics, Rockville, MD.

Lipfert, F.W., and Hammerstrom, T. (1992), Temporal Patterns in Hospital Admissions and Air Pollution, *Environ. Res.* 59:374–99.

Lutz, L.J. (1983), Health Effects of Air Pollution Measured by Outpatient Visits, *J. Fam. Practice* 16:307–13.

Martin, A.E. (1964), Mortality and Morbidity Statistics and Air Pollution, *Proc. R. Soc. Med.* 57:969–75.

Mazumdar, S., and Sussman, N. (1983), Relationships of Air Pollution to Health: Results from the Pittsburgh Study, *Arch. Env. Health* 38:17–24.

Ministry of Health (1954), Mortality and Morbidity during the London Fog of December 1952, Reports on Public Health and Medical Subjects No. 95. London, HMSO.

Muhling, P., Bory, J., and Haupt, H. (1985), Studies of Babies and Infants, The Influence of Air Pollution on Respiratory Diseases, *Staub Reinhalt. Luft* 45:35–38.

Namekata, T., Carnow, B.W., Flourno-Gill, Z., O'Farrell, E.B., and Reda, D. (1979), Model for Measuring the Health Impact from Changing Levels of Ambient Air Pollution: Morbidity Study, EPA-600/1-79-024, U.S. Environmental Protection Agency, Research Triangle Park, NC.

O'Halloran, S., and Heaf, D.P. (1989), Recurrent Accident and Emergency Department Attendance for Acute Asthma in Children, *Thorax* 44:620–26.

Pasley, B., Vernon, P., Gibson, G., McCauley, M., and Andoh, J. (1987), Geographic Variations in Elderly Hospital and Surgical Discharge Rates, New York State, *Am. J. Public Health* 77:679–84.

Plagionnakis, T., and Parker, J. (1988), *An Assessment of Air Pollution Effects on Human Health in Ontario*, Report prepared for Ontario Hydro, Toronto (Energy Economics Section, Economics and Forecast Division; Report No. 706.01 [#260]).

Ponka, A. (1991), Asthma and Low Level Air Pollution in Helsinki, *Arch. Envir. Health* 46:262–70.

Pope, C.A., III (1991), Respiratory Hospital Admissions Associated with PM_{10} Pollution in Utah, Salt Lake, and Cache Valleys, *Arch. Env. Health* 46:90–97.

Pope, C.A., III (1989), Respiratory Disease Associated with Community Air Pollution and a Steel Mill, Utah Valley. *AJPH* 79:623–28.

Ries, P.W. (1975), Current Estimates from the Health Interview Survey-1974, National Center for Health Statistics Vital and Health Statistics, series 10, no. 100, Rockville, MD.

Samet, J.M., Speizer, F.E., Bishop, Y., Spengler, J.D., and Ferris, B.G., Jr. (1981), The Relationship between Air Pollution and Emergency Room Visits in an Industrial Community, *J. APCA* 31:236–40.

Schwartz, J., Spix, C., Wichmann, H.E., and Malin, E. (1991), Air Pollution and Acute Respiratory Illness in Five German Communities, *Envir. Res.* 56:1–14.

Seskin, E.P. (1977), *Air Pollution and Health in Washington, D.C.*, EPA-600/5-77-010, U.S. Environmental Protection Agency, Corvallis, OR.

Sterling, T.D., Phair, J.J., Pollack, S.V., Schumsky, D.A., and DeGroot, I. (1966), Urban Morbidity and Air Pollution, *Arch. Env. Health* 13:158–70.

Sterling, T.D., Pollack, S.V., and Phair, J.J. (1967), Urban Hospital Morbidity and Air Pollution, *Arch. Env. Health* 15:362–74.

Sunyer, J., Anto, J.M., Murillo, C., and Saez, M. (1991), Effects of Urban Air Pollution on Emergency Room Admissions for Chronic Obstructive Pulmonary Disease, *Am. J. Epidemiology* 134:277–86.

Wilder, C.S. (1974), *Prevalence of Selected Chronic Circulatory Conditions, United States, 1972*, Vital and Health Statistics, series 10: Data from the National Health Survey, no. 94, DHEW publ. no. (HRA)75-1521, GPO.

11

Air Pollution Effects on Lung Function

We are fully conscious of the incompleteness in our knowledge of respiratory function. As there are many points still in dispute regarding the physiology of normal respiration, it is not surprising to find that there are even more deficiencies of the abnormal.

J.C. Meakins and H.W. Davies

INTRODUCTION TO LUNG FUNCTION STUDIES

In contrast to mortality and hospitalization studies, most of the lung function studies deal primarily with healthy, heterogeneous populations. (The general terms "pulmonary function," "lung function," and "respiratory function" are used interchangeably in this discussion. Specific measures were defined in Chapter 4.) Studies of asthmatics are an exception, but only a few of the studies in the literature have examined people with severely compromised cardiorespiratory systems. The task of defining the relevant effects of community air pollution on the basis of group average pulmonary function attributes is expected to be difficult, since those effects are likely to be subtle and to vary considerably among subjects. However, lung function studies also have the advantage of dealing with identified individuals, rather than with groups of people whose individual characteristics cannot be determined *a posteriori*.

Pulmonary function has been shown to be affected by a large number of factors, including genetic predisposition, weather, season, time of day, the basic respiratory health of the subject, smoking status, allergens, air pollution, and the testing procedure itself. The studies reviewed in this chapter vary considerably in their ability to control for these potential confounders in order to derive dose-response relationships for air pollution. While studies of mortality and of hospital admissions and emergency room (ER) use deal with large populations and endpoints defined with certainty (except for diagnoses, in certain instances), lung function studies tend to deal with small populations and highly variable endpoints. Since most lung function tests must be performed individually on a one-to-one basis with staff, resource limitations have generally limited these studies to either spot tests of a large number of subjects or continued followup of a smaller number of subjects.

Three kinds of lung function studies have been useful:

- Clinical laboratory exposure studies, in which animals or human volunteers are exposed to known concentrations of specific air pollutants by means of environmental chambers or direct delivery through face masks and the like.
- Epidemiological studies of transient lung function responses to naturally fluctuating levels of air pollution.
- Epidemiological studies of long-term differences in lung function among subpopulations chronically exposed to different levels of air pollution. These may include occupationally exposed groups as well as random population samples. The effects of passive smoking (environmental tobacco smoke) also fall into this category.

Each of these approaches has strengths and weaknesses, and information from all of them will be required to help define the physiological effects of air pollution on the human respiratory system.

Laboratory exposure studies are useful to compare the relative effects of different pollutants and mixtures and to observe the nature of the responses, such as the most relevant lung function metrics and whether there are lags in response or recovery. However, laboratory exposure studies suffer from lack of realism. Animal studies can also provide corroborating data on biological and morphometric changes that may have occurred as a result of exposure. However, there are inevitable questions about transferring findings from animals to humans. If human subjects are used, ethical considerations prohibit studying under stressful conditions those persons who may be most at risk, such as those with existing cardiorespiratory impairment. In most laboratory exposure tests, high ambient concentrations are used in order to generate a response that can be easily observed with a limited number of subjects. (There were some early "reverse" experiments in Los Angeles in which impaired subjects breathed purified air and their degrees of improvement relative to "normal" Los Angeles air were measured.) Practical considerations often limit the lengths of exposure that can be studied to 8 hours or less.

Epidemiological studies can also deal with acute transient responses, but with free-living populations and naturally occurring variations in air quality. This gain in realism relative to laboratory studies may be partially offset by problems in confounding; for example, during air pollution episodes, concentrations of all the pollutants present are usually elevated by similar percentages, and weather patterns may also be unusual. Thus the coincident weather effects on lung function must be considered, as well as the "natural" circadian rhythms in certain measures of lung function (Reinberg and Gervais, 1972; Kerr, 1973). For example, the slow recovery from a winter air pollution episode could be confounded with the normal seasonal cycle, which has a minimum in winter and a peak in summer (McKerrow and Rossiter, 1968). Resolving questions of the "responsible" air pollutants in natural settings will thus require comparison of acute studies conducted in different locations where the pollutant mix differs.

The underlying hypothesis being considered in this chapter is that air pollution degrades lung function, either on a transient, reversible basis or on a chronic basis due to long-term pollutant exposure. The *sequellae* of childhood respiratory disease may be important in this latter regard. Whether these respiratory effects *per se* result in increased risk of premature death is taken up in Chapter 12. There are of course no empirical data on the relationships between transient lung function decrements and sudden death; the presumption is that persons with already impaired respiratory performance will be more susceptible to further insults, even if those effects might otherwise have been reversible.

In comparing acute with chronic studies, the nature of the respiratory response is important. Pollutants that act mainly in the upper respiratory tract and large airways (large particles, SO_2, for example) may cause coughing and excessive secretion of mucus; these are acute responses, but they may have long-term consequences if repeated often enough. As will be discussed, the short-term effects of air pollution on lung function are usually reversible. Pollutants that reach the deep lung may have longer term effects, such as destruction of the alveoli and injury to proximal airways; these are essentially long-term effects, as exemplified by the effects of cigarette smoke. Long-term effects may be measured by comparing baseline lung function data among populations with differing long-term air pollution levels or by comparing rates of change (decline with age) over time among such populations. As with all cross-sectional comparisons, care must be exercised to avoid confounding from extraneous factors that may also differ among locations (see Chapter 7). In some instances, the site of action within the respiratory system may also be deduced from the specific lung function measures that show significant change; for example, peak flow has been cited as an index of large airway caliber (Vedal *et al.*, 1987). Dassen *et al.* (1986) assigned forced vital capacity (FVC) and FEV_1* as additional indices of large airways, and MMEF, MEF_{50}, and MEF_{25} as measures responding mainly to changes in small airway caliber. Frank (personal communication) points out that many measures of lung function are additionally dependent on lung volume and, thus, that changes in FVC must also be considered, independently of changes in the small airways.

This chapter begins with a review and analysis of short-term or temporal studies of the association of lung function gradients with air pollution and then takes up the question of long-term or chronic effects. The question of the

---

* Lung Function Acronyms and Abbreviations

FVC	forced vital capacity
R_{aw}	airway resistance
SR_{aw}	specific airway resistance
FEV_1	forced expiratory volume in 1 second
$FEV_{0.75}$	forced expiratory volume in 0.75 seconds
PERF, PFR	peak expiratory flow rate
$FEF_{(n)\%}$, $MEF_{(n)\%}$, $V_{max(n)}$	forced expiratory flow at (n)% vital capacity
$FEF_{25-75\%}$, MMEF, or MMFR	maximum midexpiratory flow (average slope between defined increments of the FVC).

effects of environmental tobacco smoke on lung function is then considered as an adjunct to the study of chronic effects.

TEMPORAL STUDIES OF AIR POLLUTION EFFECTS ON LUNG FUNCTION

Data on short-term and temporal lung function responses to air pollution include epidemiological studies of episodes, time-series studies including those involving repeated measurements, and laboratory exposure tests on volunteers. As discussed briefly, there are various kinds of pitfalls in each of these types of studies. It is important to understand short-term behavior in considering studies of long-term effects, since there may be interactions, depending on the experimental protocol. The reversible nature of these short-term responses is shown by studies that attempt to measure the effects of an air pollution episode by observing the subsequent recovery in lung function.

Studies of Isolated Air Pollution Episodes

Smith and Dinh (1975) tested a group of nine young adults for their responses to the traditional New Year's Eve fireworks in Honolulu. Based on nephelometer measurements, peak (ca. 15 min.) smoke level reached about 3800 $\mu g/m^3$ at this time. Although the average response of the group was not statistically significant, two of the subjects (who had a history of chronic respiratory disease) showed an average decrease in $FEV_{25-75\%}$ of 26%. They also reported previous studies by others of the responses of chronic obstructive pulmonary disease patients during fireworks displays, of which about 70% were affected. The recovery period was not measured, nor were any lag effects.

During and for about one week after the 1975 Pittsburgh air pollution episode (discussed in Chapter 5), daily FVC and $FEV_{0.75}$ measurements were made on 224 schoolchildren in the areas considered to be most polluted and in two control areas. The initial analysis of these data (Stebbings *et al.*, 1976) failed to find effects that could be attributed to changes in air quality. The working hypothesis was that pulmonary function should have improved as the episode abated and air quality improved. In a subsequent, more detailed, analysis, Stebbings and Fogleman (1979) reported that about 10 to 15% of the children in the exposed area did show an improvement in FVC of about 20 to 25%, while a smaller group in the "control" area did not. The children who showed improvements were labeled as "susceptible," but no other evidence of such susceptibility was presented, and their data for $FEV_{0.75}$ did not show improvements. The main pollutant at issue here was smoke, which reached about 700 $\mu g/m^3$ for 2 days in one local area.

Children in Missoula, Montana, who were being examined as part of a routine monitoring study, were also exposed to ash from the Mount St. Helens eruption in 1980, during which 24-hour TSP levels exceeded 11,000 $\mu g/m^3$. Lung function tests were taken the day after the peak ash exposure period ended (Johnson *et al.*, 1982); the girls showed essentially no changes beyond those expected because of seasonal variability (slight improvement), while the boys showed about a 3% decrease on average. It should be noted that the population was urged to stay indoors or to wear protective masks during this event; 10% of the boys reported being outside without wearing a protective mask near the end of the high-ash exposure period, while only 4% of the girls did so.

Additional insight into the effects of Mt. St. Helens ash on pulmonary function was gained by the studies of Buist *et al.* (1983, 1986). The first study examined 101 children who were attending a summer camp after the second eruption. The area was impacted by resuspended ash that had previously fallen; TSP levels were measured by personal samplers and were in the range of 800 to 2300 $\mu g/m^3$; respirable dust was around 170 $\mu g/m^3$. The authors used the absence of within-day and between-day effects on lung function, as determined by analysis of variance, as indicators of the lack of influence of the volcanic ash. They also reported that the camp was obviously dusty and that the dust was irritating to the throat. Subsequent lung function measurements at the camp the following year were at similar levels. Although it seems ineluctable that any effects of the ash on children's lung function must have been small, the study may have been hampered by the short duration (2 weeks), the failure to analyze daily particulate levels at various lags, and the lack of data on individuals.

The second study of Mt. St. Helens ash followed a group of about 700 exposed loggers for 4 years (Buist *et al.*, 1986). A decline in lung function (FEV_1) of about 4% was shown during the first year after the eruption; after 4 years, FEV_1 values had recovered slightly. Although the "control" group for this study was not exposed to volcanic ash, the subjects' exposure to respirable dust was not greatly different from that of the "exposed" group; for the period from June 1980 to July 1981, all the groups were exposed to 270 to 560 $\mu g/m^3$ of respirable dust. For this reason, the reliance of the authors on FEV_1 differences between "exposed" and "control" groups is probably only relevant to the specific question of volcanic ash, not to respirable dust in general.

Dassen *et al.* (1986) were able to obtain comparative spirometry on 163 Dutch schoolchildren in conjunction with the January 1985 episode in central Europe. In the area where lung function was measured, TSP, RSP, and SO_2 peaked to around 200 to 250 $\mu g/m^3$ (RSP occasionally exceeded TSP!). The children (6 to 11 years old) had been evaluated in November–December 1984 and then were remeasured at the height of the episode (January 18, 1985) and several weeks later. The statistical analysis was done for each child, accounting for growth due to aging, but the seasonal trend during the winter was not discussed. The data showed a 3 to 5% depression in lung function on January 18 that persisted at least 2 weeks but was gone after 4 weeks. No effects on lung function were observed from a later 2-day pollution increase of about 100 to 150 $\mu g/m^3$. Although these events could be explained alternatively in terms of seasonal variability, the magnitudes of the changes appear to be larger than could be explained by the seasonal cycle obtained for normal adults by McKerrow and Rossiter (1968).

Brunekreef *et al.* (1987) reexamined the January 1985 event and the Steubenville data (Dockery *et al.*, 1982) in the context of the effects of lags and of variations in response among the exposed population. They pointed out that a response persisting for a week or more after the event suggests that the operative dose should probably be defined for a lag longer than 1 day and reported that considering longer lags increased the (negative) slopes of lung function on TSP by about a factor of 3. They also questioned the use of the magnitude of lung function responses as an indicator of "susceptibility," since random variation accounts for most of this variability.

Time-Series Studies

Perhaps the earliest epidemiological study of air pollution and pulmonary function was that by Shephard *et al.* (1960); this study is also noteworthy because of its attention to subjects with impaired cardiorespiratory systems and its use of indoor air quality exposure measurements. Ten patients in Cincinnati were followed prospectively for 10 weeks from October 1, 1956. Air pollution, temperature and humidity measurements were made in each home; the patients were reported to be confined to their homes by their illnesses. Pollution data consisted of smoke measured by tape sampler and SO_2 measured by the peroxide (strong acidity) method. Pulmonary measurements were made on each patient three times each week, at the same times of day. The data on indoor versus outdoor COH levels in October showed that the high outdoor values typical of early morning and late evening were attenuated indoors, but both data sets had the same characteristic diurnal profiles. During the period plotted, indoor smoke levels were about 150 $\mu g/m^3$ in a suburban home, while outdoor downtown levels were about twice that. Outdoor gaseous acid (SO_2) levels were about 0.03 ppm during the entire survey, with a peak at 0.12 ppm. Absolute humidity was found to be significantly associated with reduced lung function: Smoke was negatively associated with some of the pulmonary function measures, particularly for one patient who lived in a more highly polluted area. The correlations for gaseous acids did not exceed those expected from just chance variation. The authors concluded that "for most of the survey period the concentration of pollutants was below the level at which even a small physiological response would occur."

Spicer *et al.* (1962) reported the daily lung function responses of seven chronic obstructive disease patients for 14 weeks and a subsequent study of 14 patients for 47 days. They reported that the patients "became better and worse together," indicating the influence of exogenous factors. Airway resistance was reported to be the most sensitive measure. Pollutants monitored nearby included SO_2, NO_2, TSP, and particles measured by the American Iron and Steel Institute (AISI) tape sampler. The statistical analysis failed to find any one pollutant associated with the daily changes. They also reported that their patients "showed striking changes in their mechanics of respiration and lung volumes within relatively short periods of time. Bearing this in mind, it is appropriate to remember how hazardous it is to assess pulmonary insufficiency on the basis of a few measurements performed on some arbitrary day in a patient's life."

I pooled the two time periods for which data were presented in Spicer *et al.*'s paper and examined alternative dose-response functions. I noted no lag effects but that airway resistance (R_{aw}) seemed to fit better against the logarithm of TSP (Figure 11-1) than with a linear plot. However, extrapolation of the regression line beyond the range of the original data, as shown, may be

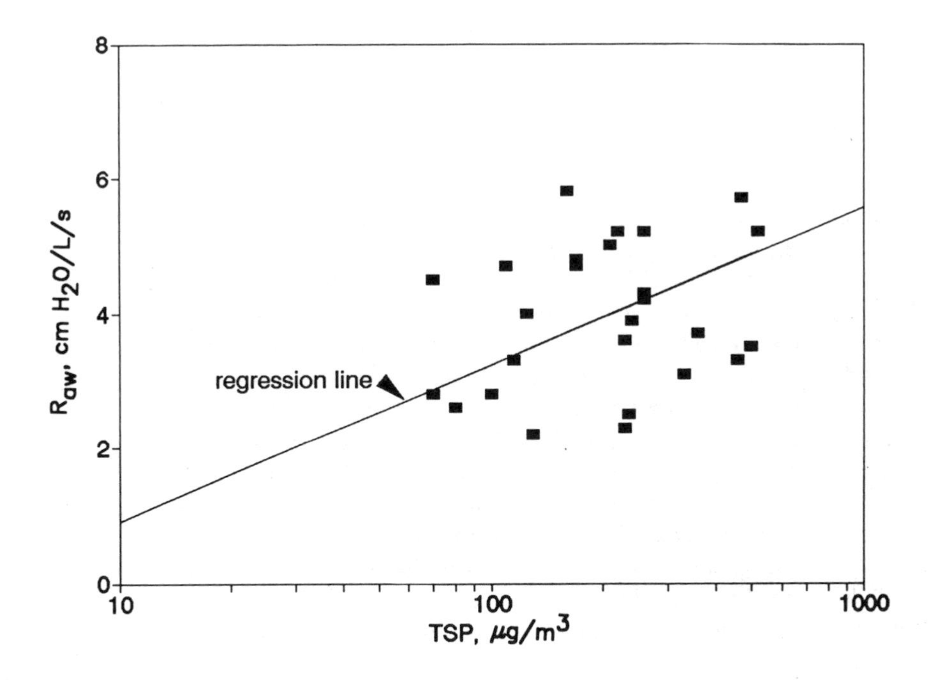

FIGURE 11-1. Daily airway resistance changes for a group of chronic obstructive lung disease patients. *Data from Spicer* et al. *(1962).*

problematic. The elasticity of this relationship is 0.26, evaluated at the mean ($p < 0.003$).

Motley and Phelps (1964) and Motley *et al.* (1959) reported on a "reverse" chamber experiment in Los Angeles, during which volunteers including emphysema patients breathed purified air, in comparison with Los Angeles smog. Lung function improvements were modest, but residual volume showed a significant decrease after 40 hours for the emphysema patients only. Nitrogen washout tests were intermediate in response.

Spodnik *et al.* (1966) studied 100 white male seminary students in Baltimore from September 1964 to June 1965, recording a number of pulmonary function and environmental variables, including TSP measured on the premises. There were 28 different days of measurements on 12 to 15 subjects each. The most significant correlations were for temperature and relative humidity. Neither same-day nor weekly average TSP values were significantly correlated with any lung function measure, even for a subset of asthmatic subjects, although the signs of the correlations suggested adverse impacts. Average TSP levels were not reported, but based on routine monitoring data, they were probably around 120 μg/m^3 (Lipfert, 1978).

Emerson (1973) followed 18 lung disease patients in London for periods varying from 12 to 82 weeks. Various pulmonary function measurements were made at intervals of a week or more, and these data were compared to air pollution data averaged over the 5 days preceding the measurements. Weather data were similarly considered. All patients lived in the area where the atmospheric measurements were made. Bivariate correlations between FEV_1 and MEFR were presented for individual patients; the most significant variable was temperature (+). One patient showed a significant positive correlation between SO_2 and both measures; another showed a positive correlation between SO_2 and MEFR. No significant correlations were shown for smoke. Emerson attributed these findings to the generally low levels of air pollution at the time (smoke from 45 to 380 $\mu g/m^3$; SO_2 from 193 to 404 $\mu g/m^3$). However, no multiple regressions were presented, and it is possible that adjustments for seasonal trend might have clarified the situation.

Kagawa and Toyama (1975) followed 21 eleven-year-old children for about 6 months, performing lung function tests weekly and comparing the results to concentrations of seven air pollutants, temperature, and humidity, measured at their school. Testing was done on Wednesdays from 1:00 to 3:00 P.M. Air concentrations were compared for the same time as the lung function tests, the preceding hour, and the preceding 24 hours. Correlations were computed for each subject, and temperature was the most significant variable for all six lung function measures, being significant for 12 to 14 of the 20 subjects. Since the correlations were essentially the same for all three averaging times, it appears that the relationship was seasonal. Air pollution was significant for no more than five subjects, and the highest correlations were for ozone, oxidant, NO, and hydrocarbons. Particulates (based on light scattering) were significant for only two subjects: The most reactive subject responded to all the other pollutants, but not always for the same lung function metric. The two most reactive subjects also had the strongest correlations between lung function and temperature; since no multiple correlations were computed, it is possible that the results for air pollution were confounded by seasonal factors. However, Kagawa and Toyama point out that the relationship between FVC and temperature is normally positive (worst in winter), but their subjects showed the opposite, with a negative correlation between FVC and temperature. It is thus possible that the lack of seasonal corrections may have obscured the true relationships with air pollution. Air pollution levels were quite modest during these experiments.

Loftsgaarden *et al.* (1981) tested groups of about 110 schoolchildren in Missoula, Montana, on days of high and low particulate air pollution. Negative gradients in lung function were seen for fine and coarse particles and for TSP. Although weather effects were not considered, seasonal variation, sex, and height of the children were controlled for. The results showed group mean declines (males, females) due to an increase of 320 $\mu g/m^3$ of TSP of 0.6%, 0.9% in FVC; 1.5%, 1.8% in FEV_1; and 1.8%, 4.6% in $FEF_{25-75\%}$. The FEV_1 changes and the changes in $FEF_{25-75\%}$ for girls were statistically significant.

Dockery *et al.* (1982) conducted an elaborate program of lung function testing on groups of about 200 elementary schoolchildren in Steubenville, OH, from 1978 to 1980. The protocol involved baseline measurements at times chosen more or less at random, measurements during high-pollution "alert" periods, and weekly follow-up measurements thereafter. Four such cycles were evaluated, three in fall and one in spring; two involved relatively modest pollution levels. The analysis used each child as his or her own control and calculated the distribution of regression-pollution slopes, using a 16-hour time lag. It was not possible to separate TSP from SO_2 during these events. The distribution of slopes showed a slight excess of negative slopes, such that the median response was less than 1% for FVC and FEV_1. Dockery *et al.* also noted that the maximum responses were observed 1 to 2 weeks after the episode, which suggests the need for a conventional time-series analysis approach to the problem. (I confirmed this finding using the tabulated group-mean averages). As a practical matter, it is diffficult to conceive of obtaining a daily time series of sufficient length on a large enough number of subjects. Lawther *et al.* (1974a,b,c; 1977) obtained such a time series for a few individuals but made no mention of any lag effects.

Vedal *et al.* (1987) reported on daily peak flow measurement on elementary schoolchildren in the Chestnut Ridge area of Pennsylvania for 9 weeks during the 1980–81 school year. Three cohorts were defined based on presence of persistent wheeze, cough or phlegm production, or neither. The air pollutants used in the study were maximum hourly concentrations of SO_2, NO_2, O_3, and COH; none was significantly associated with peak flow for any of the cohorts or with the pooled group of children. The authors cautioned that the negative findings could have resulted from the relatively low air pollution levels and from exposure misclassification, and I would add that use of the peak hourly readings may have been a contributing factor.

Borgers (1987) reported on a time-series study of three separate cohorts in West Berlin during the winter of 1982–83: 42 postmen, who worked in the most polluted part of the city; 38 fourth-grade schoolchildren; and 48 asthmatic outpatients of hospital clinic. The period of study was marked by high SO_2 concentrations (up to 489 $\mu g/m^3$), but other pollutants were not especially high (suspended particulate matter $<168\ \mu g/m^3$), and no "smog episodes" were declared during the study. The cohorts were each divided into an "exploratory" subset (38%) and a "confirmatory" subset (62%). Over 200 statistical models were evaluated, but since the confirmatory group never gave statistically significant results, Borgers concluded that the "overall interpretation was negative."

Gong (1987) studied 83 asthmatics in an area of Los Angeles where maximum hourly ozone concentrations exceeded 0.3 ppm on many occasions. Subjects kept records of respiratory symptoms and recorded their own peak flow rates in the morning and evening of each day for 230 days. Pollutants considered in this study included O_3, TSP, CO, SO_2, NO_2, NO_x, SO_4^{-2}, and hydrocarbons; in addition, a number of aeroallergens were monitored. There was no consistent statistically significant effect of any pollutant or aeroallergen on the total group; however, the responses among the group varied substantially, and statistically significant responses to ozone were shown for a subset of 63 subjects. Furthermore, eight subjects showed "clinically significant" (decrements greater than 5%) responses to ozone. Subsets of responders were not identified for any other pollutant.

Kinney *et al.* (1989) took weekly measurements of lung function of 154 schoolchildren in eastern Tennessee during February, March, and April of 1981. Their design employed repeated measures, which allows each child to serve as his or her own control. They found negative associations between FVC, $FEV_{0.75}$, MMEF, $V_{max0.75}$ and ozone (peak hour on the preceding day) and temperature, but not with particulate or sulfate levels. The maximum ozone level was about 0.08 ppm during this period; TSP data were incomplete, but the maximum value was about 72 $\mu g/m^3$, with the highest value during the coldest period. The ozone elasticities were about −0.02 for FVC, −0.028 for $FEV_{0.75}$, −0.043 for MMEF, and −0.09 for $V_{max0.75}$. There may have been some ambiguity in this study as to whether the O_3 associations were confounded by either temperature or seasonality (or both). I normalized the lung function measures with respect to their February values and regressed them against ozone, temperature, and a linear time variable (with dummy variables representing the different lung function measures). Neither time nor temperature was significant, which suggests that the ozone associations are real. A plot of the aggregated data is given in Figure 11-2 and shows a monotonic negative response to ozone for all four lung function measures.

Higgins *et al.* (1990) tested a group of 43 Los Angeles children, aged 7–13, attending a summer camp in the San Bernadino Mountains, over a period of 3 weeks. Ozone levels up to 0.25 ppm occurred there late in the day. Levels of other air pollutants, including PM_{10}, were unremarkable. Hourly ozone concentrations were used as a correlating parameter (measured 2 hours before the lung function test), and regression coefficients were computed for each child. The average elasticities were about −0.02 for both FEV_1 and FVC; the daily changes were reversible, and there was a great deal of variation among subjects, including a few individuals with positive responses. Although the ozone data were collinear with time of day, the authors argued that the effect was more likely due to ozone than to diurnal variations, in part because an analysis stratified by time of day showed that ozone remained a significant predictor for the high-ozone period. They also argued that Kerr (1973) had shown that "if ambient conditions are held constant, there is no diurnal variation in pulmonary function spirometry," while my reading of Kerr comes to

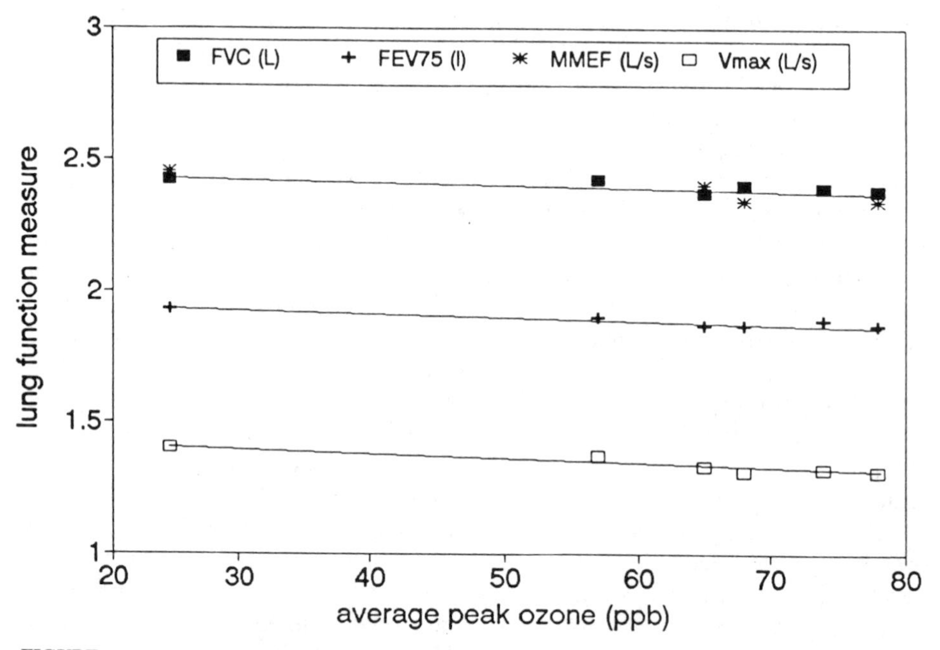

FIGURE 11-2. Children's lung function data from Eastern Tennessee, plotted vs. peak ozone. *Data from Kinney* et al. *(1989)*.

the opposite conclusion. In a multiple regression model, both fine and coarse particles were also significant predictors of the average slopes for FVC and peak flow, but the effects were positive (i.e., beneficial). It is thus difficult to accept the findings of Higgins *et al.*, especially since the tabulated data which were presented, aggregated by ranges of ozone level, do not comprise a monotonic dose-response relationship.

Berry *et al.* (1991) tested 14 children and 20 counselors at two summer camps in suburban New Jersey in 1988, a year noted for high ozone levels. They reported that only peak flow was associated (negatively) with ozone, based on 8-hour averages ($p = 0.05$). I estimated the elasticity at about -0.08. Only the children displayed a response to O_3, and the maximum response occurred when the ozone measurement was based on the 8 hours preceding the test. Two- and three-day lags displayed responses about half as large. A meta-analysis of four similar summer camp studies involving children's responses to ozone was presented by Kinney *et al.* (1988). They showed consistency of results among the four studies and also discussed the importance of considering activity patterns (exercise), which increases the dose of air pollution delivered to the lung.

The effects of wood stove air pollution on elementary schoolchildren in Oregon were studied by Heumann *et al.* (1991) during the winter of 1989–90. Two schools were selected in an area of "high" air pollution (24-hour $PM_{10} < 250\,\mu g/m^3$) and one in an area of lower PM_{10} values ($<150\,\mu g/m^3$). Demographic and social characteristics of the two areas were similar, except that there was also more tobacco exposure in the high-PM_{10} area. About 70% of the surveyed homes used wood stoves in both areas; the higher ambient (outdoor) air pollution was due to differences in topographic relief. No indoor air pollution measurements were reported. The lung function measure used was the percentage of the expected FEV_1 values, based on the height, age, and sex of each child. Spirometry was performed before, during, and after the heating season. The results showed a significantly larger decline in FEV_1 in the polluted area during the heating season (about 2%), which continued after the heating season (a further decline of 1.4% relative to the low pollution area). The last spirometry measurements were made in May. The children in the high-exposure area also had a slightly lower average baseline lung function value. In a multiple regression, outdoor air pollution exposure and use of a wood stove were associated with declines in FEV_1 during the heating season; exposure to tobacco smoke and parents' education or income were not. Thus it appears that the study assigned reduced FEV_1 to both indoor and outdoor air pollution.

Further studies on wood smoke were reported by Koenig *et al.* (1991), who tested Seattle elementary schoolchildren over two heating seasons. Spirometry was performed for each child on four occasions and compared to outdoor air pollution measurements (nephelometry) on the day before. Average PM_{10} levels were around $40\,\mu g/m^3$. Although it is difficult to estimate elasticities directly from the data presented by Koenig *et al.*, it appears that the values are around -0.07 for asthmatic children and around -0.02 for the normal children, for FVC and FEV_1. However, a subsequent analysis of these data (Koenig, 1992) found that the regressions were confounded by temperature, and, after this was taken into account, only the asthmatic children were affected by smoke. The revised elasticities for FEV_1 and FVC were about -0.025. Although data on use of wood stoves at home were obtained, the significance of this factor was not reported by Koenig *et al.*

Monthly spirometry data for 600 six-year-old Brazilian children were reported by Spektor *et al.* (1991). These children lived and attended school in an area of heavy industrial air pollution, and previous analyses had shown that a large proportion were classified as having "abnormal" spirometry. The annual PM_{10} values of five of the six school-sampling sites were in the range of about 45 to 90 $\mu g/m^3$; the sixth site had a value of about 240 $\mu g/m^3$. Average fine particle concentrations were more tightly grouped in the range of about 25 to 45 $\mu g/m^3$. All six schools showed statistically significant negative correlations between PM_{10} and peak flow and $FEF_{25-75\%}$; five of the six had significant negative relationships with FEV_1; none had a significant relationship with FVC. The authors reported that the slopes were "similar for the less exposed and more heavily exposed regions." No mention was made of seasonal factors or adjustments; the data were collected during the school year, and one might expect the normal seasonal trend to provide poorer lung function in winter, which was also the time that thermal inversions were reported. Also, the previous lung function surveys found more "abnormalities" in the winter of 1985 than in the summer of 1983, whereas the air appeared to be somewhat cleaner in 1985. The elasticity of the average FEV_1 relationship was about -0.025, but this estimate could be reduced if seasonal or temperature effects were taken into account simultaneously.

Because many of the studies that found short-term effects of air pollution on pulmonary function were based on repeated measures of each subject, who then serves as his own control, Brunekreef *et al.* (1991) examined data from three such studies for evidence of varying sensitivity, which could indicate the presence of sensitive subgroups. The authors concluded that heterogeneity of responses has been shown in the two studies implicating ozone, but not in the Steubenville study that implicated TSP. The implication was that the findings from the Steubenville study (Dockery *et al.*, 1982) could have resulted from normal sampling variability alone.

Pope *et al.* (1991) followed two cohorts of subjects in Utah County, Utah, who recorded symptoms and measured their own peak flows daily for 107 days during the winter of 1989–90. Their daily changes were then compared to 24-hour PM_{10} levels measured at three locations in the area, which averaged 46 $\mu g/m^3$ with a maximum of 195 $\mu g/m^3$. The two cohorts consisted of 34 schoolchildren, aged 9–11, who were selected based on responses to a questionnaire indicating propensity for wheezing or asthma, and 21 "patients", aged 8–72, who were receiving medical treatment for asthma. Twenty-nine percent of the first group and 90% of the second group reported using asthma medication during the study period.

Regressions were based on individual peak flow deviations from each subject's mean for the period and included variables for daily temperature and the time trend (in a personal communication, Pope reported that the results were insensitive to use of either percentage deviations or individual regression slopes). Both single-day and distributed-lag models (up to 5 days) were evaluated; for the school-based cohort, the elasticities with respect to the mean PM_{10} value were about -0.011 and -0.022, respectively, and were statistically significant. The elasticities were somewhat lower for the patient-based cohort and their responses were shifted toward greater lags (3–4 days versus 0–2 days). This study also showed increases in respiratory symptoms (school-based cohort) and use of asthma medication (both cohorts) on days of higher PM_{10} pollution. Levels of other air pollutants were generally low: SO_2 was below 0.02 ppm; NO_2 below 0.07 ppm; H^+ below 0.5 $\mu g/m^3$ as H_2SO_4. Ozone was not measured but was believed to be low, since the study period was in winter. The results were robust against the exclusion of the only pollution episode exceeding the national PM_{10} standard level that occurred during the study period. Controls for temporal trend and low temperature had only minor effects for the school-based cohort and reduced the PM_{10} effect by about 50% in the patient-based cohort.

In a follow-up study, Pope and Dockery (1992) examined two additional cohorts of 60 schoolchildren in Utah Valley during the winter of 1990–91. One cohort was asymptomatic, based on a respiratory symptom questionnaire. The other consisted of children with symptoms who were not receiving asthma medication. It appears that the "symptomatic" cohort in the second study was similar to the "school-based" cohort in the first study. The PM_{10} levels were higher in the second study. The regression coefficients in the second study were about one-fourth of those in the first study, but statistical significance was shown for 5-day moving average PM_{10} levels for both cohorts, with a larger effect for the symptomatic cohort. Controls for temporal trend and low temperature were included in these regressions. Pope and Dockery also presented data on the distributions of individual regression coefficients; the lower quartile estimates were similar to the values from the first study.

These studies are of interest for several reasons. First, they demonstrated that respiratory responses can occur at pollution levels below the national standards. However, when expressed as percentages (elasticities), the magnitudes of the responses were low compared to mortality and hospitalization responses in the same population (see Chapters 6 and 9), even though most of the subjects were specially selected for potential susceptibility and were not chosen randomly. Finally, they showed the large degree of variability in responses to be expected, even among subjects especially selected according to well-defined criteria. This variability strongly suggests that group mean responses may be inadequate in describing the results of studies among the general population.

Lung-function responses of schoolchildren in the Netherlands were studied by Hoek *et al.* (1993a,b). In the first study, peak flow was measured in 83 exercising children aged 7 to 12. The environmental variables included ozone and ambient temperature, which were found to be highly correlated. The maximum 1-hour ozone level was 236 $\mu g/m^3$. The changes in peak flow rate during exercise were positively associated with both temperature and ozone (not significant), but the peak flow rates after training were negatively associated with ozone after adjustment for temperature in a 2-stage procedure. The elasticity of this relationship was about −0.04; however, it also failed to reach statistical significance.

A larger group ($n = 533$) of similarly aged children was studied by Hoek *et al.* (1993b), and a full range of spirometric measurements was obtained. Peak flow, FVC, FEV_1, and MMEF were all significantly negatively associated with previous-day peak ozone. The elasticity for peak flow was about −0.06 ($p < 0.0001$). The maximum 1-hour ozone level was 228 $\mu g/m^3$. In both of

TABLE 11-1 Results of Temporal Studies of Lung Function

First Author	Subjects (No. tested)	Individual Slopes	Pollutants	Findings	Lag
Motley (1959, 1964)	Emphysema patients (46); normal adults (20)	Yes	LA smog	Residual volume improved	40 hrs
Shepard (1960)	Heart/lung patients (10)	Yes	Smoke	1 patient showed changes with smoke	
Spicer (1962)	COPD patients (14)	No	TSP	Elast. R_{aw}: 0.26	None
Spodnik (1966)	Young adults (100)	No	TSP	No correlations	
Smith (1973)	Young adults (9)	Yes	Fireworks	2 CRD patients showed drop in $FEV_{25-75\%}$ of 26%	None
Emerson (1973)	COPD patients (18)	Yes	SO_2	2 patients showed FEV_1 and MEFR correlations	
Kagawa (1975)	Children (21)	Yes	O_3, etc.	2–5 subjects showed correlations with air pollution	0–1 day
Stebbings (1979)	Children (224)	Yes	Smoke	10–15% showed change in FVC, no change in $FEV_{0.75}$	Not determined
Lofstgaarden (1981)	Children (110)	No	TSP	Changes in FVC, FEV_{-1}, and $FEF_{25-75\%}$	3 days
Dockery (1982)	Children (335)	Yes	TSP, SO_2	Median change = 1% (FVC, FEV_1)	1–2 week
Johnson (1982)	Children (98)	Yes	Ash	3% decrease for boys, 0 for girls	2 days
Buist (1983, 1986)	Children (64)	Yes	Ash	No significant changes	
Dassen (1986)	Children (163)	Yes	TSP, RSP, SO_2	3–5% drop in lung function	2 weeks
Buist (1986)	Loggers (712)	No	Ash	4% change in FEV_1	1 yr (?)
Vedal (1987)	Children (144)	No	SO_2, etc.	No associations with peak flow	1 day
Borgers (1987)	Postmen (42)	Yes	SO_2	No consistent changes	
	Children (38)				
	Asthmatics (48)				
Gong (1987)	Asthmatics (83)	Yes	O_3	Consistent peak flow effects only for a subset	
Kinney (1989)	Children (154)	Yes	O_3	Elast. FVC: −0.02, $FEV_{0.75}$: −0.03, MMEF: −0.04, V_{max75}: −0.09	1 day
Higgins (1990)	Children (43)	Yes	O_3	Elast. FEV_1: −0.02, FVC: −0.02	2 hr
Berry (1991)	Children (14)	Yes	O_3	Elast. peak flow: −0.08 (children only)	8 hr
	Young adults (20)				
Heumann (1991)	Children (410)	No	PM_{10}	2% drop in FEV_1, 1.4% drop later	Months
Koenig (1991)	Children (343)	Yes	Light scat	Elast. −0.03 for asthmatics only	1 day
Pope (1991)	Sensitive children (55) (10% of total)	Yes	PM_{10}	Elast. peak flow: −0.022	5 days
Spektor (1991)	Children (600)	Yes	PM_{10}	Elast. FEV_1: −0.025	1 month
Pope (1992)	Asymptomatic children (60)	Yes	PM_{10}	Elast. peak flow: −0.006	5 days
	Symptomatic children (60)	Yes	PM_{10}	Elast. peak flow: −0.009	5 days

these studies, each child served as his or her own control and systematic differences were observed among subjects. Apparently the larger study group resulted in a higher level of statistical significance.

Summary of Short-Term Relationships between Lung Function and Air Pollution

Table 11-1 summarizes the 25 studies reviewed. Of the 17 studies involving children, 14 showed statistically significant relationships with one or more air pollutants. Two of them showed significant effects only for "sensitive" children. No commonalities were apparent among the three studies that failed to reach significance; two of them involved primarily SO_2 and the other was a study of volcanic ash that employed a unique statistical approach. The two pollutants most often implicated were ozone and various measures of particulate matter. Lag periods seemed to be longer with particulates than with ozone, suggesting different mechanisms, but few studies investigated lags systematically. A consistent pattern of increasing response with increasing lag may be an indicator of causality. One must also keep in mind that all of the effects delineated in Table 11-1 were deemed to be reversible and that publication bias may have reduced the numbers of studies that did not find effects.

Of the four studies of normal adults, only the studies of loggers exposed to volcanic ash found a statistically significant relationship for normal persons. All of the studies of adults with impaired cardiorespiratory systems showed significant adverse relationships with air pollution, although there was no agreement with regard to the specific pollutants involved and there was substantial variability within the groups studied.

Repeated Measurements over the Long Term

Some of the most remarkable sets of data on repeated lung function measurements were obtained by Lawther and his colleagues in London during the 1960s (Lawther *et al.*, 1973; 1974a,b,c; 1977). The protocol involved testing each of four workers immediately after arriving at work following a walk from the railroad station, constituting an exposure of about 30 minutes during the part of the day that usually had the worst air quality. From 1960 to 1965, spirometry and peak flow measurements were made; following a pilot effort between 1964 and 1965, airway resistance measurements were obtained on three of the subjects from 1965 to 1971. This decade saw dramatic improvements in smoke levels in London, with smaller reductions in SO_2. For part of this period, daily aerosol acidity measurements were also made. In most of their analyses, Lawther *et al.* used the seven-station network daily averages of air quality. During periods of intense air pollution, such as the December 1962 episode, local hourly measurements were also used.

Several different types of analyses have been performed with these data. The spirometry values (Lawther *et al.*, 1974a) were compared with predicted values based on age and cross-sectional data from nonsmokers in London. Two of the four subjects exhibited growth in FEV_1 rather than decline, with age, during the 1961–65 period. The results for 1969 were more in line with predictions. One of the four subjects had suffered from childhood asthma and

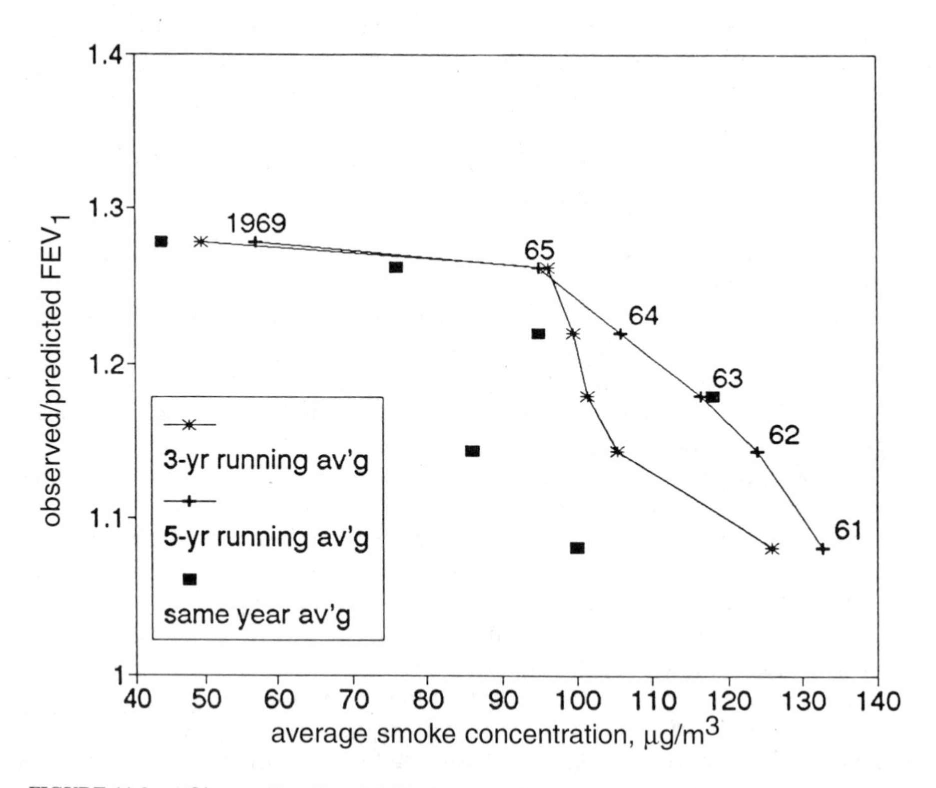

FIGURE 11-3. Observed/predicted FEV_1 for one subject, London, 1961–1969. *Data from Lawther* et al. *(1974)*.

seemed more sensitive to air pollution (as well as to medication); his long-term changes in FEV_1 relative to age-based predictions are plotted against various long-term measures of smoke in Figure 11-3. The three plots are: same-year smoke (a scatterplot of the dose-response relationship); the average of the same year and the 2 years preceding spirometry; and a 5-year running mean similarly computed (since smoke levels were improving, time runs from right to left on this plot). The latter two plots show a sharp break after 1965, when the daily measurements were interrupted. There are several possible explanations, including either a threshold in the effects of air pollution (similar to the findings of Schwartz, 1989) or a "training effect" due to the repeated spirometry measurements *per se*. However, as pointed out by Lawther *et al.* (1974a), it is difficult to imagine a training effect persisting for 5 years. There were parallel changes in other air pollutants during this period as well; all of these factors may explain the large elasticity implied by Figure 11-3 (about −0.21), in addition to the apparent heightened sensitivity of the individual involved.

The FVC measurements were less variable in the long term; the average change for the four subjects was a 1.5% reduction, compared with a 4.4% gain in FEV_1. The gain in MMEF averaged 18.6%, in part because of a 54% gain by the sensitive individual. The presence of respiratory infections was seen to exert large and relatively long-lasting effects on lung function.

Because of its apparent sensitivity, MMEF was used to analyze environmental influences on the short-term variations in spirometry. Three of the four subjects showed significant (negative) correlations between MMEF and temperature, smoke, and SO_2. Logarithmic transforms were used for the pollutant variables (because of the "log-normal distributions of these variables"). Most of the FVC correlations were positive; the sensitive subject showed negative correlations with FEV_1 and smoke and SO_2, but not temperature. Figure 11-4a presents dose-response plots for the average of three subjects for whom Lawther *et al.* (1974a) presented complete data. The logarithmic nature of the response is apparent; a unit of air pollution apparently has more of an effect at low concentrations than at high levels. This type of response was seen for two of the four subjects. The elasticities of these relationships (evaluated at the mean) were −0.03 for smoke and −0.022 for SO_2. Note that the relationships for smoke and SO_2 shown in Figure 11-4a are not independent, since high concentrations of these two pollutants tend to occur simultaneously in London. These changes in lung function were transitory and reversible. Lawther *et al.* (1977) also reported that regression models with "net particulate acid" were inferior to those employing SO_2 and relative humidity. The period of investigation includes the 1962 fog episode, during which hourly SO_2 reached 2 ppm and H^+ reached 678 $\mu g/m^3$ as H_2SO_4. Waller, who was the "sensitive" subject in the group, reported his decrement in peak flow (4%) after running for 25 minutes (outdoors) on this occasion.

The peak flow data (Lawther *et al.*, 1974b) told much the same story, except that the responses of the four subjects were all basically similar. From 1960 to 1965, the average peak flow increased by about 9%; from 1965 to 1971, it declined by about 2%. These changes are consistent with the spirometry results and support the hypothesis of an initial improvement due to the decline in smoke concentrations, followed by a period of normal decline due to aging. The subjects' ages in 1960 ranged from 30 to 41 years. The elasticity against 5-year running mean smoke was about −0.04; however, since the 1970 ratio of observed to predicted values declined in spite of improving air quality, the regression was not statistically significant. The mean seasonal effect was about 1.5%. The short-term peak flow responses were similar to those for MMEF (Figure 11-4b). The elasticities for smoke and SO_2 were −0.0135 and −0.011, respectively. Lawther *et al.* used multiple regression analysis to try to deduce which of the environmental variables (smoke, SO_2, temperature) had the main influence on peak flow varability, compared to the temporal effect alone. They concluded that time *per se* was the most important factor (rather than smoke) in addition to the logarithm of SO_2, but offered no hypotheses as to why peak flow should *in*crease as the subjects aged. A possible explanation for the failure of the smoke variable to outperform the time variable could be the use of the seven-site smoke concentration averages, rather than smoke values corresponding to the subjects' actual exposure. The SO_2 effect was estimated to be −1.9% for a concentration of 1000 $\mu g/m^3$ (24-hour average). In separate experiments during periods of intense air pollution, it was shown that exercise (walking) was required to affect peak flow readings.

In the third paper of this series (Lawther *et al.*, 1974c), the normal diurnal cycle in peak flow (increasing toward midday) was contrasted with the transient effects of air pollution. A similar SO_2 response (1.5% drop for an SO_2 increase of 1000 $\mu g/m^3$) was shown for one subject. Series of peak flow measurements

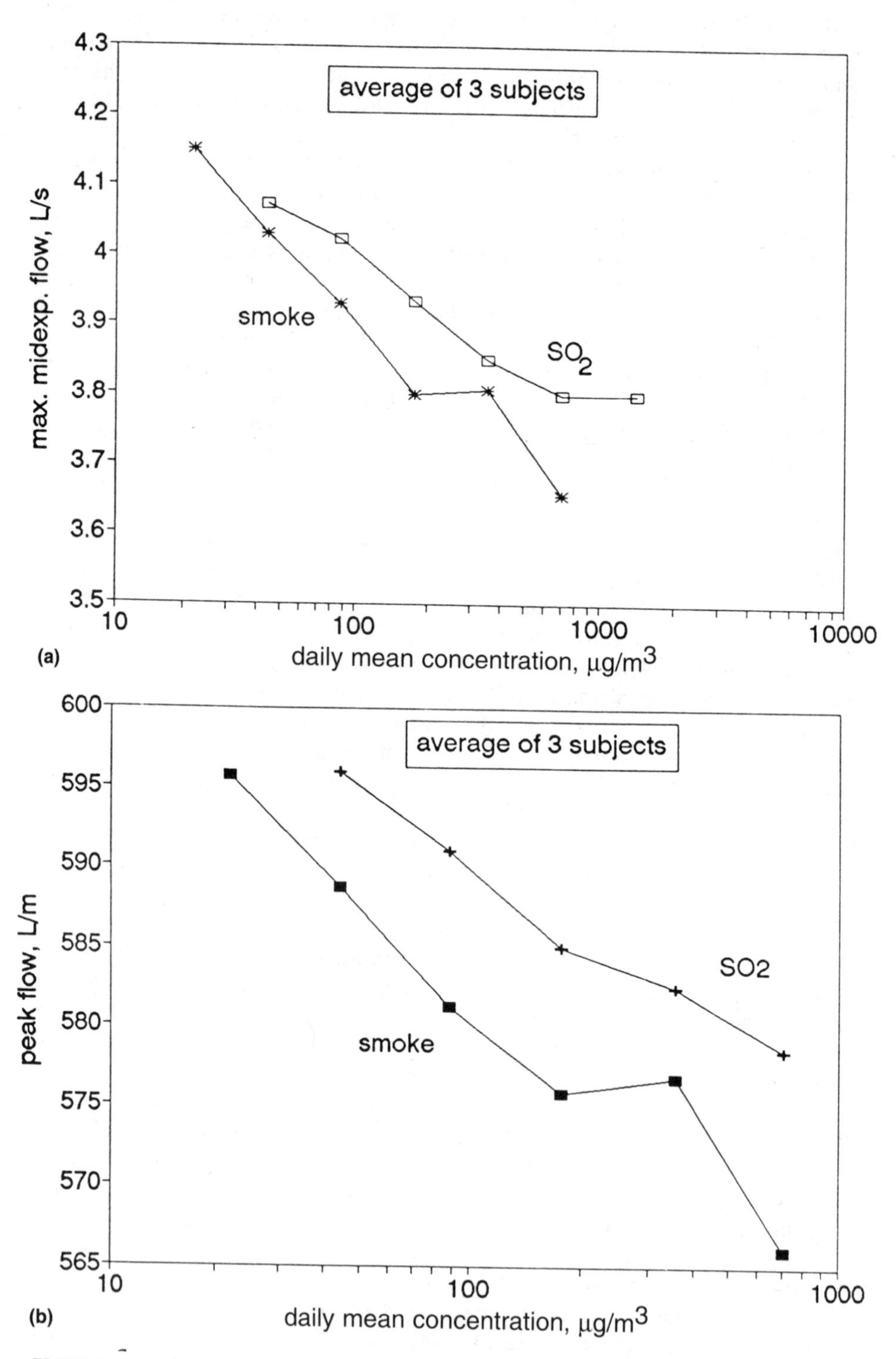

FIGURE 11-4. Lung function data for three subjects, London, 1960–65. (a) Maximum mid-expiratory flow (b) Peak flow. *Data from Lawther* et al. *(1974a, b)*.

were also obtained for two bronchitic patients; one of them showed a significant response to daily SO_2 fluctuations.

Airway resistance was found to be a sensitive indicator of air pollution effects on lung function and was used in two separate studies by Lawther *et al.* (1975, 1977). The 1977 paper was essentially an extension of the earlier spirometry studies, in which three staff members were tested upon arrival at work each day, from 1965 to 1971. Correlations were presented with temperature, relative humidity, and the logarithms of SO_2, smoke, and aerosol acidity. Although the correlations were weak (other factors, such as the presence of respiratory infections, were more important), two subjects showed significant positive correlations of R_{aw} with air pollution. Sulfur dioxide was judged to be the most important pollutant. The third subject had about half as many observations and showed nonsignificant negative correlations. A regression equation was devised for one subject,

$$SR_{aw} = 0.432 \ln SO_2 + 0.016\, RH - 1.096. \qquad [11\text{-}1]$$

The elasticity at the mean for this expression is 0.18.

The other study of airway resistance (Lawther *et al.*, 1975) involved laboratory exposures to SO_2 for up to 25 staff members, from 5 to 30 ppm. Their responses varied widely, by at least one order of magnitude. The pattern of response during repeated breathing of elevated SO_2 was one of a response plateau, from which R_{aw} eventually subsided back to the pre-exposure level, while exposure continued (even for the most sensitive subject). This pattern of apparent saturation is seen in the dose-response functions (Figure 11-5a,b), which are based on a maximum of 14 subjects. Note that the group average response approximates a straight line in log-log coordinates when the most sensitive subject is included (Figure 11-5b).

The results from the two studies (1975 and 1977) are compared in Figure 11-6 for one subject, in which the regression equation is used to represent the results of outdoor exposures, which can then be compared to the laboratory results at much higher SO_2 concentrations. The comparison is based on the percentage increase over baseline SR_{aw}, in order to remove the effect of different baselines in the two studies. Outdoor exposure is seen to create a much larger response; this may be due to the concurrent effects of temperature or the presence of other pollutants, notably smoke. (Particulate matter has been shown in animal tests to enhance the adverse effects of breathing SO_2 [Amdur, 1957].)

To summarize, the experiments of Lawther and his colleagues offer dramatic demonstration of the benefits of improved air quality, when measured by the parameters of respiratory mechanics. The actual health benefit to the individual will depend on his overall health status. The finding of a logarithmic dose-response function is consistent with the some of the analyses of air pollution on daily mortality in London (Chapter 6). This has not been observed elsewhere, but few other studies have sufficient range in their exposures to allow distinction between linear and logarithmic models. Other important lessons from these experiments include the large range in individual responses and the significant differences between responses to laboratory exposures and responses

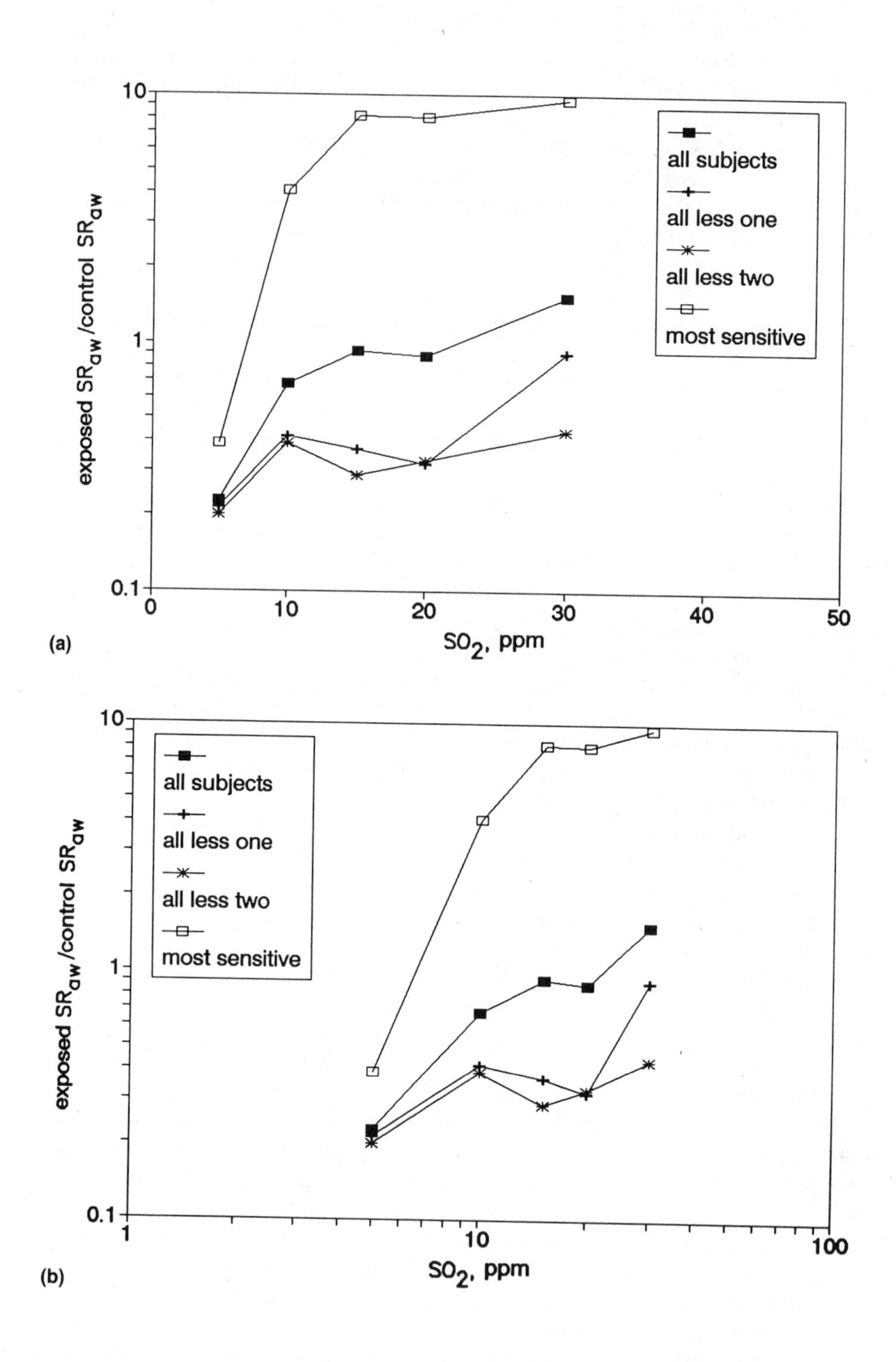

FIGURE 11-5. Airway resistance changes due to laboratory exposures to SO_2. (a) Linear SO_2 scale. (b) Logarithmic SO_2 scale. *Data from Lawther* et al. *(1975)*.

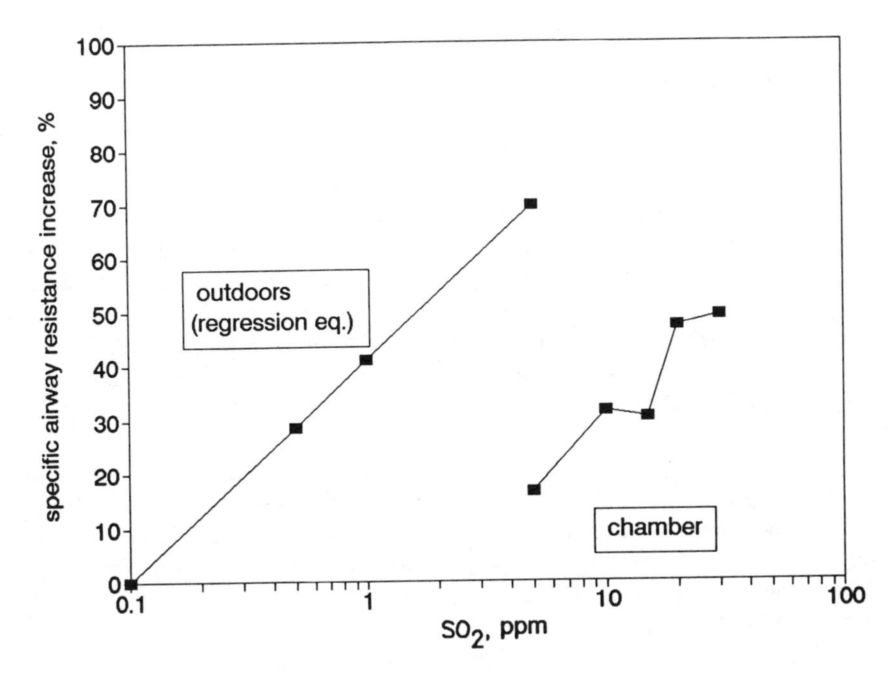

FIGURE 11-6. Comparison of effects on airway resistance of outdoor and laboratory SO_2 exposures. *Data from Lawther* et al. *(1975, 1977).*

to outdoor exposures. Both of these findings urge caution in extrapolating the results of exposing selected volunteers in laboratory settings to the general public at large.

COMPARISON OF THE EFFECTS OF DIFFERENT POLLUTANTS BASED ON LABORATORY STUDIES

The studies reviewed do not provide a clear picture of the relative importance of the different constituents of community air pollution, since mixtures were involved in most cases and one can find both significant and nonsignificant results for most of the pollutants. Laboratory exposure tests on volunteers may be useful in this regard, of which there are many in the literature, with both "normal" subjects and those with respiratory impairments such as asthma.

A useful comparison of several pollutants and combinations was performed by Stacy *et al.* (1983), by evaluating the responses of groups of 9 to 15 normal males, ages 18–40, to a variety of aerosols and gases, alone and in combination. The pollutants and concentrations used were:

clean air	
O_3	0.4 ppm
SO_2	0.75 ppm
NO_2	0.5 ppm

H_2SO_4	100 μg/m^3
NH_4HSO_4	116 μg/m^3
$(NH_4)_2SO_4$	133 μg/m^3
NH_4NO_3	80 μg/m^3

The exposure time was 4 hours and included exercise; thus the concentration levels approximated the upper end of the doses that might be expected in ambient air for a 24-hour period. Pulmonary measurements were made before exposure, after 2 hours, 4 hours, and 24 hours. A total of 19 different parameters were examined based on the spirometry; the only statistically significant responses observed were for ozone or ozone plus aerosol. None of the other gases or aerosols alone provided a significant response (group averages). Note that no carbonaceous or insoluble particles were evaluated. The authors concluded that SR_{aw}, FVC, FEV_1, and FEF_{50} were the most useful lung function metrics.

Other laboratory exposure tests have shown lung function responses of around 0.75 ppm SO_2 for exercising asthmatics (Colucci and Strieter, 1983), although the mode of administration (oral versus nasal breathing) has been shown to make a difference in response. Normal subjects are not affected by SO_2 at this level. Utell *et al.* (1990) report that 0.3 ppm for 4 hours is the minimum NO_2 dose required to elicit responses from COPD patients. Anderson *et al.* (1992) showed that neither normal nor asthmatic subjects (group averages) responded to 1-hour exposures of 200 μg/m^3 black carbon plus about 50 μg/m^3 of sulfate, with or without 100 μg/m^3 H_2SO_4. However, exposure-related decrements in lung function have been shown for occupational exposures to industrial dusts, such as coal, asbestos cement, or silica (Moore *et al.*, 1988). DuBois and Dautrebande (1958) showed reversible changes in airway resistance (200–300%) due to a few breaths of high concentrations (25–50 mg/m^3) of inert dusts ($CaCO_3$, coal dust, aluminum powder, and aerosolized India ink).

STUDIES OF LUNG FUNCTION RESPONSES TO CHRONIC AIR POLLUTION EXPOSURE

The classic method for studying the responses to chronic air pollution exposure is that of cross-sectional comparisons, as discussed in Chapters 3 and 7. There have also been a few longitudinal studies, including some involving changes due to air pollution abatement. The studies we will review are of three general types: (1) prospective, large-scale studies, covering several different countries or major sections of one country; (2) observational studies taking advantage of existing (national-level) data bases; and (3) small-scale studies involving sections of a city, metropolitan area, or state. The objective of this review is to search for consistency in both the direction and the magnitudes of the effects that have been found by these studies on various measures of lung function in response to various air pollutants.

One of the most common methodologies in the literature is the community equivalent of the case-control study, in which a polluted city or neighborhood (the case) is compared with an unpolluted area (the control). However, unlike conventional case-control studies, in which a group of individual controls can be matched with a group of individual cases, there are often fundamental differences between communities that stem directly from their differences in air pollution (presence of industry, population density, differences in socio-economic class, for example). Furthermore, a publication bias may exist with respect to the distribution of the studies that have been accepted for publication; it is difficult to conceive of a journal accepting articles in which health differences were *not* found in such community comparisons, although such findings might be equally likely. Two-community comparisons are thus of limited usefulness for the purposes of this chapter, unless the characteristics of individual subjects are considered.

The studies we review each involved at least three different locations and cover the United States (in one national study and six regional-local studies) and Western Europe. Details of the major studies are given in Table 11-2. Some of these studies lacked statistical power as presented by the original authors, for example, because separate results were presented by sex, race, smoking status, and lung function measure. This mode of data presentation is often useful for diagnostic purposes, but to derive more precise estimates of dose-response functions (i.e., the elasticity of lung function on pollution), pooling of groups can be helpful. My use of this methodology, however inelegant, has expanded the findings of the original authors in some cases. Such differences, if any, are delineated in conjunction with the reviews of each study which follow.

International Comparisons

The data of Densen *et al.* (1965) allow some comparisons to be made between various parts of the United States and England, for white males, not stratified by smoking (Figure 11-7). A reference plot of data from the first U.S. National Health and Nutrition Examination Survey (all parts of the United States, 1971–75) has been added (O'Brien and Drizd, 1981). The other plots are for New York City postmen, circa 1963; U.S. telephone workers from various parts of the country, circa 1960; postal truck drivers from rural England and London, circa 1960. More detail on the English data, which were consistent across all smoking categories, are available in Holland and Reid (1965). The effect of London residence on FEV_1 was comparable to the effect of smoking, and the urban-rural gradient was substantially worse for smokers. With the exception of the London truck drivers, the differences in FEV_1 are most pronounced for middle-aged men; the remaining plots appear to converge at age 60. In the middle-age range, the differences in FEV_1 appear to be ranked roughly in inverse proportion to air quality. Since New York had high SO_2 levels in the 1960s (and presumably before) but modest particulate levels compared to London, it is tempting to ascribe the differences in FEV_1 shown in Figure 11-7 to particulates.

The European Commission Epidemiological Survey

The Economic Community (EC) survey (Florey *et al.*, 1983) was conducted in 1975 in 19 geographical areas of six countries: Belgium, France, Germany (FRG), Ireland, Italy, and the United Kingdom. It involved children aged

TABLE 11-2 Summary of Studies of Lung Function and Chronic Air Pollution Exposure

Reference	Locations	Time Period	Subjects race, age	Subjects number	Subjects covariates	Pollutants (means, ranges)	LF metric	Pollutants elasticity	Pollutants signif. level	remarks
Chestnut *et al.* (1991)	USA (49 cities)	1971–75	all, 25–75	935	height, age, sex, smoking, obesity, occup. exp., temp., region	TSP (87)	FVC FEV_1	−0.03 to −0.056	0.001	
Schwartz (1989)	USA (44 cities)	1976–80	all, 6–24	1894 (TSP) 535 (NO_2) 1005 (O_3) 832 (SO_2)	height, age, sex, smoking, neighborhood, lung disease	TSP (62) NO_2 (62) O_3 (63) SO_2 (34)	FVC	−0.037 −0.097 −0.102 NS	0.0074 0.0332 0.0001	pollution values taken from sites within 10 mi of residence
						TSP (62) NO_2 (62) O_3 (63) SO_2 (34)	FEV_1	−0.030 −0.102 −0.103 NS	0.0004 0.0003 0.0019	
						TSP (62) NO_2 (62) O_3 (63) SO_2 (34)	PF (V_{max})	−0.070 −0.106 −0.133 NS	0.001 0.001 0.001	
Rokaw *et al.* (1980)	Los Angeles area Lancaster Burbank Long Beach	1973–72	(white, Anglo) 18–59	2433 1894 2060	age, sex, ht., wt., smoking, symptoms	av'g daily max Ox (76–228) NO_2 (60–269) SO_2 (26–134)	FVC FEV_1 $FEF_{25-75\%}$ $\dot{V}_{max}$	for pooled V_{max}, $V_{25\%}$, never smokers TSP = −0.20 NO_2 = −0.096	 0.005 0.008	data pooled from both studies
Detels *et al.* (1981)	Lancaster Glendora	1977–78	(white, Anglo) 25–59	3192 2369		geom. mean TSP (76–133)	$\dot{V}_{25\%}$ $\dot{V}_{50\%}$ $\dot{V}_{75\%}$ delta N_2	whole sample, pooled TSP = −0.19 NO_2 = −0.079 SO_2 = −0.055 Ox = −0.06	 <0.001 <0.001 0.005 0.1	

MAPSS (1980)	5 Montana cities	1977–78	white 8–10	approx. 300–400 each	ht., wt., altitude		girls, SO_2			av'g of 3 seasonal tests
						TSP (73)	FVC	−0.009	0.004	
						FP (17.2)	FEV_1	−0.012	0.004	
						SO_2 (26)	$FEF_{25-75\%}$	−0.024	0.005	
						NO_2 (26)	girls, ozone			(n = 4)
						O_3 (43)	FVC	−0.040	0.02	
							FEV_1	−0.053	0.02	
							$FEF_{25-75\%}$	−0.104	0.02	
							boys, FP			
							FVC	−0.022	0.1	
							FEV_1	−0.027	0.025	
							$FEF_{25-75\%}$	−0.058	NS	
PAARC (1982)	6 French cities (20 areas)	1974–76	white, 6–10, 25–59	19191, 2527, 16664	height, age, smoking, social class, occupation exposure	SO_2a (13–127), SO_2 (22–85)	FVC, FEV_1	for FEV_1, SO_2		multivariate analysis not done; + NO_2 result depends on Toulouse; dust & smoke highly correlated but NS on FEV_1; MMFR results reported to be the same (no data); some pollution values from prior yrs
						smoke (18–152)	men	−0.040	0.02	
						dust (75–422)	women	−0.081	0.001	
						NO_2 (12–61)	children	−0.087	<0.001	
						SO_4 (7–12)	pooled w/dummies	−0.036	0.003	
Florey *et al*. (1983)	19 areas (6 countries)	1975	6–11	22337		SO_2 (19–326)	PEFR	NS		results depend on dummy vars. for country; adjustments made to pollution data
						smoke (5–57)		−0.016	0.07	
Spinaci *et al*. (1985)	Turin, Italy (3 areas)	1980–81	white, 12	2385	age, height, smoking, SES, smoking, cent. heating	SO_2 (58–204)	FVC	NS		
						TSP (108–148)	FEV_1	NS		
							$FEF_{25-75\%}$	NS		
							$V_{max50\%}$	NS		
							pooled	SO_2 = −0.04	0.06	pooled FEV_1, V_{mx}, FEF, w/dummy vars
								TSP = −0.13	0.06	

TABLE 11-2 *Continued*

Reference	Locations	Time Period	Subjects race, age	number	covariates	Pollutants (means, ranges)	LF metric	elasticity	signif. level	remarks
Dockery	6 U.S. cities	1974–81	white	7834	age, ht.,	TSP (34–80)	FVC		NS	
et al.			6–10		parent ed.,	PM15 (20–60)	$FEV_{0.75}$		NS	
(1989)					mother smok.,	PM2.5 (12–37)	FEV_1		NS	
					gas stoves	SO_4 (3–14)	MMEF		NS	
						SO_2 (9–73)	pooled	TSP = −0.035	0.025	pooled FEV_1 & $FEV_{0.75}$
Dockery	6 U.S. cities	1974–81	white	2454	exam season,					
et al.			25–74		age, ht.,	NO_2 (12–43)	FVC		NS	
(1985)					symptoms	O_3 (35–74)	FEV_1		NS	
						H^+ (0.5–1.8)	pooled	TSP = −0.064	0.025	pooled residuals
Dodge	5 AZ smelter	1979–82	white,	780			for TSP	Anglo / Mexican–American		
(1980)	towns		8–10							
Dodge			Mex-Am,							
(1983)			8–10			SO_2 (5–84)	FVC	−0.080 / −0.19		
Dodge						TSP (37–76)	FEV_1	−0.080 / −0.18		
et al.							$\dot{V}_{max50}$	−0.010 / −0.17		
(1985)							$\dot{V}_{max75}$	−0.060 / −0.11		
							(based on 2 groups of towns)			
						SO_2 (5–102)	FEV_1	SO_2 = NS		
						TSP (35–60)		TSP = NS		
							based on 4 yrs, 4 towns			
Shenker	rural PA	1979	white	3175	sex, age, ht.,	SO_2 (66–91)	FVC		NS	other pollutants not analyzed
et al.	(3 areas)		5–14		SES, gas stove,		$FEV_{0.75}$		NS	
(1986)					parent smoking,		FEF_{25-75}		NS	
					asthma, early		$\dot{V}_{max75}$		NS	
					illness		$\dot{V}_{max90}$		NS	
Shy *et al.*	Chattanooga,	1968–69	white (?)	987	sex, height	NO_2 (81–207)	$FEV_{0.75}$	NS		pooled sexes and seasons
(1970)	TN (4 areas)		7–8			NO_3 (1.6–7.2)		NS		
						SO_4 (10–13)		−0.100	0.15	
						TSP (62–99)		−0.046	0.05	
						COH (0.8–2.1)		NS		

Notes: pollutant concentrations are in μg/m^3.
SO_2a = acidimetric method
FP = fine particles (<2.5 μm)
NS = not significant

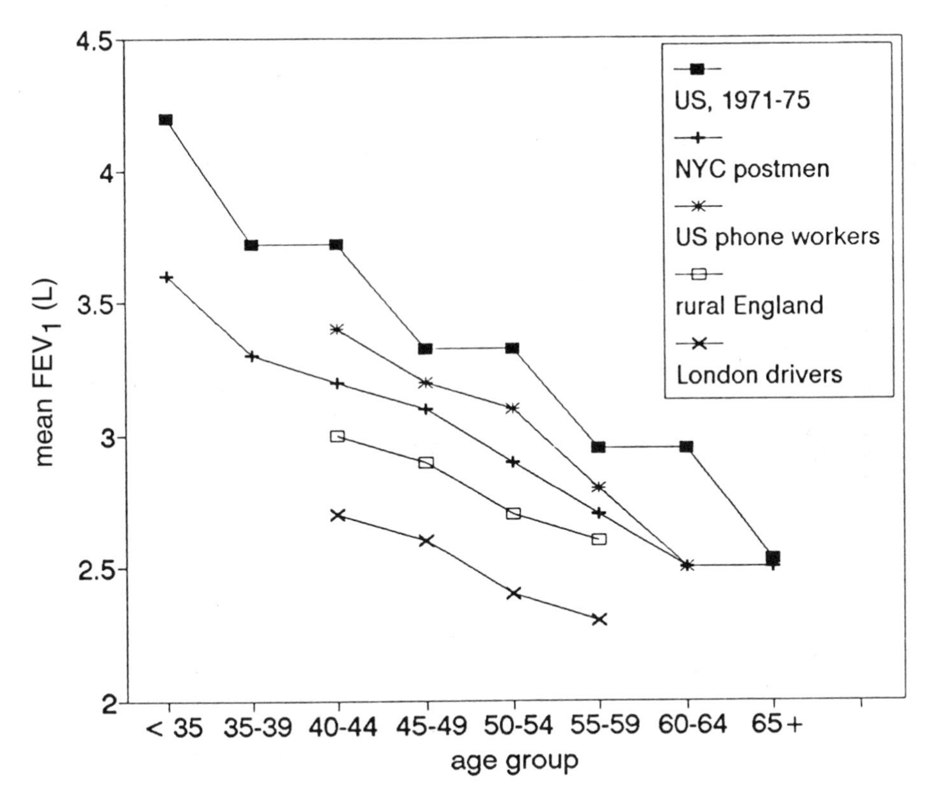

FIGURE 11-7. Comparison of white male FEV_2 values in different environmental settings. *Data from O'Brien (1981) and from Densen* et al. *(1965).*

6–11, who responded to a questionnaire on respiratory symptoms and provided spirometry data (peak flow only). The measurements were obtained between April and July 1975, so that seasonal confounding would not seem to be very important. The cross-cultural problems involved with the study design were recognized, and efforts were made to standardize all the measurements. In particular, different types of air quality measurements were made in the different areas (SO_2 and black smoke only), and correction factors were derived to adjust all data to a common basis.

Within each of the six countries, statistically significant positive and negative effects of SO_2 and black smoke were noted. The authors felt that cross-sectional studies of this type are fundamentally flawed because of the difficulty of accounting for cross-national differences unrelated to air pollution; they also expressed the opinion that consistent associations are only likely to be found at annual average levels of smoke above 140 $\mu g/m^3$ in the presence of SO_2 levels above 180 $\mu g/m^3$. The original data are plotted in Figure 11-8a; it appears that any effects of smoke are confined to within-country relationships.

In my reanalysis of these data, I pooled the data for boys and girls and included dummy variables for each country. The smoke variable nearly reached significance ($p = 0.07$); the elasticity was -0.016. Adding a variable for the latitude of each location (which was nearly significant and was intended to represent climatic differences) increased the smoke elasticity to -0.02 ($p <$

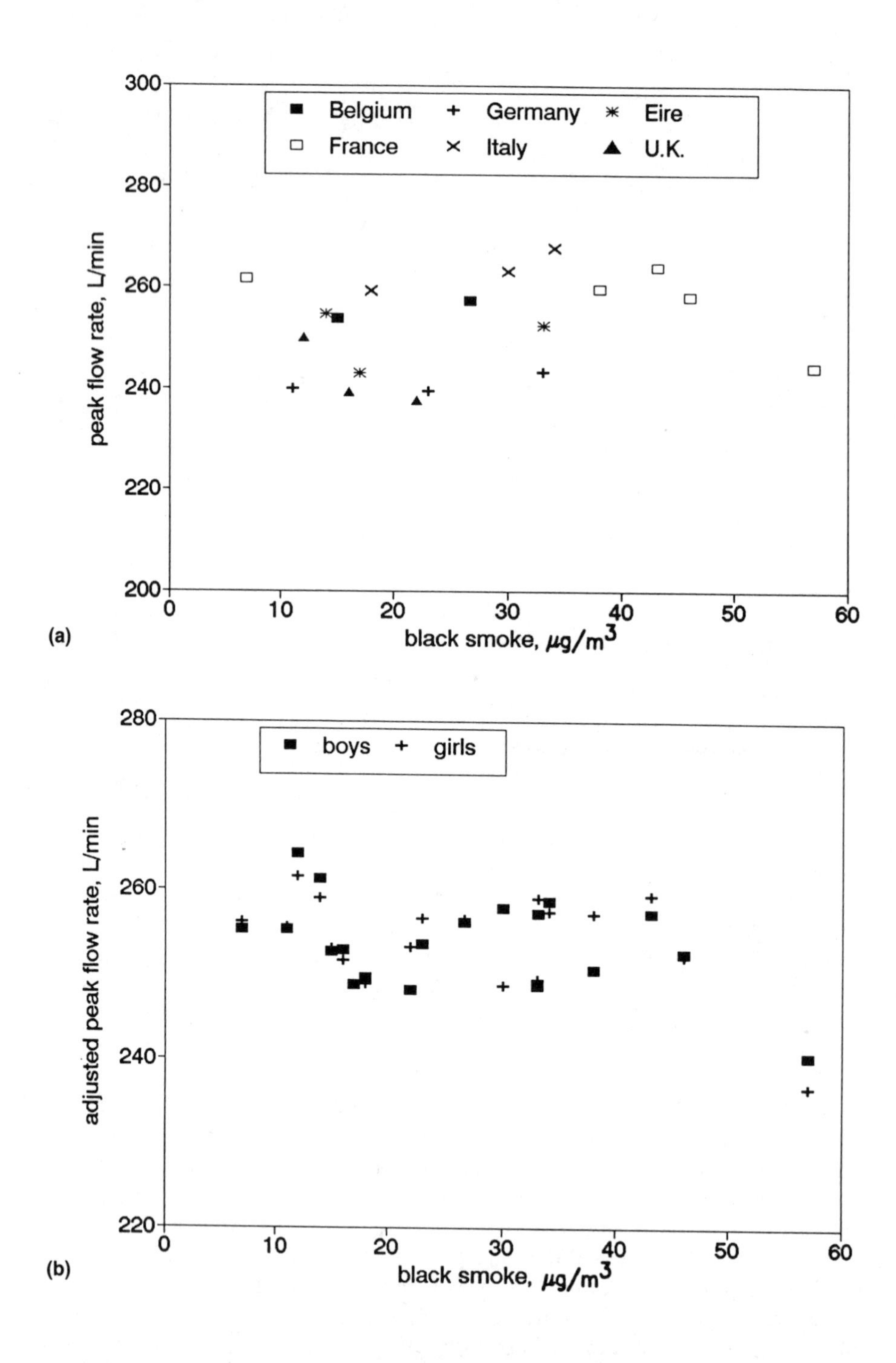

FIGURE 11-8. Children's (average of boys and girls) lung function data from the European Community Survey. (a) Peak flow versus black smoke, as observed. (b) Peak flow, adjusted for national differences, versus black smoke. *Data from Florey* et al. *(1983).*

0.03). All of the country dummy variables were significant, indicating the presence of national differences not accounted for by the adjustments for age and body size and not associated with either smoke or SO_2. Possible explanations include pollution measurement errors, unmeasured pollutants, or constitutional (genetic) differences by nationality. I "adjusted" the peak flow data using the dummy variables from the multiple regression; the results are shown in Figure 11-8b. It appears that the negative trend with black smoke is almost entirely determined by the data from the highest smoke city, which is Lyon, France. Since that location is not an outlier in the French national study (discussed later), there is no *a priori* reason to discount this observation. Nevertheless, one must conclude that any support from this study for a long-term effect of air pollution on children's lung function is weak, at best.

National U.S. Cross-Sectional Studies

Schwartz's (1989) observational study is in many ways a paradigm for deriving estimates of the effects of air pollution on lung function. The lung function data were obtained as part of the second NHANES, which was intended to be a random, national sample of the civilian, noninstitutionalized population of the United States. The NHANES II spirometry data were limited to children and young adults (ages 6–24). Care was taken to ensure standardization in the collection and processing of the data, which were acquired from 1976 to 1980. Schwartz based his analysis on the best trial for each subject and deleted data that did not meet spirometry standards. The lung function data (FVC, FEV_1, and PEFR) were fitted to regression models based on race, age, sex, number of cigarettes smoked per day, and several physiological measures. The residuals from these regressions were then regressed against air pollution, one pollutant at a time, as a two-stage procedure. The possibility of neighborhood clustering* was accounted for by using a nested random effects model similar to that of Ware *et al.* (1986).

Air pollution data were obtained from the EPA's SAROAD (now called AIRS, Aerometric Information Retrieval System) data base, using only those monitors that were located within 10 miles of the population centroid of the census tract of the subject's residence. This condition resulted in different numbers of acceptable cases for each pollutant considered, as shown in Table 11-2. Temporal averaging consisted of the 365 days preceding the examination. The ozone regressions were based on the annual average of daylight hours (8-hr averages). This statistic is about 50 to 100% greater than the annual average of all hours.

Schwartz found statistically significant negative associations between each of the three lung function metrics and NO_2, O_3, and TSP, but not for SO_2. (One of the problems with national data on SO_2 may be the variability among measurement methods in use, especially during this period.) The strongest associations were for ozone and FVC. Robustness of the results was

* Cohen (1980) and others have explored the extent to which a genetic deficiency contributes to impaired pulmonary function. To the extent that close relatives tend to live in the same neighborhood, community, or region, confounding could thus occur between local environmental effects and the effects of shared genetic defects.

established by reestimating the relationships with nonlinear pollution transforms, including thresholds; by excluding subjects with chronic respiratory conditions, smokers, and subjects not residing in the state of their birth; by including factors for family socioeconomic status, urbanization, region of the country; by using 2-year means for pollution data; and by extending the allowable monitor radius to 20 miles. Excluding subjects with chronic respiratory conditions reduced all of the pollution coefficients; excluding smokers increased them. Increasing the radius for the pollution monitors reduced only the TSP coefficients. All of these findings are consistent with a causal hypothesis. Schwartz also noted that linear dose-response models for the air pollutants appeared to be superior to logarithmic transforms.

The nonlinear dose-response relationships suggested a threshold for ozone of about 0.04 ppm for the daylight average (0.02–0.025 ppm annual average); this is a relatively high value (90th percentile in Schwartz's data), which might correspond to average daily 1-hour peaks around the federal standard of 0.12 ppm (235 $\mu g/m^3$). For TSP, a threshold was suggested around 90 $\mu g/m^3$, which exceeds the current standard of 75 $\mu g/m^3$. The NO_2 relationship was more nearly linear, with an increase in slope at about 0.04 ppm (75 $\mu g/m^3$), which is below the current federal standard. These results suggest that most of the "signal" in Schwartz's data set was derived from the locations above the 90th percentile in air pollution, for all species. The lack of significant findings for SO_2 is also noteworthy, since the 90th percentile for SO_2 was 0.019 ppm (50 $\mu g/m^3$) and there are relatively few urban locations in the United States that now exceed this value (see Table 2-8, for example). Association with SO_4^{-2} aerosol was not evaluated, but regional dummy variables were investigated that may provide some insight, given the regional distribution of this pollutant. No dummy variable was significant for FEV_1 or FVC, but peak flow was about 5% lower in the Northeast. Schwartz also reported decrements in lung function associated with central city residence, apart from the specific air pollution relationships studied.

The analytical approach taken by Chestnut *et al.* (1991) to lung function data for adults from NHANES I was similar to that used by Schwartz for youths but less comprehensive in that they considered only one air pollutant, TSP. The NHANES I examinations were conducted from 1971 to 1975 (O'Brien, 1981), and suitable air monitoring data were not available then for most other species. The population studied consisted of adults, from 25 to 75 years old, who never smoked, residing in 49 different cities across the United States. Only persons producing reproducible spirometry were included, which could have biased the results toward the healthier portion of the population. The TSP data were averaged for 3 months prior to examination, by "central urban area." This is a less stringent criterion than the distance specification used by Schwartz.

Chestnut *et al.* found statistically significant negative relationships between TSP and FVC and FEV_1 that were robust against changes in the sample and model specification. The TSP relationships were fitted to a logistic model. Other variables considered included age, physical characteristics, income, and regional dummy variables (residence in the South was associated with a decrement in lung function of about 6%). The smoothed FVC results for adults are compared to Schwartz's data for youths in Figure 11-9 and are seen to be quite

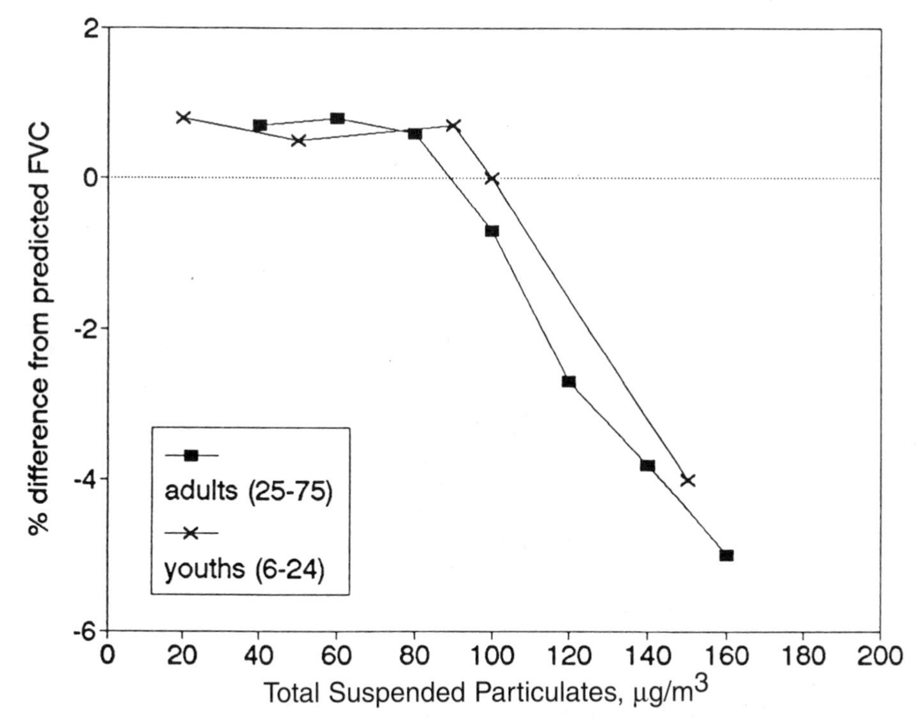

FIGURE 11-9. Comparison of the effects of TSP on forced vital capacity for youths (data from Schwartz, 1989) and adults (data from Chestnut et al. *Archives of Environmental Health* 46(3):135–44 (1991)). *Data used with permission of the Helen Dwight Reid Educational Foundation. Published by Heldref Publications, 1319 Eighteenth St., N.W., Washington, DC 20036–1802. Copyright © 1991.*

similar. The apparent threshold was slightly higher for youths, but it was not possible to test the significance of this difference. The maximum decrements in lung function were about 5% for both data sets on average. This similarity in dose-response relationships for two such different age groups suggests that lifetime exposure to air pollution is not a major factor, since adults tend to have been exposed to pollution levels (TSP) higher than current ones, in the past. It is possible, however, that the effects of TSP on adult lung function have been underestimated by Chestnut *et al.* (relative to the data for youths) because of TSP sampling errors.

Figure 11-10 shows the TSP dependence of the lower portion of the FVC distribution, which may include subjects with respiratory disease (FVC $<$ 70% of predicted). This relationship is nearly linear when the log of the prevalence is plotted (Figure 11-10b); this corresponds to the log-linear model used for mortality, and the prevalence of impairment may be thought of as analogous to a mortality rate. The slope of Figure 11-10b is quite steep; an increase of 20 $\mu g/m^3$ in TSP implies about a 30% increase in the number of people with low FVC.

The elasticity of FVC on TSP may be estimated in several different ways from the work of Chestnut *et al.* The logistic relationship is nonlinear at both high and low values of TSP; over the range of 40 to 160 $\mu g/m^3$, the elasticity

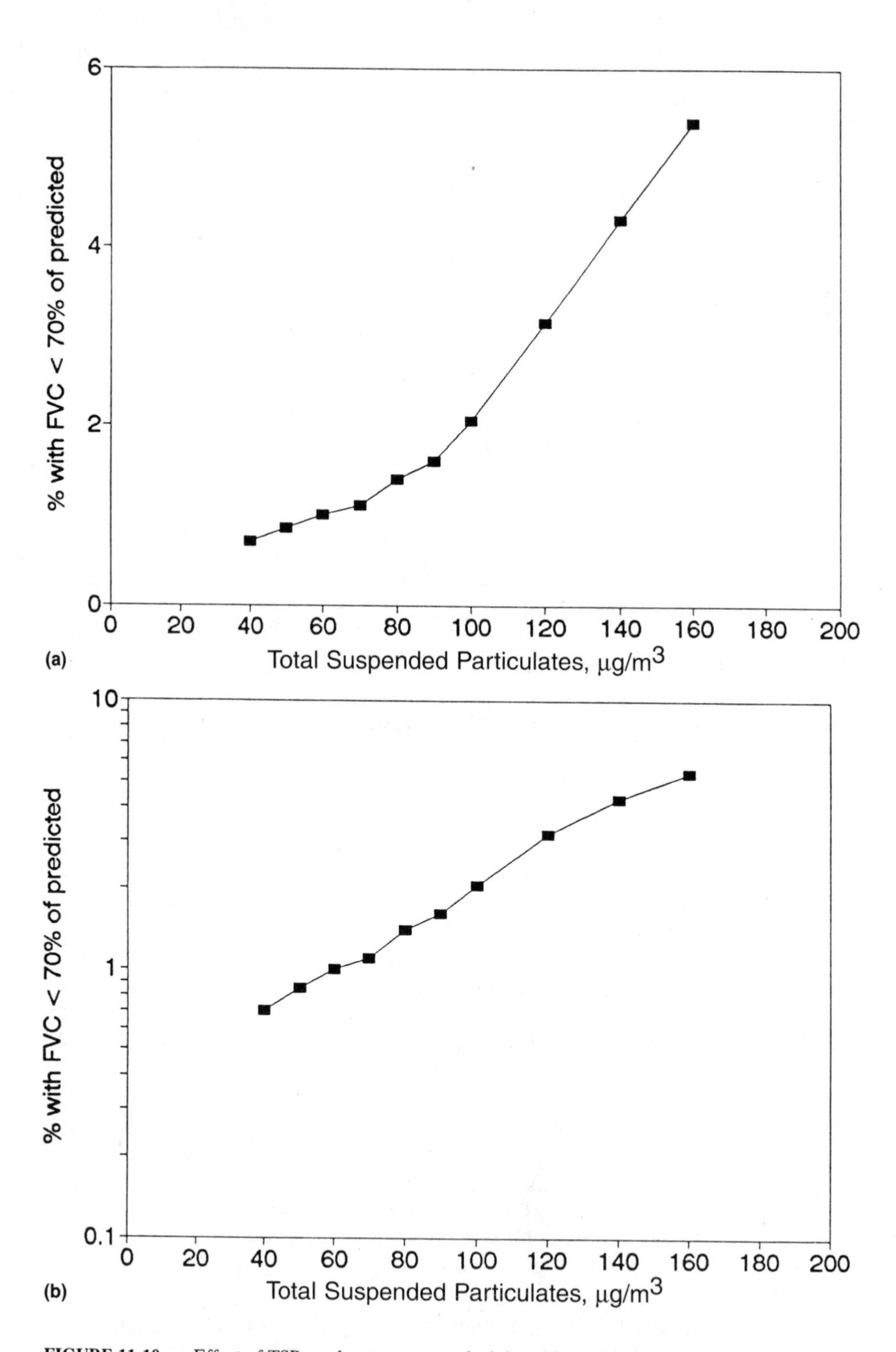

FIGURE 11-10. Effect of TSP on the percentage of adults with forced vital capacity less than 70% of predicted. After Chestnut et al. (1991). (a) Linear scale. (b) Logarithmic scale.

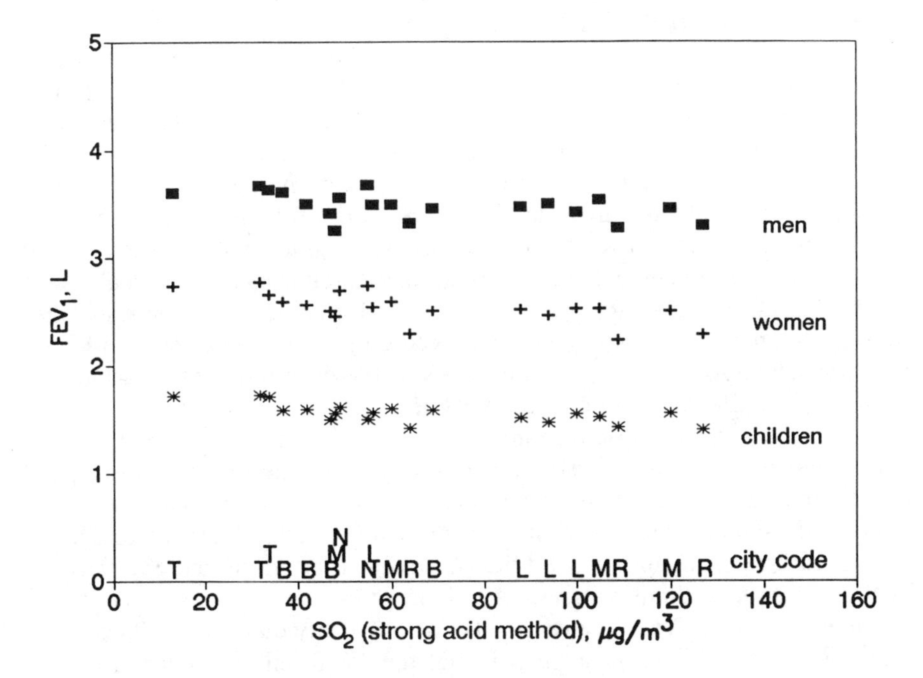

FIGURE 11-11. Lung function data from the PAARC Study for men, women, and children versus SO_2 by the strong acid method. Data from PAARC (1982b). City codes are T = Toulouse, B = Bordeaux, M = Marseille, N = Mantes, L = Lyon, R = Rouen.

was about −0.036. Based on TSP values above 80 μg/m^3, the relationship shown in Figure 11-10 gave a value of −0.056.

The PAARC Study (France)

A national survey of respiratory symptoms and lung function was conducted in France, involving children aged 6–10 and adults who were not manual workers. A number of air pollutants was measured in each of the areas studied, but the only lung function metric reported (PAARC, 1982b) was FEV_1 (referred to as VEMS in the original French text and reported in units of L/s). Data were collected from September 1974 to June 1976; no mention was made that account was taken of either seasonal or diurnal gradients. Suffer dioxide was measured in two ways:

1. Using the H_2O_2 "acidimetric," or strong acid method, labeled "SO_2a" in Table 11-2. As discussed in Chapter 2, this method is sensitive to all gases that become acidic in solution, not just to SO_2.
2. A specific method for SO_2 involving dry absorption on a zinc acetate filter was also used (Derriennic *et al.*, 1989). In all but one of the areas, the values for the first SO_2 method exceeded or equaled those for the second, sometimes by substantial amounts.

The diameters of the population zones surrounding each monitor (where the subjects resided) varied from 0.5 to 2.3 km.

The associations of lung function and air pollution were reported as correlation coefficients and scatterplots, based on FEV_1 adjusted for height, age, smoking, social class, and occupational exposure (adults). The authors reported negative associations for both measures of SO_2, positive associations for NO_2, and no association for various measures of particulates; ozone was not measured. Problems arose in the interpretation of these results because of differences in the between-city and within-city gradients. Schwartz (1989) recognized this problem and used a nested random-effects model to deal with it in his data set. The PAARC study involved 20 distinct areas but only six cities (having spirometry data). The clustering according to city can be seen in Figure 11-11, which shows that the general slopes of the dose-response functions are mainly defined by the between-city variation.

In order to derive dose-response functions, I read the values from the graphs and performed regression analysis; my correlation coefficients agreed with those originally reported (which serves as a check on the data extraction). I pooled the data for men, women, and children (and added two corresponding dummy variables) and investigated the sensitivity of the strong acid SO_2 variable to the inclusion of dummy variables for the cities (with respect to Toulouse, which was taken as the reference point). With five city dummy variables in the regression, SO_2 was nearly significant, but the coefficient was inflated by one order of magnitude because of collinearity with the city dummy variables. With dummy variables included for Bordeaux, Marseille, and Rouen (all of which showed little within-city effect of SO_2), SO_2 retained significance, with a slightly reduced coefficient ($e = -0.039, p < 0.001$). On this basis, one is led to the conclusion that a significant effect of (acidimetric) SO_2 was shown by this study (Figure 11-12a), but not according to the "specific" method for SO_2 (Figure 11-12b).

Figure 11-12c plots the children's lung function against the difference between the two SO_2 methods. This graph thus suggests that the non-SO_2 acid gases also have an effect. These might include HCl and HNO_3, for example, although the magnitude of the differences between methods suggests the presence of measurement artifacts of some kind. Alternatively, the figure suggests the dangers in making *ad hoc* inferences solely on the basis of a "good fit" to environmental data, without benefit of a physiological hypotheses as substantiation.

Regional Comparisons: The Harvard Six Cities Study

The Harvard Six Cities Study was originally conceived as a prospective cross-sectional study emphasizing sulfur oxides and related particulates, but the longitudinal findings seem to have proven more useful. The six cities cover the northeastern-northcentral quadrant of the United States (Watertown, MA; Portage, WI; St. Louis, MO; Topeka, KS; Steubenville, OH; and Kingston-Harriman, TN). Health data, including respiratory symptoms as reported by parents and spirometry, were collected in the first three cities in the fall (September–December); in the remaining three cities, health data were col-

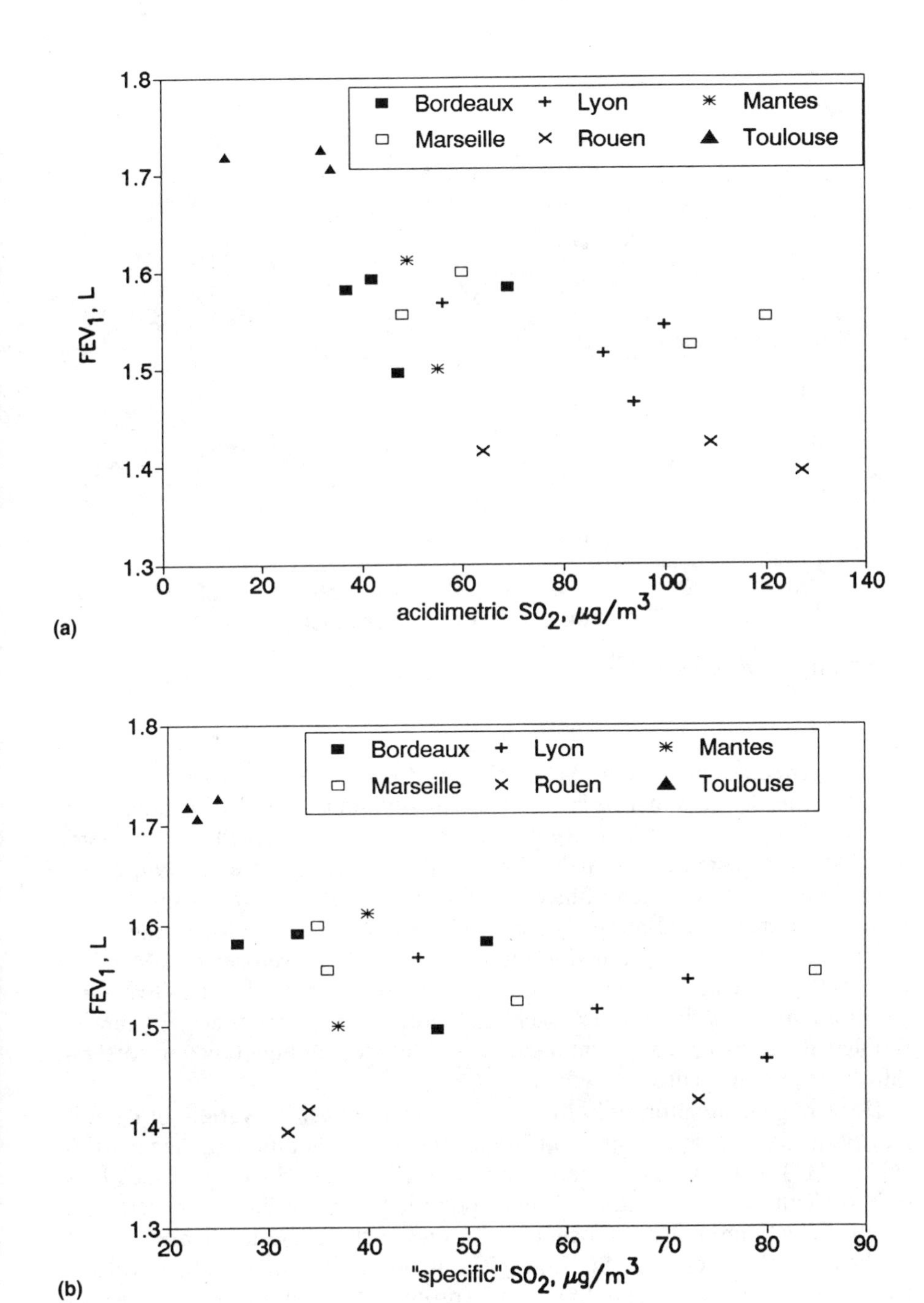

FIGURE 11-12. Children's lung function data from the PAARC Study. (a) Versus SO_2 by the strong acid method. (b) Versus SO_2 by the "specific" method. (c) Versus the difference in SO_2 methods. *Data from PAARC (1982b).*

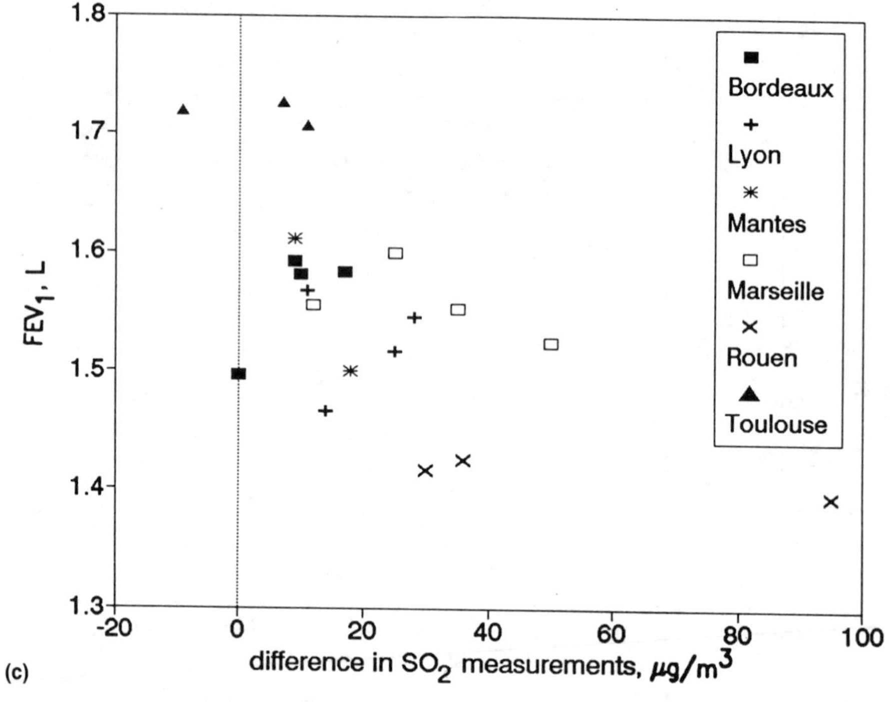

FIGURE 11-12. *Continued*

lected between January and May. This protocol raises questions about the need to consider seasonal biases that could confound the cross-sectional comparisons. To the extent that there may be seasonal biases in the health outcome variables (for example, due to patient recall [Hoppenbrouwers, 1990] or the normal annual cycle in lung function), the study has both cross-sectional and temporal sources of variation. For example, Shy *et al.* (1970) found significant differences between lung function tests conducted in November and March in the Chattanooga study, after adjustment for growth of the subjects. Given only six locations, data from which were gathered at different times, it may be problematic to assign consistent locational differences in lung function solely to differences in air quality.

Harvard's air monitoring in the six cities featured a wide variety of gaseous and particulate measurements, including the recent addition of fine particle acidity (H^+). Data were collected year round. However, with only six locations, it is difficult to define cross-sectional pollution relationships with certainty. Steubenville is an industrial location with one of the highest urban SO_2 levels in the United States, and St. Louis is a major urban area; both have high values for TSP, NO_2, and SO_2, and (probably) lower O_3 values because of titration by NO. Portage reported the highest long-term average ozone levels, probably because it is downwind from urban sources of ozone precursors.

Dockery reported lung function data for adults in 1985 and for children in 1989. The analysis for adults was limited to never-smokers not reporting

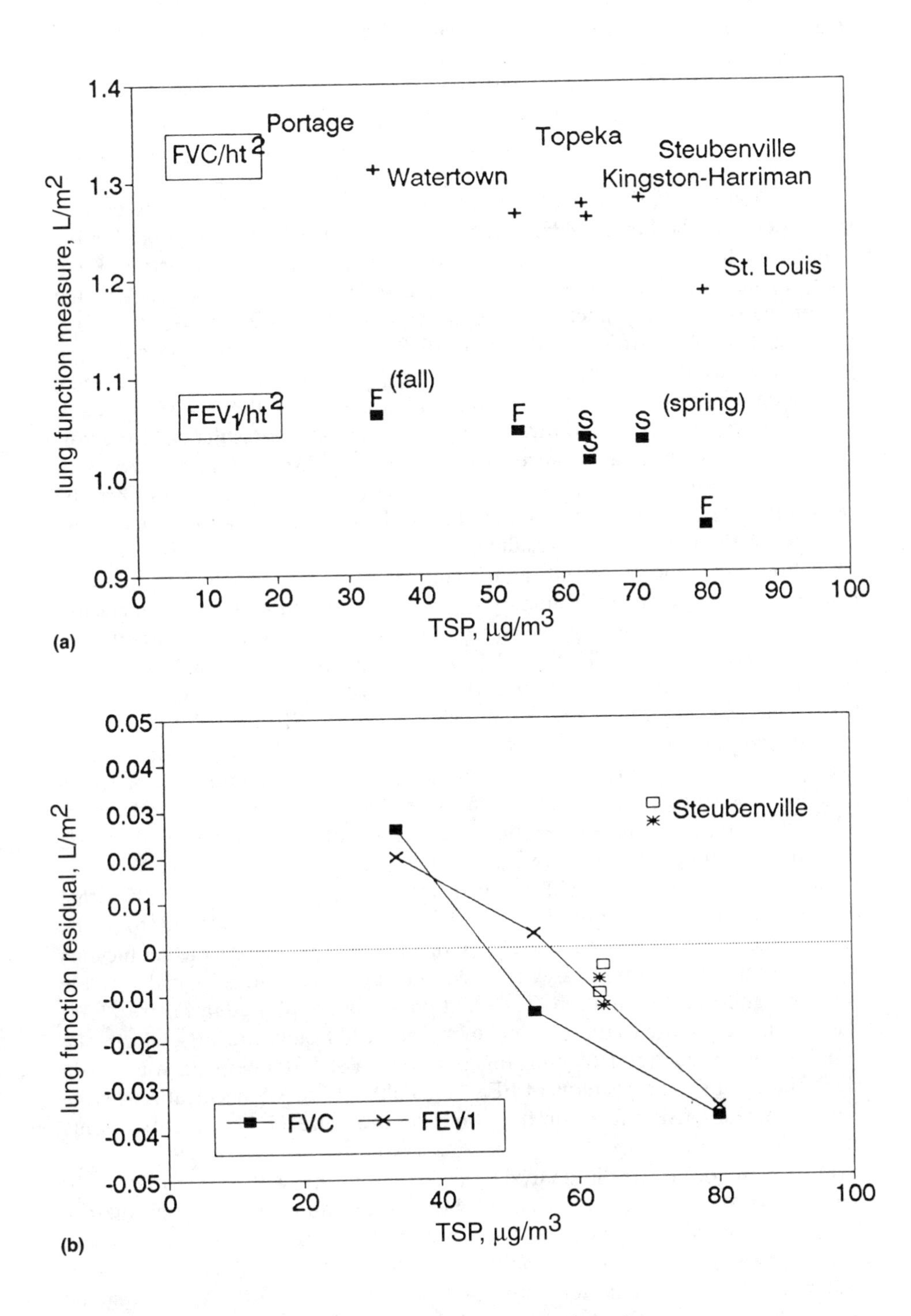

FIGURE 11-13. Lung function data from the Harvard Six City Study for adult never-smokers. (a) FEV_1/ht^2 and FVC/ht^2. (b) Residuals from multiple regression models. *Data from Dockery* et al. *(1985)*.

respiratory symptoms; the technique for spirometry was to fit the data to a regression in terms of age, sex, and height and then to examine deviations from these models for the presence of other associations such as air pollution. Dockery did not address air pollution relationships in his 1985 article; Figure 11-13a is my plot of the adult lung function data based on air pollution data reported elsewhere (annual averages). The seasonal variable suggested a slight high bias for the spring cities from this plot.

In my reanalysis, three different bases for seasonal corrections were used to try to deduce whether the Harvard Six Cities Study provides support for the hypothesis that there are chronic effects of air pollution on lung function. First, analyses were made without seasonal corrections. Then, it was assumed that the timing of the lung function "trough" was either mid-December (following the results of McKerrow and Rossiter, 1968), or, alternatively, March 1 (following the results of Spodnik *et al.*, 1966). The magnitude of any seasonal corrections were deduced from regression analysis on the Six Cities data. Finally, an *a priori* correction was imposed, based on the results of McKerrow and Rossiter on normal subjects. Correcting the MMEF data for molds or dampness (Brunekreef *et al.*, 1989) did not matter in these analyses, since the corrections were not large enough to change the order of ranking of the cities.

I pooled the lung function residual data that Dockery reported for FVC and FEV_1 for adults and regressed them against the various air pollution measures (one by one), together with dummy variables for season and type of measurement. Only TSP reached significance and only when the seasonal variable was included. As seen from Figure 11-13b, Steubenville, a "spring" city, is an outlier, but the seasonal dummy variable was of marginal significance (p = 0.10). The elasticity of lung function with respect to TSP was −0.064.

The analysis for children in the six cities followed along similar lines, except that the lung function prediction model included the effects of maternal smoking, gas stoves, and parental education. Dockery (1989) reported associations between particulate air pollution and respiratory symptoms and illness, but not lung function. However, when the two lung function measures are pooled as described above, TSP reached significance ($t = -2.7$; $p = 0.05$ with 4 degrees of freedom). Note that, although FVC and FEV_1 failed to reach significance when regressed on TSP individually, the regression coefficients were the same in all three cases (pooled and separate). The seasonal variable was not significant for children. The children's data are plotted in Figure 11-14; the two most polluted cities, Steubenville and St. Louis, are offset from the general downward trend of lung function with TSP. Steubenville was also a clear outlier in Dockery's plots of FEV_1 inhalable particle concentrations. Note that the entire "TSP effect" on lung function in Figure 11-14 amounts to only about 3% over this range.

One of the difficulties in analyzing any relationships with ozone that might exist in the Six Cities data is that of choosing the appropriate ozone metric. Only the annual averages have been reported (Dockery *et al.*, 1989), and Portage clearly has the highest value, in spite of its low values for all other pollutants. When the children's lung function data are plotted against annual ozone, five of the cities form a dose-response relationship of the expected shape; Portage is a clear outlier on this plot (Figure 11-15a). However, if some measure of peak ozone concentration were selected instead of the annual, the

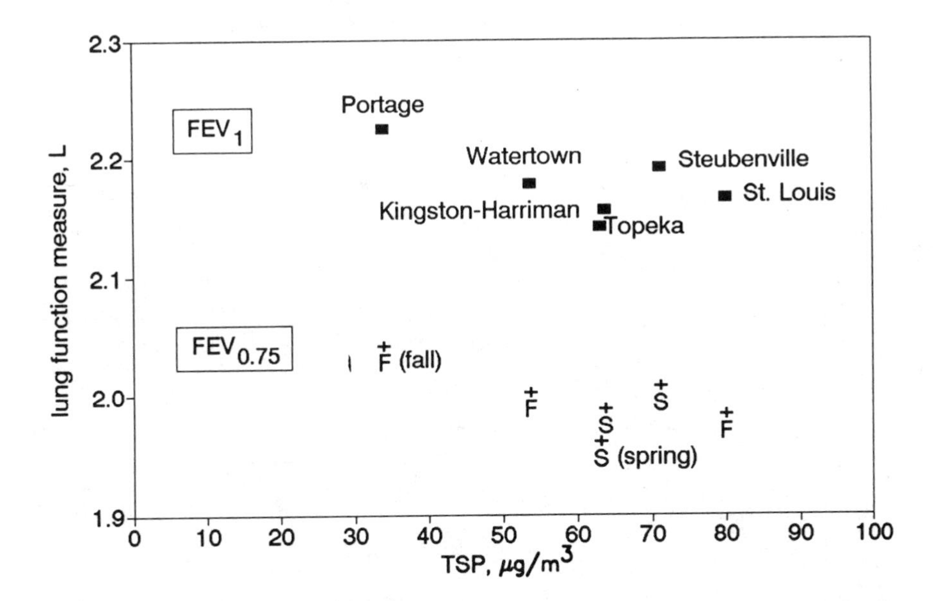

FIGURE 11-14. Children's lung function data from the Harvard Six City Study: FEV_1 and $FEV_{0.75}$ versus TSP. *Data from Dockery (1989).*

ranking of cities might very well be changed. For example, I estimated maximum 1-hour ozone values for the six locations from 1984–88 EPA SAROAD tabulations, using nearby locations where necessary; Watertown was the highest (0.15 ppm) and Portage the lowest (0.09 ppm). Using these values in a regression for the children's lung function gave an ozone elasticity of −0.036, but it failed to reach significance. These data are plotted in Figure 11-15b, with and without the seasonal adjustments.

In conclusion, the Harvard Six Cities Study provides support for Schwartz's finding of a negative association between lung function measures and TSP and suggestive support for his ozone findings.

Local Studies: The UCLA Studies of Chronic Respiratory Disease

Studies of adult respiratory health in selected communities of the Los Angeles area were begun in the mid-1970s, using a mobile laboratory that was stationed sequentially at various locations (Detels *et al.*, 1981, 1987, 1990; Rokaw *et al.*, 1980). An entire suite of respiratory measurements, including spirometry, body plethysmography, and nitrogen washout tests, was made on volunteers, 70 to 80% of whom were community residents. Air quality data were obtained from stations operated routinely by local authorities. The general procedure was to contrast (adult) residents of one or more "polluted" communities in the South Coast Air Basin with Lancaster, CA, located in the Southeast Desert Air Basin on the other side of the San Bernadino Mountains from Los Angeles. For the purpose of trying to define dose-response relationships, I pooled the lung

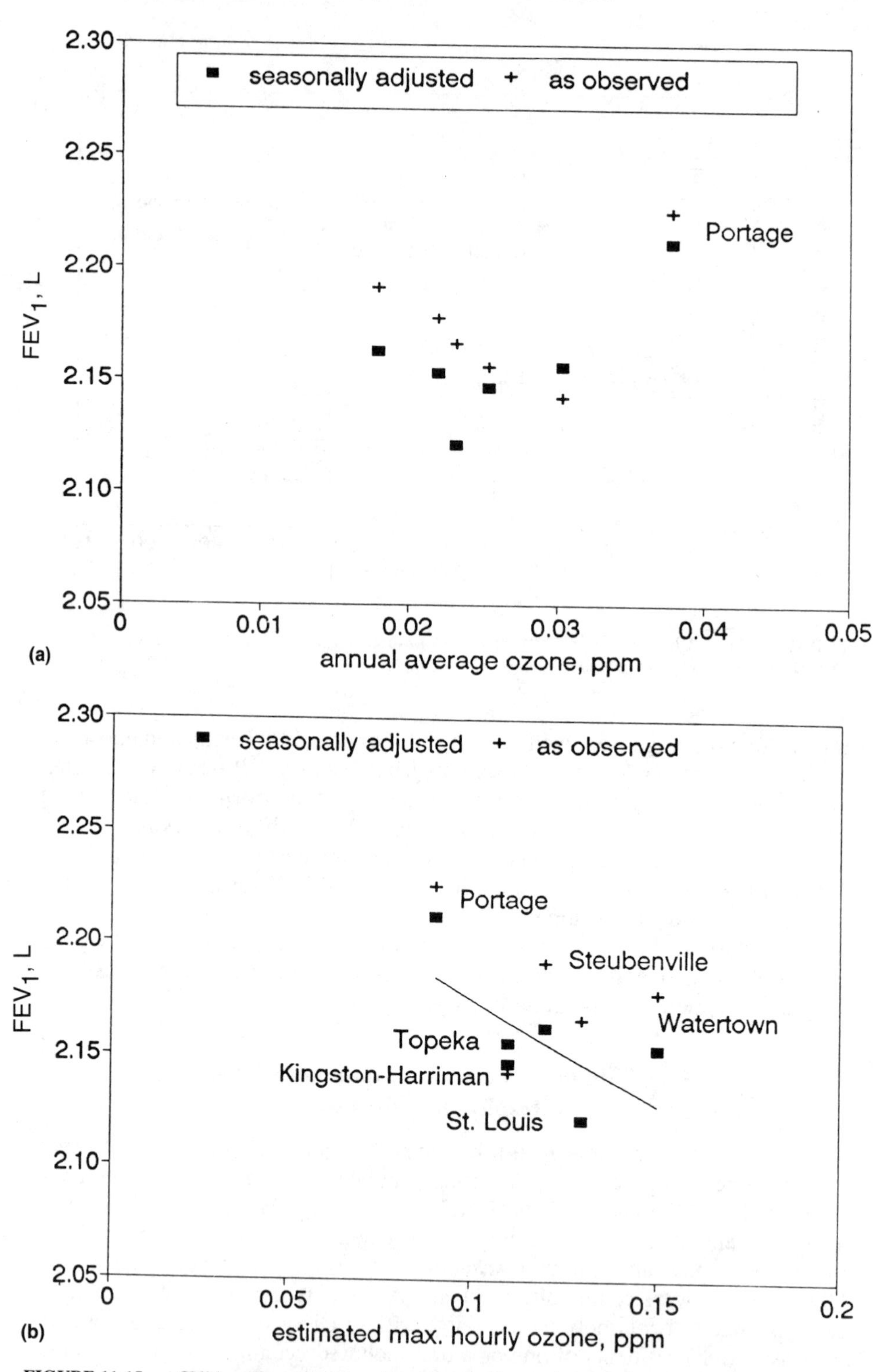

FIGURE 11-15. Children's lung function data from the Harvard Six City Study: (a) FEV_1 versus annual average ozone. (b) FEV_1 versus estimated peak ozone. *Data from Dockery (1989).*

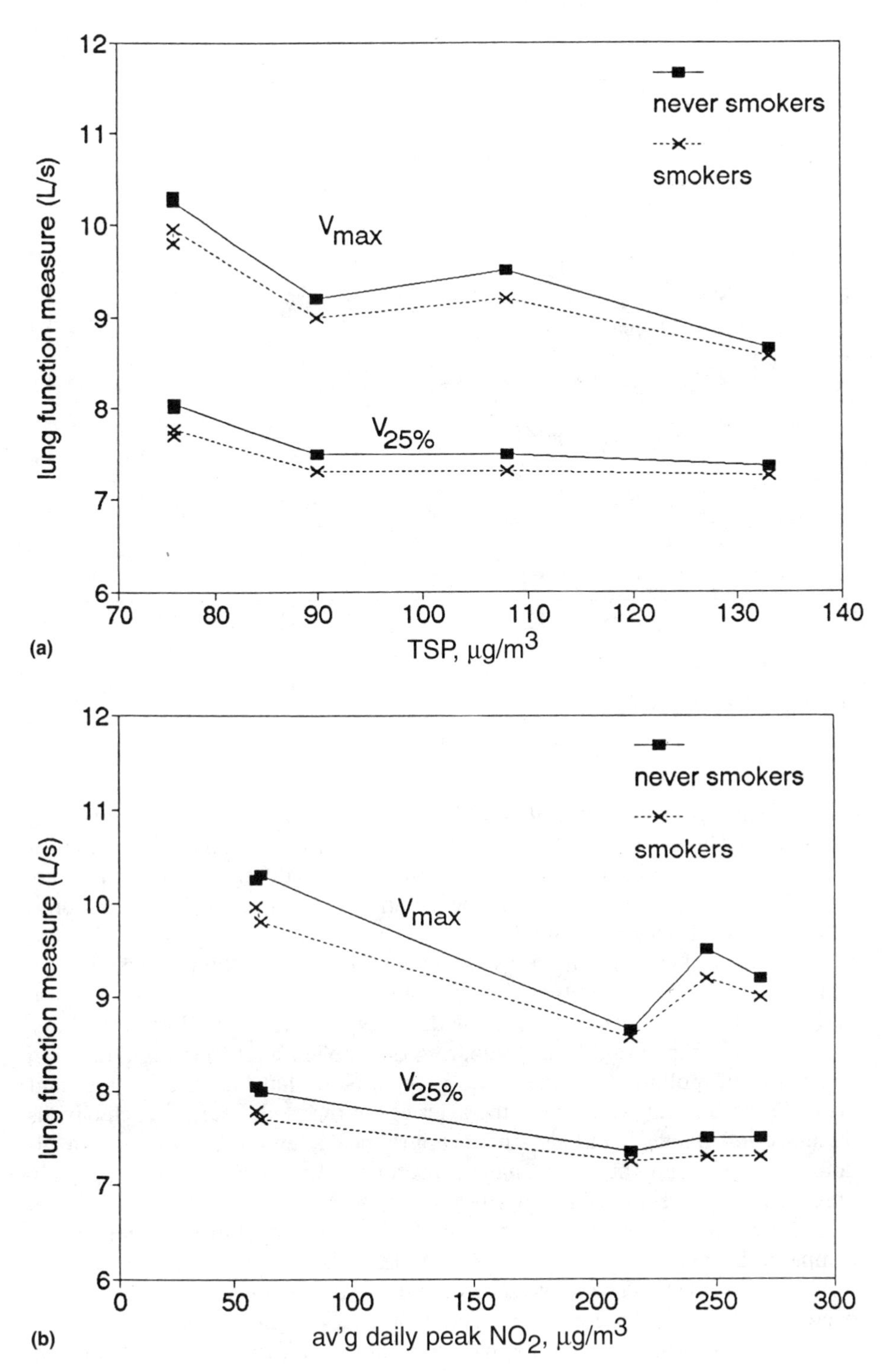

FIGURE 11-16. Lung function data from the UCLA Studies of Chronic Respiratory Disease (a) Plotted versus TSP. (b) Plotted versus NO_2. (c) Data for never-smokers plotted versus TSP. *Data from Rokaw* et al. *(1980) and Detels* et al. *(1981).*

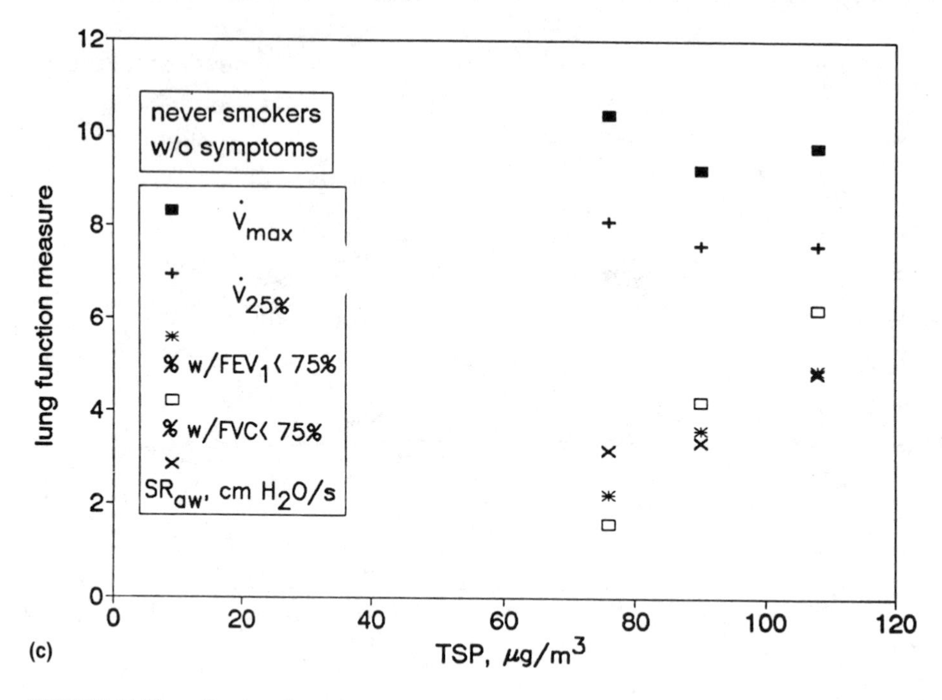

FIGURE 11-16. *Continued*

function data from Rokaw *et al.* (1980) (Lancaster, Burbank, and Long Beach) with that of Detels *et al.* (1981) (Lancaster and Glendora). Since the format of the spirometry results reported in the two studies differed somewhat, only measurements of flow rates could be compared in this way. Dose-response relationships for peak flow and flow at 25% of capacity were apparent; flows at 50% and 75% of capacity did not show such relationships. Annual average TSP and average daily peak NO_2 showed the most consistent relationships (Figure 11-16a and b); the difficulty of defining dose-response relationships on the basis of a few cities is apparent from comparing the implied intercepts (lung function values at zero pollution) of these two graphs. Note that the gaseous pollutant measures used in this study were the averages of the peak hour for each day; as discussed in Chapter 3 and elsewhere, hourly peaks tend to be noisier than 24-hour averages, and thus this analysis protocol may have served to partially obscure the true relationships, if any.

The relative TSP responses of several different lung function metrics are compared in Figure 11-16c, for Lancaster, Long Beach, and Burbank (plotted from left to right). The flow rate data (top two data sets) show a slight decrease; the other three measures show the expected increases and appear to converge on an x intercept of about 30 to 60 $\mu g/m^3$, which may be a reasonable range for the "natural" background of TSP for Los Angeles in the 1970s. The FEV_1 and FVC data are the percentages of subjects with readings below 75% of predicted values, that is, those subjects that might be classified as slightly "impaired." The lower percentage points of the distributions are more sensitive indicators than the mean values of FEV_1 or FVC; unfortunately, few studies have reported these statistics.

In summary, the UCLA studies appear to have shown that respiratory performance is adversely affected by chronic exposure to air pollution, but these studies are not helpful in identifying the responsible pollutants or the magnitudes of the effects that might be attributed solely to air pollution.

The Montana Air Pollution Study (MAPS)

Air pollution from various sources was contrasted in this cross-sectional study of five Montana cities: Anaconda (copper smelting); Billings (coal-fired power plant); Butte (copper mining); Missoula (trapping of space heating emissions, including wood smoke); and Great Falls ("clean," with good ventilation). The subjects were children in grades 3–5, who were tested during morning hours but at different times of the year; since each city was tested during the fall, winter, and spring, seasonal confounding seems unlikely. Air quality monitoring stations were established especially for this program and were integrated into the routine state-operated network; measurements included SO_2, NO_2, TSP, and the fine and coarse particle fractions from dichotomous samplers. Monthly average O_3 data were reported for four of the five cities (Great Falls was missing).

For my statistical analysis, I pooled the average values for FVC and FEV_1 and regressed the set against each pollutant for boys and girls separately (since their data seemed to follow different trends); a dummy variable was included to account for the different metrics. For boys, fine particle concentrations gave the best result; for girls, both SO_2 and O_3 yielded consistent dose-response relationships, although ozone data were missing for Great Falls. Elasticities were in the range of −0.01 to −0.06. The ozone relationship would be consistent only if Great Falls had low ozone levels, which is a reasonable assumption given its good ventilation and the tendency for ozone levels to be reduced within urban areas. The authors of this study limited their conclusions to the effects of particulates and did not comment on possible relationships with other pollutants nor on the apparent differences between sexes.

The Arizona Smelter Studies

Dodge (1980, 1783) and Dodge *et al.* (1985) reported on various aspects of a cross-sectional analysis involving five towns in Arizona, four of which were near copper smelters and were subjected to intermittent large spikes of SO_2 (up to 1000–3600 $\mu g/m^3$ 3-hour averages). The subjects were children in grades 3–5 who underwent spirometry testing and submitted questionnaires on symptom prevalence. Air monitoring data consisting of SO_2 and TSP (SO_4^{-2} was added in the last study) were obtained from state-operated stations in each town. Note that since smelters do not emit appreciable amounts of NO_x, there is no reason to expect gradients in NO_2 or O_3 among the five towns. In the first study, health outcomes were reported only for two groups of cities having "high" (72–76 $\mu g/m^3$) and "low" (37–54 $\mu g/m^3$) annual average TSP levels; gradients were not examined according to SO_2 levels, although this grouping of towns results in an average SO_2 gradient opposing the TSP gradient (12 versus 47 μg/m^3, respectively). The results showed a lung function gradient for all four metrics (FVC, FEV_1, $\dot{V}_{max_{50}}$, and $\dot{V}_{max_{75}}$), suggesting either adverse effects of TSP or beneficial effects of SO_2. By combining data from the other two papers, measurements could be compared among four of the towns for each of 4 years; neither SO_2 nor TSP showed a significant FEV_1 dose-response relationship for this data set. Since lung function data were reported only as percentages of expected values in the first paper and as adjusted absolute values in the second and third, it was not possible to reconcile the differences in conclusions. Curiously, the authors of the later papers did not refer to the first paper.

The Chattanooga Schoolchildren Study

Chattanooga, TN, offered an opportunity to study the effects of high ambient levels of nitrogen oxides, which were emitted from an explosives plant there (Shy *et al.*, 1970). The case-control method was used, based on four elementary schools (second-grade children). Cases were a high NO_2 location (with probably high HNO_3 as well) and a high particulate area. Nitrogen dioxide measurements were obtained by the Jacobs-Hochheiser method, but attempts were made to correct for the known errors by calibration. Two "controls" were selected; however, the total range in average pollutant concentrations among these four locations was not large (about a factor of 2). Lung function measurements were taken at each school in November and March; the differences according to month were significant at the 0.01 level. I pooled the data by sex and month and found no significant relationship with NO_2. However, the decline versus TSP was just significant, with an elasticity of −0.046.

The Chestnut Ridge, PA, Studies

Schenker *et al.* (1986) examined spatial gradients in a 700 km^2 area of central Pennsylvania impacted by several large coal-fired power plants. The subjects were children in grades 1–6, who were tested in a random sampling scheme, to avoid seasonal confounding. Air quality data for TSP and SO_2 were obtained from an existing network of 17 stations. The children were grouped into three areas stratified by average SO_2 level (group mean annual averages of 65, 70, and 91 μg/m^3). Respiratory symptoms in adults were analyzed in an earlier paper (Schenker *et al.*, 1983), which used a slightly different spatial grouping and presented data on the TSP gradient (annual averages from 64 to 80 μg/m^3). None of the five lung function measures showed consistent gradients with air pollution. If the elasticity were about −0.05, the lung function gradient corresponding to the pollution gradients in this study would be about −1%; since the only statistically significant lung function difference among the groups reported was 1.8% ($p < 0.05$), it appears that the study lacked sufficient statistical power to detect differences resulting from such small gradients in air pollution.

The Turin Study

Spinaci *et al.* (1985) reported data obtained from sixth-grade children living in three areas of Turin, Italy: a central urban area, a peripheral urban area, and a suburban area. The SO_2 and TSP data were obtained from seven monitoring stations in each area; stations were selected to be no more than 0.5 km from a school. By pooling the means of the various lung function measures, I was able to estimate dose-response functions from these data for both SO_2 and TSP, neither of which quite achieved statistical significance, on an aggregated basis. The magnitudes of the slopes were consistent with the other studies summarized in Table 11-2, however. Note that the same children were evaluated 2 years later after air quality had improved (Arossa *et al.*, 1987) and it was found that the spatial gradients were no longer significant.

The longitudinal changes were also analyzed by Arossa *et al.* (1987). The SO_2 and particulate levels dropped about 25 to 50% over 2 years; no attempt was made to disaggregate the two pollutants, and SO_2 was used as an index. The suburban area, which showed much less change in air quality, was used as a control. In the urban area, 1880 children were evaluated along with 162 controls. The statistical analysis considered sex, age, height, weight, smoking and exposure to passive smoke, central heating and cooking fuels, and air pollution. For the longitudinal analysis, individual slopes of lung function versus time were computed after accounting for the listed variables; it was then assumed that the changes in air quality were responsible for the improvements not attributed to physical growth. A training effect was ruled out, for example. The slope for FEV_1 was about -0.5 ml per $\mu g/m^3$, which corresponds to an elasticity of about -0.03. If the effect were assigned to the much smaller concurrent improvement in TSP, the elasticity would have been around -0.08. Note that this study does not deal with the acute responses that might have occurred on or near the days of testing.

Other Studies of Long-Term Gradients in Lung Function

Xu *et al.* (1991) tested 1440 nonsmoking adults in three areas of Beijing, China, for FVC and FEV_1 in relation to outdoor air pollution and use of coal stoves for domestic heating. Air pollution data consisted of averages of SO_2 and TSP, which ranged from 18 to 128 $\mu g/m^3$ and 261 to 449 $\mu g/m^3$, respectively. In multiple regression studies, use of coal stoves was consistently associated with reduced lung function, as were higher ambient levels of both SO_2 and TSP, thus implicating both indoor and outdoor air pollution. Elasticities were higher for FVC than for FEV_1 by a factor of 3 and higher by about a factor of 6 for TSP relative to SO_2. It was not possible to separate the two pollutants in these multiple regressions, and, given the use of only three areas, the pollution elasticities may not be reliable. The slopes of lung function on either TSP or SO_2 (ml/[$\mu g/m^3$]) were typically substantially higher than the values reported in time-varying studies.

Humphreys and Carr-Hill (1991) addressed the problem of the ecological fallacy for lung function and other health outcomes, by comparing the gradients due to social class within wards and between wards in the United Kingdom. The top social class was labeled "rich" and the bottom, "poor." Five clusters of electoral wards were defined according to the percentage of poor within each ward. For the outcome variable involving a self-assessment of health, there was a gradient within wards as well as between wards. However, for FEV_1 the only differences were within wards ("poor" having consistently lower FEV_1 for all five ward clusters). They also ranked 10 geographic regions in terms of FEV_1; Wales, Greater London, and the North West were the lowest; the East Midlands, the North, and the South East were the best. The authors concluded that the major source of unexplained variation in health outcomes was between individuals, although a "ward effect" was still present.

Brooks and Waller (1972) sampled peak flow rates for about 2900 adults who visited a public health exhibition in London in April 1970. Data were stratified by residence area, age, sex, and smoking status and standardized for height. The effect of smoking increased with age. The maximum gradient due to residence was about 2%; those born outside Greater London had higher peak flows than those born in Greater London. There was no evidence of an effect due to the month or season of birth.

Van der Lende and colleagues have studied respiratory health in two small Dutch towns since the 1960s. A "polluted" city, having annual mean SO_2 values around $250\,\mu g/m^3$ and smoke levels around $50\,\mu g/m^3$ in the 1960s, decreasing to much lower levels in the 1980s, was contrasted with a rural "clean" city. The first survey, taken in the late 1960s, found differences in respiratory symptoms but not in FEV_1; neither could an effect of smoking on FEV_1 be found (van der Lende *et al.*, 1973). After 15 years of followup (van der Lende *et al.*, 1986), significant differences in the rate of decline of FEV_1 could not be shown, although it was reported that the polluted city had a lower baseline.

Goren *et al.* (1990) compared children's lung function among three different areas of Haifa, Israel. The highest polluted area had monthly SO_2 level up to about 70 to $100\,\mu g/m^3$. Although the lung function data showed higher FVC and FEV_1 values (by 2–4%) in the low-pollution area, the authors concluded that "there is no consistent trend of reduced pulmonary function which characterizes any residential area."

Other related studies of lung function include the work of Krzyżanowski *et al.* (1990a) on the effects of residential exposure to formaldehyde on peak flow. They showed that asthmatic children were affected more than normal children at low exposures and that adults were affected even less (effects seen mainly in smokers). Studies of tunnel and turnpike workers (Tollerud *et al.*, 1983) failed to link exhaust exposure to pulmonary function, although this was not the case with earlier studies in New York City (Ayres *et al.*, 1973), which are discussed in detail in Appendix 11A. Similarly, studies of firefighters in Boston (Musk *et al.*, 1977a, 1977b) show no differences in lung function related to smoke exposure, but small differences among retirees. The authors explained these results in terms of selection factors (those with respiratory problems may be more apt to retire).

Summary of Long-Term Studies

The main findings from these studies and my reanalyses are summarized in Table 11-2. The important findings have been enclosed in boxes, to highlight their importance and to delineate what combinations of subjects, lung function measures, and pollutants have been found to be significant. Only three studies evaluated adults; the elasticities of all the studies were mostly in the range of −0.01 to −0.10. There was suggestive evidence that the elasticities for children might be higher than for adults. The pollutants found to be significant in the various studies include SO_2, TSP, O_3, and NO_2. Very few studies had enough power to evaluate the various lung function measures separately; there was some indication that peak flow might be a more sensitive indicator. In general, the long-term effects of community air pollution were small in comparison to other sources of variation.

COMPARISONS WITH THE EFFECTS OF TOBACCO SMOKE

Direct Effects of Smoking versus Air Pollution Effects

Some insights into the differences between acute and chronic responses to air pollution may be inferred from studies on the effects of cigarette smoking.

Smoking is known to reduce lung function progressively with age, amounting to an average difference in FVC of about 8% over the years 25–74 for white males (O'Brien and Drizd, 1981). Dockery *et al.* (1988) report a mean FEV_1 loss for males of 7.4 ml per pack-year smoked (a pack-year is defined as one pack per day over a year's time), which corresponds to about 0.2%, over and above the effects of aging *per se*. Based on a total inhaled volume of 15 m^3 per day and assuming that 50% of the smoke produced by a cigarette is inhaled, one pack per day corresponds to a 24-hour average air concentration of about 20,000 $\mu g/m^3$. When considered in relation to this estimate of the mass of particulate matter inhaled, the rate of lung function loss attributed to smoking is seen to be much smaller than the losses typically ascribed to community air pollution, which is difficult to comprehend.

However, the mean acute responses of 20 individuals immediately after smoking a single cigarette (either with or without filter) are given in Table 11-3 (Da Silva and Hamosh, 1980). The standard deviations of these mean responses were generally large, and only the figures marked "*" were statistically significant ($p < 0.01$). The cigarettes used in these experiments were standardized and were smoked to a constant butt length. The filtered cigarette provided about 29 mg of particulate matter (23 mg tar) and the nonfiltered cigarette, 39 and 33 mg, respectively. For comparison, at a ventilation rate of 15 m^3 per day and assuming that 50% of the smoke was inhaled, these figures would correspond to airborne concentrations of about 965 $\mu g/m^3$ and 1300 $\mu g/m^3$, on a 24-hour basis (but to much higher concentrations during the actual periods of smoking). The lung function measures most often used in air pollution studies, FVC and FEV_1, are thus seen to be relatively unresponsive to at least some of the types of environmental insults known to cause chronic injury. The experiments of Da Silva and Hamosh also showed highly significant increases in heart rate after smoking a single cigarette, of either type. Although most of the respiratory capability lost by long-term smokers may be irreversible, stopping smoking results in restoration of part of the loss (Dockery *et al.*, 1988). It is commonly assumed that all of the short-term lung function changes caused by air pollution are reversible.

TABLE 11-3 Mean Acute Responses to Smoking a Single Cigarette

	Ratio of Measure to Presmoking Value	
Measure	Unfiltered	Filtered
R_{aw}	1.27*	1.22*
FVC	0.988	0.988
FEV_1	0.992	0.990
$\dot{V}_{max50}$	0.97	0.99
$\dot{V}_{max25}$	0.98	1.046*

* Significant values ($p < 0.01$).

Source: DaSilva and Hamosh, 1980.

Effects of Environmental Tobacco Smoke on Lung Function

The health effects of passive smoking or breathing environmental tobacco smoke (ETS) have been the centerpiece of the campaign to restrict smoking in public places and workplaces. The literature on the various aspects of this topic is conflicting; a 1990 symposium (Ecobichon and Wu, 1990) sounded a very cautionary note with respect to the use of much of this literature as a basis for health policy. Other reviews have been less cautious (Rubin and Damus, 1988; Byrd *et al.*, 1989). In addition to thousands of specific compounds, ETS contains fine particles, NO_2, and CO. Carbon monoxide concentrations are 2.5 times higher in sidestream than mainstream smoke, and the particles tend to have smaller diameters. However, the additional dilution inherent in ETS exposure reduces the exposure of passive smokers by orders of magnitude, in most situations. The existence of any effects of ETS on lung function would buttress the case that low concentrations of these pollutants can cause chronic effects, in general. Many of the ETS studies have emphasized respiratory system cancer; this aspect of the problem is not discussed here, because the effects of community air pollution on cancer are generally regarded as equivocal at the pollution levels currently experienced in the developed world.

A rule-of-thumb estimate for the average mass of additional fine particles present in indoor residential air from smoking is 0.8 $\mu g/m^3$ annual average for each cigarette smoked daily, that is, 16 μ/m^3 for a one pack-per-day smoker in the home (J.D. Spengler, personal communication). Outdoor fine particle concentrations can be in the range of 20 to 40 $\mu g/m^3$ annual average. Depending on the rate of penetration into the home, the incremental effect of smoking on indoor air quality may be of equivalent magnitude to the effects of outdoor air quality. Thus the presence of a regular smoker in the home may roughly constitute a doubling of the indoor fine particle concentration. Crowded public spaces with poor ventilation and many smokers can have fine particle concentration levels near the mg/m^3 level.

As is the case with community air pollution studies, one of the problems with many ETS studies is precise characterization of the dose. Often, categorical comparisons are made on the basis of presence or absence of smoking, the number of smokers, or the average amount smoked. This is analogous to trying to relate outdoor pollution effects to the amount of pollutant emitted, rather than on the basis of ambient air quality. Since homes and other occupied spaces vary greatly in terms of volume, air circulation, and air exchange with the outside environment, ETS studies of this type are likely to have substantial uncertainties in their independent variables, which usually give rise to an underestimation of the regression coefficients.

Studies on Adults' Lung Function

The 1986 Surgeon General's Report (U.S. Dept. of Health and Human Services, 1986) reviewed five studies of pulmonary function effects in adults due to passive smoking and concluded that the small changes found were unlikely to cause a previously healthy person to develop lung disease solely as a result of exposure to ETS. The report did not comment on whether persons with impaired lung function were more common among those exposed to ETS, or how such a subgroup might be affected by ETS.

The NRC report (National Research Council, 1986) discussed the effects of both active and passive smoking on lung function. One of the primary findings in active smokers is a more rapid decline in lung function with age. Although one of the six studies of adults reviewed found the passive smoking effect to be similar to that of (actively) smoking 1 to 10 cigarettes per day, the reviewers felt that such a finding could also have resulted from misclassification of subjects.

P. Witorsch (1990) cited 11 ETS epidemiological studies from 1977 to 1989 involving lung function measurements in adults. All of them used presence or absence of smokers in the home as the index of ETS exposure. Seven of them reported a significant decrease in one or more lung function parameters; no improvements were reported. He commented that many of the changes reported were of doubtful clinical significance, that some contained internal inconsistencies or were "implausible." Examples included differences in findings by age or sex, failure to account for socioeconomic status, confounding by variations in types of cooking fuels, and misclassification by failing to recognize the presence of ex-smokers. Witorsch also reported five laboratory studies of short-term ETS exposure to asthmatic adults; three of these involving 37 subjects found adverse effects on lung function, and two of them involving 23 subjects did not. In addition to calling for more research, Witorsch concluded that the results were too variable to "permit a conclusion concerning an association between long-term ETS exposure and impaired respiratory health or pulmonary function in non-smoking adults."

ETS Studies on Children's Lung Function

The 1986 Surgeon General's Report (U.S. Dept. of Health and Human Services, 1986) reviewed 18 studies of the effects of ETS on children's lung function and concluded that an effect of maternal smoking on reduced children's lung function had been demonstrated. Although the average lung function decrement was small (1–5%), the authors felt that small differences might be more important for susceptible children and later in life.

R.J. Witorsch (1990) used the 1986 National Research Council book as a

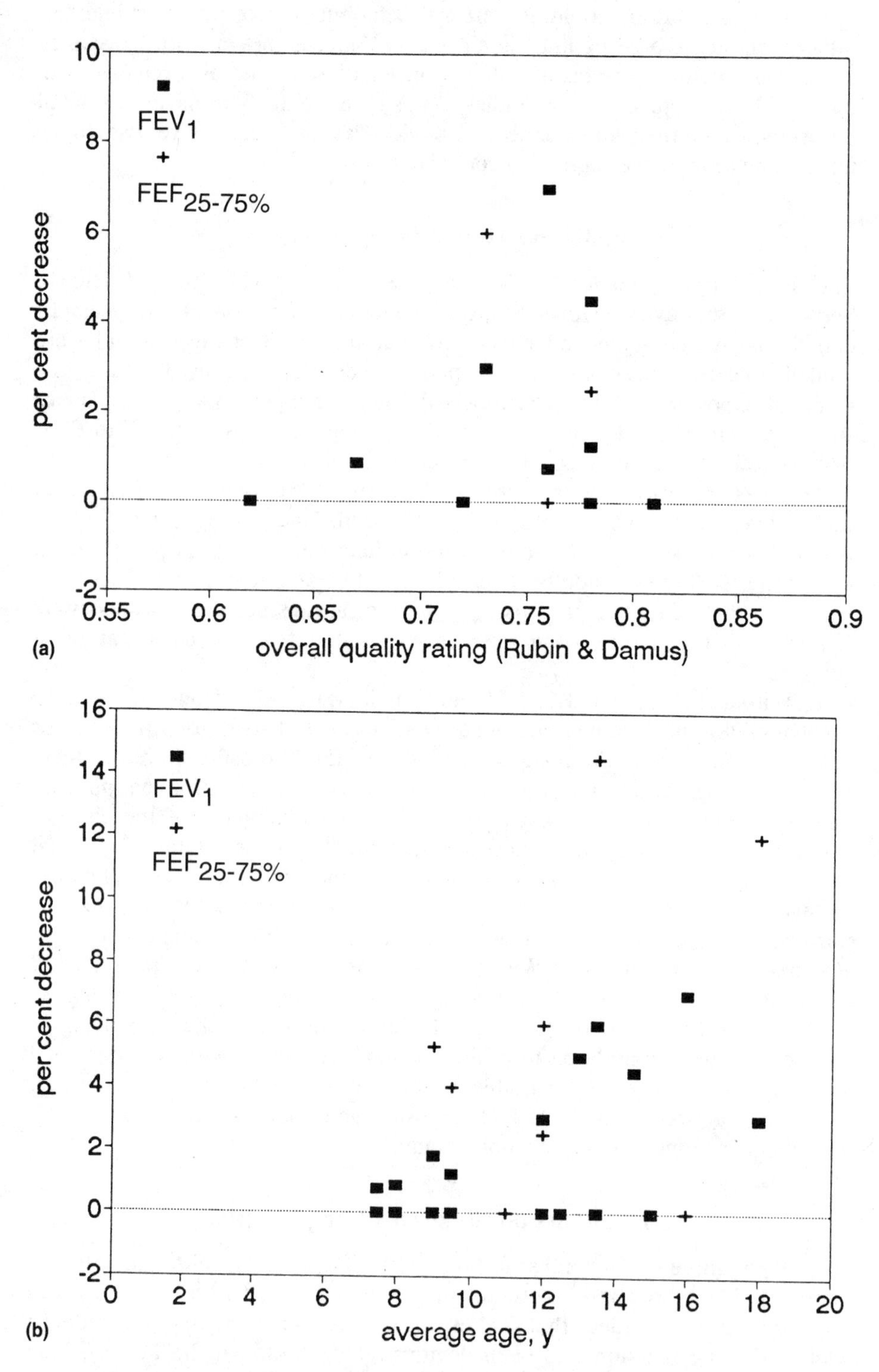

FIGURE 11-17. Comparison of effects on children's lung function found by various studies. (a) By overall quality rating of each study. (b) By average age of the children in each study. *Data from Rubin and Damus (1988) and Witorsch (1990).*

point of departure and cited 26 studies of parental smoking and children's pulmonary function. Twelve of these showed significant decrements in lung function, varying from less than 1% to 15%. No improvements in lung function were shown by any of the studies, suggesting that the differences in findings were not purely random. Rubin and Damus (1988) evaluated 30 studies of passive smoking effects on children; Figure 11-17a plots the findings of the studies reviewed by Witorsch that were also evaluated by Rubin and Damus. There is a tendency for the higher rated studies to exhibit statistically significant detrimental effects (nonsignificant studies are plotted at zero). Rubin and Damus also concluded that "while a few well-designed studies demonstrate a significant effect . . . , most studies had significant design problems that prevent reliance on their conclusions."

Among the studies reviewed by Witorsch, the weighted average decrement in FEV_1 was about 1%, counting the nonsignificant findings as zero. The statistic for those studies reporting data on $FEF_{25-75\%}$ was 3.1%. Thus, if the increase in exposure to fine particles were 100%, the average effect of ETS on children's lung function would have elasticities in the range of −0.01 to −0.03. Witorsch remained rather skeptical about these effects, in part because "the mechanism for this association remains unexplained." However, in a previous section of the paper, he hypothesized that differences in ambient air quality could be responsible for confounding studies of this type.

The possibility that the effect of maternal smoking on young children represents an *in utero* effect rather than an air quality effect deserves serious consideration (NRC, 1986). One indicator would be a decline in significance with the age of the child. As shown in Figure 11-17b, the opposite is true for the group of studies reviewed by Witorsch. Figure 11-17b seems to imply that the effect increases with age and thus with the duration of exposure. However, lung function studies in very young children are rare, so that it might be more practical to test this hypothesis with data on symptom prevalence.

Teenage athletes were examined by Tsimoyianis *et al.* (1987), who considered all sources of exposure to ETS and found that sources outside the home were also important. They found that about two-thirds of the group of 209 boys and girls had been exposed to ETS, and that among the exposed group, low FEF_{25-75}% (<70% of predicted) and cough were significantly more common. Since the absolute numbers of athletes affected were small, it is doubtful that the difference in mean lung function values would have been significant in this study.

Conclusions on Passive Smoking Effects on Lung Function

The NRC (1986) concluded that "the weight of evidence is that there are clearly observable effects of ETS on the respiratory system." My reading of this literature (and that of others; see Byrd *et al.*, 1989, for example) concurs with this finding, including detrimental effects on lung function. However, questions remain as to how the toxicity of ETS compares to nonspecific outdoor fine particles (which often contain a high percentage of sulfates), and how small average changes to a population mean should be interpreted in terms of possible clinical significance to individuals.

CONCLUDING DISCUSSION

The short- and long-term effects of air pollution appear to be reasonably consistent, in that children seem to be more sensitive, the effects are small, of the order of a few percent, and that O_3 and particulates are implicated most often. There is also evidence that suggests that even some components of the long-term effects may be reversible, for example, as suggested by the studies in London and Turin and by the observation that smoking cessation has major health benefits relative to not stopping (U.S. HEW, 1979). If so, it is possible that long-term effects are mainly the aggregate of short-term effects. Long-term, longitudinal studies are needed to better understand the chain of events by which childhood exposure becomes a determinant of adult respiratory health.

Clinical Significance of Changes in Lung Function

A change of a few percent in lung function applied uniformly to a healthy population will have little real health impact, if any. However, some of the studies reviewed above clearly show a large range of responses within populations, which suggests that real populations are not uniformly healthy. Average responses of a few percent seen in a large group are most likely the result of much responses experienced by only a fraction of the group. For example, the 25th and 75th percentiles of the effect of episodic air pollution in Steubenville on children's FVC were −2% and +1.5%, respectively (Dockery *et al.*, 1982). Lawther *et al.* (1975) stated that "any increase in airways resistance could, in some diseased lungs, further disturb the relationship between ventilation and perfusion with undesirable clinical consequences." A further factor for consideration is that most lung function studies exclude subjects who are unable to produce repeatable trials (usually 5–10% of the cohort); these subjects may in fact have the worst respiratory health, and their exclusion from a largely healthy cohort may be a significant source of bias (Eisen *et al.*, 1984).

Persistence of Effects on Children

Many of the studies of air pollution effects on lung function were conducted on children, and the evidence (however sparse) suggests that children may be more sensitive than adults, even in indoor environments. To use this information to address the question of the effects of impaired lung function on (adult) mortality thus requires longitudinal information on the persistence of such effects. A few such studies are available from the literature.

Watkins *et al.* (1986) followed a group of London children from birth to age five. Ventilatory capacity (peak flow) at age 5 was shown to be adversely affected by respiratory illness in the first year of life. The group with a history of more than one lower respiratory illness had about 9% lower peak flow, after adjustment for height, than the healthy group ($p < 0.05$). This study failed to find an effect of parental smoking, but the effect of early illness on subsequent peak flow was found to be confined to those children whose fathers had manual occupations.

Speizer and Tager (1979), pointed out that adults with diagnosed respiratory problems were more likely to have reported significant childhood respiratory illness and that early childhood illness can lead to increased chronic respiratory symptoms later on. However, insufficient follow-up has been done to establish a definitive chain of sequellae.

Acute versus Chronic Effects

It is difficult to compare acute and chronic lung function effects, since the former are reversible. Studies of long-term changes (for example, the Turin and Lawther studies) imply that long-term may also be reversible, although this does not seem to entirely be the case with cigarette smoking (ex-smokers may not recover all of their pulmonary function loss). The study of bridge and tunnel workers by Ayres *et al.* (1973) suggests that chronic exposure may result in depression of baseline lung function values with respect to "normals," and that acute exposures result in further reversible depressions (see Appendix 11A).

It is unfortunate that better follow-up studies linking childhood sensitivity to air pollution with adult respiratory health have not been done. Burrows *et al.* reported data in 1977 supporting the hypothesis that pediatric respiratory illness could lead to adult obstructive airway disease; in 1983, Samet *et al.* reviewed the literature and found the epidemiological evidence to be conflicting. Since then, Tager *et al.* (1983) reported that children's FEV_1 decrements attributed to maternal smoking persisted up to age 5, and Strope *et al.* (1991) found that two or more preschool episodes of lower respiratory disease with wheezing were associated with a child's subsequent (age 6 to 18) lung function decrement. However, in 1993 Tager *et al.* concluded that "preexisting abnormalities" were important determinants of infant lung function and that causal factors might include *in utero* exposure to tobacco smoke or familial factors, which could be either genetic or environmental. Krzyżanowski *et al.* (1990) reported that acute respiratory illness or pneumonia reduced lung function in adults as long as several years, while Jaakola *et al.* (1993) found associations in young adults between development of respiratory symptoms and loss of ventilatory function that could have been either genetic or environmental. One is thus tempted to speculate that a sensitive child may become a sensitive adult and finally an impaired adult, but the evidence to support this hypothesis appears ambivalent.

However, it appears from the reviewed studies that a given percentage change in air pollution has more effect on chronic responses, if concentrations are high enough, than on transient responses. The best examples of this may be the experiments of the Lawther group, who experienced changes of only a few percent under episodic conditions unlikely to be seen there ever again, and of the Los Angeles emphysema patients of Motley *et al.* (1959). Motley *et al.*'s patients were severely impaired individuals, yet they required 40 hours of exposure to clean air to register significant improvement. Data from the U.S. EPA Criteria Document for ozone show that for an average loss in FEV_1 of 5% at light exercise, a 2-hour exposure at nearly 0.4 ppm is required. However, Schwartz's response curves imply that a 5% drop in lung function may result from increasing annual average TSP from 100 to 160 $\mu g/m^3$, NO_2 from 0.04 to 0.06 ppm, or annual average O_3 from 0.04 to 0.06 ppm. These chronic responses are so different from the acute responses that one would suspect that the responsible mechanisms are also quite different (as they certainly are with smoking). The conclusion follows that further investigations of chronic lung responses are needed and that studies should be conducted where there is a large contrast in air quality.

An example of the need to recognize this latter requirement is found in the work of Bouhuys *et al.* (1978, 1979), who compared lung function averages among urban and rural Connecticut residents. The only major differences in air pollution were that TSP and NO_2 were both about 60% higher in the urban setting they studied, compared to the rural area. Ozone was reported to be about the same in all the areas. If the elasticity were −0.05, then a 60% increment in air pollution would yield a 3% change in lung function. The data reported by Bouhuys *et al.* showed 7.7% higher FEV_1 for "lifetime rural" residents, but this difference was not statistically significant in this study. Bouhuys *et al.* also pointed out that such small increments are not clinically significant but was unable to identify whether there were sensitive individuals in the study group. While one cannot argue with his basic premise, that controlling outdoor air pollution will not "prevent" lung disease, one cannot dismiss the effects of air pollution as trivial, either.

Finally, Glindmeyer *et al.* (1992) noted that there appears to have been a long-term improvement in male vital capacity over the last century, after considering differences in test methods and correcting for height. While it is difficult to ascribe any portion of this change to changes in air pollution, the direction of change is consistent with the improvements in urban air quality that resulted from use of cleaner fuels for space heating.

REFERENCES

Amdur, M.O. (1957), The Influence of Aerosols on the Respiratory Response of Guinea Pigs to Sulfur Dioxide, *Am. Ind. Hyg. Assoc. Q.* 18:149–55.

Anderson, K.R., Avol. E.L., Edwards, S.A., Shamoo, D.A., Peng, Ry-Chuan, Linn, W.S., and Hackney, J.D. (1992), Controlled Exposures of Volunteers to Respirable Carbon and Sulfuric Acid Aerosols, *J. Air & Waste Mgmt. Assoc.,* 42:770–76

Arossa, W., Spinaci, S., Bugiani, M., Natale, P., Bucca, C., and de Candussio, G. (1987), Changes in Lung Function of Children after an Air Pollution Decrease, *Arch. Env. Health* 42:170–74

Ayres, S.M., Evans, R., Licht, D., Griesbach, J., Reimold, F., Ferrand, E.F., and Criscitiello, A. (1973), Health Effects of Exposure to High Concentrations of Automotive Emissions, *Arch. Envir. Health* 27:168–78.

Berry, M., Lioy, P.J., Gelperin, K., Buckler, G., and Klotz, J. (1991), Accumulated Exposure to Ozone and Measurement of Health Effects in Children and Counselors at Two Summer Camps, *Envir. Res.* 54:135–50.

Borgers, D. (1987), Daily Symptoms and Lung Function in Relation to Air Pollution: A Study in Berlin (West) 1982/83, in *Health Effects of Air Pollution and the Japanese Compensation Law*, T. Namekata and C. du. V. Florey, eds., Battelle Press, Columbus, OH, pp. 29–45.

Bouhuys, A., Beck, G.L., and Schoenberg, J. (1978), Do Present Levels of Air Pollution Outdoors Affect Respiratory Health? *Nature* 276:466–71.

Bouhuys, A., Beck, G.J., and Schonenberg, J. (1979), Epidemiology of Environmental Lung Disease, *Yale J. Biol. Med.* 52:191–210.

Brooks, A.G.F., and Waller, R.E. (1972), Peak Flow Measurements among Visitors to a Public Health Exhibition, *Thorax* 27:557–62.

Brunekreef, B., Ware, J.H., Dockery, D., Speizer, F.E., Spengler, J.D., and Ferris, B.G., Jr. (1987), Pulmonary Function Changes Associated with Air Pollution Episodes: A Re-Analysis of the Steubenville and Ijmond Alert Studies, Paper 87-33.7, presented at the 80th Annual Mtg, APCA, Air Pollution Control Association, New York, NY.

Brunekreef, B., Dockery, D., Speizer, F.E., Ware, J.H., Spengler, J.D., and Ferris, B.G., Jr. (1989), Home Dampness and Respiratory Morbidity in Children, *Am. Rev. Resp. Dis.* 140:1363–67.

Brunekreef, B., Kinney, P.L., Ware, J.H., Dockery, D., Speizer, F.E., Spengler, J.D., and Ferris, B.G., Jr. (1991), Sensitive Subgroups and Normal Variation in Pulmonary Function Response to Air Pollution Episodes, *Env. Health Perspectives* 90:189–93.

Buist, A.S., Johnson, L.R., Vollmer, W.M., Sexton, G.J., and Kanarek, P.H. (1983), Acute Effects of Volcanic Ash from Mount Saint Helens on Lung Function in Children, *Am. Rev. Resp. Dis.* 127:714–19.

Buist, A.S., Vollmer, W.M., Johnson, L.R., Bernstein, R.S., and McCamant, L.E., (1986), A Four-Year Prospective Study of the Respiratory Effects of Volcanic Ash from Mt. St. Helens, *Am. Rev. Resp. Dis.* 133:526–34.

Burrows, B., Knudson, R.J., and Lebowitz, M.D. (1977), The Relationship of Childhood Respiratory Illness to Adult Obstructive Airway Disease, *Am. Rev. Respir. Dis.* 115:751–60.

Byrd, J.C., Shapiro, R.S., and Schiedermayer, D.L. (1989), Passive Smoking: A Review of Medical and Legal Issues, *Am. J. Public Health* 79:209–15.

Chestnut, L.G., Schwartz, J., Savitz, D.A., and Burchfiel, C.M. (1991), Pulmonary Function and Ambient Particulate Matter: Epidemiological Evidence from NHANES I, *Arch. Env. Health* 46:135–44.

Cohen, B.H. (1980), Chronic Obstructive Pulmonary Disease: A Challenge in Genetic Epidemiology, *Am. J. Epidemiology* 112:274–88.

Colucci, A.V., and Strieter, R.P. (1983) Dose Considerations in the SO_2-Exposed Exercising Asthmatic, *Envir. Health Perspectives* 52:221–32.

Da Silva, A.M., and Hamosh, P. (1980), The Immediate Effect on Lung Function of Smoking Filtered and Nonfiltered Cigarettes, *Am. Rev. Resp. Dis.* 122:794–97.

Dassen, W., Brunekreef, B., Hoek, G., Hofschreuder, P., Staatsen, B., de Groot, H., Schouten, E., and Biersteker, K. (1986), Decline in Children's Pulmonary Function during an Air Pollution Episode, *J. APCA* 36:1223–27.

Densen, P.M., Jones, E.W., and Bass, H.E. (1965), Respiratory Symptoms and Tests among New York City Postmen, *Arch. Env. Health* 10:370–72.

Derriennic, F., Richardson, S., Mollie, A., and Lellouch, J. (1989), Short-Term Effects of Sulphur Dioxide Pollution on Mortality in Two French Cities, *Int. J. Epidemiology* 18:186–97.

Detels, R., Sayre, J.W., Coulson, A.H., Rokaw, S.N., Massey, F.J., Jr., Tashkin, D.P., and Wu, M.-M. (1981), The UCLA Population Studies of Chronic Obstructive Respiratory Disease. IV. Respiratory Effect of Long-Term Exposure to Photochemical Oxidants, Nitrogen Dioxide, and Sulfates on Current and Never Smokers, *Am. Rev. Respir. Dis.* 124:673–80.

Detels, R., Tashkin, D.P., Sayre, J.W., Rokaw, S.N., Coulson, A.H., Massey, F.J., Jr., and Wegman, D.H. (1987), The UCLA Population Studies of Chronic Obstructive Respiratory Disease. IX. Lung Function Changes Associated with Chronic Exposure to Photochemical Oxidants; A Cohort Study among Never-Smokers, *Chest* 92:594–603.

Detels, R., Tashkin, D.P., Sayre, J.W., Rokaw, S.N., Massey, F.J., Coulson, A.H., and Wegman, D.H. (1990), The UCLA Population Studies of Chronic Obstructive

Respiratory Disease. X. A Cohort Study of Changes in Respiratory Function Associated with Chronic Exposure to SO_x, NO_x, and Hydrocarbons, *Am. J. Public Health* 80:1–9.

Dockery, D.W. (1989), Effects of Inhalable Particles on Respiratory Health of Children, *Am. Rev. Respir. Dis.* 139:587–94.

Dockery, D.W., Ware, J.H., Ferris, B.G., Jr., Speizer, F.E., Cook, N.E., and Herman, S.M. (1982), Change in Pulmonary Function in Children Associated with Air Pollution Episodes, *J. APCA* 32:937–42.

Dockery, D.W., Ware, J.H., Ferris, B.G., Jr., Glickberg, D.S., Fay, M.E., Spiro, A., III, and Speizer, F.E. (1985), Distribution of Forced Expiratory Volume in One Second and Forced Vital Capacity in Healthy, White, Adult Never-Smokers in Six U.S. Cities, *Am. Rev. Respir. Dis.* 131:511–20.

Dockery, D.W., Speizer, F.E., Ferris, B.G., Jr., Ware, J.H., Louis, T.A., and Spiro, A. (1988), Cumulative and Reversible Effects of Lifetime Smoking on Simple Tests of Lung Function in Adults, *Am. Rev. Resp. Dis.* 137:286–92.

Dodge, R. (1980), The Respiratory Health of Schoolchildren in Smelter Communities, *Am. J. Industr. Med.* 1:359–64.

Dodge, R. (1983), The Respiratory Health and Lung Function of Anglo-American Children in a Smelter Town, *Am. Rev. Respir. Dis.* 127:158–61.

Dodge, R., Solomon, P., Moyers, J., and Hayes, C. (1985), A Longitudinal Study of Children Exposed to Sulfur Oxides, *Am. J. Epidem.* 121:720–36.

DuBois, A.B., and Dautrebande, L. (1958), Acute Effects of Breathing Inert Dust Particles and of Carbachol Aerosol on the Mechanical Characteristics of the Lungs in Man. Changes in Response After Inhaling Sympathomimetic Aerosols, *J. Clin. Invest.* 37:1746–55.

Ecobichon, D.J., and Wu, J.M., eds. (1990), *Environmental Tobacco Smoke, Proceedings of the International Symposium at McGill University, 1989*, Lexington Books, Lexington, MA.

Eisen, E.A., Robins, J.M., Greaves, I.A., and Wegman, D.H. (1984), Selection Effects of Repeatability Criteria Applied to Lung Spirometry, *Am. J. Epidemiology* 120: 734–42.

Emerson, P.A. (1973), Air Pollution, Atmospheric Conditions, and Chronic Airway Obstruction, *J. Occup. Med.* 15:635–38.

Florey, C. du V., Swan, A.V., van der Lende, R., Holland, W.W., Berlin, A., and di Ferrente, E., eds. (1983), *Report of the EC Epidemiological Survey on the Relationship between Air Pollution and Respiratory Health in Primary Schoolchildren*, Commission of the European Research Communities, Environmental Research Programme, Brussels.

Glindmeyer, H.W., Diem, J.E., Jones, R.N., and Weill, H. (1982), Noncomparability of Longitudinally and Cross-Sectionally Determined Annual Change in Spirometry, *Am. Rev. Respir. Dis.* 125:544–48.

Gong, H., Jr. (1987), *Relationship between Air Quality and the Respiratory Status of Asthmatics in an Area of High Oxidant Pollution in Los Angeles County*, final report submitted to California Air Resources Board, Contracts A1-151-53 and A4-135-33.

Goren, A.I., Hellman, S., Brenner, S., Egoz, N., and Rishpon, S., (1990), Prevalence of Respiratory Conditions among Schoolchildren Exposed to Different Levels of Air Pollutants in the Haifa Bay Area, Israel, *Env. Health Perspectives* 89:225–31.

Higgins, I.T.T., D'Arcy, J.B., Gibbons, D.I., Avol, E.L., and Gross, K.B. (1990), Effect of Exposures to Ambient Ozone on Ventilatory Lung Function in Children, *Am. Rev. Resp. Dis.* 141:1136–46.

Heumann, M., Foster, L.R., Johnson, L., and Kelly, L. (1991), Woodsmoke Air Pollution and Changes in Pulmonary Function among Elementary School Children,

Paper 91-136.7, presented at the 84th Annual Meeting of the Air & Waste Management Association, Vancouver, BC.

Hoek, G., Brunekreef, B., Kosterink, P., van den Berg, R., and Hofshreuder, P. (1993a), Effect of Ambient Ozone on Peak Expiratory Flow of Exercising Children in the Netherlands, *Arch. Env. Health* 48:27–32.

Hoek, G., Fischer, P., Brunekreef, B., Lebret, E., Hofshreuder, P., and Mennen, M.G. (1993b), Acute Effects of Ambient Ozone on Pulmonary Function of Children in the Netherlands, *Am. Rev. Respir. Dis.* 147:111–17.

Holland, W.W., and Reid, D.D. (1965), The Urban Factor in Chronic Bronchitis, *Lancet* 1:445–48.

Hoppenbrouwers, T., (1990), Airways and Air Pollution in Childhood: State of the Art, *Lung* (Suppl.) 335–46.

Humphreys, K., and Carr-Hill, R. (1991), Area Variations in Health Outcomes: Artefact or Ecology, *Int. J. Epidemiology* 20:251–58.

Jaakola, M.S., Jaakola, J.J.K., Ernst, P., and Becklake, M.R. (1993), Respiratory Symptoms in Young Adults Should Not Be Overlooked, *Am. Rev. Respir. Dis.* 147:359–66.

Johnson, K.G., Loftsgarden, D.O., and Gideon, R.A., (1982), The Effects of Mount St. Helens Volcanic Ash on the Pulmonary Function of 120 Elementary Schoolchildren. *Am. Rev. Resp. Dis.* 126:1066–69.

Kagawa, J., and Toyama, T. (1975), Photochemical Air Pollution—Its Effects on Respiratory Function of Elementary Schoolchildren, *Arch. Env. Health* 30:117–29.

Kerr, H.D. (1973), Diurnal Variation of Respiratory Function Independent of Air Quality, *Arch. Env. Health* 26:144–52.

Kinney, P.L., Ware, J.H., and Spengler, J.D. (1988), A Critical Evaluation of Acute Ozone Epidemiology Results, *Arch. Env. Health* 43:168–73.

Kinney, P.L., Ware, J.H., Spengler, J.D., Dockery, D.W., Speizer, F.E., and Ferris, B.G., Jr. (1989), Short-Term Pulmonary Function Change in Association with Ozone Levels, *Am. Rev. Respir. Dis.* 139:56–61.

Koenig, J.Q., Larson, T.V., Hanley, Q.S., Rebolledo, V., Dumler, K., Checkoway, H., Wang, S.-Z., Lin, D., and Pierson, W.E. (1991), Pulmonary Function Changes in Children Associated with Particulate Matter Air Pollution from Wood Smoke, Paper 91-136.3, presented at the 84th Annual Meeting of the Air & Waste Management Association, Vancouver, BC.

Krzyżanowski, M., Quackenboss, J.J., and Lebowitz, M.D. (1990a), Chronic Respiratory Effects of Indoor Formaldehyde Exposure, *Envir. Res.* 52:117–25.

Krzyżanowski, M., Sherrill, D.L., and Lebowitz, M.D. (1990b), Longitudinal Analysis of the Effects of Acute Lower Respiratory Illnesses on Pulmonary Function in an Adult Population, *Am. J. Epidemiology* 131:412–22.

Lawther, P.J., Lord, P.W., Brooks, A.G.F., and Waller, R.E. (1973), Air Pollution and Pulmonary Resistance: A Pilot Study, *Env. Res.* 6:424–35.

Lawther, P.J., Brooks, A.G.F., Lord, P.W., and Waller, R.E. (1974a), Day-to-Day Changes in Ventilatory Function in Relation to the Environment. Part I. Spirometric Values, *Env. Res.* 7:37–40.

Lawther, P.J., Brooks, A.G.F., Lord, P.W., and Waller, R.E. (1974b), Day-to-Day Changes in Ventilatory Function in Relation to the Environment. Part II. Peak Expiratory Flow Values, *Env. Res.* 7:41–53.

Lawther, P.J., Brooks, A.G.F., Lord, P.W., and Waller, R.E. (1974c), Day-to-Day Changes in Ventilatory Function in Relation to the Environment. Part III. Frequent Measurements of Peak Flow, *Env. Res.* 8:119–30.

Lawther, P.J., Macfarlane, A.J., Waller, R.E., and Brooks, A.G.F. (1975), Pulmonary Function and Sulphur Dioxide, Some Preliminary Findings, *Env. Res.* 10:355–67.

Lawther, P.J., Brooks, A.G.F., Lord, P.W., and Waller R.E. (1977), Air Pollution and Pulmonary Airways Resistance: A 6-Year Study with Three Individuals, *Env. Res.* 13:478–92.

Lipfert, F.W. (1978), "The Association of Human Mortality with Air Pollution: Statistical Analyses by Region, by Age, and by Cause of Death," Ph.D. Dissertation, Union Graduate School, Cincinnati, Ohio. Available from University Microfilms.

Loftsgaarden, D.O., Gideon, R.A., and Johnson, K.G. (1981), *Acute Effects of Suspended Particulates on Children's Pulmonary Function Tests*, Interdisciplinary Series No. 13, Mathematics Dept., Univ. of Montana, Missoula.

McKerrow, C.B., and Rossiter, C.E. (1968), An Annual Cycle in the Ventilatory Capacity of Men with Pneumoconiosis and of Normal Subjects, *Thorax* 23:340–49.

Montana Air Pollution Study Staff (MAPS) (1980), *Montana Air Pollution Study*, Air Quality Bureau, Department of Health & Environmental Sciences, Helena, MT.

Moore, E., Martin, J., Muir, D.C.F., and Edwards, A.C. (1988), Pulmonary Function in Silicosis, *Ann. Occup. Hyg.* 32:705–11 (Suppl. 1).

Motley, H.L., and Phelps, H.W. (1964), Pulmonary Function Impairment Produced by Atmospheric Pollution, *Dis. Chest* 45:154–62.

Motley, H.L., Smart, R.H., and Leftwich, C.I. (1959), Effect of Polluted Los Angeles Air (Smog) on Lung Volume Measurements, *JAMA* 171:1469–77.

Musk, A.W., Peters, J.M., and Wegman, D.H. (1977a), Lung Function in Fire Fighters. I: A Three Year Follow-up of Active Subjects, *Am. J. Public Health* 67:626–29.

Musk, A.W., Peters, J.M., and Wegman, D.H. (1977b) Lung Function in Fire Fighters. II: A Five Year Follow-up of Retirees, *Am. J. Public Health* 67:630–33.

National Research Council (1986), *Environmental Tobacco Smoke, Measuring and Assessing Health Effects*, National Academy Press, Washington, DC.

O'Brien, R.J., and Drizd, T.A. (1981), *Basic Data on Spirometry in Adults 25–74 Years of Age*, DHHS Publ. No. (PHS) 81-1672, National Center for Health Statistics, Hyattsville, MD.

PAARC (Groupe Coopératif) (1982a), Pollution Atmosphérique et Affections Respiratores Chroniques ou à Répétition. I. Méthodes et Sujets, *Bull. Europ. Physiopath. Resp.* 18:87–99.

PAARC (Groupe Coopératif) (1982b), Pollution Atmosphérique et Affections Respiratories Chroniques ou à Répétition. II. Résultats et Discussion, *Bull. Europ. Physiopath. Resp.* 18:101–16.

Pope, C.A., III, and Dockery, D.W. (1992) Acute Health Effects of PM_{10} Pollution on Symptomatic and Asymptomatic Children, *Am. Rev. Resp. Dis.* 145:1123–28.

Pope, C.A., III, Dockery, D.W., Spengler, J.D., and Raizenne, M.E. (1991), Respiratory Health and PM_{10} Pollution: A Daily Time Series Analysis, *Am. Rev. Resp. Dis.* 144:668–74.

Reinberg, A., and Gervais, P. (1972), Circadian Rhythms in Respiratory Functions, with Special Reference to Human Chronophysiology and Chronopharmacology, *Bull. Physio-Path. Resp.* 8:663–75.

Rokaw, S.N., and Massey, F. (1962), Air Pollution and Chronic Respiratory Disease, *Am. Rev. Respir. Dis.* 86:703–04.

Rokaw, S.N., Detels, R., Coulson, A.H., Sayre, J.W., Tashkin, D.P., Allwright, S.S., and Massey, J.F., Jr. (1980), The UCLA Population Studies of Chronic Obstructive Respiratory Disease. 3. Comparison of Pulmonary Function in Three Communities Exposed to Photochemical Oxidants, Multiple Primary Pollutants, or Minimal Pollutants, *Chest* 78:252–62.

Rubin, D.H., and Damus, K. (1988), The Relationship between Passive Smoking and Child Health: Methodologic Criteria Applied to Prior Studies, *Yale J. Biol. Med.* 62:401–11.

Samet, J.M., Tager, I.B., and Speizer, F.E. (1983), The Relationship Between Respiratory Illness in Childhood and Chronic Air-Flow Obstruction in Adulthood, *Am. Rev. Respir. Dis.* 127:508–23.

Schenker, M.B., Speizer, F.E., Samet, J., Gruhl, G., and Batterman, S. (1983), Health Effects of Air Pollution Due to Coal Combustion in the Chestnut Ridge Region of Pennsylvania: Results of Cross-Sectional Analysis of Adults, *Arch. Env. Health* 38:325–30.

Schenker, M.B., Vedal, S., Batterman, S., Samet, J., and Speizer, F.E. (1986), Health Effects of Air Pollution Due to Coal Combustion in the Chestnut Ridge Region of Pennsylvania: Cross-Section Survey of Children, *Arch. Env. Health* 41:104–108.

Schwartz, J. (1989), Lung Function and Chronic Exposure to Air Pollution: A Cross-Sectional Analysis of NHANES II, *Env. Res.* 50:309–21.

Shephard, R.J., Turner, M.E., Carey, G.C.R., and Phair, J.J. (1960), Correlation of Pulmonary Function and Domestic Microenvironment, *J. Appl. Physiology* 15:70–76.

Shy, C.M., Creason, J.P., Pearlman, M.E., McClain, K.E., Benson, F.B., and Young, M.M. (1970), The Chattanooga Schoolchildren Study: Effects of Community Exposure to Nitrogen Dioxide. 1. Methods, Description of Pollutant Exposure, and Results of Ventilatory Function Testing, *J. APCA* 20:539–45.

Smith, R.M., and Dinh, V.-D. (1975), Changes in Forced Expiratory Flow Due to Air Pollution from Fireworks, *Env. Res.* 9:321–31.

Sparrow, D., Silbert, J.E., and Weiss, S.T. (1982), The Relationship of Pulmonary Function to Copper Concentrations in Drinking Water, *Am. Rev. Respir. Dis.* 126:312–15.

Sparrow, D., Rosner, B., Cohen, M., and Weiss, S.T. (1983), Alcohol Consumption and Pulmonary Function, *Am. Rev. Respir. Dis.* 127:735–38.

Speizer, F.E., and Tager, I.B. (1979), Epidemiology of Chronic Mucuc Hypersecretion and Obstructive Airways Disease, *Epidemiologic Reviews* 1:124–42.

Spektor, D.M., Hofmeister, V.A., Artaxo, P., Brague, J.A.P., Echelar, F., Nogueira, D.P., Hayes, C., Thurston, G.D., and Lippmann, M. (1991), Effects of Heavy Industrial Pollution on Respiratory Function in the Children of Cubatao, Brazil: A Preliminary Report, *Env. Health Perspectives* 94:510–54.

Spicer, W.S., Jr., Storey, P.B., Morgan, W.K.C., Kerr, H.D., and Standiford, N.E. (1962), Variations in Respiratory Function in Selected Patients and Its Relation to Air Pollution, *Am. Rev. Respir. Dis.* 86:705–19.

Spinaci, S. (1985), The Effects of Air Pollution on the Respiratory Health of Children: A Cross-Sectional Study, *Pediatr. Pulmonol.* 1:262–66.

Spodnik, M.J., Jr., Cushman, G.D., Kerr, D.H., Blide, R.W., and Spicer, W.S., Jr. (1966), Effects of Environment on Respiratory Function, *Arch. Env. Health* 13:243–54.

Stacy, R.W., Seal, E., Jr., House, D.E., Green, J., Roger, L.J., and Raggio, L. (1983), A Survey of the Effects of Gaseous and Aerosol Pollutants on Pulmonary Function of Normal Males, *Arch. Env. Health* 38:104–15.

Stebbings, J.H., Jr., and Fogleman, D.G. (1979), Identifying a Susceptible Subgroup: Effect of the Pittsburgh Air Pollution Episode upon Schoolchildren, *Am. J. Epidemiology* 110:27–40.

Stebbings, J.H., Fogleman, D.G., McClain, K.E., and Townsend, M.C. (1976), Effect of the Pittsburgh Air Pollution Episode upon Pulmonary Function in Schoolchildren, *J. APCA* 26:547–53.

Strope, G.L., Stewart, P.W., Henderson, F.W., Ivins, S.S., Stedman, H.C., and Henry, M.M. (1991), Lung Function in School-Age Children Who Had Mild Lower Respiratory Illnesses in Early Childhood, *Am. Rev. Respir. Dis.* 144:655–62.

Tager, I.B., Weiss, S.T., Munoz, A., Rosner, B., and Speizer, F.E. (1983), Longitudinal Study of the Effects of Maternal Smoking on Pulmonary Function in Children,

N. Engl. J. Med. 309:699–703.

Tollerud, D.J., Weiss, S.T., Elting, E., Speizer, F.E., and Ferris, B. (1983), The Health Effects of Automobile Exhaust. VI. Relationship of Respiratory Symptoms and Pulmonary Function in Tunnel and Turnpike Workers, *Arch. Env. Health* 38:334–39.

Tsimoyianis, G.V., Jacobson, M.S., Feldman, J.G., Antonio-Santiago, M.T., Clutario, B.C., Nussbaum, M., and Shenker, R.I. (1987), Reduction in Pulmonary Function and Increased Frequency of Cough Associated with Passive Smoking in Teenage Athletes, *Pediatrics* 80:32–36.

Ulfvarson, U., and Alexandersson, R. (1990), Reduction in Adverse Effect on Pulmonary Function after Exposure to Filtered Diesel Exhaust, *Am. J. Ind. Hyg.* 17:341–47.

U.S. Department of Health, Education and Welfare (1979), *Smoking and Health, A Report of the Surgeon General*, DHEW Publ. No. (PHS) 79-50066, Office on Smoking and Health, Rockville, Md.

U.S. Department of Health and Human Services (1986), *The Health Consequences of Involuntary Smoking*, Office on Smoking and Health, Rockville, MD.

Utell, M.J., Morrow, P.E., and Bauer, M.A. (1990), Effects of Inhaled Nitrogen Dioxide on Respiratory Function: Controlled Clinical Studies, Paper 90-147.3, presented at the 83rd Annual Meeting of the Air & Waste Management Association, Pittsburgh, PA.

van der Lende, R., Visser, B.F., Wever-Hess, J., Tammeling, G.J., de Vries, K., and Orie, N.G.M. (1973), Epidemiological Investigations in the Netherlands into the Influence of Smoking and Atmospheric Pollution on Respiratory Symptoms and Lung Function Disturbances, *Pneumonologie* 149:119–26.

van der Lende, R., Schouten, J.P., Rijcken, B., and van der Meulen, A. (1986), Longitudinal Epidemiological Studies on Effects of Air Pollution in the Netherlands, in *Aerosols: Research, Risk Assessment, and Control Strategies* S.D. Lee, T. Schneider, L.D. Grant, and P.J. Verkerk, eds., Lewis Publishers, Chelsea, MI, pp. 731–42.

Vedal, S., Schenker, M.B., Munoz, A., Samet, J.M., Batterman, S., and Speizer, F.E. (1987), Daily Air Pollution Effects on Children's Respiratory Symptoms and Peak Expiratory Flow, *Am. J. Public Health* 77:694–98.

Ware, J.H., Ferris, B.G., Jr., Dockery, D.W., Spengler, J.D., Stram, D.O., and Speizer, F.E. (1986), Effects of Ambient Sulfur Oxides and Suspended Particles on Respiratory Health of Preadolescent Children, *Am. Rev. Respir. Dis.* 133:534–42.

Watkins, C.J., Sittampalam, Y., and Bartholomew, J. (1986), Outcome of Respiratory Illness Occurring in the First Year of Life, *Brit. Med. J.* 293:925–28.

Witorsch, P. (1990), Effects of ETS Exposure on Pulmonary Function and Respiratory Health in Adults, in *Environmental Tobacco Smoke, Proceedings of the International Symposium at McGill University, 1989*, D.J. Ecobichon and J.M. Wu, eds., Lexington Books, Lexington, MA, pp. 169–86.

Witorsch, R.J. (1990), Parental Smoking and Respiratory Health and Pulmonary Function in Children: A Review of the Literature, in *Environmental Tobacco Smoke, Proceedings of the International Symposium at McGill University, 1989*, D.J. Ecobichon and J.M. Wu, eds., Lexington Books, Lexington, MA, pp. 205–26.

Xu, X., Dockery, D.W., and Wang, L. (1991), Effects of Air Pollution on Adult Pulmonary Function, *Arch. Env. Health* 46:198–206.

APPENDIX 11A A STUDY OF BRIDGE AND TUNNEL WORKERS IN NEW YORK CITY

A thorough study of New York City bridge and tunnel workers was reported by Ayres *et al.* (1973). These men were exposed to very high levels of traffic-related pollutants, especially those who worked in tunnels. Thirty-day averages were reported outside the Queens Midtown Tunnel as 63 ppm for CO, 1.4 ppm for NO_x (mostly as NO), 64 $\mu g/m^3$ RSP, 200 $\mu g/m^3$ TSP (COH = 1.7), and 30 $\mu g/m^3$ lead. There were no oxidants present; SO_2 was not reported but was probably also low, since motor fuel in New York is predominantly gasoline and the sulfur content of diesel fuel is usually low, of the order of 0.1%. Lung function data were taken and respiratory symptoms recorded, by smoking habit. Blood chemistry was also determined. Acute responses were deduced by comparing bridge workers with tunnel workers immediately after a shift. Chronic responses were deduced by obtaining detailed spirometry on a nonwork day and comparing with nonexposed workers. The largest acute responses seemed to be in flow rates; FEV_1 and FVC did not differ between the two groups of workers. Table 11A-1 compares these findings.

These data indicate both acute and chronic effects, although it is possible that one day's respite was not sufficient for lung function values to recover. This does not seem to be the case, since the maximum flow rates (MEFR) were not depressed during the day-off examination. It would also appear that the acute differences between bridge and tunnel workers are small compared to the deviations from normal values. Ayres *et al.* (1973) cite the large deviations in the flow rates after expiration of half capacity as evidence of chronic obstructive pulmonary disease. Data on closing volumes were also cited as evidence of small airway disease. Interestingly, respiratory symptom prevalence (cough, wheeze) differed significantly between bridge and tunnel workers only for nonsmokers. This study provides a clear indication that severe exposure to vehicle exhaust, in the absence of oxidants, can result in both acute and chronic respiratory dysfunction.

TABLE 11A-1 Lung Function Data for Bridge and Tunnel Workers

	After a Shift*		On Day Off		
Measure	Tunnel	Bridge	Nonsmokers	Smokers	Normal
FVC(L)	3.90	3.84	4.19	3.98	4.60
MEFR (L/min)	322	336	470	457	445
MMFR (L/min)	177	205	234	231	350
R_{aw} cm $H_2O/L/s$			2.85	2.86	1.70†

*The regression equations presented by Ayres *et al.* apparently contain typographical errors, since they do not yield the mean predicted values for the reference conditions (44 yrs of age, 175 cm tall). I assumed that the mean values, which also appear in the text, were correct.

†Average of smokers and nonsmokers.

Source: Ayres *et al.* (1973).

REFERENCE

Ayres, S.M., Evans, R., Licht, D., Griesbach, J., Reimold, F., Ferrand, E.F., and Criscitiello, A. (1973), Health Effects of Exposure to High Concentrations of Automotive Emissions, *Arch. Envir. Health* 27:168–78.

PART IV

Synthesis and Conclusions

If there has to be a choice between trusting observation or reason, it is surely better to trust observation; for reason is so often based on incomplete information on transfer routes, doses, and responses. The argument that what cannot be explained cannot occur is weak, for one can never exclude the explanation that has not been considered. In estimating risks, direct observations of evident health effects should take priority over theoretical expectations.

G. Rose, Dept. of Epidemiology and Population Sciences,
London School of Hygiene and Tropical Medicine

12

Respiratory Function as a Predictor of Mortality

These data show without question that the heart of the problem is in the lungs.

H.A. Menkes *et al.* (1984)
on the role of pulmonary dysfunction in the
evolution of extrapulmonary disease

In Parts II and III, excess rates of mortality and hospitalization and changes in lung function were shown to be significantly associated with air pollution. Since most of these studies were ecological by design (personal air pollution exposures were not measured), causality cannot be proven from their statistical fiindings, which could be regarded as only circumstantial. The case for causality would be strengthened if plausible physiological mechanisms were identified that could be consistent with the observed effects. Both qualitative and quantitative results are required to make this case.

For many years, there has been interest in developing individual predictive measures for mortality and morbidity, as well as in identifying risk factors for specific diseases. Various measures of respiratory function have been among the many factors used for this purpose. Most of these studies have been longitudinal in design, with specific account taken of individual, as opposed to group, risk factors. In this type of study, the individual serves as his or her own control as was the case with many of the lung function studies reviewed in Chapter 11. The reservations expressed in Chapter 3 about the "ecological fallacy" thus do not apply to this chapter.

The studies discussed in this chapter identified various measures of lung function as independent risk factors for heart disease morbidity and mortality, as well as for all-cause mortality. Since Chapter 11 established the association of reduced lung function with air pollution, combining the two sets of relationships could provide an independent estimate of the plausible order of magnitude of effects of air pollution on mortality for causes other than respiratory disease. However, such a synthetic approach to the problem must also show consistency with the direct evidence, both qualitatively and quantitatively, in order to help "bridge the gap." The question underlying this chapter is whether reduced lung function serves as a marker for increased risk of death, even in the absence of diagnosed disease or impairment.

Since several different measures of respiratory function have been used by the various authors, it is convenient to use elasticities for comparisons, following the examples of Sections II and III. To repeat the definition: Elasticity is defined as the percentage change in an independent variable (risk factor) associated with a given percentage change in a dependent variable, usually defined at the mean. A 1:1 proportional relationship thus has an elasticity value of 1.0. In some cases, it was necessary to estimate elasticities from categorical data, in which the mean values were not explicitly stated but could be estimated from the distributional data that were given.

Some of the studies reviewed in this chapter used the Cox proportional hazards model, which derives best-fit coefficients as powers of *e*, expressed for unit values of each independent variable (Cox, 1972). For an independent variable given in units of percent, the relationship between the proportional hazards coefficient (B_{pr}) and the elasticity is direct:

$$e = 100B_{pr}. \qquad [12\text{-}1]$$

REVIEWS AND SUMMARIES OF EPIDEMIOLOGICAL STUDIES OF LUNG FUNCTION AS A PREDICTOR OF PREMATURE MORTALITY

Although lung function (vital capacity) had been recognized as a diagnostic tool many years ago (McClure and Peabody, 1917; Wilson and Edwards, 1922), Higgins and Keller (1970) were among the first to recognize its predictive value (FEV_1) for all-cause mortality, not just for respiratory disease. The population studied was about 5000 persons over age 16 in Tecumseh, MI (a rural location northwest of Toledo, OH), who were examined in 1959–60. However, Higgins and Keller's analysis treated smoking as an alternative, rather than a joint, risk factor. For men, the mortality risk due to smoking was much greater than for low lung function (<2.0 L); for women, it was slightly less. Their analysis was based on 179 deaths; the numbers of low lung function cases were low, but mortality ratios for cardiovascular causes were in the range of 1.1 to 1.4.

The Seven Countries Study (Keys, 1980; Keys *et al.*, 1972) involved more than 12,000 men drawn from 16 cohorts in Europe, Japan, and the United States, with 5- and 10-year follow-ups and emphasis on coronary heart disease (CHD). All cohorts were aged 40–59, and special attention was paid to the cohort that was free of heart disease at entry to the study. The 1972 report dealt with the cohorts from Finland, Italy, Greece, and Yugoslavia, and U.S. railroad workers from the northeastern quadrant of the country. The analysis was limited to quintile analyses of the relationship between lung function and the development of CHD after 5 years. Both CHD diagnosis and "hard CHD" (death or definite heart attack) were considered. The latter analysis was limited by the small numbers of cases, but a multiple logistic regression for 261 cases of any CHD symptom (U.S. and Finland pooled) failed to find significance for FVC after 5 years of follow-up. Tibblin *et al.* (1975) also noted differences in the significance of risk factors, including lung function (peak expiratory flow), depending on whether fatal or nonfatal events are considered.

By 1980, lung function data were available for FVC for 10 cohorts of the Seven Countries Study, but $FEV_{0.75}$ was only available for eight of the cohorts. Differences in the instruments used for various cohorts were noted. The additional cohorts were Yugoslav groups from the University of Belgrade and three rural areas. For the purposes of lung function analysis, the cohorts were pooled into five country groups. Of these five groups, all but the Greek cohort showed a significantly lower mean ratio of FVC to height for those who died from all causes. This was also the case for CHD deaths for the U.S. and rural Italy; in Greece and Yugoslavia, there were probably too few CHD deaths for meaningful analysis. Using $FEV_{0.75}$ as the metric yielded qualitatively similar results, except that the $FEV_{0.75}$ difference was more significant for all-cause deaths than FVC and there was a significant difference in Finland for CHD deaths.

Lung function data from the Framingham study, a whole-community cohort study in Massachusetts begun in 1948, have been reported by Ashley *et al.* (1975) and by Kannel *et al.* (1983). All-cause mortality was examined in the first study using bivariate regression analysis for various subgroups. On a bivariate basis with no height adjustments to the spirometric data, smoking had a small (barely significant) effect of FVC, which had a large effect on mortality ($e = -0.92$). A multiple logistic regression was discussed (but no quantitative results were given) in which FVC/height remained significant in combination with a number of risk factors, including smoking. The authors expressed the opinion that reduced FVC was a manifestation of heart disease, rather than a consequence of smoking or "other environmental insults."

The second Framingham report (Kannel *et al.*, 1983) examined the relationship of FVC to cardiovascular risk factors and mortality, as well as noncardiovascular mortality. "Vital capacity index" (FVC/height) was used as the lung function metric, for all subjects and for nonsmokers. A subgroup of the population, "nonsmokers free from pulmonary disease and cardiac failure" was singled out for separate analysis. Figure 12-1 plots the cardiovascular mortality rates against categorical values of the vital capacity index. (The slopes of these graphs differed from the regression coefficients given in the paper, presumably because the graph weights each category or "bin" equally, whereas the actual data will be unequally distributed among the bins and were computed according to the logistic model.) For bivariate regressions of age-adjusted cardiovascular mortality on FVC/height, elasticities (based on the categorical data) of -2.0 were found for nonsmokers (men and women), -1.25 for all men, and -2.5 for all women. The effects were weaker for all-cause mortality and for cardiovascular morbidity and were reduced by 20 to 30% when age was entered as an additional independent variable. Kannel *et al.* presented another table of correlations not found elsewhere: vital capacity index versus right-hand grip strength. These values suggested that vital capacity may be an index of overall physical vitality; it was not correlated with physical activity. Kannel *et al.*

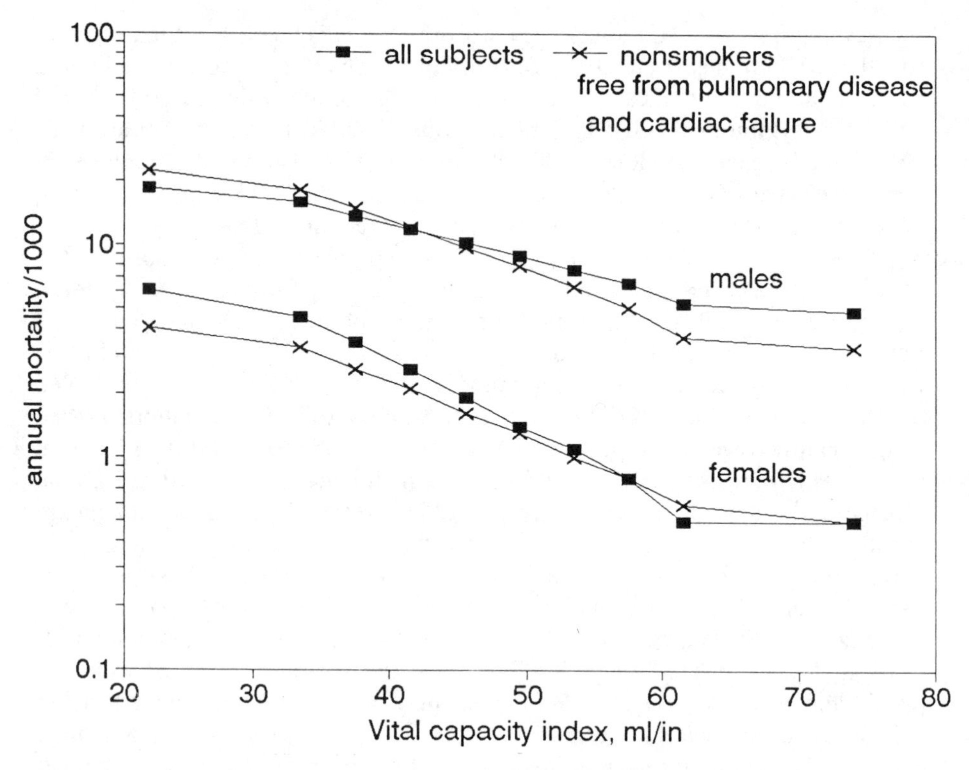

FIGURE 12-1. Risk of mortality versus vital capacity index (FVC/height) from the Framingham study. *Data from Kannel* et al. *(1983)*.

recommended that vital capacity be given a "prominent place among cardiovascular risk factors," but also noted that they knew of little evidence to suggest that vital capacity was a "modifiable factor."

Friedman *et al.* (1976) used a cohort of heart attack patients and selected controls in the San Francisco Bay area to examine the role of lung function as a predictor of heart attack or sudden cardiac death. The average time between lung function measurement and attack was about 17 months. The elasticity of heart attack on the ratio of actual to expected FVC was about −0.7, in a multiple regression model with blood pressure, cholesterol, and smoking categories as additional independent risk factors. The ratio of observed to predicted FVC was only weakly correlated with smoking. The differences in FVC and FEV_1 between heart attack–sudden death groups, compared to controls, were in the range of 0.1 to 0.3 L (5–10%).

De Hamel and de Souza (1978) followed 1566 New Zealand men for 15 years beginning in 1958 and compared the lung functions of the deceased (nonaccidental deaths) with the survivors, using regressions on weight and age, and found significantly lower values ($p < 0.0001$) for the deceased. Many of these men had worked as coal miners. The mean differences, for age 60 and 70 kg body weight, were 7.5% in FVC and 11.6% in $FEV_{0.75}$. The differences were also significant when stratified by smoking habit, for never smokers, ex-smokers, and moderate smokers, but not for heavy smokers. The magnitude of these lung function differences is substantially larger than the typical effects

TABLE 12-1 Statistical Significance of Lung Function in Predicting Mortality in a Western Australian Cohort (Cullen *et al.*, 1983)

Cause of Death	Men (40–59)	Men (60–74)	Women (40–59)	Women (60–74)
All	%FEV_1, $p < 0.001$	%FEV_1, $p < 0.001$	NS	FVC, $p < 0.05$
CVD (ICD 390–458, 746–7)	%FEV_1, $p < 0.001$	%FEV_1, $p < 0.001$	NS	%FEV_1, $p < 0.05$
CHD (ICD 410–14)	NS	NS	NS	NS
Cancer (ICD 140–209)	$p < 0.01$	NS	NS	NS

Note: NS = nonsignificant.

seen due to air pollution (Chapter 11). This study is of interest because of the length of follow-up time, between lung function measurement and death.

Beaty *et al.* (1982) studied the mortality experience of the control population which had been defined for a study of chronic obstructive pulmonary disease in Baltimore. Percent of predicted FEV_1 was used as the lung function metric, as associated with a number of risk factors, with the Cox proportional hazards model. The elasticities for men and women were each about −2.7, when entered into a regression along with age, race, and two categorical smoking variables. According to this regression, a single percentage point in FEV_1 had the same effect on risk of death as 4–5 months in age (ages 20–60+).

Cullen *et al.* (1983) followed a cohort in Western Australia for 13 years. During that time, 180 men and 68 women died. Multiple regressions for cardiovascular (CVD) or coronary heart disease (CHD) deaths were performed against lung function and several other risk factors. The lung function parameters were either FVC or percentage of predicted FEV_1 (% FEV_1); the statistical significance of these measures in predicting mortality is given in Table 12-1. The authors presented their findings in terms of standardized regression coefficients; it was thus not possible to estimate elasticities. The nonsignificant findings could have resulted from the small numbers of deaths in these subgroups (32 or fewer).

A 24-year follow-up of 874 healthy men was available from the Baltimore Longitudinal Study of Aging (Beaty *et al.*, 1985). The distributions of the 239 deaths by underlying cause of death are plotted in Figure 12-2, as subdivided according to "normal" or subnormal FEV_1 (<80% of predicted). The total rates for these two subgroups for the 24 years were 23% and 32%, respectively. The percentages of the two groups dying during the 24 years are shown in Figure 12-2, by cause of death. The low lung function group had relatively more deaths from heart disease, chronic obstructive pulmonary disease (COPD), infections diseases, and stroke, but fewer for lung cancer, external causes (accidents, etc.), and other circulatory causes.

Foxman *et al.* (1986) studied a male industrial population in Staveley, England, from 1957 to 1977. During these 20 years, 69% of the men who were aged 55–64 at first examination died. The authors compared mortality rates by lung function and symptom level and concluded that lung function was a significant predictor of mortality. However, they did not account for any

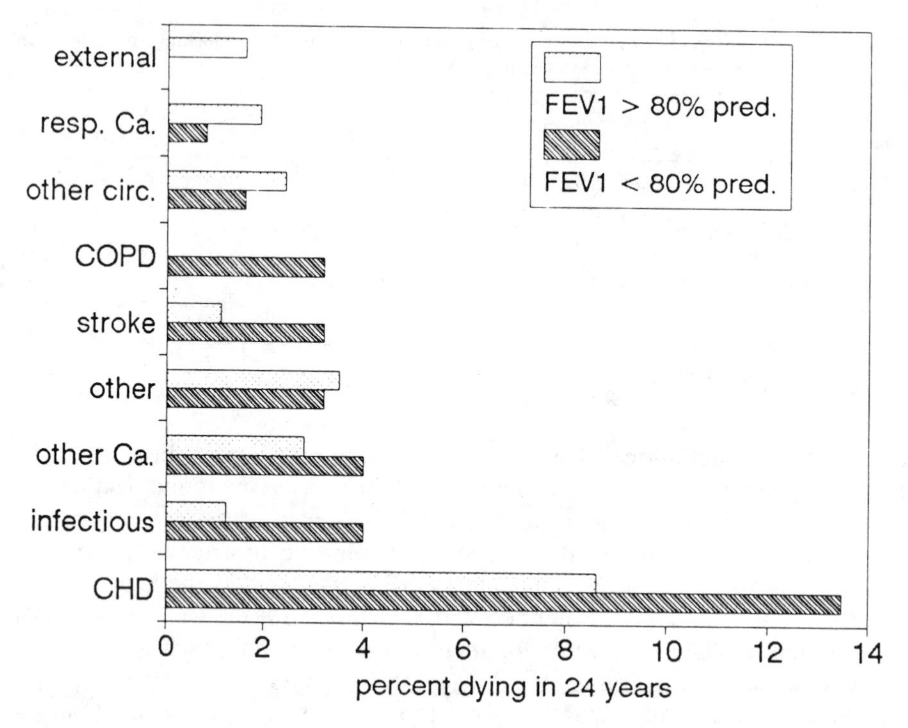

FIGURE 12-2. Comparison of distributions of causes of death by FEV_1. *Data from Beaty* et al. *(1985)*.

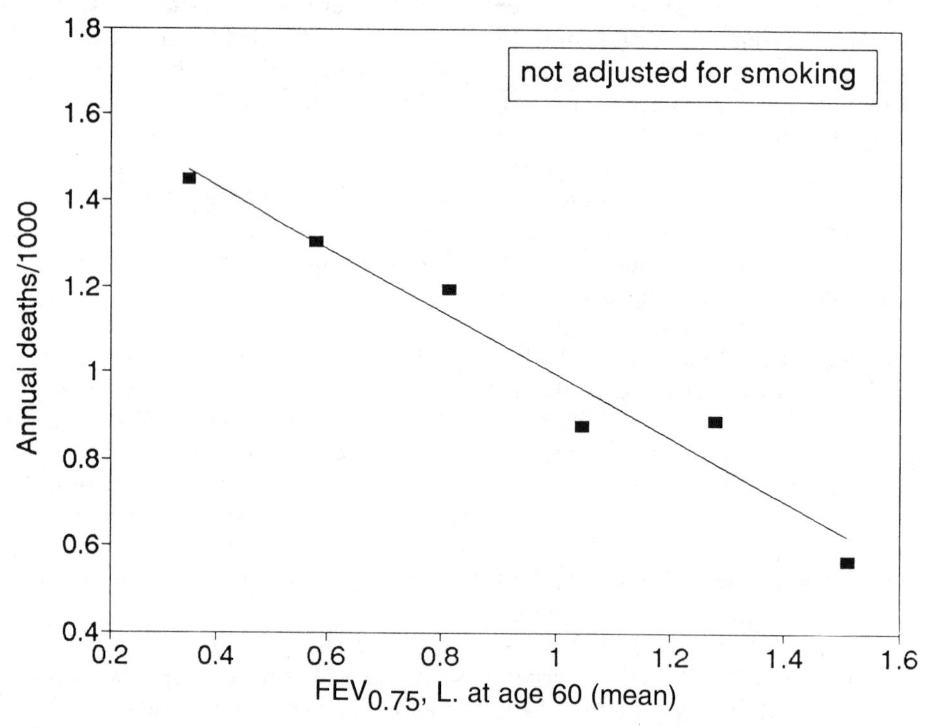

FIGURE 12-3. Excess mortality versus $FEV_{0.75}$ for British men, ages 60–80. *Data from Foxman* et al. *(1986)*.

confounding due to differences in smoking or physical stature. The data of Foxman *et al.* are plotted in Figure 12-3 and form a reasonably linear dose-response function. These are no clear indications that the excess mortality risk might be confined to the lowest lung function cases.

Krzyżanowski and Wysocki (1986) followed (1986) followed about 3000 men and women in Cracow, Poland, for 13 years, during which 523 deaths occurred. They used FEV_1 and the ratio of FEV_1 to FVC as the measures of ventilatory performance. For men, the approximate elasticity for all-cause mortality was −1.5; for women, −0.7. Multiple risk factors were not accounted for in these estimates, but in separate multiple logistic regressions, a FEV_1:FVC ratio below 0.65 was a significant risk factor for men and for women who also had heart disease. The other risk factors accounted for in the multiple regressions were age, smoking, health status, weight/height^2 (men) and age, smoking, heart disease, and chronic phlegm production (women).

A group of Swedish women in Gothenburg aged 38–60 were followed for 12 years by Persson *et al.* (1986). The lung function measure used was peak expiratory flow (PEF), in both bivariate and multivariate logistic regression analyses. Other variables considered included age, body height and mass, adipose tissue distribution, chest deformity, history of pulmonary disease, smoking, cholesterol and triglycerides, blood pressure, diabetes, and physical activity. Insufficient data were given in this paper to calculate a precise value of elasticity; my estimate of the elasticity of total mortality on peak flow was about −0.9; the corresponding value for heart attacks was about −1.2. About

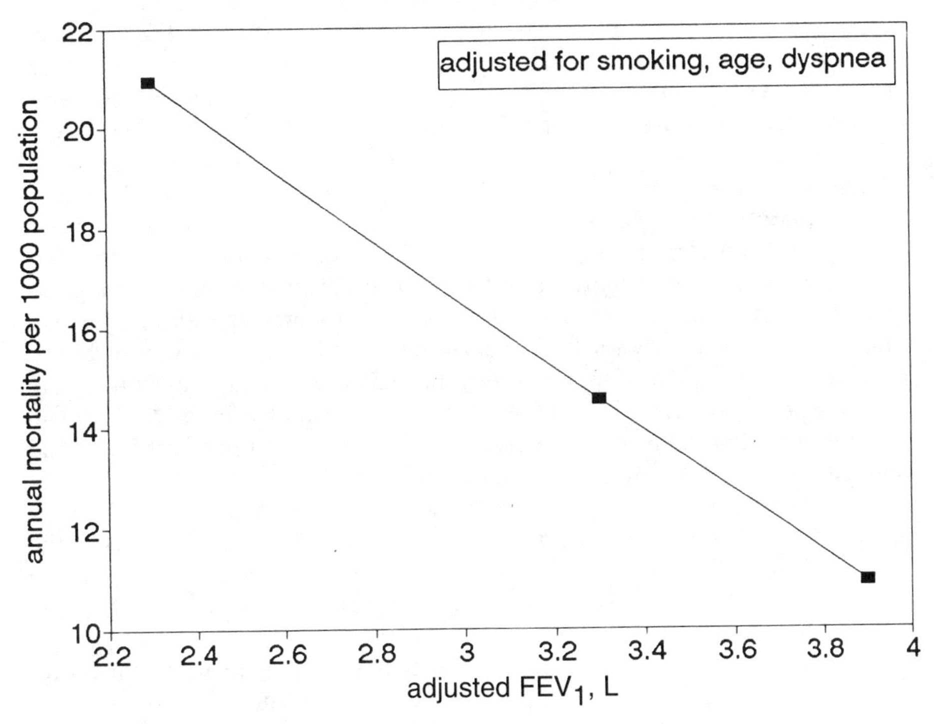

FIGURE 12-4. Annual mortality rate versus adjusted FEV_1 for Swedish males, ages 50–60. *Data from Olofson* et al. *(1987).*

40% of the group were smokers. This analysis was based on 23 heart attacks and 75 deaths.

Olofson *et al.* (1987) studied 607 Swedish men ages 50 and 60 for 11 years, during which 109 deaths occurred. The measure of lung function used was FEV_1; Figure 12-4 presents the trend of all-cause mortality, after adjusting for smoking, age and shortness of breath; the elasticity is about −1.1. Although there are only three points, there is no hint of nonlinearity.

About 18,000 British male civil servants aged 40–64 were studied by Ebi-Kryston (1988), with follow-up 10 years after examination. The measure of lung function used was the observed percentage of FEV_1 predicted on the basis of the subject's age and height. Using Cox proportional hazards regression models with variables for age, smoking status, employment grade, blood pressure, use of hypertension medication, cholesterol, diabetes, EKG changes, and presence of myocardial ischemia, the elasticities for mortality on lung function were −6 for chronic respiratory disease, −0.6 for cardiovascular disease, and −1.1 for all causes. This analysis was based on 1670 deaths. In a follow-up report (Ebi-Kryston *et al.*, 1989), three other cohorts were checked for consistency with the British civil servants; breathlessness was a better predictor of cardiovascular disease (CVD) mortality than impaired lung function ($FEV_1 <$ 65% predicted). This measure remained significant in all four studies even with other CVD risks included in the models. The range of the relative risks of mortality due to breathlessness was 1.9 to 2.5.

Cook and Shaper (1988) emphasized the role of "breathlessness" (dyspnea) and height-standardized FEV_1 in their report from the British Regional Heart Study, in which 7735 British men aged 40–59 from 24 towns in England, Wales, and Scotland were followed for 7.5 years, on average. The endpoint used was "major ischemic heart disease (IHD) event," defined as sudden cardiac death or heart attack (fatal or nonfatal). After adjusting for age, smoking, blood pressure, cholesterol, and the existence of a prior diagnosis of IHD, the rates per thousand men per year increased with the degree of breathlessness and decreased with FEV_1; 443 such events occurred. In this paper, the contrast between reduced lung function as a *consequence* of and as a *predictive* tool for IHD is discussed. The authors concluded that lung function testing has value as a diagnostic tool for IHD. From the regression data given for excess IHD risk as a function of FEV_1, I estimated the elasticity to be about −1.7; the elasticity for fatal events only would be lower. With regard to the choice of lung function measure, the 3.5% of the group with severe breathlessness had a risk for IHD of 1.3 with 95% confidence limits of 0.9 and 2.0, after adjustment for the other risk factors. The 20% with lowest FEV_1 had a risk of 1.6 (1.1, 2.3), after similar adjustment. It thus appears that FEV_1 is a slightly better index of risk. I plotted the relative odds ratios for an IHD "event" (fatal or nonfatal heart attack) against the estimated midpoints of the FEV_1 midpoints (Figure 12-5). Although the confidence limits for all but the lowest lung function quintile include 1.0, the mean values form a linear dose-response function that nearly reaches significance. One would have to pool all the individual data in order to test the linearity of such a model with confidence.

The role of lung function as a predictor of the incidence of coronary heart disease was studied by Marcus *et al.* (1989) as part of the Honolulu Heart Program. The population was a cohort of Japanese-American men who were

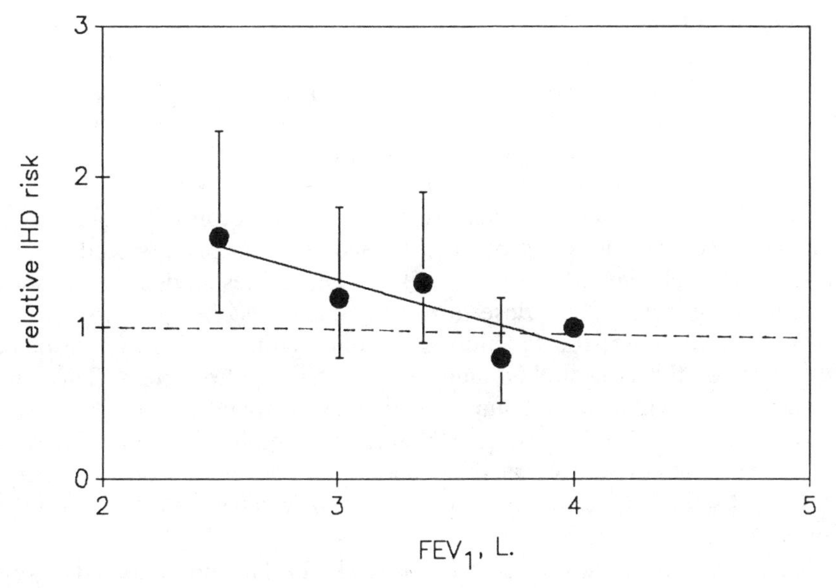

FIGURE 12-5. Relative risk of ischemic heart disease mortality versus FEV_1 for British men, ages 40–59. *Data from Cook and Shaper (1988).*

initially free of heart disease; they were followed for 15–18 years. During this time, 484 coronary heart disease (CHD) cases developed, defined as either CHD death or nonfatal heart attack. Percent of predicted FEV_1 (% $PFEV_1$) based on a healthy subgroup was used as the lung function metric, with a quintile analysis used to examine the relationships. Smoking was handled in two separate ways: first, the entire cohort was analyzed using the Cox proportional hazards model with % $PFEV_1$ and smoking status as independent variables. On this basis, lung function was highly significant, and the estimated elasticity was −0.82, including adjustment for other risk factors (pack-years of smoking, age, blood pressure, cholesterol, skinfolds, glucose, alcohol, and physical activity). The second method of analysis was by subgroups defined according to smoking status. Current and past smokers showed significant effects due to lung function, but never-smokers did not ($p = 0.85$). Further analysis of the entire cohort showed a significant interaction between lung function and smoking. The authors thus concluded that cigarette smoking was the operative variable and not lung function.

This conclusion was challenged by Cook and Shaper (1990), who felt that age should not be included as an explanatory variable along with age-adjusted lung function. They also pointed out that there would be missing risk factors (i.e., unexplained risk) for the ex-smoking group if lung function were excluded. In his reply, Marcus (1900) reported subsequent analysis of never-smokers using "raw" FEV_1 instead of percent of the predicted values; FEV_1 became a significant predictor on this basis. However, Marcus *et al.* explained the difference between their two findings in term of confounding factors, such as a history of previous illness or abnormal electrocardiograms, which would have been avoided by using predicted FEV_1 values obtained from a healthy population.

Cook *et al.* (1989, 1991) found that sex-age-height-adjusted peak flow rate was a strong predictor of all-cause mortality in an over-65 population in East Boston, MA. By smoking status, the relative risks were 1.27 ± 0.09 (for a decrease of 100 L/min) for the entire group, 1.24 ± 0.13 for never-smokers, 1.29 ± 0.13 for ex-smokers, and 1.26 ± 0.16 for current smokers. The effect persisted after including, in a proportional hazards model, factors for smoking, respiratory symptoms, cardiovascular risk factors, socioeconomic status, cognitive function scores, measures of physical activity, and self-assessed state of health, but the relative risk dropped to 1.16. This corresponds to an elasticity of 0.56. An estimate of the dose-response function is shown in Figure 12-6 based on quartile data; the function is reasonably linear, and the elasticities were −0.86 for the non-smokers and −1.20 for the entire group. This study provides strong evidence that lung function is independent of other health status indicators as a predictor of mortality (and by implication, not a surrogate for the overall health status that they represent). However, a change in peak flow of 100 L/min is quite large in the context of changes associated with air pollution.

Other supporting studies include: Ferris *et al.* (1971), who identified a group of mortality risk factors (which could not be separated) that included $FEV_1/FVC < 0.6$; Eisen *et al.* (1987), who found that inability to perform a satisfactory lung function test, because of excessive variability, was a strong predictor of all-cause mortality in six U.S. cities; Ferris *et al.* (1986), who reported the mortality risks by FEV_1 quartile for the successful tests from the same program (Figure 12-7); Kanner *et al.* (1983), who reported that survival risks for patients with chronic airflow limitation were predicted by their FEV_1/FVC ratios;

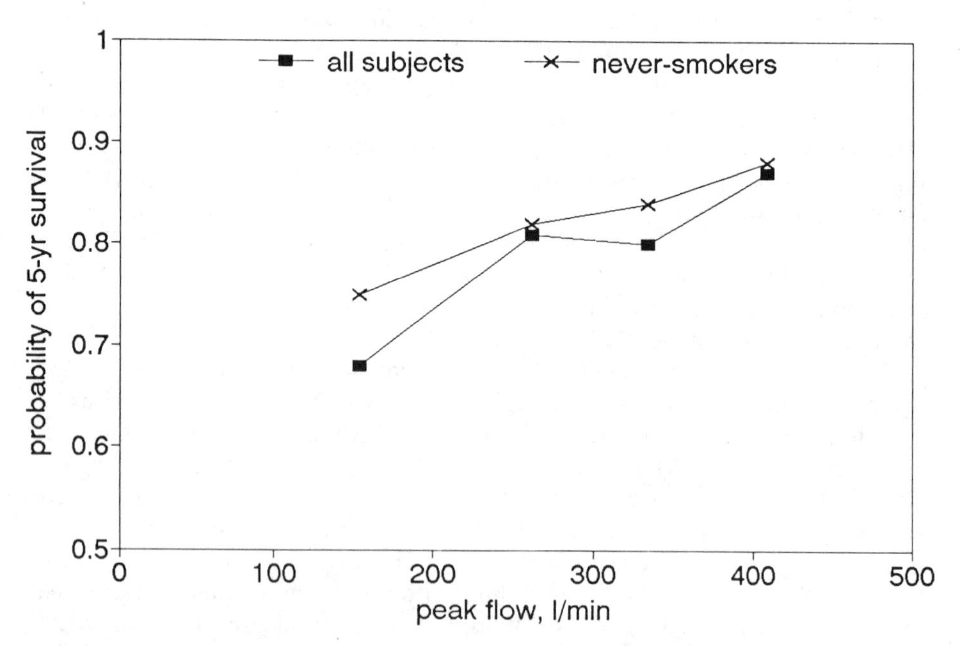

FIGURE 12-6. Probability of 5-year survival versus peak flow for persons over 65 in Boston. *Data from Cook* et al. *(1991).*

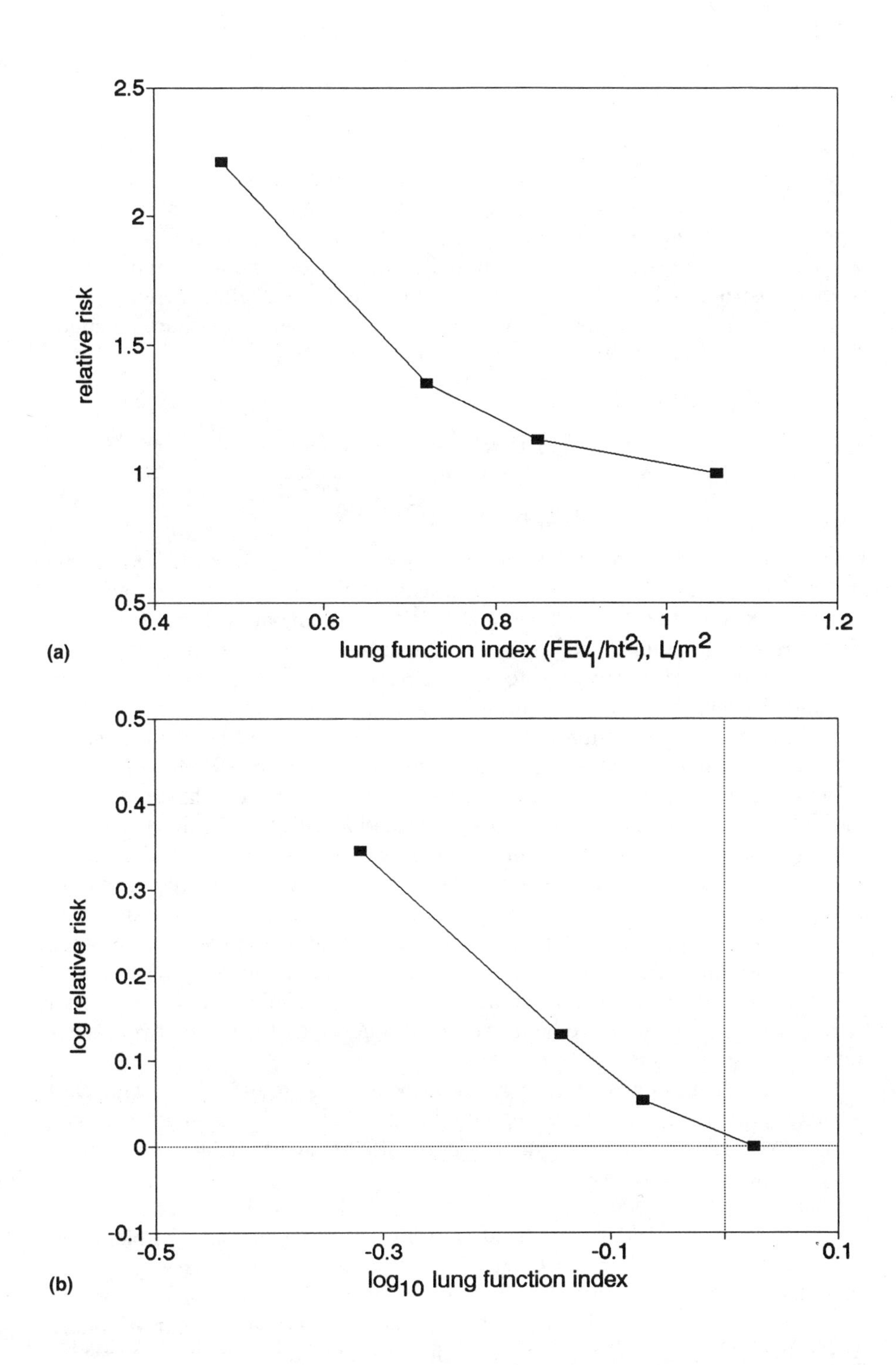

FIGURE 12-7. Relative all-cause mortality risk by quartiles of FEV_2/ht^2, from the Six-City Study. (a) Linear coordinates. (b) Logarithmic coordinates. *Data from Ferris* et al. *(1986).*

Sobol *et al.* (1974), who reported data showing that in a group of 51 patients, cigarette smoking, and not heart disease, was the source of pulmonary abnormalities; and Tibblin *et al.* (1975), who showed the risk factors for fatal ischemic heart disease (IHD) differed from those for nonfatal ischemic heart disease and that peak flow was predictive of fatal IHD.

Deutscher and Higgins (1970) studied parental longevity in relation to lung function of sons and found significant differences: Parents who lived longer had sons with better lung function, after adjustment for age and height and consideration of smoking. They speculated that constitutional familial factors might be involved. Cohen (1980) has identified some of the genetic components of chronic obstructive lung disease. To the extent that close relatives tend to live in the same city or region, it will be difficult to separate shared genetic characteristics from shared environmental characteristics.

SYNTHESIS OF STUDIES ON MORTALITY VERSUS LUNG FUNCTION

Twenty studies were summarized in Tables 12-2 and 12-3, totaling about 6000 deaths. The population groups studied included healthy adults, random samples of men and women, occupationally exposed men, ages from 20 to over 65, heart attack victims, and persons with impaired respiration. Geographic coverage includes Eastern, Western, and Southern Europe, Australia, New Zealand, Hawaii, both coasts of the United States and cities in the midwestern United States. Lung function is found to be a significant predictor of all-cause and cardiovascular mortality, with elasticities ranging from −0.7 to −2.7. The median value is about −1.1. Elasticities for respiratory deaths tend to be higher than for other causes; for heart disease, elasticities are somewhat lower; and for cancer, much lower. Lung function is less satisfactory as a predictor for heart disease morbidity. A variety of statistical analysis techniques was used in deriving these findings, including adjusting for smoking in various ways. Note that use of discrete categories rather than continuous data for smoking will tend to underestimate the true effect. Although publication bias may always be a factor in assessing the agreement among similar studies, it should be noted here that the agreement extends in large part to the relative magnitudes of the effect, not just to its existence.

Virtually all measures of lung function have been found to be significant predictors of mortality; in the one comparative study, single-breath nitrogen washout was clearly the best (Menkes *et al.*, 1985); however, elasticities from this study were not available.

These studies provide strong support for the hypothesis of Cohen (1980) that impaired lung function can lead to a host of other medical problems, in part because of compromised defenses (Figure 12-8). Rarely has a group of such diverse studies found such unanimity of results. The one discordant finding (Marcus *et al.*, 1989) applied to the analysis of cardiovascular *morbidity* (as did the 1972 paper by Keys *et al.*), which is one of the weaker relationships found. Also respiratory morbidity tends to depend on the statistical method selected to account for simultaneous effects of smoking. Many of the other

Author	Years	Location	Subjects	Death Cause	LF metric	No. Deaths	*Elasticities* All	Never-Smokers	Smokers	Control for Smoking	Analysis Method	Remarks
Beaty (1985)	1958–82	Baltimore	Healthy males	All	FEV_1	239	−1.0	0.0	−0.06	Discrete	Cox, Mult.	24-year follow-up of 874 males
Foxman (1986)	1957–77	Staveley, UK	Working males	All	$FEV_{0.75}$	267	−0.7			None	Delta	Ages 55–64, occupationally exposed
Menkes *et al.* (1985)		Baltimore	Nonpatient adlt	All	N_2 washout	84				Discrete	Mult. log.	Regression coeffs. not reported, only F's
Beaty (1982)	1971–81	Baltimore	Nonpatient adlt	All	FEV_1	100	−2.7			Discrete	Log. regr.	
Ashley *et al.* (1975)	1956–66?	Framingham, MA	Cohort study	All	FVC FEV/FVC	325	−0.9			No	Log. regr.	Means based on age 52
Olofson *et al.* (1987)	1973–84	Gothenburg	Swedish men, random	All	FEV_1	107	−1.1			Discrete	Mult. log.	Larger FEV_1 effect in heavy smokers
Higgins and Keller (1970)	1959–?	Rural Mich.	Cohort study	All	FEV_1	179				None		No account of height or smoking
Keys (1980)		7 countries	Cohort study	All	FVC $FEV_{0.75}$	580				None	Delta	Comparing dead and living LF
Cook *et al.* (1989)	1982–88	Boston, MA	Elderly	All	Peak flow	789	−0.6	−0.8	−0.8	Discrete	Regression	Peak flow effect reduced 50% in multiple regression model
de Hamel and de Souza (1978)	1958–72	New Zealand	Coal workers	All (non-acc.)	FVC $FEV_{0.75}$	209				Discrete	Regression	Regressions of LF on age and weight compared
Kannel (1983)		Framingham	Cohort study	CVD, all	FVC/ht	327		−2.0		Discrete	Regression	Still significant using "cigs" as a multiple regression variable
Cullen *et al.* (1983)		W. Australia	Cohort study	CVD, all	FEV_1	72				Discrete	Mult. regr.	Only significance levels given
Persson *et al.* (1986)	1968–80	Gothenburg	Swedish women	CVD, all	PEF	75	−0.9			Continuous	Mult. log	
Krzyzanowski and Wysocki (1986)	1968–81	Cracow, Pol.	M + F, random	All, circ., cancer	FEV_1 FEV_1/FVC	328	M = −1.6 F = −0.7			Discrete	Mult. log	Elasticities based on bivariate analysis
Ebi-Kryston (1988)	1968–78	London	Male civil servants	All CVD CRD	FEV_1	1670 889 96	−1.1 −0.6 −6.0			Continuous	Cox, mult.	Mean based on age 52 Healthy cohort
Friedman (1976)	1964–70	Oakland, CA	Check-up patients	Sudden cardiac	FVC FEV_1	197	−0.7			Discrete	Mult. regr.	Sample drawn from heart attack patients, mean age 59.4
Cook and Shaper (1988)		Britain	Men, 40–59	(CHD incid.)	FEV_1		−1.3			Continuous	Regression	No analysis of mortality
Marcus (1989)	1965–83	Honolulu	Japanese-American men	(CHD incid.)	FEV_1		−0.8	−0.1	−1.2	Disc. + cont.	Cox, mult.	Ages 45–65, reanalysis found effect on never-smokers

TABLE 12-3 Control of Possible Confounding Factors in Lung Function Mortality Studies

First Author	Higgins (1970)	Keys (1972, 1980)	Kannel (1974)	Ashley (1975)	Friedman (1976)	de Hamel (1978)	Beaty (1982)	Cullen (1983)	Kannel (1983)	Menkes (1984)	Beaty (1982)	Menkes (1985)	Foxman (1986)	Persson (1986)	Krzyzanowski (1986)	Olofson (1987)	Cook (1988)	Ebi-Kryston (1988)	Marcus (1989)	Cook (1991)
Age	X	X	X	X	X	X	X	X	X	X	X	X	X	X	X	X	X	X	X	X
Sex	X	X	X	X	X	X	X	X	X	X	X	X	X	X	X	X	X	X	X	X
Race						X	X	X			X		X	X	X	X	X		X	
Height, weight		X	X	X	X	X	X	X	X	X	X	X	X	X	X	X	X	X	X	X
Smoking		X	X	X	X	X	X	X	X	X	X	X	X	X	X	X	X	X	X	X
Socio-econ. status							X			X					X					X
Employment													X					X		
Blood pressure		X	X					X	X					X			X	X	X	X
Use of medication																		X		
Cholesterol		X						X	X					X			X	X	X	
Diabetes														X				X		X
Pulse, heart rate		X	X						X											
ECG changes			X															X		
Prior heart disease															X			X		X
Self-assessment of health															X					X
Functional ability														X					X	X
Blood glucose			X					X												
Diet (coffee, etc.)							X				X								X	
Alcohol							X	X			X									
Blood type							X				X								X	
Genetic markers							X				X									

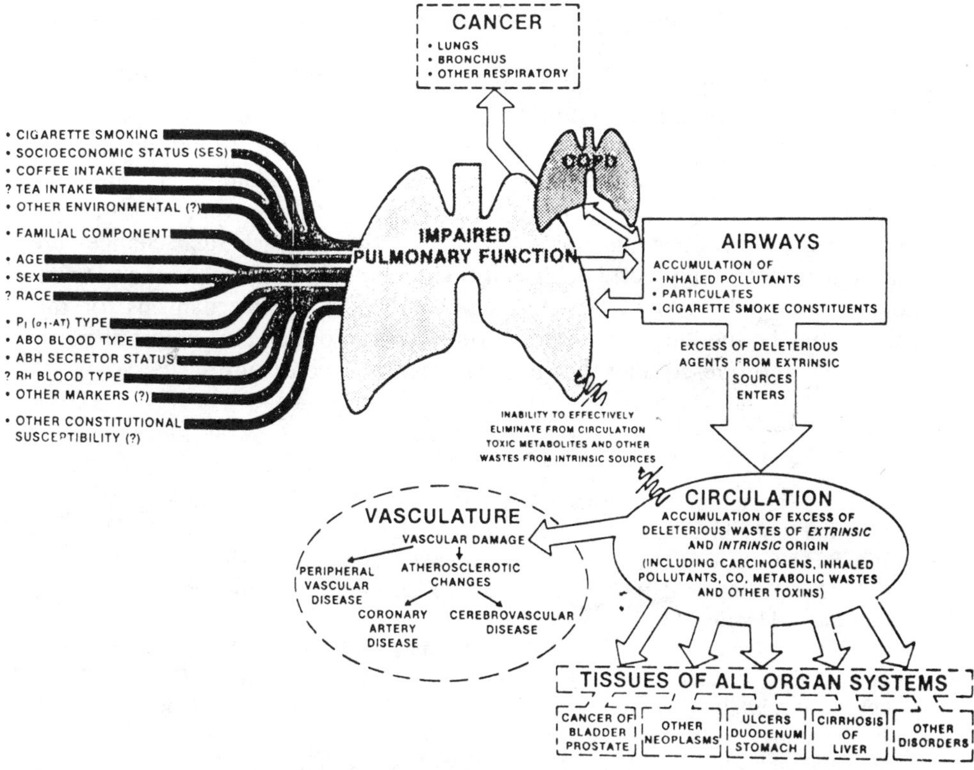

FIGURE 12-8. Diagram illustrating the hypothesis that impaired lung function can lead to many physiological disorders. *Source: Cohen (1980), used with permission.*

studies found that that the effect of smoking on lung function as substantially weaker than the effect of lung function on mortality. Also, Marcus *et al.* studied the only non-Caucasian cohort among this group of studies, so that it is not clear how widely applicable their findings might be to the general U.S. population. Keys (1980) noted that the average difference in lung function between the deceased and the survivors was only 9% and felt that it was thus unlikely to be truly independent; he speculated that this difference could be caused by the other CHD risk factors such as blood pressure, cholesterol, and physical activity. However, he recommended more study of the effects of respiratory function. Several subsequent multiple regression studies specifically included the cofactors mentioned by Keys; the effects of lung function persisted. Subsequent multifactoral studies such as Cook *et al.* (1991) would appear to have laid this issue to rest; even in Keys's own work, the correlation between lung function parameters and other CHD risk factors were quite small, of the order of 0.1 to 0.2. Finally, the most recent studies on this topic seem to come down firmly in favor of reduced lung function as an antecedent of other health problems, not just a marker for other factors.

Menkes *et al.* (1984) pointed out the interdependence of risk factors that vary cross-sectionally and thus constitute the characteristics of a population, including socioeconomic status, education, smoking, genetic makers, and familial pulmonary disease. Obstruction was a strong predictor of subsequent mortality, including deaths other than pulmonary disease, and these cross-sectional factors accounted for only a "modest" portion of the lung function decline.

I interpret these findings and those of the rest of the chapter as suggesting a causal pathway for air pollution to influence mortality for causes of death other than through respiratory disease *per se*, especially for cardiovascular causes. If air pollution can cause loss of lung function (Chapter 11), and lung function has a direct effect on mortality, then it follows that air pollution can also have an effect on mortality. The extent to which such changes are transient or essentially permanent will then depend on the frequency distributions of air quality. Menkes suggested that one of the mechanisms for this linkage may be the lung's role as the primary organ of defense against airborne toxins; a compromised lung thus implies compromised defenses.

REFERENCES

Ashley, F., Kannel, W.B., Sorlie, P.D., and Masson R. (1975), Pulmonary Function: Relation to Aging, Cigarette Habit, and Mortality, *Ann. Int. Med.* 82:739–45.

Beaty, T.H., Cohen, B.H., Newill, C.A., Menkes, H.A., Diamond, E.L., and Chen, C.J. (1982), Impaired Risk Function as a Risk Factor for Mortality, *Am. J. Epidemiology* 116:102–13.

Beaty, T.H., Newill, C.A., Cohen, B.H., Tockman, M.S., Bryant, S.H., and Spurgeon, H.A. (1985), Effects of Pulmonary Function on Mortality, *J. Chron. Dis.* 38:703–10.

Cohen, B.H. (1980), Chronic Obstructive Pulmonary Disease: A Challenge in Genetic Epidemiology, *Am. J. Epidemiology* 112:274–88.

Cook, D.G., and Shaper, A.G. (1988), Breathlessness, lung function and the risk of heart attack, *Eur. Heart J.* 9:1215–22.

Cook, D.G., and Shaper, A.G. (1990), Pulmonary Function as a Predictor of Coronary Heart Disease (letter), *Am. J. Epidemiology* 132:587.

Cook, N., *et al.* (1989), Peak Expiratory Flow Rate in an Elderly Population, *Am. J. Epidemiology* 130:66–78.

Cook, N., *et al.* (1991), Peak Expiratory Flow Rate and 5-Year Mortality in an Elderly Population, *Am. J. Epidemiology* 133:784–94.

Cox, D.R. (1972), Regression Models and Life Tables, *J. Roy. Stat. Soc.* B 34:187–220.

Cullen, K., Stenhouse, N.S., Wearne, K.L., and Welborn, T.A. (1983), Multiple Regression Analysis of Risk Factors for Cardiovascular Disease and Cancer Mortality in Busselton, Western Australia—13-Year Study, *J. Chron. Dis* 36:371–77.

de Hamel, F.A., and de Souza, P. (1978), *Lung Function and Chronic Bronchitis in New Zealand*, Dept. of Preventive and Social Medicine, University of Otago, Dunedin, New Zealand.

Deutscher, S., and Higgins, M.W. (1970), The Relationship of Parental Longevity to Ventilatory Function and Prevalence of Chronic Nonspecific Respiratory Disease among Sons, *Am Rev. Respir. Dis.* 102:180–89.

Ebi-Kryston, K.L. (1988), Respiratory Symptoms and Pulmonary Function as Predictors of 10-Year Mortality from Respiratory Disease, Cardiovascular Disease, and All Causes in the Whitehall Study, *J. Clin. Epidemiology* 41:251–60.

Ebi-Kryston, K.L., *et al.* (1989), Breathlessness, Chronic Bronchitis and Reduced Pulmonary Function as Predictors of Cardiovascular Mortality among Men in England, Scotland, and the United States, *Int. J. Epidemiology* 18:84–88.

Eisen, E.A., Dockery, D.W., Speizer, F.E., Fay, M.E., and Ferris, B.G., Jr. (1987), The Association Between Health Status and the Performance of Excessively Variable Spirometry Tests in a Population-Based Study in Six U.S. Cities, *Am. Rev. Respir. Dis.* 136:1371–76.

Ferris, B.G., Jr., Speizer, F.E., Worcester, J., and Chen, H.Y. (1971), Adult Mortality in Berlin, NH, from 1961 to 1967, *Arch. Env. Health* 23:434–39.

Ferris, B.G., Jr., Ware, J.H., Spengler, J.D., Dockery, D.W., and Speizer, F.E. (1986), The Harvard Six-Cities Study, in *Aerosols: Research, Risk Assessment and Control Strategies, Proceedings of the 2nd US-Dutch International Symposium*, S.D. Lee, T. Schneider, L.D. Grant and P.J. Verkek, eds., Williamsburg, VA, pp, 721–30.

Foxman, B., Higgins, I.T.T., and Oh, M.S. (1986), The Effects of Occupation and Smoking on Respiratory Disease Mortality, *Am. Rev. Resp. Dis.* 134:549–652.

Friedman, G.D., Klatsky, A.L., and Siegelaub, A.B. (1976), Lung Function and Risk of Myocardial Infarction and Sudden Cardiac Death, *N. Engl. J. Med.* 294:1071–75.

Higgins, M.W., and Keller, J.B. (1970), Predictors of Mortality in the Adult Population of Tecumseh, *Arch. Env. Health* 21:418–24.

Kannel, W.B., Hubert, H., and Lew, E.A. (1983), Vital capacity as a predictor of cardiovascular disease: The Framingham Study, *Am. Heart J.* 105:311–15.

Kannel, W.B., Seidman, J.M., Fercho, W., and Castelli, W.P. (1974), Vital Capacity and CHD: The Framingham Study. *Circulation* 49:1160–66.

Kanner, R.E., Renzetti, A.D., Stanish, W.M., Barkman, H.W., and Klauber, M.R. (1983), Predictors of Survival in Subjects with Chronic Airflow Limitation, *Am. J. Medicine* 74:249–55.

Keys, A. (1980), *Seven Countries*, Harvard University Press, Cambridge, MA.

Keys, A., *et al.* (1972), Lung Function as a Risk Factor for Coronary Heart Disease, *Am. J. Public Health* 62:1506–11.

Krzyżanowski, M., and Wysocki, M. (1986), The Relation of Thirteen-Year Mortality to Ventilatory Impairment and Other Respiratory Symptoms: The Cracow Study, *Int. J. Epidemiology* 15:56–64.

Marcus, E.D., Curb, J.D., MacLean, C.J., Reed, D.M., and Yano, K. (1989), Pulmonary Function as a Predictor of Coronary Heart Disease, *Am. J. Epidemiology* 129:97–104.

Marcus, E.B. (1990), Response to Letter, *Am. J. Epidemiology* 132:588.

McClure, C.W., and Peabody, F.W. (1917), Relation of Vital Capacity of Lungs to Clinical Condition of Patients with Heart Disease, *JAMA* 69:1954–59.

Menkes, H.A., Cohen, B.H., Beaty, T.H., Newill, C.A., and Khoury, M.J. (1984), Risk Factors, Pulmonary Function, and Mortality, in *Genetic Epidemiology of Coronary Heart Disease: Past, Present, and Future*, D.C. Rao *et al.*, eds., A.R. Liss, New York, pp. 501–21.

Menkes, H.A., Beaty, T.H., Cohen, B.H., and Weinman, G. (1985), Nitrogen Washout and Mortality, *Am. Rev. Resp. Dis.* 132:115–19.

Olofson, J., Skoogh, B.-E., Bake, B., and Svardsudd, K. (1987), Mortality related to smoking habits, respiratory symptoms and lung function, *Eur. J. Resp. Dis.* 71:69–76.

Persson, C., Bengtsson, C., Lapidus, L., Rybo, E., Thiringer, G., and Wedel, H. (1986), Peak Expiratory Flow and Risk of Cardiovascular Disease and Death. *Am. J. Epidemiology* 124:942–48.

Sobol, B.J., Herberts, W.H., and Emirgil, C. (1974), The High Incidence of Pulmonary Functional Abnormalities in Patients with Coronary Artery Disease, *Chest* 65:148–51.

Tibblin, G., Wilhemsen, L., and Werko, L. (1975), Risk Factors for Myocardial Infarction and Death Due to Ischemic Heart Disease and Other Causes, *Am. J. Cardiology* 35:514–22.

Wison, M.G., and Edwards, D.J. (1922), Diagnostic Value of Determining Vital Capacity of Lungs of Children, *JAMA* 78:1107–10.

13

Summary, Conclusions, and Implications

Our future is not written, it is not certain: we have awakened from a long sleep, and we have seen that the human condition is incompatible with certainty.

Primo Levi, "Eclipse of the Prophets"

Studies published during about the past 40 years on the effects of air pollution on community health and on individual respiration have been reexamined in detail in the preceding chapters. Effects on mortality, hospitalization, and respiratory mechanics were emphasized because these are the most severe health effects and their measures tend to be objective. In this final chapter, the original themes of the book and the primary conclusions from each chapter are revisited, and their causal implications are examined. Uncertainties in the findings are considered along with their potential impacts on conclusions, and information gaps and research needs are discussed.

THE BASIC THEMES, REVISITED

To set the stage for this concluding discussion, we first revisit the basic themes of the book that were presented in Chapter 1.

The Use of Observational, Ecological Studies

Most studies of air pollution and community health are *ecological*, in that effects on individuals are inferred from data on the behavior of groups. Ecological studies have long been controversial, and some of the studies reviewed have indeed offered grounds for skepticism, primarily because of poorly measured variables and the omission of potential confounders in the analyses.

However, as long as the effects of air pollution on rates of mortality or incidence of disease are of interest and we continue to lack sufficient detailed information on individuals to allow prospective studies (including personal air pollution exposures), it will be necessary to use observational studies of the ecological type. The limitations of ecological studies apply to time-series as well as to cross-sectional studies, although this similarity in shortcomings has rarely been recognized.

Rose (1985) discussed the distinctions between studying "sick individuals and sick populations" in his analysis of applications of the case-control method. He pointed out that the basic purpose of a case-control study is to identify differences between sick and healthy individuals. (Note that in a mortality study, everyone is "sick," because mortality studies are only concerned with the members of the population who died.) If the cohorts of cases and controls are properly matched, differences in their average properties may be used to identify risk factors and their confidence limits. Lacking data on personal exposures to air pollution, the case-control method cannot be used to study the effects of community air pollution, because, within a given community, it is necessary to assume that everyone is exposed to the same air pollution. As Rose pointed out, the effects of air pollution would then be completely dependent on differences in individual susceptibility. If some subsets of individuals were not exposed to the same degree (because they worked outdoors or remained indoors, for example), the circumstances that changed their exposure would be considered confounding variables. Such circumstances could include differences in age, health, life-style, or occupation, all of which could also affect their susceptibility. A pragmatic way of dealing with this source of uncertainty is to combine the two "unknowables"—individual susceptibility and the details of individual exposures—into a joint distribution; one must then assume that the victims of air pollution in an ecological study are those individuals who were both susceptible and exposed. Until means are devised for studying large numbers of individuals, there will be no way of verifying this hypothesis. Note that the two relatively large prospective studies that were reviewed in Chapter 7 did not present findings that could not be reconciled in terms of the appropriate parallel ecological studies.

Substituting different communities for individuals in the case-control protocol inevitably leads to the multivariate cross-sectional regression, since it is virtually impossible to adequately "match" whole communities, especially those cities that are large enough to have stable mortality rates. In order to assign between-community health differences to air pollution, one must adequately account for differences in average individual susceptibility and exposure. Cross-sectional regressions must therefore encompass all the attributes that might affect a community's pollution characteristics, the health responses of its individuals, and their susceptibilities. The degree to which this has been accomplished lies at the heart of the differences in findings among the many cross-sectional regressions that have appeared in the literature.

Time-series studies fare only a little better with regard to the challenge of the ecological fallacy. In a time-series study of mortality, the individual serves as his or her own control, since only two end states are considered: alive and dead. These studies have found that deaths tend to increase on or after high-pollution days, but cannot tell us anything about the health of the decedents on preceding days (which may have had higher or lower pollution), or why a given day was lethal to a particular individual. Similarly, most time-series studies of hospitalization do not track individuals; thus, we have no way of knowing whether an individual may have gone to the emergency room (ER) on other days or been previously admitted for the same diagnosis. However, even if time-series analyses cannot track individuals through time, it is possible to examine individual days, as was done by Katsouyanni *et al.* (1990) in Athens (see Chapter 6). By defining groups of days that were alike in all respects except the concentrations of certain pollutants, a type of case-control design can be derived. Furthermore, such a study design would be amenable to considering the durations of high-pollution episodes and to searching for additional relevant variables.

In general, the ecological assumption for a time-series study requires that the available fixed-location air pollution monitors must represent the community exposure accurately in the same way on each day, which is a difficult requirement because of varying meteorology. The case-control protocol could incorporate synoptic weather classifications as an additional means of categorizing groups of days (Kalkstein, 1991).

Rose offered a number of other useful concepts in his 1985 paper. As discussed above, to analyze effects on prevalence and incidence rates, we need to study populations, not individuals. The question of genetic effects has been raised; lung function deficiency has a genetic component, for example (Cohen, 1980). According to Rose, genetic factors tend to dominate individual susceptibility, but genetic heterogeneity is smaller *between* populations than *within* populations (presumably because of the central limit theorem). He pointed out that determinants of incidence rates are still unknown for a "remarkably large number of our major non-infectious diseases," and that often "the best predictor of future major disease is the presence of existing minor disease." He used lung function as an example of the latter concept, since persons currently having poor respiratory performance are likely to suffer increased rates of future lung function decline. It also seems logical to assume that such persons are the most likely victims of air pollution, regardless of whether prior exposure to air pollution was responsible for all or part of their poor respiratory performance.

The Importance of Environmental Data

While in most cases, it was necessary to accept the published air quality data as given in the various studies reviewed, three of the newer analyses highlight the need to carefully consider the air quality data used as independent variables. In the first analysis, the mortality responses during the eight major London episodes were found to fit a log-linear model based on the total dose of smoke and SO_2 averaged over the monitoring network. The maximum local air pollution values (ca. 4000 $\mu g/m^3$ of smoke), which are most often cited with regard to the 1952 London episode, were found to be less relevant than the network averages, which were substantially lower. Furthermore, the log-linear model derived essentially the same regression slope as has been derived from current daily time-series studies in the United States.

In general, temporal studies have often been guilty of ignoring the possible simultaneous contributions of unmeasured or poorly measured pollutants (such as carbon monoxide) and of not considering the effects of measurement errors on the outcome of competition between highly correlated pollutants in multiple regressions (see Appendix 3B). The "winner" in such a competition may be the pollutant whose population exposure has been measured with least error, instead of the most lethal species. This concept was shown in the second case, consisting of time-series analyses of hospital admissions in Southern Ontario, for which the 24-hour averages yielded higher bivariate correlations than the peak hour measures (Lipfert and Hammerstrom, 1992). Some of the other studies reviewed used different averaging times for various pollutants in the same regression, which can create a bias in favor of the longer averaging times.

Finally, in terms of cross-sectional air quality data, the notion that an entire SMSA might be represented by a single air monitoring station, which was introduced by Lave and Seskin in 1970, spawned a whole host of similar studies that appeared to ignore the consequences of the errors that this assumption entails. The problem of errors in the independent variables that can affect cross-sectional studies can also affect time-series studies; pollutants with localized distributions will appear to have less effect as the size of the areas around the monitoring stations (used as units of observation in cross-sectional studies) increases or as the averaging time increases. This concept was demonstrated when TSP was averaged over all the monitors in each SMSA, and the statistical significance of TSP as a predictor of mortality was found to improve greatly (Lipfert, 1992).

The Distinction between Exacerbation and Cause of Disease

At current community levels, air pollution appears more likely to exacerbate existing disease than to create new cases. Since it is difficult to detect new cases of disease (and a large percentage of heart disease tends to remain undiagnosed [Reid *et al.*, 1974]), it may never be possible to "prove" this point, but there is strong circumstantial supporting evidence. After the air pollution disasters of the 1930s–1960s, investigators were baffled that lethal effects had occurred at air concentration levels far below the toxicological limits known at the time (some of which may have been based on occupational standards). They then resorted to synergism arguments: If none of the individual pollutants known to be present reached a concentration thought to be lethal, then it must have been the combination that was somehow responsible. They apparently failed to consider models other than those of classical toxicology, for which dose limits have been determined either from animal tests or from basically healthy human volunteers. Real urban populations consist of individuals spanning a large range of exposures, susceptibilities, and disease states, including subclinical illness (Sherwin and Richters, 1991). The fatalities during the extreme and lengthy environmental stress levels of these air pollution disasters could not have been predicted from classical toxicology; neither could most of the findings of the studies reviewed in this book.

Further support for this hypothesis may be found from the shapes of the dose-response curves that have been developed in the course of this review. For mortality and hospitalization, these curves tend to be linear (or log-linear) over modest ranges of pollution levels, with no hints of the thresholds that are central to the concepts of toxicology. For some pollutants, notably SO_2, when larger ranges of pollution are involved, logarithmic transforms tend to fit, suggesting that the response tapers off at the higher concentration levels. Based on toxicological principles, this seems counterintuitive, because the concentrations involved are far below those at which a substantial portion of a (uniform) population might respond. Assuming that only one pollutant were active, such a mortality response may be understood in terms of a small group of susceptible "responders" whose numbers are limited at any given time. At some later time, others may take their place by virtue of future debilitating illness or by aging.

As discussed in Chapter 10, it is somewhat more difficult to apply the exacerbation model to the studies of hospitalization, because repeat users of hospital services are usually not identified in these purely statistical studies. However, in the absence of data that assigns patterns of cardiorespiratory disease prevalence to (long-term) patterns of air pollution, one is led to the assumption that studies of daily admissions and ER visits are concerned with acute incidents, especially since the elasticities for ER visits tend to exceed those for admissions. Almost by definition, acute responses to air pollution are most likely to occur to those persons who are either especially sensitive or have existing impairments. This conforms to the exacerbation hypothesis.

This scenario is also supported by studies of heat-wave mortality. For example, Ellis (1972) pointed out that heat-wave deaths during the first of 4 years of severe episodes (1952–55) far exceeded those in the 3 subsequent years (in spite of equivalent heat wave intensities), which could have either been the result of acclimatization or of "harvesting" of the most susceptible individuals. This finding also supports the assumption that previously healthy individuals are less likely to be affected by heat waves. Ellis also cited earlier studies by Gover (1938) in which it was claimed that the first heat wave of the season was usually the most lethal, even when followed by more severe heat waves. Kutschenreuter (1960) found that the second of two closely spaced episodes was less lethal than the first. However, an alternative explanation in the case of community air pollution might involve actions by more than one pollutant.

However, some of the lung function analyses do in fact suggest thresholds when the entire population is considered (see Figure 11-9). This implies that at least a portion of the population is not affected by pollution levels less than some limit (within the limitations of the particular lung function measure involved). To understand this apparent conflict, one must consider the differences in the populations being analyzed in these different types of studies. *Mortality studies deal with the dead, hospitalization studies deal with the sick, and (most* of the) lung function studies deal with the healthy*. It is clearly unreasonable to expect the same kinds of responses from such different segments of the population.

---

* Some lung function and symptoms studies have also been carried out on groups of asthmatics and persons with other impairments, but the point being made here is that most of the available lung function data pertain to individuals going about their normal daily routines, especially children.

This is a fundamental distinction. Use of mortality as the endpoint automatically selects the most sensitive subset of the population; these are the least healthy people in the society, by definition. During the air pollution disasters, the high mortality rates observed imply that larger segments of the population were affected, even if they did not all seek hospitalization. During the 1952 London episode, for example, not all the victims were known to be previously ill (contrary to popular opinion; see the 1954 Ministry of Health report and the discussion in Chapter 5).

This simplified scenario constitutes one of the fundamental differences between the toxicological (causal) model and the exacerbation model. Toxicological data are developed from experiments on populations that tend to be considerably more uniform than the general public (and which are sometimes selected for uniformity). As the dose is increased, the rate of response continues to increase, because the experiment is being conducted at high enough doses that the bulk of the (animal) population eventually becomes affected. In contrast, the epidemiology of community air pollution deals with the tail of the distribution, where the supply of "responders" at a given concentration level may well be limited.

The link between the well-being of the lung and the health of the rest of the organism was established by studies of lung function as a predictor of mortality for causes other than respiratory (Chapter 12). Yet, it is widely recognized that lung function measures tend to be "blunt instruments" with regard to small airway disease (Peto *et al.*, 1983), and the small airways are where the lung fulfills its basic gas exchange mission. The finding of small but consistent decrements in lung function associated with air pollution might thus be viewed as the sighting of the "tip of the iceberg." This may mean that there are many more cases of subclinical disease that have been affected and perhaps caused by air pollution, but this can only be speculation. An alternative causal pathway may be that those persons whose natural (genetically determined) lung function is deficient may be more susceptible to air pollution because the natural defenses of their lungs have been compromised (Cohen, 1980; Menkes *et al.*, 1984). Also, one cannot rule out that some fraction of the population developed respiratory impairments as a result of repeated exposures to air pollution in the past, although the evidence for this scenario based on lung function data is limited (Chapter 11).

The prospective studies of Abbey *et al.* (1991) in California are important in this light. Because all-cause mortality was not associated with cumulative air pollution exposure, while associations with acute responses were found by others (Chapter 6), we conclude that the exacerbation model is supported. Abbey *et al.* did not examine acute effects, but they reported associations of female cancer with cumulative exposure to TSP, lagged 5 years. Their study is unique in this regard, both with respect to the care taken to estimate retrospective pollution exposure and the findings of associations with cancer. Because similar associations were not reported for males, replication is required before the findings of Abbey *et al.* (1991) can be accepted at face value. Lipfert (1978) found only limited associations between cancer mortality and (same year) air pollution in his cross-sectional ecological regressions, but did not test lagged pollution data.

Consideration of Alternative Dose-Response Models

Linear dose-response models are the most common among the studies reviewed. As the range of pollutant concentrations increases, there is support for a log-linear model for mortality endpoints. This model postulates an exponential response to increasing air concentrations, that is, at higher concentrations, the natural defenses of a larger fraction of the population become overwhelmed. It was shown that a model whose slope increases as concentration *decreases* (such as shown by Ostro, 1984) is an indication of a surrogate relationship or of a missing variable. A few studies were shown to support a logarithmic pollution response, which suggests a saturation effect. Only a few studies, which were mostly inconclusive, made formal efforts to define pollution thresholds.

Even if all of the possible combinations of pollutants, lags, averaging times, and models were investigated, we are generally lacking suitable criteria for defining which model is "best." Small differences in statistical performance of alternative models, as measured by t or R^2 values, may in fact not be significant and are seldom formally tested. Tests involving the distributions of residuals probably offer the best hope, but in general the research community has not even recognized this problem, let alone agreed on a solution.

The Separation of Pollutant Species

Much of the effort expended on environmental epidemiology has been directed towards the goal of identifying "responsible" pollutants—which may well be an impossible goal. This book set out to review studies by endpoint rather than by pollutant; in many cases, a clear separation could not be made among various pollutants. This may result from collinearity among pollutants; it could also result from the heterogeneity of community populations. A few of the studies reviewed involved mainly a single pollutant (usually particulate matter), which establishes airborne particles as a sufficient (but not necessary) condition for adverse effects.

Toxicology can study one pollutant at a time and learn valuable information about mechanisms, but it can seldom bridge the relevance gap. Real populations are exposed to mixtures of species that vary in space and time. The studies reviewed have found effects attributed to all of the criteria pollutants, and it may well be that large populations will include responders to each of them, for a given endpoint. The sickest individuals may respond to weather effects as well, and thus there may be less discriminatory power at low doses than at higher levels. The ability to distinguish the effects of individual components of the urban pollutant mixture will be compromised by differences in the reliability of exposure estimates and by the structure of the models selected by the investigator. Lag times may differ by species, and there may be differences in the functional forms of the models most appropriate for each species.

For example, if a population health response is in fact due to a mix of pollutants but only one has been measured or considered, the resulting apparent dose-response function may have a counterintuitive shape (see Figure 6-9). Considering additional species or several pollutants jointly may help in such cases. This is a possible alternative explanation for the success of the logarithmic model in explaining London daily mortality, since CO has never been formally considered in this context and may be responsible for a part of the association between air pollution and mortality.

Chronic versus Acute Responses

Both types of relationships were examined for all types of studies, and the elasticity was used as a means of comparing across pollutants and models. Since most of the cross-sectional (long-term) and time-series (short-term) studies of mortality derived elasticities around 0.05, it is difficult to make a case for a substantial independent chronic effect on mortality. However, one must also consider that if cross-sectional studies based on annual average pollutant levels are interpreted as representing the annual sum of acute effects, errors may result from the failure to consider air quality data in a time sequence based on mortality data, and such errors would likely bias the association toward the null (Chapter 3). Also, if one accepts the finding by Abbey *et al.* (1991) of an association between cancer mortality and long-term air pollution exposure, this relationship will translate to a partial (chronic) relationship with all-cause mortality.

Only a few long-term studies of hospitalization were found, and their elasticities were in the same range as the time-series studies of hospitalization. In general, the hospitalization studies displayed a wider range of results than the mortality studies; this may be due to the wider variety in protocols and differences in the diagnoses included. Most of the lung-function findings appeared to be reversible and, thus, acute by definition. The weight of the evidence thus points to acute effects as the more prevalent mechanism, as discussed.

However, it is possible that the similarity in time-series and cross-sectional elasticities could be fortuitous; not all of the evidence fits this neat pattern of coincidence. The intraurban cross-sectional studies and the prospective study of Pope *et al.* (1993) found much larger elasticities, of the order of 0.20; these findings could be interpreted as evidence of chronic effects of air pollution on mortality. Some of the cross-sectional studies identified fine particles and sulfates; recent time-series studies tend to find coarser particles (TSP and PM_{10}) and, to a lesser extent, ozone. Although we caution against assigning too much credibility to the identification of specific pollutants in many of these studies, the finding of different pollutants could be an indication of different physiological mechanisms.

SUMMARY OF THE ASSOCIATIONS BETWEEN AIR POLLUTION AND MORTALITY

It was one of the purposes of this book to attempt to cast the four different types of air pollution–mortality studies into a common framework, both qualitatively and quantitatively. These categories are: (1) episodes or disasters (Chapter 5); (2) daily-weekly time-series studies (Chapter 6); (3) cross-sectional studies (Chapter 7); and (4) temporal studies for longer time periods (Chapter 8). The quantitative findings are important in order to compare the absolute levels of risk involved for various air quality policy considerations and also for use in environmental cost-benefit analyses (see for example, Ottinger *et al.*, 1990, or Krupnick and Portney, 1991).

Air Pollution Disasters

The most severe of the air pollution disasters in terms of the increases in death rates (approximately a factor of 10) were those of the Meuse Valley (Belgium) and of Donora, PA. No air monitoring was in place during either event, and a wide variety of industrial emissions was present, including acid mists and trace metals. Also, the health responses in both cases seemed to be closely tied to threshold concentrations in that they abated quickly. Only during these two episodes were substantial numbers of domestic animal deaths noted. These two events have thus been separated from the remaining events involving community air pollution. It is noteworthy that no similar events have since occurred, unless one considers poison gas disasters such as the one at Bhopal, India, in 1984.

The remaining community air pollution episodes were found to be influenced primarily by the total pollution dose (concentrations averaged over the duration of abnormal levels). There was also a suspicion, based on the 1952 London episode, that the mortality "yield" of an episode could be enhanced by the simultaneous presence of infectious disease epidemics in the population. The slopes of the responses for various episodes were in the range of 0.01 to 0.08% excess deaths per $\mu g/m^3$ for smoke and 0.03 to 0.05% excess deaths per $\mu g/m^3$ for SO_2. The product of smoke $\times$ SO_2 appeared to fit the major London episodes best, using a log-linear model. There was a tendency for excess mortality to persist longer for the more severe episodes. Since the concept of elasticity as a means to express the comparative strength of relationships is based on small perturbations about the mean, it was not used to characterize disasters.

Time-Series Studies

The results of time-series studies of mortality and air pollution were found to be influenced by the methods used to make seasonal corrections and the ways in which other potentially confounding variables were handled, including weather, day-of-week differences, holidays, and epidemics. Results may be compared in two different ways: the regression coefficients are directly comparable to the episode dose-response functions in terms of relative risk; elasticities are more useful for comparing the risks among pollutants and cities and for comparing with cross-sectional results. When logarithmic models are used, the elasticity is the only practical measure of slope.

In London, Mazumdar *et al.* (1982) reported 0.025% excess deaths per $\mu g/m^3$ for smoke and 0.0012% for SO_2, which compares well with the episode coefficients given. However, when individual years or groups of years were considered, the linear regression coefficients were found to increase as the pollution levels decreased (as in a logarithmic model); the slopes tended to bracket the findings from episodes. Ostro (1984) reported 0.017 and 0.047% excess deaths per μg/m^3 for the high and low smoke values, respectively. Shumway *et al.* (1983) reported an elasticity of 0.076 based on a logarithmic model for SO_2, which was somewhat higher than others reported for linear smoke models but which compared quite well with results in New York (Hodgson, 1970; Schimmel *et al.*, 1974), Philadelphia, and Los Angeles. The elasticities estimated from present-day pollution levels tended to be somewhat higher (0.04–0.07) than many of the older studies in London and New York, which is not consistent with the expectation of smaller effects at lower pollution levels. Differences in methodology and treatment of lag effects may be partially responsible for the apparent discrepancy.

Elasticities for heart disease mortality tend to be slightly higher than those for all causes, and elasticities for respiratory deaths (0.12–0.23) tend to be substantially higher. This suggests that persons with existing respiratory disease (most of whom are likely to be smokers) are the most sensitive to acute episodes of air pollution.

No thresholds were identified by these time-series studies, and several of them suggested that a logarithmic model provided the best fit. The logarithmic form was also suggested by the time-history plots of some of the episodes and thus should be seriously considered. However, systematic tests of alternative model specifications were not presented by any of the time-series studies reviewed.

Cross-Sectional Studies

Results of cross-sectional regression analyses tended to vary according to the size of the geographic areas considered. This is roughly analogous to the choice of averaging time (daily, weekly, monthly) in a time-series analysis. The intraurban studies tended to find much stronger effects but were limited in the degree to which potentially confounding variables were accounted for. Elasticities for all-cause mortality were in the range of 0.20 to 0.40, and regression coefficients were substantially higher than those typically found in temporal studies. Part of this difference is due to the separate age groups analyzed in the intraurban studies; it is not clear whether the remaining differences resulted from better estimates of exposure or from neglect of important confounding variables, such as smoking, use of alcohol, or other life-style parameters. One could expect sharper gradients in these parameters by neighborhood than when averaged over an entire city or county, and my personal judgment is that the intraurban studies are likely to have been confounded.

Studies of U.S. city mortality found TSP to be a significant predictor of all-cause mortality but may have suffered from monitoring that was inadequate to fully characterize each city. The results for sulfate were highly dependent on the degree of control for socioeconomic factors.

Cross-sectional results for counties and SMSAs tend to focus on regional pollutants such as sulfate aerosol, in contrast to more localized pollutants such as TSP and SO_2. To the extent that suburban and rural residents are usually only exposed to regional pollutants, these findings could be important. However, the best known of the SMSA studies suffered from the omission of potentially confounding variables such as smoking, migration, and drinking water hardness. When these variables were introduced into regression models for 1969–70 SMSA mortality, the sulfate coefficients dropped considerably. I concluded that these SMSA results were model-specific.

The most recent cross-sectional analysis of U.S. SMSAs (Lipfert, 1992) found elasticities in the 0.005–0.06 range for a variety of pollutants, including ozone and size-classified particles. Ozone was found to be associated with major cardiovascular deaths (elasticity = 0.04) and TSP with chronic obstructive pulmonary disease deaths (elasticity = 0.23). The finding of all-cause elasticity levels around 0.05 in independent U.S., British, and Canadian cross-sectional studies provides support for a causal interpretation.

The two recent prospective mortality studies reviewed were interpreted as providing support for the ecological studies. Abbey *et al.* (1991) found no association between lagged cumulative exposure to TSP and all natural-cause mortality, while Pope *et al.* (1993) found much stronger associations between coincident exposures to particulates and all-cause mortality than seen previously in either time-series or in most of the cross-sectional studies. In addition, Lipfert (1993) showed that the individual risk factors that Pope *et al.* developed for smoking and education were confirmed quantitatively in ecological studies, and that consideration of additional socioeconomic variables in their prospective study might have brought their air pollution risk factors more in line with those developed from ecological studies.

Long-Term Temporal Studies

Several different types of studies involving longer-term temporal variations were reviewed. Such studies form an important link between cross-sectional and time series types, since they address the question of the degree of prematurity of the daily mortality effects. The combined cross-sectional mortality study of nine Ontario counties over 7 years dealt with most of the potential confounders and derived elasticities of 0.04–0.06 for all-cause mortality for either SO_2 or SO_4^{-2}; TSP was not significant (Plagiannakos and Parker, 1988). For respiratory mortality, all three pollutants were significant (when taken one at a time) with elasticities from 0.10 to 0.18. These findings bridge the pure time-series and cross-sectional results very nicely. One of the other types investigated was a time-series analysis of New York City mortality based on 60 bimonthly periods (Lipfert and Wyzqa, 1992). The variables were SO_2 and smoke (measured at only one station), and temperature, which was used for seasonal adjustment. The results were also consistent with previous daily time-series results, in that the elasticity for all-cause mortality was about 0.082 (mostly due to smoke) and 0.32 for respiratory mortality. However, if all the seasonal effects were not accounted for by the temperature variable, the estimated air pollution effects could be biased high, because of the strong seasonal dependence of air quality in New York at that time.

Reconciliation and Implications of the Mortality Studies

A high degree of quantitative consistency has been shown among many studies of varying designs. The slope of the dose-response function relating the logarithm of the percentage of daily excess deaths to the level of air pollution is essentially constant from present levels (below the current ambient standards) up to the levels present in London in the 1950s (about 10 times the ambient standards). Similar results have been found in various countries, climates, and time periods. These findings are supported by cross-sectional studies of annual mortality rates and by longitudinal studies of periods of long-term abatement in air pollution. The results for mortality are supported by similar findings for hospital usage and for effects on respiratory function (discussed later).

Some previous authors (for example, Lippmann and Lioy, 1985) considered that studies of daily mortality were "useful only for evaluating the presence of acute respiratory disease responses to short-term periods of elevated exposure" and that daily excess mortality was shown in the episodes of 25 years ago largely for those having advanced chronic lung disease. The results discussed in Part II require a much broader interpretation of these studies. The comparison of episodes (excluding the Meuse Valley and Donora) with time-series results indicates that both types of studies are part of the same response continuum. In that sense, the alarms that sounded over 40 years ago are still ringing.

For major episodes, special care must be taken to account for the duration of effects after the pollution has abated, but the similarity of slopes suggests that the same basic response mechanism persists over the entire range of concentrations. Many different causes of death were involved, and the victims of these severe episodes were not limited to those known to be previously ill. Some of the uncertainties remaining (discussed below) include the types of regression models used and whether other pollutants that have been neglected in the analyses (such as CO) may have an important effect. Note that use of the total episode dose (concentration × duration) is consistent with the use of annual averages in cross-sectional studies and suggests that cumulative or distributed lags should be considered in time-series studies.

Although the consistency of such studies had been noted previously (Özkaynak and Spengler, 1985, for example), quantitative comparisons of daily time-series and cross-sectional findings have not been made and may be more controversial. Aside from an insightful paper in 1984 (Evans *et al.*), few authors had previously considered the implications of such comparisons, but some of the most recently published studies appear to reinforce the comparison. It now appears that the best estimates of elasticities from time-series and cross-sectional studies overlap, for both total mortality and for chronic obstructive pulmonary disease deaths. When one considers the likely ranges of errors in each type of study, the conclusion follows that what had previously been regarded as "chronic" effects (i.e., the results from cross-sectional regressions) are more likely to reflect mainly the annual sums of acute effects. Time-series studies find elasticities in the 0.01–0.08 range with a median value probably around 0.04–0.05. Because many of these studies used inadequate air monitoring networks, and because it is difficult to account for cumulative lag effects, these estimates may be biased low. The cross-sectional studies that tried to account for a wide range of potential confounding variables tend to find elasticities only slightly higher, in the 0.04–0.06 range, but since it is so difficult to eliminate all of the confounding variables from cross-sectional studies, these estimates might be biased high. It is unclear to what extent any high biases might be compensated by effects of errors in the exposure estimates.

An alternative reconciliation scenario could involve a different interpretation of the time-series results. If the deaths measured by daily analyses were in fact premature by only a few weeks or months, at most, and did not represent an "excess" for the year, then we would be forced to interpret the cross-sectional findings as truly chronic (or else confounded). However, the longer-term temporal studies reviewed in Chapter 8 do not support this scenario, although more studies of this type are needed. It is difficult to envision the low concentration levels typical of annual averages creating new disease cases (chronic effects), while diminishing the importance of the acute effects at higher concentrations. For these reasons, I reject this scenario.

SUMMARY OF THE ASSOCIATIONS BETWEEN AIR POLLUTION AND HOSPITALIZATION

The hospitalization studies found time-series methods (Chapter 9) to be the most useful, since factors such as bed supply and physicians' practices can influence cross-sectional comparisons (Chapter 10). The effects of air pollution on hospitalization admissions were small, representing a few percent of the total. However, they emphasized respiratory diseases, and often children's admissions played a prominent role. The findings with respect to children suggest that adverse health effects are associated with *current* air pollution levels, even when the ambient standards are met, since children would not have been affected by the much higher air pollution levels of the past. It was not possible to identify specific "responsible" pollutants with certainty, although ozone and particles figured prominently in several studies. The elasticities varied widely, in part because of differences in study designs. Lag effects were shown to be important, as well as definitions of the diseases included in the analyses.

Hospitalization studies were useful in hypothesizing differences in responses to various types of air pollution events. For the disaster at Donora, for example, many more people sought treatment than died, which is consistent with a "poisoning" model. This classification is also supported by the presence of a variety of industrial emissions, deaths of animals that occurred there, and the fact that responses stopped quickly after the pollution began to abate. For the community air pollution events in London, New York, and elsewhere, the identified morbidity responses were of the same order as the deaths, and both continued many days after the pollution abated. This suggests exacerbation of existing disease.

SUMMARY OF THE ROLES OF CHANGES IN RESPIRATORY FUNCTION

Associations between Air Pollution and Changes in Lung Function

Studies of lung function (Chapter 11) provide connections from the purely statistical to the physiological domains and from ecological studies to data on individuals. Most of the studies of lung function involve healthy individuals, and, thus, the finding of effects due to air pollution at current levels has important implications, even if the average effects are small, reversible, and not "clinically significant." The effects found mirror those of the mortality and hospitalization studies: a few percent in magnitude, affecting children and adults, associated with various pollutants, even within the same cohort. Effects of environmental tobacco smoke on children and adults appear to support the existence of chronic air pollution effects, but the degree of reversibility of these effects is not known.

Lung Function as a Predictor of Mortality

Lung function was shown to be a robust predictor of mortality (Chapter 12), even after many different other factors were controlled for. The relationships applied to initially healthy people who died from cardiovascular and respiratory causes, as well as to deaths from all causes. The more recent studies of this association implied that the relationship was indeed predictive and not just a marker for existing disease; lung function may represent an index of general physical robustness.

The relationship was approximately 1:1; a 10% increase in risk of death was associated with lung function performance about 10% below normal. The chain rule of partial derivatives might be applied to formally combine the two types of lung function effects, thus deriving an independent estimate of the effects of air pollution on mortality:

$$\partial M/\partial P = (\partial M/\partial L)(\partial L/\partial P) \qquad [13\text{-}1]$$

where M = mortality, L = lung function, and P = air pollution. Since $\partial M/\partial L = -1$, and $\partial L/\partial P = 0.05–0.10$, then $\partial M/\partial P = 0.05–0.10$, which is consistent with the results from ecological studies of air pollution and mortality and thus could be viewed as "bridging the gap" between ecological and individual studies. This model implicitly assumes linear dose-response functions, that is, that Eq. 13-1 is valid for all levels of mortality rate, lung function performance, or air pollution. It also requires that changes in lung function have a dynamic effect, that impairment can move an individual from a healthy state to one of excess risk of dying, all other factors being equal. This last hypothesis may not seem credible for small changes in lung function, hence the doubt that air pollution "causes" disease at current levels. However, it is possible that small changes may have important acute effects on severely impaired individuals, which leads to the disease exacerbation model.

QUANTITATIVE RECONCILIATION OF FINDINGS ON MORTALITY, HOSPITALIZATION, AND LUNG FUNCTION

Bates (1992) emphasized the importance of considering the coherence among studies relating various health endpoints to air pollution. His paper was mainly limited to qualitative coherence; in this section we attempt to address the topic quantitatively. As an example, we consider a hypothetical city in the northeastern United States with a population of 1,000,000 in 1985, 13% of which are 65 or older (Table 13-1). The following data sources were used: mortality, *Vital Statistics of the United States, 1985*; air pollution-mortality regression coefficients, Schwartz and Dockery (1992); hospitalization statistics, Graves (1987); air pollution-hospitalization regressions, Burnett *et al.* (1993); baseline lung function data, Enright *et al.* (1993); air pollution effects on lung function, Chestnut *et al.* (1991), Schwartz (1989), and Dockery *et al.* (1985). The analysis considers TSP as an index of all community air pollution and assumes equivalent elasticities in order to convert to a common basis (TSP). The numerical values are intended to be illustrative and their confidence limits are not considered for this purpose.

The assumed average TSP level of 50 $\mu g/m^3$ leads to an estimate of 45 deaths per year associated with air pollution for those under 65, and 285 deaths for those 65 and over. For the total population, 54 COPD and pneumonia deaths would be associated with air pollution, with 203 deaths from major cardiovascular causes and 73 deaths from other causes. If another level of air pollution were assumed, the pollution-related deaths would change proportionately. Over a 75-year life, the population risk of premature death due to air pollution is about 0.0274, which is 274 times the cancer risk level considered noteworthy for hazardous air pollutants. Almost 90% of this risk is incurred after age 64. To place these estimates in the context of common everyday risks: If the 1985 national averages applied to this hypothetical city, there would be fewer than 1 death from lightning, 24 from poisoning, 55 from falls, 91 homicides, 137 suicides, and 209 deaths from motor vehicle accidents. Thus, the estimated total mortality associated with air pollution is about the same as the sum of suicides and motor vehicle accidents.

The corresponding data on hospitalization show that there would be about as many annual respiratory admissions associated with air pollution as there were deaths, but that the age distribution is shifted toward younger patients, especially children. The data thus imply that, either many of the air pollution-related deaths for older people occur out of hospital, or there must be other admission diagnoses associated with air pollution. Cardiac diagnoses are a distinct possibility. Further increases in admissions associated with air pollution would result from consideration of cumulative lag effects. All told, the estimates of annual hospital admissions associated with air pollution appear to be roughly consistent with the corresponding estimates of mortality. Note that about 60% of all deaths take place in hospital nationwide, but fewer than 3% of hospital patients are discharged dead.

As the final step in this reconciliation and to place these figures in perspective with respiratory health as determined by lung function testing, we draw on the spirometry data of Enright *et al.* (1993), which were based on about 5000 ambulatory participants, aged 65 and over. Using the ratio of FEV_1 to FVC as a measure of impairment and assuming a normal distribution, we estimate the numbers of elderly having specified levels of FEV_1/FVC. Values less than 0.70 are often considered "impaired" (see Chapter 4); values below 0.50 would thus constitute relatively severe impairment.* Table 13-1 shows that almost 3500 people would fit this definition under baseline conditions. We assume an

TABLE 13-1 Estimated Annual Health Risks for a Hypothetical Population of 1,000,000

	Baseline			Estimated effects of air pollution (based on TSP = 50 μg/m^3)		
	All age	0–64	65+	All ages	0–64	65+
Population	1,000,000	870,000	130,000	1,000,000	870,000	130,000
Deaths, all causes	9,500	3,200	6,300	330	45	285
Deaths, respiratory causes	688	93	595	54		
Hospital admissions						
All causes	143,000	96,000	47,000			
Respiratory causes	5,900	2,400	3,500	342	192	150
Impaired lung function						
$FEV_1/FVC < 0.60$			22,000			9,350
$FEV_1/FVC < 0.50$			3,500			1,960
$FEV_1/FVC < 0.40$			480			210

elasticity of 0.05 for air pollution and estimate its effects on the distribution of FEV_1/FVC; the table shows that even this small amount of air pollution has a large effect on the numbers of impaired individuals. Enright *et al.* (1993) also report that another 6% of their sample (7800 people in our hypothetical city) were unable to perform satisfactory tests; Eisen *et al.* (1987) showed that a subgroup defined in this way would also be at increased risk of premature mortality.

These figures show that a small increase in air pollution may create a much larger (about 50%) increase in the respiratory-impaired portion of the population, which we assume comprises a large part of the pool of susceptible individuals. As a result, it may not be unreasonable to find of the order of several hundred excess deaths per year resulting from episodic swings in air pollution, especially if the perturbations happen to coincide with acute illness. We thus conclude that the quantitative estimates of acute deaths and hospitalizations may be consistent with the estimated pool of persons having impaired respiratory function. This would also be the case if the mortality estimates had included the implied chronic effects shown by Pope *et al.* (1993). However, the extreme tail portion of the distribution of lung function is involved in these estimates, and the assumption of normality may not be entirely justified.

* Cotes (1979) gives a table of relative FEV_1 values in conjunction with a scale of degrees of physical impairment. The lowest FEV_1 value is about 10% of the value for an unimpaired person: "(patient) is living: needs help with feeding." At 35%: "can walk 100 m, sing, climb 8 stairs. At 50%: "can walk 400 m." At 68%: "can walk unlimited distance at slow pace." Peto *et al.* (1983) give FEV_1 values of 10 and 30% of the FEV_1 values for age 25 as corresponding to death and disability, respectively; these correspond to about 17 and 50% of the normal values for age 65–74. These descriptions give an idea of the degree of physical impairment that could be associated with the most respiratory-impaired segment of the population.

CAUSAL IMPLICATIONS BASED ON HILL'S CRITERIA

We have shown that the quantitative estimates developed in this book are self-consistent, and that this criterion is necessary but insufficient for a causal interpretation. Chapter 3 discussed the nine criteria that Hill proposed in 1965 for assessing the likelihood that a statistical association might be causal. This book has discussed many aspects of the relationship between community health and air pollution; while each of these aspects may not satisfy all nine criteria, we now consider whether the problem in its entirety might do so.

Strength of the association. Most of the health effects of community air pollution would be judged to be "weak" associations at current levels. However, this was definitely not the case for the air pollution episodes and disasters of 20 to 40 years ago. To the extent that current studies find the same or similar dose-response functions, this criterion might be judged to have been met.

Consistency of the association. The associations between air pollution and mortality, hospitalization, and lung function have been found to be robust over space, time, and investigator.

Specificity of the association. This criterion is only partially met in that a variety of air pollutants have been shown to be associated with a specific set of endpoints. In general, respiratory diagnoses display the strongest relationships, followed by heart disease. Control diagnoses have generally not been associated with air pollution. It has not been possible to assign specific pollutants to specific diseases with confidence.

Temporality. Time-series studies have shown that the response follows the dose and not vice versa.

Biological gradient (dose-response relationships). For both time-series and cross-sectional analyses, larger doses led to stronger responses. However, in some cases, responses were attenuated at the higher levels, such that a logarithmic relationship seemed to fit better than a linear relationship. The biological implications of this behavior in a heterogeneous population are unclear.

Biological plausibility. The physiological pathway between dose and response was shown by the finding of effects of air pollution on respiratory function and of respiratory function on mortality.

Coherence. Coherence was shown by the comparability of effects on mortality, hospitalization, and lung function.

Experimental manipulation. The case studies of pollution clean-up campaigns discussed in Chapter 8 help satisfy this criterion. In addition, changes in respiratory mechanics have been induced by experimental changes in air pollution (in either direction).

Analogy. The daily changes in mortality associated with air pollution are analogous to the effects of temperature extremes, especially heat waves. The common factor may be the failure of impaired individuals to be able to maintain homeostasis (Stern and Tuck, 1986).

All told, I conclude that Hill's criteria have essentially been met, but that this conclusion depends on the coherence of the entire body of work. Doubts with regard to the validity of any of the elements could seriously weaken the credibility of the whole.

UNCERTAINTIES IN THE ANALYSIS

This reconciliation section has tried to make the case that most of the evidence points to a causal relationship between ill health and air pollution, at levels below current standards and not definitively tied to any one or group of specific pollutants. However, there are other factors to consider.

Indoor Air Pollution. One of the important factors is indoor air pollution, which can include tobacco smoke, non-specific particulates, NO_2 and other combustion products, formaldehyde, as well as the components of outdoor air pollution that infiltrate buildings. In general, neither SO_2 nor aerosol acidity is important indoors, because of absorption on surfaces in the first case and neutralization for the latter. Most people spend the overwhelming majority of their time indoors, of the order of 80% or more. Sick and impaired people may spend even higher fractions of time indoors and may be more likely to have air conditioning. Given these circumstances, how can outdoor air pollution be important?

There are at least two possible scenarios. First, some outdoor pollution does enter buildings; this was documented directly during some of the more severe episodes, for example, and has been explored in correlation studies (Wyzga and Lipfert, 1992). Second, indoor air pollution adds to the respiratory burden from outdoor air. It has been shown that living with a smoker increases a nonsmoker's risk of death by about 17% (Sandler *et al.*, 1989); presumably, this is a chronic effect. If the average smoker smoked one pack per day at home, this would add about $20\,\mu g/m^3$ of fine particles to the indoor air (J.D. Spengler, personal communication), and the corresponding dose-response function would be about 0.8% excess deaths per $\mu g/m^3$, a figure that is somewhat less than that of the findings of Pope *et al.* (1993) in their prospective study of outdoor air pollution. However, it is also possible that sidestream tobacco smoke is more toxic than the mixed aerosols typically found outdoors. Dockery *et al.* (1992) found a slope of 0.17% per $\mu g/m^3$ for the acute effects of fine particles in St. Louis, for example. The effects of passive smoking would appear to apply only to chronic effects; use of cigarette consumption, as opposed to percentage of smokers, in cross-sectional studies is intended to help capture this effect independently from the effects of outdoor air pollution (Lipfert *et al.*, 1988).

We have no evidence of acute mortality responses to either active or passive smoking, but then perhaps such studies have not yet been done. It is possible that the transition from a healthy state to an impaired state is hastened by both active and passive smoking and that, for a few individuals, an air pollution event administers the *coup de grâce*. It is also possible that many of the victims of air pollution are also smokers; for example, most chronic onstructive pulmonary disease patients and many cardiac patients are present or former smokers.

Two other points should be made about the role of indoor air quality. The degree of penetration of outdoor air depends on the air exchange rate and the degree of treatment of outdoor make-up air. In an automobile, air exchange rates tend to be very high and the local air quality tends to be worse than off the road. Given the increasing amounts of time many people spend in cars, this may be an important portion of the daily dose of pollutants. Second, little is known about the treatment of outdoor air in hospitals and nursing homes. Many of them have high air exchange rates in order to flush out pathogens. If outdoor air pollution is not removed, sick patients may suffer.

Air Pollution Metrics and Thresholds. With the exception of the studies of Abbey *et al.* (1991), little effort has been devoted to developing the most suitable air pollution metric. For acute studies using time-series methods, the possible role of the duration of elevated concentrations has not been explored. Alternative models incorporating thresholds have not been tested against linear models with statistical tests of the improvements in fit. Seasonal differences in responses that might reflect differences in the likelihood of outdoor exposure have not been explored with the more recent data sets.

Combinations of Errors. The preceding analysis relies heavily on similar findings by different types of (independent) studies. It is possible that each type has been confounded by a different source of error and that their agreement is merely fortuitous. For time-series analyses, the degree of prematurity of death is crucial, for example. Only a few long-term time-series studies have been done; more effort is clearly needed on this topic.

Cross-sectional studies have the potential for many more problems. Occupational exposures may confound, since many of the more polluted cities have heavy industry. Analysis by gender may help sort out this factor (Lipfert, 1984), as well as deletion of certain occupationally related diseases such as pneumoconiosis. When I disaggregated the 1970 city mortality study by age, I found that most of the pollution effects became nonsignificant (Lipfert, 1978). However, this could have been due to the less robust statistics of smaller population age groups. None of the cross-sectional studies of more recent mortality data have attempted to disaggregate by age; such a study should probably be done for a 5-year period of record in order to increase the reliability of the estimates of mortality rates for smaller groups.

Identification of Individuals. There are undoubtedly skeptics for whom the ecological study design will always be suspect; they would require identification of individuals and their medical, pollution exposure, and life-style histories to become convinced of causality. A retrospective case-control method could be used here, by contrasting the characteristics of those who died after low-pollution days with those who died after high-pollution days. This would require interviews with surviving relatives and would probably be extremely labor-intensive. Further, such a study would have to be done for a very recent year to get credible data, and a large number of cases would be required in order to find the small effect that air pollution seems to engender.

Unmeasured Pollutants and the "Unknown Confounder." Critics of ecological studies often raise the issues of unmeasured pollutants and unspecified confounding variables. While it is surely not possible to measure all the contaminants in urban air adequately enough for the types of studies we have reviewed, associations have been found for all of the major species and types of pollutants. The most common findings seem to be various measures of particulate matter and O_3; SO_2 has been important only in a few European studies. Questions remain about findings for NO_x and hydrocarbons, species that have not previously been associated with human health effects. Additional studies will be required to shed additional light on this issue and the role of CO.

The problem of the "unknown confounder" may be more philosophical in nature, and is most often applied to cross-sectional studies. Critics have a basic feeling that there must be another factor, yet unidentified, that is correlated with both air pollution and mortality and constitutes the true cause of the health effect differentials in question. Cross-sectional studies have improved over the years in their ability to control for additional socioeconomic and life-style factors, and additional data are becoming available. What may be required to convince the skeptics is a comprehensive look at spatial patterns and their changes over time, from, say, 1960 to 1990, using the most recent statistical methods.

Types of Regression Models. Although many different types of regression models (linear, log-linear, logarithmic) were used by the ensemble of studies, few authors provided comparisons of results for different models with a given data set. The question of pollution thresholds has received only limited attention and is probably the most important policy-oriented issue. Underlying this issue is the general question of rigorous statistical tests of the significance of differences between models.

SUMMARY OF CONCLUSIONS AND THEIR IMPLICATIONS

Conclusions. Despite the uncertainties, I feel that certain conclusions are ineluctable:

1. Air pollution can impair breathing, send people to hospital, and shorten lives, even at concentration levels below the current ambient air quality standards. At current urban levels, the risks to an individual from air pollution are comparable to other common risks such as accidents.
2. Although particulate matter and ozone figure prominently in the studies reviewed, each of the criteria pollutants (except lead, which has rarely been studied in this context) has been significantly associated with adverse community health effects. This group of studies thus provides no unequivocal mandate for increased controls on any specific pollutant. Note also that, in many locations in the United States, natural sources are responsible for an appreciable portion of existent ozone and airborne particles, on the basis of long-term averages.
3. The individuals most likely to be affected by air pollution cannot be identified *a priori*, but they are assumed to be those persons with preexisting disease who are actually exposed to outdoor air pollution.
4. Because in a large heterogeneous population there may be at least one individual who is sufficiently sensitive to a given pollutant, a no-threshold dose-response relationship cannot be ruled out.

Implications. What are the implications of these findings with respect to the regulation and control of air pollution in the United States? The studies reviewed imply that there may be no entirely "safe" levels of air pollution. The Clean Air Act requires that ambient air quality standards be set to protect the most sensitive individuals, with "an adequate margin of safety." Elaborate systems of command and control have been developed to limit emissions in accordance with these requirements. However, this requirement is incompatible with the existence of natural sources of air pollution that are not amenable to controls, including ozone from stratospheric intrusion, sulfur from biogenic sources, and airborne particles from crustal dust and wildfires. Even if thresholds cannot be defined on the basis of adverse health effects, nature provides its own thresholds of minimum risk by virtue of these natural sources and the actions of weather in creating elevated concentration levels.

Are draconian control measures justified to achieve an incremental reduction in risk that may be just over the natural threshold? In a democratic society, the public has a right to know what benefits will be purchased by its air pollution control expenditures. The public should not be misled into believing that achievement of the present ambient air quality standards, however difficult this may turn out to be, will totally eliminate their risks of illness or premature death due to air pollution.

Should the ambient air quality standards be tightened, or should the entire regulatory system be changed? There is little question that less pollution is healthier than more pollution, but perhaps the ambient standards should ultimately be regarded as guidelines, in which case-specific decisions about the degree of involuntary risk that the public is willing to tolerate would be required. This is the same question that Frank posed in 1975, "not 'What is a safe level of pollution?' but rather, 'What is a tolerable or acceptable level of sickness from pollution?'"

Although environmental quality and the basic health of the U.S. population have improved greatly over the past 20 years or so (age-adjusted mortality has dropped by more than 20%), studies still find associations between air pollution and mortality or hospitalization. What then is the outlook for the future? First, increased longevity implies the presence of more frail elderly, who may comprise a large portion of the susceptible population. Second, respiratory disease and asthma are on the increase, for reasons that are not fully understood. On the other hand, as the prevalence of smoking diminishes, there should be fewer future cases of chronic obstructive pulmonary disease.

A CALL TO ACTION

I believe that the conclusions of this review constitute a clear call to action, on three fronts. First, we must continue to seek the most cost-effective controls on community air pollution. Next, we should increase our level of understanding of community health effects, including the physiological mechanisms involved, the effects of specific pollutants on specific diseases, the types of individuals most at risk, and the ability to forecast and warn against high pollution days. Finally, we should not be afraid to seek fundamental changes in the Clean Air Act that would reflect the realities of this new knowledge, that ambient air quality thresholds may not exist and that it may not be possible to define the "responsible" pollutants with certainty.

Many of the studies reviewed in this book have been in the literature for years, including the accounts of the most severe air pollution disasters. Yet, few serious attempts have been made to draw quantitative results from them as a group. Similarly, few attempts have been made to estimate the benefits that have accrued from the great strides in air pollution control that have been made since the 1950s and 1960s. Just as in the rest of our cultural history, we must learn from the past in order to better the future. A rough calculation of the benefits of reductions in airborne particles since 1955 yields a figure in excess of one million lives that may have been extended (assuming that particles are in fact the operative agent). Unfortunately, our present state of knowledge does not permit an estimate of the actual improvement in life expectancy, and the historical monitoring record is insufficient to allow such estimates to be made for other pollutants such as ozone.

The first step toward an increased understanding of the effects of community air pollution is the recognition of the problem. This area of research has been ignored by the regulatory and health research agencies long enough, perhaps because the findings may have been inconvenient in the context of the Clean Air Act in its inability to clearly identify thresholds for specific pollutants. Studies of mortality, hospitalization, and long-term changes in respiratory function should be added to the federal agenda for environmental epidemiology. The data of Table 13-1 should convince skeptics that the current risks of air pollution are not trivial, in the context of environmental regulation.

Previous studies of community health have varied substantially in quality and in the robustness of their findings. Some, but not all, of those few community health studies sponsored by government agencies may have fared somewhat better. However, the topic has barely been scratched, considering the magnitude of the economic costs and benefits involved. Information gaps include:

- Survey data on diet and smoking habits, at the city level
- Computerized hospitalization records, accessible for research
- Population-based air quality monitoring data for all criteria pollutants
- Studies of place of death (home versus institution)
- Long-term data on lung function, for trend analysis
- Robust statistical methods of comparing alternative regression models, including the effects of errors in independent variables

Mortality records are complete, but doubts remain as to the appropriateness of underlying causes of death, as opposed to contributing causes. The full contributions of respiratory causes may be understated, especially for deaths of frail elderly (C.A. Pope, personal communication). Time-series studies have been done for several years and several locations, and we seem to be developing an understanding of this aspect of the problem. However, convincing explanations have yet to be developed for the excess mortality rates of the Midwest and parts of the South, in addition to explanations for the geographic variability in the declines in heart disease mortality that began in the late 1960s and early 1970s. Ultimately, parallel time-series studies should be done in enough locations across the nation to constitute a long-term cross-sectional study and for a time period sufficient to consider long-term lags. Such a study might be able to finally resolve questions about chronic versus acute effects.

The full faculties of the federal government and its medical establishment should be tapped for future studies of community health. This recommendation is especially timely, given the present emphasis on controlling the nation's health care costs. Community health studies require census data, vital statistics, health examination data, and environmental data; these should be provided in a coordinated and supportive effort—rather than at the convenience of what often appear to be disinterested agencies. Perhaps a test community should be selected for detailed, long-term, environmental health studies, in the manner of the Framingham (MA) studies on cardiovascular risks (Kannel *et al.*, 1974).

Finally, it is important to understand actual pollutant exposures and the roles of indoor air quality. Does air conditioning actually protect against outdoor air pollution, as hypothesized? Can the effects of indoor air pollution sources be assessed? Can the days with potential for increased mortality be forecast with sufficient accuracy to provide protection to sensitive individuals? Would such a policy be cost-effective?

These recommendations amount to a reordering of priorities for environmental (public) health, emphasizing effects on people. For example, after the risks of community air pollution are thoroughly understood, it may be appropriate to reconsider the risks involved in population exposures to hazardous air pollutants, which appear to be much smaller (J. Padgett, personal communication). The studies analyzed in this book show that the risks of ordinary community air pollution remain orders of magnitude greater than the guidelines for toxic air pollutants (which are often thought to be more dangerous). Whether reordering of regulatory priorities is appropriate may depend on many other factors, but at least the question should be addressed.

The paradox posed by Rose (1985) seems fitting: "A preventative measure which brings much benefit to the population offers little to each participating individual." Although a useful beginning has come about in the understanding of air pollution effects on community health, by way of reviewing and re-evaluating the literature, it is my hope that a more serious and coordinated epidemiological research effort can now begin.

REFERENCES

Abbey, D.E., Mills, P.K., Petersen, F.F., and Beeson, W.L. (1991), Long-Term Ambient Concentrations of Total Suspended Particulates and Oxidants As Related to Incidence of Chronic Disease in California Seventh-Day Adventists, *Env. Health Perspect.* 94:43–50.

Bates, D.V. (1992), Health Indices of the Adverse Effects of Air Pollution: The Question of Coherence, *Env. Res.* 59:336–49.

Burnett, R.T., Dales, R.E., Raizenne, M.E., Krewski, D., Summers, P.W., Roberts, G.R., Raad-Young, M., Dann, T., and Brooke, J. (1993), Effects of Low Ambient Levels of Ozone and Sulphates on the Frequency of Respiratory Admissions to Ontario Hospitals, *Env. Res.* (submitted).

Chestnut, L.G., Schwartz, J., Savitz, D.A., and Burchfiel, C.M. (1991), Pulmonary Function and Ambient Particulate Matter: Epidemiological Evidence from NHANES I, *Arch. Env. Health* 46:135–44.

Cohen, B.H. (1980), Chronic Obstructive Pulmonary Disease: A Challenge in Genetic Epidemiology, *Am. J. Epidemiology* 112:274–88.

Cotes, J.E. (1979), *Lung Function. Assessment and Applications in Medicine*, Blackwell, Oxford. p. 399.

Dockery, D.W., Schwartz, J., and Spengler, J.D. (1992), Air Pollution and Daily Mortality: Associations with Particulates and Acid Aerosols, *Envir. Res.* 59:362–73.

Eisen, E.A., Dockery, D.W., Speizer, F.E., Fay, M.E., and Ferris, B.G., Jr. (1987), The Association Between Health Status and the Performance of Excessively Variable Spirometry Tests in a Population-Based Study in Six U.S. Cities, *Am. Rev. Respir. Dis.* 136:1371–76.

Ellis, F.P. (1972), Mortality from Heat Illness and Heat-Aggravated Illness in the United States, *Env. Res.* 5:1–58.

Enright, P.L., Kronmal, R.A., Higgins, M., Schenker, M., and Haponik, E.F. (1993), Spirometry Reference Values for Women and Men 65 to 85 Years of Age, *Am. Rev. Respir. Dis.* 147:125–33.

Evans, J.S., Kinney, P.L., Koehler, J.L., and Cooper, D.W. (1984), Comparison of Cross-Sectional and Time-Series Mortality Regressions, *J. APCA* 34:551–53.

Frank, R. (1975), Biologic Effects of Air Pollution, in *Energy and Human Welfare—A Critical Analysis. Vol I. The Social Costs of Power Production*, B. Commoner, H. Boksenbaum, and M. Corr, eds., Macmillan, New York, pp. 17–27.

Gover, M. (1938), Mortality During Periods of Excessive Temperature, *Pub. Health Rep.* 53:1122–43.

Graves, E.J. (1987), Diagnosis-Related Groups Using Data From the National Hospital Discharge Survey: United States, 1985. NCHS Advance Data No. 137. National Center for Health Statistics, Hyattsville, MD.

Hill, A.B. (1965), The Environment and Disease: Association or Causation? *Proc. Roy. Soc. Med.*, Section on Occupational Medicine 58:295–300.

Hodgson, T.A., Jr. (1970), Short-Term Effects of Air Pollution on Mortality in New York City, *Env. Sci. Tech.* 4:589–97.

Kalkstein, L.S. (1991), A New Approach to Evaluate the Impact of Climate on Human Mortality, *Envir. Hlth. Persp.* 96:145–50.

Kannel, W.B., Seidman, J.M., Fercho, W., Castelli, W.P. (1974), Vital Capacity and CHD: The Framingham Study. *Circulation* 49:1160–66.

Katsouyanni, K., Karakatsani, A., Messari, I., Touloumi, G., Hatzakis, A., Kalandidi, A., and Trichopolous, D. (1990), Air Pollution and Cause Specific Mortality in Athens, *J. Epid. Comm. Health* 44:321–24.

Krupnick, A.J., and Portney, P.R. (1991), Controlling Urban Air Pollution: A Benefit-Cost Assessment, *Science* 252:522–28.

Kutschenreuter, P.H. (1960), A Study of the Effect of Weather on Mortality in New York City. A thesis (M.S.) submitted to the Graduate School of Rutgers, The State University, New Brunswick, NJ.

Lave, L.B., and Seskin, E.P. (1970), Air Pollution and Human Health, *Science* 169:723–33.

Lipfert, F.W. (1978), The Association of Human Mortality with Air Pollution: Statistical Analyses by Region, by Age, and by Cause of Death, Ph.D. Dissertation, Union Graduate School, Cincinnati, Ohio. Available from University Microfilms.

Lipfert, F.W. (1984), Air Pollution and Mortality: Specification Searches Using SMSA-based Data, *J. Env. Econ. & Mgmt.* 11:208–43.

Lipfert, F.W. (1992), Community Air Pollution and Mortality: Analysis of 1980 Data from U.S. Metropolitan Areas. Report prepared for U.S. Department of Energy, Brookhaven National Laboratory.

Lipfert, F.W., and Wyzga, R.E. (1992), Observational Studies of Air Pollution Health Effects: Are the Temporal Patterns Consistent? Paper IU-21A.09, presented at the 9th World Clean Air Conference, Montreal, Canada, September 1992.

Lipfert, F.W. (1993), A Comparison of Prospective and Ecological Studies of Mortality. Report to Electric Power Research Institute, Palo Alto, CA.

Lipfert, F.W., Malone, R.G., Daum, M.L., Mendell, N.R., and Yang, C-C. (1988), A Statistical Study of the Macroepidemiology of Air Pollution and Total Mortality. BNL Report 52122, to U.S. Dept. of Energy.

Lipfert F.W., and Hammerstrom, T. (1992), Temporal Patterns in Air Pollution and Hospital Admissions, *Environmental Research* 59:374–99.

Lippmann, M., and Lioy, P.J. (1985), Critical Issues in Air Pollution Epidemiology, *Env. Health Persp.* 62:243–58.

Mazumdar, S., Schimmel, H., and Higgins, I. (1982), Relation of Daily Mortality to Air Pollution: An Analysis of 14 London Winters, *Arch. Env. Health* 37:213–20.

Menkes, H.A., Cohen, B.H., Beaty, T.H., Newill, C.A., and Khoury, M.J. (1984), Risk Factors, Pulmonary Function, and Mortality, in *Genetic Epidemiology of Coronary Heart Diseases: Past, Present, and Future*, D.C. Rao, *et al.*, eds., A.R. Liss, Inc., New York, pp. 501–21.

Ministry of Health, Mortality and Morbidity During the London Fog of December 1952, Reports on Public Health and Medical Subjects No. 95. HMSO, London, 1954.

Ostro, B. (1984), A Search for a Threshold in the Relationship of Air Pollution to Mortality: A Reanalysis of Data on London Winters, *Env. Health Persp.* 58:397.

Ottinger, R.L., Wooley, D.R., Robinson, N.A., Hodas, D.R., and Babb, S.E. (1990), *Environmental Costs of Electricity*, Oceana Publications, New York.

Özkaynak, H., and Spengler, J.D. (1985), Analysis of Health Effects Resulting from Population Exposures to Acid Precipitation Precursors, *Env. Health Persp.* 64:45.

Peto, R., Speizer, F.E., Cochrane, A.L., Moore, F., Fletcher, C.M., Tinker, C.M., Higgins, I.T.T., Gray, R.G., Richards, S.M., Gilliland, J., and Norman-Smith, B. (1983), Relevance in Adults of Air-Flow Obstruction, but not of Mucus Hypersecretion, to Mortality from Chronic Lung Disease, *Am. Rev. Resp. Dis.* 128: 491–500.

Plagiannakos, T., and Parker, J. (1988), An Assessment of Air Pollution Effects on Human Health in Ontario, Report No. 706.01 (#260), Energy Economics Section, Economics and Forecast Division, Ontario Hydro, Toronto.

Pope, C.A., Dockery, D.W., Xu, X., Speizer, F.E., Spengler, J.D., and Ferris, B.G. (1993), Mortality Risks of Air Pollution: A Prospective Cohort Study, presented at *Aerosols in Medicine*, 9th ISAM Congress, Garmisch–Partenkirchen, Germany, March 30–April 4, 1993.

Reid, D.D., Hamilton, P.J.S., Keen, H., Brett, G.Z., Jarrett, R.J., and Rose, G. (1974), Cardiorespiratory Disease and Diabetes among Middle-aged Male Civil Servants, *Lancet* 1:469–73.

Rose, G. (1985), Sick Individuals and Sick Populations, *Int. J. Epidemiology* 14:32–38.

Sandler, D.L., Comstock, G.W., Helsing, K.J., and Shore, D.L., (1989), Deaths from All-Causes in Non-Smokers Who Lived with Smokers, *Am. J. Public Health* 79:163–67.

Schimmel, H., Murawski, T.J., and Gutfield, N. (1974), Relation of Pollution to Mortality, presented at the 67th Annual Mtg. APCA, APCA Paper 74–220.

Schwartz, J. (1989), Lung Function and Chronic Exposure to Air Pollution: A Cross-Sectional Analysis of NHANES II, *Env. Res.* 50:309–21.

Schwartz, J., and Dockery, D.W. (1992a), Increased Mortality in Philadelphia Associated with Daily Air Pollution Concentrations, *Am. Rev. Resp. Dis.* 145:600–604.

Sherwin, R.P., and Richters, V. (1991), Chronic Bronchitis in Youths (Coroner Cases): Glandular Inflammation and Alterations, presented at the Annual Conference, Society for Occupational and Environmental Health, Crystal City, VA.

Shumway, R.H., Tai, R.Y., Tai, L.P., and Pawitan, Y. (1983), Statistical Analysis of Daily London Mortality and Associated Weather and Pollution Effects, California Air Resources Board, Sacramento, CA.

Stern, N., and Tuck, M.L. (1986), Homeostatic Fragility in the Elderly, *Cardiology Clinics*, 4:201–11.

Wyzga, R.E., and Lipfert, F.W. (1992), Exposure Assessment for Particulates: A Pilot Study of a Group of Susceptibles, Paper 92-65.01, presented at the Annual Meeting of the Air & Waste Management Association, Kansas City, MO.

Index